Cytokines and the Brain

Neuroimmune Biology, Volume 6

Neuroimmune Biology

Series Editors

I. Berczi, A. Szentivanyi

Cytokines and the Brain

Volume Editors

Christopher Phelps
Elena Korneva

*Department of Anatomy, University of South Florida,
Tampa, FL, USA
and
Department of General Pathology & Pathophysiology,
Institute of Experimental Medicine,
St. Petersburg, Russia*

ELSEVIER

AMSTERDAM • BOSTON • HEIDELBERG • LONDON • NEW YORK • OXFORD
PARIS • SAN DIEGO • SAN FRANCISCO • SINGAPORE • SYDNEY • TOKYO

Elsevier
Radarweg 29, PO Box 211, 1000 AE Amsterdam, The Netherlands
Linacre House, Jordan Hill, Oxford OX2 8DP, UK

First edition 2008

Library of Congress Cataloging-in-Publication Data
A catalog record for this book is available from the Library of Congress

British Library Cataloguing in Publication Data
A catalogue record for this book is available from the British Library

ISBN: 978-0-444-53041-7
ISSN: 1567-7443

For information on all Elsevier publications
visit our website at books.elsevier.com

Printed and bound in the United Kingdom
Transferred to Digital Print 2011

Contents

Foreword

Observations suggesting the interaction of the neuroendocrine and immune systems date back to the eighteenth century. However, for a long time only sporadic experiments were performed with regard to the interaction of these complex systems. Neither the knowledge nor the research tools were available to produce sufficient evidence that would convince the scientific community of the biological significance of neuroimmune regulation in higher animals and man. The situation has gradually turned around during the past three decades, and today this research area is advancing very well [1,2].

Cytokines were discovered first in the immune system and hence were called *interleukins*. Later it became clear that these mediators function also as tissue hormones outside of the immune system and for this reason they were renamed *cytokines* [3]. It is common knowledge that the main function of cytokines is the regulation of the immune system and of inflammation. Most investigations of cytokines deal with these areas. However, cytokines also play a role in neuroimmune interaction, which is a novel and developing area. It is just being uncovered that some of the cytokines, which are well characterized in immunology, play an important role in brain development, function and in pathology, as it is presented in this volume. Clearly, the evidence presented here testifies for the fundamental importance of cytokines in the brain, which may come from glia cells, astrocytes, or neurons and play essential roles in brain physiology and pathophysiology [4–9]. Sleep is also regulated by cytokines [10]. Moreover, the blood–brain barrier has the machinery to transport cytokines [11]. Another group of tissue hormones called chemokines have also been recognized recently as important regulatory molecules in the central nervous system [12].

Immune-derived cells and mediators are part of the neuroimmune regulatory equation [13–16]. The immune cells show enhanced activity after immunization, infection, or stress [17–19] and under various pathological conditions, which include depression and various neurodegenerative disorders (e.g., multiple sclerosis [MS], Alzheimer's disease, Parkinson's disease, and stroke). These conditions share many elements of inflammation and autoimmunity. Cytokines and chemokines play an important role in initiating and propagating the inflammatory/immune response in these pathologies. It has been established that in MS there is a continuous realignment and redundancy in the inflammatory and immune responses that take place [20–23]. Cytokines also influence behavior [24,25].

Neuroimmune interaction has been predicted repeatedly from early experiments [1,26–28]. But there is much more to these systems than sharing cytokines and other mediators and communicating. It is now apparent that the neuroendocrine and immune systems do not only interact, but rather, *rely on each other* for mutual support both in health and disease There is developmental relationship and antigenic cross-reactivity. It has been known for a long time that the thymus develops from the *neural crest*, which also gives rise to the central nervous system. Moreover, glia cells that represent roughly 50% of brain cells are related to the monocyte–macrophage lineage and are bone marrow derived [29]. The new and somewhat surprising information discussed in this volume is that neurons themselves may differentiate from bone marrow derived stem cells [30,31]. If this will stand the test of scientific scrutiny, it will mean that the brain relies on the bone marrow for rejuvenation and healing ("plasticity"). Indeed, recent evidence indicates that inflammatory cells and cytokines exert a neuroprotective effect during traumatic brain injury [32]. Immunologists discovered a long time ago that immunization of animals with foreign brain tissue gave rise to T-lymphocyte-specific antibodies [33] and such antibodies were used in immunology to identify T cells until more sophisticated reagents became available. Today we know that the brain shares adhesion molecules and numerous cell surface receptors with lymphocytes. It is now also clear that cell-to-cell and cell-to-matrix interactions play important roles in brain physiology and pathology as it is also discussed in this volume [34,35].

Until recent times the immune system has been considered as an autonomous system equipped with sophisticated receptors for antigen and inner regulatory circuits, which defends the host from pathogenic insults [36]. However, on the basis of common developmental origin, shared stem cells, receptors, and mediators and mutual interdependence, it is now apparent that *the nervous-, endocrine- and immune systems are integrated parts of a united neuroimmune system, which acts as a systemic regulatory network in higher animals and man.* Thus lymphocytes with their dominant regulatory function within the immune system could be considered to be analogous to the neuron within the central nervous system. Lymphocytes, like neurons, are sensory cells with the capacity to recognize chemical structure and to distinguish self from nonself. They store such information and show a memory response. Lymphocytes are also capable of conveying information on chemical (antigenic) abnormalities to the brain through cytokine signals. Immune cells are essential for defending the body from foreign invading pathogenic organisms as well as for the elimination of aberrant cells from the host. It is now apparent that immune cells are also involved in normal physiological regulation.

In conclusion, *the neuroendocrine and immune systems form an inter-dependent unit, which is equipped with sensory capacity, the ability to process and store information and to regulate the host organism in homeostasis and harmony with the external and internal environment.* This neuroimmune system plays a fundamental regulatory role for the entire life cycle of higher animals and of man.

As it is obvious from the papers presented in this volume, the advancement of *Neuroimmune Biology* provides new and important insights continuously into the physiology/pathophysiology of higher organisms, which revolutionize our thinking, inspire new approaches for our research, and help to rationalize patient care.

Istvan Berczi, Andor Szentivanyi

REFERENCES

1. Berczi I, Szentivanyi A. The history of Neuroimmune Biology. In Neuroimmune Biology, Volume 6: Cytokines and the Brain. Korneva HA, Phelps C, Eds; Berczi I, Szentivanyi A, Series Eds; Amsterdam: Elsevier, 2007 (in this volume); pp. 3–18.
2. Sternberg EM. The Balance Within: The Science Connecting Health and Emotions. New York: WH Freeman and Company, 2000.
3. Berczi I, Szentivanyi A. Cytokines and chemokines. In Neuroimmune Biology, Volume 3: The Immune-Neuroendocrine Circuitry. History and Progress. Berczi I, Szentivanyi A, Eds; Amsterdam: Elsevier, 2003; pp. 191–220.

4. Conti B, Tabarean I, Behrens M, Bartfai T. Cytokine receptors in the Brain. In Neuroimmune Biology, Volume 6: Cytokines and the Brain. Korneva HA, Phelps C, Eds; Berczi I, Szentivanyi A, Series Eds; Amsterdam: Elsevier, 2007 (in this volume); pp. 21–38.
5. Takao T, Hashimoto K, De Souza EB. Interleukin-1 and corticotropin releasing factor receptors in the hypothalamic-pituitary-adrenal axis. In Neuroimmune Biology, Volume 6: Cytokines and the Brain. Korneva HA, Phelps C, Eds; Berczi I, Szentivanyi A, Series Eds; Amsterdam: Elsevier, 2007 (in this volume); pp. 39–54.
6. Konsman JP. Brain interleukin-1β expression and action. In Neuroimmune Biology, Volume 6: Cytokines and the Brain. Korneva HA, Phelps C, Eds; Berczi I, Szentivanyi A, Series Eds; Amsterdam: Elsevier, 2007 (in this volume); pp. 55–78.
7. Rybakina EG, Korneva EA. Interleukin 1β signal transduction via the sphingomyelin pathway in brain cells. In Neuroimmune Biology, Volume 6: Cytokines and the Brain. Korneva HA, Phelps C, Eds; Berczi I, Szentivanyi A, Series Eds; Amsterdam: Elsevier, 2007 (in this volume); pp. 79–92.
8. Spengler RN. Cytokines in synaptic function and behavior. In Neuroimmune Biology, Volume 6: Cytokines and the Brain. Korneva HA, Phelps C, Eds; Berczi I, Szentivanyi A, Series Eds; Amsterdam: Elsevier, 2007 (in this volume); pp. 111–44.
9. Prinz M, Hanisch U-K. Interleukin-2 as a neuroregulatory cytokine. In Neuroimmune Biology, Volume 6: Cytokines and the Brain. Korneva HA, Phelps C, Eds; Berczi I, Szentivanyi A, Series Eds; Amsterdam: Elsevier, 2007 (in this volume); pp. 145–66.
10. Krueger JM. Cytokines and sleep. In Neuroimmune Biology, Volume 6: Cytokines and the Brain. Korneva HA, Phelps C, Eds; Berczi I, Szentivanyi A, Series Eds; Amsterdam: Elsevier, 2007 (in this volume); pp. 213–40.
11. Banks WA. Blood-brain barrier transport of cytokines. In Neuroimmune Biology, Volume 6: Cytokines and the Brain. Korneva HA, Phelps C, Eds; Berczi I, Szentivanyi A, Series Eds; Amsterdam: Elsevier, 2007 (in this volume); pp. 93–108.
12. Florio T, Schettini G. Chemokines, their receptors and significance in brain function. In Neuroimmune Biology, Volume 6: Cytokines and the Brain. Korneva HA, Phelps C, Eds; Berczi I, Szentivanyi A, Series Eds; Amsterdam: Elsevier, 2007 (in this volume); pp. 243–74.
13. Penkowa M, Hidalgo J, Aschner M. Immune and inflammatory responses in the CNS: modulation by astrocytes. In Neuroimmune Biology, Volume 6: Cytokines and the Brain. Korneva HA, Phelps C, Eds; Berczi I, Szentivanyi A, Series Eds; Amsterdam: Elsevier, 2007 (in this volume); pp. 277–88.
14. Suzumura A. Immune response in the brain: glial response and cytokine production. In Neuroimmune Biology, Volume 6: Cytokines and the Brain. Korneva HA, Phelps C, Eds; Berczi I, Szentivanyi A, Series Eds; Amsterdam: Elsevier, 2007 (in this volume); pp. 289–304.
15. Kannan Y, Moriyama M, Nakamura Y. Lymphocytes and adrenergic sympathetic nerve system: the role of cytokines. In Neuroimmune Biology, Volume 6: Cytokines and the Brain. Korneva HA, Phelps C, Eds; Berczi I, Szentivanyi A, Series Eds; Amsterdam: Elsevier, 2007 (in this volume); pp. 307–36.
16. Goehler L. Cytokines in neural signaling to the brain. In Neuroimmune Biology, Volume 6: Cytokines and the Brain. Korneva HA, Phelps C, Eds; Berczi I, Szentivanyi A, Series Eds; Amsterdam: Elsevier, 2007 (in this volume); pp. 337–52.
17. Korneva EA, Kazakova TB. Interleukin-2 gene expression in the CNS cells after stress and antigen application. In Neuroimmune Biology, Volume 6: Cytokines and the Brain. Korneva HA, Phelps C, Eds; Berczi I, Szentivanyi A, Series Eds; Amsterdam: Elsevier, 2007 (in this volume); pp. 353–72.

18. Katafuchi T. Involvement of brain cytokines in stress-induced immunosuppression. In Neuroimmune Biology, Volume 6: Cytokines and the Brain. Korneva HA, Phelps C, Eds; Berczi I, Szentivanyi A, Series Eds; Amsterdam: Elsevier, 2007 (in this volume); pp. 391–401.
19. Phelps C, Chen L-T. Brain response to endotoxin. In Neuroimmune Biology, Volume 6: Cytokines and the Brain. Korneva HA, Phelps C, Eds; Berczi I, Szentivanyi A, Series Eds; Amsterdam: Elsevier, 2007 (in this volume); pp. 435–54.
20. Clarkson AN, Rahman R, Appleton I. Inflammation and autoimmunity as a central theme in neurodegenerative disorders: fact or fiction? Curr Opin Investig Drugs 2004;5(7):706–13.
21. Ketlinsky SA, Kalinina NM. Cytokines in demyelinating disease. In Neuroimmune Biology, Volume 6: Cytokines and the Brain. Korneva HA, Phelps C, Eds; Berczi I, Szentivanyi A, Series Eds; Amsterdam: Elsevier, 2007 (in this volume); pp. 455–84.
22. Dunn A. Cytokines and depression. In Neuroimmune Biology, Volume 6: Cytokines and the Brain. Korneva HA, Phelps C, Eds; Berczi I, Szentivanyi A, Series Eds; Amsterdam: Elsevier, 2007 (in this volume); pp. 485–506.
23. William K. Summers clinical relevance: cytokines in Alzheimer's disease. In Neuroimmune Biology, Volume 6: Cytokines and the Brain. Korneva HA, Phelps C, Eds; Berczi I, Szentivanyi A, Series Eds; Amsterdam: Elsevier, 2007 (in this volume); pp. 507–26.
24. Aubert A. Cytokines and immune-related behaviours. In Neuroimmune Biology, Volume 6: Cytokines and the Brain. Korneva HA, Phelps C, Eds; Berczi I, Szentivanyi A, Series Eds; Amsterdam: Elsevier, 2007 (in this volume); pp. 529–48.
25. Neveu PJ. The production and effects of cytokines depend on brain lateralization. In Neuroimmune Biology, Volume 6: Cytokines and the Brain. Korneva HA, Phelps C, Eds; Berczi I, Szentivanyi A, Series Eds; Amsterdam: Elsevier, 2007 (in this volume); pp. 549–64.
26. Nagy E, Berczi I, Wren GE, Asa SL, Kovacs K. Immunomodulation by bromocriptine. Immunopharmacology 1983;6:231–43.
27. Blalock JE. The immune system as a sensory organ. J Immunol 1984;132(3):1067–70.
28. Besedovsky HO, del Rey AE, Sorkin E. Immune-neuroendocrine interactions. J Immunol 1985;135(2 Suppl):750s–54s.
29. Szentivanyi A, Berczi I, Nyanteh H, Goldman A. Some evolutionary morphoregulatory and functinal aspects of the immune-neuroendocrine circuitry. In Neuroimmune Biology, Volume 3: The Immune-Neuroendocrine Circuitry. History and Progress. Berczi I, Szentivanyi A, Eds; Amsterdam: Elsevier, 2003; pp. 31–61.
30. Gottfried-Blackmore A, Croft GF, Karen Bulloch K. Sex hormones and cytokines in CNS pathology and repair. In Neuroimmune Biology, Volume 6: Cytokines and the Brain. Korneva HA, Phelps C, Eds; Berczi I, Szentivanyi A, Series Eds; Amsterdam: Elsevier, 2007 (in this volume); pp. 373–90.
31. Stewart R, Przyborski S. Non-neural adult stem cells: tools for brain repair? Bioessays 2002;24(8):708–13.
32. Correale J and colleagues. Neuroprotection by inflammation. In Neuroimmune Biology, Volume 6: Cytokines and the Brain. Korneva HA, Phelps C, Eds; Berczi I, Szentivanyi A, Series Eds; Amsterdam: Elsevier, 2007 (in this volume); pp. 403–32.
33. Thiele HG, Stark R, Keeser D. Antigenic correlations between brain and thymus. I. Common antigenic structures in rat and mouse brain tissue and thymocytes. Eur J Immunol 1972;2:424–29.
34. Berczi I, Szentivanyi A. Adhesion molecules. In Neuroimmune Biology, Volume 3: The Immune-Neuroendocrine Circuitry. History and Progress. Berczi I, Szentivanyi A, Eds; Amsterdam: Elsevier, 2003; pp. 99–115.

35. Wiranowska M, Plaas A. Cytokines and extracellular matrix remodeling in the central nervous system. In Neuroimmune Biology, Volume 6: Cytokines and the Brain. Korneva HA, Phelps C, Eds; Berczi I, Szentivanyi A, Series Eds; Amsterdam: Elsevier, 2007 (in this volume); pp. 167–198.
36. Paul WE, Ed. Fundamental Immunology. New York: Lippincott-Raven, 1999.

Preface

Cytokines have been discovered within the immune system and originally were named interleukins because these mediators were perceived to mediate chemical messages amongst leukocytes of the immune system. Later it was recognized that other somatic cells are also capable of lymphokine production, and the term *cytokine* has been adapted to describe these mediators. A more recent term describing mediators of chemotactic and pro-inflammatory potential is *chemokine*. Virtually every cell type produces these mediators all over the body, especially when suffering from various insults. So far the participation of cytokines and chemokines in immune and inflammatory reactions has been studied extensively [1].

This volume discusses current information on the role of cytokines and chemokines in brain physiology and pathophysiology. It is clear from the available evidence that these mediators fulfil physiological functions as well as they are heavily involved in host defence under pathological conditions. Brain cells produce their own cytokines, which is subject to abrupt changes under the influence of stress and of various pathological stimuli. Under pathological conditions the immune system penetrates the brain and contributes in a major way to the cytokine response in this organ.

Current experimental data and their interpretation are presented on the role of cytokines/chemokines in the brain physiology by several authors. Of the pathological conditions stress, depression, CNS pathology and repair, endotoxin shock, demyelinating diseases, and Alzheimer's disease are discussed. Behavior is also altered by cytokines, which is presented. Recently the inflammatory response was proven to exert a protective effect on the nervous system. This new information is presented in a separate chapter.

The Editors of this book hope that Basic and Clinical Scientists will find it to be useful for acquiring information on the subject area and also that this book will serve as a valuable source of reference.

Istvan Berczi

REFERENCE

1. Berczi I, Szentivanyi A. Cytokines and chemokines. In Neuroimmune Biology, Volume 3: The Immune-Neuroendocrine Circuitry. History and Progress. Berczi I, Szentivanyi A, Editors, Elsevier, Amsterdam, 2003; pp. 191–220.

In Memoriam – Christopher P. Phelps, PhD

Christopher P. Phelps, PhD, one of the editors of this volume, died unexpectedly on January 27, 2008. He was 64. He is remembered by friends and colleagues for his leadership, professionalism, and grace under pressure.

Chris joined the department in the University of South Florida College of Medicine as Assistant Professor in 1976. He rose to the rank of Professor in the department, and served as its Interim Chair from 1998 to 2001, and then as Chair from 2001 through 2006. He received his PhD in endocrinology-zoology from Rutgers University and was a Postdoctoral Fellow in Anatomy at the UCLA School of Medicine and Brain Research Institute.

His long-standing interest in the hypothalamic–pituitary–adrenal axis began when he was an undergraduate at Lafayette College in Pennsylvania, where he wrote an honors thesis on effects of overcrowding on ovaries of the Mexican swordtail fish, *Xiphophorus helleri.* In Charles Sawyers' laboratory at UCLA, Chris examined the effects of hypothalamic deafferentation on the release of luteinizing hormone, ovulation, and sexual behavior. He continued this line of research at USF, where he also made contributions examining Neuropeptide Y and the hypothalamic control of feeding behavior. Two decades ago, Chris became interested in the interaction of cytokines and the hypothalamic–pituitary–adrenal axis, and in Neuroimmune Biology, which he pursued until his death.

Samuel Saporta

Obituary

Andor Szentivanyi (1925–2005)

Dr. Andor Szentivanyi passed away on October 22nd 2008. Fifty six years ago Andor Szentivanyi and colleagues were the first to document with exact scientific methodology in animal experiments that the nervous system has a sweeping regulatory power over immune reactions. At that time Szentivanyi was a resident at the Medical School of Debrecen in Hungry, when he observed that adrenaline was ineffective to inhibit an asthmatic attack in a patient. This clinical observation inspired him and his colleagues to do animal experiments using anaphylactic shock as a model system. In these experiments it was observed that hypothalamic lesions inhibited the development of anaphylactic shock in immunized animals [1]. Hypothalamic tuberal cinereum lesions (TBL) inhibited anaphylaxis in pre-immunized guinea pigs and in later experiments also in rabbits. Anaphylaxis was elicited in immunized animals by the intravenous injection of the immunizing antigen. Antibody production was also inhibited if TBL was done prior to immunization. Such lesions did not affect the reaction of antibodies with the specific antigen, nor did the release of tissue materials mediating anaphylaxis. Hypothalamic lesions temporarily increased the resistance of the animals to histamine and inhibited the anaphylactic reaction even when the animals were provided with passively transferred antibodies, which elicited lethal shock in control animals. The Schultz-dale test, which was performed with small pieces of intestine *in vitro,* was also inhibited by TBL. The Arthus reaction, turpentine-induced inflammation, and the Sanarelli–Schwartzmann phenomenon were unaffected. Lesions of other areas of the hypothalamus or of the central nervous system were ineffective in modulating immune phenomena. Furthermore, electrical stimulation of the mammillary region of the hypothalamus had an inhibitory effect on the anaphylactic response and increased the resistance of animals to histamine [2–4]. In 1964

Korneva and Khai [5] confirmed that hypothalamic lesions in rabbits, guinea pigs, and rats inhibited the production of complement fixing antibodies.

Szentivanyi devoted his entire career of 56 years to research on the pathomechanisms of allergy and asthma. His animal experiments pointed to the importance of the beta adrenergic receptor (bADR) in anaphylactic reactions [6]. In 1968 Szentivanyi had synthesized the contemporary knowledge and all his findings in a review article [7]. He concluded that bronchial asthma, whether it is due to "extrinsic" or "intrinsic" causes, is ultimately elicited by the same mediators, such as histamine, serotonin, catecholamines, slow reactive substances plus cytokines. These are released during asthmatic reactions and should be considered as an additional group of mediators in many tissues and in most species. Glucocorticoids are natural inhibitors of inflammation. He proposed that the fundamental mechanism underlying the atopic abnormality in asthma is the abnormal function of the beta-adrenergic system, irrespective of what triggered the reaction. He concludes:

> *The beta adrenergic theory regards asthma not as an 'immunological disease, but as a unique pattern of bronchial hypersensitivity to a broad spectrum of immunological, psychic, infectious, chemical and physical stimuli. This gives to the antigen-antibody interaction the same role as that of a broad category of non-specific stimuli, which function only to trigger the same defective homeostatic mechanism in the various specialized cells of bronchial tissue [7].*

This article became a citation classic and has been published again as a classical article of lasting value in "Milestones in Allergy," L Berrens, ed., Mosby/Doyma Libros, Barcelona, Spain, in 1995 [8]. Indeed this paper has popularized the dominant role of the central nervous system in inflammatory diseases and of the importance of multi-disciplinary approaches for the understanding of such diseases.

Szentivanyi remained faithful to the idea of beta-adrenergic malfunction in atopy and asthma for his entire scientific career. This was the common thread that connected the numerous papers, reviews, book chapters, and books he published. He studied alpha and beta-adrenergic receptors; adenylate cyclase, cyclic-AMP, and signal transduction; isolated, characterized, and pharmacologically modulated phosphodiesterase; observed the systemic effect of immunization and of endotoxin on the adrenergic and cholinergic systems, on metabolism and on immune inflammatory mediators; performed clinical studies on asthma and related conditions. His major observations are

1. Beta-adrenergic sub-sensitivity exists in patients with atopic dermatitis who never received adrenergic medication. This indicates that therapeutic desensitization cannot account for the dysfunction of the beta-adrenergic system [9].
2. The beta-adrenergic reactivity of lung tissue of lymphocytes and bronchocytes from patients with atopic asthma was found to be abnormal and various patterns of drug versus disease-induced sub-sensitivity could be recognized [10–15].
3. Bronchial hyper-reactivity to cholinergic agents in asthma was not mediated through cholinergic mechanisms, but it was caused by the adrenergic abnormality, which was due to the so called "denervation super-sensitivity" [16–19].
4. Lymphocytes of asthmatic patients showed a significant decrease in adrenaline binding to beta-adrenergic receptors, which was independent of therapy [11,12,15].

Szentivanyi also studied the effects of inflammation on beta-adrenergic receptors. He studied glucocorticoid (GC) action in disease and regarded GC as important inhibitory hormones [20–24]. The pro-inflammatory cytokine, interleukin-1α (IL-1α) was found to bind to human bronchial epithelial cells and induced the production of beta adrenoreceptor mRNA [25]. Beta adrenoreceptor had an upregulatory effect on IL-1 α secretion by lymphocytes *in vitro*, which showed cell and species specificity [26]. It was discovered that immunoglobulin E (IgE) and beta adrenergic stimulation interacted in the regulation of cyclic AMP concentrations in A549 human pulmonary epithelial cells [27]. IgE decreases endothelin-1 production in A549 cells [28].

Szentivanyi's work and ideas were a major source of inspiration for me. In 1992 the World Congress of Immunology was held in Budapest, and I was invited to organize a satellite symposium on Psychoneuroimmunology. I organized this meeting with Judit Szelényi in Budapest, and we invited Szentivanyi as a Guest of Honor to deliver the keynote lecture to the symposium [29]. This was the beginning of our collaboration and friendship, which intensified over the years [30]. In 2000 we have initiated the book series, entitled *Neuroimmune Biology* and worked together on nine volumes till his passing away [31–66]. He was to write a Preface to Volume 9 as yet, but this commitment could not be fulfilled. He was very proud of *Neuroimmune Biology* and anticipated a bright future for these books.

Szentivanyi lists nearly 500 publications and 29 books to his credit. Throughout his lifetime he received numerous awards and distinctions and has been advisor/consultant to many scientific and public organizations. He was one of the founders of the Faculty of Medicine at the University of South Florida and built and organized many departments and Institutions associated with this School. He served this University with full devotion for over 35 years.

Andor was a humanist, a pacifist, a tireless champion of the poor and disadvantaged; he fought against social and political injustice and was always ready to take risks and sacrifices for his principles. His passing is a great loss to modern biomedical science, and indeed, to mankind as well.

Istvan Berczi

REFERENCES

1. Filipp G, Szentivanyi A, Mess B. Anaphylaxis and nervous system. Acta Med Hung 1952;2:163–73.
2. Szentivanyi A, Filipp G. Anaphylaxis and the nervous system. Part II. Ann Allergy 1958;16:143–51.
3. Filipp G, Szentivanyi A. Anaphylaxis and the nervous system. Part III. Allergy 1958;16:306–11.
4. Szentivanyi A, Szekely J. Effect of injury to, and electrical stimulation of, hypothalamic areas on anaphylactic and histamine shock of the guinea pig. A preliminary report. Ann Allergy 1956;14:259–61.
5. Korneva EA, Khai LM. Effect of destruction of hypothalamic areas on immunogenesis. Fed Proc 1964;23:T88.
6. Townley RG, Trapani IL, Szentivanyi A. Sensitization to anaphylaxis and to some of its pharmacological mediators by blockade of the beta adrenergic receptors. J Allergy 1967;39(3):177–79.
7. Szentivanyi A. The beta adrenergic theory of the atopic abnormality in bronchial asthma. J. Allergy 1968;42:203–32.
8. Szentivanyi, A. The beta-adrenergic theory of the atopic abnormality in bronchial asthma. Reprinted In Milestones in Allergy. Berrens L, Ed.; Barcelona, Spain: Mosby/Doyma Libros, 1995; pp. 69–102.

9. Szentivanyi A, Szentivanyi J, Wagner H. Measurement of numbers of adrenoceptors in lymphocytes and lung tissue of patients with reversible obstructive airways disease. Clin Pharmacol Ther 1980;27:193–206.
10. Szentivanyi A. The conformational flexibility of adrenoceptors and the constitutional basis of atopy. Triangle 1979;18(4):109–15.
11. Szentivanyi A, Heim O, Schultze P. Changes in adrenoceptor densities in membranes of lung tissue and lymphocytes from patients with atopic disease. Ann NY Acad Sci 1979;332:295–98.
12. Szentivanyi A. La flexibilite de conformation des adrenocepteurs et la base constitutionelle du terrain allergique. Rev Franc Allergol 1979;19:205–14.
13. Szentivanyi A, Fitzpatrick DF. The altered reactivity of the effector cells to antigenic and pharmacological influences and its relation to cyclic nucleotides. II. Effector reactivities in the efferent loop of the immune response. In Pathomechanismmus und Pathogenese Allergischer Reaktionen. Filipp G, Ed.; Werk-Verlag Dr. Edmund Banachewski, Grafelfing bei Munchen 1980; pp. 511–80.
14. Szentivanyi A, Polson JB, Krzanowski JJ. The altered reactivity of the effector cells to antigenic and pharmacological influences and its relation to cyclic nucleotides. I. Effector reactivities in the efferent loop of the immune response. In Pathomechanismus und Pathogenese Allergischer Reaktionen. Filipp G, Ed.; Werk-Verlag, Dr. Edmund Banachewski, Grafelfing bei Munchen, 1980; pp. 460–510.
15. Szentivanyi A, Krzanowski JJ, Polson JB, Anderson WH. Evolution of research strategy in the experimental analysis of the beta adrenergic approach to the constitutional basis of atopy. In Advances in Allergology and Clinical Immunology. Oehling A, Mathov E, Glazer I, Arbesman C, Eds; Oxford: Pergamon Press, 1980; pp. 301–8.
16. Szentivanyi A. Effect of bacterial products and adrenergic blocking agents on allergic reactions. In Textbook of Immunological Diseases. Samter M, Talmage DW, Rose B, Sherman WB, Vaughan JH, Eds; Boston, MA: Little, Brown and Co., 1971; pp. 356–74.
17. Szentivanyi A, Krzanowski JJ, Polson JB. The autonomic nervous system: structure, function, and altered effector responses. In Allergy: Principles and Practice. Middleton E, Reed CE, Ellis EF, Eds; St. Louis, MO: The CV Mosby Co, 1978; pp. 256–300.
18. Szentivanyi A, Williams JF. The constitutional basis of atopic disease. In Allergic Diseases of Infancy, Childhood, and Adolescence. Bierman CW, Pearlman DS, Eds; Philadelphia, PA: WB Saunders Co., 1980; pp. 173–210.
19. Szentivanyi A. Adrenergic and cholinergic receptor studies in human lung and lymphocytic membranes and their relation to bronchial hyperreactivity in asthma. In: Patient Care Publications. Darien, CT, 1982; pp. 175–92.
20. Hackney JF, Szentivanyi A. The specificity of glucocorticoids in the relaxation of respiratory smooth muscle in vitro. J Allergy Clin Immunol 1975;55:123 (abstract).
21. Hackney JF, Szentivanyi A. The unique action of glucocorticoid succinates on respiratory smooth muscle in vitro. Pharmacologist 1975;17:271 (abstract).
22. Lowitt S, Szentivanyi A, Williams JF. Endotoxin inhibition of dexamethasone induction of tryptophan oxygenase in suspension culture of isolated rat parenchymal cells. II. Effect of in vivo pretreatment of rats with endotoxin. Biochem Pharmacol 1982;31:3403–6.
23. Szentivanyi A, Szentivanyi J. Mechanisms of action of corticosteroids. In Proceedings of the International Symposium on Allergy and Immunology. Lima, Peru, 1985; pp. 531–34.
24. Szentivanyi A, Szentivanyi J. Mechanisms of action of corticosteroids. In Proceedings of the International Symposium on Allergy and Immunoloty. Lima, Peru, 1985; pp. 227–41.

25. Robicsek S, Szentivanyi, A, Calderón, EG, Heim, O, Schultze, P, Wagner, H, Lockey, RF, and Dwornik, JJ. Concentrated IL-1a derived from human T-lymphocytes binds to a specific single class surface receptor on human bronchial epithelial cells and induces the production of beta-adrenocep tor mRNA via an associated or separate receptor-linked signalling pathway. J Allergy Clin Immunol 1992;89:212.
26. Szentivanyi A, Calderón EG, Heim O, Schultze P, Wagner H, Zority J, Lockey RF, Dwornik JJ, Robicsek S. Cell- and species-specific dissociation in the beta-adrenoceptor upregulating effects of IL-1a derived from lymphocyte conditioned medium and cortisol. J Allergy Clin Immunol 89:274;1992.
27. Szentivanyi A, Ali K, Calderón EG, Brooks SM, Coffey RG, Lockey RF. The in vitro effect of immunoglobulin E (IgE) on cyclic AMP concentrations in A549 human pulmonary epithelial cells with or without beta-adrenergic stimulation. J Allergy Clin Immunol 1993;91:379.
28. Stewart, GE, Calderón, E, Szentivanyi, A, Lockey, RF. IgE decreases endothelin-1 production in A549 pulmonary epithelial cells. J Allergy Clin Immunol 1994;93:244.
29. Szentivanyi A, Abarca C. The immune-neuroendocrine circuitry – the next, and possibly, the last frontier of vertebrate immunity. In Advances in Psychoneuroimmunology, Berczi I, Szelényi J, Eds; New York: Plenum Press, 1994; pp. 41–74.
30. Berczi I, Szentivanyi A. The pituitary gland, psychoneuroimmunology and infection. In Psychoneuroimmunology, Stress, and Infection. Friedman H, Klein TW, Friedman A, Eds; Boca Raton, FL: CRC Press, 1996; pp. 71–98.
31. Berczi I, Szentivanyi A, Series editors, Berczi I, Gorczynski R, Eds. Neuroimmune Biology Volume 1. Amsterdam: New Foundation of Biology, Elsevier, 2001.
32. Berczi I, Szentivanyi A. Why neuroimmune biology? Editorial. In Neuroimmune Biology Volume 1: New foundation of Biology. Berczi I, Gorczynski R, Eds; Elsevier, 2001; pp. vii–xii.
33. Szentivanyi A, Berczi I, Pitak D, Goldman A. Studies of the hypothalamic regulation of histamine synthesis. In Neuroimmune Biology Volume 1: New foundation of Biology I. Berczi I, Gorczynski R, Eds; Elsevier, 2001; pp. 47–55.
34. Berczi I, Szentivanyi A, Series Editors, Matera L, Rapaport R, Eds. Neuroimmune Biology Volume 2: Growth and Lactogenic Hormones. Amsterdam: Elsevier, 2002.
35. Berczi I, Szentivanyi A, Series Editors and Editors. Neuroimmune Biology, Volume 3: The Immune-Neuroendocrine Circuitry. History and Progress. Amsterdam: Elsevier, 2003.
36. Berczi I, Szentivanyi A, Eds. Foreword. In Neuroimmune Biology, Volume 3: The Immune-Neuroendocrine Circuitry. History and Progress. Amsterdam: Elsevier, 2003; p. v.
37. Berczi I, Szentivanyi A, Eds. Preface. In Neuroimmune Biology, Volume 3: The Immune-Neuroendocrine Circuitry. History and Progress. Amsterdam: Elsevier, 2003; pp. vii–viii.
38. Szentivanyi A, Berczi I, Nyanteh H, Goldman A. Introduction. In Neuroimmune Biology, Volume 3: The Immune-Neuroendocrine Circuitry. History and Progress. Berczi I, Szentivanyi A, Eds; Amsterdam: Elsevier, 2003; p. 5.
39. Szentivanyi A, Berczi I, Nyanteh H, Goldman A. History. In Neuroimmune Biology, Volume 3: The Immune-Neuroendocrine Circuitry. History and Progress. Berczi I, Szentivanyi A, Eds; Amsterdam: Elsevier, 2003; pp. 7–14.
40. Szentivanyi A, Berczi I, Nyanteh H, Goldman A. The discovery of immune neuroendocrine circuitry – A generation of progress. In Neuroimmune Biology, Volume 3: The Immune-Neuroendocrine Circuitry. History and Progress. Berczi I, Szentivanyi A, Eds; Amsterdam: Elsevier, 2003; pp. 15–18.

41. Szentivanyi A, Berczi I, Nyanteh H, Goldman A. Altered effector responses. In Neuroimmune Biology, Volume 3: The Immune-Neuroendocrine Circuitry. History and Progress. Berczi I, Szentivanyi A, Eds; Amsterdam: Elsevier, 2003; pp. 21–29.
42. Szentivanyi A, Berczi I, Nyanteh H, Goldman A. Some evolutionary morphoregulatory and functinal aspects of the immune-neuroendocrine circuitry. In Neuroimmune Biology, Volume 3: The Immune-Neuroendocrine Circuitry. History and Progress. Berczi I, Szentivanyi A, Eds, Amsterdam: Elsevier, 2003; pp. 31–61.
43. Szentivanyi A, Berczi I, Nyanteh H, Goldman A. Virus associated immune and pharmacologic mechanisms in disorders of respiratory and cutaneous atopy. In Neuroimmune Biology, Volume 3: The Immune-Neuroendocrine Circuitry. History and Progress. Berczi I, Szentivanyi A, Eds; Amsterdam: Elsevier, 2003; pp. 63–95.
44. Berczi I, Szentivanyi A. Adhesion molecules. In Neuroimmune Biology, Volume 3: The Immune-Neuroendocrine Circuitry. History and Progress. Berczi I, Szentivanyi A, Eds; Amsterdam: Elsevier, 2003; pp. 99–115.
45. Berczi I, Szentivanyi A. Immunoglobulins. In Neuroimmune Biology, Volume 3: The Immune-Neuroendocrine Circuitry. History and Progress. Berczi I, Szentivanyi A, Eds; Amsterdam: Elsevier, 2003; pp. 117–27.
46. Berczi I, Szentivanyi A. Growth and lactogenic hormones, insulin-like growth factor and insulin. In Neuroimmune Biology, Volume 3: The Immune-Neuroendocrine Circuitry. History and Progress. Berczi I, Szentivanyi A, Eds; Amsterdam: Elsevier, 2003; pp. 129–53.
47. Berczi I, Szentivanyi A, Eds. The hypothalamus-pituitary-adrenal axis and opioid peptides. In Neuroimmune Biology, Volume 3: The Immune-Neuroendocrine Circuitry. History and Progress. Amsterdam: Elsevier, 2003; pp. 155–74.
48. Berczi I, Szentivanyi A, Eds. The hypothalamus-pituitary-thyroid axis. In Neuroimmune Biology, Volume 3: The Immune-Neuroendocrine Circuitry. History and Progress. Amsterdam: Elsevier, 2003; pp. 175–80.
49. Berczi I, Szentivanyi A, Eds. Nerve growth factor, leptin and neuropeptides. In Neuroimmune Biology, Volume 3: The Immune-Neuroendocrine Circuitry. History and Progress. Amsterdam: Elsevier, 2003; pp. 181–89.
50. Berczi I, Szentivanyi A, Eds. Cytokines and chemokines. In Neuroimmune Biology, Volume 3: The Immune-Neuroendocrine Circuitry. History and Progress. Amsterdam: Elsevier, 2003; pp. 191–220.
51. Berczi I, Nagy E, Baral E, Szentivanyi, A. Sterid Hormones. In Neuroimmune Biology, Volume 3: The Immune-Neuroendocrine Circuitry. History and Progress. Berczi I, Szentivanyi A, Eds; Amsterdam: Elsevier, 2003; pp. 221–70.
52. Berczi I, Szentivanyi A, Eds. Immunocompetence. In Neuroimmune Biology, Volume 3: The Immune-Neuroendocrine Circuitry. History and Progress. Amsterdam: Elsevier, 2003; pp. 281–99.
53. Berczi I, Szentivanyi A, Eds. Antigen presentation. In Neuroimmune Biology, Volume 3: The Immune-Neuroendocrine Circuitry. History and Progress. Amsterdam: Elsevier, 2003; pp. 301–13.
54. Berczi I, Szentivanyi A, Eds. Immune reactions. In Neuroimmune Biology, Volume 3: The Immune-Neuroendocrine Circuitry. History and Progress. Amsterdam: Elsevier, 2003; pp. 315–77.
55. Berczi I, Szentivanyi A, Eds. The acute phase response. In Neuroimmune Biology, Volume 3: The Immune-Neuroendocrine Circuitry. History and Progress. Amsterdam: Elsevier, 2003; pp. 463–94.

56. Berczi I, Szentivanyi A, Eds. Autoimmune disease. In Neuroimmune Biology, Volume 3: The Immune-Neuroendocrine Circuitry. History and Progress. Amsterdam: Elsevier, 2003; pp. 495–536.
57. Berczi I, Szentivanyi A, Eds. Immunodeficiency. In Neuroimmune Biology, Volume 3: The Immune-Neuroendocrine Circuitry. History and Progress. Amsterdam: Elsevier, 2003; pp. 537–58.
58. Berczi I, Szentivanyi A. The immune-neuroendocrine circuitry. In Neuroimmune Biology, Volume 3: The Immune-Neuroendocrine Circuitry. History and Progress. Amsterdam: Elsevier, 2003; pp. 561–92.
59. Berczi I, Szentivanyi A, Series editors, Straub RH, Mocchegiani E, Eds. Neuroimmune Biology, Volume 4; The Neuroendocrine Immune Network in Ageing. Amsterdam: Elsevier, 2004.
60. Berczi I, Szentivanyi A. Concluding remarks and future directions. In The Neuroendocrine Immune Network in Ageing. Straub RH, Mocchegiani E, Eds; Neuroimmune Biology Volume 4. Berczi I, Szentivanyi A, Series Eds; Amsterdam: Elsevier, 2004; pp. 409–27.
61. Bertok L, Chow DA, Eds. Natural Immunity. Neuroimmune Biology Volume 5. Berczi I, Szentivanyi A, Series Eds; Amsterdam: Elsevier, 2005.
62. Berczi I, Szentivanyi A. Foreword. In Cytokines and the Brain. Phelps C, Korneva E, Eds, Neuroimmuine Biology, Volume 6. Berczi I, Szentivanyi A, Series Eds; Amsterdam: Elsevier, 2007 (in this volume); pp. ix–xiii.
63. Berczi I, Szentivanyi A. Preface. In The Hypothalamus-Pituitary-Adrenal Axis. DelRey A, Chrousos G, Besedovsky H, Eds, Neuroimmune Biology, Volume 7. Berczi I, Szentivanyi A, Series Eds; Amsterdam: Elsevier, 2006 (submitted); pp. i–400.
64. DelRey A, Chrousos G, Besedovsky H, Eds. The Hypothalamus-Pituitary-Adrenal Axis. Neuroimmune Biology, Volume 7. Berczi I, Szentivanyi A, Series Eds. Amsterdam: Elsevier, 2006 (submitted); pp. i–400.
65. Berczi I, Szentivanyi A. Preface. In The Hypothalamus-Pituitary-Adrenal Axis. DelRey A, Chrousos G, Besedovsky H, Eds, Neuroimmune Biology, Volume 7. Berczi I, Szentivanyi A, Series Eds. Amsterdam: Elsevier, 2006 (submitted); pp. i–400.
66. Jancso G. Neurogenic inflammation in health and disease. In Neuroimmune Biology, Volume 8. Berczi I, Szentivanyi A, Series Eds; Amsterdam: Elsevier, 2006 (submitted).
67. Berczi I, Szentivanyi A. Foreword: Neurogenic Inflammation Coming of Age. In Neuroimmune Biology, Volume 8. Berczi I, Szentivanyi A, Series Eds; Amsterdam: Elsevier, 2006 (submitted).
68. Berczi I, Arnason B, Buckingham J, Eds. The brain and host defense. In Neuroimmune Biology, Volume 9, Berczi I, Szentivanyi A, Series Eds; Amsterdam: Elsevier (in preparation).

List of Contributors

Michael Aschner
Department of Pediatrics, Vanderbilt University Medical Center, Nashville, TN 37232, USA

Arnaud Aubert
Faculté des Sciences Parc de Grandmont, Tours, France

William A. Banks
GRECC, Veterans Affairs Medical Center, St. Louis and Saint Louis University School of Medicine, Division of Geriatrics, Department of Internal Medicine, University at Buffalo, The State University of New York, Buffalo, NY, USA

Tamas Bartfai
The Harold L. Dorris Neurological Research Center, Department of Molecular and Integrative Neuroscience, The Scripps Research Institute, La Jolla, CA, USA

Margarita Behrens
The Harold L. Dorris Neurological Research Center, Department of Molecular and Integrative Neuroscience, The Scripps Research Institute, La Jolla, CA, USA

Istvan Berczi
Department of Immunology, Faculty of Medicine, the University of Manitoba, Winnipeg, Manitoba, Canada

Sara Brownell
The Harold L. Dorris Neurological Research Center, Department of Molecular and Integrative Neuroscience, The Scripps Research Institute, La Jolla, CA, USA

Karen Bulloch
The Rockefeller University, Lab. of Neuroendocrinology, New York, NY, USA

Léa Chaskiel
PsychoNeuroImmunology, Nutrition and Genetics, UMR CNRS 5526/UMR INRA 1286/ University Bordeaux 2, UFR Pharmacy, University Victor Ségalen Bordeaux 2, 146, rue Léo Saignat, 33076 Bordeaux Cedex, France

Li-Tsun Chen
Department of Pathology and Cell Biology and the Neuroscience Program, College of Medicine, University of South Florida, Tampa, FL, USA

Lynn Churchill
Program in Neuroscience, Washington State University, Pullman, WA, USA

Bruno Conti
The Harold L. Dorris Neurological Research Center, Department of Molecular and Integrative Neuroscience, The Scripps Research Institute, La Jolla, CA, USA

Jorge Correale
Department of Neurology, Raúl Carrea Institute for Neurological Research, FLENI; School of Biological Sciences, Austral University, Buenos Aires, Argentina

Gist F. Croft
The Rockefeller University, Lab. of Neuroendocrinology, New York, NY, USA

Christopher N. Davis
The Harold L. Dorris Neurological Research Center, Department of Molecular and Integrative Neuroscience, The Scripps Research Institute, La Jolla, CA, USA

Errol B. De Souza
Archemix Corp., Cambridge, MA, USA

Adrian J. Dunn
Department of Pharmacology, Toxicology and Neuroscience, Louisiana State University Health Sciences Center, Shreveport, LA, USA

Marcela Fiol
Department of Neurology, Raúl Carrea Institute for Neurological Research, FLENI; School of Biological Sciences, Austral University, Buenos Aires, Argentina

Tullio Florio
Section of Pharmacology, Department of Oncology, Biology and Genetic, University of Genoa, Genoa, Italy

Lisa E. Goehler
Department of Psychology, University of Virginia, Charlottesville, VA, USA

Andrés Gottfried-Blackmore
The Rockefeller University, Lab. of Neuroendocrinology, Box 165,1230 York Ave, New York, New York 10021, USA

Uwe-Karsten Hanisch
Department of Neuropathology, University of Göttingen, Göttingen, Germany

Kozo Hashimoto
Department of Endocrinology, Metabolism and Nephrology, Kochi Medical School, Nankoku, Japan

Juan Hidalgo
Institute of Neurosciences and Department of Cellular Biology, Physiology and Immunology, Animal Physiology Unit, Faculty of Sciences, Autonomous University of Barcelona, Bellaterra, Barcelona, Spain

Tracey A. Ignatowski
Department of Pathology and Anatomical Sciences and Department of Anesthesiology, School of Medicine and Biomedical Sciences, University at Buffalo, The State University of New York, Buffalo, NY, USA

Natalia M. Kalinina
All-Russian Center of Emergency and Radiation Medical Research, St. Petersburg, Russia

Yukiko Kannan-Hayashi
Laboratory of Integrative Physiology, Division of Veterinary Sciences, Graduate School of Life and Environmental Sciences, Osaka Prefecture University, Sakai, Department of Human Life Sciences, Baika Junior College, 2-19-5, Shukunosho, Ibaraki, Osaka 567-8578, Japan

Toshihiko Katafuchi
Department of Integrative Physiology, Graduate School of Medical Sciences, Kyushu University, Fukuoka, Japan

T. B. Kazakova
State Organization "Institute for Experimental Medicine of Russian Academy of Medical Sciences," Department of General Pathology and Pathophysiology, St. Petersburg, Russia

Sergey A. Ketlinsky
State Research Center of Highly Pure Biopreparations, FMBA, St. Petersburg, Russia

Jan Pieter Konsman
PsychoNeuroImmunology, Nutrition and Genetics, UMR CNRS 5526/UMR INRA 1286/ University Bordeaux 2, UFR Pharmacy, University Victor Ségalen Bordeaux 2, 146, rue Léo Saignat, 33076 Bordeaux Cedex, France

Elena A. Korneva
Department of General Pathology and Pathophysiology, State Organization Institute for Experimental Medicine of Russian Academy of Medical Science, Saint Petersburg, Russia

James M. Krueger
Program in Neuroscience, Washington State University, Pullman, WA, USA

Mitsuaki Moriyama
Laboratory of Integrative Physiology, Division of Veterinary Sciences, Graduate School of Life and Environmental Sciences, Osaka Prefecture University, Sakai, Department of Human Life Sciences, Baika Junior College, 2-19-5, Shukunosho, Ibaraki, Osaka 567-8578, Japan

Yoichi Nakamura
Laboratory of Integrative Physiology, Division of Veterinary Sciences, Graduate School of Life and Environmental Sciences, Osaka Prefecture University, Sakai, Department of Human Life Sciences, Baika Junior College, 2-19-5, Shukunosho, Ibaraki, Osaka 567-8578, Japan

Pierre J. Neveu
Neurobiologie Integrative, Institut François Magendie, Rue Camille Saint-Saëns, Bordeaux, France

Yutaka Oomura
Department of Integrated Physiology, Kyushu University, Faculty of Medicine, Japan

Christopher Phelps
Department of Pathology and Cell Biology and the Neuroscience Program, College of Medicine, University of South Florida, Tampa, FL, USA

Anna Plaas
Department of Internal Medicine (Rheumatology) and Department of Biochemistry, Rush University Medical Center, Chicago, IL, USA

Milena Penkowa
Department of Medical Anatomy, The Panum Institute, University of Copenhagen, Copenhagen, Denmark

Marco Prinz
Department of Neuropathology, University of Göttingen, Göttingen, Germany

David M. Rector
Program in Neuroscience, Washington State University, Pullman, WA, USA

Julien Renault
Faculté des Sciences Parc de Grandmont, Tours, France

Elena G. Rybakina
Department of General Pathology and Pathophysiology, State Organization Institute for Experimental Medicine of Russian Academy of Medical Science, Saint Petersburg, Russia

Manuel Sanchez-Alavez
The Harold L. Dorris Neurological Research Center, Department of Molecular and Integrative Neuroscience, The Scripps Research Institute, La Jolla, CA, USA

Gennaro Schettini
Pharmacology, School of Medicine, University of Genova, Viale Benedetto XV2, 16100 Genova, Italy

Robert N. Spengler
Department of Pathology and Anatomical Sciences and Department of Anesthesiology, School of Medicine and Biomedical Sciences, University at Buffalo, The State University of New York, Buffalo, NY, USA

William K. Summers
Solo Research Ltd. Alzheimer's Corporation

Andor Szentivanyi
Department of Internal Medicine, College of Medicine, the University of South Florida, Tampa, FL, USA

Akio Suzumura
Department of Neuroimmunology, Research Institute of Environmental Medicine, Nagoya University, Chikusa, Nagoya, Japan

Iustin Tabarean
The Harold L. Dorris Neurological Research Center, Department of Molecular and Integrative Neuroscience, The Scripps Research Institute, La Jolla, CA, USA

Toshihiro Takao
Division of Community Medicine, Department of Community Nursing, Kochi Medical School, Nankoku, Japan

Denise Van Rossum
Department of Neuropathology, University of Göttingen, Göttingen, Germany

Andrés Villa
Department of Neurology, José María Ramos Mejía Hospital, School of Medicine, Buenos Aires University, Buenos Aires, Argentina

Marzenna Wiranowska
Departments of Pathology and Cell Biology and Internal Medicine, University of South Florida, Tampa, FL, USA

I. HISTORY

Cytokines and the Brain
Edited by C. Phelps and E. Korneva

The History of Neuroimmune Biology

ISTVAN BERCZI[1] and ANDOR SZENTIVANYI[2]

[1]*Department of Immunology, Faculty of Medicine, the University of Manitoba, Winnipeg, Manitoba, R3E OW3, Canada*
[2]*Department of Internal Medicine, College of Medicine, the University of South Florida, Tampa, FL 33612, USA*

ABSTRACT

It is known since prehistoric times that a healthy body and healthy mind are fundamental to general well-being. Scientific thinking in this area was initiated first during the late eighteenth century, when the role of the nervous system in allergic reactions had been proposed. At about the same time pathologists described that changes in thymus size were associated with serious emotional events and with diseases that alter systemic hormone levels. Hans Selye described in 1936 that various noxious agents activated the ACTH–adrenal axis and caused thymus and lymphoid involution. Szentivanyi[†] and colleagues (1951) observed that hypothalamic lesions suppress the anaphylactic response and of other immune reactions in laboratory animals. Similar observations were made by Korneva and Khai (1965). The immunoregulatory role of various classical hormones has emerged gradually and gained acceptance after systematic studies had been performed in hypophysectomized animals. In 1975 Wannemacker and coworkers discovered an immune-derived molecule, called leukocyte endogenous mediator, which caused fever. This molecule was later identified as interleukin-1 (IL-1), which was demonstrated to regulate pituitary ACTH release in several laboratories. Today it is clear that cytokines as well as nerves (e.g., vagus, sensory nerves) serve as feedback signals toward the central nervous system (CNS) from the sites of immune and inflammatory reactions. The systemic activation of the innate immune system leads to the massive release of tumor necrosis factor-alpha (TNF-α), IL-1, and IL-6 into the circulation and causes fever, the activation of the hypothalamus–pituitary–adrenal axis and induces sympathetic outflow. These alterations cause thymus involution and stimulate the production of natural antibodies by $CD5^+$ B lymphocytes and of acute-phase proteins in the liver, which amplify the natural immune system profoundly within 24–48 h. This acute-phase response represents immunoconversion from adaptive immune defense to the amplification of natural immune host defense. Today a compelling body of experimental evidence indicates that the neuroendocrine and immune systems form a Super System, which plays a fundamental regulatory role in health and disease for the entire life cycle of higher animals and of man.

[†]Dr. Szentivanyi has deceased on October 22nd, 2005

1. INTRODUCTION

That the central nervous system (CNS) has a fundamental role in the maintenance of health has been recognized since prehistoric times, and this is referred to in proverbs of many languages. The healing power of mind and faith provides one of the important foundations of religion and is included into religious texts. The placebo effect has been established in modern medicine to have a beneficial influence on cancer and on other diseases. It has been demonstrated repeatedly by exact scientific methodology that patients treated with placebo in controlled medical trials do in fact show significant improvement clinically in the absence of effective treatment. In ancient Persia, Egypt, and in the Roman Empire fever has been regarded as a reaction with healing power. This view was maintained until modern times. During the early 1900s pyrogenic substances have been developed for fever therapy [1–3].

It is generally assumed that the recognition of the role of the nervous system in immune/inflammatory reactions had been discovered in the 1960s. In actual fact extensive work was published on these matters in the European literature during the late nineteenth and early twentieth century. The earliest investigations can be traced back to the work of Solomonsen and Madsen in 1898 [4], who developed a significant interest in allergy research. This was followed by the publication of the "*vagotonia*" book of Appinger and Hess [5]. This book suggested that the phenomena of anaphylaxis and diseases of atopic allergy are due to an abnormal increase in the rate of firing of the vagus nerve [6–16]. However, this view could not be maintained because of the realization that the relationship between cholinergic and adrenergic tones is in a continuous transition. In other words, the sympathetic and parasympathetic innervation of organs is functionally by no means a constant; it is permanently shifting and seeking adjustments, as a compromise to functional exigencies.

The nature of neural influences in immune and anaphylactic processes has also been investigated in *denervation studies* [17–20]. Hyperergic inflammatory reactions were observed in *denervated* tissues. However, two features of the denervated structures handicapped the validity of these studies: (1) neural influences cannot be entirely eliminated in denervated structures because neural influences can be exercised both through neurohumoral action and by local axonal reflexes and (2) denervated tissues rapidly develop an increased reactivity to the natural chemical mediators of immune reactions as well as to other stimulating substances [21,22]. At the same time a group of Pavlov's former students (Speranski, Brucke, Alpern, Rosle, etc.) introduced the conditional reflex methodology for the analysis of the role of the nervous system in anaphylactic reactions. It is very unfortunate that both the methodology and the interpretation of these studies were completely rejected by other workers [21,22].

In these early experiments there was no attempt to trace back the observed neural influence to a specific site of the brain. This fact influenced Szentivanyi and his associates between 1949 and 1952 to identify the structure responsible for the anti-anaphylactic effect in the hypothalamus, more precisely in the *tuber cinereum*, and the anterior part of the guinea pig hypothalamus [23–26]].

In 1964 Korneva and Khai [27] also described that hypothalamic lesions in laboratory rodents (e.g., rabbits, guinea pigs, rats) inhibited the production of complement fixing antibodies. In 1949 Miklos (Nicholas) Jancso and co-workers discovered that capsaicin is a sensory irritant and that repeated local or systemic administration to rats, mice, and guinea pigs causes desensitization of sensory nerves, which involves interference with pain receptors. Systemic pretreatment of animals with capsaicin or repeated local applications prevented the inflammatory response, indicating the involvement of the nervous system in inflammation. This was confirmed by experiments on denervated tissues. These observations demonstrated the existence of a distinct form of inflammation, which depends on sensory nerve innervation. The stimulation of C-fibers was necessary to induce this inflammatory response. Subsequently, the neurogenic inflammatory response was also demonstrated in man [28,29].

Metalnikov and Chorine first proposed the behavioral modification of the immune response in 1926 [30]. In 1933, Smith and Salinger [31] observed that asthmatic attacks could be provoked in some patients with visual stimuli in the absence of the allergen. That immune reactions may be conditioned in the Pavlovian sense was later demonstrated by Ader [32], MacQueen et al. [33] and by Gorczynski et al. [34]. Various cells in the immune system were also shown to produce classical hormones and neurotransmitters. Smith and Blalock [35], Montgomery et al. [36] and DiMathia et al. [37] pioneered these observations.

2. NEURAL REGULATION OF ALLERGY AND ASTHMA

The analysis of anaphylaxis for the purpose of determining the nature of an atopic cause in all human phenomena of atopic allergy was entirely inappropriate for the following reason: anaphylaxis is an immunological phenomenon, whereas diseases of atopy [bronchial asthma, allergic rhinitis, atopic dermatitis (AD)] all have, in addition to the immunological abnormality, highly increased sensitivity to recognized neurotransmitters and neurohumors in humans (acetylcholine, histamine, serotonin, etc.). Thus, it soon became recognized that *Bordetella pertussis*-treated mice and rats show a critically enhanced reactivity to these neurohumors. Consequently, the pertussis-sensitized mouse and rat represent a far more appropriate model for the investigation of the nature of atopy in men.

Parfentjev and Goodline [38] reported first an increased sensitivity of certain strains of mice to histamine following injection of dead *Bordetella pertussis* organisms. More recently this hypersensitivity has been shown to apply also to serotonin (5-hydroxytryptamine) and to passive anaphylaxis elicited by unrelated antigen–antibody systems, reaching a maximum to all these stimuli at approximately 5 days following the injection of the pertussis organisms. The hypersensitivity then declines and reactivity returns to a normal state.

Pertussis-induced hypersensitivity to histamine and serotonin is a unique experimental model of an acquired hypersensitivity not due to classical immunologic mechanisms. The hypersensitivity is relatively transient; no anamnestic response is apparent, and the histamine and serotonin sensitivity is not transferable by lymphoid cells or serum, the classical agents of the immune response. Some investigators [39,40] have suggested that the pertussis-induced sensitivity might be the result of an impairment in the normal rapid detoxification of histamine and serotonin. In our experiments, however, no change in the excretion pattern of C-14 serotonin or C-14 histamine or their metabolic products could be detected. The urinary elimination pattern of normal mice following the injection of radioactive histamine and serotonin was compared to that of pertussis-treated animals. The absence of gross differences between these two groups suggested that the handling of histamine and serotonin by the body as a whole is not altered in the highly sensitive pertussis-vaccinated mouse. For this reason, it was concluded that the hypersensitivity is the result of a localized hyperactivity of the effector cells to normally nontoxic concentrations of histamine and serotonin. Attempts were made by various neurohumoral agents known to react with the receptors of neuroeffector cells.

Acute exposure to progressively increasing amounts of histamine and serotonin desensitized the pertussis-treated animals to these amines and the desensitization persisted for at least 24 h.

Complete cross-protection against serotonin was also obtained with histamine pretreatment, and cross-protection was incomplete with serotonin pretreatment against histamine. Desensitization attempts with another toxic amine (agmatine), which is not known to function as a neurohumor and whose toxicity for mice was not modified by pertussis sensitization, demonstrated that the resistance of either normal or pertussis-treated mice to the same amine was not altered. This procedure did not have any apparent value in protecting the latter animals against histamine.

The agmatine disparity further emphasized the critical significance of neurohumors in the pertussis-induced sensitive state and led to the administration of various autonomic drugs, which were shown to be protective in the pertussis-treated animals. Dibenzyline (α-adrenergic blocking agent) was also protective in pertussis-sensitized animals, whereas a β-adrenergic blocking agent, dichloroisoproterenol, duplicated the histamine and serotonin hypersensitivity in normal animals.

On the basis of these observations, the hypothesis was advanced that pertussis-induced hypersensitivity is the result of a functional imbalance between two types of adrenergic receptors or in the neural pathways leading to them.

Thus, the pertussis-induced hypersensitivity in mice put into focus, a possible adrenergic abnormality in manifestations of diseases of atopic allergy and the possible nature of the atopic abnormality itself [41]. At this point we will have to raise this discussion in a historical frame of reference. In 1931, F. M. Rackeman wrote about atopic disease the following lines: "The situation is somewhat analogous to that of a loaded gun. A good deal of knowledge is being obtained about the great variety of triggers (extrinsic and intrinsic causes) which fire the charge but why is the gun loaded? And what constitutes this load?" [42].

Only a minority of the total population shows some form of allergic reactivity despite that, by and large, identical conditions of antigenic exposure must be presumed to exist for all members of the same population [43,44]. The nature of the atopic abnormality in disorders of atopic allergy, which determines that only a relatively confined segment of any given population shows atopic reactivity (bronchial, cutaneous, nasal, etc.), that is, affected pathologically, is as yet unexplained, but it has been traditionally approached through immunological concepts. The association of a physicochemically and biologically distinct class of antibody, immunoglobulin E (IgE) (or reagin), with these disorders, and the production and unusual reactivity of this type of antibody is thought to account for much of the atopic abnormality in these manifestations [45]. However, as it has been extensively discussed earlier [46,47], immunological concepts cannot account for many of the significant facts surrounding these disorders. On the basis of model experiments and other considerations, an alternative to the classical immunologic concept, which came to be known as the β-adrenergic theory of the atopic abnormality, has been postulated [48]. This theory regards disorders of atopic allergy not as immunological diseases but as unique patterns of effector reactivities. In addition to the antigen–antibody interactions, atopic episodes are known to be triggered by a large variety of stimuli such as infection, various synthetic and natural chemicals, conditioned reflexes, psychic stimuli, changes in atmospheric pressure, exposure to thermal changes, nonantigenic dust, fumes, and other irritants.

Any molecular interpretation of hypersensitivity to such a large variety of unrelated stimuli would appear to necessitate the postulation that the primary lesion be connected with a final common pathway operating through a biologically unusually broad messenger or signaling system. The adenylcyclase 3′,5′-AMP system is such a messenger system capable of responding to a wide variety of neural, humoral, and hormonal agents, subserving homeostasis, especially to β-adrenergic receptor activation.

Five phases can be distinguished in the experimental analysis of this theory, both chronologically and in research strategy. In the early presentations of the theory (1962–1972), manifestations of reduced asthmatic responsiveness to catecholamines as measured by various systemic parameters of β-adrenergic reactivity were found [48,49]. These observations were highly compatible with the β-adrenergic theory. However, the interpretation of these results was handicapped by the limitations of *in vivo* studies involving complex homeostatic regulations. Importantly in the second phase of our experimental analysis (1972–1975), the same pattern of reduced β-adrenergic responsiveness was demonstrable in *in vitro* preparations of isolated cells derived from asthmatic individuals. In these studies, leukocytes and lymphocytes from atopic donors exhibited a reduced cAMP response to β-adrenergic agonists [50].

During the third phase (1975–1978), methodological and theoretical problems emerged casting doubt on the validity of interpretations of the findings on leukocytes and lymphocytes [51], as well as on the supportive value of these findings for the *de facto* existence of the postulated β-adrenergic abnormality.

Our analysis entered its fourth phase when the commercial radioactive adrenergic ligands became available [52] (1979–1986). A significant and mutually reciprocal shift was demonstrated in the numbers of adrenoceptors from β to α in lung tissues of patients with reversible airways obstruction with the aid of ^{3}H-dihydroergocryptine (^{3}H-DHE;α receptor ligand) and ^{3}H-dihydroalprenolol (^{3}H-DHA; β receptor ligand). Lymphocytes of patients with AD showed the same receptor shift. In addition, saturation curves for the binding of ^{3}H-DHA and ^{3}H-DHE to adrenergically desensitized lymphocytic membranes from patients with asthma or AD shed reduced receptor numbers of both receptors but the abnormal ratio between α and β receptors was preserved. This indicated that adrenergic desensitization, even when it contributes to the overall β-adrenergic subsensitivity, is not the cause of the original adrenergic abnormality in lymphocytes of atopic individuals [51,53].

In the emergence of the lymphocyte as a neuroendocrine cell in the past 15 years (fifth phase) [54,55] provided a new approach in the continuing analysis of adrenergic mechanisms in atopy. It was shown in mammals that the lymphocytes are able to both synthesize and store virtually all the known neurotransmitters and neurohormones. Lymphocytes were shown to contain two sets of adrenergically active proteins: (1) a secretory variant of β-arrestin and an interleukin-1α (IL-1α) antagonist, both of which downregulate β_2-adrenergic receptors in A549 human lung epithelial cells, and (2) a mixture of IL-1α and IL-1β that upregulates β_2-adrenergic receptors in the same cell line [56,57]. In subsequent studies, lymphocytes obtained from patients with asthma and AD were shown to possess significantly reduced amounts of the adrenergically upregulating components, but normal amounts of the adrenergically downregulating lymphocytic substances [56]. It has been demonstrated that the β-adrenoreceptor-mediated relaxation is attenuated in cholinergically stimulated airway smooth muscle from asthmatic patients [58,59]. Reduced β-adrenergic responsiveness has also been reported in airway smooth muscle isolated from antigen-sensitized animals [60,61]. Finally, IgE itself can enhance cholinergic neurotransmission in human airways [46].

Early and more recent reports have demonstrated the involvement of chromosome 5q31-33 in asthmatic individuals, and also in patients with AD. This region contains the β_2-adrenoceptor and steroid receptor genes. The IL-4 cluster has been identified also in the same region. IL-4 enhances B-cell proliferation and the expression of FC receptors for IgE (FCεR) [59 ,62].

Very recently, it was demonstrated that antigen-induced aggregation of FCεRl modulates the activity of GTP-cyclohydrolase I. This is a key enzyme for synthesis of the cofactor 6-tetrahydro-bi-opterin ($6BH_4$), followed by increased $6BH_4$ levels in murine mast cells. This effect was specific for the complete FCεRI and was absent in murine mastocytoma cells lacking the α chain [63]. The presence of $6BH_4$ de novo synthesis/recycling has been documented in human epidermal keratinocytes [64,65]. These cells also synthesize and degrade both catecholamines and acetylcholine and express high numbers of 132-adrenergic, muscarinic and nicotinic receptors [64,65].

Evidence is emerging for the functional polymorphisms in the β_2-adrenoceptor gene linked to atopy, which influence both β_2-adrenoreceptor structure and function. A linkage analysis of 303 children with severe asthma showed a defect on chromosome 5q31~33 [66]. Point mutations in the 132-adrenoceptor gene have been identified in mild to moderate (Gln^{27}-Glu^{27}) and in severe asthma (Arg^{16}-Gly^{16}) [67], and a substitution (Ala^{119}-Asp^{119}) has been shown in nine unrelated patients with atopic disease [68]. The latter mutation leads to defective β_2-adrenoceptor structure and function in both keratinocytes and lymphocytes with a sixfold decrease in the K_D for agonist binding [68]. These results strongly support the presence of β-adrenergic abnormality in asthma/atopy as does the important contribution from Hanifin's group on cAMP/phosphodiesterase malfunction in AD. These considerations are presented elsewhere in more detail [69].

3. NEUROENDOCRINE–IMMUNE INTERACTION

At the beginning of the nineteenth century, pathologists observed that acromegaly was often associated with thymic hyperplasia. Hammar [70] described that thymus involution was frequent under the influence of environmental or emotional factors. On the other hand, thymic hyperplasia was associated with castration, Grave's disease, Addison's disease, and acromegaly. Smith discovered in 1930 that in hypophysectomized (Hypox) rats the thymus regressed to less than half of that of controls, but there was no involution in partially hypophysectomized rats [71].

Hans Selye documented first in 1936 that the pituitary–adrenal–thymus axis was activated by various noxious stimuli, which led to the involution of the thymus and of the lymphoid organs [72,73]. Selye established that the bursa of Fabricius in the chicken was extremely sensitive to steroid hormones [74]. On the basis of his experiments, Selye [75] has described the theory of *general adaptation syndrome* (GAS) in 1946. He pointed out that this is a general reaction that leads to resistance of the organism to various insults. In 1949 Selye [76] discovered that the inflammatory response is regulated by corticosteroids. In his article, "Stress and Disease," he proposed that deficient host defense due to abnormalities of neuroendocrine factors may lead to disease [77]. Selye recognized the importance of mast cells in general pathology and performed numerous studies on these cells. His book on mast cells [78] is a contribution of lasting value on the subject.

At the time of Selye's contributions, the function of the thymus, lymph nodes, or the bursa of Fabricius was unknown. The function of these organs was discovered in the 1960s and early 1970s, decades after he published his seminal papers on stress. With the development of immunology it became clear that stress has a profound immunosuppressive effect and that the susceptibility of stressed animals to infectious disease is increased. These findings appeared to contradict Selye's conclusion that the response to stress is an adaptive defense reaction that leads to increased resistance to various noxious agents.

In man the immunosuppressive effects of ACTH and glucocorticoids was first described in 1949 by Hench and coworkers [79], who used these agents for the treatment of rheumatoid arthritis. These original observations have been confirmed by many investigators, and glucocorticoids are frequently used today as immunosuppressive and anti-inflammatory agents.

It has been recognized for some time that hormones, including those secreted by the pituitary gland, affect immune reactions [80]. We demonstrated that growth and lactogenic hormones (GLH) restore adaptive immunocompetence in hypophysectomized rats and also in animals that were immunosuppressed with bromocriptine. It was also shown that the ACTH–adrenal axis serves as an antagonist of immune activation and that dexamethason inhibits profoundly the production of TNF-α in mice treated with bacterial endotoxin [81–85]. We proposed that GLH maintain adaptive immunocompetence in higher animals and man [86]. Since the publication of these experiments, the role of pituitary hormones in immunoregulation has been accepted by the scientific community.

Wannemacher and co-workers isolated the leukocyte endogenous mediator (LEM) of fever in 1975 [87]. This was the first immune-derived molecule identified, which mediated feedback signals toward the CNS. Later LEM was found to be identical with IL-1. IL-1 serves as a signal for pituitary hormone release, which was shown by several investigators in the early 1980s [88–93]. Subsequently, other cytokines, especially IL-2, IL-6, TNF-α, and interferon γ were shown to regulate the secretion of pituitary hormones during systemic immune/inflammatory reactions [94]. It is also clear by now that nerves have immunoregulatory function and provide feedback signals from lymphoid organs and from sites of immune/inflammatory reactions toward the CNS [95–98].

4. NEUROIMMUNE REGULATON IN HIGHER ORGANISMS

Immune-derived cells and mediators are present in the CNS and are part of the neuroimmune regulatory equation [99–102]. The immune cells in the CNS show enhanced activity after immunization, infection, or stress [103–105] and in various pathological conditions, such as depression, and neurodegenerative disorders (e.g., multiple sclerosis [MS], Alzheimer's disease, Parkinson's disease, and stroke). These conditions are associated with many elements of inflammation and autoimmunity. Cytokines and chemokines initiate and propagate the inflammatory/immune response in these pathologies. In MS there is continuous realignment and redundancy in the inflammatory and immune responses [106–109]. Cytokines also have behavioral effects [110,111].

By now it is clear that there is much more to these systems than simply sharing cytokines and other mediators. It is now apparent that the neuroendocrine and immune systems do not only interact, but rather these organ systems rely on each other for mutual support both in health and in disease. It has been known for a long time that the thymus develops from the neural crest, which also gives rise to the CNS. Moreover, glia cells that represent roughly 50% of brain cells are related to the monocyte–macrophage lineage and are bone marrow derived [112]. The new and very important information discussed in this volume is that neurons themselves may differentiate from bone marrow-derived stem cells [113,114]. This means that the brain relies on the bone marrow for rejuvenation and healing ("plasticity"). Indeed, recent evidence indicates that inflammatory cells and cytokines exert a neuroprotective effect during traumatic brain injury [115]. We know for some time that the brain shares adhesion molecules and numerous cell surface receptors with lymphocytes. It is now also clear that cell-to-cell and cell-to-matrix interactions play important roles in brain physiology and pathology [116,117].

Until recently, the immune system has been considered as an autonomous system and that lymphocytes were equipped with sophisticated receptors for the recognition of antigen and were capable of defending the host from pathogenic insults. It was also recognized that the Immune

System was well organized and was regulated by extensive regulatory pathways [118]. However, on the basis of common developmental origin, shared stem cells, receptors and mediators, and mutual interdependence, it is now apparent that *the nervous, endocrine, and immune systems are integrated parts of a united Neuroimmune Supersystem. This Supersystem coordinates and regulates all the physiological and pathological processes in higher animals and man for their entire life cycle*. Thus lymphocytes with their dominant regulatory function within the immune system could be considered to be analogous to the neuron within the CNS. Lymphocytes, like neurons, are sensory cells with the capacity to recognize chemical structure and to distinguish self from nonself. They store such information and are capable of memory responses. Lymphocytes are also capable of conveying information on chemical (antigenic) abnormalities in host tissues to the brain through cytokine signals. Immune cells are essential for defending the body from foreign invading pathogenic organisms and for the elimination of aberrant cells from the host. It is now clear that immune cells are also involved in normal physiological regulation.

5. CONCLUDING REMARKS

Although research in the field of neuroimmune interaction has been conducted for over a century, the bulk of experimental evidence that supports the existence of the neuroimmune regulatory network in higher organisms has been produced during the past two decades. Today it should be apparent to anyone who is interested in this subject and is prepared to examine the supporting evidence that the CNS controls not only the endocrine system (hence often called neuro-endocrine), but the immune system is also integrated with the CNS (neuroimmune). It is clear that this *Neuroimmune Supersystem* forms a network with the host organism and regulates the entire life cycle of higher organisms during health and disease [119].

There is continuous communication within this regulatory network and also between individual organs and tissues of the body. This network is maintained with the aid of shared cytokines, neurotransmitters, neuropeptides, hormones, and also by recirculating cells of the immune system. Growth and lactogenic hormones (GLH), or equivalent hormones of more restricted specificity, and insulin-like growth factor (IGF), provide fundamental (competence) signals for growth as well as for function in each organ and tissue in the body, including the immune system. These are the hormones of immunocompetence. The hypothalamus–pituitary–adrenal axis (HPA) serves as an antagonist of the GLH–IGF-I axis and exerts a suppressive influence on the adaptive immune system as wells as on many other tissues and organs. However, the HPA axis stimulates natural immune defense during acute phase reactions (APR). Thyroid hormones, estrogens, and androgens are nuclear regulators of immune function as well as the function of other cells. Sex hormones are able to induce modulations/shifts in immune responses as in other physiological processes [120].

During acute febrile illness, a systemic activation of the natural immune system takes place, which leads to a profound increase of IL-1, TNF-α, and IL-6 in the blood stream. These cytokines act on the CNS, which results in fever, the activation of the HPA axis, and there is also "sympathetic outflow." These immune-neuroendocrine alterations lead to a profound suppression of the thymus-derived (T) lymphocyte-dependent adaptive immune system, whereas the production of natural antibodies (Nab) by $CD5^+$ B lymphocytes and of APP in the liver is rapidly increased. For instance, the level of C-reactive protein in the blood may rise up to 1000 times within 24–48 h. Nab, CRP, endotoxin-binding protein, and mannose-binding proteins have the capacity to recognize highly preserved, homologous epitopes (homotopes) on microbes as well as on altered self-components during infection and injury. These recognition molecules serve as poly-specific activators of the natural immune system during the APR. Therefore, the APR should be regarded as a highly coordinated acute and poly-specific defense reaction based on the amplification of natural immunity. The adaptive immune system is inhibited during APR. The high energy and metabolic requirement of APR is assured by catabolism in most bodily tissues, except the CNS, the liver, bone marrow, and leukocytes, which are activated. Therefore, febrile illness leads to immunoconversion through profound neuroendocrine and metabolic alterations, which mobilize all the available resources in the body in the interest of host defense. The events of APR illustrate clearly that the ultimate immunoregulator is the CNS, which has the power to rapidly suppress adaptive immune responses and to amplify natural immunity. More recent experiments indicate that the healing process that follows acute illness is also regulated by the CNS. Here vasopressin has emerged as a key neuropeptide that promotes adaptive immunocompetence and stimulates the anabolic hormones, growth hormone, and prolactin. These observations imply that vasopressin is involved with recovery and healing from acute illness [120].

Evidence is mounting rapidly that the immune system is involved in physiological processes such as mammalian reproduction, the function of the CNS, and of endocrine organs, the gastrointestinal system, and indeed, of all tissues and organs in the body. In these organs, much of this physiological role is exerted by stromal lymphoid elements (e.g., cells of the monocyte/macrophage series, tissue mast cells, T and B lymphocytes, and polymorphonuclear cells). In case of imminent danger, a local inflammatory response is initiated, which involves the rapid homing of immune-derived inflammatory cells to the site of injury/irritation. It is also clear that inflammatory cells do not only fight the pathogen but also protect host tissues form injury and contribute significantly to regeneration and healing [114].

In conclusion, *the neuroendocrine and immune systems form a Super System, which is equipped with sensory capacity, the ability to process and store information and to regulate the host organism in homeostasis and harmony with the external and internal environment.* This Neuroimmune Supersystem plays a fundamental regulatory role for the entire life cycle of higher animals and of man.

REFERENCES

1. Sternberg EM. The Balance Within: The Science Connecting Health and Emotions. New York: WH Freeman and Company, 2000.
2. Thompson WG. Placebos: A review of the placebo response. Am J Gastroenterol 2000;95(7):1637–43.
3. Nowotny A. In search of active sites in endotoxin. In: Beneficial effects of endotoxins. Nowotny A, Ed.; New York: Plenum Press, 1993; pp. 1–55.
4. Salomonsen CJ, Madsen T. Influence de quelques poisons sur le pouvoir antitoxiquen du sang. Comt Rend Acad Sci 1898;125:1229–33.
5. Eppinger H, Hess L. Die Vagotonie. Röntgenslogische Befunde der Sella turcica bei hypoplastischen weiblichem Genitale. Mitt. D. Geselesch. F. inn. Med. U. Kindehr in Wien. Hirschwald. Berlin. 1910, ix, 26.
6. Auer J. The prophylactic action of atropin in immediate anaphylaxis of guinea pigs – third communication. Am J Physiol 1910;26:439–52.
7. Bijisma UB. Hat Einspritzung von Adrenalin einen Einfluss auf die Fähigkeit zur Antikörsperbildung. Centralbl F Bakteriol 1921;36:246.

8. Friedberger E, Oshikawa K. Beziehungen zwischen Antigen und Antikörperbildung (Der Einfluss des parenterden Antigendepots auf die Antikörperbildung). Z Immunitötsfschg 1922;33:306–16.
9. Joachimoglu G, Wada U. Ueber den einfluss von Atropin und Pilokarpin auf die Antikörperbildung bei Kanichen Arch. F exper Path U Pharmakol 1922;93:269–84.
10. Tonietti F. Le fonti storiche della chimica e dell alchimia in Italia del dott. Giovanni Carbonelli. Rassegna di clin. Terap. [etc.], Roma. 1925;xxiv,248–50.
11. Bogendorfer L. Uber den Agglutination. Arch Exp Einfluss des Sympaticus zum Zustendekommen der Path Pharmacol 1928;133:107–10.
12. Fischer E, Kayserling H. Die Experimentelle Lymphogene Allergisch-hyperergische Appendicitis. Virchows Arch 1936;297:146–76.
13. Kayserling H. Untersuchungen zur Frage der Beziehunkendes Nervensystems zur Allergisch-hyperegischen Entzündung. Virchows Arch 1937;299:253–69.
14. Leupold E. Der Einfluss der Nervendurschschneidung auf dieResorption von Serumelwissam Kaninschenohr bei normergischen und hyperergischen tieren. Frankfurt Z Path 1938;52:392–406.
15. Lasovsky JM, Viropajev DN. Sur les observations morphologiques concernant le cours de l'inflammation hyperergique dans le tissue musculaire atrophie, par suite de "l'inactivité" en rapport avec le problème de la participation du systheme nerveaux aux reaction allergiques des tissues. Ann d'Anat Path. 1938;15:357–70.
16. Meltzer SJ, Meltzer Auer C. Studies on the "paradoxical" pupil-dilatation caused by adrenalin. I. The effect of subcutaneous injections and instillations of adrenalin upon the pupils of rabbits. J Physiol 1904;11:28–36; – II – On the influence of subcutaneous injections of adrenalin upon the eyes of cats after the removal of the superior cervical ganglion. J Physiol 1904;11:37–39; III – A discussion of the nature of the paradoxical pupil dilation caused by adrenalin. J Physiol 1904;11:49–51.
17. Dale HH, Gasser HS. The pharmacology of denervated mammalian muscle. J Pharmacol Exp Ther 1926;28:287–315.
18. Luco VJ. The sensitization of inhibited structures by denervation. Am J Physiol 1937;120:179–83.
19. Bard P. Studies on the cerebral cortex. I. Localized control of placing and hopping reactions in the cat and their normal management by small cortical remnants. Arch Neurol Psychiatry 1943;30:44–74.
20. Cannon WB. The Supersensitivity of Denervated Structures. New York: MacMillan, 1949.
21. Parfentjev IA, Goodline MA. Histamine Shock in mice sensitized with Hemophilus pertussis vaccine. J Pharmacol Exp Therap 1948;92:411–13.
22. Kind LS. Sensitivity of pertussis inoculated mice to serotonin. Proc Soc Exp Biol Med 1957;95:200–1.
23. Filipp G, Szentivanyi A, Mess B. Anaphylaxis and nervous system. Acta Med Hung 1952;2:163–73.
24. Szentivanyi A, Filipp G. Anaphylaxis and the nervous system. Part II. Ann Allergy 1958;16:143–51.
25. Filipp G, Szentivanyi A. Anaphylaxis and the nervous system. Part III. Ann Allergy 1958;16:306–11.
26. Szentivanyi A, Szekely J. Effect of injury to, and electrical stimulation of, hypothalamic areas on anaphylactic and histamine shock of the guinea pig. A preliminary report. Ann Allergy 1956;14:259–61.

27. Korneva EA, Khai LM. Effect of destruction of hypothalamic areas on immunogenesis. Fed Proc 1964;23:T88.
28. Jancso N. Role of the nerve terminals in the mechanism of inflammatory reactions. Bull Millard Fillmore Hosp (Buffalo NY) 1960;7–53.
29. Jancso N. Desensitization with capsaicin and related acylamides as tool for studying the function of pain receptors. In Pharmacology of Pain. RKS Lim, Ed.; Oxford: Pergamon Press, 1968.
30. Metal'nikov S, Chorine V. Roles des reflexes conditionels dans l'immunite. Ann Inst Pasteur 1926;40:893–900.
31. Smith GH, Salinger R. Hypersensitivity and the conditioned reflex. Yale J Biol Med 1933;5:387–402.
32. Ader R. Letter: Behaviorally conditioned immunosuppression. Psychosom Med 1974; 36(2):183–84.
33. MacQueen G, Marshall J, Perdue M, Siegel S, Bienenstock J. Pavlovian conditioning of rat mucosal mast cells to secrete rat mast cell protease II. Science 1989;243:83–85.
34. Gorczynski RM, Macrae S, Kennedy M. Conditioned immune response associated with allogeneic skin grafts in mice. J Immunol 1982;129:704–09.
35. Johnson HM, Smith EM, Torres BA, Blalock JE. Regulation of the *in vitro* antibody response by neuroendocrine hormones. Proc Nat Acad Sci USA 1982;79:4171–74.
36. Montgomery DW, Zukoski CF, Shah GN, Buckley AR, Pacholczyk T, Russell DH. Concanavalin A-stimulated murine splenocytes produce a factor with prolactin-like bioactivity and immunoreactivity. Biochem Biophys Res Commun 1987;145(2): 692–98.
37. DiMattia GE, Gellersen B, Bohnet HG, Friesen HG. A human B-lymphoblastoid cell line produces prolactin. Endocrinology 1988;122:2508–17.
38. Parfentjev IA, Goodline MA. Histamine shock in mice sensitized with *Hemophulus pertussis* vaccine. J Pharmacol Exp Ther 1948;92:411–13.
39. Kind LS. Sensitivity of perussis inoculated mice to serotonin. Proc Soc Exp Biol Med 1957;95:200–1.
40. Sanyal RK, West GB. Sensitizing properties of *Haemophilus pertussis*. Int Arch Allergy 1959;14:241–48.
41. Fishel CW, Szentivanyi A. The absence of epinephrine induced hyperglycemia in pertussis-sensitized mice. Fed Proc 1962;21:271.
42. Rackeman FM. Clinical Allergy. New York: Macmillan, 1931; p. 31. (Book abstract).
43. Szentivanyi A, Fillip G, Legeza I. A dohany-allergia mint ipari artolom. Orv Hetil 1951;44:1-15.
44. Szentivanyi A, Fillip G, Legeza I. Investigations on tobacco sensitivity. Acta Sci Hung 1952:3:175–84.
45. Szentivanyi A, Maurer PH, Janicki BW (Eds). Antibodies: Structure, Synthesis, Function, and Immunologic Intervention in Disease. New York: Plenum Press, 1987; pp. 1–215.
46. Szentivanyi A, Fishel CW. Die Amin-Mediatorstoffe der Allergischen Reaktion und die Reaktionsfahigkeit ihrer Erfolgszellen. In Pathogenese und Therapie Allergischer Reaktionen. Filipp G and Szentivanyi A, Eds; Stuttgart, Germany: Grundlagenforschung und Klinik. Ferdinand Enke Verlag, 1966; pp. 588–683.
47. Pearlman DS, Szentivanyi A. Excessive reactivity of defense mechanisms – allergy. In The Biologic Basis of Pediatric Practice. Cooke RE, Ed.; New York: McGraw Hill, 1968; pp. 536–46.

48. Szentivanyi A. The beta-adrenergic theory of the atopic abnormality in bronchial asthma. J Allergy Clin Immunol 1968;42:203–32.
49. Szentivanyi A, Katsh S, Townley RG. Effect of pertussis sensitization and Pharmacological beta-adrenergic blockade on the *in vitro* uptake of D-glucose-2-C^{14} into glycogen of diaphragm and upon the formation of $C^{14}O_2$ from glucose – $1C^{14}$ by adipose tissue. J Allergy Clin Immunol 1968;41:107–8.
50. Szentivanyi A, Krzanowski JJ, Polson JB. The autonomic nervous system and altered effector responses. In Allergy Principles and Practice, 3rd Edn. Middleton E, Reed CE and Ellis EP, Eds; St. Louis: Mosby, 1988; pp. 461–93.
51. Szentivanyi A, Heim O, Schultze P. Changes in adrenoceptor densities in membranes of lung tissue and lymphocytes from patients with atopic disease. Ann NY Acad Sci 1979; 332:295–98.
52. Szentivanyi A. The radioligand binding approach in the study of lymphocytic adrenoceptors and the constitutional basis of atopy. J Allergy Clin Immunol 1980;65:5–11.
53. Szentivanyi A, Heim O, Schultze P, Szentivanyi J. Adrenoreceptor binding studies with [(3)H] dihydroergocryptine on membranes of lymphocytes from patients with atopic disease. Acta Dermato-Venereol 1980;S92:19–21.
54. Szentivanyi A. Adrenergic regulation. In Bronchial Asthma – Mechanisms and Therapeutics, 3rd Edn. Weiss EB and Stein M, Eds; Boston: Brown & Co, 1993; pp. 165–91.
55. Szentivanyi A. The immune-neuroendocrine circuitry and its relation to asthma. In Bronchial Asthma – Mechanisms and Therapeutics, 3rd Edn. Weiss EB and Stein M, Eds; Boston, 1993; pp. 421–38.
56. Szentivanyi A, Schultze P, Heim O. The elution profile of the AS49 beta adrenergic (BAR) regulating activity of lymphocyte conditioned medium (LCM) of IM9 cells developed by DEAE ion exchange HPLC. Int J Immunopharmacol 1991;13:68 (Abstract).
57. Szentivanyi A, Robicsek S, Heim O. The nature of the adrenergically active constituents of lymphocyte conditioned medium (LCM) of IM9 cells. Int J Immunopharmacol 1991;13:70 (Abstract).
58. Hakonarson G, Herrick DI, Grunstein MM. Mechanism of impaired beta-adrenoreceptor responsiveness in atopic sensitized smooth muscle. Am J Physiol 1995;269:L652–54.
59. Howard M, Farrar M, Hilfiker M. Identification of T cell-derived B cell growth factor distinct from interleukin 2. J Exp Med 1982;155:914–23.
60. Ichinose M, Miura M, Tomaki M, et al. Incubation with IgE increases cholinergic neurotransmission in human airways *in vitro*. Am J Respir Crit Care Med 1996;154:1272–76.
61. Candell LM, Yun SH, Iran LLP, Ehlert PJ. Different coupling of subtypes of the muscarinic receptor to adenylate cyclase and phosphoinositidine hydrolysis in the longitudinal muscle of the rat ileum. Mol Pharmacol 1990;38:689–97.
62. Hudak SA, Gollnick SO, Conrad H, Kehry MR. Murine B cell stimulatory factor 1 (interleukin 4) increases expression of the Fc receptor for IgE on mouse B cells. Proc Natl Acad Sci USA 1987;84:4606–10.
63. Hesslinger C, Ziegler I, Kremmer E, Hultner L. IgE – mediated signal transduction regulates the tetrahydrobiopterin synthesis in mast cells: a model system for posttranslational modulation of GIP-cyclohydrolase I. In Chemistry and Biology of Pteridines and Folates. Pfleiderer W and Rokos H, Eds; Berlin, Vienna: Blackwell Science, 1997; pp. 559–64.
64. Schallreuter KU. Epidermal adrenergic signal transduction as part of the neuronal network in the human epidermis. J Invest Dermatol Symp Proc 1997;2:37–40.

65. Grando SA. Biological functions of kerationocyte cholinergic receptors. J Invest Dermatol Symp Proc 1997;2:41–48.
66. Postma DS, Bleeker ER, Amelling PS, et al. Genetic susceptibility to asthma. N Eng J Med 1995;333:894–900.
67. Schallreuter KU. Genetic aspects of atopic eczema, In Dermatology at the Millenium. Proceedings of the 19th World Congress of Dermatology Sydney, Australia, Parthenon Publishing; June 1997.
68. Hanifin JM, Chan SC. Role of cyclic nucleotide metabolism in the patho-physiology of atopic eczema In Handbook of Atopic Eczema. Ruzicka T and Ring J. Przybilla B, Eds; Berlin, Heidelberg: Springer-Verlag, 1991; pp. 232–34.
69. Rocken M, Schallreuter K, Renz H, Szentivanyi A. What exactly is atopy. Exp Dermatol 1998;7:97–104.
70. Hammar JA. The new views as to the morphology of the thymus gland and their bearing on the problem of the function of the thymus. Endocrinology 1921;5:543–73,731–60.
71. Smith PE. Effects of hypophysectomy upon the involution of the thymus in the rat. Anat Rec 1930;47:119–29.
72. Selye H. A syndrome produced by diverse nocuous agents. Nature (Lond) 1936;138:32.
73. Selye H. Thymus and adrenals in the response of the organism to injuries and intoxication. Br J Exp Pathol 1936;17:234–48.
74. Selye H. Morphological changes in the fowl following chronic overdosage with various steroids. J Morphol 1943;73:401–21.
75. Selye H. The general adaptation syndrome and the diseases of adaptation. J Clin Endocrinol 1946;6:117–230.
76. Selye H. Effect of ACTH and cortisone upon an "anaphylactoid reaction". Can Med Assoc J 1949;61:553–56.
77. Selye H. Stress and disease. Science 1955;122:625–31.
78. Selye H. The Mast Cells. Washington: Butterworth, 1965.
79. Hench PS, Kendall EC, Slocumb CH, Polley HF. The effects of a hormone of the adrenal cortex (17-hydroxy-11-dehydrocorticosterone; compound E) and of pituitary adrenocorticotrophic hormone on rheumatoid arthritis; preliminary report. Ann Rheum Dis 1949;8: 97–104.
80. Dougherty TF. Effect of hormones on lymphatic tissue. Physiol Rev 1952;32:379–407.
81. Nagy E, Berczi I. Immunodeficiency in hypophysectomized rats. Acta Endocrinol 1978; 89:530–37.
82. Berczi I, Nagy E, Kovacs K, Horvath E. Regulation of humoral immunity in rats by pituitary hormones. Acta Endocrinol 1981;98:506–13.
83. Nagy E, Berczi I, Friesen HG. Regulation of immunity in rats by lactogenic and growth hormones. Acta Endocrinol 1983;102:351–57.
84. Nagy E, Berczi I, Wren GE, Asa SL, Kovacs K. Immunomodulation by bromocriptine. Immunopharmacology 1983;6:231–43.
85. Berczi I (author/editor). Pituitary Function and Immunity (with 12 contributing authors). Monograph published by CRC Press, Inc., Boca Raton, FL, 1986.
86. Berczi I. The role of the growth and lactogenic hormone family in immune function. Neuroimmunomodulation 1994;1:201–16.
87. Wannemacher RW Jr, Pekarek RS, Thompson WL, et al. A protein from polymorphonuclear leukocytes (LEM) which affects the rate of hepatic amino acid transport and synthesis of acute-phase globulins. Endocrinology 1975;96:651–61.

88. Besedovsky HO, del Rey A, Sorkin E. Lymphokine-containing supernatants from con A-stimulated cells increase corticosterone blood levels. J Immunol 1981;126:385–87.
89. Berkenbosch F, van Oers J, del Rey A, Tilders F, Besedovsky H. Corticotropin-releasing factor-producing neurons in the rat activated by interleukin-1. Science 1987;238:524–26.
90. Sapolsky R, Rivier C, Yamamoto G, Plotsky P, Vale W. Interleukin-1 stimulates the secretion of hypothalamic corticotropin-releasing factor. Science 1987;238:522–24.
91. Nakamura H, Motoyoshi S, Kadokawa T. Anti-inflammatory action of interleukin 1 through the pituitary-adrenal axis in rats. Eur J Pharmacol 1988;141:67–73.
92. Uehara A, Gottschall PE, Dahl RR, Arimura A. Interleukin-1 stimulates ACTH release by an indirect action which requires endogenous corticotropin releasing factor. Endocrinology 1987;121:1580–82.
93. Bernton EW, Beach JE, Holaday JW, Smallridge RC, Fein HG. Release of multiple hormones by a direct action of interleukin-1 on pituitary cells. Science 1987;238:519–21.
94. McCann SM, Karanth S, Kamat A, Les Dees W, Lyson K, Gimeno M, Rettori V. Induction by cytokines of the pattern of pituitary hormone secretion in infection. Neuroimmunomodulation 1994;1:2–13.
95. Pletsityi DF, Averi'anova LL. The influence of vagotomy on the development of experimental endocarditis. Dokl Akad Nauk SSSR 1966;167:238–40 (in Russian).
96. Williams JM, Petersen RG, Shea PA, Schemdtje JF, Bauer DC, Felten DL. Sympathetic innervation of murine thymus and spleen: evidence for a functional link between the nervous and immune systems. Brain Res Bull 1981;6:83–94.
97. Nance DM, Burns J. Innervation of the spleen in the rat: evidence for absence of afferent innervation. Brain Behav Immun 1989;3:281–90.
98. Fleshner M, Goehler LE, Hermann J, Relton JK, Maier SF, Watkins LR. Interleukin-1 beta induced corticosterone elevation and hypothalamic NE depletion is vagally mediated. Brain Res Bull 1995;37:605–10.
99. Penkowa M, Hidalgo J, Aschner M. Immune and inflammatory responses in the CNS: modulation by astrocytes. In Neuroimmune Biology, Volume 6: Cytokines and the Brain. Korneva HA and Phelps C, Eds; Berczi I and Szentivanyi A, Series Eds; Amsterdam: Elsevier, 2007 (in this volume).
100. Suzumura A. Immune response in the brain: glial response and cytokine production. In Neuroimmune Biology, Volume 6: Cytokines and the Brain. Korneva HA and Phelps C, Eds; Berczi I and Szentivanyi A, Series Eds; Amsterdam: Elsevier, 2007 (in this volume).
101. Kannan Y, Moriyama M, Nakamura Y. Lymphocytes and adrenergic sympathetic nerve system: the role of cytokines. In Neuroimmune Biology, Volume 6: Cytokines and the Brain. Korneva HA and Phelps C, Eds; Berczi I and Szentivanyi A, Series Eds; Amsterdam: Elsevier, 2007 (in this volume).
102. Goehler L. Cytokines in neural signaling to the brain. In Neuroimmune Biology, Volume 6: Cytokines and the Brain. Korneva HA and Phelps C, Eds; Berczi I and Szentivanyi A, Series Eds; Amsterdam: Elsevier, 2007 (in this volume).
103. Korneva EA, Kazakova TB. Interleukin-2 gene expression in the CNS cells after stress and antigen application. In Neuroimmune Biology, Volume 6: Cytokines and the Brain. Korneva HA and Phelps C, Eds; Berczi I and Szentivanyi A, Series Eds; Amsterdam: Elsevier, 2007 (in this volume).
104. Katafuchi T. Involvement of brain cytokines in stress-induced immunosuppression. In Neuroimmune Biology, Volume 6: Cytokines and the Brain. Korneva HA and Phelps C, Eds; Berczi I and Szentivanyi A, Series Eds; Amsterdam: Elsevier, 2007 (in this volume).

105. Phelps C, Chen L-T. Brain response to endotoxin. In Neuroimmune Biology, Volume 6: Cytokines and the Brain. Korneva HA and Phelps C, Eds; Berczi I and Szentivanyi A, Series Eds; Amsterdam: Elsevier, 2007 (in this volume).
106. Clarkson AN, Rahman R, Appleton I. Inflammation and autoimmunity as a central theme in neurodegenerative disorders: fact or fiction? Curr Opin Investig Drugs 2004;5(7):706–13.
107. Ketlinsky SA, Kalinina NM. Cytokines in demyelinating disease. In Neuroimmune Biology, Volume 6: Cytokines and the Brain. Korneva HA and Phelps C, Eds; Berczi I and Szentivanyi A, Series Eds; Amsterdam: Elsevier, 2007 (in this volume).
108. Dunn A. Cytokines and depression. In Neuroimmune Biology, Volume 6: Cytokines and the Brain. Korneva HA and Phelps C, Eds; Berczi I and Szentivanyi A, Series Eds; Amsterdam: Elsevier, 2007 (in this volume).
109. Summers WK. Clinical relevance: cytokines in Alzheimer's disease. In Neuroimmune Biology, Volume 6: Cytokines and the Brain. Korneva HA and Phelps C, Eds; Berczi I and Szentivanyi A, Series Eds; Amsterdam: Elsevier, 2007 (in this volume).
110. Aubert A. Cytokines and immune-related behaviors. In Neuroimmune Biology, Volume 6: Cytokines and the Brain. Korneva HA and Phelps C, Eds; Berczi I and Szentivanyi A, Series Eds; Amsterdam: Elsevier, 2007 (in this volume).
111. Neveu PJ. The production and effects of cytokines depend on brain lateralization. In Neuroimmune Biology, Volume 6: Cytokines and the Brain. Korneva HA and Phelps C, Eds; Berczi I and Szentivanyi A, Series Eds; Amsterdam: Elsevier, 2007 (in this volume).
112. Szentivanyi A, Berczi I, Nyanteh H, Goldman A. Some evolutionary morphoregulatory and functional aspects of the immune-neuroendocrine circuitry. In Neuroimmmune Biology, Volume 3: The Immune-Neuroendocrine Circuitry. History and Progress. Berczi I and Szentivanyi A, Eds; Amsterdam: Elsevier, 2003; pp. 31–61.
113. Gottfried-Blackmore A, Croft GF, Karen Bulloch K. Sex hormones and cytokines in CNS pathology and repair. In Neuroimmune Biology, Volume 6: Cytokines and the Brain. Korneva HA and Phelps C, Eds; Berczi I and Szentivanyi A, Series Eds; Amsterdam: Elsevier, 2007 (in this volume).
114. Stewart R, Przyborski S. Non-neural adult stem cells: tools for brain repair? Bioessays 2002;24(8):708–13.
115. Correale J, Marcela F, Villa A. Neuroprotection by inflammation. In Neuroimmune Biology, Volume 6: Cytokines and the Brain. Korneva HA and Phelps C, Eds; Berczi I and Szentivanyi A, Series Eds; Amsterdam: Elsevier, 2007 (in this volume).
116. Berczi I, Szentivanyi A. Adhesion molecules. In Neuroimmmune Biology, Volume 3: The Immune-Neuroendocrine Circuitry. History and Progress. Berczi I and Szentivanyi A, Eds; Amsterdam: Elsevier, 2003; pp. 99–115.
117. Wiranowska M, Plaas A. Cytokines and extracellular matrix remodeling in the central nervous system. In Neuroimmune Biology, Volume 6: Cytokines and the Brain. Korneva HA and Phelps C, Eds; Berczi I and Szentivanyi A, Series Eds; Amsterdam: Elsevier, 2007 (in this volume).
118. Paul WE, Ed. Fundamental Immunology. New York: Lippincott-Raven, 1999.
119. Berczi I and Szentivanyi A, Series Eds and Eds. Neuroimmmune Biology, Volume 3: The Immune-Neuroendocrine Circuitry. History and Progress. Amsterdam: Elsevier, 2003.
120. Berczi I, Quintanar-Stephano A, Kovacs K. Chapter 14. Immunoconversion in the acute phase response. In Cytokines, Stress and Immunity. Nicholas P. Plotnikoff, Robert E. Faith, Anthony J. Murgo, and Robert A. Good, Eds; Boca Raton, FL: CRC Press, Taylor & Francis Group, 2006; pp. 215–54.

II. CYTOKINES IN THE BRAIN

A. CYTOKINES, THEIR RECEPTORS AND SIGNAL TRANSDUCTION IN THE BRAIN

Cytokines and the Brain
Edited by C. Phelps and E. Korneva

Cytokine Receptors in the Brain

BRUNO CONTI, IUSTIN TABAREAN, MANUEL SANCHEZ-ALAVEZ, CHRISTOPHER DAVIS, SARA BROWNELL, MARGARITA BEHRENS, and TAMAS BARTFAI

The Harold L. Dorris Neurological Research Center, Department of Molecular and Integrative Neurosciences, The Scripps Research Institute, 10550 North Torrey Pines Road, La Jolla, CA 92037, USA

ABSTRACT

Cytokines, once thought to be specialized molecules of the immune system, are now being investigated also for their synaptic and inflammatory action on the central nervous system (CNS). These proteins and their receptors can be synthesized in the brain by glial and neuronal cells and contribute to two main types of action: modulation of neuronal excitability and local inflammatory responses. We present an overview of the distribution, regulation, and function of neuronal cytokine receptors in the CNS, specifically focusing on the interleukin-1 receptor (IL-1R), tumor necrosis factor receptor (TNFR), interleukin-6 receptor (IL-6R), and interleukin-10 receptor (IL-10R), because their ligands are the most studied pro- and anti-inflammatory cytokines in the brain.

1. BASAL AND INDUCED LEVELS OF CYTOKINES IN THE BRAIN

Cytokines represent a growing class of secreted protein that act as messengers among cells of the immune system. It was recognized early on that cells of the nervous system (neurons and glia) as well as certain endocrine cells are capable of synthesizing cytokines. Furthermore, the presence of cytokine receptors on specific subpopulations of neurons and microglia support the idea that cytokines act as important messengers in the nervous system. The expression of cytokines and their receptors on immune competent cells and cells of the nervous system makes them ideal signal substances to connect the nervous and immune responses. Thus, cytokines have come to play a central role in neuroimmunology, a discipline that studies the interactions between the nervous and the immune system, often mediated by neuroendocrine and endocrine signals (neuroendocrineimmunology).

Cytokines represent a highly inducible class of proteins. Their biosynthesis and secretion can be induced by a multitude of pathophysiological states including stress, infection, inflammation, tissue injury, burn, seizures, and pain. As a result genes, cytokine concentrations can increase locally and systemically 10–1000-fold within a few hours, changing the receptor occupancy dramatically. The picomolar cytokine concentrations found in healthy brain can be elevated by trauma or infection to 10–20 nanomolar concentrations. The number of cytokine receptors also changes in response to stimuli such as trauma or infection, although only a two to threefold change, a mere tenth or hundredth the inducibility of the ligands.

Another important aspect of the expanding studies on cytokines is the recognition of the cross-inducibility of many cytokines and the "self limiting" nature of the cytokine response. Cross-inducibility is the concept that cytokines induce the production of other cytokines, creating an inflammatory cascade of a number of different pro-inflammatory cytokines. Inflammation, trauma, stress, or infection can cause a rapid induction of pro-inflammatory cytokines such as interleukin-1β (IL-1β), tumor necrosis factor-α (TNF-α), interleukin-6 receptor (IL-6), and monocyte chemoattractant protein-1 (MCP1). These cytokines can also limit their synthesis and/or action by inducing the biosynthesis of anti-inflammatory cytokines such as the IL-1 receptor antagonist (IL-1RA), interleukin-4 (IL-4) and interleukin-10 (IL-10).

Additional anti-inflammatory responses are achieved endogenously by several mechanisms such as the (1) presence of decoy cytokine receptors that bind the ligand but fail to initiate a cellular response (e.g., IL-1R type II acts as a "sink" for the agonist IL-1β as occupancy of this soluble receptor is not coupled to a cellular response); (2) binding of the ligand to binding proteins that sequester the ligand (e.g., IL-18-binding protein binds IL-18 and prevents it from stimulating its membrane-bound IL-18 receptor); (3) induction of silencing elements intracellularly (e.g., SOC-2) that prevent the transcriptional activation of other pro-inflammatory genes.

The effects of cytokines on the nervous system have been studied during development, under conditions of normal neuronal function in the adult animal and during infection and inflammation. This review will focus on the effects of cytokines in the brain of the adult healthy animal, as well as in animals with inflammation or infection that lead to significantly elevated cytokine levels both locally and systemically. We will also examine some cytokine effects on the hippocampal, hypothalamic, and pituitary function, looking at the "neuro–immuno–endocrine" interaction. Rather than listing the numerous examples of the expression and effects of each and every cytokine and their receptors in different brain areas, most of this chapter will focus on IL-1β and TNF-α as the prototypic pro-inflammatory cytokines and interleukin-1 receptor antagonist (IL-1ra) and IL-10 as the prototypic anti-inflammatory cytokines. Therefore, we will focus our attention on the TNF-α receptor (TNFR), IL-1 receptor (IL-1R), and IL-10 receptor (IL-10R). Using these representative examples, we will describe the roles played by cytokines in the modulation of the nervous system when these cytokines are present in femtomolar–picomolar (normal) and nanomolar (elevated) concentrations (inflammatory conditions). Additionally, although cytokine receptors are expressed on both neurons and glia, this chapter will focus on the neuronal effects of cytokine action.

2. DISTRIBUTION OF CYTOKINE RECEPTORS

The main source of cytokine synthesis in the brain under noninflammatory conditions is local and mostly neuronal. Cytokine effects in the brain under noninflammatory conditions are exerted through neuronal or glial cytokine receptors in an autocrine or paracrine manner. The basal cytokine levels in the healthy adult brain are femtomolar–picomolar range (Table 1).

Cytokine receptor levels are also low (femtomoles/mg protein) in noninflammatory conditions and show clear regional differences, indicating that certain neuronal types are selectively expressing these receptors and suggesting that cytokine receptors participate in normal signaling in the absence of inflammation.

Three regions of the brain have been investigated in depth with respect to cytokine receptor expression levels, distribution, and function: the hippocampus, the hypothalamus, and the

Table 1 Regional levels of some selected pro-inflammatory (in pg/mg of protein)

	IL-1β	TNF-α	IL-6
Hippocampus	~1 [107]	~200 [109]	~5 [109]
Hypothalamus	~1 [107]	~250 [107]	~7 [107]
Pituitary	~1 [110]		

pituitary gland. In these brain regions, cytokine effects on synaptic, normal neurosecretory function have been demonstrated in the absence of trauma, infection, or neurodegeneration.

The cytokine receptors IL-1R1, TNFR p55, IL-6R, and IL-10R are expressed on specific neuronal populations such as pyramidal and granule cells in the hippocampus, large neurosecretory neurons in the hypothalamus, and somatotrophes and lactotrophes in the pituitary [1–10]. In addition, cytokine receptors are present and functional in astrocytes and microglia in these brain regions. In fact, the lack of TNFR on microglia suppresses its activation, indicating that proper cytokine receptor expression is important for microglia function [11]. Finally, cytokine receptors are expressed on endothelial cells of the blood–brain barrier and the organum vasculosum laminae terminalis (OVLT). These receptors on the blood–brain barrier are highly important in cytokine signaling from peripheral immune cells to the brain. Introcerebroventricular injection of IL-1β has been shown to induce the infiltration of leukocytes into the brain through the blood–brain barrier, but this effect is abrogated in IL-1R1-deficient mice, illustrating the importance of IL-1R1 in immune cell recruitment to the brain [12,13]. It is important to note that cytokine receptors are also abundantly expressed in the peripheral nervous system and are induced in response to various stimuli. Sensory nerves and the excitable neuron-like chromaffin-type neurosecretory cells of the adrenal medulla express both cytokines and cytokine receptors.

The expression level of cytokines dramatically changes during seizures in the hippocampus and following neurotrauma, neuroinflammation, and neurodegeneration in all other brain regions. The receptor levels are unaltered or show small twofold changes following such insults. Thus, the dynamics of cytokine signaling is overwhelmingly a result of changes in the ligand concentrations, not receptor concentrations. This is not to say that the cytokine receptor signaling mechanisms do not show any plasticity, such as desensitization, or supersensitivity. An increase in cytokine receptor mRNA in the brain has been demonstrated by *in situ* hybridization for IL-1R [1,2,14] and IL-6R [15,16] following agonist exposure, seizures, and inflammation.

Table 2 shows the relative expression levels of some cytokine receptors in the rodent brain. Under normal, unstimulated conditions, the expression of cytokine receptors is mostly neuronal.

Table 2 Presence of some cytokine receptors in different brain regions

	IL-1R1	TNFR	IL-6R	IL-10
Hippocampus	+ [52]	+ [56]	+ [56]	+ [56]
Hypothalamus	++ [52, 111]	+ [111]	+ [111]	+ [9]
Pituitary	+ [112]	+ [111]	+ [111]	

Following stimulation, activated microglia and astrocytes increase the expression not only of the cytokine ligands but also of the cytokine receptors, changing the distribution to mostly non-neuronal expression of cytokine receptors. Therefore, it seems as though a shift from neuronal to non-neuronal expression of cytokine receptors occurs in response to an inflammatory stimulus, although an exception is the IL-10R that is mostly found in astrocytes and microglia even under normal conditions [9,17,18].

3. SIGNALING THROUGH CYTOKINE RECEPTORS IN THE BRAIN

The molecular components of the classical signaling events activated through cytokine receptors in the cells of the immune system and leading to transcriptional modifications are also found in the brain. So, it is considered that the same transduction pathways are mediating cytokine effects in central nervous system (CNS). However, cytokines in the brain have been shown to have a rapid action to modulate neuronal activity, suggesting the existence of other fast-acting transcriptional independent mechanisms. One such example is the ceramide-mediated action of IL-1β in the hypothalamus.

3.1. Transcription-dependent signaling

Cytokines bind with very high affinity (K_D = nanomolar-picomolar) to their membrane bound, signaling receptors. Much of previous research on cytokine signaling has been done using immune-competent cells from the periphery, which can be obtained in larger quantities than neurons and can often be sorted by FACS analysis. The signaling cascades of these pro- and anti-inflammatory cytokines were thus initially defined in non-neuronal cell types outside of the brain. Although it is assumed that signaling pathways through the same cytokine receptors are identical in non-neuronal and neuronal cells, some considerations may be important. Neurons are excitable cells with large and rapid changes in the membrane potential during action potential firing activity. Furthermore, neurons can exhibit extreme and repetitive activity such as that occuring during seizures. In addition, the major substrates for the cytokine-activated protein kinases (MAP kinases) may be different in neuronal and non-neuronal cells. Therefore, it is important to determine the specific effects of cytokines at their receptors in the brain and not just rely on previous research done in peripheral immune cells.

Cytokine receptors have been classified as belonging to different protein families (Fig. 1). These include the Toll-like receptor family (IL-1 and IL-18 receptors and receptor accessory protein), the TNF/NGFR family (TNF p55/60, p75/80 receptors), the interferon receptor (IFN-R) family (IFNγR, IL-10R), and the hematopoetic growth factor receptor family (IL-4R, IL-6R, gp130). These receptor types differ both in their structure and in their signaling cascades.

Receptors belonging to the Toll-like receptor family (i.e., IL-1R1, IL-18R) are dimers composed of a receptor chain binding the ligand and of a receptor-associated protein required for signal transduction. Both the receptor and the accessory protein are single transmembrane proteins carrying a toll interleukin receptor (TIR) recognition domain in their intracellular portion, which mediates the binding of the cytosolic adaptor protein MyD88 and initiates a cytosolic signaling cascade that eventually activates MAP kinases and nuclear factor-κB (NF-κB)-dependent transcription. The soluble form of the IL-1 receptor, the IL-1R type II, binds the IL-1 ligand but does not mediate signaling, thus acting as a decoy receptor [19]. However, some evidence suggests that some actions of IL-1β are independent of the IL-1R1

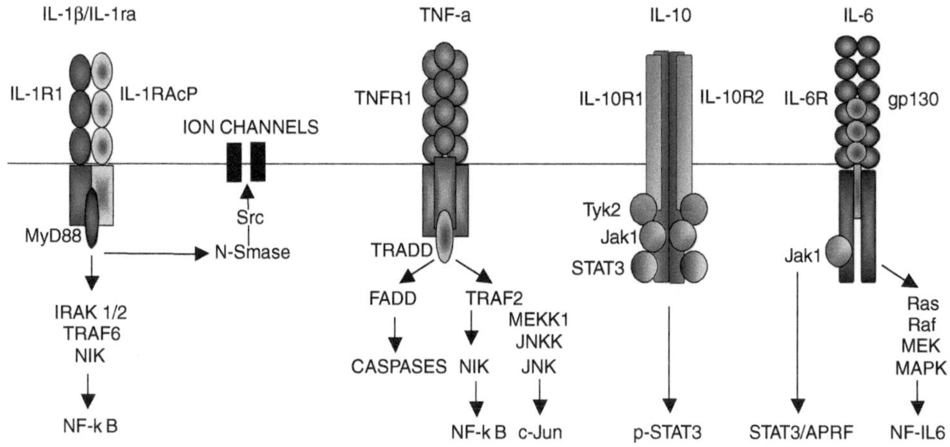

Figure 1. **Schematic representation of the four main cytokine receptor types and their signaling.** Receptors for IL-1 (IL-1R1/IL-1RacP), TNF (TNFR1), IL-10 (IL-10R1/IL-10R2), and IL-6 (IL-6R/gp130) and the transduction pathways they activate are represented. Both transcription-dependent and the ceramide-mediated transcription-independent pathwyspathways activated by IL-1 are shown. See main text for details.

and the classical IL-1 signaling pathways. It has been proposed that additional functional IL-1 receptors are expressed that could mediate IL-1R1-independent IL-1 signaling [20]. TNF-α binds to at least two receptor classes: TNFR-1 (p55/60) and TNFR-2 (p75/80). Both receptor classes are type 1 transmembrane glycoproteins. For both TNFR-1 and TNFR-2, trimeric TNF-α binds three receptor molecules, and this leads to the activation of the transcription factors c-Jun and NF-κB through different signaling cascades. Additionally, TNFR-1 possesses intracellular death domains that activate caspases that mediate apoptosis [21,22]. Mice lacking both TNF receptors have altered levels of proteins involved in signal transduction, stress response, protein folding, glucose and amino acid metabolism, vesicle trafficking, and cytoskeletal arrangement [23].

The receptor for IL-10 (IL-10R) belongs to the IFN-R family. These receptors are tetramers composed of two identical heterodimers of two different receptor components: IL-10R1 and IL-10R2. The tyrosine kinase Tyk2 constitutively associated with the intracellular domain of IL-10R2 interacts with the tyrosine kinase Jak1, which is associated with the intracellular domain of IL-10R1. The binding of IL-10 to this receptor complex induces tyrosine phosphorylation and activation of the transcription factor STAT3 [24].

IL-6R and IL-4R belong to the hematopoetic growth factor receptor family that uses the gp130 protein dimer for signal transduction. Once IL-6 or IL-4 binds its respective transmembrane receptor, gp130 dimerizes and leads to the activation of STAT3 through the Jak family protein tyrosine kinases or to the nuclear factor IL-6 (NFIL-6) through Ras and MAPK [25–27]. The receptors for IL-1, IL-6, and TNF-α are expressed on both glial and neuronal cells (Fig. 1). The strong antipyretic effects of centrally applied IL-10 would also suggest the presence of IL-10R on neuronal cells because fever has been shown to be mediated by hypothalamic neurons [9,28]. Additional cytokines and chemokines, not shown in the figure, bind to the G-protein-coupled receptor (GPCR)-type chemokine receptor and to the TGF-β receptor family members.

3.2 Fast nontranscriptional mediated action of IL-1 in the central nervous system

The biological action of IL-1 through the "classical" signaling cascade described above requires transcription-mediated changes. These changes take place in a time frame that is not likely to mediate the often observed fast action of IL-1 in the CNS. The existence of an additional IL-1 signaling pathway parallel to the one leading to the activation of NF-κB was recently discovered in the hypothalamic neurons of the preoptic area (POA). Although not fully characterized, this pathway is rapid, transcriptionally mediated, and enzymatic. It occurs when IL-1-bound to IL-1R/IL-1 receptor accessory protein (IL-1RAcP) activates the enzyme neutral sphingomyelinase (N-Smase) through MyD88, leading to the production of ceramide, potentially a second messenger. This pathway leads to the activation of the protein tyrosine kinase Src that can lead to the modification of ion channels with consequent changes in neuronal excitability [29,30] (Fig. 1). This pathway has been demonstrated to mediate the fast action (0–30 min) of IL-1β on fever and to modulate the IL-1β-mediated synaptic inhibition in POA neurons [31]. The fast action of IL-1β was also demonstrated in the hippocampus where IL-1β induced seizure by enhancing N-methyl-D-aspartate (NMDA) receptor excitability (discussed below). Additionally, a role of ceramide in IL-1β-induced seizure has been hypothesized.

4. FUNCTIONAL ROLES OF CYTOKINE AND CYTOKINE RECEPTORS IN THE BRAIN

The best evidence for the presence and the functional role of cytokine receptors in the brain comes from studies in two major groups: (1) cytokine effects on neuronal signaling functions including long-term potentiation (LTP), seizures, and homeostasis and (2) cytokine effects during fever and inflammation with particular attention to neuronal damage and neuronal cell death.

4.1. Long-term potentiation

LTP is a hippocampal circuit phenomenon widely studied as a cellular/network model of learning [32]. Studies on LTP are carried out either *in vivo* or more frequently *in vitro* using hippocampal slices for the stimulation and recording of electrical activity. The systematic study of hippocampal slices has shown that the process of preparation of hippocampal tissue slices leads to a 100–1000-fold induction of proinflammatory cytokines that lasts for several hours, temporarily preventing the examination of cytokine effects on LTP. Once proinflammatory cytokine levels return to normal levels, it has been demonstrated that IL-1β can impair the formation of LTP [4,14,33]. This inhibition can be reversed by IL-1ra [34], indicating that the IL-1RI plays a role in LTP induction. Additionally, the anti-inflammatory cytokine IL-10 [35] can reverse the IL-1β-mediated inhibition of LTP, and IL-18 has been shown to be involved in the inhibition of LTP [36] (authors' unpublished observations). Although IL-1β has been shown to inhibit LTP at pathological inflammatory conditions, there is evidence that it could actually have a dual role in being required for the formation of LTP under physiological conditions [37]. IL-1R1 null mice show impaired hippocampal learning, which indicates that IL-1 signaling is an important contributor to the formation of LTP [38].

These studies demonstrate that IL-1β and IL-18, both utilizing the Toll receptor family and activating MAP kinases, may modulate neuronal functions. Specifically, they may modulate the normal synaptic transmission that underlies memory formation and other cognitive functions by affecting the levels of adenosine and glutamate, both of which exert positive feedback on the IL-1 system. IL-1β acts on IL-1R1 to induce the release of adenosine, which binds to P2X receptors and stimulates the production of more IL-1β [39]. IL-1β also acts on IL-1R1 to induce the release of glutamate, which binds to NMDA channels and induces IL-1β [40]. The fact that cytokine levels are highly inducible by glutamate and adenosine, two excitatory neurotransmitters that are released in massive amounts during injury, including the mechanical slicing of the hippocampus, is of particular interest because seizures, traumatic brain injury, and stroke lead to massive glutamate release and will be accompanied by induction of cytokines in the brain.

4.2. Seizures

One of the best examples of functional changes following the elevation of cytokine levels in the brain is the pro-convulsive action of IL-1β mediated by IL-1R1 in the hippocampus during seizure activity.

Before seizure activity, IL-1α and IL-1β are found at very low levels in the hippocampus (femtomolar). Similarly, the level of IL-1ra, the anti-inflammatory antagonist of IL-1, is below the detection limit in most cases, and the level of the decoy receptor IL-1R type II is also low. Gabellec [41] showed that hippocampal injury rapidly induces the upregulation of IL-1α/β, IL-1ra, and IL-1 receptors.

Upon seizure onset, the levels of both IL-1α and IL-1β rapidly rise in both neurons and microglia, reaching 50–100-fold elevation within 30–90 min as demonstrated by radioimmunological as well as by immunohistochemical methods [42,43]. The levels of IL-1ra begin rising to a similar extent 30–60 min after the induction of IL-1α and IL-1β. In addition, exogenous (intracerebroventricular) application of IL-1β reduced seizure latency and enhances the time spent in clonus and tonus during seizures, whereas intracerebroventricular application of IL-1ra prolongs the latency of seizures and reduces the time spent in clonus and tonus. These experiments demonstrate the role of IL-1R in seizures, which has been confirmed through studies on IL-1RI null mice that do not develop seizures upon injection of IL-1β [42–44]. The functional significance of the dynamic changes in the ratio of proinflammatory IL-1β and anti-inflammatory IL-1ra has been studied in great detail. IL-1β has been shown to act as a powerful proconvulsant agent by enhancing NMDA receptor-mediated mechanisms; this action is inhibited by IL-1ra, and IL-1ra has been shown to act as a potent anticonvulsant. An interesting aspect of these findings is that the IL-1 system may play a role in febrile seizures that may occur during high fever in children and that is not responsive to conventional anticonvulsant therapy [45]. Additional evidence for a role of IL-1β in seizures comes from a correlational study looking at the polymorphisms in the gene encoding for IL-1β and the incidence of temporal lobe epilepsy [46].

IL-1 could be exerting its effects on seizure activity through the functional interactions between the IL-1RI and the NMDA receptors, which are co-expressed on hippocampal neurons (Fig. 2). IL-1RI might contribute to excitotoxic glutamate effects exerted at NMDA receptors through the activation of similar MAPK pathways, thus synergistically contributing to glutamate-mediated effects. Alternatively, IL-1RI may act directly by influencing the phosphorylation state of NMDA receptor through the activation of Src (Figs 2 and 3), thus enhancing the effects of glutamate by altering its receptors, and thereby affecting the duration and amplitude of the excitatory postsynaptic potential [40].

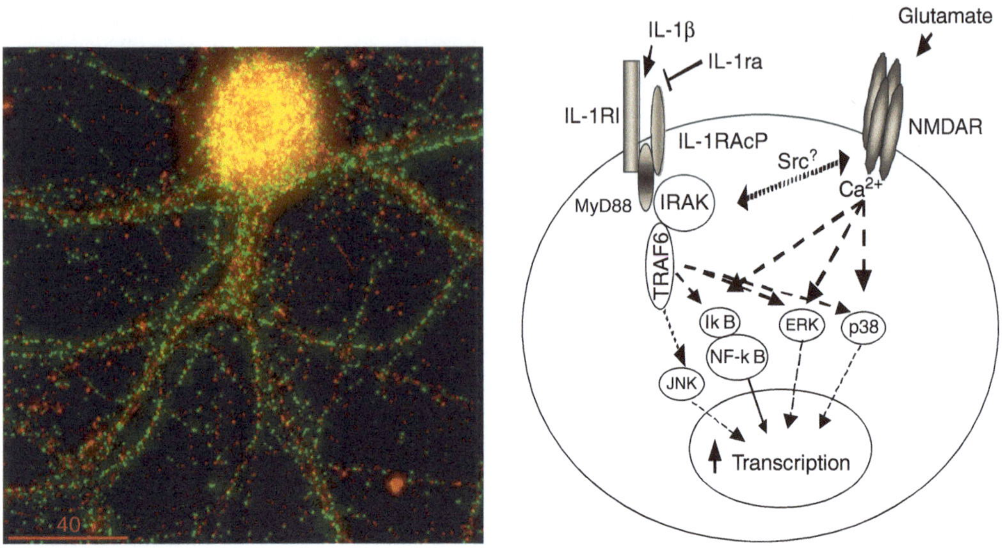

Figure 2. **Co-expression of IL-1RI and glutamate receptor on hippocampal neurons. Right.**: Dissociated murine hippocampal neurons were grown for 21 days as described [108]. After fixation in 4% paraformaldehyde, IL-1RI and the NR1 subunit of NMDA receptors were detected by immunocytochemistry using specific antibodies (anti-IL-1RI: JAMA-147, and anti-NR1 both from Pharmingen), followed by incubation with AlexaFluor conjugated secondary antibodies (594-red: IL-1RI, 488-green: NR1, Molecular Probes). Images were obtained with a confocal microscope (Olympus). **Left.**: Schematic representation of the proposed functional interaction between IL-1RI and NMDAR. Leading to IL-1R1-mediated potentiation of NMDA-mediated signaling and neurotoxicity.

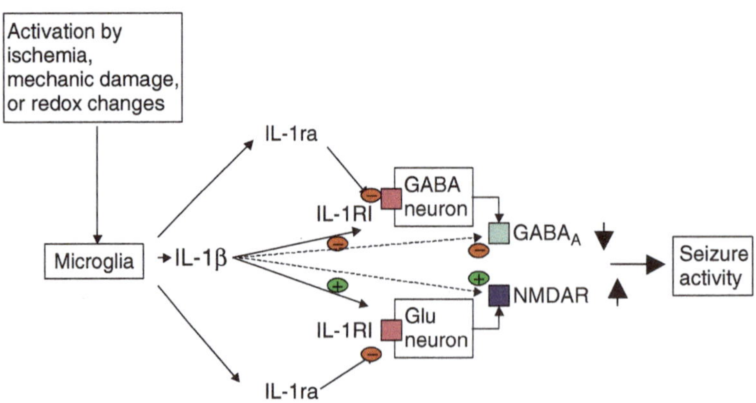

Figure 3. **Schematic representation of the effects of IL-1 system on neuronal excitability in seizures**. Activated microglia produce Il-1β, which acts to activate excitatory glutamate neurons and inhibit inhibitory GABA neurons, thus leading to an overall excitatory tone and increased seizure activity. See main text for details.

4.3. Neuroendocrine/behavioral effects

The role of IL-1 in the mechanisms of homeostasis extends beyond its effects on fever to the modulation of the hypothalamic–pituitary–adrenal axis (HPA axis) through IL-1RI in the hypothalamus and in the pituitary gland, probably the most tangible link between immune, nervous, and endocrine systems. IL-1β is a potent stimulator of CRF secretion in the hypothalamus and ACTH production and release from the pituitary gland [47]. In addition, TNF-α [48], IL-6, and IL-10 have been identified in the pituitary, explaining the effects of these cytokines on the HPA axis. Thus, local inflammation, seizures, or peripheral signaling leading to an elevation of hypothalamic IL-1β causes a strong activation of the HPA axis indistinguishable from the stimulation by stressors [49].

The behavioral effects of pro-inflammatory cytokines are mediated through cytokine receptors in the brain [50,51] and are collectively known as the sickness syndrome. These include loss of appetite [52], induction of sleep, and initiation of fever [53]. These effects are induced by IL-1β and TNF-α through IL-1RI and p55 receptors, respectively, as shown by antagonistic studies [54,55]. Cytokine receptors such as p55 and IL-1RI expressed in noradrenergic neurons are involved in anorexia [56,57] through central mediation of increased sympathetic tonus at the level of the *locus coeruleus* [58]. Cognitive impairment has also been shown upon IL-1ra and IL-6 injection in the brain [59].

Finally, cytokine levels in the cerebrospinal fluid of depressed patients are altered, and this may affect cytokine receptor occupancy in the brain [60].

4.3.1. Fever

Fever is the elevation of the core body temperature due to the alteration of the set-point located in the POA of the hypothalamus. The pro-inflammatory cytokines IL-1α, IL-1β, TNF-α, and IL-6 have been shown to be pyrogenic when administered systemically, intracerebroventricularly (ICV), or directly to the hypothalamus. Lipopolysaccharide (LPS) is a commonly used exogenous pyrogen that is obtained from the cell wall of Gram-negative bacteria. The hierarchy and specific roles of cytokine receptors in mediating the fever response to LPS has been studied by examining the fever response in mouse strains carrying null mutations of the IL-1 receptor type 1 (IL-1 R1), IL-RAcP, and the TNF-α receptors (p55/60 and p75/80) (Figs 4 and 5).

These studies have shown that LPS (acting at the CD14-LPS-binding protein Toll receptor-4 complex) can induce the synthesis and release of IL-1α/β and TNF-α. These pro-inflammatory cytokines can induce the synthesis and release of each other through the IL-1R1-IL-1RAcP complex and the trimeric p55/60 TNF receptors, respectively. LPS acts to increase the expression of IL-1R1, but this effect is abrogated by IL-10 [61,62].

The delineation of the signaling cascade for this and other cytokine responses has benefited from the availability of transgenic animals with the null mutation of cytokine genes and cytokine receptor genes. Cytokines often cross-induce each other and express an overlapping spectrum of effects (upregulated IL-1α, IL-1β, TNF-α), so it is often not easy to find clear phenotypic changes when a single cytokine is null mutated. The most instructive mutations have been the study of transgenic mice with null mutation of a given cytokine receptor or receptor subtype. This has been particularly important to define which receptor subtypes are mediating certain effects. The understanding that nonsignaling "shedded" decoy receptor are a key part of the cytokine network has led to the development of important anti-inflammatory agents such as the soluble TNFR for the treatment of rheumatoid arthritis. Additionally, the realization that IL-1ra knockout mice in a BALB/c genetic background develop spontaneous arthritis has led to the successful development of Kineret, a synthetic form of IL-1ra used to treat rheumatoid

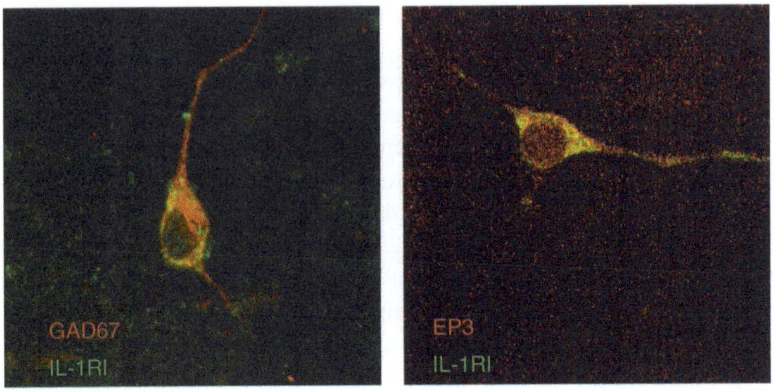

Figure 4. **GABAergic neurons in the hypothalamic culture express receptors for the pyrogens: IL-1β and PGE2 (EP3receptor).** Primary anterior hypothalamic neurons. Cultures (DIV 14) were fixed in 4% paraformaldehyde and detection of markers was obtained by incubation with antibodies against IL-1RI (Pharmingen), MAP-2 (Upstate), GAD67 (Chemicon), and EP3 (Alpha Diagnostic). AlexaFluor 594 and 488 were used as secondary antibodies.

arthritis [63]. Additional research has shown that the administration of IL-1Ra protects against cerebral ischemia in rodent models and in humans in a phase II clinical trial, indicating that it could potentially be used as a therapeutic for strokes [64,65].

Anti-inflammatory cytokines, such as IL-1ra, and IL-10, have been shown to attenuate the fever response. IL-1ra reduces the fever response to IL-1β, but not the fever produced in response to LPS or IL-6 [66]. IL-10 appears to attenuate the fever response to LPS, as well as IL-1β and thus must act in a more general manner than IL-1ra [67,68].

Hypothalamic neurons express IL-1RI as well as receptors for other pyrogens, such as EP3, the prostanoid receptor for prostaglandin E2 (PGE2) (Fig. 4). Many of these neurons are temperature sensitive and alter their firing rates in response to warming or cooling. Several groups have reported the effects of IL-1 on changing the firing properties of warm sensitive neurons [69–71]. It is assumed that the temperature-sensitive neurons are part of the circuitry that governs the set point and that the alteration of their firing rate by the endogenous pyrogen IL-1 is a direct way for this pyrogen to adjust the set point upward during fever [72].

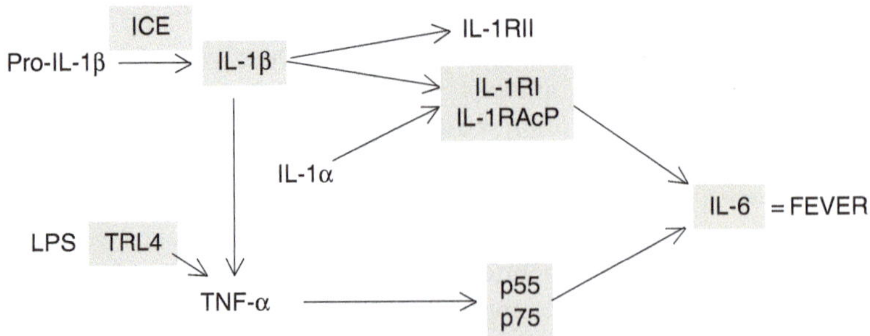

Figure 5. **Schematic representation of the cytokine cascade involved in the LPS and IL-1β-mediated fever response.** The cascade was determined using null animals for the genes ancodingencoding for some of the component molecules (highlighted) including the receptors for IL-1 and TNF-α.

4.3.2. Neuroinflammation and neurodegeneration

Studies on traumatic brain injury and stroke in humans, as well as in rodent models, have demonstrated that cytokine levels rise dramatically following neurotrauma [73–77]. This elevation of cytokine level is characterized by two distinct phases of clinical importance: (1) An early phase in which the rise of pro-inflammatory cytokines results from the local activation of microglia. During this phase, cytokines are produced within the injured brain with the blood–brain barrier still intact and impermeable to systemic cytokines except for the minimal contribution of the cytokine transporter [78]. (2) A late phase usually 12–24 h after injury or stroke, characterized by damage of the blood–brain barrier either through pathophysiological mechanisms including high cranial pressure and bradykinin release or neurosurgical intervention. During this phase, peripherally produced cytokines and immunocompetent cells, also capable of producing and responding to cytokines, can enter the brain and substantially contribute to the rise in cytokine levels and to their overall effect [79].

A key biochemical event following mechanical or ischemic neurotrauma is the massive release of the excitotoxic, depolarizing neurotransmitter glutamate. Glutamate can induce large changes in intracellular Ca^{2+} concentrations by acting at the NMDA and AMPA receptors, with subsequent transcriptional and redox changes. Cytokine genes respond rapidly and potently to glutamate overstimulation; neurons and microglia start a sustained production of IL-1β and TNF-α, which in turn induce IL-6, IL-1ra, IL-4, and IL-10. Following ischemic insults or seizures, IL-1β is the first cytokine to be induced, whereas in mechanical trauma, the levels of TNF-α rise faster than IL-1β. However, as IL-1β and TNF-α stimulate their reciprocal production through cross-inducibility, these imbalances rapidly disappear over time.

TNF-α appears to potentiate AMPA-induced excitotoxicity through TNFR1 (p55) [80]. It can act through TNFR1 to cause a reduction in the number of surface gamma-aminobutyric acid (GABA) receptors and therefore decreased inhibitory synaptic strength, thus potentially exacerbating excitotoxicity [81]. IL-1β acts through IL-1R1 and appears to cause cortical neuron death through ERK phosphorlyation [82]. It has also been shown in mice that IL-1ra administration after a traumatic brain injury reduces both the lesion volume and improves performance on cognitive tests [83]. As excessive activation of the NMDA receptor leads to excitotoxicity, IL-1 can synergize with glutamate in killing neurons with the underlying molecular mechanism being the enhancement of NMDA receptor-mediated signaling to allow large Ca^{2+} influx into the neurons. Conversely, endogenous IL-6 seems to protect neurons in cerebral ischemia. An IL-6 receptor antibody that blocked IL-6 signaling resulted in an increased infarct size and number of apoptotic cells [84]. Specifically, IL-6 has been shown to be important in protecting neurons from glutamate-induced neurotoxicity [85].

Other strongly depolarizing neurotransmitters, including substance P, are capable of inducing cytokine biosynthesis, an effect that may be of importance in neuronal injury and the development of neuropathic pain. The ability to counteract the initial surge of pro-inflammatory cytokines after stroke or traumatic brain injury with a strong induction of the anti-inflammatory cytokines IL-1ra [86], IL-4, and IL-10 could be critical for the clinical outcome and survival. A study by Bartfai and Danielsson has shown that patients who could not increase anti-inflammatory cytokine levels within 2 days of traumatic brain injury had significantly worse neurological outcomes than patients with similar injuries who could increase the production of IL-1ra within 24–36 h after the injury (unpublished observations). Thus, the individual ability to cope with secondary damage after neurotrauma may depend on polymorphism in the genes encoding for cytokines or for their receptors. A large number of polymorphisms in these genes have been associated with increased risk for different inflammatory diseases [87–105]. While major epidemological work has been devoted to determining the association between cytokine polymorphisms and large inflammatory diseases such as rheumatoid arthritis [88] and Crohn's disease [101], an association between cytokine gene polymorphisms and the outcome of neuronal injury and neurodegenerative diseases might be established as soon as a sufficient number of neurotrauma patients are genotyped.

IL-1 has also been shown to play a role in Alzheimer's disease. IL-1 acting at IL-1RI can induce the secretion of the amyloidogenic species Aβ1-40/42, which itself may stimulate microglia activation and IL-1 production, thus generating a cycle leading to the production and deposition of Aβ1-40/42 [106,107].

5. CONCLUSIONS

In summary, there is increasing evidence for cytokine receptor involvement in neuronal signaling in the CNS during noninflammatory conditions. Reports on direct neuronal effects of cytokines have been observed electrophysiologically and biochemically, suggesting that neuronal cytokine receptors are functionally active. Hippocampal, pituitary, locus coeruleus, and amygdala neurons have shown altered activity upon cytokine activation. Additionally, it is clear that cytokine receptors play a critical role in cytokine signaling in inflammatory, injured, and diseased states. Elevated cytokine receptors are associated with acute sickness syndrome and are known to contribute to neurodegeneration.

This chapter has focused on neuronal cytokine receptors. Nevertheless, it is clear that astrocytes and microglia not only produce cytokines but also express cytokine receptors. It is likely that future reports and reviews on cytokine receptors in the CNS will focus on the balance between neuronal and glial contribution of cytokine receptor function in healthy and diseased brains.

REFERENCES

1. Ban E, Milon G, Prudhomme N, Fillion G, Haour F. Receptors for interleukin-1 (alpha and beta) in mouse brain: mapping and neuronal localization in hippocampus. Neuroscience 1991;43(1):21–30.
2. Ban EM. Interleukin-1 receptors in the brain: characterization by quantitative in situ autoradiography. Immunomethods 1994;5(1):31–40.
3. Bartfai T, Schultzberg M. Cytokines in neuronal cell types. Neurochem Int 1993;22(5):435–44.
4. Bellinger FP, Madamba S, Siggins GR. Interleukin 1 beta inhibits synaptic strength and long-term potentiation in the rat CA1 hippocampus. Brain Res 1993;628(1–2):227–34.
5. Cornfield LJ, Sills MA. High affinity interleukin-6 binding sites in bovine hypothalamus. Eur J Pharmacol 1991;202(1):113–15.
6. Ericsson A, Liu C, Hart RP, Sawchenko PE. Type 1 interleukin-1 receptor in the rat brain: distribution, regulation, and relationship to sites of IL-1-induced cellular activation. J Comp Neurol 1995;361(4):681–98.
7. Frost P, Barrientos RM, Makino S, Wong ML, Sternberg EM. IL-1 receptor type I gene expression in the amygdala of inflammatory susceptible Lewis and inflammatory resistant Fischer rats. J Neuroimmunol 2001;121(1–2):32–39.
8. Katsuura G, Gottschall PE, Arimura A. Identification of a high-affinity receptor for interleukin-1 beta in rat brain. Biochem Biophys Res Commun 1988;156(1):61–67.

9. Ward H, Vigues S, Poole S, Bristow AF. The rat interleukin 10 receptor: cloning and sequencing of cDNA coding for the alpha-chain protein sequence, and demonstration by western blotting of expression in the rat brain. Cytokine 2001;15(5):237–40.
10. Wong ML, Licinio J. Localization of interleukin 1 type I receptor mRNA in rat brain. Neuroimmunomodulation 1994;1(2):110–15.
11. Sriram K, Matheson JM, Benkovic SA, Miller DB, Luster MI, O'Callaghan JP. Deficiency of TNF receptors suppresses microglial activation and alters the susceptibility of brain regions to MPTP-induced neurotoxicity: role of TNF-alpha. Faseb J 2006;20(6):670–82.
12. Ching S, He L, Lai W, Quan N. IL-1 type I receptor plays a key role in mediating the recruitment of leukocytes into the central nervous system. Brain Behav Immun 2005;19(2):127–37.
13. Lazovic J, Basu A, Lin HW, Rothstein RP, Krady JK, Smith MB, et al. Neuroinflammation and both cytotoxic and vasogenic edema are reduced in interleukin-1 type 1 receptor-deficient mice conferring neuroprotection. Stroke 2005;36(10):2226–31.
14. Cunningham ET Jr, Wada E, Carter DB, Tracey DE, Battey JF, De Souza EB. In situ histochemical localization of type I interleukin-1 receptor messenger RNA in the central nervous system, pituitary, and adrenal gland of the mouse. J Neurosci 1992;12(3):1101–14.
15. Gadient RA, Otten U. Differential expression of interleukin-6 (IL-6) and interleukin-6 receptor (IL-6R) mRNAs in rat hypothalamus. Neurosci Lett 1993;153(1):13–16.
16. Zubelewicz B, Braczkowski R, Renshaw D, Harbuz MS. Central injection of morphine stimulates plasma corticosterone and interleukin (IL)-6 and IL-6R mRNAs in the pitui tary and adrenals in adjuvant-induced arthritis. J Biol Regul Homeost Agents 1999;13(2):103–9.
17. Hulshof S, Montagne L, De Groot CJ, Van Der Valk P. Cellular localization and expression patterns of interleukin-10, interleukin-4, and their receptors in multiple sclerosis lesions. GLIA 2002;38(1):24–35.
18. Ledeboer A, Breve JJ, Wierinckx A, van der Jagt S, Bristow AF, Leysen JE, et al. Expression and regulation of interleukin-10 and interleukin-10 receptor in rat astroglial and microglial cells. Eur J Neurosci 2002;16(7):1175–85.
19. Mantovani A, Locati M, Vecchi A, Sozzani S, Allavena P. Decoy receptors: a strategy to regulate inflammatory cytokines and chemokines. Trends Immunol 2001;22(6):328–36.
20. Andre R, Moggs JG, Kimber I, Rothwell NJ, Pinteaux E. Gene regulation by IL-1beta independent of IL-1R1 in the mouse brain. GLIA 2006;53(5):477–83.
21. Grell M, Zimmermann G, Gottfried E, Chen CM, Grunwald U, Huang DC, et al. Induction of cell death by tumour necrosis factor (TNF) receptor 2, CD40 and CD30: a role for TNF-R1 activation by endogenous membrane-anchored TNF. Embo J 1999;18(11):3034–43.
22. McDermott MF. TNF and TNFR biology in health and disease. Cell Mol Biol 2001;47(4):619–35.
23. Pejovic V, Soskic V, Pan W, Kastin AJ. Brain proteome of mice lacking the receptors for tumor necrosis factor alpha. Proteomics 2004;4(5):1461–64.
24. Donnelly RP, Dickensheets H, Finbloom DS. The interleukin-10 signal transduction pathway and regulation of gene expression in mononuclear phagocytes. J Interferon Cytokine Res 1999;19(6):563–73.
25. Akira S. IL-6-regulated transcription factors. Int J Biochem Cell Biol 1997;29(12):1401–18.
26. Hibi M, Nakajima K, Hirano T. IL-6 cytokine family and signal transduction: a model of the cytokine system. J Mol Med 1996;74(1):1–12.

27. Hirano T, Ishihara K, Hibi M. Roles of STAT3 in mediating the cell growth, differentiation and survival signals relayed through the IL-6 family of cytokine receptors. Oncogene 2000;19(21):2548–56.
28. Conti P, Kempuraj D, Frydas S, Kandere K, Boucher W, Letourneau R, et al. IL-10 subfamily members: IL-19, IL-20, IL-22, IL-24 and IL-26. Immunol Lett 2003;88(3):171–74.
29. Davis CN, Tabarean I, Gaidarova S, Behrens MM, Bartfai T. IL-1beta induces a MyD88-dependent and ceramide-mediated activation of Src in anterior hypothalamic neurons. J Neurochem 2006;98(5):1379–89.
30. Sanchez-Alavez M, Tabarean IV, Behrens MM, Bartfai T. Ceramide mediates the rapid phase of febrile response to IL-1beta. Proc Natl Acad Sci USA 2006;103(8): 2904–8.
31. Tabarean IV, Korn H, Bartfai T. Interleukin-1beta induces hyperpolarization and modulates synaptic inhibition in preoptic and anterior hypothalamic neurons. Neuroscience 2006;141(4):1685–95.
32. Martin SJ, Grimwood PD, Morris RGM. Synaptic plasticity and memory: an evaluation of the hypothesis. Annu Rev Neurosci 2000;23(1):649–711.
33. Katsuki H, Nakai S, Hirai Y, Akaji K, Kiso Y, Satoh M. Interleukin-1 beta inhibits long-term potentiation in the CA3 region of mouse hippocampal slices. Eur J Pharmacol 1990;181(3):323–26.
34. Cunningham AJ, Murray CA, O'Neill LA, Lynch MA, O'Connor JJ. Interleukin-1 beta (IL-1 beta) and tumour necrosis factor (TNF) inhibit long-term potentiation in the rat dentate gyrus in vitro. Neurosci Lett 1996;203(1):17–20.
35. Kelly A, Lynch A, Vereker E, Nolan Y, Queenan P, Whittaker E, et al. The anti-inflammatory cytokine, interleukin (IL)-10, blocks the inhibitory effect of IL-1 beta on long term potentiation. A role for JNK. J Biol Chem 2001;276(49):45564–72.
36. Curran B, O'Connor JJ. The pro-inflammatory cytokine interleukin-18 impairs long-term potentiation and NMDA receptor-mediated transmission in the rat hippocampus in vitro. Neuroscience 2001;108(1):83–90.
37. Ross FM, Allan SM, Rothwell NJ, Verkhratsky A. A dual role for interleukin-1 in LTP in mouse hippocampal slices. J Neuroimmunol 2003;144(1–2):61–67.
38. Avital A, Goshen I, Kamsler A, Segal M, Iverfeldt K, Richter-Levin G, et al. Impaired interleukin-1 signaling is associated with deficits in hippocampal memory processes and neural plasticity. Hippocampus 2003;13(7):826–34.
39. Ferrari D, Pizzirani C, Adinolfi E, Lemoli RM, Curti A, Idzko M, et al. The P2X7 receptor: a key player in IL-1 processing and release. J Immunol 2006;176(7):3877–83.
40. Viviani B, Bartesaghi S, Gardoni F, Vezzani A, Behrens MM, Bartfai T, et al. Interleukin-1beta enhances NMDA receptor-mediated intracellular calcium increase through activation of the Src family of kinases. J Neurosci 2003;23(25):8692–700.
41. Gabellec MM, Jafarian-Tehrani M, Griffais R, Haour F. Interleukin-1 receptor accessory protein transcripts in the brain and spleen: kinetics after peripheral administration of bacterial lipopolysaccharide in mice. Neuroimmunomodulation 1996;3(5):304–09.
42. Vezzani A, Moneta D, Conti M, Richichi C, Ravizza T, De Luigi A, et al. Powerful anticonvulsant action of IL-1 receptor antagonist on intracerebral injection and astrocytic overexpression in mice. Proc Natl Acad Sci USA 2000;97(21):11534–39.
43. Vezzani A, Moneta D, Richichi C, Aliprandi M, Burrows SJ, Ravizza T, et al. Functional role of inflammatory cytokines and antiinflammatory molecules in seizures and epileptogenesis. Epilepsia 2002;43(Suppl 5):30–35.

44. Dube C, Vezzani A, Behrens M, Bartfai T, Baram TZ. Interleukin-1beta contributes to the generation of experimental febrile seizures. Ann Neurol 2005;57(1):152–55.
45. Gatti S, Vezzani A, Bartfai T. Putative Role of IL-1/IL-1ra in Febrile Seizures. San Diego: Academic Press; 2002.
46. Kanemoto K, Kawasaki J, Miyamoto T, Obayashi H, Nishimura M. Interleukin (IL)1beta, IL-1alpha, and IL-1 receptor antagonist gene polymorphisms in patients with temporal lobe epilepsy. Ann Neurol 2000;47(5):571–74.
47. Stepien H, Zerek-Melen G, Mucha S, Winczyk K, Fryczak J. Interleukin-1 beta stimulates cell proliferation in the intermediate lobe of the rat pituitary gland. J Endocrinol 1994;140(3):337–41.
48. Benigni F, Faggioni R, Sironi M, Fantuzzi G, Vandenabeele P, Takahashi N, et al. TNF receptor p55 plays a major role in centrally mediated increases of serum IL-6 and corticosterone after intracerebroventricular injection of TNF. J Immunol 1996;157(12):5563–68.
49. Dunn AJ. Cytokine activation of the HPA axis. Ann NY Acad Sci 2000;917:608–17.
50. Bluthe RM, Beaudu C, Kelley KW, Dantzer R. Differential effects of IL-1ra on sickness behavior and weight loss induced by IL-1 in rats. Brain Res 1995;677(1):171–76.
51. Dantzer R, Bluthe RM, Gheusi G, Cremona S, Laye S, Parnet P, et al. Molecular basis of sickness behavior. Ann NY Acad Sci 1998;856:132–38.
52. Ilyin SE, Gayle D, Flynn MC, Plata-Salaman CR. Interleukin-1beta system (ligand, receptor type I, receptor accessory protein and receptor antagonist), TNF-alpha, TGF-beta1 and neuropeptide Y mRNAs in specific brain regions during bacterial LPS-induced anorexia. Brain Res Bull 1998;45(5):507–15.
53. Bluthe RM, Dantzer R, Kelley KW. Effects of interleukin-1 receptor antagonist on the behavioral effects of lipopolysaccharide in rat. Brain Res 1992;573(2):318–20.
54. Bluthe RM, Pawlowski M, Suarez S, Parnet P, Pittman Q, Kelley KW, et al. Synergy between tumor necrosis factor alpha and interleukin-1 in the induction of sickness behavior in mice. Psychoneuroendocrinology 1994;19(2):197–207.
55. Goujon E, Parnet P, Cremona S, Dantzer R. Endogenous glucocorticoids down regulate central effects of interleukin-1 beta on body temperature and behaviour in mice. Brain Res 1995;702(1–2):173–80.
56. Ilyin SE, Plata-Salaman CR. An approach to study molecular mechanisms involved in cytokine-induced anorexia. J Neurosci Methods 1996;70(1):33–38.
57. Plata-Salaman CR. Cytokine-induced anorexia. Behavioral, cellular, and molecular mechanisms. Ann NY Acad Sci 1998;856:160–70.
58. De Sarro G, Gareri P, Sinopoli VA, David E, Rotiroti D. Comparative, behavioural and electrocortical effects of tumor necrosis factor-alpha and interleukin-1 microinjected into the locus coeruleus of rat. Life Sci 1997;60(8):555–64.
59. Bianchi M, Ferrario P, Clavenna A, Panerai AE. Interleukin-6 affects scopolamine-induced amnesia, but not brain amino acid levels in mice. Neuroreport 1997;8(7):1775–78.
60. Reichenberg A, Yirmiya R, Schuld A, Kraus T, Haack M, Morag A, et al. Cytokine-associated emotional and cognitive disturbances in humans. Arch Gen Psychiatry 2001;58(5):445–52.
61. Lynch AM, Lynch MA. The age-related increase in IL-1 type I receptor in rat hippocampus is coupled with an increase in caspase-3 activation. Eur J Neurosci 2002;15(11): 1779–88.
62. Lynch MA. Interleukin-1 beta exerts a myriad of effects in the brain and in particular in the hippocampus: analysis of some of these actions. Vitam Horm 2002;64:185–19.

63. Horai R, Saijo S, Tanioka H, Nakae S, Sudo K, Okahara A, et al. Development of chronic inflammatory arthropathy resembling rheumatoid arthritis in interleukin 1 receptor antagonist-deficient mice. J Exp Med 2000;191(2):313–20.
64. Emsley HC, Smith CJ, Georgiou RF, Vail A, Hopkins SJ, Rothwell NJ, et al. A randomised phase II study of interleukin-1 receptor antagonist in acute stroke patients. J Neurol Neurosurg Psychiatry 2005;76(10):1366–72.
65. Mulcahy NJ, Ross J, Rothwell NJ, Loddick SA. Delayed administration of interleukin-1 receptor antagonist protects against transient cerebral ischaemia in the rat. Br J Pharmacol 2003;140(3):471–76.
66. Chai Z, Alheim K, Lundkvist J, Gatti S, Bartfai T. Subchronic glucocorticoid pretreatment reversibly attenuates IL-beta induced fever in rats; IL-6 mRNA is elevated while IL-1 alpha and IL-1 beta mRNAs are suppressed, in the CNS. Cytokine 1996;8(3):227–37.
67. Zetterstrom M, Lundkvist J, Malinowsky D, Eriksson G, Bartfai T. Interleukin-1-mediated febrile responses in mice and interleukin-1 beta activation of NFkappaB in mouse primary astrocytes, involves the interleukin-1 receptor accessory protein. Eur Cytokine Netw 1998;9(2):131–38.
68. Zetterstrom M, Sundgren-Andersson AK, Ostlund P, Bartfai T. Delineation of the proinflammatory cytokine cascade in fever induction. Ann NY Acad Sci. 1998;856:48–52.
69. Nakashima T, Hori T, Mori T, Kuriyama K, Mizuno K. Recombinant human interleukin-1 beta alters the activity of preoptic thermosensitive neurons in vitro. Brain Res Bull 1989;23(3):209–13.
70. Shibata M, Blatteis CM. Differential effects of cytokines on thermosensitive neurons in guinea pig preoptic area slices. Am J Physiol 1991;261(5 Pt 2):R1096–103.
71. Vasilenko VY, Petruchuk TA, Gourine VN, Pierau FK. Interleukin-1beta reduces temperature sensitivity but elevates thermal thresholds in different populations of warm-sensitive hypothalamic neurons in rat brain slices. Neurosci Lett 2000;292(3):207–10.
72. Boulant JA. Role of the preoptic-anterior hypothalamus in thermoregulation and fever. Clin Infect Dis 2000;31 (Suppl 5):S157–61.
73. Leker RR, Shohami E. Cerebral ischemia and trauma-different etiologies yet similar mechanisms: neuroprotective opportunities. Brain Res – Brain Res Rev 2002;39(1):55–73.
74. Rothwell N, Allan S, Toulmond S. The role of interleukin 1 in acute neurodegeneration and stroke: pathophysiological and therapeutic implications. J Clin Invest 1997;100(11):2648–52.
75. Rothwell NJ, Luheshi GN. Interleukin 1 in the brain: biology, pathology and therapeutic target. Trends Neurosci 2000;23(12):618–25.
76. Shohami E, Ginis I, Hallenbeck JM. Dual role of tumor necrosis factor alpha in brain injury. Cytokine Growth Factor Rev 1999;10(2):119–30.
77. Touzani O, Boutin H, Chuquet J, Rothwell N. Potential mechanisms of interleukin-1 involvement in cerebral ischaemia. J Neuroimmunol 1999;100(1–2):203–15.
78. Banks WA, Plotkin SR, Kastin AJ. Permeability of the blood-brain barrier to soluble cytokine receptors. Neuroimmunomodulation 1995;2(3):161–65.
79. Pan W, Cain C, Yu Y, Kastin AJ. Receptor-mediated transport of LIF across blood-spinal cord barrier is upregulated after spinal cord injury. J Neuroimmunol 2006;174(1–2): 119–25.
80. Bernardino L, Xapelli S, Silva AP, Jakobsen B, Poulsen FR, Oliveira CR, et al. Modulator effects of interleukin-1beta and tumor necrosis factor-alpha on AMPA-induced excitotoxicity in mouse organotypic hippocampal slice cultures. J Neurosci 2005;25(29):6734–44.

81. Stellwagen D, Beattie EC, Seo JY, Malenka RC. Differential regulation of AMPA receptor and GABA receptor trafficking by tumor necrosis factor-alpha. J Neurosci 2005;25(12):3219–28.
82. Lu KT, Wang YW, Wo YY, Yang YL. Extracellular signal-regulated kinase-mediated IL-1-induced cortical neuron damage during traumatic brain injury. Neurosci Lett 2005;386(1):40–45.
83. Jones NC, Prior MJ, Burden-Teh E, Marsden CA, Morris PG, Murphy S. Antagonism of the interleukin-1 receptor following traumatic brain injury in the mouse reduces the number of nitric oxide synthase-2-positive cells and improves anatomical and functional outcomes. Eur J Neurosci 2005;22(1):72–78.
84. Yamashita T, Sawamoto K, Suzuki S, Suzuki N, Adachi K, Kawase T, et al. Blockade of interleukin-6 signaling aggravates ischemic cerebral damage in mice: possible involvement of Stat3 activation in the protection of neurons. J Neurochem 2005;94(2): 459–68.
85. Peng YP, Qiu YH, Lu JH, Wang JJ. Interleukin-6 protects cultured cerebellar granule neurons against glutamate-induced neurotoxicity. Neurosci Lett 2005;374(3):192–96.
86. Tehranian R, Andell-Jonsson S, Beni SM, Yatsiv I, Shohami E, Bartfai T, et al. Improved recovery and delayed cytokine induction after closed head injury in mice with central overexpression of the secreted isoform of the interleukin-1 receptor antagonist. J Neurotrauma 2002;19(8):939–51.
87. Fernandez-Real JM, Broch M, Vendrell J, Richart C, Ricart W. Interleukin-6 gene polymorphism and lipid abnormalities in healthy subjects. [see comments]. J Clin Endocrinol Metab 2000;85(3):1334–39.
88. Fishman D, Faulds G, Jeffery R, Mohamed-Ali V, Yudkin JS, Humphries S, et al. The effect of novel polymorphisms in the interleukin-6 (IL-6) gene on IL-6 transcription and plasma IL-6 levels, and an association with systemic-onset juvenile chronic arthritis. J Clin Invest 1998;102(7):1369–76.
89. Francis SE, Camp NJ, Dewberry RM, Gunn J, Syrris P, Carter ND, et al. Interleukin-1 receptor antagonist gene polymorphism and coronary artery disease. Circulation 1999;99(7):861–66 [see comments].
90. Grove J, Daly AK, Bassendine MF, Day CP. Association of a tumor necrosis factor promoter polymorphism with susceptibility to alcoholic steatohepatitis. Hepatology 1997;26(1):143–46 [see comments].
91. Higuchi T, Seki N, Kamizono S, Yamada A, Kimura A, Kato H, et al. Polymorphism of the 5′-flanking region of the human tumor necrosis factor (TNF)-alpha gene in Japanese. Tissue Antigens 1998;51(6):605–12.
92. Hohjoh H, Nakayama T, Ohashi J, Miyagawa T, Tanaka H, Akaza T, et al. Significant association of a single nucleotide polymorphism in the tumor necrosis factor-alpha (TNF-alpha) gene promoter with human narcolepsy. Tissue Antigens 1999;54(2):138–45.
93. Hohler T, Kruger A, Gerken G, Schneider PM, Meyer zum Buschenfelde KH, Rittner C. Tumor necrosis factor alpha promoter polymorphism at position -238 is associated with chronic active hepatitis C infection. J Med Virol 1998;54(3):173–77.
94. Huang D, Pirskanen R, Hjelmstrom P, Lefvert AK. Polymorphisms in IL-1beta and IL-1 receptor antagonist genes are associated with myasthenia gravis. J Neuroimmunol 1998;81(1–2):76–81.
95. Huang DR, Pirskanen R, Matell G, Lefvert AK. Tumour necrosis factor-alpha polymorphism and secretion in myasthenia gravis. J Neuroimmunol 1999;94(1–2):165–71.

96. Keen RW, Woodford-Richens KL, Lanchbury JS, Spector TD. Allelic variation at the interleukin-1 receptor antagonist gene is associated with early postmenopausal bone loss at the spine. Bone 1998;23(4):367–71.
97. Linker-Israeli M, Wallace DJ, Prehn JL, Nand R, Li L, Klinenberg JR. A greater variability in the 3′ flanking region of the IL-6 gene in patients with systemic lupus erythematosus (SLE). Autoimmunity 1996;23(3):199–209.
98. Liu ZH, Cheng ZH, Yu YS, Tang Z, Li LS. Interleukin-1 receptor antagonist allele: is it a genetic link between Henoch-Schönlein nephritis and IgA nephropathy? Kidney Int 1997;51(6):1938–42.
99. McGarry F, Walker R, Sturrock R, Field M. The -308.1 polymorphism in the promoter region of the tumor necrosis factor gene is associated with ankylosing spondylitis independent of HLA-B27. J Rheumatol 1999;26(5):1110–16.
100. Morse HR, Bidwell JL, Raveche ES. A poly(C) repeat polymorphism in the promoter of the IL-10 gene in NZB mice. Eur J Immunogenet 1999;26(5):377–78.
101. Negoro K, Kinouchi Y, Hiwatashi N, Takahashi S, Takagi S, Satoh J, et al. Crohn's disease is associated with novel polymorphisms in the 5′-flanking region of the tumor necrosis factor gene. Gastroenterology 1999;117(5):1062–68.
102. Roy S, McGuire W, Mascie-Taylor CG, Saha B, Hazra SK, Hill AV, et al. Tumor necrosis factor promoter polymorphism and susceptibility to lepromatous leprosy. J Infect Dis 1997;176(2):530–32.
103. Seki N, Kamizono S, Yamada A, Higuchi T, Matsumoto H, Niiya F, et al. Polymorphisms in the 5′-flanking region of tumor necrosis factor-alpha gene in patients with rheumatoid arthritis. Tissue Antigens 1999;54(2):194–97.
104. Terry CF, Loukaci V, Green FR. Cooperative influence of genetic polymorphisms on interleukin 6 transcriptional regulation. J Biol Chem 2000;275(24):18138–44.
105. Vinasco J, Beraun Y, Nieto A, Fraile A, Mataran L, Pareja E, et al. Polymorphism at the TNF loci in rheumatoid arthritis. Tissue Antigens 1997;49(1):74–78.
106. Griffin WS, Mrak RE. Interleukin-1 in the genesis and progression of and risk for development of neuronal degeneration in Alzheimer's disease. J Leukocyte Biol 2002;72(2):233–38.
107. Tha KK, Okuma Y, Miyazaki H, Murayama T, Uehara T, Hatakeyama R, et al. Changes in expressions of proinflammatory cytokines IL-1beta, TNF-alpha and IL-6 in the brain of senescence accelerated mouse (SAM) P8. Brain Res 2000;885(1):25–31.
108. Goslin K, Banker G. Experimental observations on the development of polarity by hippocampal neurons in culture. J Cell Biol 1989;108(4):1507–16.
109. Agnello D, Villa P, Ghezzi P. Increased tumor necrosis factor and interleukin-6 production in the central nervous system of interleukin-10-deficient mice. Brain Res 2000;869(1–2):241–43.
110. Goujon E, Parnet P, Laye S, Combe C, Dantzer R. Adrenalectomy enhances proinflammatory cytokines gene expression, in the spleen, pituitary and brain of mice in response to lipopolysaccharide. Brain Res Mol Brain Res 1996;36(1):53–62.
111. Utsuyama M, Hirokawa K. Differential expression of various cytokine receptors in the brain after stimulation with LPS in young and old mice. Exp Gerontol 2002;37(2–3):411–20.
112. Bristulf J, Bartfai T. Interleukin-1 beta and tumour necrosis factor-alpha stimulate the mRNA expression of interleukin-1 receptors in mouse anterior pituitary AtT- 20 cells. Neurosci Lett 1995;187(1):53–56.

Cytokines and the Brain
Edited by C. Phelps and E. Korneva

Interleukin-1 and Corticotropin-Releasing Factor Receptors in the Hypothalamic–Pituitary–Adrenal Axis

TOSHIHIRO TAKAO[1], KOZO HASHIMOTO[2], and ERROL B. DE SOUZA[3]

[1]*Division of Community Medicine, Department of Community Nursing, Kochi Medical School;*
[2]*Department of Endocrinology, Metabolism and Nephrology, Kochi Medical School, Nankoku 783-8505, Japan; and*
[3]*Archemix Corp. 300 Third Street, Cambridge, MA 02142, USA*

ABSTRACT

Interleukin-1 (IL-1) receptors were localized in mouse brain and pituitary using [^{125}I]IL-1α and [^{125}I]IL-1ra as radioligands. Receptor autoradiography and *in situ* hybridization studies demonstrated high densities and a discrete localization of IL-1 receptors and receptor mRNA, respectively, in the dentate gyrus of the hippocampus, choroid plexus, and anterior pituitary. Ether-laparotomy stress in mice resulted in a significant increase in [^{125}I]IL-1α binding in the pituitary with no significant alterations observed in the brain; in contrast, [^{125}I]oCRF binding in the pituitary was significantly decreased after the ether-laparotomy stress. The upregulation of IL-1 receptors in the mouse pituitary gland following ether-laparotomy stress was attenuated in a dose-dependent manner by systemic administration of corticotropin-releasing factor (CRF) receptor antagonist D-Phe^{12}-Nle^{21,38} human CRF(12-41)NH2. Moreover, i.p. injection of r/h CRF resulted in a dramatic increase in [^{125}I]IL-1α binding in the pituitary at 2 and 6 h after the injection although it did not affect [^{125}I]IL-1α binding in the hippocampus. Pretreatment with the nonpeptide, type 1 selective CRF antagonist, CRA 1000 significantly decreased ether-laparotomy stress-induced increases of IL-1R1 mRNA levels in the pituitary. Moreover, ether-laparotomy caused a significant increase of IL-1R1 mRNA in the pituitary of wild-type mice, and this increment of IL-1R1 mRNA in the pituitary was abolished in the CRF knockout (KO) mouse group. The treatment of AtT-20 mouse pituitary adenoma cells for 24 h with neuroendocrine mediators of stress such as CRF and catecholamine receptor (β_2 adrenergic) agonists produced a dose-dependent increase in cAMP and [^{125}I]IL-1α binding.

1. INTRODUCTION

Interleukin-1 (IL-1) and corticotropin-releasing factor (CRF) represent two candidates involved in coordinating the brain-endocrine-immune responses to stress. IL-1-like activity is present in the cerebrospinal fluid (CSF) [1,2], and IL-1 mRNA has been detected in rodent brain [3]. Immunohistochemical studies have identified neurons positive for IL-1β-like immunoreactivity in both hypothalamic [4,5] and extrahypothalamic [5] sites in human brain. Central as well as peripheral administration of IL-1 has potent neuroendocrine actions including stimulation of the hypothalamic–pituitary–adrenocortical (HPA) axis [6–9]. These effects of IL-1 are presumably mediated through actions of the cytokine at specific high-affinity receptors. Previous studies have identified at least two types of IL-1 receptors that are differentially expressed on the surface of certain types of immune cells and human- and murine-derived cell lines [10,11]. Recombinant human IL-1α and IL-1β bind to both type I receptors that are essential for all IL-1-mediated signaling [12,13] and type II receptors on various B-cell lines, including the Raji human B-cell lymphoma line [14,15]. More recent studies using several IL-1β mutants that differ in their affinities for the IL-1 type I receptor but have similar affinities for the IL-1 type II receptor have shown that IL-1β-induced adrenocorticotropin hormone (ACTH), corticosterone, and IL-6 production is mediated by IL-1 type I receptors [16].

CRF is a 41-amino-acid polypeptide in a large variety of mammalian species [17]. CRF mRNA and protein are abundantly distributed in the central nervous system (CNS) with major sites of expression in the paraventricular nucleus (PVN) of the hypothalamus, cerebral cortex, cerebellum, and the amygdalar–hippocampal complex, an area important for stress adaptation, learning, and memory [18]. CRF coordinates the adaptive behavioral and physical changes that occur during stress. However, when hypersecreted chronically, CRF causes symptoms pertaining to cognition, appetite, sleep, and anxiety [19]. Signals from CRF and CRF-related peptides are transduced across cell membranes through activation of two types of CRF receptor, R1 and R2, encoded by different genes [20–22]. The two CRF receptors belong to the class II G-protein-coupled receptor superfamily and share 70% homology at the amino acid level [23]. The CRF receptor type 1α (CRF-R1α) is a 415-amino acid protein, containing seven hydrophobic α-helices that are predicted to span the plasma membrane. CRF-R1α is widespread within both the CNS and the periphery [20]. Recently, small molecular weight CRF-R1 antagonists that could cross blood–brain barrier and could be taken orally have been developed for the treatment of stress-related disorders such as anxiety and depression [24,25]. In this chapter, we summarize some of the data from our recent studies using binding studies and *in situ* hybridization studies to localize IL-1 receptors in the mouse hypothalamus and pituitary. Then, we describe the modulation of IL-1 receptors and type 1 IL-1 receptor mRNA (IL-1R1 mRNA) by CRF using iodine-125-labeled recombinant human IL-1 ($[^{125}I]$IL-1)α binding and semi-quantitative reverse transcription–polymerase chain reaction (RT–PCR) to clarify interactions between IL-1 and CRF receptors in modulating the HPA axis function.

2. LOCALIZATION OF IL-1 RECEPTORS IN HYPOTHALAMUS

2.1. Distribution of $[^{125}I]$IL-1-binding sites: homogenate binding and receptor autoradiographic studies

The regional distribution of binding sites for $[^{125}I]$IL-1α was examined in homogenates of discrete areas of mouse CNS. The highest density of binding sites in mouse CNS was present in the hippocampus [26,27]. Progressively lower, but significant densities of binding sites were detected in hypothalamus.

$[^{125}I]$IL-1α or $[^{125}I]$IL-1 receptor antagonist (IL-1ra) were utilized as radioligands for autoradiographic studies. Overall, very low densities of $[^{125}I]$IL-1α- and $[^{125}I]$IL-1ra-binding sites were present throughout the brain. Very high densities and a discrete localization of IL-1 receptors were evident in the hippocampal formation and in the choroid plexus [26,27]. There was an absence of specific $[^{125}I]$IL-1ra or $[^{125}I]$IL-1α binding in the hypothalamus and other brain areas.

2.2. Distribution of type I IL-1 receptor mRNA: *in situ* hybridization studies

In situ hybridization histochemistry was used to investigate the distribution of cells expressing type I IL-1 receptor mRNA in the CNS. The strongest autoradiographic signal in the forebrain was found in the hippocampal formation [28]. The autoradiographic signal in the hypothalamic PVN and most aspects of median eminence were comparable to background [29]. An intense autoradiographic signal was found over endothelial cells of postcapillary venules throughout the CNS including the hypothalamic area, both in the parenchyma and at the pial surface [29].

2.3. Physiological and pharmacological actions

IL-1 might exert at least some of its central effects on the hypothalamic–pituitary axis at the level of the hippocampus, as has been postulated for glucocorticoids in the regulation of the HPA axis [30,31]. This possibility seems particularly appealing given the relative absence of [^{125}I]IL-1 binding and type I IL-1 receptor mRNA in ether the PVN or the median eminence although IL-1 has been shown to stimulate release of CRF from perfused rat ushypothalami [32,33] and to produce an increase in plasma ACTH following microinjection into the median eminence *in vivo* [34]. Another possibility is that IL-1 receptors in endothelial cells of postcapillary venules close to hypothalamic area may be involved in the regulation of HPA axis as it was reported that IL-1 in the brain is involved in the stress response and that stress-induced activation of monoamine release and the HPA axis were inhibited by IL-1ra administration directly into the rat hypothalamus [35]. A recent study demonstrated that IL-1R1s are expressed in endothelial cells of brain venules and suggested that vascular IL-1R1 distribution is an important factor determining blood–brain barrier prostaglandin-dependent activation of brain structures during infection [36]. However, the possibility that the radioligands used in the present study (recombinant human [^{125}I]IL-1α and [^{125}I]IL-1ra) only label a subtype of these receptors that is present in the hippocampus but is absent in the hypothalamus could not be excluded.

3. LOCALIZATION OF IL-1 RECEPTORS IN PITUITARY

3.1. Distribution of [^{125}I]IL-1 binding sites: receptor autoradiographic studies

There was a homogeneous distribution of [^{125}I]IL-1 α- and [^{125}I]IL-1ra-binding sites throughout the anterior lobe, suggesting that IL-1 may modulate the release of multiple anterior pituitary hormones as well as ACTH. No specific [^{125}I]IL-1ra or [^{125}I]IL-1α binding was present in the intermediate and posterior lobes of the pituitary [26].

3.2. Distribution of type I IL-1 receptor mRNA: *in situ* hybridization studies

A dense and homogeneously distributed autoradiographic signal was observed over the entire anterior lobe of the pituitary [29]. The autoradiographic signal over the posterior and intermediate lobes was comparable to background. The quality of counterstaining following proteinase K treatment of frozen sections did not allow for localization to specific cell types within the anterior lobe.

3.3. Physiological and pharmacological actions

A number of studies have demonstrated that IL-1 directly induces proopiomelanocortin (POMC)-derived peptide secretion from pituitary cells [37,38] and from AtT-20 mouse pituitary tumor cells [39,40] as well as inducing POMC-derived peptide secretion through stimulation of hypothalamic CRF [41]. IL-1 can also modify the release of other pituitary hormones such as luteinizing hormone, follicle-stimulating hormone, thyrotropin, growth hormone, and prolactin [42]. The present study demonstrating a homogeneous distribution of [^{125}I]IL-1-binding sites and type I IL-1 receptor mRNA in the anterior pituitary lobe further substantiates a generalized action of the cytokine to alter anterior pituitary hormone secretion.

4. MODULATION OF IL-1 RECEPTORS BY CRF IN THE HPA AXIS

4.1. Effect of ether-laparotomy stress on IL-1 and CRF receptors

In order to further investigate IL-1 receptors and CRF receptors in the HPA axis, we used ether-laparotomy stress. Male C57BL/6 mice were laparotomized under ether anesthesia (approximately 20 s exposure), and the small intestine was pulled for 10 s, and then the abdominal incision was sutured. Ether-laparotomy stress significantly increased plasma ACTH and corticosterone at 2 h after the onset of stress; plasma ACTH and corticosterone levels returned to nonstressed basal levels at 6 h after the stress [43] (Fig. 1). Iodine-125-labeled ovine CRF ([^{125}I]oCRF) binding in the pituitary was unchanged at 2 h but significantly decreased at 6 h after the ether-laparotomy stress (Fig. 2). Ether-laparotomy stress did not affect [^{125}I]oCRF binding in the frontal cortex both at 2 h and 6 h after stress when compared with nonstressed control mice [43] (Fig. 2). The demonstration that ether-laparotomy stress reduced CRF binding in the anterior lobe of the pituitary is in agreement with previous studies showing the

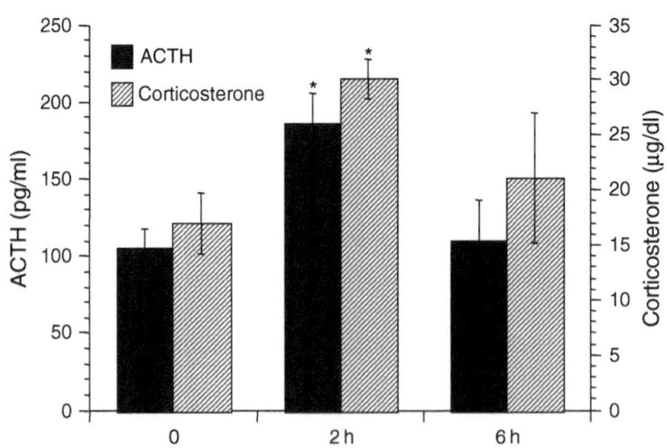

Figure 1. Plasma ACTH and corticosterone levels following ether-laparotomy stress. Mice were laparotomized under ether anesthesia, and the small intestine was pulled, and then the abdominal incision was sutured. The mice were decapitated at 2 h or 6 h after the onset of the anesthesia. Data represent the mean ± SEM (ACTH: control mice, $n = 15$, stressed mice, $n = 8$; corticosterone: control mice, $n = 13$, stressed mice, $n = 9$). *Significant alterations at $p < 0.05$ when compared to nonstressed controls. Reproduced with permission from Elsevier [43].

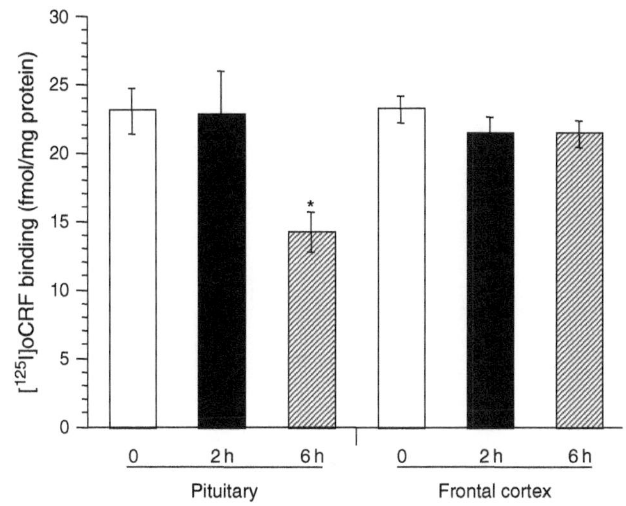

Figure 2. Regulation of [^{125}I]oCRF binding in the pituitary and frontal cortex following ether-laparotomy stress. The mice were decapitated at 2 h or 6 h after the onset of the anesthesia. The pituitary data represent the mean of five pooled pituitaries ± SEM (2 h and 6 h: $n = 3$; 0 h: $n = 6$). Data in the frontal cortex represent the mean ± SEM ($n = 5$; control $n = 10$). *Significant alterations at $p < 0.05$ when compared to nonstressed controls. Reproduced with permission from Elsevier [43].

downregulation of CRF receptors in the pituitary after immobilization stress [44] and footshock-induced stress [45]. A previous study demonstrated that ether-laparotomy increased plasma ACTH levels and decreased CRF concentrations in median eminence at 2 h after the stress [46], suggesting increased portal CRF levels which, in turn, downregulated pituitary CRF receptor. In contrast to the marked reduction of CRF binding in the pituitary, CRF receptor densities in the brain were not affected following ether-laparotomy stress, which is consistent with previous observations [44,47]. The mechanism in differential regulation of CRF receptors between brain and pituitary may be related to differences in the nature of hormone–receptor interactions in secretory and neural tissue [44].

In order to evaluate whether IL-1 receptors as well as CRF receptors are involved in the elevation of plasma ACTH and HPA axis modulation after stress, we investigated [^{125}I]IL-1α binding in the pituitary, hippocampus, spleen, and testis. [^{125}I]IL-1α binding in the pituitary was significantly increased at 2 h following ether-laparotomy stress and tended to be higher than non-stressed levels at 6 h after the stress, but was not statistically significant [43] (Fig. 3). In contrast, [^{125}I]IL-1α binding in the hippocampus, spleen, and testis was unchanged at both 2 h and 6 h after the onset of ether-laparotomy stress (Fig. 3).

As the measurement of IL-1 binding has been obtained using a ligand concentration in excess of the K_D value (50–100 pM), it is likely that the stress-induced increase in [^{125}I]IL-1α binding in the pituitary most likely represents an increase in receptor density; however, changes in receptor affinity cannot be excluded. These results are keeping with the observation by Ban et al. that IL-1 receptor number increased within 5–6 h after restraint stress in pituitary but not in the hippocamus [48]. The differences in the peak increases in IL-1 receptors between the studies may possibly be attributed to the use of different stressors, as they studied restraint stress and we utilized ether-laparotomy stress. Furthermore, it has been reported that sound, cold exposure or footshock stress did not modify IL-1-binding levels in the hippocampus [49],

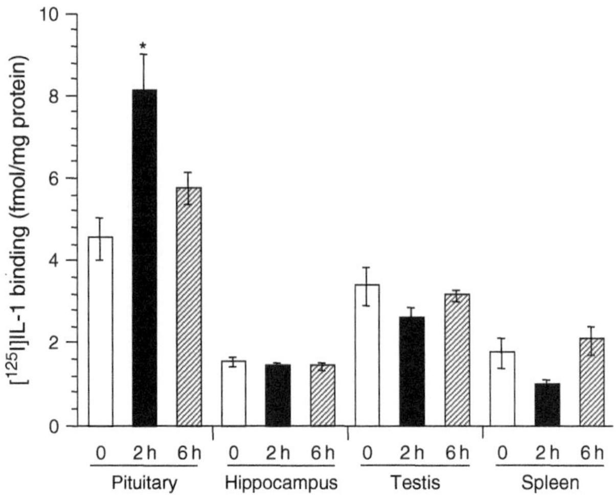

Figure 3. Regulation of [^{125}I]IL-1α binding in the pituitary, hippocampus, spleen, and testis following ether-laparotomy stress. The mice were decapitated at 2 h or 6 h after the onset of the anesthesia. The pituitary data represent the mean of five pooled pituitaries ± SEM (2 h and 6 h: $n = 3$; 0 h: $n = 6$). Data in the hippocampus, spleen, and testis represent the mean ± SEM ($n = 5$; 0 h: $n = 10$). *Significant alterations at $p < 0.05$ when compared to nonstressed controls. Reproduced with permission from Elsevier [43].

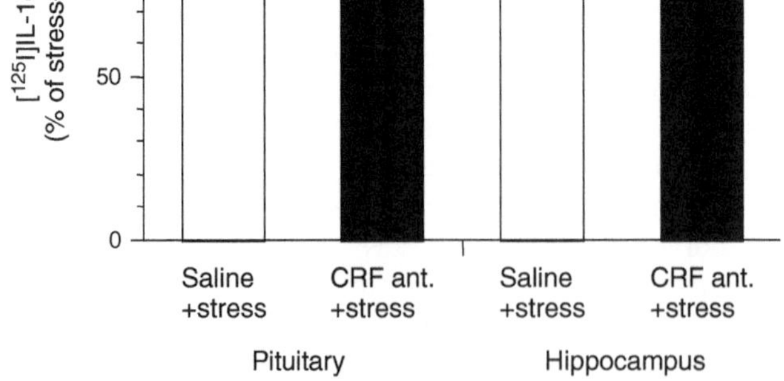

Figure 4. Effect of i.p. injection of 800 μg/kg of [D-Phe^{12}-Nle^{21,38} human CRF(12-41)NH_2 (D-Phe CRF[12-41]), a nonselective CRF receptor antagonist (CRF ant.), or saline on ether-laparotomy stress-induced increase of [^{125}I]IL-1α binding in the pituitary and hippocampus. The mice were decapitated at 2 h after the onset of the stress. Data are expressed as a percent of stress-free controls. The pituitary data represent the mean of five pooled pituitaries ± SEM ($n = 3$). Data in the hippocampus represent the mean ± SEM ($n = 6$). *Significant alterations at $p < 0.05$ when compared to saline-injected stress-free group. Reproduced with permission from Elsevier [50].

suggesting that IL-1 involvement in stress response is confined to the pituitary. The upregulation of IL-1 receptors in the mouse pituitary gland following ether-laparotomy stress was attenuated in a dose-dependent manner by systemic administration of the selective, high-affinity CRF receptor antagonist D-Phe^{12}-Nle^{21,38} human CRF(12-41)NH2, further suggesting a role for endogenous CRF in stress-induced modulation of IL-1 receptors *in vivo* [50] (Fig. 4).

4.2. Effect of CRF treatment on IL-1 receptors

In order to further characterize the mechanisms involved in stress-induced regulation of IL-1 receptors, we examined [^{125}I]IL-1α binding following treatment with CRF (40 μg/kg/0.2 ml, i.p.) and/or the potent glucocorticoid analog dexamethasone (DEX, 1 mg/kg/0.2 ml of 4% ethanol-saline), two major regulators of the body's response to stress. Intraperitoneal injection of CRF significantly induced plasma ACTH and corticosterone elevation at 2 h after the injection and returned to basal levels at 6 h after the injection. [^{125}I]IL-1α binding in the pituitary was significantly increased at 2 h and 6 h following i.p. injection of CRF (Fig. 5). In contrast, [^{125}I]IL-1α binding in the hippocampus, spleen, and testis was unchanged at both 2 h and 6 h after the CRF injection [51] (Fig. 5). A single i.p. injection of DEX alone did not alter [^{125}I]IL-1α binding in the pituitary and hippocampus. However, i.p. injection of DEX inhibited CRF-induced increase of [^{125}I]IL-1α binding at 2 h after injection in the pituitary, but DEX did not affect CRF-induced increase of [^{125}I]IL-1α binding in the hippocampus [51].

The stress-induced alterations in IL-1 binding in the pituitary most likely contribute to alterations in the HPA axis. It has been demonstrated that CRF can sensitize the pituitary gland *in vivo* to the direct ACTH-releasing activity of IL-1 [52], suggesting that the CRF-induced changes in IL-1 receptors have functional consequences.

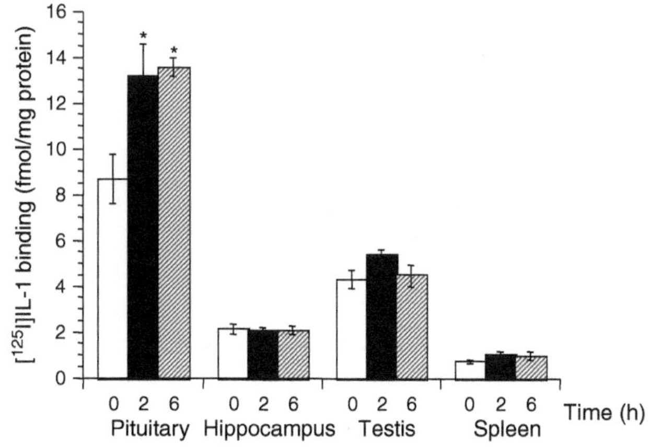

Figure 5. Effect of i.p. injection of rat/human CRF (40 μg/kg/0.2 ml of saline) on [^{125}I]IL-1α binding in the pituitary, hippocampus, testis, and spleen. The mice were decapitated at 2 h or 6 h after the onset of the injection. The pituitary data represent the mean of five pooled pituitaries ± SEM (2 h and 6 h: $n = 3$; 0 h: $n = 6$). Data in the hippocampus, spleen, and testis represent the mean ± SEM (2 h and 6 h: $n = 5$; 0 h: $n = 6$). *Significant alterations at $p < 0.05$ when compared to 0 h controls. Reproduced with permission from Elsevier [51].

4.3. Effect of nonpeptide CRFR1 antagonist, CRA 1000 on IL-1 receptors

Recently, nonpeptide CRF receptor antagonists were developed, which (by their chemical structure) offer the advantages of selectivity (for CRFR1 versus CRFR2), oral administration, long duration of action, and activity within the CNS [53,54]. Using these nonpeptidic CRFR1 antagonists, IL-1β-induced fever and anxiogenic-like behavior in the elevated plus-maze test were found to be mediated by CRFR1 in rats [53–55]. In order to evaluate the role of CRFR1 on modulation of IL-1 receptors by ether-laparotomy stress, we used the nonpeptide CRFR1 antagonist, CRA 1000 {2-[N-(2-methylthio)-4-isopropylphenyl]-N-ethylamino-4-[4-(3-fluorophenyl)-1,2,3,6-tetrahydropyridin-1-yl]-6-methylpyrimidine} and examined the effect of CRA 1000 on ether-laparotomy stress-induced IL-1 receptors and type I IL-1 receptor (IL-1R1) mRNA in the pituitary utilizing [^{125}I] IL-1α binding and semi-quantitative RT–PCR.

Ether-laparotomy stress resulted in a robust increase in [^{125}I] IL-1α binding in the pituitary at 2 h after the onset of the stress in the Vehicle-Stress group compared with the vehicle pretreatment without stress group. CRA 1000 pretreatment did not affect ether-laparotomy stress-induced increases of [^{125}I] IL-1α binding in the pituitary [56] (Fig. 6). Ether-laparotomy stress induced significant increases in IL-1R1 mRNA levels in the pituitary at 2 h after the onset of the stress in the Vehicle-Stress group compared with the vehicle pretreatment without stress group (Fig. 7). CRA 1000 pretreatment significantly decreased ether-laparotomy stress-induced increases of IL-1R1 mRNA levels in the pituitary compared with the vehicle pretreatment with stress group [56] (Fig. 7).

CRA 1000 did not affect ether-laparotomy induced increase of [^{125}I] IL-1α binding in the pituitary in our study. These discrepancies between CRA 1000 and D-Phe CRF (12-41) are somewhat puzzling; however, the differences in CRF receptor subtype selectivity between CRA 1000 (CRFR1 selective, nonpeptide antagonist) and D-Phe CRF (12-41) (CRFR nonselective,

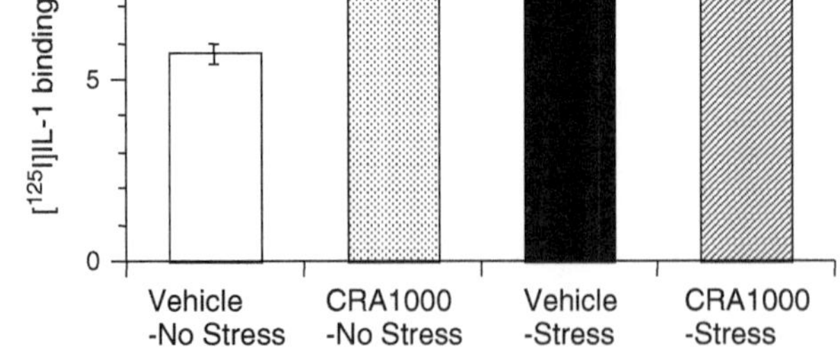

Figure 6. Effect of CRA 1000, a nonpeptide selective CRF receptor antagonist, on [^{125}I] IL-1α binding in the pituitary following ether-laparotomy stress in C57/BL6 mice. The mice were administered p.o. with CRA1000 (10 mg/kg) or 0.2 ml of vehicle (saline with 0.3% Triton-X) in the morning. Thirty minutes later, the mice were laparotomized while under ether anesthesia. The animals were sacrificed at 2 h after ether-laparotomy stress, and the tissues were dissected. Data represent the mean ± SEM ($n = 3$). * and ** represent significant alterations at $p < 0.05$ and $p < 0.01$ when compared to Vehicle-No Stress group. Reproduced with permission from Elsevier [56].

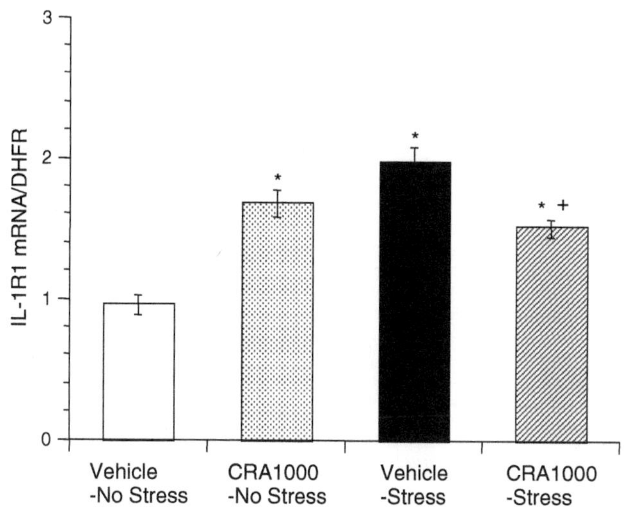

Figure 7. Effect of CRA 1000 on stress-induced alterations in IL-1R1 mRNA in the pituitary gland of C57/BL6 mice. The mice were administered p.o. with a nonpeptide selective CRF receptor antagonist CRA1000 (10 mg/kg) or 0.2 ml of vehicle (saline with 0.3% Triton-X) in the morning. Thirty minutes later, the mice were laparotomized while under ether anesthesia. The animals were sacrificed at 2 h after ether-laparotomy stress and the tissues were dissected. Data represent the mean ± SEM ($n = 5$). * and $^+$ represent significant alterations at $p < 0.01$ when compared to Vehicle-No Stress group and Vehicle-Stress group, respectively. Reproduced with permission from Elsevier [56].

peptide antagonist) may be responsible. CRF receptor blockade by CRA 1000 at the pituitary level may be less effective than D-Phe CRF (12-41). The reason that CRA 1000 inhibited ether-laparotomy stress-induced increases of IL-1R1 mRNA levels but not available receptor is unclear but may relate in part to the time of sacrifice which may have been insufficient for changes in mRNA levels to be reflected in altered receptor protein levels. It is reported that the IL-1-binding site was not identical with the type 1 IL-1 receptor but closely resembled the blood–brain barrier transporter [57], speculating that CRA 1000 may have differential effects to the transporter and the type 1 receptor. Moreover, IL-1 signaling and IL-1R-associated kinase activation may be involved in the discrepancy of modulation between IL-1R1 mRNA levels and available receptors [58].

4.4. Modulation of IL-1 receptors in the CRF knockout mouse

The role of CRF on IL-1 receptors during stress was further examined in CRF-deficient (knockout, KO) mice. We used male CRF-deficient (KO) mice, first, to evaluate the role of CRF deficiency on ether-laparotomy stress-induced activation of the HPA axis. Second, we examined the possible role of CRF in stress-induced modulation of IL-1R1 mRNA expression in the hippocampus, pituitary, and adrenal, utilizing semi-quantitative RT–PCR. The adult heterozygous CRF-deficient mice originally generated by Muglia and collaborators [59] were chosen as the first-generation parents. The heterozygous CRF-deficient male and female mice were mated to produce the F2 generation. To obtain a sufficient number of CRF KO mice within the same age range, the following pairs of F2 were mated: female heterozygous with male homozygous for the null allele. Wild-type (WT) mice were generated by interbreeding WT offspring from mating heterozygous parents [60].

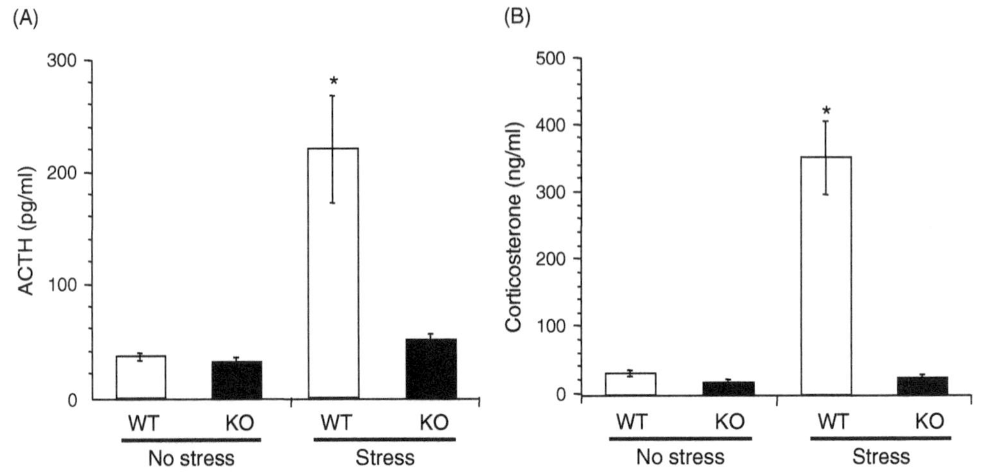

Figure 8. Effect of ether-laparotomy stress on plasma ACTH (A) and corticosterone (B) levels in wild-type (WT) and CRF knockout (KO) mice. The plasma was collected 2 h after ether-laparotomy stress and assayed for ACTH and corticosterone. Data represent the mean ± SEM ($n = 5$). *Significant alterations at $p < 0.01$ when compared to WT/No stress groups.

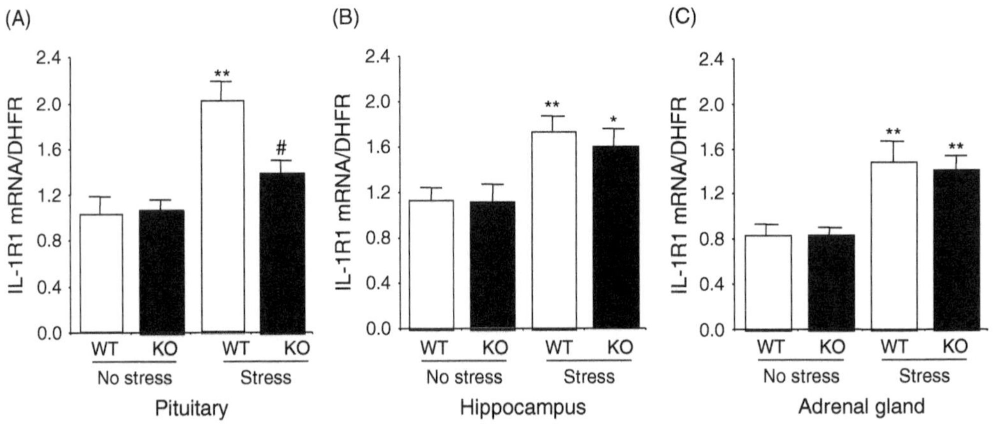

Figure 9. Effect of ether-laparotomy stress on IL-1R1 mRNA levels in the pituitary, hippocampus, and adrenal gland in wild-type (WT) and CRF knockout (KO) mice. The animals were sacrificed at 2 h after ether-laparotomy stress and the tissues were dissected. Data represent the mean ± SEM ($n = 5$). * and ** represent significant alterations at $p < 0.05$ and $p < 0.01$ when compared to no stress groups in each genotype, respectively. # represents significant alterations at $p < 0.01$ when compared to WT/Stress group.

Ether-laparotomy stress significantly increased plasma ACTH (Fig. 8A) and corticosterone (Fig. 8B) levels in WT/Stress group compared with WT/no stress group. In contrast, plasma ACTH and corticosterone levels were unchanged between stress and no stress groups in CRF KO mice. Ether-laparotomy caused a significant increase of IL-1R1 mRNA levels in the pituitary of WT mice (Fig. 9A). The increment of IL-1R1 mRNA in the pituitary was abolished in the KO/Stress group when compared with the WT/Stress group. In contrast, comparable increases of hippocampal (Fig. 9B) and adrenal (Fig. 9C) IL-1R1 mRNA levels were seen after ether-laparotomy stress both in the WT/Stress and KO/Stress groups. No significant differences were observed in basal (no stress) levels of IL-1R1 mRNA between genotypes in pituitary, hippocampus, and adrenal gland.

The elevation of plasma ACTH and corticosterone levels were abolished in the CRF KO mice. These data are in agreement with previous studies demonstrating absent or impaired ACTH and corticosterone responses following restraint stress and fasting in CRF KO mice [59,61,62]. In contrast, it was demonstrated that lipopolysaccharide (LPS) administration induced increases in plasma ACTH and corticosterone levels in KO mice, but the increases in hormone levels were smaller than in WT mice, suggesting that the role of CRF on HPA axis activation may depend on the type of stressor [63].

Ether-laparotomy stress caused a significant increase in IL-1R1 mRNA expression in the pituitary of WT mice; the increases in IL-1R1 mRNA expression in the pituitary was abolished in CRF KO mice, further suggesting that CRF plays a role in stress-induced IL-1R1 mRNA elevation in the pituitary although the mechanism(s) of the tissue-specific differences in IL-1R1 mRNA regulation followed by ether-laparotomy stress remains unclear.

4.5. Modulation of IL-1 receptors in AtT-20 mouse pituitary tumor cells

The mechanisms involved in the increased [^{125}I]IL-1α binding in the pituitary following stress are unknown. The alterations may be a consequence of increased synthesis of IL-1 receptors, unmasking of cryptic receptors and/or alterations in internalization of IL-1 receptors. The evidence of upregulation of IL-1 receptors in AtT-20 mouse pituitary adenoma cells following treatment with agents that stimulate cAMP production (e.g., CRF, isoproterenol, and forskolin) suggest that the increase in pituitary [^{125}I]IL-1α binding is secondary to stress-induced effects on CRF secretion and possibly other mediators including the catecholamines [64,65]. Conversely, treatment of AtT-20 cells with agents that inhibit adenylate cyclase activity such as somatostatin, decreased IL-1 binding, further suggesting that modulation of IL-1 receptors at the pituitary is most likely mediated through cAMP [64]. The CRF-induced increase in [^{125}I]IL-1α binding in AtT-20 cells appears to be mediated through specific membrane receptors for CRF because the CRF receptor antagonist, α-helical ovine CRF(9-41), blocked the CRF-induced increase in IL-1 receptors without producing any change in [^{125}I]IL-1α binding by itself. [^{125}I]IL-1α saturation assays were performed in CRF-treated and control cell cultures in order to determine whether the increase in [^{125}I]IL-1α binding following CRF treatment was related to changes in the affinity and/or concentration of IL-1 receptors on AtT-20 cells. Scatchard analysis of the saturation data indicated that the K_D values in the control- and CRF-treated cells were similar, while the density of receptors in the CRF-treated cultures was significantly higher than in the control-treated cells. The stress-induced alterations in IL-1 binding in the pituitary likely contribute to alterations in ACTH. Taken together, the data from ether-laparotomy stress and CRF treatment studies suggest that IL-1 may, in part, act at the pituitary level to modify and/or maintain ACTH under stress conditions. Additional studies will be required to determine the precise nature of the interaction between CRF and IL-1 receptors on the HPA axis by the endogenous ligands during stress.

5. SUMMARY AND CONCLUSIONS

IL-1 receptors were localized in mouse brain and pituitary using [^{125}I]IL-1α and [^{125}I]IL-1ra as radioligands. Receptor autoradiography and *in situ* hybridization studies demonstrated high densities and a discrete localization of IL-1 receptors and receptor mRNA, respectively, in the dentate gyrus of the hippocampus, choroid plexus, and anterior pituitary. Ether-laparotomy stress in mice resulted in a significant increase in [^{125}I]IL-1α binding in the pituitary with no significant alterations observed in the brain; in contrast, [^{125}I]oCRF binding in the pituitary was significantly decreased after the ether-laparotomy stress. The upregulation of IL-1 receptors in the mouse pituitary gland following ether-laparotomy stress was attenuated in a dose-dependent manner by systemic administration of CRF receptor antagonist D-Phe^{12}-Nle^{21,38} human CRF(12-41)NH2. Moreover, i.p. injection of r/h CRF resulted in a dramatic increase in [^{125}I]IL-1α binding in the pituitary at 2 h and 6 h after the injection although it did not affect [^{125}I]IL-1α binding in the hippocampus. Pretreatment with the nonpeptide, type 1 selective CRF antagonist, CRA 1000 significantly decreased ether-laparotomy stress-induced increases of IL-1R1 mRNA levels in the pituitary. Moreover, ether-laparotomy caused a significant increase of IL-1R1 mRNA in the pituitary of WT mice, and this increment of IL-1R1 mRNA in the pituitary was abolished in the CRF KO mouse group. The treatment of AtT-20 mouse pituitary adenoma cells for 24 h with neuroendocrine mediators of stress such as CRF and catecholamine receptor (β_2 adrenergic) agonists produced a dose-dependent increase in cAMP and [^{125}I]IL-1α binding.

These data provide further support for a role for IL-1 in coordinating HPA responses to stress and a role for CRF in modulating IL-1 receptors during stress conditions.

ACKNOWLEDGEMENTS

We are extremely grateful to Dr. Robert C. Newton and Ms. Maryanne Covington for providing recombinant human. IL-1s and Dr. Joseph A. Majzoub for providing the heterozygous CRH KO mice. The data presented in this chapter involved collaborative studies with Drs. Elizabeth L. Webster, Daniel E. Tracey, Klaus D. Dieterich, Tatsuya Nishioka, Koichi Asaba, Hossein Pournajafi Nazarloo, and Ms. Mitsuko Nakatukasa. We thank them for their contributions.

REFERENCES

1. Brisby H, Olmarker K, Larsson K, Nutu M, Rydevik B. Proinflammatory cytokines in cerebrospinal fluid and serum in patients with disc herniation and sciatica. Eur Spine J 2002;11:62–66.
2. Singhal A, Baker AJ, Hare GM, Reinders FX, Schlichter LC, Moulton RJ. Association between cerebrospinal fluid interleukin-6 concentrations and outcome after severe human traumatic brain injury. J Neurotrauma 2002;19:929–37.
3. Johnson AB, Bake S, Lewis DK, Sohrabji F. Temporal expression of IL-1beta protein and mRNA in the brain after systemic LPS injection is affected by age and estrogen. J Neuroimmunol 2006;174:82–91.
4. Breder CD, Dinarello CA, Saper CB. Interleukin-1 immunoreactive innervation of the human hypothalamus. Science 1988;240:321–24.
5. Breder CD, Saper CB. Interleukin 1b-like immunoreactive innervation in the human central nervous system. Soc Neurosci Abstr 1989;15:715.
6. DeKeyser FG, Leker RR, Weidenfeld J. Activation of the adrenocortical axis by surgical stress: involvement of central norepinephrine and interleukin-1. Neuroimmunomodulation 2000;7:182–88.

7. Mulla A, Buckingham JC. Regulation of the hypothalamo-pituitary-adrenal axis by cytokines. Baillieres Best Pract Res Clin Endocrinol Metab 1999;13:503–21.
8. Sapolsky R, Rivier C, Yamamoto G, Plotsky P, Vale W. Interleukin-1 stimulates the secretion of hypothalamic corticotropin-releasing factor. Science 1987;238:522–24.
9. Gadek-Michalska A, Bugajski J. Nitric oxide mediates the interleukin-1beta- and nicotine-induced hypothalamic-pituitary-adrenocortical response during social stress. J Physiol Pharmacol 2005;56:491–503.
10. Bomsztyk K, Sims JE, Stanton TH, Slack J, McMahan CJ, Valentine MA, Dower SK. Evidence for different interleukin 1 receptors in murine B- and T-cell lines. Proc Natl Acad Sci USA 1989;86:8034–38.
11. Chizzonite R, Truitt T, Kilian PL, Stern AS, Nunes P, Parker KP, Kaffka KL, Chua AO, Lugg DK, Gubler U. Two high-affinity interleukin 1 receptors represent separate gene products. Proc Natl Acad Sci USA 1989;86:8029–33.
12. Cullinan EB, Kwee L, Nunes P, Shuster DJ, Ju G, McIntyre KW, Chizzonite RA, Labow MA. IL-1 receptor accessory protein is an essential component of the IL-1 receptor. J Immunol 1998;161:5614–20.
13. Labow M, Shuster D, Zetterstrom M, Nunes P, Terry R, Cullinan EB, Bartfai T, Solorzano C, Moldawe rLL, Chizzonite R, McIntyre KW. Absence of IL-1 signaling and reduced inflammatory response in IL-1 type I receptor-deficient mice. J Immunol 1997;159:2452–61.
14. Dripps DJ, Verderber E, Ray KN, Thompson RC, Eisenberg SP. Interleukin-1 receptor antagonist binds to the type II interleukin-1 receptor on B cells and neutrophils. J Biol Chem 1991;266:20311–15.
15. Liu C, Hart RP, Liu XJ, Clevenger W, Maki RA, De Souza EB. Cloning and characterization of an alternatively processed human type II interleukin-1 receptor mRNA. J Biol Chem 1996;271:20965–72.
16. Van Dam AM, Malinowsky D, Lenczowski MJ, Bartfai T, Tilders FJ. Interleukin 1 (IL-1) type I receptors mediate activation of rat hypothalamus-pituitary-adrenal axis and interleukin 6 production as shown by receptor type selective deletion mutants of IL-1beta. Cytokine 1998;10:413–17.
17. Vale W, Spiess J, Rivier C, Rivier J. Characterization of a 41-residue ovine hypothalamic peptide that stimulates secretion of corticotropin and beta-endorphin. Science 1981;213:1394–97.
18. Bittencourt JC, Sawchenko PE. Do centrally administered neuropeptides access cognate receptors? an analysis in the central corticotropin-releasing factor system. J Neurosci 2000;20:1142–56.
19. Arzt E, Holsboer F. CRF signaling: molecular specificity for drug targeting in the CNS. Trends Pharmacol Sci 2006;27(10):531–38.
20. Chen R, Lewis KA, Perrin MH, Vale WW. Expression cloning of a human corticotropin-releasing-factor receptor. Proc Natl Acad Sci USA 1993;90:8967–71.
21. Perrin M, Donaldson C, Chen R, Blount A, Berggren T, Bilezikjian L, Sawchenko P, Vale W. Identification of a second corticotropin-releasing factor receptor gene and characterization of a cDNA expressed in heart. Proc Natl Acad Sci USA 1995;92: 2969–73.
22. Chen A, Perrin M, Brar B, Li C, Jamieson P, Digruccio M, Lewis K, Vale W. Mouse corticotropin-releasing factor receptor type 2alpha gene: isolation, distribution, pharmacological characterization and regulation by stress and glucocorticoids. Mol Endocrinol 2005;19:441–58.

23. Grammatopoulos DK, Chrousos GP. Functional characteristics of CRH receptors and potential clinical applications of CRH-receptor antagonists. Trends Endocrinol Metab 2002;13:436–44.
24. Gannon RL, Millan MJ. The corticotropin-releasing factor (CRF)(1) receptor antagonists CP154,526 and DMP695 inhibit light-induced phase advances of hamster circadian activity rhythms. Brain Res 2006;1083:96–102.
25. Overstreet DH, Keeney A, Hogg S. Antidepressant effects of citalopram and CRF receptor antagonist CP-154,526 in a rat model of depression. Eur J Pharmacol 2004;492:195–201.
26. Takao T, Culp SG, Newton RC, De Souza EB. Type I interleukin-1 (IL-1) receptors in the mouse brain-endocrine-immune axis labelled with ^{125}I-recombinant human IL-1 receptor antagonist. J Neuroimmunol 1992;41:51–60.
27. Takao T, Tracey DE, Mitchell WM, De Souza EB. Interleukin-1 receptors in mouse brain: characterization and neuronal localization. Endocrinology 1990;127:3070–78.
28. Cunningham ET Jr, Wada E, Carter DB, Tracey DE, Battey JF, De Souza EB. Localization of interleukin-1 receptor messenger RNA in murine hippocampus. Endocrinology 1991;128:2666–68.
29. Cunningham ET Jr, Wada E, Carter DB, Tracey DE, Battey JF, De Souza EB. *In situ* histochemical localization of type I interleukin-1 receptor messenger RNA in the central nervous system, pituitary and adrenal gland of the mouse. J Neurosci 1992; 12:1101–14.
30. Jacobson L, Sapolsky R. The role of the hippocampus in feedback regulation of the hypothalamic-pituitary-adrenocortical axis. Endocr Rev 1991;12:118–34.
31. Keller-Wood ME, Dallman MF. Corticosteroid inhibition of ACTH secretion. Endocr Rev 1984;5:1–24.
32. Navarra P, Tsagarakis S, Faria MS, Rees LH, Besser M, Grossman AB. Interleukin-1 and -6 stimulate the release of corticotropin-releasing hormone-41 from rat hypothalamus *in vitro* via the eicosanoid cyclooxygenase pathway. Endocrinology 1991;128:37–44.
33. Tsagarakis S, Gillies G, Rees LH, Besser M, Grossman A. Interleukin-1 directly stimulates the release of corticotropin releasing factor from rat hypothalamus. Neuroendocrinology 1989;49:98–101.
34. Sharp BM, Matta SG, Peterson PK, Newton R, Chao C, Mcallen K. Tumor necrosis factor-alpha is a potent ACTH secretagogue: comparison to interleukin-1 beta. Endocrinology 1989;124:3131–33.
35. Shintani F, Nakaki T, Kanba S, Kato R, Asai M. Role of interleukin-1 in stress responses. A putative neurotransmitter. Mol Neurobiol 1995;10:47–71.
36. Konsman JP, Vigues S, Mackerlova L, Bristow A, Blomqvist A. Rat brain vascular distribution of interleukin-1 type-1 receptor immunoreactivity: relationship to patterns of inducible cyclooxygenase expression by peripheral inflammatory stimuli. J Comp Neurol 2004;472:113–29.
37. Bernton EW, Beach JE, Holaday JW, Smallridge RC, Fein HG. Release of multiple hormones by a direct action of interleukin-1 on pituitary cells. Science 1987; 238:519–21.
38. Kehrer P, Turnill D, Dayer J-M, Muller AF, Gaillard RC. Human recombinant interleukin-1beta and -alpha, but not recombinant tumor necrosis factor alpha stimulate ACTH release from rat anterior pituitary cells in vitro in a prostaglandin E_2 and cAMP independent manner. Neuroendocrinology 1988;48:160–66.
39. Katahira M, Iwasaki Y, Aoki Y, Oiso Y, Saito H. Cytokine regulation of the rat proopiomelanocortin gene expression in AtT-20 cells. Endocrinology 1998;139:2414–22.

40. Kovalovsky D, Paez Pereda M, Labeur M, Renner U, Holsboer F, Stalla GK, Arzt E. Nur77 induction and activation are necessary for interleukin-1 stimulation of proopiomelanocortin in AtT-20 corticotrophs. FEBS Lett 2004;563:229–33.
41. MohanKumar SM, MohanKumar PS.: Systemic interleukin-1beta stimulates the simultaneous release of norepinephrine in the paraventricular nucleus and the median eminence. Brain Res Bull 2005;65:451–56.
42. Rothwell NJ. The endocrine significance of cytokines. J Endocrinol 1991;128:171–73.
43. Takao T, Tojo C, Nishioka T, Kurokawa H, Takemura T, Hashimoto K, De Souza EB. Reciprocal modulation of corticotropin-releasing factor and interleukin-1 receptors following ether-laparotomy stress in the mouse. Brain Res 1994;660:170–74.
44. Hauger RL, Millan MA, Lorang M, Harwood JP, Aguilera G. Corticotropin-releasing factor receptors and pituitary adrenal responses during immobilization stress. Endocrinology 1988;123:396–405.
45. Anderson SM, Kant GJ, De Souza EB. Effects of chronic stress on anterior pituitary and brain corticotropin-releasing factor receptors. Pharmacol Biochem Behav 1993; 44:755–61.
46. Nishioka T, Iyota K, Nakayama T, Suemaru S, Numata Y, Hashimoto K. Effects of ether-laparotomy and water immersion-restraint stress on CRH concentration in the hypothalamus, extrahypothalamic tissues and peripheral blood. Endocrine J 1993;40:213–20.
47. Hauger RL, Lorang M, Irwin M, Aguilera G. CRF receptor regulation and sensitization of ACTH responses to acute ether stress during chronic intermittent immobilization stress. Brain Res 1990;532:34–40.
48. Ban E, Marquette C, Sarrieau A, Fitzpatrick F, Fillion G, Milon G, Rostene W, Haour F. Regulation of interleukin-1 receptor expression in mouse brain and pituitary by lipopolysaccharide and glucocorticoids. Neuroendocrinology 1993;58:581–87.
49. Betancur C, Lledo A, Borrell J, Guaza C. Corticosteroid regulation of IL-1 receptors in the mouse hippocampus: effects of glucocorticoid treatment, stress, and adrenalectomy. Neuroendocrinology 1994;59:120–28.
50. Takao T, Tojo C, Nishioka T, Makino S, Hashimoto K, De Souza EB. Stress-induced upregulation of pituitary interleukin-1 receptors is mediated by corticotropin releasing factor. Life Sci 1996;59:PL165–68.
51. Takao T, Tojo C, Nishioka T, Hashimoto K, De Souza EB. Corticotropin-releasing factor treatment upregulates interleukin-1 receptors in the mouse pituitary: reversal by dexamethasone. Brain Res 1995;688:219–22.
52. Payne LC, Weigent DA, Blalock JE. Induction of pituitary sensitivity to interleukin-1: a new function for corticotropin-releasing hormone. Biochem Biophys Res Commun 1994; 198:480–84.
53. Lundkvist J, Chai Z, Teheranian R, Hasanvan H, Bartfai T, Jenck F, Widmer U, Moreau JL. A non peptidic corticotropin releasing factor receptor antagonist attenuates fever and exhibits anxiolytic-like activity. Eur J Pharmacol 1996;309:195–200.
54. Schulz DW, Mansbach RS, Sprouse J, Braselton JP, Collins J, Corman M, Dunaiskis A, Faraci S, Schmidt AW, Seeger T, Seymour P, Tingley Fr, Winston EN, Chen YL, Heym J. CP-154,526: a potent and selective nonpeptide antagonist of corticotropin releasing factor receptors. Proc Natl Acad Sci USA 1996;93:10477–82.
55. Baram TZ, Chalmers DT, Chen C, Koutsoukos Y, De Souza EB. The CRF1 receptor mediates the excitatory actions of corticotropin releasing factor (CRF) in the developing rat brain: in vivo evidence using a novel, selective, non-peptide CRF receptor antagonist. Brain Res 1997;770:89–95.

56. Pournajafi Nazarloo H, Takao T, Nanamiya W, Asaba K, De Souza EB, Hashimoto K. Effect of non-peptide corticotropin-releasing factor receptor type 1 antagonist on adrenocorticotropic hormone release and interleukin-1 receptors followed by stress. Brain Res 2001;902:119–26.
57. Banks W. Characterization of interleukin-1alpha binding to mouse brain endothelial cells. J Pharmacol Exp Ther 1999;291:665–70.
58. Ito A, Takii T, Matsumura T, Onozaki K. Augmentation of type I IL-1 receptor expression and IL-1 signaling by IL-6 and glucocorticoid in murine hepatocytes. J Immunol 1999;162:4260–65.
59. Muglia L, Jacobson L, Dikkes P, Majzoub JA. Corticotropin-releasing hormone deficiency reveals major fetal but not adult glucocorticoid need. Nature 1995;373:427–32.
60. Pournajafi Nazarloo H, Takao T, Taguchi T, Ito H, Hashimoto K. Modulation of type I IL-1 receptor and IL-1beta mRNA expression followed by endotoxin treatment in the corticotropin-releasing hormone-deficient mouse. J Neuroimmunol 2003;140:102–8.
61. Jacobson L, Muglia LJ, Weninger SC, Pacák K, Majzoub JA. CRH deficiency impairs but does not block pituitary-adrenal responses to diverse stressors. Neuroendocrinology 2000;71:79–87.
62. Jeong KH, Jacobson L, Pacak K, Widmaier EP, Goldstein DS, Majzoub JA. Impaired basal and restraint-induced epinephrine secretion in corticotropin-releasing hormone-deficient mice. Endocrinology 2000;141:1142–50.
63. Venihaki M, Majzoub JA. Animal models of CRH deficiency. Front Neuroendocrinol 1999;20:122–45.
64. Takao T, Dieterich KD, Tracey DE, De Souza EB. Cyclic AMP-dependent modulation of interleukin-1 receptors in the mouse AtT-20 pituitary tumor cell line. Brain Res 1994;656:177–81.
65. Webster EL, Tracey DE, De Souza EB. Upregulation of interleukin-1 receptors in mouse AtT-20 pituitary tumor cells following treatment with corticotropin-releasing factor. Endocrinology 1991;129:2796–98.

Cytokines and the Brain
Edited by C. Phelps and E. Korneva

Brain Interleukin-1β Expression and Action in the Absence of Neuropathology

LÉA CHASKIEL and JAN PIETER KONSMAN

PsychoNeuroImmunology, Nutrition and Genetics, UMR CNRS 5526/UMR INRA 1286/University Bordeaux 2, UFR Pharmacy, University Victor Ségalen Bordeaux 2, 146, rue Léo Saignat, 33076 Bordeaux Codex, France

ABSTRACT

The brain parenchyma has long been considered immunologically privileged based on the lack of a true lymphoid system. It is, however, clear now that brain infection or injury elicits innate immune responses including the release of cytokines, such as interleukin-1β (IL-1β). Interestingly, low levels of IL-1β are already present in the developing and adult brain in the absence of any infection or injury, indicating that this cytokine plays a role in the modulation of central nervous system functioning. The aim of this chapter is to provide some insight into the different roles IL-1β may play in the brain in the absence of neuropathology and to outline the difficulties and pitfalls associated with the study of brain IL-1β expression and action. Brain IL-1β expression during postnatal development has been hypothesized to regulate neuronal survival in a nerve growth factor (NGF)-dependent way. It is becoming clear that the constitutive low expression of IL-1β in the adult hippocampus plays a role in spatial learning and memory processes, possibly through its role in maintaining long-term potentiation. The circadian variation in expression of IL-1β in the adult forebrain is proposed to play a role in the physiological regulation of sleep, probably as a circulating factor in the cerebrospinal fluid. Constitutive IL-1β expression in neuroendocrine hypothalamic nuclei might act as a true neuropeptide and mediate neuroendocrine responses in response to physical stressors. The peripheral administration of bacterial lipopolysaccharides induces *de novo* IL-1β synthesis at the blood–brain interface without disrupting the blood–brain barrier. This transient IL-1β production at the blood–brain interface in response to peripheral immune stimuli is thought to constitute a signal mediating behavioral changes during infectious and inflammatory diseases. In conclusion, it is recognized increasingly that IL-1β production and action in the brain is involved in various neurobiological responses that are distinct from its classical role in mediating cellular responses to infection and injury.

1. INTRODUCTION

Neuroimmunology is the scientific discipline that studies interactions between the immune and the nervous systems. Despite its relative youth, it has a very respectable record in challenging prevailing dogmas. The brain was long considered an immunologically privileged organ based on the presence of a blood–brain barrier (BBB) and the lack of professional antigen-presenting cells as well as of a true lymphoid system in the brain parenchyma. Indeed, it has been argued that the appearance of the brain during evolution predates that of the adaptive immune system, which exerted certain constraints on the "colonization" of the brain parenchyma by lymphatic vessels and antigen-presenting cells, such as dendritic cells [1]. This view is evolving now that dendritic cells have been found in circumventricular organs, meninges, and choroid plexus where the BBB is nonfunctional as well as associated with vessels making up the BBB [2,3]. Interestingly, increased number of dendritic cells have been observed in multiple sclerosis or ischemic brain [3,4]. Moreover, it has recently been shown that augmenting the number of dendritic cells increases the clinical severity in an animal model of multiple sclerosis [3]. Furthermore, despite the absence of a true lymphoid system in the brain parenchyma, there is evidence indicating that antigens present in the brain interstitial fluid drain through peri-arterial spaces and cerebrospinal fluid to lymph nodes [5,6]. In addition, it has recently been shown that antigen-loaded dendritic cells injected into the brain can migrate to cervical lymph nodes and initiate specific T-cell homing to the central nervous system (CNS) [7]. These findings indicate that migration of dendritic cells throughout perivascular brain compartments plays an important role in immune invasion of the CNS in T-cell-dependent neuropathologies.

During the early stages of brain evolution, the brain acquired all innate immune effectors, including cells of mononuclear lineage and their soluble products in the form of cytokines and chemokines [1]. And when antigens are introduced carefully into the brain parenchyma, they do elicit transient innate immune responses, including the release of cytokines, and the recruitment of monocytes and neutrophils [1]. Research in this fast-moving field has now convincingly shown that, albeit different from the periphery, inflammatory, and immune responses can be mounted in the brain [8,9].

As one of the first reports on cytokine induction in the CNS described interleukin-1 (IL-1) expression in glial cells of the brains from patients afflicted with Alzheimer's disease [10], many of the subsequent studies have focused on the role of this pro-inflammatory cytokine in neuropathology. This focus on IL-1 expression in the CNS has also generated findings that could not be explained within the context of neuropathology. First, IL-1 activity has been found in the developing brain and suggested to act as a growth factor [11]. Second, IL-1β-like immunoreactivity has been observed in neurons of the adult human and rat brain in the absence of brain damage [12,13]. This together with the finding that IL-1 activity increases rapidly after exposure of animals to physical stressors [14,15] indicates that IL-1β may act as a neuromodulator. Finally, peripheral immune stimulation by administration of bacterial lipopolysaccharides (LPS) was found to induce *de novo* IL-1β synthesis at the blood–brain interface, suggesting that brain IL-1β induction is important in immune-to-brain signaling [16–19]. The aim of this chapter is to provide some insight into the different roles IL-1β may play in the healthy brain.

2. BRAIN IL-1β EXPRESSION AND ACTION: CHARACTERISTICS AND PITFALLS

2.1. IL-β expression and release

Before addressing the specific role of IL-1 in the brain, it is important to point out some of the general characteristics of IL-1. IL-1 is part of the family of cytokines, which can be defined as regulatory proteins controlling survival, growth and effector functions of tissue cells. Cytokines are produced by many different cell types and generally require *de novo* synthesis before release, as they are often not stored. The CNS seems to be an exception to this rule as we will see below. Once released, cytokines mostly act locally in an autocrine or paracrine manner at low concentrations. As various cell types express their receptors, cytokines have a wide variety of effects (pleiotropy). Finally, different cytokines can provoke the same biological effects (redundancy). These general features of cytokines bear consequences for the study of IL-1 expression and action in the brain.

The IL-1 family has three well-known endogenous ligands, two agonists: IL-1α, IL-1β, and one IL-1 receptor antagonist (IL-1ra). Some recently discovered molecules related to the classical IL-1 family members have been found to be expressed in the brain or brain-derived cells. However, as relatively few studies have addressed the role of these new IL-1 family members in brain pathology, the present review will mostly focus on the role of IL-1α and IL-1β.

Although IL-1α and IL-1β mostly induce similar effects when administered exogenously, their physiological roles probably differ. IL-1α and IL-1β are produced as immature propeptides upon transcription of different genes but differ in their maturation and secretion. Both propeptides lack hydrophobic leader sequences and remain in the cytosol [20]. Pro-IL-1α can be cleaved by calcium-dependent membrane proteases to yield mature IL-1α [21]. However, IL-1α rarely appears in extracellular biological fluids, and when it does, it is thought to originate from lysed cells and cleavage by extracellular proteases [22]. As pro-IL-1α is just as active as mature IL-1α and appears to remain intracellular or on the surface of mononuclear cells, IL-1α is thought to act intracellularly or through cell-to-cell contact [23,24].

Most of proIL-1β is present in the cytosol, but a fraction moves into specialized secretory lysosomes, where it colocalizes with procaspase-1 [25]. During the initiation of IL-1β synthesis, procaspase-1 is cleaved to caspase-1 or IL-1 converting enzyme (ICE), which, in turn cleaves pro-IL-1β to yield mature IL-1β [26]. The exact route by which IL-1β is secreted is still a matter of debate, but involves activation of the purinergic P2X7 receptor [27].

2.2. IL-1 type 1 receptor signaling

IL-1 signals through the IL-1 type 1 receptor (IL-1R1) [28] and its action is negatively regulated by IL-1ra and soluble IL-1 receptors that are processed by extracellular matrix proteins induced by IL-1 [29] and other pro-inflammatory cytokines. The type 2 IL-1 receptor does not contain an intracellular domain and is, therefore, thought to function as a decoy receptor limiting IL-1β's actions [30]. To induce biological responses after IL-1β binding, the IL-1R1 has to associate with the IL-1R accessory protein (AcP) to recruit the adapter protein MyD88 [31]. MyD88 subsequently binds the IL-1 receptor-associated kinase-1 (IRAK-1) [31], which, once phosphorylated, transiently binds the tumor necrosis factor receptor-associated factor 6 (TRAF6). IRAK-1 and TRAF6 then leave the IL-1R1 and interact at the membrane with a pre-existing complex of the transforming-growth factor β-activated kinase 1 (TAK1) and its binding proteins and provoke their phosphorylation [31]. IRAK-1 mediates the cytosolic translocation of the TRAF6-TAK1-binding protein complex after which IRAK-1 is degraded. The TRAF6-TAK1-binding proteins complex then forms a multiprotein complex with other cytosolic proteins and stimulates TAK1 kinase activity. TAK1, in turn, phosphorylates inhibitory factor κB kinase (IKK) and mitogen-activated protein kinases (MAPKK) and thus leads to activation of the transcription factor nuclear factor-κB (NFκB) and Jun terminal/p38/p42 MAPKs [31]. Interestingly, a splice variant of IRAK lacking kinase activity, and, therefore, thought to act as a dominant negative regulator, is the predominant form of IRAK found in the brain [32,33], suggesting that IL-1 signaling in the brain is tightly regulated.

2.3. Potential pitfalls when studying IL-1β expression and release in the brain

In vivo measurements of brain IL-1 production, for example, with push-pull probes, have to be interpreted with caution, because IL-1β and other pro-inflammatory cytokines are inevitably produced after a couple of hours as a consequence of the tissue damage created by the stab wound [34]. The same goes for *ex vivo* brain tissue blocks or slice preparations [35] and primary brain cultures [36]. Even though no more IL-1β is detected at a time point of study, one has to keep in mind that the animal, tissue explant, slice, or culture preparation has been subject to the effects of IL-1β and, therefore, does not necessarily reflect the situation in the healthy CNS. For example, IL-1β has been shown to upregulate expression of the IL-1R1 in brain cells both *in vitro* and *in vivo* [37,38], suggesting that animals with intracerebral cannulas, brain tissue explants, or cell cultures are more sensitive to the effects of IL-1β. These confounding effects may be restricted by placing guide cannulas at some distance of the brain site of interest through which an injection cannula ending at the site of interest is later inserted at the time of the actual experiment. In the case of tissue explants, it is recommended to study IL-1 release as soon as possible as to avoid picking up IL-1β production induced by the damage exerted on the tissue.

Given that cytokines are active at low concentrations, IL-1β synthesis may often be below the detection limit of anatomical techniques, such as immunocytochemistry and *in situ* hybridization [39]. In this respect, it is of interest to note that glucocorticoids inhibit both the expression and the action of IL-1β in the CNS [40,41]. Therefore, stress to the experimental animal should be limited to a strict minimum when studying brain IL-1β expression and action. Finally, when IL-1-like immunoreactivity (ir) is detected in the CNS, it is important to carry out Western blot or in situ hybridization control experiments, because other cytokines, such as acidic fibroblast growth factor, share amino acid sequences with IL-1 [42].

In most cases, the only available evidence in favor of IL-1 production in a brain structure is the detection of its immunoreactivity or mRNA in tissue homogenates. Before the advent of these techniques, the presence of IL-1 in tissue homogenates was based on the induction of IL-1-dependent biological responses in culture. Although bioassays are potentially sensitive, the fact that different cytokines can induce the same biological response represents a potential confounding factor. However, the main confounding factor associated with approaches involving brain tissue homogenates is the risk of false-positive signals due to presence of blood cells in brain parenchyma or meninges, because animals are often not perfused with saline prior to sacrifice. Finally, an inherent drawback of analyses performed on tissue homogenates is the inherent lack of anatomical detail. Hopefully, other approaches, including Laser Capture Microdissection allowing polymerase chain reaction (PCR) analysis on defined cell populations, will provide both the sensitivity and the anatomical detail required to further our understanding of IL-1 synthesis in the healthy CNS.

Despite technical difficulties and pitfalls, several groups have convincingly shown that IL-1β is present in healthy nervous tissue of several species, ranging from protochordates [43] through fish [44] and amphibians [45] to mammals, including rat [13,46,47], pig [48], and man [12,49]. This conserved expression of IL-1β in nervous tissue throughout evolution indicates that IL-1β plays important roles in nervous tissue.

2.4. Potential pitfalls when studying brain IL-1 receptors

As activation of a few dozen IL-1R1s can be sufficient to induce biological responses, many of the issues discussed above are also valid for the detection of IL-1R1 in the CNS by anatomical techniques, such as autoradiography, *in situ* hybridization, and immunohistochemistry. Faced with these difficulties, many researchers have changed strategies and studied behavioral and (electro)physiological responses after local IL-1 injection into brain structures or IL-1 application onto brain slices under the assumption that changes in the response are due to the presence of IL-1 receptors in the brain parenchyma of the structure of interest. Although similar approaches may have proven valid for neurotransmitters, it is important to point out that IL-1R1s are present on brain blood vessels and perivascular glia (see section 4.2) and that IL-1 induces the expression of enzymes synthesizing diffusing mediators, such as prostaglandins and nitric oxide, in these cells [38,50]. It is, therefore, impossible to conclude from the effects of IL-1 administration alone that IL-1 receptors are present in the parenchyma of a brain structure. Moreover, the electrophysiological effects of IL-1 concentrations above 1 nM are not blocked by the IL-1ra [51], indicating that it acts on receptors other than the IL-1R1. Given that the backbone structure of the putative receptor binding loop of the FGF is very similar to that of IL-1β [52,53] and that FGF receptors are expressed in the rodent brain [54,55], it is conceivable that high doses of IL-1β act on FGF receptors to induce biological effects.

The availability of the recombinant form of the IL-1ra has enabled researchers not only to control for the specificity of the effect of exogenously added IL-1, but also to study the roles of endogenous brain IL-1. However, given that low concentrations of endogenous IL-1β are sufficient to induce a biological response, high doses of IL-1ra are often necessary to prevent IL-1β's effects through the IL-R1. One has to keep in mind in this respect that high doses of cytokines administered into the CNS are rapidly cleared from cerebrospinal fluid and found at higher concentrations in plasma than after intravenous cytokine administration [56]. As IL-1β may exert effects at the periphery, which, in turn, alter brain functioning, it is important to monitor plasma levels of both IL-1ra and IL-1β in order to ascertain that intracerebral IL-1ra antagonizes only the action of endogenously produced brain IL-1β.

We have recently developed a new approach to study the presence of functional IL-1 receptors in the brain in an indirect manner and to test the role of IL-1 receptor-bearing nervous cells in a biological response. This approach consists of the local injection of a conjugate molecule of IL-1β coupled to the ribosome-inactivating plant protein saporin. Saporin is not toxic extracellularly, but rapidly kills cells once it has entered these [57]. Given saporin's characteristics, it is feasible to design receptor ligand-saporin conjugates that can be used as Trojan horses if the conjugate is internalized after ligand binding to the membrane receptor. Using an IL-1β-saporin conjugate, we have recently been able to kill glia and neurons proposed to express functional IL-1 receptors (see section 4.2).

3. BRAIN IL-1β AND RECEPTOR EXPRESSION DURING DEVELOPMENT

As mentioned, numerous studies have shown increased IL-1 expression by glial cells in response to CNS trauma or disease [10,58–60]. Interestingly, at the time of the first studies on IL-1 induction during neuropathology, IL-1 bioactivity was shown in embryonic and postnatal mouse brain [11]. The finding that microglia cultured from early postnatal brain produce high levels of IL-1 and show morphological and cytochemical similarities to activated or "amoeboid" microglia (based on its characteristic shape) observed in developing and pathological CNS [61] has led to the idea that IL-1 is produced by amoeboid microglia during CNS development [1]. It was only 10 years later that this hypothesis was tested and found to be wrong at least concerning postnatal development. Using IL-1β- and IL-1α-specific primers and

PCR amplification, IL-1 expression in cultured microglia was confirmed, but no IL-1 expression was observed in postnatal rat cortex known to contain amoeboid microglia [62]. Although oligodendrocytes have been shown to express IL-1β mRNA by double-labeling in situ hybridization on postnatal rat brain sections [63], no IL-1β mRNA was found in postnatal corpus callosum by reverse transcriptase (RT)–PCR [62]. The reasons for this discrepancy remain obscure, but it is at least clear from these studies that amoeboid microglia are not a predominant *in vivo* source of IL-1 during postnatal CNS development.

High levels of IL-1β are present during postnatal development of the rat olfactory bulb [64] and hippocampus [62]. IL-1β-ir in the postnatal olfactory bulb is found in fiber-rich layers and is dramatically reduced after sensory deprivation by naris closure [64], a procedure known to cause neuronal cell death. The expression of IL-1β-ir in the olfactory bulb also occurs in developing *Xenopus leavis* using antibodies raised against human IL-1β recognizing a 17 kDa frog protein [65]. This observation indicates that IL-1β expression in the developing olfactory bulb constitutes an evolutionary conserved phenomenon. In addition, IL-1β- and IL-1R1-ir are found in some hypothalamic cells, the fifth, seventh, and ninth cranial ganglia, primary motoneurons, and dorsal root ganglion cells of the developing frog nervous system [65]. As several IL-1β- and IL-1R1-expressing nervous cells are known to be long-lived neurons, these findings indicate that IL-1β and its signaling receptor are involved in the maintenance of cell survival during development.

Interestingly, the reduction in IL-1β-ir and increased neuronal death in the developing rat olfactory bulb after naris closure is accompanied by an increase in NGF p75 receptor-ir, suggesting that IL-1β interacts with NGF. NGF is produced in the olfactory bulb and hippocampus [66,67] during postnatal development and promotes the survival of mostly cholinergic neurons in the brain [68]. The observation that IL-1β, but not IL-1α, increases NGF production in astrocytes both *in vivo* and *in vitro* [46,69–71] indicates that IL-β regulates neuronal survival in an NGF-dependent manner in the postnatal brain. Given that the olfactory bulb and the dentate gyrus of the hippocampus are the main target of newly generated neurons in the adult nervous system [72] and that IL-1β mRNA in these structures is expressed in the same layers as NGF [46], IL-1β-induced NGF production may play a role in favoring survival of these neurons.

4. CONSTITUTIVE BRAIN EXPRESSION OF IL-1β AND ITS RECEPTORS

4.1. Brain IL-1β expression and production

IL-1β expression is not limited to brain development and occurs at low levels in the adult hippocampus, cerebellum, and hypothalamus as shown by RT–PCR [73–76]. Although these studies were performed on nonperfused brain tissue, the observation that expression levels in forebrain as opposed to hindbrain varied throughout the day [76] argues against the possibility that a false-positive signal due to cells present in blood or meninges alone accounts for the results obtained in these studies.

The low expression levels and diurnal variations may *a posteriori* explain why some *in situ* hybridization studies have shown IL-1β mRNA expression in the olfactory bulb, hippocampus [46], cerebellum, and hypothalamus of the healthy rat brain [77], whereas others have not [19,39]. Immunohistochemical stains revealed IL-1β-ir in cell bodies of the rat olfactory bulb and hippocampus after intracerebroventricular injection of colchicine [13], a toxin that blocks axonal transport and thus results in accumulation of peptide in neuronal cell bodies. It is important to note that colchicine has been shown to induce IL-1 expression in some cells [78] and that the above observations may, therefore, represent artifacts. However, as the immunohistochemical observations after colchicine treatment are in accordance with the results obtained by in situ hybridization on untreated brain tissue, it can be argued that they do not represent a nonspecific effect of colchicine administration. In addition, IL-1β-ir has been observed in neuronal cell bodies of the paraventricular, supraoptic, and accessory neuroendocrine nuclei of the human, pig, and rat hypothalamus without colchicine administration [47–49]. IL-1β-ir fibers are found in the olfactory bulb, hippocampus, bed nucleus of the stria terminalis, the suprachiasmatic nucleus and descend into the hypothalamic pituitary tract in the rat, pig, and human brain [12,47–49]. The presence of IL-1β in the olfactory bulb, hippocampus, and hypothalamus of the adult brain thus constitute exceptions to the general rule that IL-1β is not stored intracellularly and indicates that IL-1β might act as a neuromodulator in these neuronal circuits.

For IL-1β to be active, its proform need to be cleaved by ICE/caspase-1 to yield mature IL-1β. The mRNA coding for ICE/caspase-1 is constitutively expressed at low levels in the mouse and rat brain [79,80]. Moreover, ICE/caspase-1 is active in the hypothalamus and hippocampus as shown with a model peptide substrate [80], indicating that IL-1β can be cleaved in the healthy CNS. IL-1 is, indeed, bioactive in the healthy adult brain, because cell-free supernatant as well as cell lysate of brainstem, cortex, diencephalon, and hippocampus of normal rat contains IL-1 activity as measured with a sensitive T-lymphocyte cell proliferation assay [81]. As mentioned, bioassays are highly sensitive to detect IL-1, but can be compromised by the presence of molecules other than IL-1. Aware of this potential bias, the authors used a monoclonal antibody against the IL-1 receptor to show that the assay was specific for IL-1 [81]. It thus seems safe to say that IL-1 is released and bioactive in the healthy brain.

4.2. Brain IL-1 receptor expression

To induce biological effects, IL-1β has to act on the signaling IL-1R1. An initial study using radioactive recombinant murine IL-1α revealed binding sites in the rat olfactory bulb, hippocampus, choroid plexus, cerebellum, and to a lesser extent in the hypothalamus and anterior dorsal thalamus [82]. However, with radioactive recombinant rat IL-1β as a ligand, binding was observed in the rat choroid plexus only [83,84]. In the mouse brain, hippocampus and choroid plexus were labeled when using recombinant human IL-1α or IL-1β [85–87]. These observations indicate that the results obtained in IL-1-binding studies vary with the ligand and species used and render any firm conclusion as to the localization of IL-1 receptors problematic.

In situ hybridization studies have revealed IL-1R1 mRNA expression in the choroid plexus and postcapillary vennules of both rat and mouse brain [88–91]. IL-1R1 mRNA is strongly expressed in the mouse hippocampus, whereas its presence in the rat hippocampus has been subject to some controversy [89–91]. The olfactory bulb has only been investigated in the rat and was found to express IL-1R1 mRNA [89]. In addition, IL-1R1 mRNA was observed in rat, but not mouse, basomedial hypothalamus, basolateral amygdala, and in several brain stem motor nuclei [89–91]. Conversely, IL-1R1 mRNA expression has been reported to outline the brain stem raphe nuclei in mice [88], whereas in rats only vascular labeling has been described in this structure [89,91]. Although these findings suggest the existence of species differences in CNS expression of the IL-1R1, other factors may play a role. For example, IL-1R1 mRNA in the basolateral amygdala of the Fisher rat, which is characterized by high stress-induced corticosterone responses, is not detectable, whereas IL-R1 mRNA is strongly expressed in the same structure of the Lewis rat known for its attenuated corticosterone responses [92]. As important differences exist in hypothalamus–pituitary–adrenal axis reactivity between rats and mice [93], the failure to detect IL-1R1 mRNA in the mouse amygdala may be due to inhibition of its expression by glucocorticoids.

Although IL-1R1 mRNA in the rat and mouse brain is expressed in choroid plexus and postcapillary venules, the presence of IL-1R1 protein has until recently not been established in these structures. Using an antiserum raised against the extracellular portion of the rat recombinant IL-1R1, we have observed IL-1R1-immunoreactivity in brain postcapillary venules, perivascular macrophages, meninges, circumventricular organs, and choroid plexus, where it colocalized with IL-1R1 mRNA [94]. The IL-1 type 2 receptor is not found in the brain under basal conditions, but can be rapidly upregulated in blood vessels in response to focal brain inflammation [95]. With a well-characterized monoclonal antibody, IL-1R1-ir has been shown in neuronal soma of the mouse hippocampus and the cerebellum as well as in epithelial cells of the choroid, but not in postcapillary vennules, of the mouse brain [96]. In the rat CNS, only very weak IL-1R1-immunoreactivity has been described in the rat hippocampus [59]. The reasons for the discrepancies between studies are not yet clear, but may include differences in processing techniques.

Given the absence of IL-1 binding in the rat hippocampus, the weak IL-1R1-immunoreactivity [59] and the controversy with respect to neuronal IL-1R1 mRNA expression in this structure [89,91], the presence of functional IL-1 receptors in the rat hippocampus can be questioned. To address this issue, we injected IL-1β coupled to the intracellular toxin saporin into the dorsal hippocampus of the rat. As explained above, saporin is not toxic extracellularly, but rapidly kills cells once it has entered these [57]. As IL-1 is internalized after binding to its receptor [97], cells that bear functional IL-1 receptors will be killed after IL-1β-saporin binding to membrane IL-1 receptors. Following this approach, we have shown neuronal cell death in the dentate gyrus and CA1-3 regions of the rat hippocampus after injection of the IL-1β-saporin conjugate, but not after administration of equimolar quantities of nonconjugated IL-1 and saporin.

Strong neuronal staining of the rat paraventricular and supraoptic hypothalamus has been reported using antisera raised against peptide 156–170 from mouse IL-1R1 [98] (and our own unpublished observations). However, as this peptide only contains one sequence of three aligned antigenic amino acids of rat IL-1R1 and no IL-1R1 mRNA other than associated with blood vessels is present in the rat paraventricular and supraoptic hypothalamus [89–91,94], the capacity of these antisera to detect rat IL-1R1 can be questioned. Moreover, recent electrophysiological data show that the excitatory effects of IL-1β on supraoptic neurons are mimicked and mediated by prostaglandins through their action on EP4 receptors [99]. Altogether, these observations indicate that IL-1β acts on endothelial or perivascular cells of the supraoptic hypothalamus to induce the synthesis of prostaglandins, which, in turn, act on EP4 receptors to excite magnocellular neurons.

5. ROLE OF CONSTITUTIVE BRAIN IL-1B IN LEARNING AND SLEEP REGULATION

What can be the physiological roles of IL-1β in the adult brain? In many cases characterization of the nervous cell types expressing a given molecule or its receptor has provided hints regarding its functions.

5.1. Hippocampal IL-1, long-term potentiation, and spatial learning

In the hippocampus IL-1β may play a role in long-term potentiation (LTP), one of the mechanisms proposed to underlie learning and memory. Interestingly, LTP induces IL-1β mRNA and protein in the rat brain hippocampus both *in vitro* and *in vivo* [100]. Moreover, intracerebroventricular administration of IL-1ra impairs LTP [39] as well as hippocampus-dependent learning and memory in rats [101]. Finally, IL-1R1-deficient mice do not show hippocampal LTP and perform worse than wild-type mice in hippocampus-dependent behavioural learning tasks, including the Morris water maze and contextual fear conditioning [102]. These observations indicate that hippocampal IL-1β plays a role in spatial learning and memory processes by acting on neuronal IL-1R1s, possibly through its role in maintaining LTP.

5.2. Preoptic and brainstem IL-1 action and sleep regulation

IL-1β mRNA levels in the hypothalamus, hippocampus, and cortex are higher during the inactive as compared to the active daily phase of rats [103]. Interestingly, intracerebroventricular administration or cortical application of IL-1β increases nonrapid-eye-movement (NREM) sleep in animals [104–106]. As IL-1R1-deficient mice sleep less than wild-type controls [107], IL-1 can be considered an endogenous mediator of sleep regulation. Moreover, intracerebroventricular administration of anti-IL-1β just before the inactive phase reduced NREM sleep during the subsequent 12-h period [108]. Sleep deprivation further increases IL-1β mRNA in the hypothalamus, hippocampus, and cortex [103]. Under these conditions, intracerebroventricular administration of anti-IL-1β or cortical application of the soluble form of the IL-1 receptors attenuates NREM sleep rebound after sleep deprivation [106,108]. Taken together, these findings clearly demonstrate that brain IL-1β plays a role in the physiological regulation of NREM sleep.

Although the cellular source of IL-1β in the regulation of sleep remains unknown, it is interesting to note that the rat pineal gland, known to play an important role in diurnal rhythms, constitutively expresses IL-1β mRNA [109]. Just as in other central nervous structures, pineal IL-1β mRNA levels are higher during the inactive as compared to the active phase [109]. Given that the pineal gland sits atop of the cortex and that, at best, only a few scattered cells express IL-1β mRNA in the cortex [46], it is possible that the diurnal variation in "cortical" IL-1β expression is at least partly due to pineal IL-1β.

As melatonin produced by the pineal gland enters the cerebrospinal fluid of the third ventricle through the pineal recess [110], IL-1β can be proposed to follow the same route. Mimicking IL-1β action in the ventricular system by intracerebroventricular injection increases cerebrospinal fluid prostaglandin-D synthase levels [111]. Subsequent experiments have shown that IL-1β injected into the ventricular and subarachnoid spaces of the preoptic hypothalamus promotes NREM-sleep in a prostaglandin-dependent manner [112]. It has recently been shown that cultured leptomeninges, which express IL-1R1s *in vivo* [91,94], express prostaglandin-D synthase after stimulation with IL-1β [113], indicating that these tissues are one of the main targets of brain IL-1β involved in sleep regulation. As prostaglandins are easily diffusing mediators, prostaglandin-D can be proposed to mediate the reported effects of IL-1β application on sleep- and wake-related discharge patterns of neurons in the ventrolateral preoptic area [114], a structure known to be important role in NREM sleep regulation [115].

However, prostaglandin-D receptors are predominantly present on the arachnoid trabecular cells of the leptomeninges in the basal forebrain and not found in the parenchyma of the preoptic area [116,117]. When prostaglandin-D is injected into this area, it induces NREM sleep and adenosine release in wild-type mice, but not in prostaglandin-D receptor-deficient mice [117]. Interestingly, adenosine agonist mimics the somnogenic effects of prostaglandin-D, while adenosine A2 receptor antagonists attenuate prostaglandin-D-induced sleep [118]. This indicates that adenosine mediates the sleep-regulating effects of prostaglandin-D. Moreover, injection of an adenosine A2 receptor agonist or prostaglandin-D into the subarachnoid space induces the expression of the cellular activation marker c-Fos in the ventrolateral preoptic area [119,120]. Finally, adenosine, just like IL-1β (see above), excites a subset of sleep-promoting neurons through A2A receptors in the ventrolateral preoptic nucleus [121]. It is, therefore, likely that the effects of IL-1β on sleep- and wake-related discharge patterns of neurons in the ventrolateral preoptic area in freely moving animals [114] are mediated by adenosine in response to prostaglandin-D release. Prostaglandin-D is produced as a result of interaction of IL-1β with IL-1R1s on meningeal and perivascular cells.

Interestingly, the injection of IL-1 into the aqueduct of Sylvius also leads to an increase in NREM sleep [122], suggesting that IL-1 may act in the brainstem to regulate NREM sleep. Serotonin has a long-standing record in the regulation of sleep–wake cycles [123] and intracerebroventricular administration of IL-1 enhances serotonin release in the preoptic area as well as in the dorsal raphe nuclei containing the cell bodies of serotoneric neurons [123]. Moreover, serotonin depletion or administration of a serotonin receptor antagonist attenuates the increase in NREM sleep after intracerebroventricular injection of IL-1β [124,125]. As dorsal raphe injections of IL-1β at doses 5–10 times lower than those administered intracerebroventricularly enhance NREM sleep, it is likely that this nucleus contains IL-1-responsive cells mediating its effects on NREM sleep.

In situ hybridization for the IL-1R1 mRNA show positive reactivity in the midline raphe nuclei in mice [88]. It is conceivable that this receptor is expressed by neurons and that IL-1β in this species acts directly on serotonergic neurons in the dorsal raphe to promote NREM sleep. However, in the rat, IL-1R1 mRNA in the raphe nuclei is associated with the vasculature [91,94], indicating that brain IL-1β exerts its sleep-regulating effects through the induction of diffusible messengers, such as prostaglandins, in (peri)vascular cells. The sites and mechanisms mediating the NREM sleep-promoting effects of IL-1β in the brain stem thus await further study.

6. BRAIN IL-1β PRODUCTION AND ACTION DURING STRESS

6.1. Brain IL-1β production in response to physical stressors

As mentioned above, IL-1β-ir is found in neuronal cell bodies of the paraventricular, supraoptic, and accessory neuroendocrine nuclei of the human, pig, and rat hypothalamus [47–49]. Double labeling experiments revealed that IL-1β-ir in these brain nuclei as well as in fibers within the median eminence and neurohypophysis is present in neurons containing oxytocin or vasopressin [47,49]. Moreover, IL-1β in the neurohypophysis seems to be released in response to osmotic challenge as drinking of hypertonic salt solutions and lactation reduce the quantity of immunoreactive IL-1β terminals in the neurohypophysis [47].

Several observations indicate that pre-formed neuronal IL-1β can be released in the hypothalamus. First of all, rat hypothalamic explants release immunoreactive IL-1β in a ICE/caspase-1-dependent way in response to depolarizing stimuli or nitric oxide [126–129] within a time window too short to allow *de novo* IL-1β synthesis. Moreover, glucocorticoids and IL-10, which are known to inhibit IL-1β synthesis, do not affect IL-1β release from hypothalamic explants [126]. Second, some physical stressors, such as foot shock and immobilization, result in a rapid increase in immunoreactive IL-1β and IL-1 activity in the hypothalamus only [14,130,131]. However, metabolic stressors, such as hypoglycemia, and more psychological stressors, including restraint and social isolation, do not alter concentrations of immunoreactive IL-1β in the hypothalamus [132–134].

Interestingly, physical stressors including, immobilization [135,136], tailshocks [134], and formalin injection into hind paws [137], but not psychological stressors, such as predator exposure [138], also induce IL-1β mRNA expression in the hypothalamus. Although IL-1β mRNA induction occurs in the paraventricular nucleus, where IL-1β-immunoreactive neurons have been described, it seems to occur in glia cells at least in response to formalin injection [137]. Whether IL-1β synthesized in these glia cells somehow replenishes IL-1β-ir neurons in the paraventricular hypothalamus remains an outstanding question. Taken together, the data available in the literature now provide ample evidence to make the case that hypothalamic IL-1β is rapidly released upon physical stressors, including tailshock and immobilization, and may, therefore, play a role as a neuropeptide.

Stress-induced increases in immunoreactive IL-1β have been described in other brain structures than the hypothalamus, but were indistinguishable from rises in plasma IL-1β and occurred only in adrenalectomized rats with basal corticosterone levels maintained by subcutaneous pellets [14]. It can, therefore, not be excluded that the increases in brain IL-1β other than in the hypothalamus reflect rises in plasma IL-1β due to stress-induced bacterial translocation, which are normally prevented by corticosterone.

6.2. Brain IL-1 action in response to physical, metabolic, and psychological stressors

Only a handful of studies have addressed the role of brain IL-1 action in biological responses to stressors. Intrahypothalamic administration of IL-1ra just before, but not right after, immobilization inhibits subsequent rises in hypothalamic indol- and catecholamines as well as in plasma adrenocorticotropin hormone [130]. So, IL-1β seems to be rapidly released in the hypothalamus in response to physical stressors and to contribute to activation of the hypothalamo–pituitary–adrenal axis. Interestingly, antagonizing endogenous IL-1 action also attenuates endocrine responses to metabolic and psychological stressors, such as hypoglycemia and restraint. Indeed, IL-R1-deficient mice do not show the rise in circulating corticosterone levels observed in wild-type mice in response to restraint stress or hypoglycemia [139]. However, given that IL-R1-deficient mice lack IL-1R1 not only in the nervous system, but in all bodily tissues, and that restraint and hypoglycemia do not increase brain immunoreactive IL-1 (see above), it is currently not clear whether the results obtained are due to lack of endogenous IL-1 receptors in the brain. In this respect, it is important to note that overexpression of IL-1ra in the brain lowers brain monoamine responses to isolation stress [140]. This observation indicates that, even though no increase in brain IL-1 could be detected in response to isolation stress [133,134], endogenous brain IL-1 does play a role in mediating neurochemical changes to some psychological stressors.

7. BRAIN IL-1β INDUCTION AND ACTION DURING THE ACUTE PHASE RESPONSE

Peripheral tissue injury or infection triggers a complex series of reactions mounted by the host to prevent further damage, isolate, and destroy infectious organisms and initiate repair processes. These responses collectively known as the acute-phase response are rapidly activated by pro-inflammatory cytokines, including IL-1β, that are synthesized after detection of microbial products by tissue macrophages [141]. Peripheral tissue injury or infection is often accompanied by elevations in body temperature, activation of the hypothalamo–pituitary–adrenal axis and behavioral changes such as increased sleep, loss of appetite, and reduced interest in usual activities as well as cognitive impairment [142–145]. As these responses involve the CNS, but peripheral cytokines cannot passively cross the BBB that separates the majority of the brain vessels from nervous tissue, this gives to rise the question as to how the immune system signals the brain during the acute phase response. In view of the finding that brain IL-1β plays a role in the physiological regulation of sleep, our group hypothesized that IL-1β acts in the brain to induce the physiological and behavioral changes characteristic of the acute phase response.

To test this hypothesis, we administered IL-1ra either into the lateral brain ventricle or into the peritoneal cavity before the intraperitoneal injection of IL-1β. The fever response induced by intraperitoneal injection of IL-1β is abrogated only by administration of IL-1ra through the same route and not after intracerebroventricular administration [146]. These results can now be explained by the fact that IL-1β acts peripherally to induce IL-6 [147] or at the BBB to induce the prostaglandin-synthesizing enzyme COX-2 [94,148], which are both indispensable for the fever response and known to act in the CNS [149–151]. In contrast, both intraperitoneal and intracerebroventricular administration of IL-1ra attenuate the reduction in social interaction and food-motivated behavior induced by intraperitoneal injection of IL-1β [146].

Although useful when it comes to identifying potential mechanisms of action, pharmacological experiments with bolus injections of IL-1 are, however, unlikely to mimic levels of this cytokine during disease. Instead, administration of bacterial LPS induces the secretion of pro-inflammatory cytokines, including IL-1β, in a pattern similar to that seen in natural infection [152]. Importantly, peripheral LPS administration provokes IL-1β induction in the brain without disrupting the BBB [75]. As brain meninges, circumventricular organs, and choroid plexus are structures that are situated at the blood–brain interface without a BBB [153] and express LPS-binding molecules CD14 and Toll-like receptor 4 [154], it seems likely that circulating bacterial LPS acts in these brain structures to induce IL-1β synthesis. Indeed, IL-1β mRNA and IL-1β-ir are observed in resident cells of the meninges, circumventricular organs, and choroid plexus as early as 2 h after intraperitoneal injection of bacterial LPS [19,155]. Double-labeling studies identified IL-1β-ir cells as phagocytic cells in these organs [155–157]. However, at this early time point, LPS-induced brain IL-1β is not yet bioactive as measured by the IL-1 receptor-dependent thymocyte proliferation assay [158]. At later time points, IL-1β-ir cells are observed in brain parenchyma adjacent to circumventricular organs and at the ventricular side of the choroid plexus [155]. Moreover, bioactive IL-1 can be found in the brain extracellular and ventricular CSF at similar time points [158]. These findings indicate that after peripheral injection of LPS, IL-1β produced in circumventricular organs and choroid plexus spreads throughout the brain's extracellular spaces [159].

The finding that two-thirds of peripheral LPS-induced brain IL-1β originated from the meninges and choroid plexus [157] is in accordance with the hypothesis that these structures play an important role in CNS IL-1β during sickness. Whether the remaining third represents circumventricular organs or still other sources is still open for debate. Regardless of the exact source, brain IL-1β plays a role in suppressing behavioral activity and food intake during the acute phase response, because intracerebroventricular administration of IL-1ra attenuates anorexia and behavioral depression after LPS administration, without altering peripheral IL-1β production and action [160,161]. Collectively, these observations indicate that IL-1β induction and action within the CNS constitutes an immune-to-brain signaling pathway mediating sickness-associated behavioral changes.

The central nervous sites involved in IL-1β's effects on behavior and cognition during sickness remain elusive. However, it is tentative to speculate that IL-1β induction in the meninges covering the preoptic hypothalamus is responsible for increased NREM sleep, as IL-1β injected into the subarachoid space of this structure promotes sleep [112]. Interestingly, hippocampal IL-1β concentration is increased while LTP induction is reduced after peripheral LPS administration with both effects being attenuated by intracerebroventricular administration of a caspase-1 inhibitor [162]. These findings suggest that contrary to the role of IL-1 in maintaining LTP in healthy animals [39], higher concentrations of hippocampal IL-1β, as may occur during peripheral infection, reduce LTP induction. As LTP is one of the mechanism proposed to underlie learning and memory, cognitive impairments seen during sickness may be due to IL-1β's effects on LTP in the hippocampus.

8. CONCLUSIONS

Despite the technical difficulties and pitfalls associated with the study of IL-1β expression and action in the healthy brain, it is now clear that the constitutive low expression of IL-1β in the brain plays a role in the physiological regulation of sleep and most probably in hippocampus-dependent learning processes. Neuronal IL-1β expression in neuroendocrine hypothalamic nuclei might act as a neuropeptide and mediate neuroendocrine responses in response to physical stressors. Peripheral infection mimicked by administration of bacterial LPS provokes *de novo* synthesis of IL-1β in phagocytic cells of brain circumventricular organs, choroid plexus, and meninges, which is thought to mediate behavioral changes; including reduced food intake and exaggerated sleep, typically associated with infectious diseases. These findings are interesting given the fact that physical stressors do not induce sickness behavior and sleepiness and can be interpreted to suggest that IL-1β expression and action in the brain, like in the periphery, may be compartmentalized. In conclusion, IL-1β appears to play various roles in the brain that are not related to central nervous trauma or infection. It is to be hoped that future research will elucidate the mechanisms and sites of IL-1 action in the brain in the absence of neuropathology.

REFERENCES

1. Lowenstein PR. Dendritic cells and immune responses in the central nervous system. Trends Immunol 2002;23(2):70.
2. McMenamin PG. Distribution and phenotype of dendritic cells and resident tissue macrophages in the dura mater, leptomeninges, and choroid plexus of the rat brain as demonstrated in wholemount preparations. J Comp Neurol 1999;405(4):553–62.

3. Greter M, Heppner FL, Lemos MP, Odermatt BM, Goebels N, Laufer T, et al. Dendritic cells permit immune invasion of the CNS in an animal model of multiple sclerosis. Nat Med 2005;11(3):328–34.
4. Reichmann G, Schroeter M, Jander S, Fischer HG. Dendritic cells and dendritic-like microglia in focal cortical ischemia of the mouse brain. J Neuroimmunol 2002;129 (1–2):125–32.
5. Weller RO. Pathology of cerebrospinal fluid and interstitial fluid of the CNS: significance for Alzheimer disease, prion disorders and multiple sclerosis. J Neuropathol Exp Neurol 1998;57(10):885–94.
6. Johnston M, Zakharov A, Papaiconomou C, Salmasi G, Armstrong D. Evidence of connections between cerebrospinal fluid and nasal lymphatic vessels in humans, non-human primates and other mammalian species. Cerebrospinal Fluid Res 2004;1(1):2.
7. Karman J, Ling C, Sandor M, Fabry Z. Initiation of immune responses in brain is promoted by local dendritic cells. J Immunol 2004;173(4):2353–61.
8. Perry VH, Bell MD, Brown HC, Matyszak MK. Inflammation in the nervous system. Curr Opin Neurobiol 1995;5(5):636–41.
9. Perry VH, Anthony DC, Bolton SJ, Brown HC. The blood-brain barrier and the inflammatory response. Mol Med Today 1997;3(8):335–41.
10. Griffin WS, Stanley LC, Ling C, White L, MacLeod V, Perrot LJ, et al. Brain interleukin 1 and S-100 immunoreactivity are elevated in Down syndrome and Alzheimer disease. Proc Natl Acad Sci USA 1989;86(19):7611–15.
11. Giulian D, Young DG, Woodward J, Brown DC, Lachman LB. Interleukin-1 is an astroglial growth factor in the developing brain. J Neurosci 1988;8(2):709–14.
12. Breder CD, Dinarello CA, Saper CB. Interleukin-1 immunoreactive innervation of the human hypothalamus. Science 1988;240(4850):321–24.
13. Lechan RM, Toni R, Clark BD, Cannon JG, Shaw AR, Dinarello CA, et al. Immunoreactive interleukin-1 beta localization in the rat forebrain. Brain Res 1990;514(1): 135–40.
14. Nguyen KT, Deak T, Will MJ, Hansen MK, Hunsaker BN, Fleshner M, et al. Timecourse and corticosterone sensitivity of the brain, pituitary, and serum interleukin-1beta protein response to acute stress. Brain Res 2000;859(2):193–201
15. Shintani F, Nakaki T, Kanba S, Kato R, Asai M. Role of interleukin-1 in stress responses. A putative neurotransmitter. Mol Neurobiol 1995;10(1):47–71.
16. Gatti S, Bartfai T. Induction of tumor necrosis factor-alpha mRNA in the brain after peripheral endotoxin treatment: comparison with interleukin-1 family and interleukin-6. Brain Res 1993;624(1–2):291–94.
17. Van Dam AM, Bauer J, Tilders FJ, Berkenbosch F. Endotoxin-induced appearance of immunoreactive interleukin-1 beta in ramified microglia in rat brain: a light and electron microscopic study. Neuroscience 1995;65(3):815–26.
18. Buttini M, Boddeke H. Peripheral lipopolysaccharide stimulation induces interleukin-1 beta messenger RNA in rat brain microglial cells. Neuroscience 1995;65(2): 523–30.
19. Quan N, Whiteside M, Herkenham M. Time course and localization patterns of interleukin-1beta messenger RNA expression in brain and pituitary after peripheral administration of lipopolysaccharide. Neuroscience 1998;83(1):281–93.
20. Andersson J, Bjork L, Dinarello CA, Towbin H, Andersson U. Lipopolysaccharide induces human interleukin-1 receptor antagonist and interleukin-1 production in the same cell. Eur J Immunol 1992;22(10):2617–23.

21. Kavita U, Mizel SB. Differential sensitivity of interleukin-1 alpha and -beta precursor proteins to cleavage by calpain, a calcium-dependent protease. J Biol Chem 1995;270(46): 27758–65.
22. Wakabayashi G, Gelfand JA, Jung WK, Connolly RJ, Burke JF, Dinarello CA. Staphylococcus epidermidis induces complement activation, tumor necrosis factor and interleukin-1, a shock-like state and tissue injury in rabbits without endotoxemia. Comparison to Escherichia coli. J Clin Invest 1991;87(6):1925–35.
23. Maier JA, Voulalas P, Roeder D, Maciag T. Extension of the life-span of human endothelial cells by an interleukin-1 alpha antisense oligomer. Science 1990;249(4976): 1570–74.
24. Weitzmann MN, Savage N. Nuclear internalisation and DNA binding activities of interleukin-1, interleukin-1 receptor and interleukin-1/receptor complexes. Biochem Biophys Res Commun 1992;187(2):1166–71.
25. Andrei C, Dazzi C, Lotti L, Torrisi MR, Chimini G, Rubartelli A. The secretory route of the leaderless protein interleukin 1beta involves exocytosis of endolysosome-related vesicles. Mol Biol Cell 1999;10(5):1463–75.
26. Burns K, Martinon F, Tschopp J. New insights into the mechanism of IL-1beta maturation. Curr Opin Immunol 2003;15(1):26–30.
27. Buell G, Chessell IP, Michel AD, Collo G, Salazzo M, Herren S, et al. Blockade of human P2X7 receptor function with a monoclonal antibody. Blood 1998;92(10): 3521–28.
28. Sims JE, Gayle MA, Slack JL, Alderson MR, Bird TA, Giri JG, et al. Interleukin 1 signaling occurs exclusively via the type I receptor. Proc Natl Acad Sci USA 1993; 90(13):6155–59.
29. Cui X, Rouhani FN, Hawari F, Levine SJ. Shedding of the type II IL-1 decoy receptor requires a multifunctional aminopeptidase, aminopeptidase regulator of TNF receptor type 1 shedding. J Immunol 2003;171(12):6814–19.
30. Colotta F, Dower SK, Sims JE, Mantovani A. The type II 'decoy' receptor: a novel regulatory pathway for interleukin 1. Immunol Today 1994;15(12):562–66.
31. Janssens S, Beyaert R. Functional diversity and regulation of different interleukin-1 receptor-associated kinase (IRAK) family members. Mol Cell 2003;11(2):293–302.
32. Rao N, Nguyen S, Ngo K, Fung-Leung WP. A novel splice variant of interleukin-1 receptor (IL-1R)-associated kinase 1 plays a negative regulatory role in Toll/IL-1R-induced inflammatory signaling. Mol Cell Biol 2005;25(15):6521–32.
33. Su J, Richter K, Zhang C, Gu Q, Li L. Differential regulation of interleukin-1 receptor associated kinase 1 (IRAK1) splice variants. Mol Immunol 2007;44(5):900–5.
34. Woodroofe MN, Sarna GS, Wadhwa M, Hayes GM, Loughlin AJ, Tinker A, et al. Detection of interleukin-1 and interleukin-6 in adult rat brain, following mechanical injury, by in vivo microdialysis: evidence of a role for microglia in cytokine production. J Neuroimmunol 1991;33(3):227–36.
35. Coltman BW, Ide CF. Temporal characterization of microglia, IL-1 beta-like immunoreactivity and astrocytes in the dentate gyrus of hippocampal organotypic slice cultures. Int J Dev Neurosci 1996;14(6):707–19.
36. Giulian D, Baker TJ. Characterization of ameboid microglia isolated from developing mammalian brain. J Neurosci 1986;6(8):2163–78.
37. Friedman WJ. Cytokines regulate expression of the type 1 interleukin-1 receptor in rat hippocampal neurons and glia. Exp Neurol 2001;168(1):23–31.

38. Cao C, Matsumura K, Shirakawa N, Maeda M, Jikihara I, Kobayashi S, et al. Pyrogenic cytokines injected into the rat cerebral ventricle induce cyclooxygenase-2 in brain endothelial cells and also upregulate their receptors. Eur J Neurosci 2001;13(9):1781–90.
39. Schneider H, Pitossi F, Balschun D, Wagner A, del Rey A, Besedovsky HO. A neuromodulatory role of interleukin-1beta in the hippocampus. Proc Natl Acad Sci USA 1998;95(13):7778–83.
40. Goujon E, Parnet P, Cremona S, Dantzer R. Endogenous glucocorticoids down regulate central effects of interleukin-1 beta on body temperature and behaviour in mice. Brain Res 1995;702(1–2):173–80.
41. Goujon E, Parnet P, Laye S, Combe C, Kelley KW, Dantzer R. Stress downregulates lipopolysaccharide-induced expression of proinflammatory cytokines in the spleen, pituitary, and brain of mice. Brain Behav Immun 1995;9(4):292–303.
42. Bird S, Zou J, Wang T, Munday B, Cunningham C, Secombes CJ. Evolution of interleukin-1beta. Cytokine Growth Factor Rev 2002;13(6):483–502.
43. Pestarino M, De Anna E, Masini MA, Sturla M. Localization of interleukin-1 beta mRNA in the cerebral ganglion of the protochordate, Styela plicata. Neurosci Lett 1997;222(3): 151–54.
44. Engelsma MY, Stet RJ, Schipper H, Verburg-van Kemenade BM. Regulation of interleukin 1 beta RNA expression in the common carp, Cyprinus carpio L. Dev Comp Immunol 2001;25(3):195–203.
45. Zou J, Bird S, Minter R, Horton J, Cunningham C, Secombes CJ. Molecular cloning of the gene for interleukin-1beta from Xenopus laevis and analysis of expression in vivo and in vitro. Immunogenetics 2000;51(4–5):332–38.
46. Bandtlow CE, Meyer M, Lindholm D, Spranger M, Heumann R, Thoenen H. Regional and cellular codistribution of interleukin 1 beta and nerve growth factor mRNA in the adult rat brain: possible relationship to the regulation of nerve growth factor synthesis. J Cell Biol 1990;111(4):1701–11.
47. Watt JA, Hobbs NK. Interleukin-1beta immunoreactivity in identified neurons of the rat magnocellular neurosecretory system: evidence for activity-dependent release. J Neurosci Res 2000;60(4):478–89.
48. Molenaar GJ, Berkenbosch F, van Dam AM, Lugard CM. Distribution of interleukin 1 beta immunoreactivity within the porcine hypothalamus. Brain Res 1993;608(1):169–74.
49. Huitinga I, van der Cammen M, Salm L, Erkut Z, van Dam A, Tilders F, et al. IL-1beta immunoreactive neurons in the human hypothalamus: reduced numbers in multiple sclerosis. J Neuroimmunol 2000;107(1):8–20.
50. Proescholdt MG, Chakravarty S, Foster JA, Foti SB, Briley EM, Herkenham M. Intracerebroventricular but not intravenous interleukin-1beta induces widespread vascular-mediated leukocyte infiltration and immune signal mRNA expression followed by brain-wide glial activation. Neuroscience 2002;112(3):731–49.
51. Desson SE, Ferguson AV. Interleukin 1beta modulates rat subfornical organ neurons as a result of activation of a non-selective cationic conductance. J Physiol 2003;550(Pt 1): 113–22.
52. Eriksson AE, Cousens LS, Weaver LH, Matthews BW. Three-dimensional structure of human basic fibroblast growth factor. Proc Natl Acad Sci U S A 1991;88(8):3441–45.
53. Zhang JD, Cousens LS, Barr PJ, Sprang SR. Three-dimensional structure of human basic fibroblast growth factor, a structural homolog of interleukin 1 beta. Proc Natl Acad Sci USA 1991;88(8):3446–50.

54. Yazaki N, Hosoi Y, Kawabata K, Miyake A, Minami M, Satoh M, et al. Differential expression patterns of mRNAs for members of the fibroblast growth factor receptor family, FGFR-1-FGFR-4, in rat brain. J Neurosci Res 1994;37(4):445–52.
55. Belluardo N, Wu G, Mudo G, Hansson AC, Pettersson R, Fuxe K. Comparative localization of fibroblast growth factor receptor-1, -2, and -3 mRNAs in the rat brain: in situ hybridization analysis. J Comp Neurol 1997;379(2):226–46.
56. Di Santo E, Benigni F, Agnello D, Sipe JD, Ghezzi P. Peripheral effects of centrally administered interleukin-1beta in mice in relation to its clearance from the brain into the blood and tissue distribution. Neuroimmunomodulation 1999;6(4):300–4.
57. Wiley RG. Targeting toxins to neural antigens and receptors. Methods Mol Biol 2001;166:267–76.
58. Benveniste EN. Inflammatory cytokines within the central nervous system: sources, function, and mechanism of action. Am J Physiol 1992;263(1 Pt 1):C1–16.
59. Sairanen TR, Lindsberg PJ, Brenner M, Sirén AL. Global forebrain ischemia results in differential cellular expression of interleukin-1beta (IL-1beta) and its receptor at mRNA and protein level. J Cereb Blood Flow Metab 1997;17(10):1107–20.
60. Pearson VL, Rothwell NJ, Toulmond S. Excitotoxic brain damage in the rat induces interleukin-1beta protein in microglia and astrocytes: correlation with the progression of cell death. Glia 1999;25(4):311–23.
61. Giulian D, Baker TJ, Shih LC, Lachman LB. Interleukin 1 of the central nervous system is produced by ameboid microglia. J Exp Med 1986;164(2):594–604.
62. Hurley SD, Walter SA, Semple-Rowland SL, Streit WJ. Cytokine transcripts expressed by microglia in vitro are not expressed by ameboid microglia of the developing rat central nervous system. Glia 1999;25(3):304–9.
63. Blasi F, Riccio M, Brogi A, Strazza M, Taddei ML, Romagnoli S, et al. Constitutive expression of interleukin-1beta (IL-1beta) in rat oligodendrocytes. Biol Chem 1999;380(2): 259–64.
64. Lim JH, Brunjes PC. Activity-dependent regulation of interleukin-1 beta immunoreactivity in the developing rat olfactory bulb. Neuroscience 1999;93(1):371–74.
65. Jelaso AM, Acevedo S, Dang T, Lepere A, Ide CF. Interleukin-1beta and its type 1 receptor are expressed in developing neural circuits in the frog, Xenopus laevis. J Comp Neurol 1998;394(2):242–51.
66. Whittemore SR, Ebendal T, Larkfors L, Olson L, Seiger A, Stromberg I, et al. Development and regional expression of beta nerve growth factor messenger RNA and protein in the rat central nervous system. Proc Natl Acad Sci USA 1986;83(3): 817–21.
67. Hashimoto Y, Furukawa S, Omae F, Miyama Y, Hayashi K. Correlative regulation of nerve growth factor level and choline acetyltransferase activity by thyroxine in particular regions of infant rat brain. J Neurochem 1994;63(1):326–32.
68. Korsching S. The neurotrophic factor concept: a reexamination. J Neurosci 1993;13(7): 2739–48.
69. Spranger M, Lindholm D, Bandtlow C, Heumann R, Gnahn H, Naher-Noe M, et al. Regulation of nerve growth factor (NGF) synthesis in the rat central nervous system: comparison between the effects of interleukin-1 and various growth factors in astrocyte cultures and in vivo. Eur J Neurosci 1990;2(1):69–76.
70. Friedman WJ, Thakur S, Seidman L, Rabson AB. Regulation of nerve growth factor mRNA by interleukin-1 in rat hippocampal astrocytes is mediated by NFkappaB. J Biol Chem 1996;271(49):31115–20.

71. Juric DM, Carman-Krzan M. Interleukin-1 beta, but not IL-1 alpha, mediates nerve growth factor secretion from rat astrocytes via type I IL-1 receptor. Int J Dev Neurosci 2001;19(7):675–83.
72. Seaberg RM, van der Kooy D. Adult rodent neurogenic regions: the ventricular subependyma contains neural stem cells, but the dentate gyrus contains restricted progenitors. J Neurosci 2002;22(5):1784–93.
73. Gabellec MM, Griffais R, Fillion G, Haour F. Expression of interleukin 1 alpha, interleukin 1 beta and interleukin 1 receptor antagonist mRNA in mouse brain: regulation by bacterial lipopolysaccharide (LPS) treatment. Brain Res Mol Brain Res 1995;31(1–2): 122–30.
74. Higgins GA, Olschowka JA. Induction of interleukin-1 beta mRNA in adult rat brain. Brain Res Mol Brain Res 1991;9(1–2):143–48.
75. Pitossi F, del Rey A, Kabiersch A, Besedovsky H. Induction of cytokine transcripts in the central nervous system and pituitary following peripheral administration of endotoxin to mice. J Neurosci Res 1997;48(4):287–98.
76. Taishi P, Bredow S, Guha-Thakurta N, Obál F Jr, Krueger JM. Diurnal variations of interleukin-1 beta mRNA and beta-actin mRNA in rat brain. J Neuroimmunol 1997;75 (1–2):69–74.
77. Yabuuchi K, Minami M, Katsumata S, Satoh M. In situ hybridization study of interleukin-1 beta mRNA induced by kainic acid in the rat brain. Brain Res Mol Brain Res 1993;20 (1–2):153–61.
78. Manie S, Schmid-Alliana A, Kubar J, Ferrua B, Rossi B. Disruption of microtubule network in human monocytes induces expression of interleukin-1 but not that of interleukin-6 nor tumor necrosis factor-alpha. Involvement of protein kinase A stimulation. J Biol Chem 1993;268(18):13675–81.
79. Nett MA, Cerretti DP, Berson DR, Seavitt J, Gilbert DJ, Jenkins NA, et al. Molecular cloning of the murine IL-1 beta converting enzyme cDNA. J Immunol 1992;149(10): 3254–59.
80. Tingsborg S, Zetterström M, Alheim K, Hasanvan H, Schultzberg M, Bartfai T. Regionally specific induction of ICE mRNA and enzyme activity in the rat brain and adrenal gland by LPS. Brain Res 1996;712(1):153–58.
81. Quan N, Zhang Z, Emery M, Bonsall R, Weiss JM. Detection of interleukin-1 bioactivity in various brain regions of normal healthy rats. Neuroimmunomodulation 1996;3(1): 47–55.
82. Farrar WL, Kilian PL, Ruff MR, Hill JM, Pert CB. Visualization and characterization of interleukin 1 receptors in brain. J Immunol 1987;139(2):459–63.
83. Takao T, Newton RC, De Souza EB. Species differences in [125I]interleukin-1 binding in brain, endocrine and immune tissues. Brain Res 1993;623(1):172–76.
84. Marquette C, Van Dam AM, Ban E, Laniece P, Crumeyrolle-Arias M, Fillion G, et al. Rat interleukin-1 beta binding sites in rat hypothalamus and pituitary gland. Neuroendocrinology 1995;62(4):362–69.
85. Cunningham ET Jr, Wada E, Carter DB, Tracey DE, Battey JF, De Souza EB. Localization of interleukin-1 receptor messenger RNA in murine hippocampus. Endocrinology 1991;128(5):2666–68.
86. Ban E, Milon G, Prudhomme N, Fillion G, Haour F. Receptors for interleukin-1 (alpha and beta) in mouse brain: mapping and neuronal localization in hippocampus. Neuroscience 1991;43(1):21–30.

87. Parnet P, Amindari S, Wu C, Brunke-Reese D, Goujon E, Weyhenmeyer JA, et al. Expression of type I and type II interleukin-1 receptors in mouse brain. Brain Res Mol Brain Res 1994;27(1):63–70.
88. Cunningham ETJ, Wada E, Carter DB, Tracey DE, Battey JF, De Souza EB. In situ histochemical localization of type I interleukin-1 receptor messenger RNA in the central nervous system, pituitary, and adrenal gland of the mouse. J Neurosci 1992;12(3): 1101–14.
89. Yabuuchi K, Minami M, Katsumata S, Satoh M. Localization of type I interleukin-1 receptor mRNA in the rat brain. Brain Res Mol Brain Res 1994;27–36.
90. Wong ML, Licinio J. Localization of interleukin 1 type I receptor mRNA in rat brain. Neuroimmunomodulation 1994;1(2):110–15.
91. Ericsson A, Liu C, Hart RP, Sawchenko PE. Type 1 interleukin-1 receptor in the rat brain: distribution, regulation, and relationship to sites of IL-1-induced cellular activation. J Comp Neurol 1995;361(4):681–98.
92. Frost P, Barrientos RM, Makino S, Wong ML, Sternberg EM. IL-1 receptor type I gene expression in the amygdala of inflammatory susceptible Lewis and inflammatory resistant Fischer rats. J Neuroimmunol 2001;121(1–2):32–39.
93. Armario A, Castellanos JM. A comparison of corticoadrenal and gonadal responses to acute immobilization stress in rats and mice. Physiol Behav 1984;32(4):517–19.
94. Konsman JP, Vigues S, Mackerlova L, Bristow A, Blomqvist A. Rat brain vascular distribution of interleukin-1 type-1 receptor immunoreactivity: relationship to patterns of inducible cyclooxygenase expression by peripheral inflammatory stimuli. J Comp Neurol 2004;472(1):113–29.
95. Docagne F, Campbell SJ, Bristow AF, Poole S, Vigues S, Guaza C, et al. Differential regulation of type I and type II interleukin-1 receptors in focal brain inflammation. Eur J Neurosci 2005;21(5):1205–14.
96. French RA, VanHoy RW, Chizzonite R, Zachary JF, Dantzer R, Parnet P, et al. Expression and localization of p80 and p68 interleukin-1 receptor proteins in the brain of adult mice. J Neuroimmunol 1999;93(1–2):194–202.
97. Praast G, Hofmeister R, Grube K, Hans W, Kühlwein T, Krammer PH, et al. The internalized interleukin-1 activation complex in fibroblasts localizes to the Golgi apparatus. J Inflamm 1995;46(3):125–38.
98. Diana A, Van Dam AM, Winblad B, Schultzberg M. Co-localization of interleukin-1 receptor type I and interleukin-1 receptor antagonist with vasopressin in magnocellular neurons of the paraventricular and supraoptic nuclei of the rat hypothalamus. Neuroscience 1999;89(1):137–47.
99. Chakfe Y, Zhang Z, Bourque CW. IL-1beta directly excites isolated rat supraoptic neurons via upregulation of the osmosensory cation current. Am J Physiol Regul Integr Comp Physiol 2006;290(5):R1183–90.
100. Layé S, Parnet P, Goujon E, Dantzer R. Peripheral administration of lipopolysaccharide induces the expression of cytokine transcripts in the brain and pituitary of mice. Brain Res Mol Brain Res 1994;27(1):157–62.
101. Yirmiya R, Winocur G, Goshen I. Brain interleukin-1 is involved in spatial memory and passive avoidance conditioning. Neurobiol Learn Mem 2002;78(2):379–89.
102. Avital A, Goshen I, Kamsler A, Segal M, Iverfeldt K, Richter-Levin G, et al. Impaired interleukin-1 signaling is associated with deficits in hippocampal memory processes and neural plasticity. Hippocampus 2003;13(7):826–34.

103. Taishi P, Chen Z, Obál F Jr, Hansen MK, Zhang J, Fang J, et al. Sleep-associated changes in interleukin-1beta mRNA in the brain. J Interferon Cytokine Res 1998;18(9): 793–98.
104. Opp MR, Krueger JM. Interleukin 1-receptor antagonist blocks interleukin 1-induced sleep and fever. Am J Physiol 1991;260(2 Pt 2):R453–57.
105. Opp MR, Obal F Jr, Krueger JM. Interleukin 1 alters rat sleep: temporal and dose-related effects. Am J Physiol 1991;260(1 Pt 2):R52–58.
106. Yasuda T, Yoshida H, Garcia-Garcia F, Kay D, Krueger JM. Interleukin-1beta has a role in cerebral cortical state-dependent electroencephalographic slow-wave activity. Sleep 2005;28(2):177–84.
107. Fang J, Wang Y, Krueger JM. Effects of interleukin-1 beta on sleep are mediated by the type I receptor. Am J Physiol 1998;274(3 Pt 2):R655–60.
108. Opp MR, Krueger JM. Anti-interleukin-1 beta reduces sleep and sleep rebound after sleep deprivation in rats. Am J Physiol 1994;266(3 Pt 2):R688–95.
109. Tsai SY, McNulty JA. Interleukin-1beta expression in the pineal gland of the rat. J Pineal Res 1999;27(1):42–48.
110. Tricoire H, Malpaux B, Moller M. Cellular lining of the sheep pineal recess studied by light-, transmission-, and scanning electron microscopy: Morphologic indications for a direct secretion of melatonin from the pineal gland to the cerebrospinal fluid. J Comp Neurol 2003;456(1):39–47.
111. Ishizaka M, Ohe Y, Senbongi T, Wakabayashi K, Ishikawa K. Inflammatory stimuli increase prostaglandin D synthase levels in cerebrospinal fluid of rats. Neuroreport 2001;12(6):1161–65.
112. Terao A, Matsumura H, Saito M. Interleukin-1 induces slow-wave sleep at the prostaglandin D2-sensitive sleep-promoting zone in the rat brain. J Neurosci 1998;18(16):6599–7.
113. Muraki T, Fujimori K, Ishizaka M, Ohe Y, Urade Y, Okajima F, et al. Effects of interleukin-1beta and prostaglandin E2 on prostaglandin D synthase production in cultivated rat leptomeningeal cells. J Cereb Blood Flow Metab 2004;24(4):409–18.
114. Alam MN, McGinty D, Bashir T, Kumar S, Imeri L, Opp MR, et al. Interleukin-1beta modulates state-dependent discharge activity of preoptic area and basal forebrain neurons: role in sleep regulation. Eur J Neurosci 2004;20(1):207–16.
115. Saper CB, Cano G, Scammell TE. Homeostatic, circadian, and emotional regulation of sleep. J Comp Neurol 2005;493(1):92–98.
116. Gerashchenko D, Beuckmann CT, Kanaoka Y, Eguchi N, Gordon WC, Urade Y, et al. Dominant expression of rat prostanoid DP receptor mRNA in leptomeninges, inner segments of photoreceptor cells, iris epithelium, and ciliary processes. J Neurochem 1998;71(3):937–45.
117. Mizoguchi A, Eguchi N, Kimura K, Kiyohara Y, Qu WM, Huang ZL, et al. Dominant localization of prostaglandin D receptors on arachnoid trabecular cells in mouse basal forebrain and their involvement in the regulation of non-rapid eye movement sleep. Proc Natl Acad Sci USA 2001;98(20):11674–79.
118. Satoh S, Matsumura H, Suzuki F, Hayaishi O. Promotion of sleep mediated by the A2a-adenosine receptor and possible involvement of this receptor in the sleep induced by prostaglandin D2 in rats. Proc Natl Acad Sci USA 1996;93(12):5980–84.
119. Scammell T, Gerashchenko D, Urade Y, Onoe H, Saper C, Hayaishi O. Activation of ventrolateral preoptic neurons by the somnogen prostaglandin D2. Proc Natl Acad Sci USA 1998;95(13):7754–59.

120. Scammell TE, Gerashchenko DY, Mochizuki T, McCarthy MT, Estabrooke IV, Sears CA, et al. An adenosine A2a agonist increases sleep and induces Fos in ventrolateral preoptic neurons. Neuroscience 2001;107(4):653–63.
121. Gallopin T, Luppi PH, Cauli B, Urade Y, Rossier J, Hayaishi O, et al. The endogenous somnogen adenosine excites a subset of sleep-promoting neurons via A2A receptors in the ventrolateral preoptic nucleus. Neuroscience 2005;134(4):1377–90.
122. Walter JS, Meyers P, Krueger JM. Microinjection of interleukin-1 into brain: separation of sleep and fever responses. Physiol Behav 1989;45(1):169–76.
123. Ursin R. Serotonin and sleep. Sleep Med Rev 2002;6(1):55–69.
124. Imeri L, Bianchi S, Mancia M. Muramyl dipeptide and IL-1 effects on sleep and brain temperature after inhibition of serotonin synthesis. Am J Physiol 1997;273(5 Pt 2): R1663–68.
125. Imeri L, Mancia M, Opp MR. Blockade of 5-hydroxytryptamine (serotonin)-2 receptors alters interleukin-1-induced changes in rat sleep. Neuroscience 1999;92(2):745–49.
126. Tringali G, Mirtella A, Mancuso C, Guerriero G, Preziosi P, Navarra P. The release of immunoreactive interleukin-1 beta from rat hypothalamic explants is modulated by neurotransmitters and corticotropin-releasing hormone. Pharmacol Res 1997;36(4): 269–73.
127. Mancuso C, Tringali G, Grossman A, Preziosi P, Navarra P. The generation of nitric oxide and carbon monoxide produces opposite effects on the release of immunoreactive interleukin-1beta from the rat hypothalamus in vitro: evidence for the involvement of different signaling pathways. Endocrinology 1998;139(3):1031–37.
128. Tringali G, Dello Russo C, Vairano M, Preziosi P, Navarra P. Depolarization and a NO-donor stimulate interleukin-1beta release from the rat hypothalamus via a mechanism involving caspase 1. Neurosci Lett 1999;276(2):119–22.
129. Tringali G, Mancuso C, Mirtella A, Pozzoli G, Parente L, Preziosi P, et al. Evidence for the neuronal origin of immunoreactive interleukin-1 beta released by rat hypothalamic explants. Neurosci Lett 1996;219(3):143–46.
130. Shintani F, Nakaki T, Kanba S, Sato K, Yagi G, Shiozawa M, et al. Involvement of interleukin-1 in immobilization stress-induced increase in plasma adrenocorticotropic hormone and in release of hypothalamic monoamines in the rat. J Neurosci 1995;15 (3 Pt 1):1961–70.
131. O'Connor KA, Johnson JD, Hansen MK, Wieseler Frank JL, Maksimova E, Watkins LR, et al. Peripheral and central proinflammatory cytokine response to a severe acute stressor. Brain Res 2003;991(1–2):123–32.
132. Pugh CR, Nguyen KT, Gonyea JL, Fleshner M, Wakins LR, Maier SF, et al. Role of interleukin-1 beta in impairment of contextual fear conditioning caused by social isolation. Behav Brain Res 1999;106(1–2):109–18.
133. Hennessy MB, Deak T, Schiml-Webb PA, Wilson SE, Greenlee TM, McCall E. Responses of guinea pig pups during isolation in a novel environment may represent stress-induced sickness behaviors. Physiol Behav 2004;81(1):5–13.
134. Deak T, Bordner KA, McElderry NK, Barnum CJ, Blandino P Jr, Deak MM, et al. Stress-induced increases in hypothalamic IL-1: a systematic analysis of multiple stressor paradigms. Brain Res Bull 2005;64(6):541–56.
135. Minami M, Kuraishi Y, Yamaguchi T, Nakai S, Hirai Y, Satoh M. Immobilization stress induces interleukin-1 beta mRNA in the rat hypothalamus. Neurosci Lett 1991;123(2): 254–56.

136. Suzuki E, Shintani F, Kanba S, Asai M, Nakaki T. Immobilization stress increases mRNA levels of interleukin-1 receptor antagonist in various rat brain regions. Cell Mol Neurobiol 1997;17(5):557–62.
137. Yabuuchi K, Maruta E, Minami M, Satoh M. Induction of interleukin-1 beta mRNA in the hypothalamus following subcutaneous injections of formalin into the rat hind paws. Neurosci Lett 1996;207(2):109–12.
138. Plata-Salaman CR, Ilyin SE, Turrin NP, Gayle D, Flynn MC, Bedard T, et al. Neither acute nor chronic exposure to a naturalistic (predator) stressor influences the interleukin-1beta system, tumor necrosis factor-alpha, transforming growth factor-beta1, and neuropeptide mRNAs in specific brain regions. Brain Res Bull 2000;51(2):187–93.
139. Goshen I, Yirmiya R, Iverfeldt K, Weidenfeld J. The role of endogenous interleukin-1 in stress-induced adrenal activation and adrenalectomy-induced adrenocorticotropic hormone hypersecretion. Endocrinology 2003;144(10):4453–58.
140. Oprica M, Zhu S, Goiny M, Pham TM, Mohammed AH, Winblad B, et al. Transgenic overexpression of interleukin-1 receptor antagonist in the CNS influences behaviour, serum corticosterone and brain monoamines. Brain Behav Immun 2005;19(3):223–34.
141. Baumann H, Gauldie J. The acute phase response. Immunol Today 1994;15(2):74–80.
142. Elmquist JK, Scammell TE, Saper CB. Mechanisms of CNS response to systemic immune challenge: the febrile response. Trends Neurosci 1997;20(12):565–70.
143. Krueger JM, Majde JA. Microbial products and cytokines in sleep and fever regulation. Crit Rev Immunol 1994;14(3–4):355–79.
144. Pugh R, Fleshner M, Watkins LR, Maier SF, Rudy JW. The immune system and memory consolidation: a role for the cytokine IL-1beta. Neurosci Biobehav Rev 2001;25(1): 29–41.
145. Konsman JP, Parnet P, Dantzer R. Cytokine-induced sickness behaviour: mechanisms and implications. Trends Neurosci 2002;25(3):154–59.
146. Kent S, Bluthe RM, Dantzer R, Hardwick AJ, Kelley KW, Rothwell NJ, et al. Different receptor mechanisms mediate the pyrogenic and behavioral effects of interleukin 1. Proc Natl Acad Sci USA 1992;89(19):9117–20.
147. Cartmell T, Poole S, Turnbull AV, Rothwell NJ, Luheshi GN. Circulating interleukin-6 mediates the febrile response to localised inflammation in rats. J Physiol 2000;526 (Pt 3):653–61.
148. Rivest S. What is the cellular source of prostaglandins in the brain in response to systemic inflammation? Facts and controversies. Mol Psychiatry 1999;4(6):500–7.
149. Li S, Wang Y, Matsumura K, Ballou LR, Morham SG, Blatteis CM. The febrile response to lipopolysaccharide is blocked in cyclooxygenase-2(–/–), but not in cyclooxygenase-1(–/–) mice. Brain Res 1999;825(1–2):86–94.
150. Li S, Ballou LR, Morham SG, Blatteis CM. Cyclooxygenase-2 mediates the febrile response of mice to interleukin-1beta. Brain Res 2001;910(1–2):163–73.
151. Chai Z, Gatti S, Toniatti C, Poli V, Bartfai T. Interleukin (IL)-6 gene expression in the central nervous system is necessary for fever response to lipopolysaccharide or IL-1 beta: a study on IL-6-deficient mice. J Exp Med 1996;183(1):311–16.
152. Chen TY, Lei MG, Suzuki T, Morrison DC. Lipopolysaccharide receptors and signal transduction pathways in mononuclear phagocytes. Curr Top Microbiol Immunol 1992;181:169–88.
153. Gross PM. Circumventricular organ capillaries. Prog Brain Res 1992;91:219–33.

154. Laflamme N, Rivest S. Toll-like receptor 4: the missing link of the cerebral innate immune response triggered by circulating gram-negative bacterial cell wall components. Faseb J 2001;15(1):155–63.
155. Konsman JP, Kelley K, Dantzer R. Temporal and spatial relationships between lipopolysaccharide-induced expression of Fos, interleukin-1beta and inducible nitric oxide synthase in rat brain. Neuroscience 1999;89(2):535–48.
156. Vernet-der Garabedian B, Lemaigre-Dubreuil Y, Delhaye-Bouchaud N, Mariani J. Abnormal IL-1beta cytokine expression in the cerebellum of the ataxic mutant mice staggerer and lurcher. Brain Res Mol Brain Res 1998;62(2):224–27.
157. Garabedian BV, Lemaigre-Dubreuil Y, Mariani J. Central origin of IL-1beta produced during peripheral inflammation: role of meninges. Brain Res Mol Brain Res 2000; 75(2):259–63.
158. Quan N, Sundar SK, Weiss JM. Induction of interleukin-1 in various brain regions after peripheral and central injections of lipopolysaccharide. J Neuroimmunol 1994;49 (1–2): 125–34.
159. Konsman JP, Tridon V, Dantzer R. Diffusion and action of intracerebroventricularly injected interleukin-1 in the CNS. Neuroscience 2000;101(4):957–67.
160. Konsman JP. Immune-to-Brain Communication: A Functional Neuroanatomical Approach. Groningen: University of Groningen; 2000.
161. Layé S, Gheusi G, Cremona S, Combe C, Kelley K, Dantzer R, et al. Endogenous brain IL-1 mediates LPS-induced anorexia and hypothalamic cytokine expression. Am J Physiol 2000;179:R93–98.
162. Vereker E, Campbell V, Roche E, McEntee E, Lynch MA. Lipopolysaccharide inhibits long term potentiation in the rat dentate gyrus by activating caspase-1. J Biol Chem 2000;275(34):26252–58.

Cytokines and the Brain
Edited by C. Phelps and E. Korneva

Interleukin-1β Signal Transduction via the Sphingomyelin Pathway in Brain Cells

ELENA G. RYBAKINA and ELENA A. KORNEVA

Department of General Pathology and Pathophysiology, State Organization Institute for Experimental Medicine of Russian Academy of Medical Science, Saint Petersburg, Russia

ABSTRACT

Cytokine signaling plays an important role in the communication of the neuroendocrine and immune systems. It is of special interest how the cytokine signals are transmitted in nerve and immune-competent cells. Interleukin-1β (IL-1β) is one of the key cytokines that regulate host defense, and it is an important mediator of neuroimmune interaction. However, the pathways of signal transduction by IL-1β have not been fully elucidated. Over the last decade, a new sphingomyelin signal transduction pathway has been described for IL-1β, tumor necrosis factor-α, and γ-interferon. This pathway is initiated by the membrane-bound enzyme, neutral sphingomyelinase (nSMase). nSMase was shown to play a key role in the sphingomyelin cascade. Recent studies have demonstrated that IL-1 receptor subtype I and an accessory protein are involved in the activation of nSMase by IL-1β. In turn, nSMase activation initiates intracellular signaling through the sphingomyelin pathway in nerve cells. It was suggested that the pathways for transmission of the IL-1β signal into nerve cells, astrocytes, and immune-competent cells are similar and include the activation of nSMase. Further experiments suggested that IL-1β is involved in the stress reaction and that cytokine signaling through the sphingomyelin pathway plays a role in the stress response of both nerve and immune cells. The data reviewed in the present chapter provide compelling evidence that glucocorticoid hormones and short immunomodulatory peptides are able to modify the sphingomyelin pathway of IL-1β signaling in cells of both the neuroendocrine and the immune systems.

1. INTRODUCTION

Metalnikoff [1,2] studied the interactions of the immune and neuroendocrine systems during the first-third of the twentieth century. This field was eventually developed into a scientific discipline (named as psychoneuroimmunology, neuroimmunomodulation, or immunophysiology) as the result of classical research performed during the last-third of the same century [3–12]. At present the term Neuroimmune Biology integrates data concerning the biology of neuroimmune interactions. Modern experimental approaches are used in this scientific field to study fine mechanisms of neuroimmune interactions, such as the clarification of receptor signaling.

One of the main mediators of neural–immune interactions is the cytokine, interleukin-1 (IL-1) – the first interleukin discovered in 1948 as the endogenous pyrogen, which induced fever [13,14] and the acute-phase response [15]. During the last decade it has been finally confirmed that IL-1 is the key endogenous regulator of host defense, which are exerted by the adaptive and innate immune reactions. IL-1 is released by a wide variety of cells, mainly mononuclear phagocytes. In the central nervous system (CNS) this cytokine is produced by nerve cells, microglia, and macroglia cells [16–19]. Changes in brain IL-1 levels are involved in the production of its central effects. The binding of IL-1 to specific receptors and transmission of its signal into the cell results in alterations of the cell's metabolism and synthetic activity.

According to current concepts, the IL-1 family consists of three ligands: two agonists IL-1α, IL-1β, and one IL-1 receptor antagonist (IL-Ra). IL-1α is involved mainly in intracellular interactions, IL-1β predominantly acts on distant receptors, and IL-1Ra binds to IL-1 receptors and blocks cytokine signaling [18–21]. Two isoforms of IL-1Ra have been described: secreted IL-1Ra (sIL-1Ra) is actively released from the cells and intracellular IL-1Ra (icIL-1Ra) is retained in the cytoplasm [22,23].

Peripheral activities of IL-1 are mainly immunomodulatory, proinflammatory, and hemopoietic [19–21]. The balance between IL-1 and IL-1Ra has been shown to play an important role in controlling the inflammatory responses as well as in the susceptibility to and severity of many diseases [20,23]. In the CNS IL-1β is the predominant form of the cytokine, and it was demonstrated to induce numerous biological effects including fever, slow-wave sleep analgesia, anorexia, *c-fos* expression, and activation of the hypothalamic–pituitary axis (HPA) [12,18,19,24].

Membrane IL-1 receptors belong to the immune-globulin super-family and include IL-1 type I and type II receptors and an accessory protein – the second subunit of IL-1 receptor type I complex. The two receptors differ mainly in their intracellular part, which is much shorter in IL-1 receptor type II [19,25].

It is well known that practically all cell types of the host, including brain cells, are targets for IL-1 action. Receptors for IL-1 are found in virtually all parts of the brain [18,19,26]. Central and peripheral IL-1β action proceeds through type I IL-1 receptors and by an accessory protein, which initiates the passage of the cytokine signal into the target cell [25,26]. The type II IL-1 receptor is known to act as a regulated decoy target, which does not signal the cell, rather it interferes with the biological activity of IL-1 [27]. Circulating soluble type II IL-1 receptor binds IL-1; a soluble form of the IL-1 receptor accessory protein increases the affinity of binding of human IL-1α and IL-1β to the soluble human type II IL-1 receptor by approximately 100-fold, while leaving unaltered the low binding affinity of IL-1Ra [28]. These findings suggest that the soluble form of the IL-1 receptor accessory protein contributes to the antagonism of IL-1 action by the type II decoy receptor.

IL-1 is pleiotrophic – it is capable of inducing a wide spectrum of effects in numerous cell types. For this reason the mechanisms of IL-1 cellular signaling are of particular interest. Transduction of IL-1 signal into the cell is the most important functional component involved in producing its physiological effects. A large number of studies of IL-1β appeared in the past decade, which led to conflicting conclusions and failed to clarify which signal transduction pathway is involved in the central effects of IL-1β. Here we examine one of the possible pathways of IL-1β intracellular signaling in brain cells and compare it with peripheral immune cells. The role of this pathway in the stress reaction is also examined.

2. THE SPHINGOMYELIN PATHWAY OF INTERLEUKIN-1β SIGNALING

Since 1992 several groups obtained evidence, indicating that IL-1β, tumor necrosis factor (TNF-α), and γ-interferon rapidly induced membrane sphingomyelin turnover in various types of cells and produce phosphorylcholine and ceramide – as second messenger molecule in an intracellular signaling cascade [29–31]. Ceramide is formed in this reaction by the action of plasma membrane-associated enzyme, Mg^{2+}-dependent neutral sphingomyelinase (nSMase). Ceramide was shown to convert into ceramide-1-phosphate and to potentiate the effects of these agents on cell growth, differentiation, and apoptosis [32–35]. It is well known now that sphingomyelin hydrolysis and ceramide production are key events in cellular regulation [34,36].

In cells ceramide has been shown to modulate protein phosphorylation, the activity of serine-threonine protein kinases, the levels of the *c-myc* proto-oncogene, the nuclear factor-kB (NF-kB), the activity of phospholipase A2 and prostaglandin release, resulting in cellular and biological changes [34,37,38]. When using the model of fibroblasts from patients with type A Niemann–Pick disease, the participation of nSMase in IL-1/TNF-induced NF-kB activation was suggested [39].

Over the last decade the world literature has accumulated enough data to establish the new sphingomyelin signal transduction pathway for IL-1β, TNF-α, and γ-interferon. This pathway is initiated by the activation of the membrane-bound enzyme, nSMase, which seems to play a key role in the sphingomyelin cascade [40–45]. The steps of this cytokine signal transduction through the sphingomyelin pathway up to NF-kB activation has been described (Fig. 1). Present data favor the view that the sphingomyelin pathway of cytokine signaling is one of the principle signaling mechanisms that mediate most biological effects of IL-1β and TNF-α [38–43,46].

The sphingomyelin cytokine signal-transduction pathway has been mostly studied on cell lines derived from lymphocytes and fibroblasts. The possibility that cytokine signals might be

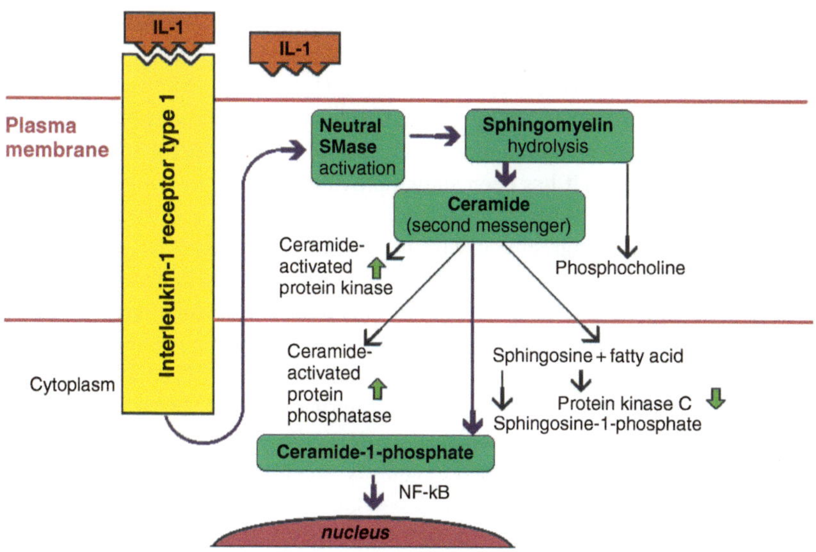

Figure 1. A schematic representation of the sphingomyelin pathway of IL-1β signal transduction. ↑ – activation; ↓ – inhibition.

transmitted through the sphingomyelin pathway in brain cells has not been definitively established. Furthermore, there are only a few reports on the sphingomyelin pathway in IL-1β and TNF-α signaling in the cells of the intact host [47–49]. Occasional reports have demonstrated that IL-1β and TNF-α activate nSMase in isolated rat brain myelin to form [^{3}H]ceramide [50,51]. Increased levels of brain cortical nSMase have also been seen in P/10 mice with accelerated senescence, which show premature aging [52]. This suggests the involvement of this enzyme in various antiproliferative processes, including apoptosis, cell cycle inhibition, and brain aging.

Finally, it was incompletely understood whether the cytokine binding to their specific receptors (in our case – IL-1β binding to IL-1 type I receptor) was critical for initiation of the sphingomyelin pathway for signaling.

3. ACTIVATION OF NEUTRAL SPHINGOMYELINASE AND THE NECESSITY OF INTERLEUKIN-1β RECEPTOR TYPE I FOR SIGNALING NERVE CELLS

In order to confirm that IL-1β signaling in the nervous system proceeds through the sphingomyelin pathway, it is reasonable to measure IL-1β-induced changes by the activity of nSMase – the initial step of the sphingomyelin cascade. For this, the effect of IL-1β on nSMase activity was studied in brain synaptosomes from Wistar rats and in the P2 membrane fractions of brain cerebral cortex from wild-type mice and mice deficient in IL-1 receptor type I (IL-1 receptor type I knockout mice).

It was shown that pre-incubation of rat brain synaptosomes (37°C, 15–60 min) with various concentrations of recombinant IL-1β (Stockholm University, Sweden) led to changes in nSMase activity. Stimulation was most effective when synaptosomes were pre-incubated for 30 min with 10–30 nmoles of IL-1β [53].

In order to reveal the effect of IL-1β on nSMase in cortical cell membranes of mice, the activity of the enzyme was analyzed after incubation of P2 fractions with IL-1β (10^{-7}–10^{-12} M, 37°C, 30 min). IL-1β stimulated the activity of the enzyme in P2 fractions from the cortex of wild-type mice in a dose-dependent manner (Fig. 2). Maximum stimulation of the enzyme was observed at 10^{-8}–10^{-10} M IL-1β concentrations [54–57].

Thus, the stimulatory effect of IL-1β on nSMase activity in membranes of rat and mouse brain has been demonstrated. These results support the hypothesis that IL-1β signals nerve cells through the sphingomyelin pathway, as it has been shown previously for immunocompetent cells.

IL-1β stimulates nSMase activity in dose-dependent manner in P2 cerebral cortex membrane fraction from mice. This is in agreement with the effect of this cytokine on rat brain synaptosomes and on the release of [^{3}H]ceramide from isolated myelin [50]. We observed a dose- and time-dependent stimulation by IL-1β of nSMase activity in plasma membranes of mouse thymocytes [54,56]. These data suggest that the pathways for transmission of the IL-1β signal into nerve and immunocompetent cells are similar and include nSMase activation.

In general signal transduction involves the binding of the ligand (here IL-1β) to its specific receptor. So far the literature contains no direct evidence for this principle in the case of IL-1β. Three experimental models have been used to analyze this question: (1) the incubation of IL-1β with membranes from nerve cells lacking type I receptors because of genomic mutations; (2) the incubation of these membranes with IL-1β simultaneously with its receptor antagonist, IL-Ra; (3) the employment of antibody 4C5, which is specific to an accessory protein, which serves as a subunit of the type I receptor complex [58].

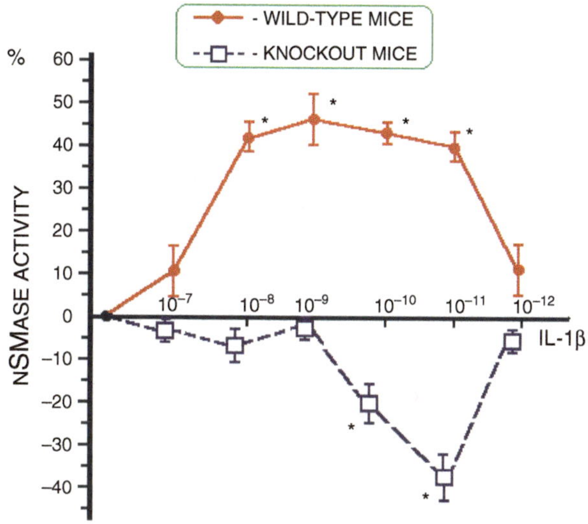

Figure 2. The dose-dependent effects of IL-1β on neutral sphingomyelinase (nSMase) activity in cerebral cortex membrane fractions from wild-type mice and mice with a genetic defect in the expression of the type I IL-1 receptor. The data here and on Fig. 3 are expressed as means ± SEM. Comparisons between values are made using the Student's t-test. The abscissa shows the IL-1β concentration, M; the ordinate shows specific enzyme activity, %. *$p < 0.05$ as compared to the basal level of enzyme activity.

IL-1β did not stimulate nSMase activation in the P2 fraction from the cortex of IL-1 receptor type I knockout mice [55,57]. On the contrary, the activity of nSMase in cortical membranes of such mice was inhibited when IL-1β was used at concentrations of 10^{-10}–10^{-11} M (Fig. 2). This inhibition might be due to IL-1β binding to type II receptors, which operate as regulatory "traps" for this cytokine, in the absence of type I receptors.

These data suggest that the function of IL-1 type II receptors are not restricted to the uptake of IL-1 without signaling, as demonstrated previously [27]. It seems that these receptors mediate the inhibitory effect of IL-1 on metabolism and cell functioning. The result of binding to the "inhibitory" receptor could be the suppression of nSMase activity at low IL-1β concentration [57]. IL-1β did not stimulate nSMase when membrane preparations from the cortex of wild-type mice and mice deficient in IL-1 type I receptor were incubated simultaneously with IL-1β (10^{-8} and 10^{-11} M) and IL-1Ra (10^{-7} and 10^{-10} M, respectively) [55,57].

The fundamental importance of an accessory protein in IL-1β signaling is of interest. It was shown previously that another type of cellular SMase – acid sphingomyelinase (aSMase) – was strongly linked with the internalization of IL-1 receptor type I, which is mediated by IL-1 receptor accessory protein. aSMase and nSMase are activated by different pathways [59]. Anti-accessory protein antibody 4C5 (Sclavo, Italy), at 10^3–10^5 ng/ml, inhibited nSMase activity in the P2 cerebral cortex membrane fraction from mice and also prevented the stimulatory effect of IL-1β (10^{-11} M) on nSMase activity.

It can be concluded that IL-1β has to be bound to its type I receptor for nSMase activation in nerve cells. Furthermore, IL-1 receptor must also form a complex with its accessory protein for the initiation of the sphingomyelin pathway of IL-1β signal transduction in nerve cells.

4. CHANGES IN NEUTRAL SPHINGOMYELINASE ACTIVITY IN MEMBRANES OF NERVE AND IMMUNE-COMPETENT CELLS UNDER STRESS

One of the most important directions in which immunophysiology is presently advancing involves the question of stress effects on host defense functions. The mechanisms by which the body resists unfavorable situations include both innate and acquired immune reactions. Altered immune regulation has a fundamental role in the host's response to any type of stress at a very early stage. It is clear now that stress exerts not only negative (suppressive) but also activating effects on host defense [10,60–65].

The role of cytokine signal transduction pathways in stress reactions is of special interest. The sphingomyelin pathway of cytokine signaling has mostly been investigated on cell lines *in vitro*. In these models the role of this pathway in "cellular" oxidative stress has been studied in detail [35–38]. In animal experiments the effect of various forms of stress was analyzed first on the sphingomyelin pathway of cytokine signaling of nerve and immune cells.

Mice (CBAxC57BL6)F1 were used to explore two models of experimental stress. The following experiments were done: (1) rotation stress (rotation of animals at 78 rpm for 1 h) and (2) combined stress (cooling in metal containers to 4–5°C for 2 h followed by immobilization in the same containers at room temperature for 18 h). Mild rotation stress did not inhibit the antibody response and even had a tendency to enhance it. On the contrary, severe prolonged combined stress produced significant suppression of the antibody response [10,63,66]. Both types of stress led to altered interactions of IL-1β with its receptor on lymphoid target cells. This correlated with changes in the antibody response [63]. On the basis of these results, we were interested to study whether or not the sphingomyelin pathway of IL-1β signaling is involved in the stress reaction, especially in the brain.

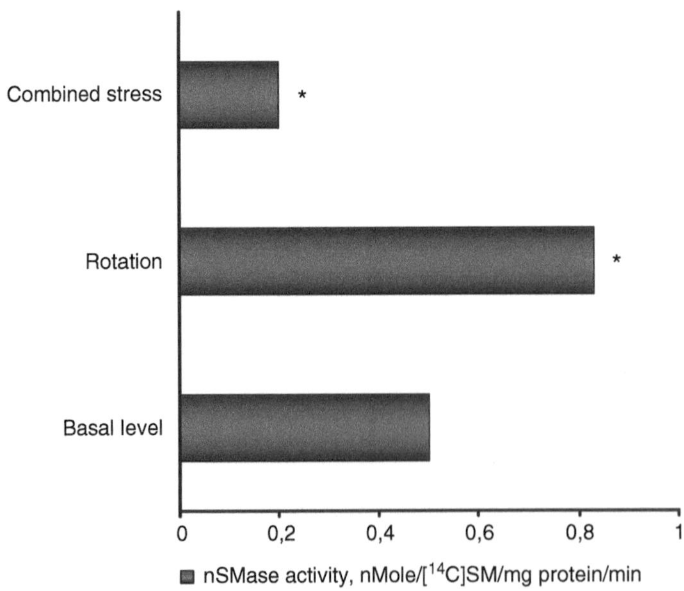

Figure 3. The effects of rotation and combined stress on the activity of neutral sphingomyelinase (nSMase) in cerebral cortex membrane fractions from mice. SM – sphingomyelin. *$p < 0.05$ as compared to the basal level of nSMase activity.

These experiments showed that combined stress, which induced marked immunosuppression, also led to the inhibition of nSMase activity in the P2 fraction of the brain cortex of mice, as compared with nonstressed animals. Rotation stress was immunostimulatory and induced an increase in brain nSMase activity (Fig. 3).

IL-1β at concentrations 10^{-7}–10^{-11} M induced similar rises on nSMase activity in membrane P2 fractions from the cortex of intact mice but not of mice subjected to rotation stress. One may suggest that the increased amount of endogenous IL-1 produced during the rotation stress reaction is enough for nSMase activation and the addition of cytokine preparation into the culture medium does not change the intensity of this reaction [66]. Additional experiments using the same stress models showed similar alterations in nSMase activity in thymocyte membrane fractions obtained from mice after exposure to rotation or combined stress.

In conclusion, these results suggest that the sphingomyelin pathway of IL-1β signaling is involved in the conduction of the stress effect to nerve tissue and to immune cells. Indeed, this signaling pathway plays a critical role in the stress response of nerve and immune cells.

5. MODULATORY EFFECTS OF GLUCOCORTICOID HORMONES AND SHORT PEPTIDES ON THE ACTIVITY OF NEUTRAL SPHINGOMYELINASE

A reasonable approach to study the influence of stress on IL-1β intracellular signaling is to investigate the effects of glucocorticoids on nSMase activity in the membranes of nerve and immune cells. There are practically no published data on this problem. We employed two experimental models to study the effects of blood corticosterone concentrations on nSMase activity: (1) adrenalectomy in mice (CBAxC57BL6)F1, which leads to near zero serum corticosterone level and to increased serum level of IL-1α, and (2) intraperitoneal injection of hydrocortizone in two doses (0.3 mkg/g and 50 mkg/g) to mice of the same line. This resulted in increased concentration of corticosterone and IL-1α in serum. Both procedures stimulated nSMase activity in membranes of brain cortex cells and of thymocytes.

These findings indicate that glucocorticoids modulate the sphingomyelin pathway of cytokines signaling in immune and nerve cells. These results support the functional interaction of glucocorticoids and IL-1, as it has been postulated [12,67], and show that this interaction also affects the level of IL-1 signal transduction.

Further *in vitro* experiments showed that a synthetic glucocorticoid, dexamethazone, stimulated nSMase, when added directly to aliquots of the P2 membrane fraction of the brain cortex from mice. This finding also suggested the involvement of the sphingomyelin cascade in glucocorticoid signaling of nerve cells. This has been previously proposed [68], but not supported experimentally.

Similar experimental approach was used for studying the effects of biologically active peptides on IL-1β signal transduction through the sphingomyelin pathway in membranes of nerve and immune cells. We used short synthetic peptides, namely Vilon (Lys-Glu) and Epitalon (Ala-Glu-Asp-Gly), which are experimental medications with immunomodulatory activities [69]. These drugs have been designed at the St. Petersburg Institute of Bioregulation and Gerontology, North-Western Branch of Russian Academy of Medical Science. The information used for the creation of these agents was the amino acid sequence of complex peptide substances extracted from the thymus and from the pineal gland, respectively. Recent results suggest that such short peptides, administered intramuscularly or intravenously, do not only modulate immune reactions but also influence brain functions and consequently participate in the interactions of the immune and neuroendocrine systems [70].

These studies were the first to show that Vilon and Epitalon modified cytokine signal transduction through the sphingomyelin pathway, as assessed by alterations of nSMase activity in cerebral cortex and thymocyte membranes [71]. The combined application of these peptides with IL-1β enhanced the stimulation of nSMase. These data show that these short synthetic peptides are effective bioregulators.

It is important to examine the effects of these peptides on nSMase activity and on the antibody response of animals exposed to stresses. In mice exposed to immunostimulatory rotation stress, both peptides (0.1–100 ng/ml) normalized nSMase activity in thymocyte membranes and P2 fraction of the brain cortex. Similar results were obtained with the P2 fraction of mice exposed to immunosuppressive combined stress [71].

Our aim was to restore nSMase activity impaired stress to baseline levels. We confirmed the protective effects of Vilon and Epitalon against stress-induced impairment. These novel data are in agreement with earlier reported stress-protective effects of natural immunoregulatory proteins and peptides, such as IL-1 and defensins [10,61].

6. INTERLEUKIN-1β SIGNALING THROUGH THE SPHINGOMYELIN PATHWAY IN ASTROCYTES

Over the last decade our knowledge of the physiological role of brain macroglial cells, called astrocytes, in CNS function has been extended significantly. Evidence is increasing that, in addition to performing their classical supportive functions for neurons and participating in the blood–brain barrier, astrocytes are immunocompetent cells. They are capable of antigen presentation and initiate and maintain immune responses and inflammation during CNS disorders. Astroglial cells can produce both pro- and anti-inflammatory cytokines, including IL-1α and IL-1β, and they express receptors for numerous cytokines [72–74].

The question was asked whether or not these electrically nonexcitable glial cells are able to transduce the IL-1β signal through the sphingomyelin pathway. For these experiments, astrocytes were cultivated and differentiated *in vitro* from precursor cells, obtained from the cerebellum of 7-day-old Wistar rats. It was found for the first time that recombinant IL-1β (Sigma) stimulated the activity of nSMase in membrane fractions from astrocytes in a dose-dependent manner and in the same concentration range, what has been previously shown for rat brain synaptosomes and the P2 cerebral cortex membrane fractions from mice. Maximum stimulation of the enzyme was observed at IL-1β concentrations of 10^{-8}–10^{-10} M. These results indicate that in astrocytes at least one of the IL-1β signal transduction pathways involves the sphingomyelin cascade and is initiated by nSMase activation.

Prenatal stress exposure (cooling –20°C, 20 min, for three circles) led to increased nSMase activity in membranes of astrocytes, cultivated from precursor cells of 7-day-old rats. The pre-incubation of these membranes with IL-1 for 30 min resulted in additional increases of nSMase activity as compared to those measured after incubation of membranes without IL-1β.

These results show that in different cell systems of the brain – neurons and astrocytes – IL-1β signal transduction occurs through the sphingomyelin pathway. nSMase activity in membranes of cultivated astroglial cells is sensitive to prenatal stress exposure.

7. CONCLUSIONS

One may conclude on the basis of the evidence presented in this chapter that IL-1β signaling of neurons is mediated by type I IL-1 receptors and by the sphingomyelin pathway for signal transduction. IL-1β signaling of neurons, astrocytes, and immune-competent cells are similar as the activation of nSMase is involved, which is the key enzyme of the sphingomyelin cascade. These findings confirm the critical role of IL-1β in coordinating neuroendocrine and immunological responses during health and disease.

Our results show that IL-1β signaling by the sphingomyelin pathway is activated during the stress reaction. The activity of nSMase is altered in membranes of neurons and thymocytes during stress. The intensity of sphingomyelin patway activation might serve as an informative index for stress-induced disorders. Furthermore, the membrane-bound enzyme, nSMase, might be a potential target for the pharmacological correction of disturbed immune–neuroendocrine interaction.

Glucocorticoid hormones and short immunomodulatory peptides were shown to modify IL-1β signal transduction through the sphingomyelin pathway, as indicated by alterations of nSMase activity. Of particular importance are the data concerning the protective effects of synthetic peptides against stress. Vilon and Epitalon regulate IL-1β signals and may be useful for clinical application for the correction of impaired host defense.

It should be noted that the conclusions of the present work are relevant only to IL-1β but not to IL-1α signaling. Published data suggest that IL-1α and IL-1β induce different signals by binding to separate sites on type I IL-1 receptor [75].

The results presented here open up new possibilities for analyzing the role of immunoregulatory cytokines in brain function.

ACKNOWLEDGMENTS

The authors are greatly indebted to Prof. T. Bartfai for initiation and guidance of the study concerning IL-1β signaling in cells of brain cerebral cortex. This work was also supported in part by the Russian Fund for Basic Research (grants No. 03-04-49236 and No. 06-04-48609).

REFERENCES

1. Metalnikov S, Chorine V. Roles des reflexes conditionels dans I'immunite. Ann Inst Pasteur 1926;40:893–900.
2. Metalnikov S. Role du systeme nerveux et des facteurs biologiques et psychique daus l'immunite. Paris, 1934.
3. Korneva EA, Khai LM. Effect of destruction of hypothalamic areas on immunogenesis. Fed Proc 1964;23:T88.
4. Solomon GF. Emotions, stress, the central nervous system and immunity. Ann NY Acad Sci 1969;164:335–43.
5. Pierpaoli W, Besedovsky HO. Role of the thymus in programming of neuroendocrine functions. Clin Exp Immunol 1975;20:323–28.
6. Blalock JE, Smith EM. A complete regulatory loop between the immune and neuroendocrine system. Fed Proc 1985;101:472–78.

7. Jankovic BD, Spector NH. Effects on the immune system of lesioning and stimulation of the nervous system: neuroimmunomodulation. In Enkephalins and Endorphins: Stress and the Immune System. Plotnikoff NP, Faith RE, Murgo AJ, Good RA, Eds; New York: Plenum Publishing Corporation, 1986; pp. 189–220.
8. Blalock JE. A molecular basis for bidirectional communication between the immune and neuroendocrinesystems. Physiol Rev 1989;69:1–32.
9. Blalock JE The syntax of immune-neuroendocrine communication. Immunol Today 1994; 15:504–11.
10. Korneva EA, Rybakina EG, Orlov DS, Shamova OV, Shanin SN, Kokryakov VN. Interleukin-1 and defensins in thermoregulation, stress and immunity. Ann NY Acad Sci 1997;813:465–73.
11. Korneva EA, Barabanova SV, Golovko OI, Nosov MA, Novikova NS, Kazakova TB. C-fos and gene expression in rat brain cells and splenic lymphocytes after nonantigenic and antigenic stimuli. In: Conti A, Maestroni GJM, McCann SM, Sternberg EM, Lipton JM, Smith CC (Eds). Neuroimmunomodulation. Perspectives at the new Millenium. Ann NY Acad Sci 2000;917:197–209.
12. Turnbull AV, Rivier CL. Regulation of the hypothalamic-pituitary-adrenal axis by cytokines: actions and mechanisms of action. Physiol Rev 1999;79:1–71.
13. Beeson PB. Temperature-elevating effect of a substance obtained from polymorphonuclear leucocytes. J Clin Invest 1948;27:524–34.
14. Bennett IL, Beeson PB. Studies on the pathogenesis of fever. II. Characterization of fever producing substances from polymorphonuclear leukocytes and from the fluid of sterile exudates. J Exp Med 1953;98:493–508.
15. Kampschmidt RF. Leucocytic endogenous mediator. J Reticuloendothel Soc 1978; 23: 278–97.
16. Bartfai T, Schultzberg M. Cytokines in neuronal cell types. Neurochem Int 1993; 22:435–44.
17. Hopkins SJ, Rothwell NJ. Cytokines and the nervous system. I: Expression and recognition. Trends Neurosci 1995;18:83–88.
18. Dinarello CA. Induction of interleukin-1 and interleukin-1 receptor antagonist. Sem Oncol 1997;3:S9–81;S9–93.
19. Dinarello CA. Interlekin 1: a proinflammatory cytokine. In Inflammation: Basic Principles and Clinical Correlates, 3rd Ed. Gallin JI, Snydrman R, Eds; Philadelphia: Lappincott Williams & Wilkins, 1999; pp. 443–61.
20. Arend WP. The balance between IL-1 and IL-1Ra in disease. Cytokine Growth Factor Rev 2002;13:323–40.
21. Dinarello CA. Unraveling the NALP-3/IL-1β inflammasome: a big lesson from a small mutation. Immunity 2004;20:243–46.
22. Holtkamp GM, Devos AF, Kijlstra A, Peek R. Expression of multiple forms of IL-1 receptor antagonist (IL-1 Ra) by human retinal pigment epithelial cells: identification of a new IL-1Ra exon. Eur J Immunol 1999;29:215–22.
23. Redlitz KH, Yamshcikov VF, Cominelli F. Differential contribution of IL1Ra isoforms to allele-specific IL-1Ra mRNA accumulation. J Interferon Cytokine Res 2004;24:253–60.
24. Van Dam A-M, Malinowsky D, Lenczowski JP, Bartfai T, Tilders FJH. Interleukin 1 (IL-1) type I receptors mediate activation of rat hypothalamus-pituitary-adrenal axis and interleukin 6 production as shown by receptor type selective deletion mutants of IL-1β. Cytokine 1998;10:413–17.

25. Sims JE, Dower SK. Interleukin-1 receptors. Eur Cytokine Netw 1994;5:539–46.
26. Malinowsky D, Chai Z, Bristulf J, Simoncsits A, Bartfai T. The type I interleukin-1 receptor mediates fever in the rat as shown by interleukin-1 receptor subtype selective ligands. Neurosci Lett 1995;201:33–36.
27. Colotta F, Re F, Muzio M, Bertini R, Polentarutti N, Sironi M, Giri JG, Dower SK, Sims JE, Mantovani A. Interleukin 1 type II receptor: a decoy target for IL-1 that is regulated by IL-4. Science 1993;261:472–75.
28. Smith DE, Hanna R, Friend D, Moore H, Chen H, Farese AM, MacVittie TJ, Virca GD, Sims JE. The soluble form of IL-1 receptor accessory protein enhances the ability of soluble type II IL-1 receptor to inhibit IL-1 action. Immunity 2003; 18:87–96.
29. Dressler KA, Mathias S, Kolesnick RN. Tumor necrosis factor-alpha activates sphingomyelinsignal transduction pathway in a cell-free system. Science 1992;255:1715–18.
30. Mathias S, Jounes A, Kan CC, Orlow I, Joseph C, Kolesnick RN. Activation of the sphingomyelin signalling pathway in intact EL4 cells and in a cell-free system by IL-1 beta. Science 1993;259:519–22.
31. Dbaibo GS, Obeid LM, Hannun YA. Tumor necrosis factor-alfa (TNF alfa) signal transduction through ceramide. J Biol Chem 1993;268:17762–66.
32. Hannun YA. Sphingomyelin cycle and the second messenger function of ceramide. J Biol Chem 1994;269:3125–28.
33. Testi R. Sphingomyelin breakdown and cell fate. TIBS 1996;21:468–71.
34. Kolesnick R. The therapeutic potential of modulating the ceramide/sphingomyelin pathway. J Clin Invest 2002;110:3–8.
35. Andrieu-Abadie N, Levade T. Sphingomyelin hydrolysis during apoptosis. Biochimica et Biophysica Acta 2002;1585:126–34.
36. Hannun YA, Luberto C. Ceramide in the eukaryotic stress. Trends Cell Biol 2000; 10:73–80.
37. Liu B, Obeid LM, Hannun YA. Sphingomyelin in cell regulation. Sem Cell Dev Biol 1997;8:311–22
38. Mathias S, Pena LA., Kolesnick RN. Signal transduction of stress via ceramide. Biochem J 1998;335:465–80.
39. Kuno K., Sukegawa K, Ishikawa Y, Orii T, Matsushima K. Acid sphingomyelinase is not essential for the IL-1 and tumor necrosis factor receptor signalling pathway leading to NF-kB activation. Int Immunol 1994;6:269–72.
40. Adam D, Wiegmann K, Adam-Klages S, Ruff A, Kronke M. A novel cytoplasmic domain of the p55 tumor necrosis factor receptor initiates the neutral sphingomyelinase pathway. J Biol Chem 1996;271:14617–22.
41. Gamard CJ, Dbaibo GS, Lin B, Obeid LM, Hannun YA. Selective involvement of ceramide in cytokine-induced apoptosis. Ceramide inhibits phorbol ester activation of nuclear factor kappa B. J Biol Chem 1997;272:16474–16481.
42. Kolesnick RN. Signal transduction through the sphingomyelin pathway. Mol Chem Neuropathol 1994;21:287–97.
43. Kolesnick RN, Golde DW. The sphingomyelin pathway in tumor necrosis factor and interleukin-1 signaling. Cell 1994;77:325–28.
44. Jung SY, Suh JH, Park HJ, Jung KM, Kim MY, Na DS, Kim DK. Identification of multiple forms of membrane-associated neutral sphingomyelinase in bovine brain. J Neurochem 2000;75:1004–14.

45. Goni F, Alonso A. Sphingomyelinases: enzymology and membrane activity. FEBS Lett 2002;531:38–46.
46. Ballou LR, Laulederkind SJF, Rosloniec EF, Raghow R. Ceramide signalling and the immune response Biochim Biophys Acta 1996;1301:273–87.
47. Kaszkin M, Huwiler A, Scholz K, van den Bosch H, Pfeilschifter J. Negative regulation of interleukin-1beta-activated neutral sphingomyelinase by protein kinase C in rat mesangial cells. FEBS Lett 1998;440:163–66.
48. Hinkovska-Galcheva V, Kjeldsen L, Mansfield PJ, Boxer LA, Shayman JA, Suchard SJ. Activation of a plasma membrane-associated neutral sphingomyelinase and concomitant ceramide accumulation during IgG-dependent phagocytosis in human polymorphonuclear leukocytes. Blood 1998;91:4761–69.
49. Campbell V, Lynch MA. The role of ceramide in the modulation of intracellular Ca^{2+} levels by interleukin 1β in rat cortical synaptosomes. Cytokine 2000;12:487–90.
50. Ziemba S, Chakraborty G, Ledeen R. Evidence of the presence of Mg^{2+}-dependent neutral sphingomyelinase in CNS myelin. J Neurochem 1996;66:S46C.
51. Chakraborty G, Ziemba S, Drivas A, Ledeen RW. Myelin contains neutral sphingomyelinase activity that is stimulated by tumor necrosis factor-alpha. J Neurosci Res 1997; 50:466–76.
52. Kim SS, Kang MS, Choi YM, Suh YH, Kim DK. Sphingomyelinase activity is enchanced in cerebral cortex of senescence-accelerated mouse-P/10 with advancing age. Miochem Biophys Res Comm 1997;237:583–87.
53. Nalivaeva NN, Rybakina EG, Shanin SN, Kozinets IA, Pivanovich IY. Ganglioside GMI potentiates the effect of IL-1β on neutral sphingomyelinase activity in rat brain synaptosomes. Biochem Soc Trans 1997;25:214S.
54. Rybakina EG, Nalivaeva NN, Kozinets IA, Shanin SN, Pivanovich IY. Involvement of the sphingomyelin pathway in interleukin 1 signaling in murine immune-competent and nerve cells. Immunol Lett 1997;56:67.
55. Nalivaeva NN, Rybakina EG, Pivanovich IY, Kozinets IA, Shanin SN, Bartfai T. Activation of neutral sphingomyelinase by IL-1beta requires the type I Interleukin 1 receptor. Cytokine 2000;12:229–32.
56. Rybakina EG, Pivanovich IY, Kozinets IA, Shanin SN, Korneva EA. The sphingomyelin pathway of Interleukin-1β signaling in neural-immune interactions. Int J Immunorehabilitation 2001;3:70.
57. Rybakina EG, Nalivaeva NN, Pivanovich IY, Shanin SN, Kozinets IA, Korneva EA. The role of neutral sphingomyelinase in Interleukin-1β signal transduction in mouse cerebral cortex cells. Neurosci Behav Physiol 2001;31:439–44.
58. Greenfeder SA, Nunes P, Kwee L, Labow M, Chizzonite RA, Ju G. Molecular cloning and characterization of a second subunit of the interleukin 1 receptor complex. J Biol Chem 1995;270:13757–65.
59. Hofmeister R, Wiegmann K, Korherr C, Bernardo K, Kronke M, Falk W. Activation of acid sphingomyelinase by interleukin-1 (IL-1) requires the IL-1 receptor accessory protein. J Biol Chem 1997;272:27730–36.
60. Dunn AJ. Role of cytokines in infection-induced stress. Ann NY Acad Sci 1993; 697:189–202.
61. Rybakina EG, Shanin SN, Kozinets IA, Fomicheva EE, Korneva EA. Cellular mechanisms of cold stress-related immunosuppression and the action of interleukin-1. Int J Tiss Reac 1997;19:135–40.

62. Lauc G, Dabelic S, Dumic J, Flogel M. Stressing and natural killer cell activity in professional soldiers. In Stress of Life from Molecules to Man. Csermely P, Ed. Ann NY Acad Sci 1998;851:526–30.
63. Korneva EA, Shanin SN, Rybakina EG. The role of interleukin-1 in stress-induced changes in immune system functions. Neurosci Behav Physiol 2001;31:431–37.
64. Heuser I, Lammers C-H. Stress and the brain. Neurobiol Aging 2003;24:S69–76.
65. Shanin SN, Rybakina EG, Novikova NS, Kozinets IA, Rogers VJ, Korneva EA. Natural killer cell cytotoxic activity and c-Fos protein synthesis in rat hypothalamic cells after painful electric stimulation of the hind limbs and EHF irradiation of the skin. Med Sci Monit 2005;11:BR309–15.
66. Rybakina EG. Interleukin 1 and its signal transduction mechanisms in realization of stress reaction. Pathophysiology 1998;5:146.
67. Besedovsky HO, Del Rey A, Sorkin E, Dinarello CA. Immunoregulatory feedback between Interleukin 1 and glucocorticoid hormones. Science 1986;223:652–54.
68. Buckingham JC. Stress and the neuroendocrine-immune axis: the pivotal role of glucocorticoids and lipocortin 1.Br J Pharmacol 1996;118:1–19.
69. Khavinson VK, Kvetnoi IM. Peptide bioregulators inhibit apoptosis. Bull Exp Biol Med 2000;130:1175–76.
70. Khavinson VK, Goncharova N, Lapin B. Synthetic tetrapeptide Epitalon restors disturbed neuroendocrine regulation in senescent monkeys. Neuroendocrinol Lett 2001;22:251–54.
71. Khavinson VK, Korneva EA, Malinin VV, Rybakina EG, Pyvanovich IY, Shanin SN. Effect of Epitalon on interleukin-1β signal transduction and the reaction of thymocyte blast transformation under stress. Neuroendocrinol Lett 2002;23:411–16.
72. Benveniste EN. Cytokines: influence on glial cell gene expression and function. Chem Immunol 1997;69:31–75.
73. Xiao BG, Link H. Immune regulation within the central nervous system. J Neurol Sci 1998;157:1–12.
74. Dong Y, Benveniste EN. Immune function of astrocytes. Glia 2001;36:180–90.
75. Clark BD, Ikejima T, Mancilla J, Orencole SF, Sirko SP, Ishii N, Okuda K, Dinarello CA. An antibody to a 17 amino acid synthetic peptide for the type I interleukin 1 receptor preferentially blocks interleukin-1 beta binding. J Interferon Cytokine Res 1996;16:1079–88.

Blood–Brain Barrier Transport of Cytokines

WILLIAM A. BANKS

GRECC, Veterans Affairs Medical Center-St. Louis and Saint Louis University School of Medicine, Division of Geriatrics, Department of Internal Medicine

ABSTRACT

Cytokine interactions with the blood–brain barrier (BBB) and blood–cerebrospinal fluid (B-CSF) barrier are an integral part of neuroimmune interactions. These barriers constitute potential limitations or restrictions against the free exchange of humoral message-bearing molecules residing in blood, CSF, and extracellular fluid (ECS) within brain and non-neural body fluid compartments. Many cytokines are transported across the BBB, and there are unique transporters for groups and families of cytokines that enable movements from blood to brain and the converse. This chapter reviews these transport movements in either direction for their behavioral effects, their species specificity and the alterations in transport that may occur or can be demonstrated in health and after pathological processes. The rate at which these cytokine transporters act is shown to not be static and is known to vary diurnally, among different strains of animals and with different brain locations. Cytokines are shown to affect the transport of other substances across the BBB, including viruses and regulatory proteins such as insulin. Secretion of cytokines by cells that comprise the BBB is also reviewed in order to illustrate the range of cytokine interactions with the barriers.

1. INTRODUCTION

Interactions between cytokines and the blood–brain barrier (BBB) are an intimate part of neuroimmunology. The BBB prevents, limits, or impairs the exchange between the central nervous system (CNS) and blood of most soluble substances and so contributes to the compartmentalization of the immune system and CNS. But the cells that comprise the BBB, like the immune and many types of brain cells, have receptors for cytokines and respond to them. These receptors affect many aspects of BBB function that are important to brain–immune interactions, including the trafficking of immune cells and the transport of substances with direct or indirect immune functions. In extreme cases, some cytokines can induce disruption of the BBB, which must be viewed as a pathological event. Most BBB–cytokine interactions, however, are likely important to the normal functioning of the neuroimmune axis. Many cytokines are transported across the BBB, others are presented on the membrane of the brain endothelial cells, which comprise the BBB, and the BBB itself secretes cytokines. Cytokine–BBB interactions are not static but are dynamic interactions that respond to physiological and pathological events.

This review will concentrate on what is known about the transport of cytokines across the BBB. However, it will also address to a limited extent other cytokine–BBB interactions, especially as they relate to cytokine transport.

2. BLOOD-TO-BRAIN TRANSPORT

Numerous cytokines have been investigated for their abilities to cross the BBB. Table 1 lists most of these and characterizes them with regard to whether they can cross the BBB in either the brain-to-blood or blood-to-brain direction and whether any such passage is saturable or nonsaturable.

2.1. Interleukin-1

Interleukin-1α (IL-1α) was the first cytokine to be shown to cross the BBB by a saturable system [1]. It shares [2] transporter activity with IL-1β and IL-1 receptor antagonist (IL-1ra).

Table 1. BBB permeability of representative cytokines

Representative cytokines and related substances	Permeability	Reference
Brain-derived neurotrophic factor	T	[120]
Ciliary neurotrophic factor	T	[121,122]
Cytokine-induced neutrophil chemoattractant-1	NS	[10]
Epidermal growth factor	T	[123]
Epogen	NS	[124,125]
Fibroblast growth factor	T	[126]
Glial cell line-derived neurotrophic factor	NT	[127]
Interferons	T	[94,128]
Interleukin-1α	T	[1,2,129]
Interleukin-1β	T	[2]
Interleukin-1 receptor anatagonist	T	[130]
Interleukin-2	NT	[11,12]
Interleukin-2	E	[12]
Interleukin-6	T	[4,131]
Interleukin-6	NE	[22]
Interleukin-10	NT	[132]
Leptin	T	[72]
Leukemia inhibitory factor	T	[133]
MIP's	NT	[134]
Nerve growth factor	T	[95,135]
Neurotrophin 3	T	[95,121]
Soluble receptors	NT	[136]
Transforming growth factor-α	T	[137]
Transforming growth factor-β	NT	[138]
Tumor necrosis factor-α	T	[5,84]
Tumor necrosis factor-α	NE	[5,21,86]

T = Saturable blood-to-brain transport; E = Saturable brain-to-blood transport; NS = Nonsaturable blood-to-brain transport; NE = Nonsaturable brain-to-blood transport; NT = No blood-to-brain transport.

Like most cytokines, its saturable transport is only in the blood-to-brain direction; it has no saturable component in the brain-to-blood direction. This means that its transporter is likely not facilitated diffusion but an energy requiring system [3]. Many of the classic saturable transport systems of the BBB, such as the one for glucose, are facilitated diffusion systems [3]. That is, they transport substances from the side of highest concentration to lowest regardless of whether that is in the brain-to-blood or blood-to-brain direction. Hence, these systems are bidirectional. Facilitated diffusion systems are also characterized as energy independent. Unidirectional systems, such as that described for IL-1α, are usually active transporters; that is, they are energy requiring.

2.2. Interleukin-6 and TNF

The transporter for IL-6 is distinct from that for the IL-1s [4]. It is also unidirectional. Similarly, the transporter for tumor necrosis factor-α (TNF-α) is distinguishable from those for the IL-1s and for IL-6 [5]. Transporters can be distinguished from one another by the inability of ligands to cross inhibit. Thus, it can be concluded that IL-1α and IL-1β share either a single transporter or a family of transporters with overlapping affinity because they inhibit one another's transport. In contrast, IL-6 does not inhibit the blood-to-brain transport of IL-1 or TNF, whereas TNF does not inhibit the transport of IL-1 or IL-6 [2,4,5]. Most combinations for possible cross-inhibition of cytokine transport have not been tested for the subsequently described cytokine transporters. However, it is assumed that most cytokines are transported by their own unique transport systems.

2.3. Miscellaneous considerations

One cytokine has been characterized to cross the BBB by a nonsaturable mechanism. Most substances with substantial nonsaturable transmembrane diffusion are small lipid soluble molecules [6]. However, several small peptides, such as delta sleep-inducing peptide, have been shown to cross by this mechanism in amounts capable of inducing CNS effects [7–9]. At a molecular weight of 7.8 KD, cytokine-induced neutrophil chemoattractant-1 (CINC-1) is the largest substance to date shown to cross by transmembrane diffusion [10].

A few cytokines appear to be totally excluded from crossing the BBB. In most cases of total exclusion, some factor other than just the lack of a transporter seems to be present. One of the best examples of this is IL-2, discussed in more detail below [11,12]. The total or near total lack of permeation of IL-2 across the BBB depends on more than just the lack of a blood-to-brain saturable transporter, but also on the presence of a brain-to-blood saturable transporter, binding in serum, and enzymatic activity.

Most work investigating the contribution of the BBB to the relation between brain and blood levels of cytokines have relied on acute, pharmacologic studies. Such studies can measure the rate at which a cytokine can cross the BBB but cannot be used to calculate what percent of brain levels of cytokines are derived from blood. The latter is a steady-state measure and so depends on the rates of brain-to-blood efflux and enzymatic degradation, in addition to the rate of blood-to-brain influx. One study used the finding that human IL-1α can cross the murine BBB to determine steady-state concentrations in brain of a blood-derived cytokine. Human IL-1α was infused peripherally by Alzet pump and brain and blood levels measured after 48 h of infusion [13]. The brain/blood ratio of human IL-1α was 0.126. This means that the amount of IL-1α in a gram of brain was about 12% of that found in a milliliter of serum. This is a surprisingly high number as the extracellular fluid space of the brain is about 15–20% of brain weight. If IL-1α is largely confined to this volume, then the concentrations in the extracellular fluid of the brain would be about 60–80% of serum levels.

3. BRAIN-TO-BLOOD TRANSPORT

3.1. Interleukin-2

IL-2 is to date the only cytokine that has been found to be transported from the brain to blood by a saturable efflux system [12]. This efflux system in combination with the absence of a blood-to-brain saturable transporter, serum protein binding, and either a BBB or CNS environment that is enzymatically active against IL-2, virtually excludes any blood-borne IL-2 from entering the CNS [11,12]. However, IL-2 does enter the CSF from blood at least in some clinical conditions [14]. This is very interesting in light of the profound effects that IL-2 has on behavior [15–18], producing a schizophrenic-like condition [19,20].

3.2. IL-1, TNF, IL-6

The lack of a saturable transport system in the brain-to-blood direction does not mean that cytokines produced in the CNS are unable to enter the blood. Cytokines (IL-1, TNF, IL-6) secreted into the CNS can enter the circulation with the reabsorption of the cerebrospinal fluid (CSF) [1,2,21,22]. Blood levels achieved after administering cytokines into the lateral ventricle of the brain can even exceed those achieved after administering cytokines intravenously, especially at later time points. This is because with an intravenous injection, the cytokine bolus can be rapidly cleared from blood, whereas the slower movement of cytokine from CSF to blood acts like an intravenous infusion. This nonsaturable efflux from CSF to blood may result in large increases in blood levels of cytokines and may be important in CNS infections [23].

3.3. Cerebrospinal fluid contributions

CSF enters the blood stream through two main routes. One is reabsorption directly into the venous circulation through the subarachnoid villi, and the other is indirectly through drainage at the olfactory bulbs to the cervical lymph nodes [24–26]. Substances drained by way of the lymphatic route have the ability to act as immunomodulators [27,28]. Thus, cytokines entering the blood stream from the CNS may have different effects, especially depending on which efflux route they use to enter the blood stream, than cytokines secreted directly into the blood from circulating immune cells or peripheral tissues.

4. BEHAVIORAL EFFECTS OF CYTOKINES AND THE BBB

4.1. General

Cytokines produce a wide variety of effects on CNS function after either their central or peripheral administration [29–32]. The CNS effects of cytokines can be physiologic as in the case of TNF-mediated effects on physiologic sleep [33,34], adaptive as in the case of IL-1 induction of sickness behavior [35], or pathologic as in interferon gamma induction of depression [36] or IL-2 induction of stereotypic behavior [37]. An interesting hallmark is that typically a cytokine will produce similar behaviors after peripheral or central administration. In fact, an antagonist to the cytokine, given centrally, can often block the effect of a cytokine given peripherally [38–40]. This shows that somehow peripheral cytokine administration results in an elevated level of cytokine in the CNS.

4.2. Mechanisms of behavioral effects

How peripheral cytokines induce an elevation in their CNS levels has been the topic of a great deal of investigation in neuroimmunology. Mechanisms other than directly crossing the BBB have been elucidated [41–44]. The first mechanism established was the ability of cytokines to act at vagal nerve endings to induce afferent signaling to the brain [45,46]. Such signaling can evoke many responses in the CNS, including the local release of cytokines [47]. Many of the aspects of sickness behavior, especially when invoked by the intraperitoneal administration of cytokines or lipopolysaccharide (LPS), seem to be mediated by afferent vagal fibers [40,48]. Furthermore, it appears that other nerves are capable of similar actions. For example, administration of LPS to the palate results in afferent signaling to the CNS through the glossopharyngeal nerve [49].

4.3. Circumventricular organs

Cytokines can also act at circumventricular organs (CVOs), areas of the CNS whose vasculature does not form a BBB. Circulating cytokines are able to leak into CVOs at rates of 20–160 times faster than they are transported [44,50]. Once there, the cytokines can interact with nerve bodies and endings which project from or into the CVO from other regions of the brain [51,52]. The CVOs tend to be delimited from the rest of the brain by a tanycytic barrier [53–56], but this barrier itself appears to be secretory. The increase in body temperature that results from peripheral administration of cytokines or LPS is largely mediated through prostaglandin release from the CVOs [57].

4.4. Blood–brain barrier

Cytokines can also affect the passage of other substances across the BBB, which in turn affect CNS functions. Classically, treatment with LPS or with proinflammatory cytokines disrupt the BBB [58–60], although this can be surprisingly difficult to achieve on a reliable basis [61]. Insulin, leptin, pituitary adenylate cyclase activating polypeptide, and gp120 which is the viral coat of HIV have their transport across the BBB perturbed by treatment with LPS [62–65]. In some cases, such as leptin, the effects are indirect in that the BBB transporter itself is not affected. Instead, leptin transport is altered because serum levels of leptin and triglycerides are elevated. Serum triglycerides, in turn, inhibit leptin transport [66]. In contrast, the insulin transporter is directly affected by a prostaglandin-independent pathway. It has been postulated that the cytokine-enhanced transport of insulin into the CNS may be one mechanism by which sepsis induces insulin resistance. Cytokines affect nonpermeability aspects of BBB function as well. For example, IL-1 induces expression of the mu opiate receptor on brain endothelial cells [67].

Given the multiple pathways by which cytokines can affect CNS function, it can be difficult to determine what contribution, if any, transport of a cytokine across the BBB may make to any given function. One exception to this is the effect of IL-1 on memory. Impairment of learning is one of the features of sickness behavior, an adaptive response to illness largely mediated through IL-1 release [35,68]. We have shown that much of the impairment in learning induced by the intravenous administration of human IL-1 to the mouse could be prevented by giving species-specific blocking antibodies directed against human IL-1 directly into the posterior division of the septum [69]. This showed that much of the learning impairment depended on the ability of IL-1 to cross the BBB at this region of the brain. The effects of TNF on fever as conducted by Stefferl et al. [70] also fit the pattern for the effect being dependent on TNF transport. Whereas human or murine TNF each produced fever when injected directly into the brain of mice, only murine TNF produced fever when injected peripherally. This finding is easily explained by the finding that murine, but not human, TNF is transported across the BBB of the mouse [71].

The classification of leptin as a cytokine provides another example of an important CNS effect being dependent on BBB transport. Leptin is transported across the BBB into many regions of the brain, but uptake is particularly high into the arcuate nucleus [72]. There, leptin interacts with its receptors to induce anorexia and increase thermogenesis [73]. Leptin accesses the arcuate nucleus primarily by being transported across the vascular BBB of the hypothalamus [72]. Leptin is also transported across the choroid plexus into the CSF [74]. It is likely that leptin in the CSF of the fourth ventricle is able to reach receptors in the arcuate nucleus [75]. Others have proposed that leptin can leak from the median eminence into the adjacent arcuate nucleus, but as Kastin has pointed out [76,77], there is no experimental evidence for this. The tanycytic barrier, which develops between the median eminence and the hypothalamus after birth is likely to prevent any significant leakage [54–56,78]. Barriers at other CVO have been shown to prevent the leakage of other cytokines into adjacent brain tissues [50,53].

5. SPECIES SPECIFICITY

Cytokines show a degree of homology both among chemical species and among genetic species. For example, human IL-1α is 26% homologous with human IL-1β [79] and 60–70% homologous with murine IL-1α [80]. Transport across the BBB shows variation across species in both senses of the word. For example, the mouse transports across its BBB murine IL-1α, murine IL-1β, and human IL-1α [2]. Neither rat nor mouse, however, transports human IL-1β across their BBBs [81–83]. Likewise, the mouse transports murine, but not human, TNF across its BBB [71,84,85]. Interestingly, much early work on TNF used human TNF injected into rats. As a result, nearly all those effects can be ascribed to TNF working through peripheral mechanisms [70,86].

6. PHYSIOLOGICAL ALTERATIONS

6.1. Diurnal rhythms

Both IL-1 and TNF have been shown in mouse to have diurnal rhythms in their transports across the BBB [71,85]. That the fluctuations in blood-to-brain uptake are caused by variations in the transporters, and not to some nonspecific aspect of BBB function, is shown by the findings that human TNF, a cytokine not transported across the mouse BBB, shows no circadian rhythm. The barrier at the spinal cord also transports IL-1α [87] and shows diurnal variation in this uptake as well [71]. Peak transport into brain and spinal cord both occur at about 0800 h at which time the transport rate into brain is about twice that of spinal cord. The nadir in transport for both tissues was at 2400 h (midnight) with a decrease in peak transport rate of fivefold for brain and about 10-fold for spinal cord. These patterns were different for both testis, another tissue with a blood barrier [88] that transports IL-1 [89], and muscle, which had peak uptakes at 1200 h (noon) and differences between peak and nadir transport rates of about three- and twofold, respectively.

6.2. Regional variation

One of the most striking aspects of cytokine transport across the BBB is its heterogeneity across brain regions. For example, IL-1α is not transported into the striatum or midbrain of the CD-1 mouse [90]. In the SAMP8 mouse, a strain of mouse used in Alzheimer's research [91,92], these regions as well as the hippocampus, thalamus, hypothalamus, and occipital cortex are unable to transport IL-1α. In contrast, TNF is transported throughout the brain, but transport rates vary by 10-fold [93]. Differences also occur between rates of transport into whole brain and spinal cord [94,95]. For example, IL-1 and colony-stimulating factor (GMCSF) have a transport rate into spinal cord that is about half the rate of transport into brain [71,96].

7. PATHOLOGICAL ALTERATIONS IN TRANSPORT

Transport systems for cytokines show a surprising array of responses to CNS injuries [97]. Immunologic and various types of traumatic injuries in particular have been studied. The first such study was the response of the TNF transporter during the clinical phase of experimental allergic encephalomyelitis [EAE]. The blood-to-brain transport of TNF was greatly increased [98]. The BBB was also disrupted to serum albumin, but even after correcting for this, the saturable component to the TNF transporter was increased by about two- to threefold. With clinical recovery, the TNF transporter rate returned to normal. Interestingly, the TNF transporter is not perturbed in animals treated with LPS, suggesting that the effect is not mediated by proinflammatory cytokines.

Other cytokines have altered transport across the BBB after various injuries. For example, spinal cord injury [99,100], right temporal lobe blunt trauma [83], and stroke [101] alter TNF transport. In contrast, cardiac arrest, or LPS treatment, has no effect on TNF transport [84,102], even though these events alter other aspects of BBB function. Alterations in the transport rates of cytokines across the BBB do not necessarily occur at the time or at the site of injury but can be both anatomically and temporally removed [103]. Disruptions of the BBB resulting from trauma have long been known to also show such anatomical and temporal disassociations from the trauma [104]. Interestingly, the areas and times at which the cytokine transporters are altered are not the same as those at which disruption of the BBB occurs [83,99,105].

8. OTHER CYTOKINE–BBB INTERACTIONS

As mentioned in the introduction, cytokines and the BBB interact in ways besides that of transport. One of the less explored but potentially very exciting of these interactions is the ability of brain endothelial cells to secrete cytokines. IL-1, IL-3, IL-6, IL-8, TNF, GMCSF, monocyte chemoattractant protein-1, nerve growth factor, RANTES, and endothelin-1 have all been shown to be secreted from brain endothelial cells or the choroid plexus [106–113]. Based on *in vitro* cultures of brain endothelial cells, it seems that some cytokines are constitutively released, whereas others are released with stimulation. For example, LPS and HIV-1 Tat induce the release of IL-6 [111,114,115], whereas HIV-1 gp120 or HIV-1 itself stimulates release of endothelin-1 [106].

The BBB is polarized; that is, its blood side and brain side are not identical, but differ in the kind and concentration of membrane lipids, glycoproteins, receptors, and transporters [116–119]. This raises the possibility that the cytokine secretion is also polarized; that is, that cytokines might be preferentially released from one side of the BBB or the other. This seems to be the case as shown in a recent study. The most striking example of polarized release was that of IL-6, which was released 10-fold more from the blood side than the brain side of cultured brain endothelial cells [114]. Polarization also occurs with regard to which side is most sensitive to stimuli. LPS applied to the blood side of cultured brain endothelial cells increased IL-6 release by fourfold, but LPS applied to the brain side of those cells increased IL-6 release by eightfold, a difference which was highly statistically significant. Similarly, gp120 or AIDS virus applied to the blood side of cultured endothelial cells increased the release of endothelin from the brain side of endothelial cells by several fold [106]. Therefore, the BBB has the potential to respond to signals received from one of its sides by releasing regulatory substances such as cytokines into the other.

9. CONCLUSIONS

Many cytokines are transported across the BBB. Transporters are unique for at least groups or families of cytokines. The amount transported across the BBB is sufficient to affect CNS functions. At least some of the effect of blood-borne IL-1 on memory is mediated by its crossing the BBB and interacting with receptors at the posterior division of the septum. The rate at which these transporters act is not static, but varies diurnally, among strains of animals, with brain location, and with type of cytokine. The transporters are also altered by disease and injury to the CNS. Cytokines also affect the transport of other substances across the BBB, including viruses such as HIV and regulatory proteins such as insulin, and are themselves secreted from the cells, which comprise the BBB. These various ways in which cytokines and the BBB interact illustrate that the BBB is a major link in the neuroimmune axis.

ACKNOWLEDGMENTS

Supported by VA Merit Review and DA019396.

REFERENCES

1. Banks WA, Kastin AJ, Durham DA. Bidirectional transport of interleukin-1 alpha across the blood- brain barrier. Brain Res Bull 1989;23:433–37.
2. Banks WA, Ortiz L, Plotkin SR, Kastin AJ. Human interleukin [IL] 1α, murine IL-1α and murine IL-1β are transported from blood to brain in the mouse by a shared saturable mechanism. J Pharmacol Exp Ther 1991;259:988–96.
3. Davson H, Segal MB. Special Aspects of the Blood-Brain Barrier. Physiology of the CSF and Blood-Brain Barriers. Boca Raton: CRC Press, 1996; pp. 303–485.

4. Banks WA, Kastin AJ, Gutierrez EG. Penetration of interleukin-6 across the blood-brain barrier. Neurosci Lett 1994;179:53–56.
5. Gutierrez EG, Banks WA, Kastin AJ. Murine tumor necrosis factor alpha is transported from blood to brain in the mouse. J Neuroimmunol 1993;47:169–76.
6. Cornford EM, Braun LD, Oldendorf WH, Hill MA. Comparison of lipid-mediated blood-brain-barrier penetrability in neonates and adults. Am J Physiol 1982;243:C161–68.
7. Kastin AJ, Akerstrom V. Orexin A but not orexin B rapidly enters brain from blood by simple diffusion. J Pharmacol Exp Ther 2005;289:219–23.
8. Banks WA, Kastin AJ, Coy DH. Evidence that [^{125}I]N-Tyr-delta sleep-inducing peptide crosses the blood-brain barrier by a non-competitive mechanism. Brain Res 1984;301:201–7.
9. Banks WA, Kastin AJ. Peptides and the blood-brain barrier: lipophilicity as a predictor of permeability. Brain Res Bull 1985;15:287–92.
10. Pan W, Kastin AJ. Changing the chemokine gradient: CINC1 crosses the blood-brain barrier. J Neuroimmunol 2001;115:64–70.
11. Waguespack PJ, Banks WA, Kastin AJ. Interleukin-2 does not cross the blood-brain barrier by a saturable transport system. Brain Res Bull 1994;34:103–9.
12. Banks WA, Niehoff ML, Zalcman S. Permeability of the mouse blood-brain barrier to murine interleukin-2: predominance of a saturable efflux system. Brain Behav Immun 2004;18:434–42.
13. Banks WA, Kastin AJ. Relative contributions of peripheral and central sources to levels of IL-1α in the cerebral cortex of mice: assessment with species-specific enzyme immunoassays. J Neuroimmunol 1997;79:22–28.
14. Saris SC, Rosenberg SA, Friedman RB, Rubin JT, Barba D, Oldfield EH. Penetration of recombinant interleukin-2 across the blood-cerebrospinal fluid barrier. J Neurosurg 1988; 69:29–34.
15. Zalcman SS. Interleukin-2 potentiates novelty - and GBR 12909-induced exploratory activity. Brain Res 2001;899:1–9.
16. Zalcman S, Murray L, Dyck DG, Greenberg AH, Nance DM. Interleukin-2 and -6 induce behavioral-activating effects in mice. Brain Res 1998;811:111–21.
17. Zalcman S, Green-Johnson JM, Murray L, Wan W, Nance DM, Greenberg AH. Interleukin-2 induced enhancement of an antigen-specific IgM PFC response is mediated by the sympathetic nervous system. J Pharmacol Exp Ther 1994;271:977–82.
18. Zalcman SS. Interleukin-2-induced increases in climbing behavior: inhibition by dopamine D-1 and D-2 receptor antagonists. Brain Res 2002;944:157–64.
19. Rapapport MH, Lohr JB. Serum soluble interleukin-2 receptors in neuroleptic naive schizophrenic patients and in medicated schizophrenic subjects with and without tardive dyskinesia. Acta Psychiatr Scand 1994;90:311–15.
20. Rapapport MH, Caliguiri MP, Lohr JB. An association between increased soluble interleukin-2 receptors and a disturbance in motor muscle force in schizophrenic patients. Prog Neuropsychopharmacol Biol Psychiatr 1997;21:817–27.
21. Chen G, Reichlin S. Clearance of [^{125}I]-tumor necrosis factor-α from the brain into the blood after intracerebroventricular injection into rats. Neuroimmunomodulation 1998;5:261–69.
22. Chen G, Castro WL, Chow HH, Reichlin S. Clearance of ^{125}I-labelled interleukin-6 from brain into blood following intracerebroventricular injection in rats. Endocrinology 1997;138:4830–36.
23. Romero LI, Kakucska I, Lechan RM, Reichlin S. Interleukin-6 [IL-6] is secreted from the brain after intracerebroventricular injection of IL-1β in rats. Am J Physiol 1996;270:R518–24.

24. Davson H, Segal MB. The Return of the Cerebrospinal Fluid to the Blood: The Drainage Mechanism. Physiology of the CSF and Blood-Brain Barriers. Boca Raton: CRC Press, 1996; pp. 489–523.
25. Boulton M, Flessner M, Armstrong D, Mohamed R, Hay J, Johnston M. Contribution of extracranial lymphatics and arachnoid villi to the clearance of a CSF tracer in the rat. Am J Physiol 1999;276:R818–23.
26. Kida S, Pantazis A, Weller RO. CSF drains directly from the subarachnoid space into nasal lymphatics in the rat. Anatomy, histology and immunological significance. Neuropathol Appl Neurobiol 1993;19:480–88.
27. Cserr HF, Knopf PM. Cervical lymphatics, the blood-brain barrier and the immunoreactivity of the brain: a new view. Immunol Today 1992;13:507–12.
28. Knopf PM, Cserr HF, Nolan SC, Wu TY, Harling-Berg CJ. Physiology and immunology of lymphatic drainage of interstitial and cerebrospinal fluid from the brain. Neuropathol Appl Neurobiol 1995;21:175–80.
29. Bartholomew SA, Hoffman SA. Effects of peripheral cytokine injections on multiple unit activity in the anterior hypothalamic area of the mouse. Brain Behav Immun 1993; 7:301–16.
30. Bianchi M, Panerai AE. Peripherally administered IL-1α interferes with scopolomine-induced amnesia in mice. Cogn Brain Res 1993;1:257–59.
31. Lancel M, Mathias S, Faulhaber J, Schiffelhosz T. Effect of interleukin-1 beta on EEG power density during sleep depends on circadian phase. Am J Physiol 1996;270:R830–37.
32. Venters HD, Tang Q, Liu Q, VanHoy RW, Dantzer R, Kelley KW. A new mechanism of neurodegeneration : a proinflammatory cytokine inhibits receptor signaling by a survival peptide. Proc Natl Acad Sci USA 1999;9617:9879–84.
33. Krueger JM, Obal FJ, Fang J, Kubota T, Taishi P. The role of cytokines in physiological sleep regulation. Ann NY Acad Sci 2001;933:211–21.
34. Opp MR, Obal F Jr, Krueger JM. Interleukin 1 alters rat sleep: temporal and dose-related effects. Am J Physiol 1991;260:R52–58.
35. Larson SJ, Dunn AJ. Behavioral effects of cytokines. Brain Behav Immun 2001;15: 371–87.
36. Anisman H, Merali Z, Poulter MO, Hayley S. Cytokines as a precipitant of depressive illness: animal and human studies. Curr Pharm Design 2005;11:963–72.
37. Lacosta S, Merali Z, Anisman H. Influence of acute and repeated interleukin-2 administration on spatial learning, locomotor activity, exploratory behaviors, and anxiety. Behav Neurosci 1999;113:1030–41.
38. Cremona S, Goujon E, Kelly KW, Dantzer R, Parnet P. Brain type I but not type II IL-1 receptors mediate the effects of IL-1 beta on behavior in mice. Am J Physiol 1998;274: R735–40.
39. del Rey A, Monge-Arditi G, Besedovsky HO. Central and peripheral mechanisms contribute to the hypoglycemia induced by interleukin-1. Ann NY Acad Sci 1998;840: 153–61.
40. Kent S, bret-Dibat JL, Kelley KW, Dantzer R. Mechanisms of sickness-induced decreases in food-motivated behavior. Neurosci Biobehav Rev 1996;20:171–75.
41. Banks WA. Neuroimmune networks and communication pathways: the importance of location. Brain Behav Immun 2004;18:120–22.
42. Kelley KW, Bluthe RM, Dantzer R, Zhou J-H, Shen W-H, Johnson RW, et al. Cytokine-induced sickness behavior. Brain Behav Immun 2003;17:S112-S118.

43. Wieseler-Frank J, Maier SF, Watkins LR. Immune-to-brain communication dynamically modulates pain: physiological and pathological consequences. Brain Behav Immun 2005;19:104–11.
44. Banks WA. Cytokines, CVOs, and the Blood-Brain Barrier. In: Ader R, Felten DL, Cohen N, Eds; San Diego: Academic Press, 2001; pp. 483–97.
45. Watkins LR, Maier SF, Goehler LE. Cytokine-to-brain communication: a review & analysis of alternative mechanisms. Life Sci 1995;57:1011–26.
46. Laye S, Bluthe RM, Kent S, Combe C, Medina C, Parnet P, et al. Subdiaphragmatic vagotomy blocks induction of IL-1 beta mRNA in mice brain in response to peripheral LPS. Am J Physiol 1995;268:R1327–31.
47. Laye S, Parnet P, Goujon E, Dantzer R. Peripheral administration of lipopolysaccharide induces the expression of cytokine transcripts in the brain and pituitary of mice. Brain Res 1994;27:157–62.
48. Kent S, Bluthe RM, Dantzer R, Hardwick AJ, Kelley KW, Rothwell NJ, et al. Different receptor mechanisms mediate the pyrogenic and behavioral effects of interleukin 1. Proc Natl Acad Sci USA 1992;89:9117–20.
49. Brief DJ, Davis JD. Reduction of food intake and body weight by chronic intraventricular insulin infusion. Brain Res Bull 1984;12:571–75.
50. Plotkin SR, Banks WA, Kastin AJ. Comparison of saturable transport and extracellular pathways in the passage of interleukin-1α across the blood-brain barrier. J Neuroimmunol 1996;67:41–47.
51. Ferguson AV. The area postrema: a cardiovascular control centre at the blood-brain interface? Can J Physiol Pharmacol 1991;69:1026–34.
52. Johnson AK, Gross PM. Sensory circumventricular organs and brain homeostatic pathways. FASEB J 1993;7:678–86.
53. Maness LM, Kastin AJ, Banks WA. Relative contributions of a CVO and the microvascular bed to delivery of blood-borne IL-1α to the brain. Am J Physiol 1998;275:E207–12.
54. Krisch B. The functional and structural borders between the CSF- and blood-dominated milieus in the choroid plexuses and the area postrema of the rat. Cell Tissue Res 1986; 245:101–15.
55. Rethelyi M. Diffusional barrier around the hypothalamic arcuate nucleus in the rat. Brain Res 1984;307:355–58.
56. Peruzzo B, Pastor FE, Blazquez JL, Schobitz K, Pelaez B, Amat P, et al. A second look at the barriers of the medial basal hypothalamus. Exp Brain Res 2000;132:10–26.
57. Blatteis CM. Role of the OVLT in the febrile response to circulating pyrogens. In Circumventricular Organs and Brain Fluid Environment. Ermisch A, Landgraf R, Ruhle HJ, Eds; Amsterdam: Elsevier, 1992; pp. 409–12.
58. Abraham CS, Harada N, Deli MA, Niwa M. Transient forebrain ischemia increases the blood-brain barrier permeability for albumin in stroke-prone spontaneously hypertensive rats. Cell Mol Neurobiol 2003;22:455–62.
59. Fiala M, Looney DJ, Stins M, Way DD, Zhang L, Gan X, et al. TNF-α opens a paracellular route for HIV-1 invasion across the blood-brain barrier. Mol Med 1997;3:553–64.
60. Rosenberg GA, Estrada EY, Dencoff JE, Stetler-Stevenson WG. Tumor necrosis factor-α-induced gelatinase B causes delayed opening of the blood-brain barrier: an expanded therapeutic window. Brain Res 1995;703:151–55.
61. Banks WA, Kastin AJ. The interleukins -1α, -1β, and -2 do not disrupt the murine blood-brain barrier. Int J Immunopharmacol 1992;14:629–36.

62. Banks WA, Kastin AJ, Brennan JM, Vallance KL. Adsorptive endocytosis of HIV-1gp120 by blood-brain barrier is enhanced by lipopolysaccharide. Exp Neurol 1999; 156:165–71.
63. Nonaka N, Hileman SM, Shioda S, Vo P, Banks WA. Effect of lipopolysaccharide on leptin transport across the blood-brain barrier. Brain Res 2004;1016:58–65.
64. Nonaka N, Shioda S, Banks WA. Effect of lipopolysaccharide on the transport of pituitary adenylate cyclase activating polypeptide across the blood-brain barrier. Exp Neurol 2005;191:137–44.
65. Xaio H, Banks WA, Niehoff ML, Morley JE. Effect of LPS on the permeability of the blood-brain barrier to insulin. Brain Res 2001;896:36–42.
66. Banks WA, Coon AB, Robinson SM, Moinuddin A, Shultz JM, Nakaoke R, et al. Triglycerides induce leptin resistance at the blood-brain barrier. Diabetes 2004;53: 1253–60.
67. Vidal EL, Patel NA, Wu G, Fiala M, Chang SL. Interleukin-1 induces the expression of: opioid receptors in endothelial cells. Immunopharmacology 1998;38:261–66.
68. Pugh CR, Fleshner M, Watkins LR, Maier SF, Rudy JW. The immune system and memory consolidation: a role for the cytokine IL-1β. Neurosci Biobehav Rev 2001;25:29–41.
69. Banks WA, Farr SA, La Scola ME, Morley JE. Intravenous human interleukin-1α impairs memory processing in mice: dependence on blood-brain barrier transport into posterior division of the septum. J Pharmacol Exp Ther 2001;299:536–41.
70. Stefferl A, Hopkins SJ, Rothwell NJ, Luheshi GN. The role of TNF-alpha in fever: opposing actions of human and murine TNF-alpha and interactions with IL-beta in the rat. Br J Pharmacol 1996;118:1919–24.
71. Banks WA, Kastin AJ, Ehrensing CA. Diurnal uptake of circulating interleukin-1α by brain, spinal cord, testis and muscle. Neuroimmunomodulation 1998;5:36–41.
72. Banks WA, Kastin AJ, Huang W, Jaspan JB, Maness LM. Leptin enters the brain by a saturable system independent of insulin. Peptides 1996;17:305–11.
73. Friedman JM, Halaas JL. Leptin and the regulation of body weight in mammals. Nature 1998;395:763–70.
74. Zlokovic BV, Jovanovic S, Miao W, Samara S, Verma S, Farrell CL. Differential regulation of leptin transport by the choroid plexus and blood-brain barrier and high affinity transport systems for entry into hypothalamus and across the blood-cerebrospinal fluid barrier. Endocrinology 2000;141:1434–41.
75. Maness LM, Kastin AJ, Farrell CL, Banks WA. Fate of leptin after intracerebroventricular injection into the mouse brain. Endocrinology 1998;139:4556–62.
76. Kastin AJ, Akerstrom V, Pan W. Validity of multiple-time regression analysis in measurement of tritiated and iodinated leptin crossing the blood-brain barrier: meaningful controls. Peptides 2001;22:2127–36.
77. Kastin AJ, Pan W. Intranasal leptin: blood-brain barrier bypass [BBBB] for obesity? Endocrinology 2006;147:2086–87.
78. Krisch B, Leonhardt H. Relations between leptomeningeal compartments and the neurohemal regions of circumventricular organs. Biomed Res 1989;10(Suppl 3):155–68.
79. March CJ, Mosley B, Larsen A, Cerretti DP, Braedt G, Price V, et al. Cloning, sequene and expression of two distinct human interleukin-1 complimentary DNAs. Nature 1985; 315:641–47.
80. Gray PW, Glaister D, Chen E, Goeddel DV, Pennica D. Two interleukin 1 genes in the mouse: cloning and expression of the cDNA for murine interleukin 1 beta. J Immunol 1986;137:3644–48.

81. Plotkin SR, Banks WA, Maness LM, Kastin AJ. Differential transport of rat and human interleukin-1α across the blood-brain barrier and blood-testis barrier in rats. Brain Res 2000;881:57–61.
82. Maness LM, Banks WA, Zadina JE, Kastin AJ. Selective transport of blood-borne interleukin-1α into the posterior division of the septum of the mouse brain. Brain Res 1995;700:83–88.
83. Pan W, Kastin AJ, Rigai T, McLay R, Pick CG. Increased hippocampal uptake of tumor necrosis factor alpha and behavioral changes in mice. Exp Brain Res 2003;149:195–99.
84. Osburg B, Peiser C, Domling D, Schomburg L, Ko YT, Voight K, et al. Effect of endotoxin on expression of TNF receptors and transport of TNF-alpha at the blood-brain barrier of the rat. Am J Physiol 2002;283:E899–908.
85. Pan W, Cornelissen G, Halberg F, Kastin AJ. Selected contributions: circadian rhythm of tumor necrosis factor-alpha uptake into mouse spinal cord. J Appl Physiol 2002;92:1357–62.
86. Bodnar RJ, Pasternak GW, Mann PE, Paul D, Warren R, Donner DB. Mediation of anorexia by human recombinant tumor necrosis factor through a peripheral action in the rat. Cancer Res 1989;15:6280–84.
87. Banks WA, Kastin AJ, Ehrensing CA. Transport of blood-borne interleukin-1α across the endothelial blood-spinal cord barrier of mice. J Physiol [London] 1994;479:257–64.
88. Neaves WB. The blood-testis barrier. In The Testis. Johnson AD, Gomes WR, Eds; New York: Academic Press, 1977; pp. 125–62.
89. Banks WA, Kastin AJ. Human interleukin-1α crosses the blood-testis barriers of the mouse. J Androl 1992;13:254–59.
90. Moinuddin A, Morley JE, Banks WA. Regional variations in the transport of interleukin-1α across the blood-brain barrier in ICR and aging SAMP8 mice. Neuroimmunomodulation 2000;8:165–70.
91. Morley JE, Farr SA, Kumar VB, Banks WA. Alzheimer's disease through the eye of a mouse: acceptance lecture for the 2001 Gayle A. Olson and Richard D. Olson prize. Peptides 2002;23:589–99.
92. Morley JE. The SAMP8 mouse: a model of Alzheimer's disease? Biogerontology 2002;31:57–60.
93. Banks WA, Moinuddin A, Morley JE. Regional transport of TNF-α across the blood-brain barrier in young ICR and young and aged SAMP8 mice. Neurobiol Aging 2001;22: 671–76.
94. Pan W, Banks WA, Kastin AJ. Permeability of the blood-brain barrier and blood-spinal cord barriers to interferons. J Neuroimmunol 1997;76:105–11.
95. Pan W, Banks WA, Kastin AJ. Permeability of the blood-brain barrier to neurotrophins. Brain Res 1998;788:87–94.
96. McLay RN, Kimura M, Banks WA, Kastin AJ. Granulocyte-macrophage colony-stimulating factor crosses the blood-brain and blood-spinal cord barriers. Brain 1997;120:2083–91.
97. Banks WA. Blood-brain barrier transport of cytokines: a mechanism for neuropathology. Curr Pharm Design 2005;11:973–84.
98. Pan W, Banks WA, Kennedy MK, Gutierrez EG, Kastin AJ. Differential permeability of the BBB in acute EAE: enhanced transport of TNF-α. Am J Physiol 1996;271:E636–42.
99. Pan W, Banks WA, Kastin AJ. BBB permeability to ebiratide and TNF in acute spinal cord injury. Exp Neurol 1997;146:367–73.
100. Pan W, Zhang L, Liao J, Csernus B, Kastin AJ. Selective increase in TNF alpha permeation across the blood-spinal cord barrier after SCI. J Neuroimmunol 2003;134:111–17.

101. Pan W, Ding Y, Yu Y, Ohtaki H, Nakamachi T, Kastin AJ. Stroke upregulates TNFalpha transport across the blood-brain barrier. Exp Neurol 2006;198:222–33.
102. Mizushima H, Banks WA, Dohi K, Shioda S, Matsumoto K. Effect of cardiac arrest on brain weight and the permeability of the blood-brain barrier and blood-spinal cord barrier to albumin and tumor necrosis factor-α. Life Sci 1999;20:2127–34.
103. Pan W, Kastin AJ. Increase in TNF alpha transport after SCI is specific for time, region, and type of lesion. Exp Neurol 2001;170:357–63.
104. Baldwin SA, Fugaccia I, Brown DR, Brown LV, Scheff SW. Blood-brain barrier breach following cortical contusion in the rat. J Neurosurg 1996;85:476–81.
105. Pan W, Kastin AJ, Pick CG. The staircase test in mice after spinal cord injury. Int J Neuroprotect Neuroregen 2005;1:32–37.
106. Didier N, Banks WA, Creminon C, Dereuddre-Bosquet N, Mabondzo A. Endothelin-1 production at the in-vitro blood-brain barrier during HIV infection. Neuroreport 2002; 13:1179–83.
107. Fabry Z, Fitzsimmons KM, Herlein JA, Moninger TO, Dobbs MB, Hart MN. Production of the cytokines interleukin 1 and 6 by murine brain microvessel endothelium and smooth muscle pericytes. J Neuroimmunol 1993;47:23–34.
108. Lee YW, Hennig B, Fiala M, Kim KS, Toborek M. Cocaine activates redox-regulated transcription factors and induces TNF-alpha expression in human brain endothelial cells. Brain Res 2001;920:125–33.
109. Hofman F, Chen P, Incardona F, Zidovetzki R, Hinton DR. HIV-tat protein induces the production of interleukin-8 by human brain-derived endothelial cells. J Neuroimmunol 1999;94:28–39.
110. Moser KV, Reindl M, Blasig I, Humpel C. Brain capillary endothelial cells proliferate in response to NGF, express NGF receptors and secrete NGF after inflammation. Brain Res 2004;1017:53–60.
111. Reyes TM, Fabry Z, Coe CL. Brain endothelial cell production of a neuroprotective cytokine, interleukin-6, in response to noxious stimuli. Brain Res 1999;851:215–20.
112. Simpson JE, Newcombe J, Cuzner ML, Woodrofe MN. Expression of monocyte chemoattractant protein-1 and other beta-chemokines by resident glia and inflammatory cells in multiple sclerosis. J Neuroimmunol 1998;84:238–49.
113. Vadeboncoeur N, Segura M, Al-Numani D, Vanier G, Gottschalk M. Proinflammatory cytokine and chemokine release by human brain microvascular endothelial cells stimulated by Streptococcus suis serotype 2. FEMS Immunol Med Microbiol 2003; 35:49–58.
114. Verma S, Nakaoke R, Dohgu S, Banks WA. Release of cytokines by brain endothelial cells: a polarized response to lipopolysaccharide. Brain Behav Immun 2006;20:449–55.
115. Zidovetzki R, Wang JL, Chen P, Jeyasseelan R, Hofman F. Human immunodeficiency virus Tat protein induces interleukin 6 mRNA expression in human brain endothelial cells via protein kinase C- and cAMP-dependent protein kinase pathways. AIDS Res Hum Retroviruses 1998;14(10):825–33.
116. Chikale EG, Burton PS, Borchardt RT. The effect of verapamil on the transport of peptides across the blood-brain barrier in rats: kinetic evidence for an apically polarized efflux mechanism. J Pharmacol Exp Ther 1995;273:298–303.
117. Betz AL, Goldstein GW. Polarity of the blood-brain barrier: neutral amino acid transport into isolated brain capillaries. Science 1978;202:225–27.

118. Clapham PR, Weber JR, Whitby D, McIntosh K, Dalgleish AG, Maddon PJ, et al. Soluble CD4 blocks the infectivity of diverse strains of HIV and SIV for T cells and monocytes but not for brain and muscle cells. Nature 1989;337:368–70.
119. Vorbrodt AW. Glycoconjugates and anionic sites in the blood-brain barrier. In: Nicolini M, Zatta PF, Eds; Oxford: Pergamon Press, 1994; pp. 37–62.
120. Pan W, Banks WA, Fasold MB, Bluth J, Kastin AJ. Transport of brain-derived neurotrophic factor across the blood-brain barrier. Neuropharmacology 1998;37:1553–61.
121. Poduslo JF, Curran GL. Permeability at the blood-brain barrier and blood-nerve barriers of the neurotrophic factors: NGF, CNTF, NT-3, BDNF. Mol Brain Res 1996;36:280–86.
122. Pan W, Kastin AJ, Maness LM, Brennan JM. Saturable entry of ciliary neurotrophic factor into the brain. Neurosci Lett 1999;263:69–71.
123. Pan W, Kastin AJ. Entry of EGF into brain is rapid and saturable. Peptides 1999;20: 1091–98.
124. Banks WA, Jumbe NL, Farrell CL, Niehoff ML, Heatherington A. Passage of erythropoietic agents across the blood-brain barrier: A comparison of human and murine erythropoietin and the analog Darbopoetin alpha. Eur J Pharmacol 2004;505:93–101.
125. Brines ML, Ghezzi P, Keenan S, Agnello D, de Lanerolle NC, Cerami C, et al. Erythropoietin crosses the blood-brain barrier to protect against experimental brain injury. Proc Natl Acad Sci USA 2000;97:10526–31.
126. Cuevas P, Fermamdez-Ayerdi A, Carceller F, Colin S, Mascarelli F, Munoz-Willery I, et al. Central nervous system distribution of fibroblast growth factor injected into the blood stream. Neurol Res 1996;18:267–72.
127. Kastin AJ, Akerstrom V, Pan W. Glial cell line-derived neurotrophic factor does not enter normal mouse brain. Neurosci Lett 2003;340:239–41.
128. Habif DV, Lipton R, Cantell K. Interferon crosses the blood-cerebrospinal fluid barrier in monkeys. Proc Soc Exp Biol Med 1975;149:287–89.
129. Banks WA, Kastin AJ, Gutierrez EG. Interleukin-1α in blood has direct access to cortical brain cells. Neurosci Lett 1993;163:41–44.
130. Gutierrez EG, Banks WA, Kastin AJ. Blood-borne interleukin-1 receptor antagonist crosses the blood-brain barrier. J Neuroimmunol 1994;55:153–60.
131. Luheshi GN, Gay J, Rothwell NJ. Circulating IL-6 is transported into the brain via a saturable transport mechanism in the rat. Br J Pharmacol 1994;112:637P.
132. Kastin AJ, Akerstrom V, Pan W. Interleukin-10 as a CNS therapeutic: the obstacle of the blood-brain/blood-spinal cord barrier. Mol Brain Res 2003;114:168–71.
133. Pan W, Kastin AJ, Brennan JM. Saturable entry of leukemia inhibitory factor from blood to the central nervous system. J Neuroimmunol 2000;106:172–80.
134. Banks WA, Kastin AJ. Reversible association of the cytokines MIP-1α and MIP-1β with the endothelia of the blood-brain barrier. Neurosci Lett 1996;205:202–6.
135. Loy R, Talialatela G, Angelucci L, Heyer D, Perez-Polo R. Regional CNS uptake of blood-borne nerve growth factor. J Neurosci Res 1994;39:339–46.
136. Banks WA, Plotkin SR, Kastin AJ. Permeability of the blood-brain barrier to soluble cytokine receptors. Neuroimmunomodulation 1995;2:161–65.
137. Pan W, Vallance K, Kastin AJ. TGFα and the blood-brain barrier: accumulation in cerebral vasculature. Exp Neurol 1999;160:454–59.
138. Kastin AJ, Akerstrom V, Pan W. Circulating TGF-beta1 does not cross the intact blood-brain barrier. J Mol Neurosci 2003;21:43–48.

B. CYTOKINES IN BRAIN PHYSIOLOGY

Cytokines and the Brain
Edited by C. Phelps and E. Korneva

Cytokines in Synaptic Function

TRACEY A. IGNATOWSKI and ROBERT N. SPENGLER

Department of Pathology and Anatomical Sciences and Department of Anesthesiology, School of Medicine and Biomedical Sciences, University at Buffalo, The State University of New York, 206 Farber Hall, 3435 Main Street, Buffalo, NY 14214, USA

ABSTRACT

Cytokines perform an extensive range of overlapping functions both outside and inside the central nervous system (CNS). Some of the roles for cytokines include immunologic, inflammatory, and hematopoietic functions, as well as the regulation of neuron differentiation, survival, and plasticity, and in particular the regulation of neuron synaptic function. The transmission of signals by neurons in the CNS involves multiple interactions between the neuron and numerous accessory cells. These interactions are mediated by various factors, including short-acting soluble cytokines. These polypeptides impart profound effects over synaptic responsiveness. Although cytokines have diverse actions, they all share some common properties that dictate their involvement in the nervous system. Originally characterized as either pro- or anti-inflammatory mediators primarily produced by cells of the immune system, it is now apparent that cytokines serve major roles in the nervous system. These roles involve preserving neuron homeostasis, not necessarily in the face of an ensuing inflammatory or immune response. Appropriately, the term cytokine has now been adopted as a general umbrella for many types of secreted proteins that mediate diverse biological responses including synaptic responsiveness. Belonging to the cytokine family are (1) interleukins, (2) interferons, (3) chemokines, and (4) tumor necrosis factors. For the purpose of this chapter, each of these family members will be discussed with relevance to their roles in synaptic function. Particular emphasis will be placed on the group including tumor necrosis factor-α (TNF-α). The synapse is the location for the delivery of neurotransmitter from neuron to neuron. However, our concepts of synaptic transmission are in flux, requiring constant reevaluation. With the blossoming of our knowledge of cytokine involvement in synaptic function, our concepts are continually evolving. Most synapses rely on a chemical intermediary. Moreover, the chemical synapse is characterized by great flexibility and, therefore, adapts to environmental cues. This flexibility of the synapse depends on modifications by mediators such as cytokines. It is now obvious that many cytokines are used constitutively by cells of the CNS as signaling molecules involved in intercellular communication. Therefore, this interaction of cytokines at the synapse allows flexibility in synaptic function and subsequent behavior. This leads to a complex processing of information that is necessary to modify neuronal circuits in the multitude of processes governed by the nervous system. The role of cytokines in brain signaling has, therefore, gained recent recognition as emphasized in this presentation. Neurons are the fundamental building blocks of the CNS that transmit information at the synapse. Consequently, control over synaptic function would have significant impact on the brain and spinal cord. As with any control system, numerous mechanisms interact with each other to tightly maintain synaptic regulation of neurotransmitter release. Given the complexity of synaptic regulation that exists when accounting for the tasks performed by the brain, it is not surprising that many mediators, including pleiotropic cytokines, play an integral role in this regulation. By increasing our understanding of the relationship between cytokines and synaptic functions, powerful new classes of drugs that act in the CNS will be developed. Neurological disorders are becoming increasingly common due to the considerable growth of an aging population. These disorders are costly and can arise from many diseases (e.g., AIDS). The design of drugs targeted at cytokines and/or their effects at the synapse, we believe, will improve brain function and thereby benefit individuals experiencing disorders of the CNS.

1. INTRODUCTION

The regulation of the synapse, and therefore the ability of a neuron to transmit its signal, involves multiple interactions between that neuron and numerous accessory cells. These interactions are mediated by many factors, including short-acting soluble mediators, called cytokines. These polypeptides impart profound effects over synaptic responsiveness. Most of these small proteins have a wide spectrum of effects and in response to a stimulus are quickly released by diverse populations of cells throughout the body. Although cytokines have many varied actions, they all share some common properties that dictate their involvement in the nervous system. Cytokine actions are pleiotropic, meaning that each cytokine can influence many cell types and mediate many different effects. In addition, they are also redundant, implicating the importance of the events that they regulate. Many different cytokines are produced by numerous cell types, including neurons. As neurons produce cytokines, these mediators can have an autocrine effect on the neuron that produces them. Cytokines can also affect other cells in their vicinity (paracrine effects), or they may affect many cells systemically (humoral effects). Finally, cytokines mediate their effects by binding to specific high-affinity receptors on their target cells. Originally they were known as chemical messengers imparting critical information among cells of the immune system and consequently orchestrating coherent inflammatory and immune processes. Based on the latter they were originally referred to as pro- and anti-inflammatory cytokines; however, as their role in other systems including the nervous system became apparent, a broader appreciation of cytokine actions developed. More recently, cytokines have been shown to play integral roles in normal growth and development along with the maintenance of homeostasis of the nervous system, including synaptic plasticity. The term cytokine has currently been adopted as a general umbrella designation for many types of secreted proteins that mediate diverse biological responses such as synaptic responsiveness. The groups belonging to the cytokine family include (1) tumor necrosis factors, (2) interleukins, (3) interferons, and (4) chemokines. For the purpose of this chapter, each of these groups will be briefly discussed with relevance to their roles in synaptic function. Special consideration will be given to the family including tumor necrosis factor-α (TNF).

The synapse is the primary location for the transmission of information from neuron to neuron. These unique sites on the neuron are the active zone for neurotransmitter release. However, our concepts of synaptic transmission are in flux, requiring constant reevaluation. With the blossoming of our knowledge of cytokine involvement in synaptic function, our concepts of the interactive relationships between the classical neurotransmitters and the cytokines are continually evolving. The electrical synapse is where the electrical signal from one cell can directly cross to the adjacent cell through the conducting pathways of gap junctions. However, most synapses rely on a chemical intermediary. This chemical intermediary, or transmitter, is released from the presynaptic cell in order to affect the postsynaptic cell. It is the chemical synapse that is characterized by great flexibility; therefore, different afferent signals can have diverse effects on the presynaptic response and consequent postsynaptic reply. These differences depend on the phenotype of the transmitters released, the receptors activated, the prior activity of that neuron, and in particular the prior and current modifications by mediators such as cytokines. As the presence of cytokines and their receptors became apparent in the nervous system, a role for these mediators in neuronal function was implied. It is now obvious that many cytokines are used constitutively by cells of the central nervous system (CNS) as signaling molecules involved in intercellular communication [1–3]. The identification of cytokine protein and mRNA expression in neurons and non-neuronal cells of the brain implies a functional role for these messengers in the CNS [4–6]. The localization of some cytokine receptors in the brain overlaps; however, some localization is distinct from other cytokine receptors [4–8]. This discreet receptor distribution implicates a role for cytokines in various neuron functions, such as neuronal transmission, which in turn could be responsible for the behavioral effects of cytokines. In fact, cytokines have been shown to modulate the release at acetylcholine (Ach) [9], norepinephrine (NE) [10–12], dopamine (DA), and serotonin (5-HT) [13] in the brain. Therefore, this (potential) interaction of cytokines with neurotransmitter release at the synapse may lead to a complex processing of information that is necessary to modify neuronal circuits in the multitude of processes governed by the nervous system. The role of cytokines on neuronal circuit behavior and brain signaling has gained recent attention as demonstrated in this chapter.

Information is transmitted between neurons at the synapse. Consequently, modulation of synaptic function would have significant impact on the brain and spinal cord. In light of its importance in neuron and brain function, the modulation of synaptic transmission has been extensively studied [14–16]. As with any control system, redundant mechanisms, parallel and convergent pathways, and positive- and negative-feedback loops interact with each other to tightly maintain synaptic regulation of transmission. Our understanding of how the pleiotropic, redundant mediators collectively termed cytokines regulate the complexity of synaptic function has evolved. The role(s) of the various members of the cytokine family of mediators in synaptic function will be addressed.

2. BASIC CONCEPTS AND PRINCIPLES OF SYNAPTIC TRANSMISSION

A fundamental principle of neuron function is that they secrete chemicals as a major form of communication. Neurons secrete chemical neurotransmitters across points of contact called synapses in order to activate another neuron. This interneuronal communication occurs at synapses throughout the CNS. This region of the neuron membrane is specific for neurotransmitter release, response, and catabolism. However, neurons also use other forms of communication at the synapse. These include electrical synaptic transmission, as well as autocrine, paracrine, and long-range (humoral) signaling with substances produced by both neurons and non-neuronal cells. The synapse was originally referred to as areas of close contact, specialized for effective transmission from one neuron to another. It has subsequently been redefined as a structure anatomically differentiated and functionally specialized for transmission between excitable cells. The examination of neurons revealed numerous areas of "swellings," or synaptic terminals, which provide the synaptic contact linking neurons in functional chains. This synaptic contact, although closely applied to the postsynaptic cell membrane, bears no physical contact, and the intervening space separating the two cells is referred to as the synaptic cleft.

2.1. Electrical transmission

Synapses in which the synaptic cleft is narrowed to the point of virtual obliteration are referred to as gap, or tight junctions. In this type of synapse, transmission is mediated by direct current flow as opposed to release of a chemical transmitter. Gap junction channels present a low-resistance pathway between two neurons. This allows for direct communication between the cytoplasm of two cells. There are functional advantages of this type of synapse. The speed of transmission is increased, and the synapse is practically free of harm from pharmacological or metabolic attack due to the lack of chemical transmitters. While this may appear to be advantageous for the organism, electrical transmission does not allow for a great deal of modification and adaptation to the environment. Because relatively little charge can flow through the gap junctions associated with electrical synapses in order to charge the large membrane capacitance of the postsynaptic cell, the electrical synapse is inefficient. Another functional disadvantage of electrical synapses is that they can only produce a stimulatory signal. As this type of synapse is less common in the brain, our focus of cytokine regulation of synaptic transmission will remain on chemical transmission.

2.2. Chemical transmission

Functional flexibility is a hallmark of chemical synapses. Prior activity of the presynaptic neuron may modify the chemical synapse. In fact, chemical synapses are often subject to modulation of their ion channels on the presynaptic membrane elicited by mediators released either from the postsynaptic cell or from adjacent cells. This flexibility is necessary for the neuron circuit to process complex information that ultimately leads to processes such as learning and behavior. The released neurotransmitter may diffuse across the synaptic cleft and bind to receptors on the postsynaptic cell or also feed back on the presynaptic cell and act in an autocrine fashion by activating autoreceptors. Therefore, neurotransmitters are endogenous compounds released from neurons that produce functional changes on the target cell, whether it is presynaptic or postsynaptic.

It is apparent that a large number of chemical messengers broadly qualify as intercellular transmitters. However, it is generally accepted that specific criteria must be met for a compound to be a neurotransmitter. Consequently, a very small number of compounds are recognized as true neurotransmitters. The criteria for chemical synaptic transmission are (1) a neurotransmitter must be synthesized and released by a neuron. Therefore, an enzyme system must be present in the neuron from which the chemical transmitter is synthesized, and the chemical transmitter must be stored in a sequestered, quantal form; (2) the substance must be released from the synaptic region in a chemically or pharmacologically recognizable form. In other words, an action potential excites, or depolarizes the presynaptic neuron membrane releasing an effective quantity of the chemical transmitter, and a mechanism must exist for that release; (3) the released transmitter must react in a specific manner with the postsynaptic membrane, and an exogenously administered neurotransmitter must faithfully reproduce the functional changes to the postsynaptic cell as occurs after depolarization of the presynaptic terminal; (4) a competitive antagonist should effectively block the effects of the neurotransmitter in a concentration-dependent manner; and (5) finally, active mechanisms should be in place to rapidly terminate the actions of the released transmitter. Presently, transmitters are classified as either "classical" neurotransmitters that meet the stringent criteria outlined above, or as intercellular transmitters, such as some cytokines, that broadly qualify, yet do not meet the stringent classical criteria. The interactive relationship between these different compounds is fostering exciting research. The modulation of synaptic transmission involves many mediators, including but not limited to neurotransmitters, prostaglandins, neuropeptides (substance P, neuropeptide Y), neurohormones (CRF, ACTH), as well as cytokines. These modulators may affect synaptic neurotransmission directly or through indirect mechanisms. The regulation of synaptic transmission by cytokines will be further explored.

3. TUMOR NECROSIS FACTOR-α

TNF is a pleiotropic cytokine produced by a plethora of cell types affecting numerous cellular functions. TNF is synthesized as a 26 kDa membrane-bound protein that is proteolytically processed by the metalloprotease TNFα-converting enzyme (TACE) to produce a 17 kDa soluble form of TNF [17]. The biologically active form of TNF consists of a 51 kDa trimer. TNF was originally classified as a proinflammatory cytokine that regulates both innate and adaptive immune responses. However, its role as a neuromodulator is becoming apparent. In fact, receptors for TNF (TNFR-1, p55TNFR and TNFR-2, p75TNFR) are constitutively expressed on all neuronal cell types, except for astrocytes and oligodendrocytes, which predominantly express the TNFR-1 variant [4,5,18,19]. The mRNA for TNF accumulates in neurons, and the TNF protein is expressed and released [20–25], qualifying TNF as a neuromodulator. The role of TNF in mediating synaptic transmission and function is discussed below.

3.1. Direct regulation

The cytokine, TNF, plays an important role in the nervous system by regulating the synaptic release of neurotransmitter. When applied exogenously to superfused brain tissue, TNF inhibits the stimulation (stimulations 1 and 2, S1 and S2, at 2 Hz, 120 shocks) evoked release of NE from noradrenergic axon terminals in the isolated median eminence [10,11]. In addition, NE release inhibits further release of the neurotransmitter by activating α_2-adrenergic receptors throughout the CNS [26–28]. Therefore, TNF inhibition of NE release may involve its affect on the principal regulator of neurotransmitter release, the α_2-adrenergic receptor. As the median eminence appears not to be subject to α_2-adrenergic receptor-mediated negative feedback modulation, it is presumed that TNF directly inhibits stimulated NE release in this brain region. Similarly, when TNF is applied to slices of the hippocampus, a brain region rich in α_2-adrenergic autoreceptors, the cytokine inhibits stimulated (S1 at 1 HZ and S2 at 4 Hz) NE release in a concentration- and frequency-dependent manner, similar to α_2-adrenergic inhibition of NE release [12,21,29]. The inhibition of stimulated NE release by TNF is more pronounced at lower frequencies of stimulation (0.5–2 Hz), when the α_2-adrenergic autoreceptor is functioning at a diminished capacity [12,21]. Therefore, the α_2-adrenergic receptor tempers TNF inhibition of NE release [12,29]. In both studies, it should be noted that the addition of TNF was 15–16 min prior to stimulation, indicating that TNF does not require a long exposure time to develop modulatory effects. Interestingly, TNF inhibition of stimulated NE release under physiological conditions is altered in pathophysiological conditions. For example, the inhibition of stimulated NE release by TNF is supersensitized, or increased, during conditions whereby TNF expression is enhanced in the brain (chronic pain) [30,31]. Conversely, TNF facilitates stimulated NE release during chronic antidepressant drug administration to rats and during both the natural dissipation of neuropathic pain and the theraupetic alleviation of pain, conditions that are associated with decreased TNF expression in the brain [12,22,31–35]. Therefore, the manner in which synaptic transmission is modulated by cytokines, such as TNF, depends on the environment, including prior exposure to cytokines and the sensitivity of receptors specific for classic neurotransmitters.

While it is evident that TNF plays a crucial role in synapses in the CNS, TNF similarly regulates NE release in the peripheral nervous system. The evoked release of NE from mouse isolated atria and from isolated superfused human atria is inhibited by TNF [36,37]. The inhibition of NE release by TNF in the peripheral nervous system demonstrates that mediators produced by cells of the immune system can also locally modulate sympathetic nervous system activity.

As observed in tissue preparations, TNF also modulates synaptic function of cells in culture. TNF, through activation of TNFR-1 and TNFR-2, results in the activation of cultured dorsal root ganglion (DRG) neurons and the ensuing transient Ca^{2+}-dependent inward currents [38]. This TNF-dependent increase in intracellular Ca^{2+} may be responsible for increased neuron excitability and neurotransmission. However, TNF stimulation of TRFR-1 in the absence of increased Ca^{2+} concentration increases the frequency of spontaneous miniature synaptic currents in cultured rat hippocampal neurons [39]. Therefore, while TNF directly modulates neuron functions, the complex nature of these modulations makes it hard to predict the outcome. This outcome depends on whether a particular neuron is functioning physiologically or whether it is experiencing pathophysiological conditions.

3.2. Indirect regulation

While TNF influences the response of neurons, in the majority of experiments, this influence appears to be indirect. For example, TNF indirectly inhibits evoked release of NE from longitudinal muscle myenteric plexus preparations from rat jejunum [40]. In this preparation, TNF time and concentration dependently inhibit NE release. This tissue preparation required a long-term (≥30 min) exposure to TNF to develop its modulatory effect, which is in contrast to the short exposure time needed for modulation in the median eminence and the hippocampus (see section 3.1). This modulatory effect is observed as a biphasic inhibition of NE release. This biphasic suppression of NE release is mediated through two separate mechanisms: an early TNF-induced synthesis of arachidonic acid metabolites (PGE_2) and a late TNF-mediated synthesis and release of interleukin-1 (IL-1). As both prostaglandins and IL-1β exert suppressive effects on neurotransmitter release [41,42], it is plausible that the proximal cytokine TNF initiates the cascade of events that ultimately regulates NE release. In fact, it is well established that cytokines influence each other, which results in a cascade of subsequent cytokine release (see section 4.1).

When cultured sympathetic superior cervical ganglia (SCG) neurons are depolarized with excess K^+, NE is released [43]. However, depolarization (high K^+)-induced secretion of NE from SCG neurons is decreased when those neurons are exposed to TNF. Moreover, the inhibition of NE release is absent during the first K^+-evoked NE release and occurs following a repeat depolarization. Therefore, the TNF-induced inhibition of NE release may be due to a prolonged recovery from Ca^{2+}-channel-dependent inactivation of NE secretion [43]. TNF also increases the somatic Ca^{2+} current density in SCG neurons [44]. Furthermore, regulation of intracellular Ca^{2+} plays a primary role in modulation of neurotransmitter release [45]. However, in this instance, the effect of TNF on somatic Ca^{2+} current density is unrelated to its modulatory effect on NE release [44]. Interestingly, NE secretion from SCG neurons induced by a nicotinic agonist 1,1-Dimethyl-4-phenylpiperazinium (DMPP) is enhanced by TNF in the presence of non-neuronal cells [46]. The finding that TNF enhances NE secretion from neurons only when non-neuronal cells are included in the culture indicates that TNF induces the release of a factor(s) from the non-neuronal cells that mediates the enhanced NE release. In fact, nicotinic receptor current density is increased in SCG neurons from these mixed cultures exposed to TNF [46]. Therefore, in keeping with its pleiotropic nature, TNF exerts multiple, independent effects on SCG neurons.

3.3. Long-term potentiation

In most synapses, transmission is initiated by a nerve impulse that opens presynaptic voltage-gated channels to allow Ca^{2+} to flow through [47]. Fusion of the synaptic vesicles containing the neurotransmitter with the plasma membrane in the presynaptic cell is triggered by the Ca^{2+}, resulting in the release of the neurotransmitter into the synaptic cleft. The neurotransmitter diffuses across the cleft and activates the postsynaptic cell. The membrane receptors of the postsynaptic cell are activated by neurotransmitters that cause ion channels to open and elicit either a brief excitatory (depolarization) or inhibitory (hyperpolarization) change in potential across the membrane. This event may progress to a long-term potentiation (LTP) of the postsynaptic membrane. The LTP of synaptic transmission is defined as an enhancement of the postsynaptic potential that lasts hours to weeks and is induced by high-frequency stimulation that is relatively brief (1s or less) [48]. LTP is a basic model for synaptic plasticity that encompasses changes in synaptic strength and transmission and is also considered to be the cellular basis for learning and memory formation [49,50]. In fact, the hippocampus is the primary brain region implicated in memory formation [51]. As TNF is produced in the hippocampus and is associated with memory and learning, it is not surprising that this cytokine modulates synaptic plasticity [12,21,52,53]. While LTP induced in hippocampal slices is not influenced by acute exposure (5–15 min) to TNF, chronic TNF exposure (≥50 min) inhibits LTP in a concentration-dependent manner [54]. Similar findings of TNF inhibition of LTP have been reported [55,56]. The mechanisms by which TNF inhibits LTP are under investigation by several laboratories. A number of postulates explain the delayed TNF influence (necessity of chronic exposure) on LTP. TNF may indirectly influence synaptic transmission by reducing the glutamate-induced increase in astrocyte Ca^{2+} levels [57]. Increased intracellular Ca^{2+} in astrocytes induces glutamate release that, in turn, activates *N*-methyl-D-aspartate (NMDA) receptors on adjacent neurons and increases neuronal intracellular Ca^{2+} thereby promoting neurotransmitter release [58]. Additionally, TNF inhibition of LTP occurs during the early/induction phase (1–2 h post-LTP induction) and the late/maintenance phase of LTP (hours to weeks). The inhibition of early-LTP by TNF is mediated through activation of p38 MAPK, a mitogen-activated protein (MAP) kinase that is present in the dentate gyrus [59]. In addition, targets of p38 MAPK activity induced by TNF that may ultimately inhibit early-LTP include heat-shock protein 27 or cytoplasmic phospholipase A2, an enzyme that produces arachidonic acid from membrane phospholipids [59–61]. Aside from the TNFR-1-mediated activation of p38 MAPK, the TNF inhibition of early-LTP was also found to be dependent upon metabotropic

(G-protein-coupled) glutamate receptor (mGluR) activation [62]. The concomitant activation of mGluR and TNFR-1 results in increased Ca^{2+} levels, leading to inhibition of LTP. Interestingly, TNF inhibition of late-phase LTP does not involve activation of p38 MAPK [59]. This latter inhibition may result from TNF-induced changes in protein synthesis [63,64], which likely involves NF-κB target gene transactivation or expression [55].

TNF enhances synaptic transmission. Glutamate is the main excitatory amino acid neurotransmitter in the CNS and stimulates both α-amino-3-hydroxy-5-methyl-4-isoxazolepropionic acid (AMPA) and NMDA receptors. The blockade of glutamate uptake by TNF has been shown to underlie glutamate-induced excitotoxicity [65]. Additionally, TNF significantly increases expression of ionotropic, glutamate-sensitive AMPA receptors located at synapses [66]. TNF also increases the mean frequency of miniature excitatory postsynaptic currents in neurons, thereby enhancing synaptic strength [66]. In fact, TNF stimulation of neuronal TNFR-1 not only increases the expression of AMPA receptors, but simultaneously decreases the expression of $GABA_A$ receptors, which are primarily responsible for mediating fast, inhibitory synaptic transmission [67]. The net effect of TNF is to increase synaptic efficacy by altering the balance of excitatory and inhibitory neurotransmission which results in strengthened excitatory synapses and weakened inhibitory synapses. Taken together, an increase in TNF production may play varied roles in mediating glutamate-induced excitotoxicity [68]. In fact, TNF-mediated changes in neuron excitability may be responsible for the neuropathologies associated with learning and memory deficits where elevated levels of TNF have been observed.

The endogenous TNF regulation of synaptic plasticity, in the form of long-term depression (LTD), has been demonstrated through the use of mice genetically deficient in both TNFR-1 and TNFR-2 (TNF receptor knockouts) [69]. In contrast to the high-frequency stimulation (100 Hz) used to induce LTP, low-frequency stimulation (1 Hz) is used to induce LTD of synaptic transmission. Following low frequency stimulation of hippocampal slices from mice lacking TNF receptors, the LTD response (population spike amplitude) is impaired when compared to slices from wild-type mice [69]. This finding demonstrates a critical role for TNF in the induction of LTD, a process that similar to LTP, involves long-term changes in synaptic strength and transmission that affect learning and memory. Collectively, these findings support TNF regulation of synaptic plasticity in regions of the brain associated with learning and memory, such as the hippocampus.

4. INTERLEUKINS

Many of the soluble messengers that belong to the cytokine family are referred to as "interleukin." The term "interleukin" was originally coined to categorize the soluble mediators released by leukocytes as a mechanism of intercommunication between leukocytes. However, it is now quite apparent that their source and roles go beyond this original characterization. As with most of the cytokines, interleukins are primarily secreted, but they can also be expressed on the cell surface. Characteristic of cytokines, interleukins are produced by a plethora of cell types and have multiple effects on different cells. While it is well established that interleukins have diverse effects in the immune system, their functions in the nervous system, such as directing neuron to neuron communication, is less well understood. The effects of interleukins on neurons and leukocytes emphasize the interactive relationship between the immune and the nervous system. The role of two widely investigated interleukins in modulating neurotransmission is addressed in this section.

4.1. Interleukin-1

IL-1 can be produced by virtually any leukocyte, endothelial cells, fibroblasts, and most interesting, by selective neurons. It plays a prominent role in inflammation, immunity, and development, and now apparent is its role in neuron to neuron communication. IL-1 is a cytokine with numerous, prevalent effects in the nervous system. This cytokine consists of two forms, IL-1α and IL-1β. Both forms bind and activate the same surface receptor and elicit similar biological responses. IL-1β is the primary secreted form and is the predominant IL-1 molecule expressed in brain tissue [70]. This proinflammatory, pleiotropic cytokine has redundant biological activities similar to TNF. IL-1β is an important neuroimmune modulator involved in the generation of fever, activation of the hypothalamic–pituitary–adrenal axis, sleep induction, and sickness behavior, which are responses requiring neuron involvement [71–73]. Given the importance of the CNS processes regulated by IL-1, the neuronal mechanisms involved in transmitting the signals elicited by this cytokine are under intense investigation. Whether administered systemically or directly into the hypothalamus, IL-1 increases the stimulated release of DA, NE, and 5-HT in the hypothalamus [13,74–76]. Likewise, when IL-1 is administered either systemically, intracerebroventricularly (i.c.v.), or directly into the hippocampus, this cytokine enhances hippocampal release of 5-HT and NE [77–80]. As the hypothalamus is the brain region associated with neuroendocrine and autonomic functioning, and the hippocampus, in addition to its role in learning and memory formation, is involved with maintaining homeostasis of the organism [81], it is not surprising that IL-1 affects neurotransmission in these discreet regions of the brain. In fact, these regions of the brain express increased levels of receptors for IL-1 supporting a primary role for IL-1 in regulating neurotransmission in the hypothalamus and hippocampus [7]. In addition to regulating the release of monoamines, the binding of IL-1 to its receptor in the hippocampus interferes with both cholinergic and glutamatergic neurotransmission thereby impairing working memory [82]. In light of the pivotal role for IL-1 in regulating hippocampal synaptic function, many electrophysiological studies have demonstrated direct effects of IL-1 on hippocampal neurons. For example, exposure of neurons *in vitro* to IL-1 alters neuron firing patterns [83,84], inhibits excitatory synaptic responses [85–90], and modifies inhibitory synaptic responses [91–94]. Therefore, IL-1 directs synaptic plasticity in an autocrine, paracrine, and humoral fashion, which generally results in inhibition of LTP [85,86,88]. Many of these investigations employed pathological levels of IL-1 (nanomolar). However, Schneider et al. (1998) demonstrated the importance of pathological versus physiological levels of IL-1 on neuron function. This study demonstrated that low, physiologic levels of IL-1 (femtomolar to low picomolar) serve a physiological role in the maintenance phase of LTP [95]. Similar findings support that high levels of IL-1 inhibit while low levels maintain LTP [96]. Therefore, IL-1, depending on its concentration, can either promote or inhibit synaptic plasticity demonstrating the profound effects this cytokine has on neuronal functioning.

It is evident that IL-1 is involved in synaptic responses in the CNS. However, its mechanism(s) of action in the synaptic cleft is also elucidated by investigating how it regulates neurotransmitter release in the peripheral nervous system. The evoked release of NE from mouse-isolated atria as well as from isolated superfused human atria is inhibited by IL-1β [36,37]. This IL-1β-mediated inhibition of NE release is blocked by both IL-1 receptor antagonist and the cyclooxygenase inhibitor, diclofenac. Therefore, IL-1β inhibition of NE release, at least in part, acts through the production of prostaglandins [36,37]. IL-1 also decreases NE release from stimulated rat myenteric plexus [41]. This inhibition of NE release by IL-1 is biphasic and is characterized by an early direct effect of IL-1 on sympathetic nerves, and a delayed effect mediated by the synthesis of endogenous IL-1 that, in turn, suppresses NE release. This demonstrates networking, one of the fundamental characteristics of cytokines, involving cytokine induction of the release of additional cytokines. In addition, IL-1 both depolarizes and hyperpolarizes membranes of neurons of the guinea pig pelvic plexus ganglia [97]. The inhibition of fast excitatory postsynaptic potentials by IL-1 may result from the inhibition of the prejunctional release of Ach [98]. Inhibition of neurotransmitter release by IL-1 in the peripheral nervous system demonstrates that, in addition to their potent regulatory effects on central transmission, mediators produced by cells of the immune system can locally modulate sympathetic nervous system activity.

4.2. Interleukin-6

The cytokine IL-6 is a redundant cytokine that performs inflammation and immune functions similar to IL-1. In the brain, IL-6 predominantly plays a protective role: improving survival of neurons in culture [99–101]; protecting neurons from excitotoxic and ischemic insults [102–105]; and promoting growth of axons, and consequently, the number of synapses in a region [106–109, for review]. The levels of this cytokine are oftentimes increased during disorders of the CNS (depression, Alzheimer's disease, Parkinson's disease (PD); see section 7), suggesting that IL-6 functions to promote neuron survival in the face of unfavorable conditions.

Similar to other cytokine family members, IL-6 inhibits LTP through its affect on synaptic plasticity [110,111]. This inhibition of LTP is mediated by IL-6 inhibition of glutamate release and inhibition of the spread of excitation in the cerebral cortex [112]. Furthermore, IL-6-mediated inhibition of LTP is accompanied by inhibition of the MAPK/ERK intracellular signal transduction pathway involved in the regulation of gene transcription [113]. As MAPK/ERK activation is required during LTP [114,115], its inhibition provides a causal factor in the inhibition of LTP by IL-6. As in the CNS, IL-6 inhibits neurotransmitter release in the peripheral nervous system [116,117]. In fact, exogenously administered IL-6 inhibits neuronal hyperexcitability and electrically evoked C-fiber post discharge and wind-up observed in animals receiving spinal nerve ligation (a model for neuropathic pain) [116]. These data support a neuroprotective role for IL-6, whereby the abnormal transmission of pain signals is attenuated by IL-6.

5. INTERFERONS

The family of interferons (IFN) comprises a group of cytokines produced by virus-infected cells. The term IFN, however, is used for several unrelated proteins. IFNs are pleiotropic cytokines produced by many cell types in response to virus infection, endotoxin (lipopolysaccharide, LPS), double-stranded RNA, and antigenic and mitogenic stimuli. Among the many functions of the IFNs is the ability to bind to nearby cells and induce a general antiviral state. However, these cytokines affect numerous cellular functions and therefore are plurifunctional mediators. IFNs are released during infections and predominantly regulate immune system functions; however, they also function in the nervous system. The immune system cross-talks with the nervous system allowing the individual to adapt a behavioral response. This response can be beneficial to the organism, or in some conditions the immune and inflammatory response can be detrimental. Therefore, the response of the nervous system can be either functional or dysfunctional. In fact, the production of IFNs induces behavior disturbances [118,119], because neurons express IFN receptors, which allow them to respond to IFNs in their environment. It is obvious that some members of the IFN family of cytokines engage in regulating the activity of the CNS. Consequently, IFNs participate in a network in the CNS that goes well beyond an immune and inflammatory response. Moreover, IFNs function as neuromodulators in the nervous system and in particular mediate synaptic transmission and function.

5.1. Classification

The classification of IFNs has evolved into new classes and subtypes along with our emerging knowledge of the field. The IFNs are classified into three major groups, which are characterized according to the type of leukocyte that secretes them. Activated leukocytes generally exhibit greater secretion of IFNs than inactivated cells. IFN-α and IFN-β are secreted from numerous leukocytes and inhibit viral replication. IFN-γ is secreted from T-lymphocytes and natural killer cells and enhances as well as inhibits numerous leukocyte activities.

5.2. Cellular communication

IFNs are neuromodulators that affect direct synaptic transmission and function. Neurons are indeed responsive to IFNs as they express receptors for IFN-γ located in both postsynaptic and presynaptic membranes [120]. Yet, neurophysiological studies on the effects of IFNs on neuronal functions remain sparse [121]. IFN-α was shown to enhance the excitability and spontaneous activity of cerebral and cerebellar neurons [122]. Unlike other excitatory substances, these effects occur only after 30 min and last several hours. In addition, IFN-α and -β have been reported to have excitatory effects on CA3 pyramidal neurons [119]. In early research directed at understanding cytokine involvement in synaptic function, the role of IFNs on the electrically induced potentiation of synaptic transmission in rat hippocampal slices was studied by using extracellular field potential recordings [123]. Exposure of rat hippocampal slices to rat IFN reduced the size of short-term potentiation and suppressed LTP. In addition, repeated IFN-α administration inhibits dopaminergic neuronal activity in the mouse brain [124]. The authors suggest that this inhibitory action of IFN-α in DA neurons may be involved in adverse central effects, such as Parkinsonism and depression with suicidal potential. It has subsequently become obvious that IFNs display differential effects between short-term exposure of neurons and long-term exposure [121]. The acute application of IFN-γ to differentiated neurons in culture (48 h) displays an increased frequency of AMPA receptor-mediated spontaneous excitatory postsynaptic currents. However, immunocytochemistry for this receptor showed no alteration in receptor clustering at this time point. A prolonged 2-week administration of IFN-γ results in a reduction in receptor clustering, but there were no significant differences in spontaneous excitatory postsynaptic currents. In addition, IFN-γ has been reported to reduce the number of synapses in hippocampal neurons [125]. It has recently been shown that during the development of synapses, short-term exposure to IFN-γ induces marked changes in spontaneous activity that are significant only after a latency period of weeks [126]. Thus, exposure to IFNs at the critical period of development, during peak excitatory synaptogenesis, may result in certain disorders of the nervous system that occur after a latency period. It is now clear that cytokines such as IFNs can influence the development of the nervous system, including synaptogenesis. Therefore, inflammation or an immune reaction may lead to functional abnormalities at several mechanistic points in the CNS.

6. CHEMOKINES

The family of cytokines with chemoattractant properties is referred to as chemokines. Chemokines are produced by a plethora of cell types affecting numerous cellular functions and therefore are plurifunctional mediators. Chemokines are smaller than the inflammatory cytokines and, on the whole, belong to a family of small (8–10 kD) proteins that were originally thought to act primarily as chemoattractants that were selective for specific types of leukocytes [127,128]. While they induce chemotaxis, they also stimulate tissue extravasation and change the functioning of various leukocytes during inflammation [129]. It is now apparent that the original characteristic of this group of cytokines was a vast underestimate of their wide spectrum of effects.

Many, if not all, of the cells in the CNS including neurons generate chemokines [130,131]. While there are greater than 50 different chemokines characterized, there have been about 20 corresponding receptors identified. Neurons express various chemokine receptors allowing them to respond to chemokines in their environment. Consequently, chemokines participate in a network in the CNS that goes well beyond leukocyte trafficking. Therefore, chemokines serve as neuromodulators in the nervous system. Specifically, we will address the putative role(s) of chemokines in mediating synaptic transmission and function.

6.1. Classification

The chemokines are classified into four major groups, which are characterized on the number and the arrangement of four conserved N-terminal cysteine (C) residues in the mature proteins: (1) CXC chemokines (α chemokines) have a single amino acid residue separating the first two conserved cysteine residues; (2) CC chemokines (β chemokines) have the first two conserved cysteine residues adjacent to each other; (3) C chemokines (γ chemokines, lymphotactin) lack the first and the third of the four conserved cysteine residues; and (4) CX_3C chemokines contain three amino acids between two cysteines. Fractalkine is the only member of this group, which exists in two forms.

6.2. Localization to neurons

It is obvious that some members of the chemokine family of cytokines engage in regulating the activity of the CNS under physiological conditions. The use of *in situ* and *in vitro* analysis has confirmed the presence of chemokine receptors on neurons. In addition, regional distribution of the expression of genes for various chemokines exists in the CNS, suggesting differential effects at diverse neurons [130]. For example, the CXCR2 receptor is prominent in the dentate gyrus of the hippocampus of humans [132]. The Duffy-antigen-related chemokine (DARC) receptor is found almost exclusively on Purkinje cells [133]. The receptor RNA specific for CXCR4 is highest on specific neurons such as hippocampal hilar neurons and cerebral cortical neurons [134]. Therefore, distinct classes of chemokine receptors are localized to specific types of neuronal cells. This provides the CNS with a framework upon which to discriminate the communication directed by chemokines.

6.3. Cellular communication

Neurons are indeed responsive to chemokines. Using patch-clamp experiments on Purkinje neurons, the acute application of GROβ (CXC chemokine, growth-related gene product) with and without a synthetic inhibitor of the ERK (signal-regulated kinase) pathway affects neurotransmitter release in the cerebellum [135]. This work provided initial insight on chemokine control of the signaling pathways involved in synaptic transmission in the CNS. In fact, GROβ stimulates the signaling pathway of the extracellular signal-regulated kinases and enhances both evoked and spontaneous postsynaptic currents. When cultured rat hippocampal neurons are exposed to any one of several different chemokines (SDF-1α, RANTES, fractalkine, MIP-1α, and IL-8), a myriad of functional responses occur [136]. These functional responses include a reduction in voltage-dependent Ca^{2+} currents as well as the induction of single or oscillatory Ca^{2+}-spike activity. Administration of fractalkine activates the MAP kinase signaling pathway which induces the cAMP-response-element-binding protein (CREB). Specific chemokines can modify neuronal functions directly in a fashion similar to classical neurotransmitters. In fact, fractalkine modulates synaptic transmission directly in hippocampal neurons [130]. In Purkinje cells the CXC chemokines (GROβ, IL-8, and GROα) increase neurotransmitter release and reduce the magnitude of LTP [137]. Stromal-derived cell factor-1α (SDF-1α) causes Ca^{2+} transients sensitive to glutamate receptor antagonists both in granule cells and in Purkinje neurons [138]. In response to SDF-1α stimulation, Purkinje neurons also exhibit a slow inward current and an increase in spontaneous electrical activity coupled to the release of glutamate. In fact, the coupling of CXCR2 (CXC chemokine receptor) with GluR1 (glutamate receptor) modulates glutamatergic synaptic transmission [139]. The chemokine receptor CXCR2 regulates the functional properties of AMPA-type glutamate receptor GluR1 in Purkinje neurons. Stimulation of chemokine CXC receptor 4 induces synaptic depression of evoked parallel fibers inputs onto Purkinje neurons in the cerebellum [140]. Therefore, SDF-1α depresses evoked neurotransmission reducing neurotransmitter release. In addition, acute exposure to CXC chemokine ligand 10 alters synaptic plasticity in mouse hippocampal slices [141]. Fractalkine/CX3CL1 depresses central synaptic transmission in the hippocampus [142]. Therefore, a fractalkine is a potent neuromodulator of evoked synaptic transmission. Finally, SDF-1α induces, through presynaptic mechanisms, alterations in the excitability of DA neurons as confirmed by current-clamp experiments [143]. Therefore, it is becoming more evident that chemokines have important physiological cellular roles within the adult mammalian CNS. The characteristic role for chemokines is to regulate leukocyte trafficking in the CNS during various immunoinflammatory states. However, one of the primary roles of chemokines is obviously the control of synaptic plasticity. It has now become clear that during pathological states the chemokines and their receptors are also involved in numerous processes other than chemotaxis. In neuroinflammatory disorders of diverse etiology, there is a coordinate induction or upregulation in the CNS of chemokine genes belonging to all four major groups [130].

It is now obvious that chemokines and consequently their receptors are pleiotropic molecules whose impact in the CNS goes far beyond chemotaxis. Accumulating evidence is establishing these molecules as essential regulators of cellular communication in both physiological and pathological settings. In fact, studies of the involvement of chemokines in physiology and neurological diseases have indeed broadened our understanding of these molecules beyond leukocyte chemoattraction and into novel mechanisms that control neuronal communication. It is interesting to speculate on the broader role of chemokines in neuronal function. While it is obvious that chemokines released in the vicinity of the synapse will attract leukocytes, they will also regulate the release of neurotransmitter. In turn, the neurotransmitter will act upon receptors on the leukocytes setting up a dynamic event between the synapse and the accumulating leukocytes. No doubt we will continue to discover novel events where chemokines direct neuronal communication.

7. IMPLICATIONS FOR DISEASES

Cytokine expression is found to be increased in many pathological conditions of the CNS including Alzheimer's disease, Parkinson's disease (PD), schizophrenia, depression, and neuropathic pain. Abnormalities in the properties of neuronal circuits often give rise to many neurological disorders. Through their actions on mechanisms that control neurotransmitter release, cytokines influence neuronal circuits, modifying mood states and contributing to the memory deficits associated with many neurological disorders. The evidence for cytokine over-expression in the brain as an important factor in the pathogenesis of these particular neurological disorders is presented forthwith.

7.1. Alzheimer's disease

Alzheimer's disease is a senile dementia characterized by mental deterioration. This disease is accompanied by the presence of amyloid-containing senile-plaques in affected regions of the brain and neurofibrillary tangles in dead and dying neurons [144]. Excess β-amyloid protein deposition in the brain is widely believed to be responsible for neurodegeneration and is the primary pathogenic event in Alzheimer's disease [144–146]. In fact, the enhanced production of cytokines in the brain induces an over-production and accumulation of β-amyloid protein [147–149]. Furthermore, as addressed throughout this chapter, cytokines also have profound effects on synaptic function, including the release of neurotransmitters. As cytokines affect neurotransmission and are increased during Alzheimer's disease, abnormalities in neurotransmission may in fact contribute to the pathology of this disease [150,151]. Of particular interest, the enhanced accumulation of β-amyloid feeds back to regulate the production of these cytokines. The β-amyloid protein stimulates the production of cytokines in the brain that contribute to the ongoing inflammatory cascade, synaptic dysfunction, and neuron death characteristic of this disease [152–154]. These findings point to the pivotal role of cytokine participation in the pathogenesis of Alzheimer's disease. Supporting this view is the elevated level of cytokines observed in Alzheimer patients [155–157]. As anti-cytokine therapies are being implemented for treatment of Alzheimer's disease, it is hoped that by blocking the inflammatory cascade, the neuronal damage resultant from this process will be halted if not overcome [158–159]. While neurotransmitters regulate the production of cytokines [23,160–163], it is interesting to speculate about the dynamic equilibrium existing between the over-production of cytokines, β-amyloid, and the modified release of neurotransmitters during Alzheimer's disease. For example, a change in any one of these three constituents would shift the equilibrium thereby affecting the other components.

7.2. Parkinson's disease

Selective loss of the DA producing neurons in the ventral mesencephalon substantia nigra pars compacta results in a deficiency in the neurotransmitter DA at nerve terminals in the striatum. This lack of bioavailable DA results in a disorder associated with movement abnormalities that are characteristic of PD. The enhanced production of proinflammatory cytokines as well as the inhibition of anti-inflammatory factors directs the pathology and the etiology of this neurodegenerative disease [164]. Numerous studies assessing postmortem brains and cerebrospinal fluid (CSF) from PD patients as well as the assessment of tissues taken from animal models of PD have found increased levels of proinflammatory cytokines, such as TNF, IL-1, and IL-6 [165–171]. Over-expression of these cytokines is believed to underlie the neurodegeneration characteristic of PD. Interestingly, it is purported that the disruption in neuron transmission directed to microglial cells that normally serves to keep the microglial cell quiescent, initiates activation of these cells, which results in the excess production of cytokines [172–174]. Therefore, the use of anti-cytokine strategies is justified in the design of therapeutic strategies to prevent or attenuate the progression of PD [175,176].

7.3. Schizophrenia

Schizophrenia is a psychiatric disorder in which both positive (bizarre behavior, hallucinations, delusions) and negative (loss of interest and energy) symptoms are observed [177]. A range of modified levels of cytokines are noted in schizophrenic patients, including an increase in plasma or CSF levels of TNFα, IL-6, and IL-2 [178–182]. In fact, numerous investigations have demonstrated the involvement of cytokines and catecholamines in the etiology and in the pathophysiology of schizophrenia [179,183]. While it is widely accepted that DA plays a role in schizophrenia, it is of particular interest that IL-2 stimulates DA release from rat striatal and mesencephalic cells in culture [184,185]. As IL-2 is increased in schizophrenic patients, this IL-2-induced potentiation of DA release may underlie some of the behavioral abnormalities observed in these patients. Further evidence for IL-2 involvement in schizophrenia stems from the finding that IL-2 microinfused directly into the rat locus coeruleus produces similar behavioral and electrical activity changes as that which is observed in schizophrenic patients [186].

Recently it has been proposed that cytokines, through their influence on synaptic plasticity and neurotransmission, contribute to the etiology of schizophrenia during the development of the nervous system [187]. In particular, evidence is provided for a "Two Hit Theory" whereby the combination of a genetic factor predisposing an individual to inappropriate development of the CNS, in which cytokines play a role, along with an environmental factor that initiates cytokine-mediated changes leading to impairment of neuronal functioning together help explain the etiology of this disorder [187].

The treatment of schizophrenic patients with antipsychotic agents is associated with modifications in cytokine networks [188,189]. However, findings from studies investigating both animal and human subjects, whether approached by using either *in vitro* or *in vivo* methods, have yielded inconsistent results and are therefore controversial. Yet, a preponderance of these studies demonstrates a decrease in the levels of cytokines following the treatment of subjects with antipsychotic agents [181,190–193]. Interestingly, a decrease in the production of IL-2 would support a decrease in DA release, thereby alleviating the positive symptoms of schizophrenia. Recently, it has been shown that an atypical antipsychotic drug used in the treatment of schizophrenia, while it decreases the levels of IL-2 and at the same time decreases associated DA-metabolite levels, it also increases NE release [179]. The increase in NE release appears to, at least in part, be due to inhibition of the NE transporter, thereby improving the negative symptoms of schizophrenia [179]. As many studies demonstrate that TNF inhibits NE release [10–12] and that TNF is increased in schizophrenic patients [181,194], it would follow that by decreasing TNF levels antipsychotic agents would thereby similarly result in an increase in NE release. As investigation into the role of cytokines and their influence of neurotransmission during schizophrenia continues, the dynamics of cytokine and monoamine interactions will be revealed.

7.4. Depressive disorders

Cytokines, such as IL-1, IL-6, and TNF, provoke a constellation of behavioral, neuroendocrine, and central neurotransmitter changes that have been implicated in depression [195]. Some of these changes induced in animals include but are not limited to an increase in plasma levels of glucocorticoids and adrenocorticotropic hormone, an increase in monoamine (NE, 5-HT) turnover within the brain, and provocation of behavioral symptoms such as anhedonia [78,196–198]. Similar to the animal models for depressive behavior, evidence supports the observation that cytokines are involved in the pathogenesis of depression. Persons with depression exhibit increased serum levels of TNF, IL-1, and IL-6 [199–202]. Moreover, healthy volunteers administered TNF exhibit depression [203,204]. Additionally, when TNF is administered i.v. for 24 h to cancer patients, symptoms of depression are consistently produced [205]. Likewise, symptoms of depression are correlated with the levels of proinflammatory cytokines in the plasma of cancer patients [206,207]. Of particular interest is that many antidepressant drugs have anti-inflammatory properties, including decreasing the over-expression of the proinflammatory cytokine TNF [21,22,208–210].

Cytokines have profound effects on synaptic transmission, such as regulating both serotonergic and noradrenergic systems in the brain [10–12,211]. For example, IL-1 augments the release of 5-HT and NE, two monoamines involved in the pathophysiology of depression, in selected brain regions of the rat; it also modulates the activity of three other neurotransmitters implicated in the pathophysiology of depression, Ach, DA, and γ-aminobutyric acid [13,212]. Therefore, it is suggested that aberrant regulation of neurotransmission, secondary to enhanced cytokine expression, plays a role in the etiology of depression [12,21,22,161,213]. A mechanism involved in this cytokine-induced alteration in neurotransmission is proposed for TNF and α_2-adrenergic receptor regulation of NE release in Fig. 1 and is described in Section 8.

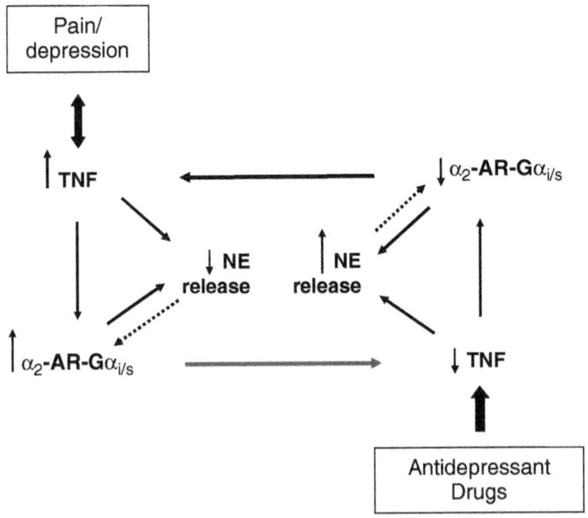

Figure 1. The postulated interactive relationship at the synapse between levels of TNF and α_2-adrenergic autoreceptor coupling to specific G-proteins directs neurotransmitter release. Changes in the production of TNF affect G-protein levels that, in turn, direct G-protein coupling of presynaptic receptors. For example, we propose that an increase in Gα_i-protein levels favors α_2-adrenergic receptor-Gα_i-protein coupling when TNF levels are high, whereas a decrease in Gα_i-protein levels favors α_2-adrenergic receptor-Gα_s-protein coupling when TNF levels are low. As both TNF and activation of the α_2-adrenergic autoreceptor regulate NE release, this interactive relationship illustrates redundancy, parallel pathways, and feedback loops that interact to tightly maintain synaptic regulation of neurotransmitter release. α_2-AR = α_2-adrenergic receptor; NE = norepinephrine, TNF = tumor necrosis factor-α.

7.5. Neuropathic pain

Neuropathic pain is a chronic pain resulting from abnormal operation of the pain sensory system after a primary lesion, or dysfunction in the nervous system [214,215]. As neuropathic pain is integrated in the brain, it is reasonable that adaptive changes in the brain are involved. In fact, neurons in the brain make the proinflammatory cytokine TNF [22–24]. Additionally, TNF is produced in excess by neurons and non-neuronal cells during the development of neuropathic pain [20,30,35,216]. This production of TNF in distinct brain regions orchestrates the development and the dissipation of this disorder [31]. In fact, considerable evidence supports the involvement of TNF in the pathogenesis of persistent pain. TNF mediates the hyperalgesia, or increased sensitivity to painful stimuli, following administration of bacterial endotoxin LPS [217]. Furthermore, i.c.v. microinjection of TNF induces hyperalgesia in naïve rats, supporting a site within the brain for its nociceptive effects [216,218]. Elevated levels of cytokines in brain and spinal cord are associated with increased pain. Increases in CNS levels of TNF occur during painful mononeuropathy [219]; drugs that inhibit this production of TNF decrease hyperalgesia [219,220]. Of particular interest, under appropriate conditions, central administration of TNF produces antinociception [31]. However, the majority of literature supports a pro-nociceptive role for TNF [217,221–223]. Regardless of the role played, TNF is constitutively expressed in the brain. Taken together, the evidence provided supports that central sites of TNF action are involved [217].

Historically, noradrenergic neuronal dysfunction is believed to be a key factor in the pathogenesis or maintenance of neuropathic pain, including complex regional pain syndrome, postherpetic neuralgia (shingles), phantom limb pain, and diabetic neuralgia [224–226]. Clearly, the CNS plays an important role in this component of persistent pain [227]. During neuropathic pain, there is a decrease in stimulated hippocampal NE release along with enhanced α_2-adrenergic inhibition of NE release. This supports the hypothesis that central processing during persistent pain is related to adaptations in synaptic availability of NE (and possibly 5-HT) [228,229]. TNF, acting possibly in an autocrine fashion, also decreases release of NE in the brain, augmenting pain perception [20,31,216]. This decreased NE release in the brain would attenuate the descending inhibitory pathway to the spinal or peripheral nerve injury, thereby maintaining persistent pain. Interestingly, drugs efficacious in treating pain (analgesics) such as tricyclic antidepressants and α_2-adrenergic agonists transform the α_2-adrenergic receptor that regulates NE release from inhibitory to facilitatory [24,29,31], through their regulation of TNF production in the brain in conjunction with managing NE bioavailability. Similarly, TNF inhibition of NE release is transformed to facilitation during effective alleviation of pain [24,31]. Therefore, research is recently targeted at cytokines and their receptors in the development of effective therapies for chronic pain [230–232].

8. G-PROTEINS

The majority of transmitter receptors throughout the CNS are coupled to guanine nucleotide-binding proteins (G-proteins). For instance, G-proteins couple adenylate cyclase to β_2- and α_2-adrenergic receptors; induction or suppression of these G-proteins has profound effects on noradrenergic neuron responses. In fact, the cytokine TNF directs this G-protein receptor system. Not only does TNF regulate NE release, but this pleiotropic cytokine also directly up-regulates $G\alpha_i$ protein expression [233–235]. Increased TNF levels increases plasma membrane G-protein activity [233]. As $G\alpha_i$ proteins couple noradrenergic receptors to adenylate cyclase, they control the potency of these receptors. Therefore, changes in NE release, at least in part, occur as a result of TNF effect on expression of G-proteins [234]. While TNF directly controls NE release, it also induces adaptive effects on this dynamic process. The α_2-adrenergic receptor is described as coupling to Gα_i; however, cell surface receptors such as these may in fact be coupled to various distinct GTP-binding proteins. Of particular interest is that the α_2-adrenergic receptor simultaneously couples to two G-proteins with opposing effects (Gα_i and Gα_s) [236–239]. Additionally, other G-protein-coupled receptors, including the β_2-adrenergic receptor, couple to Gα_s as well as to Gα_i [240,241]. As many neurological disorders are attributed to various neurochemical disturbances, such as dysregulation of NE, 5-HT, and Ach and their receptors [242–244], it is plausible to develop therapeutic strategies to regulate the G-protein repertoire because of their vast effects on numerous receptor systems. While TNF directs G-protein expression and therefore the signal elicited by adrenergic receptor activation, changes in TNF production would transform a myriad of other receptors as well producing a more effective therapeutic strategy in neurological disorders.

A model developed in our laboratory and presented in Fig. 1 depicts postulated interactions between α_2-adrenergic responses and neuron sensitivity to TNF, directed by decreased or increased TNF production. As TNF is constitutively expressed in neurons (as well as adjacent non-neuronal cells), it constantly regulates noradrenergic function. These events maintain homeostasis between the effects elicited by TNF and α_2-adrenergic receptor activation of noradrenergic activity. The α_2-adrenergic receptor (α_2-AR) normally inhibits NE release, and activation of these inhibitory autoreceptors (favored coupling to Gα_i subunit) decreases NE release from noradrenergic neurons. Additionally, activation (Fig. 1, gray arrow) of the α_2-adrenergic receptor coupled to Gα_i decreases neuron-associated TNF. Because TNF normally inhibits NE release, this reduction in levels of TNF increases NE release, thus maintaining homeostasis of NE levels. Decreases in levels of TNF, as a result of chronic administration of antidepressant drugs, transform α_2-adrenergic receptor function. The α_2-adrenergic receptor now exists predominantly as a stimulatory autoreceptor (designated as α_2-adrenergic receptor coupled to Gα_s), and its activation facilitates NE release and increases TNF production. This system remains in check because increased levels of TNF support coupling of the α_2-adrenergic receptor to the Gα_i subunit. These events continually occur within an equilibrium by which physiologic levels of TNF and normal functioning of the α_2-adrenergic receptor are preserved. A pathologic increase in TNF production, as for example during chronic pain or depression, disrupts the system's natural balance, such that increased TNF levels are sustained and NE release remains low, while the α_2-adrenergic receptor undergoes a dysfunctional adaptation, disproportionately favoring one functional form (couples to Gα_i) over the other (coupling to Gα_s) (favors "left" side of model). Conversely, antidepressant drugs perturb the system in the opposite fashion; decreasing levels of TNF, increasing NE release, and returning the α_2-adrenergic receptor to a state of functional balance. This is a mechanism by which TNF levels and the α_2-adrenergic receptor participate in a reciprocal fashion in the pathogenesis of depression and chronic pain and in the therapeutic efficacy of antidepressant drugs.

9. CONCLUDING REMARKS

Cytokines exert redundant effects within the brain, but also produce specific effects through selective activation of neurotransmitter pathways. Cytokines have a double-edged sword reputation, based on their developmental and neuroprotective effects [1,245], as well as their contribution to neurotoxicity, neuronal damage, and neurodegeneration [246,247]. The activation of neurons results in the direct upregulation and production of cytokines, which indicates that these mediators serve as neuromodulators involved in normal neuronal function [3,25]. Neurotransmitters, co-transmitters, and other mediators fine-tune cytokine production in the CNS, which translates into a dynamic control system that sustains cytokine balance and preserves homeostasis through the regulation of synaptic functions.

Significant strides have been made in our knowledge of the interactions that occur between cytokines and neurons in the CNS, however, questions remain unanswered. The complexity of the interactions between these moieties, while daunting, allows for considerable research endeavors in the future. As research continues to uncover information about the roles of cytokines in the CNS, the importance of their interactive relationship with the functioning of neurons will become more evident. Through this area of research, the involvement of cytokines as principal players in the CNS will continue to be exposed. Finally, awareness of the intercellular transactions that cytokines direct will solidify our understanding of how cytokines ultimately control synaptic plasticity.

ACKNOWLEDGEMENTS

The authors wish to thank Dr. Mary Spengler for careful critique of this manuscript. Work presented in this manuscript was supported by National Institutes of Health Grant NS-41352.

REFERENCES

1. Golan H, Levav T, Mendelsohn A, Huleihel M. Involvement of tumor necrosis factor alpha in hippocampal development and function. Cereb Cortex 2004;14:97–105.
2. Pan W, Zadina JE, Harlan RE, Weber JT, Banks WA, Kastin AJ. Tumor necrosis factor-alpha: a neuromodulator in the CNS. Neurosci Biobehav Rev 1997;21:603–13.
3. Vitkovic L, Konsman JP, Bockaert J, Dantzer R, Homburger V, Jacque C. Cytokine signals propagate through the brain. Mol Psychiatry 2000;5:604–15.
4. Bartfai T, Schultzberg M. Cytokines in neuronal cell types. Neurochem Int 1993;22:435–44.
5. Kinouchi K, Brown G, Pasternak G, Donner DB. Identification and characterization of receptors for tumor necrosis factor-α in the brain. Biol Biopharm Res Commun 1991;181:1532–38.
6. Szelenyi J. Cytokines and the central nervous system. Brain Res Bull 2001;54:329–38.
7. Ban E, Milon G, Prudhom N, Fillion G, Haour F. Receptors for interleukin-1 (α and β) in mouse brain: mapping and neuronal localization in hippocampus. Neuroscience 1991;43:21–30.
8. Takao T, Tracey DE, Mitchell WM, De Souza EB. Interleukin-1 receptors in mouse brain: characterization and neuronal localization. Endocrinology 1990;3070–78.
9. Hanisch UK, Seto D, Quirion R. Modulation of hippocampal acetylcholine release: a potent central action of interleukin-2. J Neurosci 1993;13:3368–74.
10. Elenkov IJ, Kovacs K, Duda E, Stark E, Vizi ES. Presynaptic inhibitory effect of TNF-α on the release of noradrenaline in isolated median eminence. J Neuroimmunol 1992;41:117–20.
11. Elenkov IJ, Kovacs K, Duda E, Vizi ES. Modulatory effect of TNF-α on the release of noradrenaline in isolated median eminence. Pharmacol Res 1992;25:23–24.

12. Ignatowski TA, Spengler RN. Tumor necrosis factor-α: presynaptic sensitivity is modified after antidepressant drug administration. Brain Res 1994;665:293–299.
13. Shintani F, Kanba S, Nakaki T, Nibuya M, Kinoshita N, Suzuki E, Yagi G, Kato R, Asai M. Interleukin-1β augments release of norepinephrine, dopamine and serotonin in the rat anterior hypothalamus. J Neurosci 1993;13:3574–81.
14. Ito M. Long-term depression. Annu Rev Neurosci 1989;12:85–102.
15. Nicoll RA, Kauer JA, Malenka RC. The current excitement in long-term potentiation. Neuron 1988;1:97–103.
16. Zucker RS. Short-term synaptic plasticity. Annu Rev Neurosci 1989;2:13–31.
17. Idriss HT, Naismith JH. TNFα and the TNF receptor superfamily: structure-function relationship(s). Microsc Res Tech 2000;50:184–95.
18. Benveniste EN, Benos DJ. TNF-alpha- and IFN-gamma-mediated signal transduction pathways: effects on glial cell gene expression and function. FASEB J 1995;9:1577–84.
19. Dopp JM, Mackenzie GA, Otero GC, Merrill JE. Differential expression, cytokine modulation, and specific functions of type-1 and type-2 tumor necrosis factor receptors in rat glia. J Neuroimmunol 1997;75:104–12.
20. Covey WC, Ignatowski TA, Renauld AE, Knight PR, Nader ND, Spengler RN. Expression of neuron-associated tumor necrosis factor alpha in the brain is increased during persistent pain. Reg Anesth Pain Med 2002;27:357–66.
21. Ignatowski TA, Noble BK, Wright JR, Gorfien JL, Heffner RR, Spengler RN. Neuronal-associated tumor necrosis factor (TNFα): its role in noradrenergic functioning and modification of its expression following antidepressant drug administration. J Neuroimmunol 1997;79:84–90.
22. Nickola TJ, Ignatowski TA, Reynolds JL, Spengler RN. Antidepressant drug-induced alterations in neuron-localized tumor necrosis factor-α mRNA and α_2-adrenergic receptor sensitivity. J Pharmacol Exp Ther 2001;297:680–87.
23. Renauld AE, Spengler RN. Tumor necrosis factor expressed by primary hippocampal neurons and SH-SY5Y cells is regulated by α_2-adrenergic receptor activation. J Neurosci Res 2002;67:264–74.
24. Spengler RN, Sud R, Knight PR, Ignatowski TA. Antinociception mediated by α_2-adrenergic activation involves increasing tumor necrosis factor α (TNFα) expression and restoring TNFα and α_2-adrenergic inhibition of norepinephrine release. Neuropharmacology 2007;52:576–89.
25. Vitkovic L, Bockaert J, Jacque C. "Inflammatory" cytokines: neuromodulators in normal brain? J Neurochem 2000;74:457–71.
26. Dixon WR, Mosimann WF, Weiner N. The role of presynaptic feedback mechanisms in regulation of norepinephrine release by nerve stimulation. J Pharmacol Exp Ther 1979;209:196–204.
27. Langer SZ. Presynaptic regulation of the release of catecholamines. Pharmacol Rev 1981;32:337–62.
28. Starke K. Presynaptic receptors. Annu Rev Pharmacol Toxicol 1981;21:7–30.
29. Ignatowski TA, Chou RC, Spengler RN. Changes in noradrenergic sensitivity to tumor necrosis factor-α in brains of rats administered clonidine. J Neuroimmunol 1996;70: 55–63.
30. Covey WC, Ignatowski TA, Knight PR, Spengler RN. Brain-derived TNFα: involvement in neuroplastic changes implicated in the conscious perception of persistent pain. Brain Res 2000;859:113–22.

31. Ignatowski TA, Sud R, Reynolds JL, Knight PR, Spengler RN. The dissipation of neuropathic pain paradoxically involves the presence of tumor necrosis factor-α (TNF). Neuropharmacology 2005;48:448–60.
32. Reynolds JL, Ignatowski TA, Gallant S, Spengler RN. Amitriptyline administration transforms tumor necrosis factor-alpha regulation of norepinephrine release in the brain. Brain Res 2004;1023:112–20.
33. Reynolds JL, Ignatowski TA, Sud R, Spengler RN. Brain-derived tumor necrosis factor-α and its involvement in noradrenergic neuron functioning involved in the mechanism of action of an antidepressant. J Pharmacol Exp Ther 2004;310:1216–25.
34. Reynolds JL, Ignatowski TA, Sud R, Spengler RN. An antidepressant mechanism of desipramine is to decrease tumor necrosis factor-α production culminating in increases in noradrenergic neurotransmission. Neuroscience 2005;133:519–31.
35. Sud R, Ignatowski TA, Lo CPK, Spengler RN. Uncovering molecular elements of brain-body communication during development and treatment of neuropathic pain. Brain Behav Immun 2007;21:112–24.
36. Abadie C, Foucart S, Page P, Nadeau R. Interleukin-1β and tumor necrosis factor-α inhibit the release of [^{3}H]-Noradrenaline from isolated human atrial appendages. Naunyn-Schmiedeberg's Arch Pharmacol 1997;355:384–89.
37. Foucart S, Abadie C. Interleukin-1β and tumor necrosis factor-α inhibit the release of [^{3}H]-noradrenaline from mice isolated atria. Naunyn-Schmiedeberg's Arch Pharmacol 1996;354:1–6.
38. Pollock J, McFarlane SM, Connell MC, Zehavi U, Vandenabeele P, MacEwan DJ, Scott RH. TNF-α receptors simultaneously activate Ca^{2+} mobilization and stress kinases in cultured sensory neurons. Neuropharmacology 2002;42:93–106.
39. Grassi F, Mileo AM, Monaco L, Punturieri A, Santoni A, Eusebi F. TNF-alpha increases the frequency of spontaneous miniature synaptic currents in cultured rat hippocampal neurons. Brain Res 1994;659:226–30.
40. Hurst S, Collins SM. Mechanism underlying tumor necrosis factor-α suppression of norepinephrine release from rat myenteric plexus. Am J Physiol 266:G1123-G1129.
41. Hurst SM, Collins SM. Interleukin-1β modulation of norepinephrine release from rat myenteric nerves. Am J Physiol 1993;264(Gastrointest Liver Physiol 27):G30-G35.
42. Wedel OT, Strandhoy JW. The effects of prostaglandins E_2 and F_{2a} on synaptosomal accumulation and release of ^{3}H-norepinephrine. Prostaglandins 1978;16:441–49.
43. Soliven B, Albert J. Tumor necrosis factor modulates the inactivation of catecholamine secretion in cultured sympathetic neurons. J Neurochem 1992;58:1073–78.
44. Soliven B, Albert J. Tumor necrosis factor modulates Ca^{2+} currents in cultured sympathetic neurons. J Neurosci 1992;12:2665–71.
45. Stjärne L, Msghina M, Stjärne E. "Upstream" regulation of the release probability in sympathetic nerve varicosities. Neuroscience 1990;36:571–87.
46. Soliven B, Wang N. Tumor necrosis factor-α regulates nicotinic responses in mixed cultures of sympathetic neurons and nonneuronal cells. J Neurochem 1995;64:883–94.
47. Reichardt LF, Kelly RB. A molecular description of nerve terminal function. Annu Rev Biochem 1983;52:871–926.
48. Bliss TVP, Lomo T. Long-lasting potentiation of synaptic transmission in the dentate area of the anaesthetized rabbit following stimulation of the perforant path. J Physiol 1973;232:331–56.

49. Bliss TVP, Collingridge GL. A synaptic model of memory: long-term potentiation in the hippocampus. Nature 1993;361:31–39.
50. Madison DV, Malenka RC, Nicoll RA. Mechanisms underlying long-term potentiation of synaptic transmission. Annu Rev Neurosci 1991;14:379–97.
51. Scoville WB, Milner B. Loss of recent memory after bilateral hippocampal lesions. J Neurol Neurosurg Psych 1957;20:11–21.
52. Floyd RA, Krueger JM. Diurnal variation of TNFα in the rat brain. NeuroReport 1997;8:915–18.
53. Rivier C. Influence of immune signals on the hypothalamic-pituitary axis of the rodent. Front Neuroendocrinol 1995;16:151–82.
54. Tancredi V, D'Arcangelo G, Grassi F, Tarroni P, Palmieri G, Santoni A, Eusebi F. Tumor necrosis factor alters synaptic transmission in rat hippocampal slices. Neurosci Lett 1992;146:176–78.
55. Butler MP, Moynagh PN, O'Connor JJ. Methods of detection of the transcription factor NF-κB in rat hippocampal slices. J Neurosci Meth 2002;119:185–90.
56. Cunningham AJ, Murray CA, O'Neill LA, Lynch MA, O'Connor JJ. Interleukin-1 beta (IL-1 beta) and tumor necrosis factor (TNF) inhibit long-term potentiation in the rat dentate gyrus in vitro. Neurosci Lett 1996;203:17–20.
57. Köller H, Trimborn M, von Giesen H-J, Schroeter M, Arendt G. TNFα reduces glutamate induced intracellular Ca2+ increase in cultured cortical astrocytes. Brain Res 2001;893:237–43.
58. Parpura V, Barsarsky TA, Liu F, Jeftinija K, Jeftinija S, Haydon PG. Glutamate-mediated astrocyte-neuron signaling. Nature 1994;369:744–47.
59. Butler MP, O'Connor JJ, Moynagh PN. Dissection of tumor-necrosis factor-α inhibition of long-term potentiation (LTP) reveals a p38 mitogen-activated protein kinase-dependent mechanism which maps to early-but not late-phase LTP. Neuroscience 2004;124:319–26.
60. Kramer RM, Roberts EF, Um SL, Borsch-Haubold AG, Watson SP, Fisher MJ, Jakubowski JA. p38 Mitogen-activated protein kinase phosphorylates cytosolic phospholipase A2 (cPLA2) in thrombin-stimulated platelets: evidence that proline-directed phosphorylation is not required for mobilization of arachidonic acid by cPLA2. J Biol Chem 1996;271:27723–29.
61. Rousseau S, Houle F, Landry J, Huot J. p38 MAP kinase activation by vascular endothelial growth factor mediates actin reorganization and cell migration in human endothelial cells. Oncogene 1997;15:2169–77.
62. Cumiskey D, Butler MP, Moynagh PN, O'Connor JJ. Evidence for a role for the group I metabotropic glutamate receptor in the inhibitory effect of tumor necrosis factor-α on long-term potentiation. Brain Res 2007;1136:13–19.
63. Nguyen PV, Kandel ER. A macromolecular synthesis-dependent late phase of long-term potentiation requiring cAMP in the medial perforant pathway of rat hippocampal slices. J Neurosci 1996;16:3189–98.
64. Otani S, Abraham WC. Inhibition of protein synthesis in the dentate gyrus, but not the entorhinal cortex, blocks maintenance of long-term potentiation in rats. Neurosci Lett 1989;106:175–80.
65. Zou JY, Crews FT. TNFα potentiates glutamate neurotoxicity by inhibiting glutamate uptake in organotypic brain slice cultures: neuroprotection by NFκB inhibition. Brain Res 2005;1034:11–24.

66. Beattie EC, Stellwagen D, Morishita W, Bresnahan JC, Ha BK, Von Zastrow M, Beattie MS, Malenka RC. Control of synaptic strength by glial TNFα. Science 2002;295: 2282–85.
67. Stellwagen D, Beattie EC, Seo JY, Malenka RC. Differential regulation of AMPA receptor and GABA receptor trafficking by tumor necrosis factor-α. J Neurosci 2005;25:3219–28.
68. Hermann GE, Rogers RC, Bresnahan JC, Beattie MS. Tumor necrosis factor-alpha induces cFOS and strongly potentiates glutamate-mediated cell death in the rat spinal cord. Neurobiol Dis 2001;8:590–99.
69. Albensi BC, Mattson MP. Evidence for the involvement of TNF and NF-κB in hippocampal synaptic plasticity. Synapse 2000;35:151–59.
70. Rothwell NJ. Functions and mechanisms of interleukin 1 in the brain. Trends Pharmacol Sci 1991;12:430–36.
71. Patel HC, Boutin H, Allan SM. Interleukin-1 in the brain: mechanisms of action in acute neurodegeneration. Ann NY Acad Sci 2003;992:39–47.
72. Rothwell NJ, Hopkins SJ. Cytokines and the nervous system II: actions and mechanisms of action. Trends Neurosci 1995;18:130–36.
73. Rivier C. Effect of peripheral and central cytokines on the hypothalamic-pituitary-adrenal axis of the rat. Ann NY Acad Sci 1993;697:97–105.
74. Mohankumar PS, Thyagarajan S, Quadri SK. Interleukin-1 stimulates the release of dopamine and dihydroxyphenylacetic acid from hypothalamus in vivo. Life Sci 1991;48:925–30.
75. Mohankumar PS, Thyagarajan S, Quadri SK. Interleukin-1β increases 5-hydroxyindoleacetic acid release in the hypothalamus in vivo. Brain Res Bull 1993;31:745–48.
76. Shintani F, Nakaki T, Kanba S, Sato K, Yagi G, Shiozawa M, Aiso S, Kato R, Asai M. Involvement of interleukin-1 in immobilization stress-induced increase in plasma adrenocorticotropic hormone and in release of hypothalamic monoamines in the rat. J Neurosci 1995;15:1961–70.
77. Linthorst ACE, Flachskamm C, Holsboer F, Reul JMHM. Local administration of recombinant human interleukin-1 beta in the rat hippocampus increases serotonergic neurotransmission, hypothalamic-pituitary-adrenocortical axis activity, and body temperature. Endocrinology 1994;135:520–32.
78. Linthorst ACE, Flachskamm C, Müller-Preuss P, Holsboer F, Reul JMHM. Effect of bacterial endotoxin and interleukin1-β on hippocampal serotonergic neurotransmission, behavioral activity, and free corticosterone levels: an in vivo microdialysis study. J Neurosci 1995;15:2920–34.
79. Merali Z, Lacosta S, Anisman H. Effects of interleukin-1β and mild stress on alterations of norepinephrine, dopamine and serotonin neurotransmission: a regional microdialysis study. Brain Res 1997;761:225–35.
80. Zalcman S, Green-Johnson JM, Murray L, Nance DM, Dyck D, Anisman H, Greenberg AH. Cytokine-specific central monoamine alterations induced by interleukin-1, -2 and -6. Brain Res 1994;643:40–49.
81. De Kloet ER. Brain corticosteroid receptor balance and homeostatic control. Front Neuroendocrinol 1991;12:95–64.
82. Matsumoto Y, Yoshida M, Watanabe S, Yamamoto T. Involvement of cholinergic and glutamatergic functions in working memory impairment induced by interleukin-1beta in rats. Eur J Pharmacol 2001;430:283–88.
83. Li Z, Inenaga K, Kawano S, Kannan H, Yamashita H. Interleukin-1β directly excites hypothalamic supraoptic neurons in rats *in vitro*. NeuroReport 1992;3:91–93.

84. Mo ZL, Katafuchi T, Hori T. Effects of IL-1β on neuronal activities in the dorsal motor nucleus of the vagus in rat brain slices. Brain Res Bull 1996;41:249–55.
85. Bellinger FP, Madamba S, Siggins GR. Interleukin 1β inhibits synaptic strength and long-term potentiation in the rat CA1 hippocampus. Brain Res 1993;628:227–34.
86. Cunningham AJ, Murray CA, O'Neill LA, Lynch MA, O'Connor JJ. Interleukin-1β (IL-1β) and tumor necrosis factor (TNF) inhibit long-term potentiation in the rat dentate gyrus *in vitro.* Neurosci Lett 1996;203:17–20.
87. Coogan A, O'Connor JJ. Inhibition of NMDA receptor-mediated synaptic transmission in the rat dentate gyrus *in vitro* by IL-1β. NeuroReport 1997;8:2107–10.
88. Katsuki H, Nakai S, Hirai Y, Akaji K, Kiso Y, Satoh M. Interleukin-1β inhibits long-term potentiation in the CA3 region of mouse hippocampal slices. Eur J Pharmacol 1990;181:323–26.
89. Luk WP, Zhang Y, White TD, Lue FA, Wu C, Jiang C-G, Zhang L, Moldofsky H. Adenosine: a mediator of interleukin-1β-induced hippocampal synaptic inhibition. J Neurosci 1999;19:4238–44.
90. Yang S, Liu Z-W, Wen L, Qiao H-F, Zhou W-X, Zhang Y-X. Interleukin-1β enhances NMDA receptor-mediated current but inhibits excitatory synaptic transmission. Brain Res 2005;1034:172–79.
91. Ikegaya Y, Delcroix I, Iwakura Y, Matsuki N, Nishiyama N. Interleukin-1β abrogates long-term depression of hippocampal CA1 synaptic transmission. Synapse 2003;47:54–57.
92. Pringle AK, Gardner CR, Walker RJ. Reduction of cerebral $GABA_A$ responses by interleukin-1 (IL-1) through an indomethacin insensitive mechanism. Neuropharmacology 1996;35:147–52.
93. Zeise ML, Espinoza J, Morales P, Nalli A. Interleukin-1β does not increase synaptic inhibition in hippocampal CA3 pyramidal and dentate gyrus granule cells of the rat in vitro. Brain Res 1997;768:341–44.
94. Zeise ML, Madamba S, Siggins GR. Interleukin-1β increases synaptic inhibition in rat hippocampal pyramidal neurons in vitro. Regul Pept 1992;39:1–7.
95. Schneider H, Pitossi F, Balschun D, Wagner A, Del Rey A, Besedovsky HO. A neuromodulatory role of interleukin-1β in the hippocampus. Proc Natl Acad Sci USA 1998;95:7778–83.
96. Coogan AN, O'Neill LAJ, O'Connor JJ. The P38 mitogen-activated protein kinase inhibitor SB203580 antagonizes the inhibitory effects of interleukin-1β on long-term potentiation in the rat dentate gyrus *in vitro.* Neuroscience 1999;93:57–69.
97. Lin J, Krier J. Human recombinant interleukin-1β inhibits nicotinic transmission in neurons of guinea pig pelvic plexus ganglia. Am J Physiol (Gastrointest Liver Physiol 32) 1995;269:G981-G987.
98. Main C, Blennerhassett P, Collins SM. Human recombinant interleukin 1β suppresses acetylcholine release from rat myenteric plexus. Gastroenterology 1993;104:1648–54.
99. Hama T, Kushima Y, Miyamoto M, Kubota M, Takei N, Hatanaka H. Interleukin-6 improves the survival of mesencephalic catecholaminergic and septal cholinergic neurons from post-natal, two-week-old rats in culture. Neuroscience 1991;40:445–52.
100. Kushima Y, Hama T, Hatanaka H. Interleukin-6 as a neurotrophic factor for promoting the survival of cultured catecholaminergic neurons in a chemically-defined medium from fetal and post-natal rat midbrains. Neurosci Res 1992;13:267–80.
101. Kushima Y, Hatanaka H. Interleukin-6 and leukemia inhibitory factor promote the survival of acetylcholinesterase-positive neurons in culture from embryonic rat spinal cord. Neurosci Lett 1992;143:110–14.

102. Loddick SA, Turnbull AV, Rothwell NJ. Cerebral interleukin-6 is neuroprotective during permanent focal cerebral ischemia in the rat. J Cereb Blood Flow Metab 1998;18:176–79.
103. Matsuda S, Wen TC, Morita F, Otsuka H, Igase K, Yoshimura H, Sakanaka M. Interleukin-6 prevents ischemia-induced learning disability and neuronal and synaptic loss in gerbils. Neurosci Lett 1996;204:109–12.
104. Toulmond S, Vige X, Fage D, Benavides J. Local infusion of interleukin-6 attenuates the neurotoxic effects of NMDA on rat striatal cholinergic neurons. Neurosci Lett 1992;144:49–52.
105. Yamada M, Hatanaka H. Interleukin-6 protects cultured rat hippocampal neurons against glutamate-induced cell death. Brain Res 1994;643:173–80.
106. Ihara S, Iwamatsu A, Fujiyoshi T, Komi A, Yamori T, Fukui Y. Identification of interleukin-6 as a factor that induces neurite outgrowth by PC12 cells primed with NGF. J Biochem 1996;120:865–68.
107. Ihara S, Nakajima K, Fukada T, Hibi M, Nagata S, Hirano T, Fukui Y. Dual control of neurite outgrowth by STAT3 and MAP kinase in PC12 cells stimulated with interleukin-6. EMBO J 1997;16:5345–52.
108. Wu YY, Bradshaw RA. Induction of neurite outgrowth by interleukin-6 is accompanied by activation of Stat3 signaling pathway in a variant PC12 cell (E2) line. J Biol Chem 1996;272:13023–32.
109. Gadient RA, Otten UH. Interleukin-6 (IL-6) – a molecule with both beneficial and destructive potential. Prog Neurobiol 1997;52:379–90.
110. Bellinger FP, Madamba SG, Campbell IL, Siggins GR. Reduced long-term potentiation in the dentate gyrus of transgenic mice with cerebral overexpression of interleukin-6. Neurosci Lett 1995;198:95–98.
111. Li AJ, Katafuchi T, Oda S, Hori T, Oomura Y. Interleukin-6 inhibits long-term potentiation in rat hippocampal slice. Brain Res 1997;748:30–38.
112. D'Arcangelo G, Tancredi V, Onofri F, D'Antuono M, Giovedi S, Benfenati F. Interleukin-6 inhibits neurotransmitter release and the spread of excitation in the rat cerebral cortex. Eur J Neurosci 2000;12:1241–52.
113. Tancredi V, D'Antuono M, Cafè C, Giovedì S, Buè MC, D'Arcangelo G, Onofri F, Benfenati F. The inhibitory effects of interleukin-6 on synaptic plasticity in the rat hippocampus are associated with an inhibition of mitogen-activated protein kinase ERK. J Neurochem 2000;75:634–43.
114. English JD, Sweatt JD. Activation of p42 mitogen-activated protein kinase in hippocampal long-term potentiation. J Biol Chem 1996;271:24329–32.
115. English JD, Sweatt JD. A requirement for the mitogen-activated protein kinase cascade in hippocampal long term potentiation. J Biol Chem 1997;272:19103–6.
116. Flatters SJL, Fox AJ, Dickenson AH. Spinal interleukin-6 (IL-6) inhibits nociceptive transmission following neuropathy. Brain Res 2003;984:54–62.
117. Flatters SJL, Fox AJ, Dickenson AH. Nerve injury alters the effects of interleukin-6 on nociceptive transmission in peripheral afferents. Eur J Pharmacol 2004;484:183–91.
118. Crnic LS, Segall MA. Behavioral effects of mouse interferons-α and -γ and human interferon-α in mice. Brain Res 1992;590:277–84.
119. Müller M, Fontana A, Zbinden G, Gähwiler BH. Effects of interferons and hydrogen peroxide on CA3 pyramidal cells in rat hippocampal slice cultures. Brain Res 1993;619:157–62.
120. Vikman K, Robertson B, Grant G, Liljeborg A, Kristensson K. Interferon-γ receptors are expressed at synapses in the rat superficial dorsal horn and lateral spinal nucleus. J Neurocytol 1998;27:749–60.

121. Vikman SK, Owe-Larsson B, Brask J, Kristensson KS, Hill RH. Interferon-γ-induced changes in synaptic activity and AMPA receptor clustering in hippocampal cultures. Brain Res 2001;896:18–29.
122. Calvet MC, Gresser I. Interferon enhances the excitability of cultured neurons. Nature 1979;278:558–60.
123. D'Arcangelo G, Grassi F, Ragozzino D, Santoni A, Tancredi V, Fabrizio E. Interferon inhibits synaptic potentiation in rat hippocampus. Brain Res 1991;564:245–48.
124. Shuto H, Kataoka Y, Horikawa T, Fujihara N, Oishi R. Repeated interferon-α administration inhibits dopaminergic neural activity in the mouse brain. Brain Res 1997;747:348–51.
125. Kim IJ, Beck HN, Lein PJ, Higgins D. Interferon gamma induces retrograde dendritic retraction and inhibits synapse formation. J Neurosci 2002;22:4530–39.
126. Brask J, Kristensson K, Hill RH. Exposure to interferon-γ during synaptogenesis increases inhibitory activity after a latent period in cultured rat hippocampal neurons. Eur J Neurosci 2004;19:3193–201.
127. Rossi D, Zlotnik A. The biology of chemokines and their receptors. Annu Rev Immunol 2000;18:217–42.
128. Yoshie O, Imai T, Nomiyama H. Chemokines in immunity. Adv Immunol 2001;78: 57–110.
129. Zlotnik A, Morales J, Hedrick JA. Recent advances in chemokines and chemokine receptors. Crit Rev Immunol 1999;19:1–47.
130. Asensio VC, Campbell IL. Chemokines in the CNS: plurifunctional mediators in diverse states. Trends Neurosci 1999;22:504–12.
131. White FA, Bhangoo SK, Miller RJ. Chemokines: Integrators of pain and inflammation. Nature Rev Drug Discov 2005;4:834–44.
132. Horuk R, Martin AW, Wang Z, Schweitzer L, Gerassimides A, Guo H, Lu Z, Hesselgesser J, Perez HD, Kim J, Parker J, Hadley TJ, Peiper SC. Expression of chemokine receptors by subsets of neurons in the central nervous system. J Immunol 1997;158:2882–90.
133. Horuk R, Martin A, Hesselgesser J, Hadley T, Lu Z-H, Wang Z-X, Peiper SC. The Duffy antigen receptor for chemokines: structural analysis and expression in the brain. J Leukoc Biol 1996;59:29–38.
134. Wong ML, Xin WW, Duman RS. Rat LCR1: cloning and cellular distribution of a putative chemokine receptor in brain. Mol Psychiatry 1996;1:133–40.
135. Ragozzino D, Giovannelli A, Mileo AM, Limatola C, Santoni A, Eusebi F. Modulation of the neurotransmitter release in rat cerebellar neurons by GROβ. NeuroReport 1998;9:3601–6.
136. Meucci O, Fatatis A, Simen AA, Bushell TJ, Gray PW, Miller RJ. Chemokines regulate hippocampal neuronal signaling and gp 120 neurotoxicity. Proc Natl Acad Sci USA 1998;95:14500–5.
137. Giovannelli A, Limatola C, Ragozzino D, Mileo AM, Ruggieri A, Ciotti MT, Mercanti D, Santoni A, Eusebi F. CXC chemokines interleukin-8 (IL-8) and growth-related gene product α (GROα) modulate Purkinje neuron activity in mouse cerebellum. J Neuroimmunol 1998;92:122–32.
138. Limatola C, Giovannelli A, Maggi L, Ragozzino D, Castellani L, Ciotti MT, Vacca F, Mercanti D, Santoni A, Eusebi F. SDF-1α-mediated modulation of synaptic transmission in rat cerebellum. Eur J Neurosci 2000;12:2497–504.

139. Lax P, Limatola C, Fucile S, Trettel F, Di Bartolomeo S, Renzi M, Ragozzino D, Eusebi F. Chemokine receptor CXCR2 regulates the functional properties of AMPA-type glutamate receptor GluR1 in HEK cells. J Neuroimmunol 2002;129:66–73.
140. Ragozzino D, Renzi M, Giovannelli A, Eusebi F. Stimulation of chemokine CXC receptor 4 induces synaptic depression of evoked parallel fibers inputs onto Purkinje neurons in mouse cerebellum. J Neuroimmunol 2002;127:30–36.
141. Vlkolinský R, Siggins GR, Campbell IL, Krucker T. Acute exposure to CXC chemokine ligand 10, but not its chronic astroglial production, alters synaptic plasticity in mouse hippocampal slices. J Neuroimmunol 2004;150:37–47.
142. Bertollini C, Ragozzino D, Gross C, Limatola C, Eusebi F. Fractalkine/CX_3CL1 depresses central synaptic transmission in mouse hippocampal slices. Neuropharmacology 2006;51:816–21.
143. Guyon A, Skrzydelsi D, Rovère C, Rostène W, Parsadaniantz SM, Nahon JL. Stromal cell-derived factor-1α modulation of the excitability of rat substantia nigra dopaminergic neurones: presynaptic mechanisms. J Neurochem 2006;96:1540–50.
144. Iversen LL, Mortishire-Smith RJ, Pollack SJ, Shearman MS. The toxicity in vitro of β-amyloid protein. (Review Article) Biochem J 1995;311:1–16.
145. Esler WP, Wolfe MS. A portrait of Alzheimer secretases – New features and familiar faces. Science 2001;293:1449–54.
146. Gotz J, Chen F, van Dorpe J, Nitsch RM. Formation of neurofibrillary tangles in P301L tau transgenic mice induced by Aβ42 fibrils. Science 2001;293:1491–95.
147. Del Bo R, Angeretti N, Lucca E, De Simoni G, Forloni G. Reciprocal control of inflammatory cytokines, IL-1 and IL-6 and β-amyloid production in cultures. Neurosci Lett 1995;188:70–74.
148. Forloni G, Demicheli F, Giorgi S, Bendotti C, Angeretti N. Expression of amyloid precursor protein mRNAs in endothelial, neuronal and glial cells: modulation by interleukin-1. Mol Brain Res 1992;16:128–34.
149. Vandenabeele P, Fiers W. Is amyloidogenesis during Alzheimer's disease due to an IL-1-/IL-6-mediated 'acute phase response' in the brain? Immunol Today 1991;12:217–19.
150. Behan DP, Heinrichs SC, Troncoso JC, Liu X-J, Kawas CH, Ling N, De Souza EB. Displacement of corticotrophin releasing factor from its binding protein as a possible treatment for Alzheimer's disease. Nature 1995;378:284–87.
151. Lee EHY, Hung HC, Lu KT, Chen WH, Chen HY. Protein synthesis in the hippocampus associated with memory facilitation by corticotrophin-releasing factor in rats. Peptides 1992;13:927–37.
152. Forloni G, Mangiarotti F, Angeretti N, Lucca E, De Simoni MG. β-amyloid fragment potentiates IL-6 and TNF-α secretion by LPS in astrocytes but not in microglia. Cytokine 1997;9:759–62.
153. Hu J, Akama KT, Krafft GA, Chromy BA, Van Eldik LJ. Amyloid-β peptide activates cultured astrocytes: morphological alterations, cytokine induction and nitric oxide release. Brain Res 1998;785:195–206.
154. Weisman D, Hakimian E, Ho GJ. Interleukins, inflammation, and mechanisms of Alzheimer's disease. Vitam Horm 2006;74:505–30.
155. Blum-Degen D, Müller T, Kuhn W, Gerlach M, Przuntek H, Riederer P. Interleukin 1β and interleukin-6 are elevated in the cerebrospinal fluid of Alzheimer's and de novo Parkinson's disease patients. Neurosci Lett 1995;202:17–20.

156. Fillit H, Dig W, Buee L, Kalman J, Altstiel L, Lawlor B, Wolf-Klein G. Elevated circulating tumor necrosis factor levels in Alzheimer's disease. Neurosci Lett 1991;129:318–20.
157. Griffin WST, Stanley LC, Ling C, White L, MacLeod V, Perrot LJ, White CL III, Araoz C. Brain interleukin-1 and S-100 immunoreactivity are elevated in Down syndrome and Alzheimer disease. Proc Natl Acad Sci USA 1989;86:7611–15.
158. McGeer PL, Rogers J. Anti-inflammatory agents as a therapeutic approach to Alzheimer's disease. Neurology 1992;42:447–49.
159. Tobinick E, Gross H, Weinberger A, Cohen H. TNF-alpha modulation for treatment of Alzheimer's disease: a 6-month pilot study. Med Gen Med 2006;8:25.
160. Ignatowski TA, Spengler RN. Regulation of macrophage-derived tumor necrosis factor production by modification of adrenergic receptor sensitivity. J Neuroimmunol 1995;61: 61–70.
161. Nickola TJ, Ignatowski TA, Spengler RN. Antidepressant drug administration modifies the interactive relationship between α_2-adrenergic sensitivity and levels of TNF in the rat brain. J Neuroimmunol 2000;107:50–58.
162. Severn A, Rapson NT, Hunter CA, Liew FY. Regulation of tumor necrosis factor production by adrenaline and β-adrenergic agonists. J Immunol 1992;148:3441–45.
163. Spengler RN, Chensue SW, Giacherio DA, Blenk N, Kunkel SL. Endogenous norepinephrine regulates tumor necrosis factor-α production from macrophages in vitro. J Immunol 1994;152:3024–31.
164. McDonald WM, Richard IH, DeLong MR. Prevalence, etiology, and treatment of depression in Parkinson's disease. Biol Psychiatry 2003;54:363–75.
165. Barcia C, de Pablos V, Bautista-Hernandez V, Sanchez-Bahillo A, Bernal I, Fernandez-Villalba E, Martin J, Banon R, Fernandez-Barreiro A, Herrero MT. Increased plasma levels of TNF-alpha but not of IL1-beta in MPTP-treated monkeys one year after the MPTP administration. Parkinsonism Relat Disord 2005;11:435–39.
166. Boka G, Anglade P, Wallach D, Javoy-Agid F, Agid Y, Hirsch EC. Immunocytochemical analysis of tumor necrosis factor and its receptors in Parkinson's disease. Neurosci Lett 1994;172:151–54.
167. Ferrari CC, Pott Godoy MC, Tarelli R, Chertoff M, Depino AM, Pitossi FJ. Progressive neurodegeneration and motor disabilities induced by chronic expression of IL-1beta in the substantis nigra. Neurobiol Dis 2006;24:183–93.
168. Mogi M, Togari A, Kondo T, Mizuno Y, Komure O, Kuno S, Ichinose H, Nagatsu T. Caspase activities and tumor necrosis factor receptor R1 (p55) level are elevated in the substantia nigra from parkinsonian brain. J Neural Transm 2000;107:335–41.
169. Nagatsu T, Sawada M. Inflammatory process in Parkinson's disease: role for cytokines. Curr Pharm Des 2005;11:999–1016.
170. Sriram K, Matheson JM, Benkovic SA, Miller DB, Luster MI, O'Callaghan JP. Mice deficient in TNF receptors are protected against dopaminergic neurotoxicity: implications for Parkinson's disease. FASEB J 2002;16:1474–76.
171. Sriram K, Matheson JM, Benkovic SA, Miller DB, Luster MI, O'Callaghan JP. Deficiency of TNF receptors suppresses microglial activation and alters the susceptibility of brain regions to MPTP-induced neurotoxicity: role of TNF-alpha. FASEB J 2006;20:670–82.
172. Kim YS, Joh TH. Microglia, major player in the brain inflammation: their roles in the pathogenesis of Parkinson's disease. Exp Mol Med 2006;38:333–47.
173. Neumann H. Control of glial immune function by neurons. Glia 2001;36:191–99.

174. Wei R, Jonakait GM. Neurotrophins and the anti-inflammatory agents interleukin-4 (IL-4), IL-10, IL-11 and transforming growth factor-beta1 (TGF-beta1) down-regulate T cell costimulatory molecules B7 and CD40 on cultured rat microglia. J Neuroimmunol 1999;95:8–18.
175. Greig NH, Mattson MP, Perry T, Chan SL, Giordano T, Sambamurti K, Rogers JT, Ovadia H, Lahiri DK. New therapeutic strategies and drug candidates for neurodegenerative diseases: p53 and TNF-alpha inhibitors, and GLP-1 receptor agonists. Ann NY Acad Sci 2004;1035:290–315.
176. McCoy MK, Martinez TN, Ruhn KA, Szymkowski DE, Smith CG, Botterman BR, Tansey KE, Tansey MG. Blocking soluble tumor necrosis factor signaling with dominant-negativetumor necrosis factor inhibitor attenuates loss of dopaminergic neurons in models of Parkinson's disease. J Neurosci 2006;26:9365–75.
177. Black DW, Yates WR, Andreasen NC. Schizoprenia, schizophreniform disorder, and delusional (paranoid) disorders. In The American Psychiatric Press Textbook of Psychiatry. Talbott JA, Hales RE, Yudofsky SC, Eds; Washington: American Psychiatric Press, 1988; p. 367.
178. Garver DL, Tamas RL, Holcomb JA. Elevated interleukin-6 in the cerebrospinal fluid of a previously delineated schizophrenia subtype. Neuropsychopharmacol 2003;28:1515–20.
179. Hori H, Yoshimura R, Yamada Y, Ikenouchi A, Mitoma M, Ida Y, Nakamura J. Effects of olanzapine on plasma levels of catecholamine metabolites, cytokines, and brain-derived neurotrophic factor in schizophrenic patients. Int Clin Psychopharmacol 2007;22:21–27.
180. Maes M, Bosmans E, Calabrese J, Smith R, Meltzer HY. Interleukin-2 and interleukin-6 in schizophrenia and mania: effects of neuroleptics and mood stabilizers. J Psychiatric Res 1995;29:141–52.
181. Monteleone P, Fabrazzo M, Tortorella A, Maj M. Plasma levels of interleukin-6 and tumor necrosis factor alpha in chronic schizophrenia: effects of clozapine treatment. Psychiatry Res 1997;71:11–17.
182. Schuld A, Hinze-Selch D, Pollmächer T. Cytokine network in patients with schizophrenia and its significance for the pathophysiology of the illness. Nervenarzt 2004;75:215–26.
183. Smith RS. A comprehensive macrophage-T-lymphocyte theory of schizophrenia. Med Hypoth 1992;39:248–57.
184. Alonso R, Chaudieu I, Diorio J, Krishnamurthy A, Quirion R, Boksa P. Interleukin-2 modulates evoked release of [3H]dopamine in rat cultured mesencephalic cells. J Neurochem 1993;61:1284–90.
185. Lapchak PA. A role for IL-2 in the regulation of striatal dopaminergic function. NeuroReport 1992;3:165–68.
186. De Sarro CB, Masuda Y, Ascioti C, Audino MG, Nistico G. Behavioral and ECoG spectrum changes induced by intracerebral infusion of interferons and interleukin 2 in rats are antagonized by naloxone. Neuropharmacology 1990;29:167–79.
187. Nawa H, Takahashi M, Patterson PH. Cytokine and growth factor involvement in schizophrenia - support for the developmental model. Mol Psychiatry 2000;5:594–603.
188. Drzyzga L, Obuchowicz E, Marcinowska A, Herman ZS. Cytokines in schizophrenia and the effects of antipsychotic drugs. Brain Behav Immun 2006;20:532–45.
189. Pollmächer T, Haack M, Schuld A, Kraus T, Hinze-Selch D. Effects of antipsychotic drugs on cytokine networks. J Psychiatric Res 2000;34:369–82.
190. Cazzullo CL, Sacchetti E, Galluzzo A, Panariello A, Adorni A, Pegoraro M, Bosis S, Colombo F, Trabattoni D, Zagliani A, Clerici M. Cytokine profiles in schizophrenic pateients treated with risperidone: a 3-month follow-up study. Prog Neuro-Psychopharmacol Biol Psychiatry 2002;26:33–39.

191. Leykin I, Mayer R, Shinitzky M. Short and long-term immunosuppressive effects of clozapine and haloperidol. Immunopharmacol 1997;37:75–86.
192. Szuster-Ciesielska A, Slotwinska M, Stachura A, Marmurowska-Michalowska H, Kandefer-Szerszen M. Neuroleptics modulate cytokine and reactive oxygen species production in blood leukocytes of healthy volunteers. Arch Immunol Ther Exp 2004;52:59–67.
193. Zhang XY, Zhou DF, Cao LY, Zhang PY, Wu GY, Shen YC. Changes in serum interleukin-2, -6, and -8 levels before and during treatment with risperidone and haloperidol: relationship to outcome in schizophrenia. J Clin Psychiatry 2004;65:940–47.
194. Paterson GJ, Ohashi Y, Reynolds GP, Pratt JA, Morris BJ. Selective increases in the cytokine, TNFalpha, in the prefrontal cortex of PCP-treated rats and human schizophrenic subjects: influence of antipsychotic drugs. J Psychopharmacol 2006;20:636–42.
195. Hayley S, Poulter MO, Merali Z, Anisman H. The pathogenesis of clinical depression: stressor- and cytokine-induced alterations of neuroplasticity. Neuroscience 2005;135:659–78.
196. Anisman H, Merali Z. Anhedonic and anxiogenic effects of cytokine exposure. Adv Exp Med Biol 1999;461:199–233.
197. Hayley S, Kelly O, Anisman H. Murine tumor necrosis factor α sensitizes plasma corticosterone and manifestation of shock: Modulation by histamine. J Neuroimmunol 2003;131:60–69.
198. Merali Z, Lacosta S, Anisman H. Effects of interleukin-1β and mild stress on alterations of central monoamines: a regional microdialysis study. Brain Res 1997;761:225–35.
199. Maes M. Evidence for an immune response in major depression: a review and hypothesis. Prog Neuropsychopharmacol Biol Psych 1995;19:11–38.
200. Maes M, Scharpe S, Meltzer HY, Bosmans E, Suy E. Relationships between IL-6 activity, acute phase protein, and function of the hypothalamic-pituitary-adrenal axis in severe depression. Psychiatry Res 1993;49:11–27.
201. Mikova O, Yakimova R, Bosmans E, Kenis G, Maes M. Increased serum tumor necrosis factor alpha concentrations in major depression and multiple sclerosis. Eur Neuropsychopharmacol 2001;11:203–8.
202. Raison CL, Capuron L, Miller AH. Cytokines sing the blues: inflammation and the pathogenesis of depression. Trends Immunol 2006;27:24–31.
203. Meyers CA. Mood and cognitive disorders in cancer patients receiving cytokine therapy. In Cytokines, Stress, and Depression. Danzer R, Wollman EE, Yirmiya R, Eds; New York: Kluwer Academic/Plenum Publishers, 1999; pp. 75–82.
204. Wichers M, Maes M. The psychoneuroimmuno-pathophysiology of cytokine induced depression in humans. Int J Neuropsychopharm 2002;5:375–88.
205. Spriggs DR, Sherman ML, Michie H, Arthur KA, Imamura K, Wilmore D, Frei E III, Kufe DW. Recombinant hTNF administered as a 24-hour intravenous infusion. A phase 1 and pharmacologic study. J Natl Cancer Inst 1988;80:1039–44.
206. Bower JE, Ganz PA, Aziz N, Fahey JL. Fatigue and proinflammatory cytokine activity in breast cancer survivors. Psychosomatic Med 2002;64:604–11.
207. Meyers CA, Albitar M, Estey E. Cognitive impairment, fatigue, and cytokine levels in patients with acute myelogenous leukemia or myelodysplastic syndrome. Cancer 2005;104:788–93.
208. Brustolim D, Ribeiro-dos-Santos R, Kast RE, Altschuler EL, Soares MBP. A new chapter opens in anti-inflammatory treatments: the antidepressant bupropion lowers production of tumor necrosis factor-alpha and interferon-gamma in mice. Int Immunopharmacol 2006;6:903–7.

209. Kast RE. Anti- and pro-inflammatory considerations in antidepressant use during medical illness: bupropion lowers and mirtazapine increases circulating tumor necrosis factor-alpha levels. Gen Hosp Psych 2003;25:495–96.
210. Xia Z, DePierre JW, Nassberger L. Tricyclic antidepressants inhibit Il-6, IL-1 beta, and TNF alpha release in human blood monocytes and IL-2 and IFN gamma in T cells. Immunobiology 1996;34:37–47.
211. Hayley S, Brebner K, Lacosta S, Merali Z, Anisman H. Sensitization to the effects of tumor necrosis factor-α: neuroendocrine, central monoamine, and behavioral variations. J Neurosci 1999;19:5654–65.
212. Plata-Salaman, CR. Immunoregulators in the nervous system. Neurosci Biobehav Rev 1991;15:185–215.
213. Ignatowski TA, Noble B, Wright JR, Gorfien J, Spengler RN. TNFα: A neuromodulator in the central nervous system. Adv Exp Med Biol 1996;402:219–24.
214. Merskey H, Bogduk, N (Eds). Classification of Chronic Pain, 2nd Edn. Seattle: IASP Press; 1994.
215. Roizen MF, Fleisher LA. Essence of Anesthesia Practice. Philadelphia: W.B. Saunders and Co., 1997; p. 271.
216. Ignatowski TA, Covey WC, Knight PR, Severin CM, Nickola TJ, Spengler RN. Brain-derived TNFα mediates neuropathic pain. Brain Res 1999;841:70–77.
217. Watkins LR, Wiertelak EP, Goehler LE, Smith KP, Martin D, Maier SF. Characterization of cytokine-induced hyperalgesia. Brain Res 1994;654:15–26.
218. Oka T, Wakugawa Y, Hosoi M, Oka K, Hori T. Intracerebroventricular injection of tumor necrosis factor-α induces thermal hyperalgesia in rats. Neuroimmunomodulation 1996;3:135–40.
219. Sommer C, Marziniak M. Experimental painful mononeuropathy: Inhibitors of TNFα-production induce a decrease in hyperalgesia and an increase of spinal met-enkephalin. Soc Neurosci 1996; 22:2086.
220. Sommer C, Schmidt C, George A, Toyka KV. A metalloprotease-inhibitor reduces pain-associated behavior in mice with experimental neuropathy. Neurosci Lett 1997;237:45–48.
221. Wagner R, Myers RR. Endoneurial injection of TNF-alpha produces nociceptive behaviors. NeuroReport 1996;7:2897–901.
222. Wang CX, Nuttin B, Heremans H, Dom R, Gybels J. Production of tumor necrosis factor in spinal cord following traumatic injury in rats. J Neuroimmunol 1996;69:151–56.
223. Watkins LR, Goehler LE, Relton J, Brewer MT, Maier SF. Mechanisms of tumor necrosis factor-alpha (TNF-alpha) hyperalgesia. Brain Res 1995;692:244–50.
224. Melzack R. Phantom limb pain: implications for treatment of pathologic pain. Anesthesiology 1971;35:409–19.
225. Nathan PW. On the pathogenesis of causalgia in peripheral nerve injuries. Brain 1947;70:145–70.
226. Sunderland M, Kelly A. The painful sequelae of injuries to peripheral nerves. Aust NZJ Surg 1948; 18:75.
227. Schwartzman RJ. Reflex sympathetic dystrophy. Cur Opin Neurol Neurosurg 1993;6:531–36.
228. Alojado ME, Ohta Y, Kemmotsu O. The effect of clonidine on the activity of neurons in the rat dorsal raphe nucleus *in vitro*. Anesth Analg 1994;79:257–60.
229. Desmeules JA, Kayser V, Weil-Fuggaza J, Bertrand A, Guilbaud G. Influence of the sympathetic nervous system in the development of abnormal pain-related behaviours in a rat model of neuropathic pain. Neuroscience 1995;67:941–51.

230. Iannone F, Trotta F, Montecucco C, Giacomelli R, Galeazzi M, Matucci-Cerinic M, Ferri C, Cutolo M, Maria Bambara L, Triolo G, Ferraccioli G, Valentini G, Lapadula G, GISEA (Gruppo Italiano per lo Studio delle Early Arthritis). Etanercept maintains the clinical benefit achieved by infliximab in patients with rheumatoid arthritis who discontinued infliximab because of side effects. Ann Rheum Dis 2007;66:249–52.
231. Mulleman D, Mammou S, Griffoul I, Watier H, Goupille P. Pathophysiology of disk-related low back pain and sciatica. II. Evidence supporting treatment with TNF-alpha antagonists. Joint Bone Spine: Revue du Rhumatisme 2006;73:270–77.
232. Sasaki N, Kikuchi S, Konno S, Sekiguchi M, Watanabe K. Anti-TNF-alpha antibody reduces pain-behavioral changes induced by epidural application of nucleus pulposus in a rat model depending on the timing of administration. Spine 2007;32:413–16.
233. Klein JB, Scherzer JA, Harding G, Jacobs AA, McLeish KR. TNFα stimulates increased plasma membrane guanine nucleotide binding protein activity in polymorphonuclear leukocytes. J Leukoc Biol 1995;57:500–6.
234. Reithmann C, Gierschik P, Jakobs KH, Werdan K. Regulation of adenylyl cyclase by noradrenaline and tumor necrosis factor-α in rat cardiomyocytes. Eur Heart J 1991; 12(F):139–42.
235. Scherzer JA, Lin Y, McLeish KR, Klein JB. TNF translationally modulates the expression of G_1 protein alpha(i2) subunits in human polymorphonuclear leukocytes. J Immunol 1997;158:913–18.
236. Chabre O, Conklin BR, Brandon S, Bourne HR, Limbird LE. Coupling of the alpha 2A-adrenergic receptor to multiple G-proteins. A simple approach for estimating receptor-G-protein coupling efficiency in a transient expression system. J Biol Chem 1994;269:5730–34.
237. Eason MG, Jacinto MT, Liggett SB. Contribution of ligand structure to activation of α_2-adrenergic receptor subtype coupling to G_s. Mol Pharmacol 1994;45:696–702.
238. Eason MG, Kurose H, Holt BD, Raymond JR, Liggett SB. Simultaneous coupling of α_2-adrenergic receptors to two G-proteins with opposing effects. J Biol Chem 1992;267:15795–801.
239. Pauwels PJ, Tardif S, Colpaert FC, Wurch T. Modulation of ligand responses by coupling of α_{2A}-adrenoceptors to diverse G_α-proteins. Biochem Pharmacol 2001;61:1079–92.
240. Kilts JD, Gerhardt MA, Richardson MD, Sreeram G, Mackensen GB, Grocott HP, White WD, Davis RD, Newman MF, Reves JG, Schwinn DA, Kwatra MM. β_2-Adrenergic and several other G protein-coupled receptors in human atrial membranes activate both G_s and G_i. Circ Res 2000;87:705–9.
241. Xiao R-P, Avdonin P, Zhou Y-Y, Cheng H, Akhter SA, Eschenhagen T, Lefkowitz RJ, Koch WJ, Lakatta EG. Coupling of β_2-adrenoceptor to G_i proteins and its physiological relevance in murine cardiac myocytes. Circ Res 1999;84:43–52.
242. Bartolini A, Ghelardini C, Fantetti L, Malcangio M, Malmberg-Aiello P, Giotti A. Role of muscarinic receptor subtypes in central antinociception. Br J Pharmacol 992;105:77–82.
243. Jasmin L, Tien D, Janni G, Ohara PT. Is noradrenaline a significant factor in the analgesic effect of antidepressants? Pain 2003;106:3–8.
244. Millan MJ, Bervoets K, Rivet JM, Widdowson P, Renouard A, Le Marouille-Girardon S, Gobert A. Multiple alpha-2 adrenergic receptor subtypes. II. Evidence for a role of rat R alpha-2A adrenergic receptors in the control of nociception, motor behavior, and hippocampal synthesis of noradrenaline. J Pharmacol Exp Ther 1994; 270:958–72.

245. Stoll G, Jander S, Schroeter M. Detrimental and beneficial effects of injury-induced inflammation and cytokine expression in the nervous system. Adv Exp Med Biol 2002;513:87–113.
246. Gelbard HA, Dzenko KA, DiLoreto D, del Cerro C, del Cerro M, Epstein LG. Neurotoxic effects of tumor necrosis factor alpha in primary human neuronal cultures are mediated by activation of the glutamate AMPA receptor subtype: implications for AIDS neuropathogenesis. Dev Neurosci 1993;15:417–22.
247. Meistrell III ME, Botchkina GI, Wang H, Di Santo E, Cockroft KM, Bloom O, Vishnubhakat JM, Ghezzi P, Tracey KJ. Tumor necrosis factor is a brain damaging cytokine in cerebral ischemia. Shock 1997;8:341–48.

Cytokines and the Brain
Edited by C. Phelps and E. Korneva

Interleukin-2 as a Neuroregulatory Cytokine

MARCO PRINZ, DENISE VAN ROSSUM, and UWE-KARSTEN HANISCH

Department of Neuropathology, University of Göttingen, Göttingen, Germany

ABSTRACT

Interleukin-2 (IL-2) has been widely known as a T-cell growth factor. Despite the hesitation to accept central nervous system (CNS) functions for a typical immunoregulatory factor, experimental work *in vitro*, *in situ*, and *in vivo* as well as clinical observations have been accumulating evidence for the functional expression of IL-2 receptors (IL-2R) on glial and neuronal cells. The distribution of mRNAs and proteins of the IL-2/IL-2R system in the CNS correlates with the demonstration of functional responses to IL-2 administrations. IL-2 may participate in the development of the CNS as a growth and differentiation factor. Strikingly potent effects on the release properties of certain neuronal and neurosecretory cell populations suggest neuromodulatory actions also in the mature brain. Regulating cholinergic transmission in the frontal cortex and hippocampal formation or stimulating the activity of the hypothalamic–pituitary–adrenal axis, IL-2 appears to exert regioselective and cell type-specific influences. Whether intrinsic IL-2 production is constitutively providing effective amounts of IL-2 is not yet convincingly demonstrated. Nevertheless, under certain conditions, circulating IL-2 may penetrate the blood–brain barrier. T cells invading the CNS during neuroinflammatory processes could also release IL-2. Yet some of the effects assigned to IL-2 may relate to IL-15, the cytokine sharing signal-transducing receptor subunits with IL-2. Nevertheless, the demonstration of nonredundant immune functions suggests that IL-2 and IL-15 exert individual activities in the CNS. Finally, therapeutic interventions aiming at a manipulation of IL-2 or IL-15 signaling or targeting cells by IL-2R or IL-15R molecules have to consider a CNS presence of the receptors and their coupling to glial and neuronal functions.

1. INTRODUCTION

Cytokines have originally been associated with hematopoietic cells and the physiology of the immune system. Nowadays there is no doubt as to the fact that these small signaling proteins have a much more universal spectrum of cellular sources and targets. Countless interactions between cells and tissues of the body are based on the exchange of cytokines and chemokines. They are produced and effective in many organs and tissues, and there is virtually no cell, which is not synthesizing or responding to a cytokine throughout its lifetime. It was still a longer process to accept that certain cytokines could play neurodevelopmental roles or would modulate neuronal activities in the mature central nervous system (CNS). The specific situation of the CNS in terms of its "immune privilege" – or better to say, the special control over and conditions for inflammatory processes and immune responses – certainly raised some hesitation about an intrinsic expression and day-to-day functioning of "typical" immunoregulatory peptides. Meanwhile, a large body of molecular and physiological evidence has proven the very presence of several cytokines and their cognate receptors in CNS tissues of human and laboratory animal origin.

While some cytokine systems may preferentially organize for the proliferation and differentiation of neural cell populations during development, others may, indeed, serve a signaling of neurons and glia also later in ontogeny. Some cytokines obviously participate in the neuro-immune-endocrine communication. They could act as soluble messengers that can be reciprocally exchanged between these systems – especially in situations when the body has to mount defensive measures against invasion by foreign material or neoplastic threats. Nevertheless, disturbances in the CNS homeostasis, such as trauma, infection, or degenerative processes, can apparently lead to a *de novo* or enhanced expression of critical cytokines and their receptors by both neuronal and non-neuronal cells. Pro-inflammatory cytokines and numerous chemokines are particularly responsible for the recruitment and local interactions of invading immune cells. At the same time, these factors seem to directly affect resident cells, with the consequences ranging from adequate functional adjustments to the exacerbation of destructive cascades. It thus comes also with considerable clinical interest when neurobiologists focus on the actions of cytokines.

Interleukin-2 (IL-2), one of the most intensively investigated cytokines, is mainly produced by T-helper cells upon antigenic challenges and has commonly been termed the T-cell growth factor. IL-2 thereby crucially participates in the mechanisms of host defence and immune responses. On the other hand, potent effects on neural (neuronal and glial) cells have been described as well. What is the physiological significance of a T-cell-regulating factor in the CNS? This question can only partially be answered. Even though most circulating (serum) proteins are normally denied CNS entry across the blood–brain barrier (BBB) certain cytokines, such as IL-2, can apparently penetrate the brain parenchyma in certain areas and may thus gain access to receptors on resident cell populations. IL-2 may, therefore, serve as one of the humoral mediators in neuro-immune communication. Whether transfer rates are sufficient for IL-2 to trigger substantial responses is controversial, but BBB impairment would certainly facilitate the BBB passage. Under pathophysiological conditions, infiltrating T cells could release the cytokine inside the CNS. On the other hand, molecular detection of IL-2-related mRNA and protein in neural cells and tissues points to some intrinsic production and allows to correlate inducible responses with a genuine expression of the cytokine and its receptors in selected CNS regions. Finally, the therapeutic use of IL-2 for cancer treatment as accompanied by sometimes markedly increased circulating amounts and tissue levels was reported to cause neurological and neuropsychiatric side effects.

In this chapter, we focus on evidence for the expression and function of the IL-2 system in the CNS. We also consider the closely related IL-15 system and discuss neuroregulatory features as well as the potential CNS consequences that could arise from their dysregulation and clinical manipulation. Regarding the immunobiology of IL-2 and IL-15 as well as their receptors, signalling, and diverse functions in the defence system, the reader may refer to a selection of recent reviews [1–5].

2. IL-2 AND IL-15: TWO CYTOKINES SHARING A HETEROMERIC RECEPTOR COMPLEX

Described 30 years ago, IL-2 was assigned its name according to the interleukin nomenclature in 1979 and cloned in 1983, allowing to relate several biological activities to a single protein [6–8]. As the most prominent function, IL-2 was found to serve as a T-cell growth factor during an immune response, driving the activation and clonal expansion of T cells upon antigenic stimulation, controlling their growth and death and influencing the formation of T-cell memory. While the T-cell-related functions were regarded major contributions, IL-2 also affected the growth and activities of other cell populations [9]. Indeed, IL-2 has activities for B cells, natural killer (NK) cells, lymphokine-activated killer (LAK) cells, macrophages, and monocytes as well as neutrophils. Yet the recent years reshaped the picture of IL-2 functions. Even though their spectrum seems to be more restricted in terms of the primarily affected T-cell population(s) and features, their importance for lymphoid homeostasis is not diminished [1].

Antigen- or mitogen-stimulated $CD4^+$ T-helper cells of Th1 type are the main source of IL-2, with additional signals being required for maximal production, but B cells can also synthesize significant quantities [10,11]. Parallel induction of the IL-2 receptor subunit-α (IL-2Rα) and its association with the IL-2R chains ß and γ create trimeric high-affinity IL-2 receptors through which the auto- and paracrine IL-2 effects are mediated. IL-2 itself is a 15 kD polypeptide of 133 amino acids, in the case of the human mature form [10,12]. Extensive sequence homology and structural conservation is noticed when comparing mammalian species [9,13,14]. The folding topology characterizes IL-2 as a member of the four-core α-helices bundle cytokine family [10,15–17].

The observation that IL-2 absence in respective knockout mice did not precipitate in the assumed immune deficiency stimulated speculations about an alternative factor with IL-2-like properties. The search for this factor led to the discovery of IL-15 ("IL-T"), revealing a 15 kD cytokine with the ability to bind to IL-2R [10,18–23]. Even though the primary structures are not closely related, the 3D models of IL-2 and IL-15 reveal similarities.

IL-2 and IL-15 share biological activities, especially in innate immunity [10,24]. Both also stimulate the growth of activated $CD4^+$ and/or $CD8^+$ as well as $CD4^-CD8^-$ T cells, promote cytolytic activity, including that of cytotoxic T and NK cells, and support B-cell differentiation and immunoglobulin synthesis. While overlapping in many aspects, also distinct effects have been attributed to each of these factors, occasionally causing even opposite outcomes on a given cell population [1,2,25]. Accordingly, IL-2 is rather critical for maintenance of self-tolerance through activation-induced cell death (AICD) in T cells and a control over regulatory T cells (Treg), a subpopulation of $CD4^+$ T cells. IL-2 now appears essential for the containment of T-cell activation and prevention of autoimmunity. Separation of IL-2-carried effects from those of other cytokines with a related receptor/signaling system refined its nonredundant activity repertoire. In contrast, IL-15 may exhibit rather antiapoptotic features [26]. It is important for proper $CD8^+$ memory cell and NK-cell functions, supporting homeostasis and maintenance of these populations. IL-15 was also reported to serve as a chemoattractant for T cells and to have growth factor activity for mast cells, features not obvious for IL-2 [23]. In striking contrast to IL-2, IL-15 is not synthesized by T cells. Apparently, IL-15 carries also some nonimmune activities. Demonstrations of anabolic effects in muscle tissue support such a broader involvement. Indeed, when compared to IL-2, IL-15 shows a rather different and much broader expression pattern. Accordingly, differences have been found as to the control of expression. While IL-2 expression is controlled by transcription and mRNA stabilization, the production of IL-15 is mainly regulated at a post-transcriptional level, including intracellular trafficking [10,23,27–32].

The similarities in some of the biological effects of IL-2 and IL-15 can be explained by their receptor mechanisms (Fig. 1). The principle of subunit sharing is a common theme among the cytokine receptors, but it is extensively used in the case of the IL-2R/IL-15R system. Specific receptors for IL-2 are formed by the di- and trimeric interaction of the structurally distinct

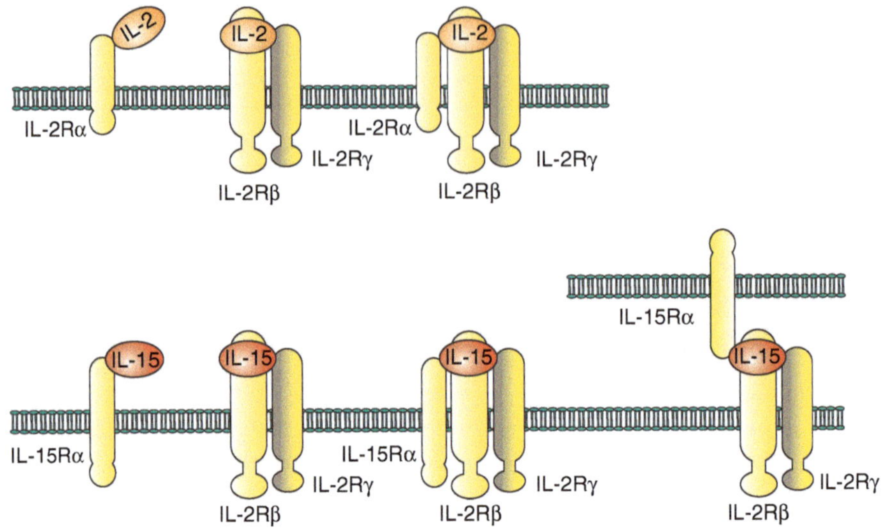

Figure 1. Schematic arrangement of the IL-2R and IL-15R. Four subunits constitute dimeric and trimeric receptor complexes for the signaling of the two cytokines. IL-2Rβ and IL-2Rγ (γ_c, common γ chain) are shared to serve the signal transduction while each cytokine has also a specific α subunit, that is, IL-2Rα and IL-15Rα. The cytosolic consequences of IL-2R and IL-15R stimulation are mediated through an array of kinases, including JAK1, JAK3, syk, lck, and routes of MAPK pathways. IL-15 may use additional systems for its signalling. Most notably, certain cells such as activated monocytes and dendritic cells may present IL-15Rα-bound IL-15 in a *trans*-configuration to cells which express the IL-2/IL-15Rβ and γ_c chain, such as $CD8^+$ T cells or NK cells. (Adopted from Ref. 10.)

IL-2R subunits α, β, and γ, which associate with additional proteins to constitute a fully functional signaling complex [9,10,12,23]. IL-2Rα (also known as p55, CD25, or Tac for "T-cell activation") binds IL-2 with only low affinity ($K_D \sim 10^{-8}$ M) and without any known cytosolic consequence. Soluble IL-2Rα version may affect the amount of circulating IL-2, while the membrane-inserted version controls the formation of high-affinity IL-2R when it joins the IL-2Rβγ heterodimer, the IL-2R of intermediate affinity ($K_D \sim 10^{-9}$ M). A number of cell types constitutively express IL-2Rß (p70/75, CD122) and IL-2Rγ (p64, CD132, γ_c for "common γ" chain as to its importance for receptors of IL-2, -4, -7, -9, -15, and -21), whereas IL-2Rα mostly depends on induction and reveals high turnover rates [9,15,23,33]. In the dimer, both IL-2Rß and IL-2Rγ seem to physically interact with IL-2, and both contribute to the signaling. Dimerization is required for effective binding as well as for the full delivery of a cytosolic signal. Trimeric IL-2Rαßγ then binds IL-2 with a K_D of about 10^{-11} M, representing a high-affinity full-function binding site. Its is known that reduced IL-2Rγ synthesis or signaling efficacy causes serious impairments of immune functions, and it should be stressed again that the γ chain is a building block of several cytokine receptors [9,10,15,17,34,35].

IL-2Rβγ accepts also IL-15 as a ligand, but recruitment of a specific ("private") IL-15Rα subunit is needed for the formation of a trimeric IL-15 receptor [23,25]. IL-15Rα, a 60 kD protein, is related to IL-2Rα but can bind IL-15 with high affinity ($K_D \sim 10^{-11}$ M) and independently of the signaling units IL-2Rβγ. IL-15Rα additionally stabilizes the IL-15·IL-2Rßγ association [36–38]. The differences in the molecular interactions between the ligands and the receptor subunits in the IL-2R *versus* IL-15R complexes as well as the presence of an intracellular domain in IL-15Rα may underlie the differences in the signaling consequences of IL-2 and IL-15 [23,39]. Indeed, the solution(s) to the question how two cytokines sharing the two signaling subunits of their receptor complexes organize for distinct signaling consequences and biological functions are far from trivial. A key to the functional diversity of IL-2 and IL-15 is seen in the differential expression patterns and regulation of IL-2Rα and IL-15Rα (namely in activated T and B cells, monocytes, and dendritic cells). Some signaling contribution of IL-15Rα could also underlie the functional diversity of IL-2 *versus* IL-15. As to their relative abundance, IL-15Rα shows a broader expression than IL-2Rα. Considering the constitutive expression of IL-2Rß and IL-2Rγ in certain cells IL-15 – itself revealing relatively widespread distribution – could probably dominate the employment of the shared receptors. However, substantial progress has been made in providing new insights into IL-15–IL-15R interactions. In a principle of *trans*-presentation, an IL-15Rα-expressing cell (e.g., monocytes, dendritic cells) can offer cell surface-bound IL-15 to a cell expressing IL-2/IL-15Rβ together with γ_c, thereby initiating signaling in the target (e.g., $CD8^+$ T cells, NK cells). IL-15 seems to be largely membrane bound, in contrast to the secreted IL-2, which binds to its di- or (preformed) trimeric receptors. Through *trans*-presentation, monocytes or dendritic cells could "hand-over" very efficiently even limited amounts of IL-15. Recycling incorporated IL-15–IL-15Rα complexes to the surface could thereby make further use of already presented ligand. In addition to the IL-2Rβγ–IL-15Rα complex, IL-15 was mentioned to signal through an alternative receptor, a 60–65-kD protein termed IL-15RX [23,40]. Discovered in mast cells, the receptor apparently allowed for IL-15 signaling in the absence of IL-2Rßγ/IL-15Rα.

Ligand binding to the IL-2R causes the phosphorylation of the β and γ subunits as well as of several other proteins [12]. As IL-2R itself lacks intrinsic enzymatic activity recruitment of protein kinases to and by the signaling subunits organizes for a very complex array of phosphorylation cascades that finally lead to multiple transcriptional events. Roles in the IL-2/IL-15 signaling have been proposed for phosphoinositide 3-kinase (PI3K), protein kinase C (PKC), Raf, Ras, mitogen-activated protein kinases $p44/42^{MAPK}$ (also ERK1/2), and soluble nonreceptor protein tyrosine kinases of the src family, such as p56lck or p72syk [9,12,41–47]. Prominent consequences of IL-2Rβγ stimulation relate to the activation of the JAK-STAT system, consisting of Janus kinases, which can phosphorylate signal transducers and activators of transcription [23,48–50]. STATs are latent transcription factors, which become activated by JAKs, subsequently dimerise and traffic to the nucleus. In the IL-2Rβγ signaling, JAK1 and JAK3 mediate an activation of STAT3 and STAT5. Interestingly, depending on the critical activation of JAK3 and the recruitment of a varying set of additional proximal signaling enzymes (such as Tyk2) and their substrates (e.g., STAT6), the receptor seems to trigger different downstream scenarios in different cell types. Still, several aspects of the complicated IL-2/IL-15R signaling remain enigmatic. Taking together, and as an emerging concept, signaling through the very same receptor subunits allows IL-2 and IL-15 to activate similar as well as still distinct cytosolic pathways which translate into individual effects on cell physiology depending on both the nature of the ligand and the type and status of a given cell [3,23].

3. EXPRESSION OF IL-2, IL-15, AND THEIR RECEPTORS IN THE CNS

Evidence for the expression of IL-2R and IL-15R in the CNS derives from both molecular detections and demonstrations of functional responses. Most of the studies revealing mRNA, protein, immunoreactive material, or binding sites were based on rodent brain tissue or cell culture models. On the other hand, clinical findings offered valuable hints as to the functional implications of IL-2 in the brain, especially when high-dose treatments of cancer patients were accompanied by neurological and neuropsychiatric side effects. As for other cytokines, IL-2 could be shown to affect the physiology and viability of neurons, astrocytes, microglia, oligodendrocytes, or endothelial cells as populations of these cells may constitutively or occasionally express the receptors.

Initial reports on the CNS expression of IL-2 and IL-2-binding structures date back to the 1980s when first studies claimed evidence for the presence of a functional IL-2/IL-2R system [51]. The very low expression levels challenged the technical limits of molecular detection at this time. Meanwhile, the molecular components required for building di- and trimeric IL-2R/IL-15R were all detected in brain tissues or neural cell culture, and receptor units have even been cloned from CNS material.

IL-2R are found in the frontal cortex, hippocampus, septum, striatum, cerebellum, and the locus coeruleus [51–54]. [^{125}I]-IL-2-binding sites are present with high density in the pyramidal cell layer of the hippocampal formation, suggestive of localization to neuronal somata. Stronger binding was also detected in the corpus callosum and the anterior commissure, association with the fibre tracts indicating binding to oligodendrocytes [54].

Immunodetection identified IL-2Rα in various brain regions, especially in the neuron-enriched layers of the hippocampus and dentate gyrus as well as the cerebellum [54,55]. Several reports point to an expression by microglia, astrocytes, oligodendrocytes, as well as neurons [56–65]. The apparently widespread and constitutive presence of message and immunoreactive material (protein) in the brain contrasts with the situation in the immune system in which the cellular expression of IL-2Rα strictly depends on induction, the only exception being Treg cells.

IL-2Rß is seen in the frontoparietal extension of the cortex, the caudate nucleus, hippocampus, amygdala, hypothalamus, thalamus, substantia nigra, cerebellum, and the corpus callosum [51,58,61,66]. Microglia, astrocytes, oligodendroglioma, and pituitary cells were reported for IL-2Rß synthesis [58,61,64,67].

IL-2Rγ appears to be present in cortical areas, the hippocampus, medulla oblongata, and the cerebellum. Microglia, oligodendroglioma cells, and neuronal populations have been mentioned as likely cellular carriers [44,58,61,68].

The action of cytokines on CNS cells expressing the respective receptors requires either their endogenous production or a transport across the BBB and blood–cerebrospinal fluid (CSF) border. Indeed, certain cytokines seem to penetrate the BBB at rates higher than those estimated from their physicochemical features. Transport mechanisms were postulated that allow circulating cytokines to enter the brain tissue even under normal conditions, when the integrity of the BBB is not impaired (for a review see [10]). Yet it is not clear whether the quantities are sufficient for causing cellular effects. However, cytokines may inundate the brain tissue when the BBB is disrupted, for example, due to mechanical damage or stroke or as a result of high circulating cytokine concentrations. In this regard, IL-2 itself has been reported to cause a vascular leak syndrome [10]. Alternatively, immune cells can invade the brain during inflammatory processes and could thereby temporarily serve as a local source. Cytokines of leukocytic origin would then not only serve the communication with endothelial or glial cells but also influence the functional properties of neurons with respective receptor expression.

A postulation of neurophysiological functions for the IL-2/IL-2R system in the healthy adult brain in general, however, depends on a demonstration of intrinsic IL-2 synthesis. An experimental proof of such a synthesis is not trivial because co-purification of immune-derived material can result from blood contamination. Early studies on rat brain material probably overestimated the tissue content of IL-2-like material [53] as became subsequently evident [10,52,56].

Immunocytochemical staining for IL-2 in rodent brains revealed the strongest signals in the frontoparietal extent of the cerebral cortex, preferentially layer IV, the caudate-putamen, lateral septum, pyramidal, and granule cell layers of the hippocampus and dentate gyrus, respectively, as well as in the arcuate nucleus/median eminence of the hypothalamus. Labeling was also identified in the interpenduncular nucleus, locus coeruleus, and the molecular layer of the cerebellum. Only weak staining was reported for the caudal regions of the cerebral cortex and the thalamus [51,54]. At a cellular level, the signal associated with perikarya [54], the neuronal staining pattern being supported by electron-microscopical and *in situ* hybridization findings made in the cerebral cortex, hippocampus, habenula, and the arcuate nucleus [51]. IL-2 mRNA was found in white and gray matter of rat, mouse, and human brain tissues and cells by several laboratories [56,69–71]. Even though endogenous cytokine synthesis by the resident populations, that is, neurons as well as glia, has thus been demonstrated in certain anatomical divisions of the adult healthy CNS *de novo* or enhanced production can also be observed upon experimental stimulation or may arise as a consequence of disturbed homeostasis. Activated glial cells, namely microglia and astrocytes, can then serve as inducible sources for a whole array of cytokines – local productions thereby covering also CNS tissue regions that normally are devoid of those molecules.

The anatomical data for the expression of IL-2 and its function-associated molecules, especially IL-2Rα, also indicated mismatches and suggested the parallel existence of related molecules or functional involvements apart from IL-2 signaling proper [10]. *Trans*-presentation principles, as now demonstrated for IL-15Rα, could explain isolated expression of a nonsignaling unit. The IL-2/IL-2R system of the nervous system may also differ in certain structural features from the much better known "immune" version, although full identity has been reported for CNS-derived cDNAs of receptor molecules (for a more detailed discussion see [10,51]). Still, some authors have reported on marked structural deviations and unusual species among transcripts and proteins. They probably result from alternative splicing, post-translational processing, or other tissue-specific modifications. IL-2 may also cross-interact with certain other receptor systems, due to structural similarities. Opioid-like actions were reported. Similarities between IL-2 and corticotropin-releasing factor (CRF) may add to the endocrine effects of the cytokine, which are well documented [10].

IL-15 and IL-15Rα are widely spread throughout the CNS, the distribution likely covering microglia, astrocytes, and neurons as cellular sources [44,58,72]. Studies on the anatomical and cellular expression of IL-15Rα, IL-2Rß, and IL-2Rγ revealed the constituents for trimeric IL-15R complexes – or their rather co-localized presence – in multiple regions of the mouse brain [58] and an alternative splice variant of IL-15 mRNA in certain microglial cells [73]. As a cell with lineage links to monocytes, microglia could represent a primary carrier for IL-15Rα-based presentation of IL-15 to potential recipients of its signaling capacity. The presence of IL-2/IL-15Rβ and γ_c, on the other hand, also points to a role as a target of IL-15 effects.

4. EFFECTS ON NEURAL CELL GROWTH, SURVIVAL, AND DIFFERENTIATION

IL-2 can support the survival and development of cortical, hippocampal, septal, striatal, cerebellar, and sympathetic neurons as well as neuroblastoma cells [59,65,74–81]. It also affects the morphology of neuronal cells *in vitro*, that is, neurite outgrowth and branching patterns.

Direct effects, such as on the morphology, can be distinguished from indirect, glia-mediated actions, namely on the viability. Reports on survival-promoting activities of IL-2 occasionally contrast with those on toxic effects, even for comparable culture models [10,74,76]. On the other hand, neurotoxicity *in vivo* likely involves contributions of (micro)glia and infiltrating immune cells [10,82–88] (Table 1).

Early observations of proliferative influences of IL-2 on neural cells concerned the inhibitory effects on [^{3}H]thymidine incorporation in oligodendrocytic progenitor cells [63]. These observations suggested a growth-controlling activity of IL-2 in conjunction with IL-1 in immature oligodendrocytes. Other authors discovered positive IL-2 effects on the proliferation and maturation of more mature oligodendrocytes [60,89,90]. Obviously, IL-2 could exert both negative and positive growth control, depending on the developmental stage of the cells. A series of studies reporting on the existence and enzymatic generation of an unusual dimeric IL-2 version also mentioned toxic effects for oligodendrocytes [1,91–93]. The authors suggested that the IL-2-like dimer could help outgrowing nerves to overcome the growth inhibition as organized by oligodendrocytes [94]. Interestingly, dimeric IL-2 had already been described earlier in samples of post-traumatic brain tissue but may not be exclusively a product of the CNS [10].

IL-2 modifies the proliferative and scar-forming activity of astroglia, a feature shared with many cytokines. This feature could be relevant for the early ontogenesis or after brain injury. Yet the literature on these IL-2 effects gives a controversial picture. Certain studies failed to detect any effect or left doubts as to the actual involvement of IL-2 [10].

Table 1. Effects of IL-2 on the growth and differentiation of neural cells, the firing activity of neuronal populations, animal behaviour, and the release of transmitters, neuropeptides and hormones

Cell growth and differentiation	Transmitter and hormone release
Survival of cortical neurons ↑	Hippocampal acetylcholine ↑/↓
Survival of hippocampal neurons ↑/↓	Frontal cortical acetylcholine ↓
Survival of septal neurons ↑	Striatal dopamine ↑
Survival of striatal neurons ↑	Dopamine in the nucleus accumbens ↓
Growth of pituitary (tumor) cells ↑/↓	Mesencephalic dopamine ↑/↓
Neurite outgrowth in hippocampal cells ↑	Hypothalamic noradrenaline ↓
Outgrowth of sympathetic neuritis ↑	Hippocampal 5-HT ↑
Axonal abnormalities and demyelination	Hypothalamic 5-HT ↑
Growth of oligodendrocytes ↑/↓	Pituitary ACTH ↑
Growth of astrocytes ↑	Plasma glucocorticoids ↑
Growth of microglia ↑	Pituitary PRL and TSH ↑
Electrophysiological and behavioral effects	Pituitary FSH, LH and GH ↓
Cortical sensory transmission ↓	Hypothalamic CRF ↑
Hippocampal Ca^{2+} currents ↓	CRF in the amygdala ↑
Hippocampal long-term potentiation ↓	Hypothalamic vasopressin ↑
Firing in the supraoptic nucleus ↑	Hypothalamic LHRH ↓
Firing in the paraventricular nucleus ↑	Hypothalamic somatostatin ↑
Locomotion and exploratory activity ↑/↓	Hypothalamic GHRH ↓
Sedation and sleep ↑	Microglial NO ↑

Arrows indicate induction or positive modulation (↑) and suppression (↓) of the respective function. For additional examples, more detail and a list of sources see Ref. 10.

Also for microglia, the macrophages of the brain, IL-2 offers growth support upon activation of the cells with lipopolysaccharide (LPS) [64]. Moreover, IL-2 seems to enhance the microglial release of NO, a feature requiring the presence of interferon-γ (IFN-γ) [95]. IL-15 was found to have similar but also partially distinct effects. It supported microglial survival but rather caused NO release suppression [58]. IL-15 may serve a kind of autocrine regulation because it could be produced by microglia while these cells express the full repertoire of receptor subunits [58]. IL-2 and IL-15 could also play some roles in a lymphocyte–microglial communication during inflammatory episodes. Microglial IL-15 could chemoattract T cells [96]. Invading T cells could then even enhance the production of IL-15 through the release of IFNγ [72]. IFNγ is a complex regulator of microglial chemokine production, which could reorganize the chemoattraction of immune cells during neuroinflammatory processes [97]. Under IFNγ, the chemoattractive pattern would shift its relative preferences from neutrophils toward monocytes/macrophages and favor Th2 over Th1 cells. IFNγ-enhanced IL-15 synthesis by microglia may fit into such a profile. IL-2 as another T-cell product could thereby act on activated microglia to further drive proliferative expansion [58,98].

5. EFFECTS ON THE FIRING ACTIVITY AND THE RELEASE PROPERTIES OF NEURONS

A whole series of experimental studies points to modulatory actions of IL-2 on several neurotransmitter systems [10,51]. The cytokine was demonstrated to control (to enhance or to suppress) the release or tissue levels of dopamine, acetylcholine, noradrenaline, as well as serotonin (5-HT) or – in some cases – their related metabolites [53,69,99–106]. Together these studies revealed strong evidence for the release-controlling functions because the demonstrated effects appeared to be specific for IL-2. Other cytokines tested in parallel did not exert the activity. The effects were found selective for the neurotransmitter type, and they were selective as to the anatomical region. Modulatory activities were found for neurons of a given brain structure, such as the frontal cortex, the hippocampus, striatum, or the hypothalamus, while populations of other areas did not respond the same way or at all. It should be emphasized that functional responses were thus far seen with topographic match to the regions either revealing expression of IL-2 or IL-2R or both. Most notably, release-regulating effects were sometimes obtained with extremely low concentrations of IL-2 (for a survey see [10]). On the other hand, IL-15 may cover similar neuronal influences although not yet being demonstrated.

Acetylcholine is probably the transmitter for which the release-controlling features of IL-2 were shown in most detail. IL-2 revealed a very potent, regioselective and receptor-dependent modulation of the release as evoked by K^+ or veratridine [53,54,101,104]. While low concentrations (<0.1 pM) caused an augmentation of transmitter release from hippocampal slices nanomolar doses clearly suppressed the evoked release, resulting in a biphasic dose–response relation. IL-2 is thus one of – or even – the most potent molecule ever reported to have an effect on the release performance of cholinergic neurons [10]. The phenomenon of a rundown is suggestive of receptor desensitisation and could result from an internalization of the ligand–receptor complex as demonstrated for non-neural cells [107,108]. For this model of hippocampal acetylcholine release, it was shown that the augmenting effects of IL-2 were likely to result from a more direct action of IL-2 on cholinergic nerve terminals while the inhibition was organized in a more indirect way which probably involved GABAergic interneurons [104]. Involvement of high-affinity IL-2R was already suggested by the subpicomolar IL-2 concentrations causing release augmentation. Substantial proof for an IL-2R-mediated effect was obtained from blocking experiments. Anti-IL-2Rα antibodies could interfere with both the negative and the positive release control by IL-2 [104]. Moreover, the overall anatomical pattern found for the release modulation indicated that IL-2 influences mainly addressed certain cholinergic projections rather than interneurons [10,101]. Interestingly, effects of IL-2 were only seen when the release of acetylcholine was primarily triggered by another agent, whereas the basal levels remained unaffected. In other words, IL-2 appears to be a modulator, rather than a stimulus of release activity. This feature was also described for other transmitter systems.

Dopamine release from mesencephalic cells *in vitro* was found to be under a similar biphasic IL-2 control [99]. Enhancement of K^+- or NMDA-evoked release was obtained with low picomolar concentrations. Low nanomolar doses led to attenuation. Noradrenaline and 5-HT are further transmitter candidates that show regional responses to IL-2 exposure. Using hypothalamic tissue preparations, IL-2 treatment resulted in a decreased noradrenaline release [109], whereas IL-2 increased 5-HT levels in the hypothalamus and hippocampus [102,105].

It is necessary to emphasize that all these studies still do not provide a complete map of the IL-2-regulated neurotransmission systems, and variations in the experimental settings do not yet allow to directly compare or to transfer findings between *in vitro* and *in vivo* situations, developmental stages or species. Nevertheless, the body of experimental proofs for a potent IL-2 action as a neuroregulatory cytokine is already too large to be ignored, even though the true physiological implications are not known yet.

Taking the IL-2-mediated acetylcholine release modulation in the hippocampus as an example, one may speculate that the cytokine could influence or even interfere with cognitive and memory functions, because this structure is known to be critical for learning and memory. There are, indeed, experimental and clinical findings in support of this notion. For example, IL-2 was reported for profound effects on electrophysiological properties of hippocampal neurons in varying animal species [10,110]. A classical study published in 1990 already described IL-2-mediated inhibition of LTP [111]. Systemic IL-2 treatment in mice caused impaired mnesic function, the decreased performance in behavioral tasks even correlating with neuronal losses in the hippocampus [12,87,112]. Some effects were mostly obtained with aged animals, pointing to an aggravation of latent functional impairment. On the other hand, chronic intracerebroventricular (i.c.v.) infusion of IL-2 in young adult rats resulted only in a transiently impaired memory task performance, which was – interestingly – accompanied by a transient alteration in cholinergic receptor systems of the hippocampus [85]. Still, IL-2 presence seems to be a prerequisite for normal hippocampal development as well as learning and memory function in adulthood [79–81,113].

The cholinergic deficit is one of the hallmarks in Alzheimer's disease (AD). In association with neurodegenerative plaques and neurofibrillary tangles, several cytokines, complement factors, and other immune-related markers have been detected in brains of AD patients, contributing to the concept of an inflammatory component in the neurodegenerative processes. IL-2(R) immunoreactive material and IL-2-binding sites are elevated in AD tissue [10]. Could an increased level of IL-2 or enhanced IL-2 signaling in the hippocampus thus contribute to a decreased hippocampal acetylcholine release? The electrophysiological, neurochemical, molecular biology, and behavioural approaches give good reason to assume interference with the cholinergic system and mnesic functions. There are also data proving for an interaction of IL-2 and cholinergic function in the cortex (see [10] for details). Moreover, the use of IL-2 in the treatment of certain types of tumours has frequently been accompanied by severe neurological and neuropsychiatric side effects, such as cognitive impairment, disorientation, or memory loss.

Systemic high-dose administrations of IL-2 thereby resulted in CSF concentrations high enough to be effective for inhibition of acetylcholine release under experimental conditions [101,114].

Parenchymal or i.c.v. injections of IL-2 provided insights into a whole variety of electrophysiological and behavioral responses. IL-2 administration into the locus coeruleus was found to induce strong sedative and soporific effects in laboratory animals [10,115–118]. Choosing the nucleus caudatus or the substantia nigra as the targets for stereotactic delivery, animals showed ipsilateral turning and asymmetric body posture, indicators for the dopaminergic component among the IL-2-affected transmitter systems [69,99,100,103]. IL-2 was even shown to trigger epileptiform episodes [115,119]. More recently, IL-2R/IL-15R gene deletion approaches have been introduced to unravel the neurobehavioral significance of the IL-2/IL-15 system [120].

Another major line of evidence for the release-controlling activities of IL-2 in the CNS relates to the neuroendocrine axis [10]. Direct cell contacts through innervation of glands and lymphoid organs and exchange of soluble factors, such as neurotransmitters, neuropeptides, hormones, and cytokines, have been demonstrated as the two communication principles by which neuro-immune-endocrine interactions are coordinated. In this communication – and largely through endocrine "chains of command" – CNS structures control and support the function of immune cells. The concept also suggests informational exchanges in the immune-to-CNS direction, namely when the brain has to be "informed" about a challenge of the defence system [121]. Indeed, during immune responses alterations in the neuronal discharge frequencies can be recorded from the hypothalamus, electrophysiological indication for the activation of vegetative and neuroendocrine centres [122].

IL-1α, IL-1ß, IL-2, IL-6, and tumor necrosis factor-α (TNFα) can regulate endocrine functions by affecting "higher" hierarchical structures, such as the hypothalamus [10,51,121]. The secretory activity of the pituitary gland is also known to be influenced by several cytokines. At both levels, cytokines may thus indirectly control the function of peripheral glands and target organs. The influences are complex because cytokines organize for regulatory networks by affecting different elements of the neuroendocrine axis simultaneously or by inducing other neuroendocrine factors. Down the road, the hypothalamic–pituitary–adrenocortical (HPA) function appears to be activated whereas functions of the thyroid gland and the gonads are inhibited.

IL-2 itself can modulate the release of the pituitary hormones and hypothalamic peptides, Met-enkephalin, ß-endorphin, adrenocorticotropic hormone (ACTH), prolactin (PRL), luteinizing hormone (LH), follicle-stimulating hormone (FSH), thyrotropic hormone (TSH), growth hormone (GH), CRF, GH-releasing hormone (GHRH), and somatostatin. IL-2 can thus influence peripheral glands through the pituitary [10,123,124], the hypothalamus, and other brain structures controlling pituitary and adrenal activities [125–129] but also directly, as in the case of the adrenal gland [130]. Mechanistic models have already been presented which nicely integrate the IL-2 system in the control of the HPA function (for detail see [10]). The work on cell cultures and tissue preparations found support in a number of animal studies. Systemic as well as direct CNS application of IL-2 demonstrated changes in the firing behavior of hypothalamic neurons in the supraoptic and paraventricular nuclei, structures involved in the control of water retention and HPA activity [10,131]. Injections or infusions of IL-2 thereby revealed not only acute responses of the endocrine systems, such as the HPA, but indicated chronic effects [10,82,83,85,102,132,133]. Plasma levels of IL-2 may occasionally increase along with an activation of the immune system. Scenarios with substantial elevations of circulating IL-2 amounts relate to transplant rejection and therapeutic infusion of IL-2 as an antitumor drug, but may also occur following peripheral nerve lesion [51]. Then, sufficient IL-2 amounts could gain access to IL-2-receptive cell populations in the pituitary as well as the hypothalamus [134,135]. What would be the endocrine impact in response to an IL-2 signal? Taking glucocorticoids as a product of the HPA activation, they would not only serve the needs of the defence system by adjusting metabolic and vegetative functions, they are also known for potent influences on immune activities [121,122]. Glucocorticoids likely affect the extent, the duration, and probably also the type of an immune reaction. Exerting an anti-inflammatory and immunosuppressive effect in general, glucocorticoids have a much more refined repertoire of suppressive and permissive influences. An endocrine feedback may help preventing hyperresponses of the immune system, would probably play a role in the termination of a response, or could also explain the phenomenon of sequential antigen competition. Already activated T cells are less affected by their suppressive action, supporting clonal selection. Activated B cells can even be stimulated. Such a discriminating control in conjunction with a circadian rhythm of HPA activity and the timed induction may have also some impact on the favoring of Th1 *versus* Th2 types of immune responses [136–138].

6. NEUROPATHOLOGICAL AND THERAPEUTIC IMPLICATIONS

Entering the brain from the circulation upon BBB breakdown or through some suggested transport, being delivered to the CNS tissue by a Trojan horse action of T lymphocytes during inflammatory or (auto)immune processes or resulting from a intrinsic production by activated glia, elevated levels of IL-2 may impact on CNS functions – and even tissue integrity. Next to its T-lymphocyte-related limb of activity (as just one of the immune effects of IL-2), direct toxicity for oligodendrocytes and disturbances of the firing mode and transmitter release in IL-2R-expressing neurons shall add to a rather complex cascade of transient or irreversible alterations. Several degenerative CNS diseases with inflammatory or (auto)immune contributions show alterations in the IL-2/IL-2R system. A link has even been drawn to psychiatric disorders. The data from studies showing cellular or *in vivo* responses to IL-2 exposure suggest that – whatever the cellular source of IL-2 would be – CNS consequences would be inevitable. As a factor in the communication between lymphocytes and glial populations and as a neuromodulatory agent, IL-2 could thus prove critical in the functional coordination of both resident neural cells as well as immune infiltrates.

Much of the understanding of the CNS facet of IL-2 pleiotropy came from its clinical use [25]. The original therapeutic employment promised positive results for the treatment of various types of tumors and was raising hope in the case of neoplasias not responding to conventional means. However, undesired side effects had to be noticed as well. Patients suffered from capillary leak syndrome, hypotension, edema, increased energy expenditure, massive protein breakdown, and various cardiovascular, hematological, hepatic, renal, gastrointestinal, pulmonary, and dermatological abnormalities [10,25]. Most notably with respect to the IL-2R on neural cells, adverse neurological effects and neuropsychiatric symptoms were listed among the common complications with occasionally treatment-limiting severity. For example, headache, motor weakness and somnolence, memory impairment, cognitive failure, confusion and disorientation, anxiety, or paranoid delusions were mentioned [51]. Some of the CNS complications did not resolve upon termination of an IL-2 treatment, including disturbances of neuroendocrine activity and memory function, or even developed into progressive brain injury [139–142]. Similar observations were made in animal models [10,82–85,87].

IL-15, by its suggested pro-inflammatory profile and inhibitory outcome on self-tolerance, might upon dysregulation play disease-promoting roles and, for example, participate in autoimmune disorders. IL-15 signaling thus represents an interesting target for therapeutic intervention. IL-2R-directed strategies are considered for a treatment of certain leukemias or were designed for interference with transplant rejection. Protocols for therapeutic manipulations of the IL-2/IL-15R system have been improved based on the experience made with the focus on IL-2, offering now novel and adapted applications. Nevertheless, still much has to be learned about the potential CNS tissue impact and the subtle neurophysiological engagement of IL-2/IL-15 signaling.

At this point, it should be at least mentioned that strategies aiming at a suppression of IL-2R signaling in immune cells by using toxic antibodies against IL-2Rα molecules have to bear in mind that brain structures could get under "friendly fire" when coincidental BBB impairment would allow for CNS inundation. Those antibody tools, that is, antibodies coupled to toxins, target at the IL-2Rα chain as the inducible subunit conferring high affinity on the IL-2Rβγ complex. This strategy could be effective in eliminating malignant T cells or T cells that drive allograft rejection. On the contrary, while being strictly dependent on induction in most immune cells, the IL-2Rα chain seems to be constitutively present in multiple brain regions. Maybe, the wide distribution of the various building bocks of the IL-2/IL-15R system in the nervous tissues will gain more attention when more of the neurophysiological and neuropathophysiological implications of IL-2 – an especially IL-15 – emerge.

ACKNOWLEDGMENT

This work was supported by the German Research Foundation (DFG, SFB507 for UKH).

REFERENCES

1. Waldmann TA. The biology of interleukin-2 and interleukin-15: implications for cancer therapy and vaccine design. Nat Rev Immunol 2006;6(8):595–601.
2. Ma A, Koka R, Burkett P. Diverse functions of IL-2, IL-15, and IL-7 in lymphoid homeostasis. Annu Rev Immunol 2006;24:657–79.
3. Budagian V, Bulanova E, Paus R, Bulfone-Paus S. IL-15/IL-15 receptor biology: a guided tour through an expanding universe. Cytokine Growth Factor Rev 2006;17(4):259–80.
4. Bulfone-Paus S, Bulanova E, Budagian V, Paus R. The interleukin-15/interleukin-15 receptor system as a model for juxtacrine and reverse signaling. Bioessays 2006;28(4):362–77.
5. Fehervari Z, Yamaguchi T, Sakaguchi S. The dichotomous role of IL-2: tolerance versus immunity. Trends Immunol 2006;27(3):109–11.
6. Morgan DA, Ruscetti FW, Gallo R. Selective in vitro growth of T lymphocytes from normal human bone marrows. Science 1976;193(4257):1007–08.
7. Devos R, Plaetinck G, Cheroutre H, Simons G, Degrave W, Tavernier J et al. Molecular cloning of human interleukin 2 cDNA and its expression in E. coli. Nucleic Acids Res 1983;11(13):4307–23.
8. Taniguchi T, Matsui H, Fujita T, Takaoka C, Kashima N, Yoshimoto R et al. Structure and expression of a cloned cDNA for human interleukin-2. Nature 1983;302(5906):305–10.

9. Gaffen SL, Goldsmith A, Greene WC. Interleukin-2 and the interleukin-2 receptor. In The Cytokine Handbook. Thompson A, Ed; SanDiego: Academic Press, 1998; pp. 73–103.
10. Hanisch UK. Effects of interleukin-2 and interferons on the nervous system. In: Ader R, Felten DL, Cohen N, Eds; San Diego: Academic Press, 2001; pp. 585–631.
11. Kakiuchi T, Tamura T, Gyotoku Y, Nariuchi H. IL-2 production by B cells stimulated with a specific antigen. Cell Immunol 1991;138(1):207–15.
12. Ellery JM, Nicholls PJ. Alternate signalling pathways from the interleukin-2 receptor. Cytokine Growth Factor Rev 2002;13(1):27–40.
13. Mai Z, Kousoulas KG, Horohov DW, Klei TR. Cross-species PCR cloning of gerbil (Meriones unguiculatus) interleukin-2 cDNA and its expression in COS-7 cells. Vet Immunol Immunopathol 1994;40(1):63–71.
14. McKnight AJ, Classon BJ. Biochemical and immunological properties of rat recombinant interleukin-2 and interleukin-4. Immunology 1992;75(2):286–92.
15. Theze J. Cytokine receptors: a combinative family of molecules. Eur Cytokine Netw 1994;5(4):353–68.
16. Bazan JF. Unraveling the structure of IL-2. Science 1992;257(5068):410–13.
17. Davies DR, Wlodawer A. Cytokines and their receptor complexes. FASEB J 1995;9(1):50–56.
18. Bamford RN, Grant AJ, Burton JD, Peters C, Kurys G, Goldman CK et al. The interleukin (IL) 2 receptor beta chain is shared by IL-2 and a cytokine, provisionally designated IL-T, that stimulates T-cell proliferation and the induction of lymphokine-activated killer cells. Proc Natl Acad Sci USA 1994;91(11):4940–44.
19. Burton JD, Bamford RN, Peters C, Grant AJ, Kurys G, Goldman CK et al. A lymphokine, provisionally designated interleukin T and produced by a human adult T-cell leukemia line, stimulates T-cell proliferation and the induction of lymphokine-activated killer cells. Proc Natl Acad Sci USA 1994;91(11):4935–39.
20. Carson WE, Giri JG, Lindemann MJ, Linett ML, Ahdieh M, Paxton R et al. Interleukin (IL) 15 is a novel cytokine that activates human natural killer cells via components of the IL-2 receptor. J Exp Med 1994;180(4):1395–1403.
21. Giri JG, Ahdieh M, Eisenman J, Shanebeck K, Grabstein K, Kumaki S et al. Utilization of the beta and gamma chains of the IL-2 receptor by the novel cytokine IL-15. EMBO J 1994;13(12):2822–30.
22. Grabstein KH, Eisenman J, Shanebeck K, Rauch C, Srinivasan S, Fung V et al. Cloning of a T cell growth factor that interacts with the beta chain of the interleukin-2 receptor. Science 1994;264(5161):965–68.
23. Waldmann TA, Tagaya Y, Bamford RN. Interleukin-2, interleukin-15, and their receptors. Int Rev Immunol 1998;16:205–26.
24. Waldmann TA, Dubois S, Tagaya Y. Contrasting roles of IL-2 and IL-15 in the life and death of lymphocytes: implications for immunotherapy. Immunity 2001;14(2):105–10.
25. Fehniger TA, Cooper MA, Caligiuri MA. Interleukin-2 and interleukin-15: immunotherapy for cancer. Cytokine Growth Factor Rev 2002;13(2):169–83.
26. Marks-Konczalik J, Dubois S, Losi JM, Sabzevari H, Yamada N, Feigenbaum L et al. IL-2-induced activation-induced cell death is inhibited in IL-15 transgenic mice. Proc Natl Acad Sci USA 2000;97(21):11445–50.
27. Bamford RN, Battiata AP, Burton JD, Sharma H, Waldmann TA. Interleukin (IL) 15/IL-T production by the adult T-cell leukemia cell line HuT-102 is associated with a human T-cell lymphotrophic virus type I region /IL-15 fusion message that lacks many upstream

AUGs that normally attenuates IL-15 mRNA translation. Proc Natl Acad Sci USA 1996;93(7):2897–902.
28. Bamford RN, Battiata AP, Waldmann TA. IL-15: the role of translational regulation in their expression. J Leukoc Biol 1996;59(4):476–80.
29. Tagaya Y, Bamford RN, DeFilippis AP, Waldmann TA. IL-15: a pleiotropic cytokine with diverse receptor/signaling pathways whose expression is controlled at multiple levels. Immunity 1996;4(4):329–36.
30. Bamford RN, DeFilippis AP, Azimi N, Kurys G, Waldmann TA. The 5′ untranslated region, signal peptide, and the coding sequence of the carboxyl terminus of IL-15 participate in its multifaceted translational control. J Immunol 1998;160(9):4418–26.
31. Onu A, Pohl T, Krause H, Bulfone-Paus S. Regulation of IL-15 secretion via the leader peptide of two IL-15 isoforms. J Immunol 1997;158(1):255–62.
32. Tagaya Y, Kurys G, Thies TA, Losi JM, Azimi N, Hanover JA et al. Generation of secretable and nonsecretable interleukin 15 isoforms through alternate usage of signal peptides. Proc Natl Acad Sci USA 1997;94(26):14444–49.
33. Theze J, Alzari PM, Bertoglio J. Interleukin 2 and its receptors: recent advances and new immunological functions. Immunol Today 1996;17(10):481–86.
34. Russell SM, Keegan AD, Harada N, Nakamura Y, Noguchi M, Leland P et al. Interleukin-2 receptor gamma chain: a functional component of the interleukin-4 receptor. Science 1993;262(5141):1880–83.
35. Noguchi M, Nakamura Y, Russell SM, Ziegler SF, Tsang M, Cao X et al. Interleukin-2 receptor gamma chain: a functional component of the interleukin-7 receptor. Science 1993;262(5141):1877–80.
36. Anderson DM, Kumaki S, Ahdieh M, Bertles J, Tometsko M, Loomis A et al. Functional characterization of the human interleukin-15 receptor alpha chain and close linkage of IL15RA and IL2RA genes. J Biol Chem 1995;270(50):29862–69.
37. de Jong JL, Farner NL, Widmer MB, Giri JG, Sondel PM. Interaction of IL-15 with the shared IL-2 receptor beta and gamma c subunits. The IL-15/beta/gamma c receptor-ligand complex is less stable than the IL-2/beta/gamma c receptor-ligand complex. J Immunol 1996;156(4):1339–48.
38. Giri JG, Kumaki S, Ahdieh M, Friend DJ, Loomis A, Shanebeck K et al. Identification and cloning of a novel IL-15 binding protein that is structurally related to the alpha chain of the IL-2 receptor. EMBO J 1995;14(15):3654–63.
39. Dubois S, Mariner J, Waldmann TA, Tagaya Y. IL-15Ralpha recycles and presents IL-15 In trans to neighboring cells. Immunity 2002;17(5):537–47.
40. Tagaya Y, Burton JD, Miyamoto Y, Waldmann TA. Identification of a novel receptor/signal transduction pathway for IL-15/T in mast cells. EMBO J 1996;15(18):4928–39.
41. Brennan P, Babbage JW, Burgering BM, Groner B, Reif K, Cantrell DA. Phosphatidylinositol 3-kinase couples the interleukin-2 receptor to the cell cycle regulator E2F. Immunity 1997;7(5):679–89.
42. Gomez J, Garcia A, Borlado L, Bonay P, Martinez A, Silva A et al. IL-2 signaling controls actin organization through Rho-like protein family, phosphatidylinositol 3-kinase, and protein kinase C-zeta. J Immunol 1997;158(4):1516–22.
43. Maslinski W, Remillard B, Tsudo M, Strom TB. Interleukin-2 receptor signal transduction: translocation of active serine-threonine kinase Raf-1 from IL-2 receptor into cytosol depends on IL-2-induced tyrosine kinase activation. Transplant Proc 1993; 25(1 Pt 1):109–10.

44. Adunyah SE, Wheeler BJ, Cooper RS. Evidence for the involvement of LCK and MAP kinase (ERK-1) in the signal transduction mechanism of interleukin-15. Biochem Biophys Res Commun 1997;232(3):754–58.
45. Hatakeyama M, Kawahara A, Mori H, Shibuya H, Taniguchi T. c-fos gene induction by interleukin 2: identification of the critical cytoplasmic regions within the interleukin 2 receptor beta chain. Proc Natl Acad Sci USA 1992;89(6):2022–26.
46. Ihle JN. Cytokine receptor signalling. Nature 1995;377(6550):591–94.
47. Kirken RA, Malabarba MG, Xu J, DaSilva L, Erwin RA, Liu X et al. Two discrete regions of interleukin-2 (IL2) receptor beta independently mediate IL2 activation of a PD98059/rapamycin/wortmannin-insensitive Stat5a/b serine kinase. J Biol Chem 1997; 272(24):15459–65.
48. Johnston JA, Bacon CM, Finbloom DS, Rees RC, Kaplan D, Shibuya K et al. Tyrosine phosphorylation and activation of STAT5, STAT3, and Janus kinases by interleukins 2 and 15. Proc Natl Acad Sci USA 1995;92(19):8705–09.
49. Lai SY, Xu W, Gaffen SL, Liu KD, Longmore GD, Greene WC et al. The molecular role of the common gamma c subunit in signal transduction reveals functional asymmetry within multimeric cytokine receptor complexes. Proc Natl Acad Sci USA 1996; 93(1):231–35.
50. Zhu MH, Berry JA, Russell SM, Leonard WJ. Delineation of the regions of interleukin-2 (IL-2) receptor beta chain important for association of Jak1 and Jak3. Jak1-independent functional recruitment of Jak3 to Il-2Rbeta. J Biol Chem 1998; 273(17): 10719–25.
51. Hanisch UK, Quirion R. Interleukin-2 as a neuroregulatory cytokine. Brain Res Brain Res Rev 1995;21(3):246–84.
52. Araujo DM, Lapchak PA. Induction of immune system mediators in the hippocampal formation in Alzheimer's and Parkinson's diseases: selective effects on specific interleukins and interleukin receptors. Neuroscience 1994;61(4):745–54.
53. Araujo DM, Lapchak PA, Collier B, Quirion R. Localization of interleukin-2 immunoreactivity and interleukin-2 receptors in the rat brain: interaction with the cholinergic system. Brain Res 1989;498(2):257–66.
54. Lapchak PA, Araujo DM, Quirion R, Beaudet A. Immunoautoradiographic localization of interleukin 2-like immunoreactivity and interleukin 2 receptors (Tac antigen-like immunoreactivity) in the rat brain. Neuroscience 1991;44(1):173–84.
55. Luber-Narod J, Rogers J. Immune system associated antigens expressed by cells of the human central nervous system. Neurosci Lett 1988;94(1–2):17–22.
56. Arzt E, Stelzer G, Renner U, Lange M, Muller OA, Stalla GK. Interleukin-2 and interleukin-2 receptor expression in human corticotrophic adenoma and murine pituitary cell cultures. J Clin Invest 1992;90(5):1944–51.
57. Arzt E, Buric R, Stelzer G, Stalla J, Sauer J, Renner U et al. Interleukin involvement in anterior pituitary cell growth regulation: effects of IL-2 and IL-6. Endocrinology 1993;132(1):459–67.
58. Hanisch UK, Lyons SA, Prinz M, Nolte C, Weber JR, Kettenmann H et al. Mouse brain microglia express interleukin-15 and its multimeric receptor complex functionally coupled to Janus kinase activity. J Biol Chem 1997;272(46):28853–60.
59. Haugen PK, Letourneau PC. Interleukin-2 enhances chick and rat sympathetic, but not sensory, neurite outgrowth. J Neurosci Res 1990;25(4):443–52.
60. Otero GC, Merrill JE. Cytokine receptors on glial cells. Glia 1994;11(2):117–28.

61. Otero GC, Merrill JE. Molecular cloning of IL-2R alpha, IL-2R beta, and IL-2R gamma cDNAs from a human oligodendroglioma cell line: presence of IL-2R mRNAs in the human central nervous system. Glia 1995;14(4):295–302.
62. Petitto JM, Huang Z. Molecular cloning of the coding sequence of an interleukin-2 receptor alpha subunit cDNA in murine brain. J Neuroimmunol 1995;59(1–2):135–41.
63. Saneto RP, Altman A, Knobler RL, Johnson HM, de Vellis J. Interleukin 2 mediates the inhibition of oligodendrocyte progenitor cell proliferation in vitro. Proc Natl Acad Sci USA 1986;83(23):9221–25.
64. Sawada M, Suzumura A, Marunouchi T. Induction of functional interleukin-2 receptor in mouse microglia. J Neurochem 1995;64(5):1973–79.
65. Shimojo M, Imai Y, Nakajima K, Mizushima S, Uemura A, Kohsaka S. Interleukin-2 enhances the viability of primary cultured rat neocortical neurons. Neurosci Lett 1993;151(2):170–73.
66. Satoh J, Kurohara K, Yukitake M, Kuroda Y. Interleukin-15, a T-cell growth factor, is expressed in human neural cell lines and tissues. J Neurol Sci 1998;155(2):170–77.
67. Petitto JM, Huang Z, Rinker CM, McCarthy DB. Isolation of IL-2 receptor-beta cDNA clones from AtT-20 pituitary cells: constitutive expression and role in signal transduction. Neuropsychopharmacology 1997;17(2):57–66.
68. Petitto JM, Huang Z, Raizada MK, Rinker CM, McCarthy DB. Molecular cloning of the cDNA coding sequence of IL-2 receptor-gamma (gammac) from human and murine forebrain: expression in the hippocampus in situ and by brain cells in vitro. Brain Res Mol Brain Res 1998;53(1–2):152–62.
69. Lapchak PA. A role for interleukin-2 in the regulation of striatal dopaminergic function. Neuroreport 1992;3(2):165–68.
70. Ashman RB, Bolitho EM, Fulurija A. Cytokine mRNA in brain tissue from mice that show strain-dependent differences in the severity of lesions induced by systemic infection with Candida albicans yeast. J Infect Dis 1995;172(3):823–30.
71. Eizenberg O, Faber-Elman A, Lotan M, Schwartz M. Interleukin-2 transcripts in human and rodent brains: possible expression by astrocytes. J Neurochem 1995;64(5):1928–36.
72. Lee YB, Satoh J, Walker DG, Kim SU. Interleukin-15 gene expression in human astrocytes and microglia in culture. Neuroreport 1996;7(5):1062–66.
73. Prinz M, Hanisch UK, Kettenmann H, Kirchhoff F. Alternative splicing of mouse IL-15 is due to the use of an internal splice site in exon 5. Brain Res Mol Brain Res 1998;63(1):155–62.
74. Awatsuji H, Furukawa Y, Nakajima M, Furukawa S, Hayashi K. Interleukin-2 as a neurotrophic factor for supporting the survival of neurons cultured from various regions of fetal rat brain. J Neurosci Res 1993;35(3):305–11.
75. Moroni SC, Rossi A. Enhanced survival and differentiation in vitro of different neuronal populations by some interleukins. Int J Dev Neurosci 1995;13(1):41–49.
76. Sarder M, Saito H, Abe K. Interleukin-2 promotes survival and neurite extension of cultured neurons from fetal rat brain. Brain Res 1993;625(2):347–50.
77. Yao H, Wang FZ, Ding AS, Liu ZW, Ling SG. [Effects of interleukin-1 and interleukin-2 on electrophysiological characteristics of rat hippocampal neurons in culture]. Sheng Li Xue Bao 1994;46(6):539–45.
78. Mennicken F, Quirion R. Interleukin-2 increases choline acetyltransferase activity in septal-cell cultures. Synapse 1997;26(2):175–83.

79. Petitto JM, McNamara RK, Gendreau PL, Huang Z, Jackson AJ. Impaired learning and memory and altered hippocampal neurodevelopment resulting from interleukin-2 gene deletion. J Neurosci Res 1999;56(4):441–46.
80. Petitto JM, Huang Z, Lo J, Beck RD, Rinker C, Hartemink DA. Relationship between the development of autoimmunity and sensorimotor gating in MRL-lpr mice with reduced IL-2 production. Neurosci Lett 2002;328(3):304–8.
81. Beck RD, Jr., King MA, Huang Z, Petitto JM. Alterations in septohippocampal cholinergic neurons resulting from interleukin-2 gene knockout. Brain Res 2002; 955(1–2):16–23.
82. Hanisch UK, Rowe W, Sharma S, Meaney MJ, Quirion R. Hypothalamic-pituitary-adrenal activity during chronic central administration of interleukin-2. Endocrinology 1994; 135(6): 2465–72.
83. Hanisch UK, Rowe W, van Rossum D, Meaney MJ, Quirion R. Phasic hyperactivity of the HPA axis resulting from chronic central IL-2 administration. Neuroreport 1996;7(18):2883–88.
84. Hanisch UK, Neuhaus J, Quirion R, Kettenmann H. Neurotoxicity induced by interleukin-2: involvement of infiltrating immune cells. Synapse 1996;24(2):104–14.
85. Hanisch UK, Neuhaus J, Rowe W, van Rossum D, Moller T, Kettenmann H et al. Neurotoxic consequences of central long-term administration of interleukin-2 in rats. Neuroscience 1997;79(3):799–818.
86. Hanisch UK. Microglia as a source and target of cytokines. Glia 2002;40(2):140–55.
87. Nemni R, Iannaccone S, Quattrini A, Smirne S, Sessa M, Lodi M et al. Effect of chronic treatment with recombinant interleukin-2 on the central nervous system of adult and old mice. Brain Res 1992;591(2):248–52.
88. Sessa M, Nemni R, Iannaccone S, Quattrini A, Confalonieri V, Canal N. In vivo modulation of myelin gene expression by human recombinant IL-2. Brain Res Mol Brain Res 1992;12(4):331–34.
89. Benveniste EN, Merrill JE. Stimulation of oligodendroglial proliferation and maturation by interleukin-2. Nature 1986;321(6070):610–13.
90. Benveniste EN, Herman PK, Whitaker JN. Myelin basic protein-specific RNA levels in interleukin-2-stimulated oligodendrocytes. J Neurochem 1987;49(4):1274–79.
91. Eitan S, Zisling R, Cohen A, Belkin M, Hirschberg DL, Lotan M et al. Identification of an nterleukin 2-like substance as a factor cytotoxic to oligodendrocytes and associated with central nervous system regeneration. Proc Natl Acad Sci USA 1992;89(12): 5442–46.
92. Eitan S, Schwartz M. A transglutaminase that converts interleukin-2 into a factor cytotoxic to oligodendrocytes. Science 1993;261(5117):106–8.
93. Eizenberg O, Kaplitt MG, Eitan S, Pfaff DW, Hirschberg DL, Schwartz M. Linear dimeric interleukin-2 obtained by the use of a defective herpes simplex viral vector: conformation-activity relationship. Brain Res Mol Brain Res 1994;26(1–2):156–62.
94. Eitan S, Solomon A, Lavie V, Yoles E, Hirschberg DL, Belkin M et al. Recovery of visual response of injured adult rat optic nerves treated with transglutaminase. Science 1994;264(5166):1764–68.
95. Sakai N, Kaufman S, Milstien S. Parallel induction of nitric oxide and tetrahydrobiopterin synthesis by cytokines in rat glial cells. J Neurochem 1995;65(2):895–902.
96. Wilkinson PC, Liew FY. Chemoattraction of human blood T lymphocytes by interleukin-15. J Exp Med 1995;181(3):1255–59.

97. Häusler KG, Prinz M, Nolte C, Weber JR, Schumann RR, Kettenmann H et al. Interferon-gamma differentially modulates the release of cytokines and chemokines in lipopolysaccharide-and pneumococcal cell wall-stimulated mouse microglia and macrophages. Eur J Neurosci 2002;16(11):2113–22.
98. Petitto JM, Huang Z, Lo J, Streit WJ. IL-2 gene knockout affects T lymphocyte trafficking and the microglial response to regenerating facial motor neurons. J Neuroimmunol 2003;134(1–2):95–103.
99. Alonso R, Chaudieu I, Diorio J, Krishnamurthy A, Quirion R, Boksa P. Interleukin-2 modulates evoked release of [3H]dopamine in rat cultured mesencephalic cells. J Neurochem 1993;61(4):1284–90.
100. Anisman H, Kokkinidis L, Merali Z. Interleukin-2 decreases accumbal dopamine efflux and responding for rewarding lateral hypothalamic stimulation. Brain Res 1996;731(1–2):1–11.
101. Hanisch UK, Seto D, Quirion R. Modulation of hippocampal acetylcholine release: a potent central action of interleukin-2. J Neurosci 1993;13(8):3368–74.
102. Pauli S, Linthorst AC, Reul JM. Tumour necrosis factor-alpha and interleukin-2 differentially affect hippocampal serotonergic neurotransmission, behavioural activity, body temperature and hypothalamic-pituitary-adrenocortical axis activity in the rat. Eur J Neurosci 1998;10(3):868–78.
103. Petitto JM, McCarthy DB, Rinker CM, Huang Z, Getty T. Modulation of behavioral and neurochemical measures of forebrain dopamine function in mice by species-specific interleukin-2. J Neuroimmunol 1997;73(1–2):183–90.
104. Seto D, Kar S, Quirion R. Evidence for direct and indirect mechanisms in the potent modulatory action of interleukin-2 on the release of acetylcholine in rat hippocampal slices. Br J Pharmacol 1997;120(6):1151–57.
105. Song C, Leonard BE. Interleukin-2-induced changes in behavioural, neurotransmitter, and immunological parameters in the olfactory bulbectomized rat. Neuroimmunomodulation 1995;2(5):263–73.
106. Zalcman S, Green-Johnson JM, Murray L, Wan W, Nance DM, Greenberg AH. Interleukin-2-induced enhancement of an antigen-specific IgM plaque-forming cell response is mediated by the sympathetic nervous system. J Pharmacol Exp Ther 1994; 271(2):977–82.
107. Fung MR, Ju G, Greene WC. Co-internalization of the p55 and p70 subunits of the high-affinity human interleukin 2 receptor. Evidence for a stable ternary receptor complex. J Exp Med 1988;168(5):1923–28.
108. Legrue SJ, Sheu TL, Chernajovsky Y. The role of receptor-ligand endocytosis and degradation in interleukin-2 signaling and T-lymphocyte proliferation. Lymphokine Cytokine Res 1991;10(6):431–36.
109. Lapchak PA, Araujo DM. Interleukin-2 regulates monoamine and opioid peptide release from the hypothalamus. Neuroreport 1993;4(3):303–6.
110. Plata-Salaman CR, ffrench-Mullen JM. Interleukin-2 modulates calcium currents in dissociated hippocampal CA1 neurons. Neuroreport 1993;4(5):579–81.
111. Tancredi V, Zona C, Velotti F, Eusebi F, Santoni A. Interleukin-2 suppresses established long-term potentiation and inhibits its induction in the rat hippocampus. Brain Res 1990;525(1):149–51.
112. Bianchi M, Panerai AE. Interleukin-2 enhances scopolamine-induced amnesia and hyperactivity in the mouse. Neuroreport 1993;4(8):1046–48.
113. Beck RD, Jr., Wasserfall C, Ha GK, Cushman JD, Huang Z, Atkinson MA et al. Changes in hippocampal IL-15, related cytokines, and neurogenesis in IL-2 deficient mice. Brain Res 2005;1041(2):223–30.

114. Saris SC, Patronas NJ, Rosenberg SA, Alexander JT, Frank J, Schwartzentruber DJ et al. The effect of intravenous interleukin-2 on brain water content. J Neurosurg 1989;71(2):169–74.
115. De Sarro G, Rotiroti D, Audino MG, Gratteri S, Nistico G. Effects of interleukin-2 on various models of experimental epilepsy in DBA/2 mice. Neuroimmunomodulation 1994;1(6):361–69.
116. De Sarro GB, Nistico G. Behavioral and electrocortical power spectrum effects after intracerebral microinfusion of interleukin-2 in rats are antagonized by naloxone. Int J Neurosci 1990;51(3–4):209–10.
117. Nistico G, De Sarro G. Is interleukin 2 a neuromodulator in the brain? Trends Neurosci 1991;14(4):146–50.
118. Nistico G, De Sarro G. Behavioral and electrocortical spectrum power effects after microinfusion of lymphokines in several areas of the rat brain. Ann NY Acad Sci 1991;621:119–34.
119. Nistico G. Communications among central nervous system, neuroendocrine and immune systems. Prog Neurobiol 1993;40:463–75.
120. Petitto JM, Huang Z, Hartemink DA, Beck R. IL-2/15 receptor-beta gene deletion alters neurobehavioral performance. Brain Res 2002;929(2):218–25.
121. Besedovsky HO, del Rey A. Immune-neuro-endocrine interactions: facts and hypotheses. Endocr Rev 1996;17(1):64–102.
122. Besedovsky HO, del Rey A. Introduction: immune-neuroendocrine network. Front Horm Res 2002;29:1–14.
123. Karanth S, McCann SM. Anterior pituitary hormone control by interleukin 2. Proc Natl Acad Sci USA 1991;88(7):2961–65.
124. Smith LR, Brown SL, Blalock JE. Interleukin-2 induction of ACTH secretion: presence of an interleukin-2 receptor alpha-chain-like molecule on pituitary cells. J Neuroimmunol 1989;21(2–3):249–54.
125. Cambronero JC, Rivas FJ, Borrell J, Guaza C. Interleukin-2 induces corticotropin-releasing hormone release from superfused rat hypothalami: influence of glucocorticoids. Endocrinology 1992;131(2):677–83.
126. Karanth S, Aguila MC, McCann SM. The influence of interleukin-2 on the release of somatostatin and growth hormone-releasing hormone by mediobasal hypothalamus. Neuroendocrinology 1993;58(2):185–90.
127. Karanth S, Lyson K, McCann SM. Role of nitric oxide in interleukin 2-induced corticotropin-releasing factor release from incubated hypothalami. Proc Natl Acad Sci USA 1993;90(8):3383–87.
128. Raber J, Bloom FE. IL-2 induces vasopressin release from the hypothalamus and the amygdala: role of nitric oxide-mediated signaling. J Neurosci 1994;14(10):6187–95.
129. Raber J, Koob GF, Bloom FE. Interleukin-2 (IL-2) induces corticotropin-releasing factor (CRF) release from the amygdala and involves a nitric oxide-mediated signaling; comparison with the hypothalamic response. J Pharmacol Exp Ther 1995;272(2):815–24.
130. Tominaga T, Fukata J, Naito Y, Usui T, Murakami N, Fukushima M et al. Prostaglandin-dependent in vitro stimulation of adrenocortical steroidogenesis by interleukins. Endocrinology 1991;128(1):526–31.
131. Bartholomew SA, Hoffman SA. Effects of peripheral cytokine injections on multiple unit activity in the anterior hypothalamic area of the mouse. Brain Behav Immun 1993;7(4):301–16.

132. McCann SM, Lyson K, Karanth S, Gimeno M, Belova N, Kamat A et al. Role of cytokines in the endocrine system. Ann NY Acad Sci 1994;741:50–63.
133. McCann SM, Karanth S, Kamat A, Les DW, Lyson K, Gimeno M et al. Induction by cytokines of the pattern of pituitary hormone secretion in infection. Neuroimmunomodulation 1994;1(1):2–13.
134. Banks WA, Kastin AJ. The interleukins-1 alpha, -1 beta, and -2 do not acutely disrupt the murine blood-brain barrier. Int J Immunopharmacol 1992;14(4):629–36.
135. Waguespack PJ, Banks WA, Kastin AJ. Interleukin-2 does not cross the blood-brain barrier by a saturable transport system. Brain Res Bull 1994;34(2):103–9.
136. Constant SL, Bottomly K. Induction of Th1 and Th2 CD4+ T cell responses: the alternative approaches. Annu Rev Immunol 1997;15:297–322.
137. Petrovsky N, Harrison LC. Th1 and Th2: swinging to a hormonal rhythm. Immunol Today 1995;16(12):605.
138. Ramierz F, Fowell DJ, Puklavec M, Simmonds S, Mason D. Glucocorticoids promote a TH2 cytokine response by CD4+ T cells in vitro. J Immunol 1996;156(7):2406–12.
139. Denicoff KD, Durkin TM, Lotze MT, Quinlan PE, Davis CL, Listwak SJ et al. The neuroendocrine effects of interleukin-2 treatment. J Clin Endocrinol Metab 1989; 69(2):402–10.
140. Meyers CA, Yung WK. Delayed neurotoxicity of intraventricular interleukin-2: a case report. J Neurooncol 1993;15(3):265–67.
141. Ravaud A, Negrier S, Cany L, Merrouche Y, Le Guillou M, Blay JY et al. Subcutaneous low-dose recombinant interleukin 2 and alpha-interferon in patients with metastatic renal cell carcinoma. Br J Cancer 1994;69(6):1111–14.
142. Vecht CJ, Keohane C, Menon RS, Punt CJ, Stoter G. Acute fatal leukoencephalopathy after interleukin-2 therapy. N Engl J Med 1990;323(16):1146–47.

Cytokines and the Brain
Edited by C. Phelps and E. Korneva

*Cytokines and Extracellular Matrix Remodeling in the Central Nervous System**

MARZENNA WIRANOWSKA[1] and ANNA PLAAS[2]

[1]*Departments of Pathology and Cell Biology and Internal Medicine*
[2]*Department of Internal Medicine (Rheumatology) and Department of Biochemistry, Rush University Medical Center, Chicago, IL, USA*

ABSTRACT

The extracellular space consists of approximately a quarter of the volume of the central nervous system (CNS), which is otherwise mostly occupied by cellular elements and blood vessels. This space is filled with various ions, transmitters, metabolites, peptides, and extracellular matrix (ECM) molecules produced by neurons and glia. The best-known ECM domain in brain was recently visualized as "interstitial clefts," consisting of astrocytic processes enveloped by basal lamina with ECM molecules detected within the clefts. Both the basal lamina and the ECM serve as depots for cytokines and growth factors and both contain proteoglycans, important molecules with the capacity to bind cytokines and growth factors. The ECM components and their ability to store cytokines and growth factors are important in CNS cell development, such as stem cell cytogenesis, synaptic plasticity but also in CNS pathology. It is postulated that most stem cell niches are composed of ECM and other noncellular components that regulate stem cell control and allow for homologous cell–cell interactions as well as stem cell migration toward a damaged tissue. Thus, ECM serves as an immediate source of cytokines and growth factors during normal CNS development, as well as during CNS injury and inflammation. In addition, recent studies have shown that an enzymatic degradation of ECM components may result in uncontrolled diffusion of cytokines and growth factors as well as the activation of the ECM molecules, for example, laminin, therefore contributing to CNS pathology. The information contained in this chapter describes a number of ECM components in the CNS and their role in both normal physiological conditions and CNS injury. The interactions of cytokines and ECM molecules during CNS injury is presented using examples of CNS disease involving either autoimmune (multiple sclerosis), bacterial (meningitis), viral (HIV dementia), or neoplastic (glioma) pathogenic stimuli.

1. INTRODUCTION

Extracellular matrix (ECM) components are highly specialized assemblies of macromolecules that provide for the structural cohesiveness and dynamic physiochemical properties of a tissue or organ. In the central nervous system (CNS) cellular elements and blood vessels consist of

*This Chapter is dedicated to the memory of Dr. Christopher Phelps

approximately 80% of the total CNS volume, and the remaining 15–25% is the extracellular space [1]. This space contains various ions, transmitters, metabolites, peptides, neurohormones, and molecules of the ECM produced by neurons and glia. While neurons can communicate and interact both by synapses and by diffusion of ions in the ECM, the glia rely entirely on the diffusion of ions in the ECM as a mode of communication. The majority of identified ECM components could belong to (1) fibrous elements, for example, collagen; (2) link proteins, for example, fibronectin or laminin; or (3) space-filling molecules, for example, glycosaminoglycans (GAGs) [2]. ECM macromolecular components of the CNS that have been studied include tissue-specific collagens, non-collagenous glycoproteins such as laminin and fibronectin link proteins, heparan sulfate (HS) and chondroitin sulfate (CS) proteoglycans, and hyaluronan. Several decades of biochemical research and more recent gene manipulation approaches in mice have resulted in acquisition of comprehensive information on the structural characteristics and tissue-specific distributions of these molecules in the CNS, as well as the functional importance of their ability to form intermolecular contacts with each other and with receptors on the cell surface [3–5].

In the CNS, as in all other tissues and organs, ECM functions to maintain *structural integrity* and *metabolic homeostasis* which includes adaptive remodeling during neuronal function. Specifically, processes such as transport of neuro-regulatory substances from the sites of production to the sites of action, provision of surfaces on which cells can migrate, adhere, form intercellular junctions, and produce myelin sheaths, provision of high affinity binding domains for growth factors and cytokines by serving as their ligands on cell surface receptors, as well as providing an environment for stem cells during their cytogenesis.

ECM components also play critical roles in the generation of CNS pathologies, such as localized inflammation and injuries due to viral, bacterial, autoimmune, neurodegenerative insults, or participate in host defense during neoplastic transformations. Thus, cytokine and growth factor-driven activation of disease-specific signal transduction pathways and their intracellular cross-talk circuits involves (1) the engagement of transmembrane receptors such as integrins or CD44 with their ECM ligands, cytokines or growth factors, (2) the direct activation of a regulatory influence on CNS cells, (3) the provision of migratory pathways for invasive neoplastic cells, and (4) the mediation of activation of stem cells in neoplastic diseases of the CNS. These pathological conditions, during which ECM molecules work in concert with cytokines, could affect glia cells and the vascular bed within the brain parenchyma that ultimately participate in CNS pathology.

In this review, we will discuss the current knowledge of ECM localization and its specific anatomical CNS domains, which play a role in either synaptic plasticity or serve as a reservoir for growth factors and cytokines, for example, as they may function in cytogenesis of stem cells (Section 2); we present the structural and functional characteristics of the main classes of ECM molecules with emphasis on those known to serve as ligands for cytokines in the adult brain (Section 3); we describe the participation of ECM molecules and cytokines in the regulation of synaptic plasticity (Section 4); and ECM involvement in remodeling and interaction of cytokines during the CNS diseases (Section 5). The examples of CNS diseases to be discussed include: autoimmune (multiple sclerosis [MS]), viral (HIV dementia), and bacterial (meningitis), as well as neoplastic disease (cerebral gliomas).

2. ECM LOCALIZATION IN BRAIN

Until recently, the most well-known ECM domain in the brain was believed to be the "*perineuronal nets*," described by Camillo Golgi in 1893 as "enwrapping the cell bodies and proximal dendrites of certain neurons in the adult mammalian CNS where they represent supportive and protective scaffolding." The application of molecular biology and immunohistochemical methods have revealed that these structures are indeed composed of ECM molecules that fill the space between the glial processes and the nerve cells that they outline.

Attempts to visualize a higher order of perineuronal net organization led to a hypothesis that some nerve cells are surrounded by three almost-concentric nets. Two of them are extracellular and composed by intercalated *astrocytic processes* in the first net and the interposed ECM molecules between astrocytic processes and the neuronal surface in the second net. However, the third net is currently believed to be intracellular and consisting of a *membrane cytoskeleton* composed of *spectrin* and *anykrin* [2,6].

The current model for the structure of perineuronal nets proposes a more heterogenous content of the CNS, including recently described "*interstitial clefts*" present within the brain. Interstitial clefts consist of astrocytic walls enveloped by basal lamina, (an ECM structure) and the ECM molecules detected within these clefts, all of which are thought to function as depots for cytokines and growth factors [7]. As described by Brightman [8], interstitial clefts vary in size, shape, and their content of ECM and basal lamina. The clefts and the configuration of their astrocytic walls may be variable depending on the brain region. For example, in normal midbrain and also in astroglial scars, astrocytic processes of the glia limitans form thin, parallel, concentric sheets, which comprise the walls of narrow interstitial clefts. These long clefts are not as branched as the reticular network of clefts characteristic of gray matter neuropil. The geometry of the clefts and the specialization of their content such as ECM and basal lamina are relevant in the movement of solutes (e.g., ions, neurotransmitters, hormones, and cytokines), along the clefts [8]. At places where the astroglial and meningeal cell processes or lamella are thicker than a critical thickness of approximately 20–30 nm, *endocytotic* or *transcytotic pits* could be formed. However, below this thickness the astrocytic cell processes do not invaginate and do not form transcytotic vesicles and large hydrophilic solutes cannot pass across the thin portion of a cell process. Consequently, solutes within the clefts may remain arrested for relatively long time and perhaps stay there indefinitely if they cannot be removed by a receptor-mediated transcytosis. The width and content of interstitial clefts found in the mature brain was reported to be different than that of the fetal brain, which permit cell migration and outgrowth of neurites. In contrast, in the mature brain the interstitial clefts are narrower, but they are still able to permit solute convection. However, cell migration through these clefts could not occur unless their ECM content is degraded enzymatically [8].

While the bulk of the normal CNS parenchyma may not provide much space for ECM, some regions of the brain are rich in ECM. For example, in the *circumventricular organs* (CVOs), which are supplied by fenestrated, permeable capillaries with no *blood–brain barrier* (BBB), the perivascular ECM is focally enlarged around arterioles and venules [8]. These vessels are surrounded by stromal connective tissue space [8] and they are lined by subepithelial and endothelial basal laminae. In addition to CVOs, the ECM in the CNS is the most abundant in several other areas such as the subarachnoid space, perivascular spaces, subependymal pockets, and also at sites of injury [8]. *Heparan sulfate proteoglycans* (HSPGs) are the common components found in the ECM and basal lamina, as well as in ependymal, astroglial, and endothelial interfaces of the CNS [8]. It has been suggested that the HSPGs of the basal lamina and ECM serve as a storage site for growth factors and cytokines [7].

The basal lamina of the brain appears as a homogenous layer, but in fact, it consists of several layers. These include (1) *pia intima* (or pia-glia), a component of the pia mater which consists of fine reticular and elastic fibers that together are regarded as the external limiting membrane of the CNS (ependyma is the internal limiting membrane); (2) *subpial astrocytic feet* (part of glia limitans) which form a continuous cytoplasmic sheath beneath the pia intima; and (3) an *osmophilic basement membrane* located between the two other mentioned layers. As blood vessels enter or leave the brain parenchyma, they carry with them pia-glia and *arachnoid* which form a cuff around the vessel. The space between the blood vessel and its "sheath" that it carries with it (pia-glia and arachnoid) is designated as *Virchow-Robin space*. As we now know, under normal conditions there is no real space between those layers (they eventually become continuous when an artery dives into the brain substance). At the capillary level of the blood vessel there are no "adventitial" elements around the capillaries, only astrocytic endfeet surrounding the basement membrane of the capillary endothelium.

Recently, two novel forms of brain basal lamina were described by Mercier et al. [9]. One form of basal lamina, which is located in the most neurogenic area of the adult brain, the subependymal layer of the lateral ventricles, was termed "*fractones*." The second form of basal lamina that was found between capillaries in dense capillary beds was called an "*anastomotic*" *basal lamina*. Fractones (at the former location) were found to consist of labyrinthine basal lamina projecting from subependymal blood vessels, terminating beneath ependyma, and being contacted directly by numerous processes of astrocytes and microglial cells. The "anastomotic" basal lamina was found to enclose extraparenchymal cells that networked with the perivascular cells coursing in the sheaths of adjacent blood vessels. The "anastomotic" basal lamina, which is also overlaid by macrophages was found to contain numerous collagen fibrils. In addition, Mercier et al. [9] found that basal lamina located at the pial surface formed labyrinthine tube-like structures enclosing numerous fibroblasts and *astrocytic endfeet*, with pouches of collagen fibrils at the interface between the two cell types. Mercier et al. suggested that cytokines and growth factors produced by connective tissue cells might concentrate in these basal lamina, where they may interact with ECM proteins promoting biological processes, for example, cytogenesis and morphological changes of the stem cells [9].

2.1. Brain ECM and stem cells

The first cellular and molecular elements of the adult brain niche for *neural stem cells* have been identified. The elements of this microenvironment, beside the stem cells, include the endothelial cells of the vasculature, ECM components, and basal lamina [10].

It was considered, however, that some stem cell niches do not contain the heterologous cells but that they are composed of only ECM and other noncellular components which regulate stem cell control and allow for homologous cell–cell interactions. The recent data suggest that there is an association between the stem cells and the endothelial cells within the niche. For example, it was reported that the soluble factors released by the endothelial cells induced activation of neural precursor receptor, Notch, leading to self-renewal of neural stem cells [11]. The interaction of stem cells with the basement membrane (alias basal lamina) includes stem cells residing on the basement membrane. In addition, the basement membrane participates in creating the specialized microenvironment by providing a possible venue for stem cell movement and shifting of their location. The subependymal basal lamina of the medial wall of the lateral ventricle (subventricular zone), the most neurogenic area of the adult brain, also known as "fractones" [9], lies in proximity to the stem cells [8]. This subependymal basal lamina which contains HSPG was suggested to sequester trophic factors and cytokines. The cytokines and growth factors could easily diffuse from there having a short distance to travel with the possibility of achieving a high local concentration before reaching the stem cells [12–14].

The subventricular zone is known to be an important germinal layer that forms during development [14]. Moreover, the subventricular zone persists into adulthood, retaining the capacity to generate both neurons and glia. The ECM, including growth factors and their receptors which are contained within the subventricular zone, regulate the primitive cells and define their niches. For example, the receptor for the epidermal growth factor (EGFR) which cooperates with two other neural progenitor cell-surface receptors, such as Notch and β-1 integrins, was suggested to play an important role in recognition of the niche that surrounds stem cells [15]. Studies have shown that EGF and basic fibroblast growth factor (bFGF), which are possibly stored locally in the ECM, have a stimulatory effect on cells residing in the subventricular zone. Thus, while situated within the ECM, growth factors could then stimulate the cells of the subventricular zone to retain their capacity for self-renewal and generation of neurons and glia [13,14]. Moreover, it has been reported that growth factors and cytokines as well as their receptors are not only stored in the ECM throughout the life of an individual (fetal to adult), but they also may be expressed by cells in the developing fetal brain and in the adult brain. For example, interneurons during neocortical development, express CXC chemokine receptor 4 (CXCR4). The same receptor is also found in adult brain and plays a role in adaptive neuronal plasticity and cerebral leukocyte recruitment following a brain lesion [16]. Under homeostatic conditions, chemokines and chemokine receptors were shown to be expressed constitutively in the brain and to be responsible for promotion of quiescence and survival of the neural progenitor cells. While induced into quiescence, these cells retained their multipotential ability to develop into neurons or astrocytes. However, the inhibitory effects of chemokines on neural progenitor cells could be regulated by ECM in the brain, for example, by HS which was shown to block these inhibitory effects of chemokines [17]. Another example of endogenously expressed signaling molecules by cells in the developing fetal brain was the recent finding of approximately 132 genes representing five receptor-defined signaling pathways differentially expressed by oligodendrocyte progenitor cells [18]. The principal components of these signaling pathways included, for example, platelet-derived growth factor receptor-α (PDGFRA), type 3 fibroblast growth factor receptor (FGFR3), receptor tyrosine phosphatase-beta/zeta (RTPZ), Notch, and syndecan 3. Differential overexpression of RTPZ was associated with that of several ECM molecules such as pleiotropin, tenascin, and CS proteoglycans. This suggested that RTPZ signaling the pathway may regulate the self-maintenance of adult human white matter progenitor cells and it can also be modulated to induce differentiation into oligodendrocytes and astrocytes [18].

It is well known that the ECM environment of the CNS plays an important regulatory role during development and adulthood, providing signals for cell growth, differentiation, and migration. Disruption of ECM interactions can cause severe developmental defects [12,19,20]. During CNS development, ECM not only defines functional boundaries for cells, but it is also involved in signaling following an injury [12,20]. These subventricular regions contain ECM of a well-defined composition [12,20], and therefore, the matrix molecules with cytokines and growth factors residing within may be a part of regulatory system for CNS stem cells. This long-time known concept of ECM regulating neural stem cells was recently supported by yet another new finding related to the role of tenasin-C affecting precursor populations. The embryonic neural stem cells following their switch from neural to glial precursor production program become localized within the subventricular region where the ECM molecules, for example, highly expressed tenasin-C, play a role in regulation of this program. For example, the transgenic mice lacking tenasin-C were shown to have altered response to some growth factors, for example, FGF2 and delayed acquisition of EGF receptor needed for stem cells development, as mentioned earlier. Also, in these mice, the total number of the CNS stem cells was altered with an increased number of stem cells destined to predominantly develop into neurons. Furthermore, it was found that tenasin-C was essential in regulation of the developmental program of oligodendrocyte precursor cells (OPCs). Therefore, it was concluded that the molecules like tenasin-C present within ECM in the subventricular zone contributes to generation of a stem cell "niche" and coordination of neural stem cell development [21–23].

It was found that certain ECM components play an important role in regulating motility and velocity of the multipotent stem/progenitor cells and their clonal structures – neurospheres. In a study by Kearns, et al. [12], laminin and fibronectin were found to stimulate the motility of these cells with CS proteoglycans being inhibitory. In addition, it was shown that tenascin R (TN-R), the ECM glycoprotein expressed in the adult mouse olfactory bulb, initiates and directs migration of neuroblasts toward that structure [24]. Interestingly, tenascin was reported to be expressed by cells of CNS tumors [25]. The neural stem cells, similarly to glioma cells, can be induced by many growth factors to migrate. In that regard, Heese et al. [26] showed that scatter factor/hepatocyte growth factor (SF/HGF) was the most powerful chemoattractant for neural stem cells in vitro. In addition, it was recently suggested that neural progenitor precursor cells can be triggered to migrate and to take part in reparative processes following activation of matrix metalloproteinases (MMPs) by pro-inflammatory cytokines. However, these progenitor cells are attenuated by interferon-β (IFN-β) [27]. These findings identifying ECM microenvironment components capable of stimulating and/or directing migration of stem cells may have important implications for the development of therapeutic strategies for brain repair, as well as in understanding the origins of invasive brain tumors.

The role(s) of EGF that may be locally stored in brain ECM and the subependymal basal lamina, along with other growth factors and cytokines has high significance. For example, it was reported that, there was a dramatic expansion of proliferating stem cells in the subventricular zone following infusion of EGF into the lateral ventricle *in vivo* [14,28] and after EGF treatment of adult rat subventricular zone neural precursors *in vitro* [29]. A similar observation was made in the case of brain tumor cells showing an expression of naturally occurring constitutively active variant type III of the epidermal growth factor receptor (EGFRvIII). EGFRvIII was shown to affect glioma cell motility through an increase in the number of cellular protrusions (lamellipodia) [30]. These observed similarities between neural stem cells and brain tumor cells, in respect to ECM/growth factors interaction, could be explained in part, by the discovery of a novel, non-neural cell, present in the developing and mature brain. These cells are often referred as oligodendrocyte progenitor cells (OPCs/NG2) because of their ability to differentiate into oligodendrocytes from oligodendrocyte/type-2 astrocyte (O-2A) progenitors [25]. The adult OPC cells were shown to continue their replication *in vivo* [10,14]. This recently recognized cell population of glial cells expresses the transmembrane CS proteoglycan, called neuro-glial 2 (NG2). The significance of NG2 as a lineage marker and a provision of cellular recognition source for some gliomas is currently being considered. The NG2 cell population interacts with laminin, tenascin, and certain collagen types, suggesting a role in the modulation of glioma cell adhesion to various ECM components. It was also found that NG2 cells negatively modulate the adhesion of melanoma cells through CD44 adhesion molecule, hyaluronic acid (HA), fibronectin, and some types of integrins [25].

Recent reports suggest a functional relationship between stem cells and tumor cells based on their interactive properties with ECM components and their coexistence within malignancies [31]. Also various *in vivo* and *in vitro* studies reported that neural stem cells had a tendency to migrate toward gliomas [26]. Only a minority of human cancer cells was shown to have the ability to form tumors in immunocompromised mice, and these few tumorigenic stem cells were identified as being positive for the HA receptor CD44. The genotypic and phenotypic connection between CNS stem cells residing in the ECM niche, which stores growth factors and cytokines, and the glioma cells which, similarly to CNS stem cells, require ECM rich environment needs to be further established.

2.2. Basal lamina, CNS vascular bed, and blood–brain barrier

The endothelial cells lining the blood vessels of the CNS contain tight junctions, and they are the primary site of the BBB. Under physiological conditions, the BBB, unlike other vascular beds, controls the exchange of water soluble compounds, selectively transports certain solutes and nutrients, while preventing a free diffusion of blood-borne molecules into the brain parenchyma and vice versa. Therefore, the brain endothelium is a dynamic biological system and not a rigid barrier, as originally thought [32]. The endothelial cells, which are at the interface between blood and brain, play a critical role in transport of nutrients, leukocyte trafficking, and receptor-mediated signaling. The structures of the brain vessels responsible for the filtering of solutes are the glycocalyx, located on the surface of the endothelial cells, and the subendothelial basal lamina [7]. It has been observed, that ions, neurotransmitters, and hormones may leave the CSF within cerebral ventricles, to enter the interstitial clefts between the cells lining the ventricles, but a fraction of these solutes may be retained by the subendothelial basal lamina. These solutes may then be released from the basal lamina in a regulated way. Consequently, basal lamina, similar to the ECM, serves as a storage site for some of these neuro-regulatory molecules [7,33].

Under pro-inflammatory conditions, BBB permeability increases and allows plasma constituents to escape into brain tissue [34]. The edema fluid infiltrates the brain parenchyma and penetrates freely in the white matter between nerve fibers where there are no extracellular junctions. In gray matter where an elaborated network of anatomical structures exists, the fluid expansion is limited. Between myelinated nerve fibers, edema fluid accumulates in separate compartments of the extracellular space [35]. The events occurring during the development of an inflammatory lesion in the CNS result from an active involvement of the endothelial cell basement membrane of the BBB. The endothelial basement membranes contain laminin 8 and/or 10 and their expression is influenced by proinflammatory cytokines. It was found that the endothelial cell laminin isoforms, laminin 8 and 10, play a decisive role in T-cell recruitment across the BBB in experimental autoimmune encephalomyelitis (EAE) [36]. When T cells migrate into the CNS during EAE, they encounter two distinct basement membranes (bm): the endothelial bm which contains laminins 8 and 10 and the parenchymal bm which contains laminins 1 and 2. However, it was found that characteristic inflammatory cuffs (accumulation of leukocytes outside blood vessels) occurred exclusively around endothelial basement membranes containing laminin 8, whereas in the presence of laminin 10 no infiltration of lymphocytes was detectable [36]. This finding clearly points to the active role of the brain vascular matrix components in the process of inflammation. Moreover, in addition to its active involvement in the development of inflammation, the matrix of the brain vascular system may also play a role in a storage and regulated release of growth factors and inflammatory cytokines.

2.3. ECM in brain as a storage place for growth factors and cytokines

It has been known for quite a while that the subendothelial ECM of the large blood vessels, for example, aorta, and that of small blood vessels, for example, capillaries (as they occur in the brain), is capable of storing growth factors [33]. For example, FGF, which is involved in autocrine growth promotion, was suggested to be synthesized and incorporated into the subendothelial ECM of blood vessels [33]. An increasing number of growth factors including insulin-like growth factor (IGF), transforming growth factor-β (TGF-β), HGF were found to be associated with ECM proteins and in particular with HS, a GAG in the ECM [37]. It was concluded that signaling by growth factors can be regulated outside of cells by ECM proteins and proteolytic enzymes. Consequently, rapid and localized changes in the activity of these growth factors are induced following their release from ECM storage. This serves to control cell proliferation/angiogenesis, differentiation, and remodeling/synthesis of the ECM by these growth factors. Therefore, ECM being a storage place for many growth factors plays a major role in the control of growth factor signaling. HSPG, a component of ECM, has been repeatedly implicated in high affinity binding of growth factors [33,38,39]. For example, it was reported that FGF is bound to HS in the basement membranes of cornea and released specifically by treatment with heparin, HS, or heparanase, but not CS or chondroitinase. Therefore, it was concluded that basement membrane and specifically the HS domain may serve as physiologic storage depots for angiogenic molecules such as FGF, making it accessible to the vascular endothelium, for example, in the brain [38].

A similar observation related to ECM being a storage site for growth factors is also true for cytokines [40]. Again, a number of cytokines were shown to be interacting strongly with HS and the exact mode and specificity of binding to HS has been further elaborated [41]. For example, it was shown that the HS molecule binds to the IFN-γ dimer. Furthermore, this binding occurs in such a way that two sulfated terminal sequences of the HS-binding domain interacted directly with two IFN-γ C-termini (a regulatory element of IFN-γ activity) and bridge the two cytokine monomers through an internal N-acetyl-rich sequence [41]. Another cytokine, interleukin-8 (IL-8), a member of the chemokine family, has been shown to bind to GAGs [40,42]. Furthermore, it has been suggested that the HS domain on endothelial cell surfaces provides specific ligand sites allowing it to retain the highly diffusible inflammatory chemokine in order to be presented to leukocytes. In addition, it has been reported that the dimeric form of IL-8 mediates binding to two sulfated domains of HS and that the HS sequence binds in a horseshoe fashion over two antiparallel-oriented helical regions on dimeric IL-8 [40,43]. Moreover, it was also reported that GAGs (heparin and HS) bind and modulate the activity of other cytokines such as IL-7, as well as IL-10 [44]. Other GAGs such as CS proteoglycans were also shown to bind chemokines, for example, RANTES [45].

Therefore, it can be concluded that ECM components of the CNS are able to bind and retain growth factors and inflammatory cytokines and consequently, serve as an immediate localized reservoir of these molecules that may be quickly utilized as an inflammatory reaction develops in the CNS. Furthermore, production of growth factors and cytokines by CNS cells and leukocytes infiltrating through the BBB occurs as the inflammatory process continues. This will be further illustrated by examples of several inflammatory CNS diseases described in Section 5 of this chapter. The structural and functional characteristics of the main classes of ECM molecules found in the CNS are discussed below.

3. CNS-ECM MAIN MOLECULAR CLASSES AND COMPOSITION

3.1. Hyaluronic acid/hyaluronan

HA is the only GAG chain not covalently bound to a protein core. It is synthesized on the inner surface of the plasma membrane by HA synthase and secreted during its elongation. It is a linear GAG polymer, (up to several million Daltons in molecular weight), which may present many sites of interaction for HA-binding proteins. The ECM of the CNS is relatively rich in chondroitin sulfate proteoglycans (CSPG) and HA, both of which are a space-filling molecules. HA, which is not sulfated, is bound noncovalently to a range of proteins, such as CSPGs including brain-enriched HA-binding protein (BEHAB/Brevican) specific to the CNS. Other HA-binding proteins include hyaluronectin, which is found in the nodes of ranvier of myelinated peripheral nerves, and glial HA-binding protein found mainly in the white matter. Glial HA-binding protein is identical to the HA-binding region of versican. Receptors for HA include CD44, involved in adhesion and endocytosis and RHAMM, implicated in the motility of astrocytes, microglia, and axon growth [2].

3.2. Adhesive glycoproteins

Early studies on the structure and function of ECM in the peripheral nervous system and CNS focused on their role in adhesion to maintain normal cytoarchitecture and to define the key spatial relationships among dissimilar cell types in axonal regeneration, cerebral edema, and cerebral neoplasia. Thus, a large number of *in vitro* and *in vivo* studies in this area have uncovered the pivotal importance of interactions of ECM glycoproteins, such as *tenascin*, *laminin*, *netrin*, and *reelin* with their CNS cell-specific receptors in normal and adaptive function of the adult brain.

The tenascins are a family of large modular glycoproteins that exist in the ECM as assemblies of several subunits, each of which has unique binding domains. Both tenascins-C [46] and R [47–50] are found in the adult brain. The latter is essentially restricted to CNS tissues where it is produced by oligodendrocytes, and the former is highly expressed in regions of inflammation in the CNS. In addition, tenasin-C is associated with reactive astrocytes that surround a site of injury where it may be serving as an anti-adhesive for extraneous monocytes. Furthermore, it was found that tenasin C is endogenously expressed by OPCs, and it was found highly expressed in the subventricular zone where it is believed to pay a role in the neural stem cells development. Tenasin-C is a major component of the ECM in developing nervous tissue, and it is expressed in zones of proliferation, migration, and morphogenesis. The data obtained from the studies using tenasin-C knockout mice emphasized the role tenasin-C plays in OPCs proliferation and migration toward zones of myelination [21–23] (Section 2.1).

TN-R can bind to ECM molecules and cell surface receptors. The C-terminal lectin domains on the CS proteoglycans brevican and neurocan represent similar binding partners in the ECM. These complexes localize to perineuronal nets (currently also known as interstitial clefts) in the adult CNS [51–54], as has been demonstrated by immunohistochemistry. These associations, and possibly others with as yet unidentified binding partners of TN-R, have an important role in the organization of the perineuronal nets/interstitial clefts and the *nodes of Ranvier*. Thus, these associations influence the propagation of action potentials at such sites, as is suggested by the phenotypic abnormalities of TN-R-null mice.

TN-R is one of the main carriers of the HNK-1 carbohydrate modification, which contains a 3′-sulfated glucuronic acid. Mice deficient in the sulfotransferase responsible for the production of this epitope show similar reduction in *long-term potentiation* (LTP) efficiency in the hippocampus as that reported for TN-R deficient animals, suggesting that its function in modulating synaptic activity is largely mediated by this carbohydrate structure [55].

Transcription factors, such as NF-κb, c Jun, and SP1, as well as the homeobox gene Prx1 have been shown to regulate the expression of tenascin C and R promoter constructs and may thus provide the intracellular link between cytokine stimulation and tenascin gene expression in the CNS [53,54,56–58].

Laminins are a family of proteins that are composed of three subunits, one heavy chain and two light chains. Five subunits in the laminin family have been cloned and sequenced, the A chain and the M chain, and three light chains, B1, B2, and S. In the brain, the α2- and M chains are present in association with neuronal fibers. They are potent stimulators of neurite outgrowth *in vitro* for various cell types, and their deposition along radial glial fibers and developing nerve pathways suggests that they also play an important role in routing the developing neurons *in vivo*. Various neuronal surface receptors have been identified for laminin. These receptors include several members of the *integrin family*, including integrins α1β1, α2β1, α3β1, and α6β1, as well as nonintegrin laminin-binding proteins such as LBP-110, the 67 kDa laminin receptor, α-dystroglycan, and β1,4 galactosyltransferase [59,60].

Netrins are another group of secreted multidomain glycoproteins in the ECM of the CNS. The four isotypes identified so far (netrins 1,2,3,4) all contain a laminin-type globular domain of the "VI" type, three laminin-type EGF repeats, and a carboxyl-terminal "netrin module." Netrin-1 and 2 is expressed in the adult brain by multiple types of neurons and in oligodendrocytes in the CNS. The protein is absent from compact myelin, but enriched in periaxonal myelin at the interface between axons and oligodendrocytes. This distribution suggests that in the adult nervous system netrin-1 may function to mediate cell–cell interactions. Furthermore, as netrin receptors are expressed in neurons following injury, it has been speculated that netrin 1 may influence axonal regeneration. Netrin-3 protein is present on motor neurons and subpopulations of neurons within sensory and sympathetic ganglia, suggesting a role in axon pathfinding and fasciculation within the peripheral nervous system. Netrin-4 is a distant relative of netrins-1–3, and its globular domain is more closely related to those of laminins than to those of other netrins. Netrin-4 is broadly expressed in both neural and non-neural tissues of embryonic and adult mice. In postnatal brain, it is selectively expressed in subsets of neurons, including cerebellar granule and hippocampal pyramidal cells [61–64].

Reelin is an ECM glycoprotein that has been implicated in control of cell–cell interactions critical for cell positioning in the brain. Recent research has shown that its expression is critical in the regulation of hippocampal layer formation by interacting with its receptors and stimulating a range of intracellular signal transduction pathways blocking neuronal migration. Reelin is a ligand for lipoprotein receptors, for α3β1 integrin, and for proteins of the cadherin-related neuronal receptors family [2].

3.3. Lecticans chondroitin sulfate proteoglycans

The *CSPG* in brain *versican, neurocan*, and *brevican* have been termed "lecticans" or "hyalecticans," as they have conserved structural motif both at their N- and C-termini that exhibit high-affinity binding for HA and other ECM protein ligands, such as Tenascins C and R, respectively. Comprehensive review articles on the structure of these proteoglycans in the CNS have appeared elsewhere [65–69].

Brevican is the most abundant proteoglycan in the terminally differentiated CNS. It is a component of perineuronal nets and also localized on the surface of large neuronal somata and primary dendrites. It is produced by neuronal and glial cells in two isoforms [65,70,71], one of which is secreted into the ECM and the other that is localized to the cell membrane through a GPI-anchor. Its localization in the perineural nets/interstitial clefts suggests a role to isolate and stabilize existing synapses and to prevent the formation of new synaptic contacts. It has also been speculated that brevican might function as an insulator, sealing the synapses and preventing loss of transmitter substances from the synaptic cleft to the periphery [72]. The efficacy of synaptic cross-talk might be dependent on the extracellular space surrounding the synapses, that is, on intersynaptic geometry and diffusion parameters [73].

The proteoglycan *neurocan* is highly expressed during modeling and remodeling phases in the developing brain and is largely synthesized by neurons. It accummulates in perineuronal nets/interstitial clefts during late embryonic development but then declines steeply during the early postnatal period. Despite its spatially and temporally restricted expression [74,75], neurocan's role in normal CNS biology has not been clarified. Thus, neurocan-deficient mice display no deficits in brain anatomy, morphology, and ultrastructure. Perineuronal nets/interstitial clefts were also normal in those animals, and only mild deficits in synaptic plasticity were seen, due to reduced late-phase hippocampal LTP activity [76]. Neurocan is strongly expressed in response to injury and is deposited in the vicinity of astrocytes, where it may act to modulate axonal regeneration after CNS injury [77].

The third member of CNS lectican (aggrecan family) is versican variant (V2), which contains only a short (α) chondrotin sulfate attachment domain [78,79]. This variant of versican is predominantly associated with the myelinated fiber tracts of the brain and in the optic nerve; it may also serve as a barrier to neurite outgrowth from central and peripheral neurons [78,79].

There is ample evidence for proteolytic processing of lecticans in the adult brain, by matrix metallo proteinases and ADAMTS proteinases. *In vivo* experiments have indicated an important role for the HA binding N-terminal fragment of brevican in the modulation of cell migration, because glioma cells overexpressing it displayed increased invasiveness [80], but it is currently unclear whether such fragments have physiological functions in normal brain.

3.4. NG2

NG2 is a CSPG that exists in three different forms: one transmembrane, full-length form of 300 kDa; and two shorter forms, one of which remains attached to the membrane and one that is secreted into ECM. *In vivo* NG2 is expressed in oligodendrocyte precursor platelet growth factor (PDGF-α) receptor-positive cells. The NG2 cells are believed to be either a pool of precursors of oligodendrocytes still present in the adult brain, or a new type of astrocyte [2]. It is known that NG2 inhibits neurite outgrowth of cereberal neurons and dorsal root ganglia in the presence of laminin, and it may be involved in the inhibition of axonal regeneration. This is supported by a study showing that NG2 is upregulated in sites of brain injury [2,81]. The molecular mechanism of NG2 action is not known, but what is known about it is that NG2 binds to the ECM molecules such as collagen, tenascin, and laminin. The recent literature considers that NG2 may be a lineage marker and provision of cellular source for some gliomas [25] (see Section 2 and 5 of this Chapter).

3.5. Cell surface heparan sulfate proteoglycans

HSPGs, as described earlier (Section 2), play a significant role as molecules involved in storage of growth factors and cytokines in the brain. They have been examined mainly in relation to their known regulatory participation in events involving growth factor receptors and the ECM that regulate oligodendrocyte progenitor proliferation, migration, and adhesion phenomena.

The *syndecan family* of cell surface HSPGs interacts through their cytoplasmic C-terminal tail with a membrane-associated kinase. Syndecans have distinct cellular distributions in adult rat brain, with syndecan-2 and -3 being the major syndecans expressed in neurons of the forebrain. At the cellular protein level, syndecan-2 and -3 are differentially localized within neurons; syndecan-3 is concentrated in axons, whereas syndecan-2 is localized in synapses [2]. Recently, it was reported that the specimens of normal brain expressed syndecan -2, -3, and -4 mRNA, but not syndecan-1 mRNA. However, invasive glioma was shown to express of all types of syndecans mRNA including syndecan-1 mRNA and with syndecan-1 protein localized on the cell surface of glioma cells [82] (Section 5.5). Transgenic expression in the hypothalamus of syndecan-1 produces mice with hyperphagia and maturity-onset obesity resembling mice with reduced action of α-melanocyte-stimulating hormone (α-MSH) [83]. In normal animals, syndecan-3 is expressed in hypothalamic regions that control energy balance. Food deprivation increases levels of syndecan-3 in the hypothalamus by several-fold. Syndecan-3 null mice have reduced reflex hyperphagia in response to food deprivation. Taken together it was postulated that regulation of syndecan-3 levels in the brain is a physiologic modulator of *feeding behavior* [83]. Recently, the gene for syndecan-3, along with number of other genes representing the principal components of several receptor-defined signaling pathways, was shown to be differentially expressed by white matter progenitor cells [18] (Section 2.1).

Glypicans 1-5 are another family of cell membrane-bound HSPGs, and glypicans 1 and 4 are the most prominent members in adult brain function. The spatiotemporal expression pattern of glypican-4 in the developing cerebral wall significantly overlaps with that of FGF2. These results suggest that glypican-4 plays a critical role in the regulation of FGF2 action [2].

Agrin is a HS proteglycan in neuromuscular junctions where it directs the formation of the postsynaptic apparatus. It is also expressed in the brain, concentrated at interneuron synapses, where its binding to a receptor induces the combined activation of the extracellular signal-regulated kinases (ERK1/ERK2) and p38 in central neurons raising the possibility that it might serve a related function at neuron-neuron synapses [84,85].

The astroglial proteoglycan *phosphocan* [86,87] is an mRNA splice variant representing the entire extracellular portion of a receptor-type protein tyrosine phosphatase. A glycoform of phosphocan (phosphocan-KS) that contains both CS and keratan sulfate is present in the postnatal rat CNS. In contrast, the concentrations of both phosphocan and phosphocan-KS rise steadily after embryonic day 20 to reach a plateau at about 2 weeks postnatally. Similar to neurocan, phosphocan can bind to a range of neural cell adhesion molecules (Ng-CAM/L1, NCAM, TAG-1/axonin-1) and to tenascin-C where it may also act as an inhibitor of neuronal and glial adhesion and neurite outgrowth [2].

4. FUNCTIONAL INVOLVEMENT OF ECM MOLECULES IN SYNAPTIC PLASTICITY

One of the main activities of an adult brain is in memory consolidation (learning) and retrieval. At the cellular level this function is believed to be mediated by structural changes of the synapses known as "*synaptic plasticity.*" Hippocampal LTP is one of the best-studied models of this process at the molecular level, where defined stimuli are applied in response to which neurons are switched to an altered level of metabolic activity over a prolonged period of time. Recently published data suggest that modulations that alter cell–matrix interactions, and thereby affect synaptic plasticity, play an important regulatory role in the induction and maintenance of LTP. Some of these are reviewed below.

4.1. Functional specificity of ECM molecules and cytokines indicated by knockout murine models

4.1.1. ECM knockout mice

Tenascin-R null mice showed impaired LTP in cornu ammon (CA1), an architectural subregion of the hippocampus, and it has been suggested that this particular function is mostly mediated by the posttranslational glycosylation on the protein, that is, the HNK-1 epitope which contains a 3′-sulfated glucuronic acid residue [88].

Tenascin-C-null mice underwent a series of well-characterized learning tests, and it was noted that abnormalities of myelination or oligodendrocyte precursor distribution were not detectable, suggesting that local concentrations of tenascin-C are not critical for the pattern of myelination in these regions of the adult CNS. Interestingly, however, the investigators reported a number of behavioral abnormalities in these mice, including hyperlocomotion and deficits in coordination during beam walking that can be ascribed to a tenascin-C deficiency [89]. Although the precise molecular disturbance underlying these observations remains to be determined, the behavioral findings may have been related to the results of another study by Garcion et al. [90] that showed tenascin-C is involved in cell proliferation and migration of oligodendrocyte precursors during development. In addition, this study showed that migration of OPCs occurred toward the zone of myelination. The data obtained from more recent studies by Garwood et al. [22], who used tenasin-C knockout mice, suggested the existence of mechanisms compensating for the absence of tenasin-C resulting in development of oligodendrocytes derived from tenasin-C knockout neurospheres.

Laminin: the dynamic nature of the laminin matrix and its amenability to modification by exogenous molecules was best illustrated by the phenotype of the brain-selective deletion mouse mutant of *dystroglycan*, a neuronal receptor for laminin [2,89,90].

Brevican null mice develop normal brain anatomy; however, perineuronal nets/interstitial clefts formed, appeared to be less prominent in mutant than in wild-type mice. The brevican deficient mice also showed significant deficits in the maintenance of hippocampal LTP, but no obvious impairment of excitatory and inhibitory synaptic transmission was found, suggesting a more complex mechanism for the LTP defect [91].

4.1.2. Cytokine knockout mice

It has been reported that *stromal cell-derived factor-1* (SDF-1) and *CXC chemokine receptor 4* (CXCR4) are key signaling molecules guiding migration of distinct neuronal subtypes during late cerebrocortical development. A deletion of genes encoding for SDF-1 or its receptor CXCR4 is *lethal* for mice immediately after birth. These mice suffer with severe abnormalities affecting many organs including hematopoietic, cardiovascular systems, and brain. In mice with a null mutation in CXCR4 or SDF-1, the interneurons were found to be severely underrepresented in the superficial layers and ectopically placed in the deep layers of the neocortex. It was concluded that SDF-1 selectively regulates migration and layer-specific integration of CXCR4-expressing interneurons during cortical development [16]. In addition, abnormal development of the hippocampal dentate gyrus has been described in mice lacking the CXCR4 chemokine receptor [92].

4.2. Functional specificity of ECM molecules indicated by *in vivo* and *in vitro* application of ECM molecule fragments

It has been reported that laminin degradation by plasmin regulates LTP. Incubation with anti-laminin antibodies prevented the degradation of laminin and impairment of LTP maintenance by plasmin. Taken together, these findings suggested that the laminin-mediated cell–ECM interaction may be necessary for the maintenance of LTP [93].

Infusion of chondroitinases ABC into the brains of normal mice reduced LTP and long-term depression (LTD) in the CA1 region of the hippocampus [94]. The treatment of brain slices of normal rats with antibodies against brevican results in acute impairment of LTP maintenance and is more characteristic for interference with signal transduction processes. It has been shown that secreted brevican can bind to cell surface sulfatides and sulfoglucuronyl glycolipid (SGGLs). Blocking these interactions might cause changes in signal transduction, which eventually result in the LTP defect [94]. In addition, the recent studies found that brevican which is known to inhibit neurite outgrowth and to stabilize synapses *in vivo* can be modulated by TGF-β *in vitro.* Therefore, this finding suggested that TGF-β plays an important role in neuronal plasticity [95].

Of other interest is the heparin-binding growth-association molecule (HB-GAM), *pleiotrophin* that binds to syndecan 3, thus linking these ECM components directly to the signaling pathways through the cytoskeleton. Injection of the purified syndecan, heparin, or heparitinase at those sites inhibited LTP [96,97]. Pleiotrophin, as well as midkine, which is another ECM protein, are used by the developing neurons. These molecules serve to support axonal growth and provide guidance for the neurons with pleiotrophin reported to promote differentiation of the neuronal stem cells. Both of these proteins form a two-member family of ECM proteins known to bind to HS. Recently, pleiotrophin was shown to contain the domains which are homologous to trombospondin type I repeats known to be present in many ECM proteins that interact with the cell surface [98]. However, each individual domain alone failed to give a significant effect on the assays of LTP and neurite outgrowth.

4.3. Metalloproteinases, cytokines, and ECM molecules

MMPs play an important role in the development of the CNS by catalyzing the normal turnover of ECM molecules. MMPs are also important players in a disease process. Members of the MMP family include beside the MMPs, also ADAMs (a disintegrin and metalloproteinase) and the ADAMTS (a disintegrin and metalloproteinase with trombospondin motif) [99]. The MMP and ADAMTS family of more than 20 members of enzymes act on a wide range of substrates *in vitro* and their tissue and cell-specific expression patterns are strictly regulated during matrix remodeling. MMPs are products of different genes, but show a high degree of homology. *In vitro* MMPs show overlapping affinities to different ECM and non-ECM substrates, and many MMPs can degrade several different classes of ECM proteins, for example, MMP-2 (gelatinase A) and MMP-9 (gelatinase B) degrade collagens I, IV, V, VII and X, gelatin, elastin, fibronectin, and proteoglycans. They function not only as effectors of tissue remodeling, but also interact with the cytokine network. For example, MMP expression and activity is modulated by cytokines such as TNF-α/β, IL-1, and IL-2. On the other hand, MMPs act as sheddases or convertases as they transform membrane-bound cytokines, cytokine receptors, and adhesion molecules into their soluble forms [100]. ADAMTSases, secretory proteases which selectively cleave lecticans were detected in cultures of neurons and astrocytes but not microglia. Their presence was associated with high levels of cleavage products of brevican, and their activity was shown to be reduced by TGF-β [95]. ADAMTS-ases were also shown to be expressed at high levels on glioblastomas [101] (Section 5.5). In neuro-inflammatory diseases, MMPs are expressed by blood-derived immune cells and/or resident CNS cells, and they are effectors of BBB opening and invasion of brain parenchyma by immune cells. MMPs' enzymatic activity can be inhibited by naturally occurring specific endogenous tissue inhibitors of metalloproteinases (TIMPs). The balance between the MMPs and the TIMPs which regulates a number of cell–matrix interactions [2,100] could be modified by cytokines and growth factors. For example, a recent study showed that, in neural precursors, the proinflammatory cytokine TNF-α can upregulate mRNA levels of MMP-9. This effect could be inhibited by IFN-β resulting in suppression of TNF-α-induced levels of secreted MMP-9 and MMP-2 followed by the enhancement of TIMP1 and TIMP2 mRNA expression [27].

4.4. Cytokine regulation of synaptic plasticity

Cytokines are proteins that were first discovered and described in terms of their pleiotrophic effects on cellular activation in the immune system and were grouped as pro-inflammatory or anti-inflammatory, depending on their activities on particular cell types. In the 1980s attention was first drawn to cytokine presence and their biological activities in the CNS and interest continues to the present, largely in the context of response to infection or injury (see Section 5 of this Chapter). However, recent work indicates that cytokines may also be constitutively expressed in active, normal adult brain. In this context, cytokines fulfill a role of modulating the functional output of individual neurons to mount an integrated response to normal physiological stimuli. The neuronal cytokine system regulates these expressions of cooperation in LTP and may also play a crucial role in the brain in a steady-state condition.

For example, a large number of investigators report detection of interleukins, IL1β and its receptor in normal mouse, rat and human brain [102]. Results of recent studies have shown a physiological role for IL1β in synaptic plasticity serving as a key mediator in the maintenance of LTP. Thus, production of the cytokine was found to be significantly increased during LTP by both neuronal and glial cells in the hippocampus, implying that its action is required downstream of the inducing event of LTP [103]. Complimentary findings that IL-1 signaling within the hippocampus plays a critical role in learning and memory processes was reported in another study, where IL-1β and IL-1 receptor antagonist (IL-1ra) were given to rats by intracerebroventricular (icv) injection immediately following a learning task and memory function testing 1–8 days later. The results from this study showed that IL-1ra caused hippocampal-dependent spatial memory impairment, whereas IL-1β had no effect. Furthermore, in the passive avoidance response, which also depends on hippocampal functioning, IL-1ra again caused memory impairment, whereas IL-1β caused memory improvement [103].

Apart from their role in the regulation of synaptic plasticity in normal adult brain, cytokines are also major players in CNS pathology. As it was earlier described (Section 2 of this Chapter), ECM molecules in the brain may serve as a reservoir and as a storage location for various growth factors and cytokines that play a role in normal development of the CNS and/or may be released upon initiation of an inflammatory state. We will now provide several examples of CNS diseases in which cytokines and ECM components play a significant role in the development of CNS pathology.

5. ECM–CYTOKINES AND GROWTH FACTORS IN CNS INJURY

It is known that ECM components are involved in signaling following injury in the CNS [12]. For example, laminin 8 and 10 play a role in T-cell recruitment across the BBB in EAE, an animal model of MS. The formation of inflammatory cuffs (accumulation of leukocytes outside and around blood vessels) in EAE is closely related to the presence of laminin 8 [36]. Brain injury is followed by a cascade of cellular and molecular mechanisms regulating scar formation [104]. It was found that astrogliosis and expression of TGF-β are key components of scar formation. A cytokine, connective tissue growth factor (CTGF), was found to mediate the effects of TGF-β and is believed to play a role in tissue regeneration and aberrant deposition of ECM. The restricted accumulation of CTGF-positive cells such as reactive astrocytes and fibroblastoid cells lining the adjacent laminin-positive basal lamina, suggests participation of both CTGF-positive astrocytes and fibroblastoid cells in scar formation. This was further supported by the finding of elevated numbers of CTGF-positive cells in injured rat brains [104]. It was also reported that expression of TGF-β2 correlates with the deposition of scar tissue (glial/collage) in the lesioned spinal cord [105]. After brain injury, in addition to showing an increased expression of TGF-β, there were also elevated levels of EGF receptor, neurocan, and phosphocan detected [106]. These CNS injury-induced increases in growth factors that regulate expression of CSPGs were suggested as a future targets in repair of CNS injury [106]. During the "wound" healing response activated microglia/macrophages and astrocytes release inflammatory cytokines such as IL-1 and IFN-γ, both of which can signal a change in ECM production in adjacent tissue cells. For example, IL-1α was shown to upregulate biosynthesis of HS proteoglycan, perlecan, after injury [107]. Similarly, IFN-γ has been shown to act upon astrocytes to alter biochemical and physical properties preventing neuronal regeneration, and this effect may be related to its inhibitory action on tenascin, laminin, and fibronectin [108]. On the other hand, it was reported that administration of IFN-γ can also have a downregulatory effect on microglial pathways involving remodeling of ECM in the brain [109]. Microglia are highly plastic cells that participate in inflammatory and injury responses within CNS and migrate after activation. Their normally weak adhesion to laminin (expressed by astrocytes) is markedly increased after exposure to inflammatory cytokines such as TNF, IFN-α, and IFN-γ, and this, together with increased production of laminin postinjury, may control the extent of microglial infiltration and activation at the site of injury [110,111].

5.1. The ECM–cytokine reservoir in disease

The binding and storage of cytokines in ECM through cytokine–ECM complexes ensures cytokine availability and local proximity to potentially responding cells. For example, it was found that TNF-α binds avidly to immobilize ECM glycoproteins such as laminin and fibronectin, which in turn are capable of regulating bioavailability and activity of TNF-α [112]. The migration of T cells is associated with secretion of various matrix-degrading enzymes; one of them is heparanase capable of degrading HS of the cell surface and HS proteoglycan of ECM. This action facilitates cell movement and migration through extravascular tissue but also causes the release of various growth factors and cytokines bound and stored by HS which, in turn, could be pro-angiogenic, further augmenting the development of inflammation. On the other hand, the inhibition of heparanase activity also inhibits ECM-degrading activity and the infiltration of T cells into the inflammatory lesion [112].

Recent findings provide further support for the importance of ECM as a cytokine and growth factors storage during disease processes. For example, radiation of stromal fibroblasts (potential ECM storage of growth factors and cytokines) in pancreatic carcinoma resulted in promotion of malignant behavior [113]. Radiotherapy, a major adjuvant treatment for many malignant tumors including brain tumors has been recently reported to activate several pathways involved in tumor invasion and metastasis acting through tumor–stromal interaction, which result in increased expression of MMP-2 [113]. Also, it was more recently reported that hypoxia enhanced heparanase cleavage of ECM heparan sulfate, a major GAG responsible for the storage of cytokines and growth factors in the ECM, which resulted in increased tumor cell invasion. This invasion was inhibited by treatment with anti-heparanase antibodies [114]. Both of these findings further support the importance of the ECM as a reservoir of cytokines and growth factors in a disease process.

5.2. Vasculature, BBB, and inflammation

Injury produced during CNS disease and inflammatory conditions results in both the ECM and the cellular components of blood vessels expressing endogenous cytokines, chemokines, matrix-degrading enzymes, and other inflammatory mediators that can participate in transduction of stimulatory signals to immune capable cells. For example, adhesion molecules on the luminal surfaces of capillary endothelial cells can be modified by an adjacent inflammatory process during systemic infection. This modification leads to presence of "activated" ECM components, which are presented to circulating leukocytes, therefore facilitating their extravasation and recruitment to the site of inflammation. The invading T lymphocytes are active participants in the dynamics of the microenvironment, for example, through production and release of a wide array of cytokines, chemokines, and matrix-degrading enzymes, for example, MMPs, into the inflammatory milieu. In turn, modulation of the ECM by leukocytes may cause a regulatory feedback of the initial immune reaction. For example, leukocytes following recognition and binding to ECM ligands through integrins (leukocyte receptors) are able to transduce signals from the environment to the inside of the cell. In monocytes, ECM binding by integrins results in upregulation of several genes encoding cytokines expression including IL-1, IL-8, and tumor necrosis factor (TNF) [112,115,116].

Both brain microvascular smooth muscle and endothelial cells can produce granulocyte macrophage colony-stimulating factor and support colony formation of granulocyte-macrophage-like cells [117]. This observation points to an active participation of both brain microvasculature cell types in acute inflammatory responses within the CNS. Recent data suggest that following brain ischemia, differential sensitivities to extracellular mediators of blood derived monocytes/macrophages and resident microglia were observed [118]. During brain injury and trauma, perilesional upregulation of *connective tissue growth factor* (CTGF) expression also suggests its potential role in BBB function and angiogenesis [104].

During health, very few leukocytes penetrate the BBB, but in response to inflammatory conditions, for example, MS, the barrier becomes compromised and intense infiltration of the CNS by T lymphocytes occurs. T lymphocytes' activity within the brain underlines the onset and progression of the disease [119].

5.3. ECM–cytokine interaction in CNS autoimmune disease

MS is an autoimmune disorder, which results in demyelination characterized by the appearance of lesions in the CNS white matter. The symptoms first develop in patients in their 20s and 30s, irregularly across the initial course of the disease, but become more extensive and widespread during chronic stages. The demyelination leads to progressive dysfunction in motor, sensory, and vegetative systems and eventually severe disability in the majority of patients [100]. Although the pathogenesis of this disease is still largely unknown, both viral infections and genetic causes are suspected. It is proposed that the structural homology between a self-protein in the CNS such as myelin basic protein (MBP), and an infectious pathogen leads to auto-sensitization and to a subsequent immune response aimed against both the self-protein and a pathogen. This immune activation leads to an inflammatory response and autoimmune tissue destruction in the CNS. Similarly to many other autoimmune diseases, MS is characterized by recurrent increased focal BBB damage, perivascular lymphocyte infiltration and localized degradation of myelin, leading to the formation of gliotic lesions –plaques, and eventually axonal disruption [100]. The breakdown of the BBB followed by extravasation of T cells and their invasion of the brain parenchyma is the most prominent characteristic in the development of the pathogenesis. In addition, an altered cytokine network has been associated with this disease [120]. Many cytokines, for example, IL-1, IL-2, TNF or chemokines, and proteases play an important role in promoting and mediating neurodegeneration [121]. On the other hand, one cytokine, IFN-β, is still the most effective therapy to date for most of the MS patients, but its mechanism of action is not completely understood [120].

There is a large body of evidence stating that an overproduction of MMPs is involved in the development of neuronal loss in MS. CNS cells such as astrocytes, microglia, and endothelial cells, as well as the cells of the immune system, including T cells and macrophages have been shown to secrete MMPs [100,122,123]. Moreover, it was recently reported that MMPs contribute to opening of the BBB [100]. *In vitro* studies have shown that some cytokines, for example, IL-2, can induce upregulation of the proteolytic enzymes, MMP-2 and MMP-9 in T cells, leading to an increase of cell migration across the BBB [100]. On the other hand, the results of *in vivo* studies using mice models deficient for this enzyme, for example, young MMP-9 knockout mice, revealed that the animals were less susceptible to the development of experimental autoimmune encephalitis (EAE), an animal model of MS [100,124]. EAE can be induced in susceptible mouse strains by immunization with myelin proteins, or adaptive transfer of myelin reactive T cells. A critical step in this induced inflammation is migration of leukocytes from the blood stream into the CNS parenchyma. Leukocytes must first pass across the endothelial cell monolayer, endothelial basement membrane, and then across the glia limitans, consisting of astrocyte endfeet and associated basement membrane [36]. The endothelial cells and parenchymal basement membranes define the perivascular space, where leukocytes accumulate and are entrapped during acute EAE. Moreover, it was shown that leukocyte parenchymal infiltration (past the glia limitans) may be prevented, indicating that progression through the astrocytic basement membrane is functionally different versus passing through the endothelial cell basement membrane. It was shown by Sixt et al., [36] that these two basement membranes have a different composition: the endothelial basement membrane either contains laminin 8 and is permissive for T-cell transmigration or laminin 10 and is restrictive to T-cell transmigration. However, penetration of the parenchymal basement membrane, which contains laminins 1 and 2 (not adhesive for T cells) would only occur if there is a disruption of this outer membrane. This may occur through proteolysis involving MMPs [36].

The proteolytic enzymes, MMP-2 and MMP-9 whose activities are directed toward ECM, were found to be at high levels in MS patients [125]. Interestingly, an evaluation of MS patients at the end of 9 months of IFN-β therapy showed that the levels of MMP-9 decreased significantly [120,126,127] or did not change [128] with concomitant increase of TIMP-2 (inhibitor of MMPs) levels. It was postulated by these authors that IFN-β modulates T-cell activities, including MMP and TIMP production [120]. Furthermore, this inhibitory effect may relate to a decreased production of MMP by the CNS cells as well as by the immune cells such as T cells, preventing T cells from migrating into the CNS. These observations may indicate that following IFN-β therapy the integrity of the BBB and its basal lamina are protected, resulting in a slowing progression of the disease [120]. Consequently, downregulation of MMP release may be the molecular basis for the beneficial effect of IFN-β in MS [100]. The results of recent studies suggested that the inflammatory response associated with EAE, an animal model of MS disease, affects neural precursor cells by activating their differentiation and migration into the areas of inflammation in white matter. The pro-inflammatory cytokines such as TNF-α and IFN-γ were shown to enhance neural progenitor cell migration accompanied by increased levels of MMPs and ECM remodeling. These pro-inflammatory cytokine-induced effects on neural precursor cells may have significance in promoting the reparative activity of these cells [27].

5.4. ECM–cytokine interaction in the bacterial and viral CNS disease

5.4.1. Bacterial meningitis

Infection of leptomeninges, soft membranes enveloping the CNS, results in leptomeningitis, abbreviated as meningitis. Viral meningitis is usually benign and resolves spontaneously in contrast to the more serious nature of bacterial meningitis, which continues to be an important clinical problem. In the latter the major clinical problems result not from the bacterial pathogen, as it was originally thought, but from the over-reactive immune response of the host. The end result may be brain damage, including hearing loss, as well as cognitive impairment, as well as the frequent occurrence of a fatal outcome. During this disease the integrity of the BBB is impaired, which is accompanied by intrathecal production of cytokines, for example, IL-1 and TNF, in addition to blood-derived leukocytes that are released into the CSF leading to neuronal injury. MMPs play a central role in the pathogenesis of this disease. For example, an increase in MMP-8, a neutrophil collagenase, is a specific feature of bacterial meningitis [100]. It had been shown previously *in vivo* in rats that bacterial collagenase disrupts ECM, allowing the BBB to open [129]. Moreover, it is believed that MMPs and TNF-α-converting enzyme (TACE) may play an important role in the pathophysiology of this disease. TACE releases proteolytically several cell-surface proteins, including the pro-inflammatory cytokine, TNF-α and its receptors. TNF-α in turn stimulates cells to produce active MMPs, which facilitates leukocyte extravasation, followed by brain edema and degradation of ECM. Furthermore, it was reported by Leib, et al. [130] that MMP-8 and MMP-9 were transcriptionally upregulated in CSF cells and in the brain of infant rats during pneumococcal meningitis, which was accompanied by high levels of TNF-α. The combined inhibition of MMPs and TACE resulted in a cure for meningitis in these rats. This finding may offer the basis for a novel clinical therapeutic strategy aimed to prevent brain injury [130].

5.4.2. HIV-associated dementia

The human immunodeficiency virus-1 (HIV-1) is the etiologic agent of AIDS, which infects and incapacitates the immune system, primarily depleting T-helper (CD4+) lymphocytes. A third of adults and half the children infected with AIDS develop neurological cognitive/motor impairments, HIV-associated dementia (HAD), which are attributed to HIV infection of the CNS. Following CNS entry HIV initiates activation of various inflammatory mediators, chemokine receptors, and ECM degrading enzymes which negatively affect neuronal and glial function [131]. The ability of T lymphocytes to invade the CNS and to secrete proinflammatory cytokines plays a crucial role in HIV pathogenesis. The neuropathology of HAD relates to activity of toxic molecules and cytokines secreted by infected immune-competent mononuclear phagocytes and also by astrocytes. This HAD neuropathology could be compared to that which occurs in Alzheimer's disease (AD). The progression of AD is associated with Lib, an astrocyte-derived molecule, a type I transmembrane protein that belongs to leucine-rich repeat superfamily. In AD, Lib mRNA was found to be expressed in activated astrocytes near the senile plaques. In addition, its gene was described as inducible and responsive to β-amyloid and proinflammatory cytokines. Interestingly, Lib was also found to promote migration of glioma cells through the ECM, and in general, this astrocyte-derived molecule may be responsible for regulation of interactions between cells and ECM in the brain [132]. It is believed that astrocytes may contribute to the development of HAD by their production of TIMPs, for example, TIMP-1. TIMPs are known to downregulate activity of MMPs, which proteolytically alter ECM affecting BBB integrity. The levels of TIMP-1, an MMP inhibitor, measured in CSF and brain of HAD patients were reduced, but MMP-2 levels were increased when compared to the normal controls. It was concluded that in the HAD patients, the ability of astrocytes to downregulate the destructive effects of MMPs through the expression of TIMP-1 is diminished [133]. It was also found that interaction of astrocytes with cytokine-producing T lymphocytes resulted in an increased production of MMP-3 and active MMP-9 by astrocytes despite TIMPs being expressed. It was concluded in this study that changes in MMP and TIMP expression were probably mediated by T-lymphocyte-released cytokines. In addition, evaluation of the CNS tissue from HIV patients detected MMP-9 expression by the neural cells within the perivascular space, which is infiltrated by mononuclear cells [134]. This further emphasizes the importance of MMP and TIMP in the virally infected CNS tissue infiltrated by cytokine-producing T lymphocytes. The fusion domain of HIV, gp41, was found to interact with T lymphocytes in a specific way, through HS located on the cell surface. This interaction was blocked by IL-8, and it was abolished when cells were pre-treated with heparatinase [135]. In addition, the presence of cell membrane HSPG in the extracellular milieu was found to be the essential receptor for HIV-1 Tat, the transactivator HIV protein during cellular internalization. Again, Tat uptake was completely inhibited by soluble heparin and GAG enzymes, which degrade HS chains [136]. The importance of sulfated GAGs in viral infection was also observed in other systems, for example, during infection of the CHO cell line with respiratory syncytial virus (RSV). It was found that the role of surface GAGs in RSV infection involves a specific GAG structural configuration that includes N-sulfate and at least 10 saccharide subunits [137]. Taken together, a collaboration of ECM molecules, matrix proteases, and cytokines in the CNS determine HAD pathology.

5.5. ECM–cytokine interaction in brain neoplasia

Malignant gliomas are among the most devastating forms of neoplasia. Locally they are highly invasive tumors but they rarely metastasize. Glioma invasion is a complex process involving the interaction of tumor cells with host cells and with the ECM. The invading cells, which migrate several millimeters or centimeters from the main lesion, initiate tumor regrowth under the control of yet unknown microenviromental factors and form recurrent tumors adjacent to the original site of lesion [138]. The microregional ECM heterogeneity in brain modulates glioma cell invasion [139]. For example, it was found that tumor cell aggressiveness correlated with the ability of CD44 adhesion molecule in glioma cells to bind to HA and the laminins in the ECM [140,141]. Recently, it has been reported that *laminin-8* of the blood vessel's basement membrane is upregulated in high-grade glioma, glioblastoma multiforme (GBM), promoting tumor cell invasion. A weak expression of laminin-9 was found in the capillary basement membrane of normal human brain. It was also observed that a "switch" from laminin-9 to laminin-8 expression, associated with the gradual increase from low-grade gliomas to high-level gliomas was associated with increased neovascularization and by that process contributing to tumor aggressiveness. It was concluded that laminin-8 may be used as a predictor of brain tumor recurrence and patient survival and may also be considered as a potential molecular target for glioma treatment [142]. This observation may be specific for the brain tissue only because it was not found in other tumor models, such as a model of Lewis lung carcinoma. On the contrary, in the model of Lewis lung carcinoma in mice, a deletion of laminin-8 was shown to result in an increased tumor neovascularization and metastasis [143]. Taken together, these data show that laminin-8 may play a critical role in tumor growth and angiogenesis.

During glioma invasion the ECM is proteolytically modified, and certain synthesized ECM components such as HA and CSPGs are expressed and/or released by glioma cells into the environment. For example, NG2-positive cells recently found within brain tumors were first identified as oligodendrocyte progenitor cells. NG2-positive cells were known to retain their proliferative capacity throughout the life of the organism, while expressing this CSPG (Section 2.1). Recently, based on a finding of an elevated expression of NG2 gene level in gliomas, when compared with normal brain, both high- and low-grade gliomas were compared to oligodendrocyte progenitor cells [144]. NG2 expressed by these cells has been shown to have a strong association with both ECM ligands and cellular ligands such as CD44 adhesion molecule and therefore implicated in the migratory behavior of glioma cells. It was reported that many established human glioma cell lines, which showed no migratory behavior *in vivo* [145], also did not express NG2 [146]. On the contrary, some of the highly migratory *in vivo* glioma cells expressed this marker (Wiranowska et al., personal communication, 2006).

The extent of the contributions of normal host brain tissue versus the contributions of glioma cells to the production of ECM components is still not well understood [147]. The molecular mechanisms underlying glioma infiltration are different from that of other cancers. Several new classes of genes associated with glioma invasion were recently reported when highly migratory glioma clones were studied. Most of these genes had not been linked to glioma invasion to date. Moreover, they included among others those genes encoding for ECM components, cytokines, and proteases [148]. It was found that advanced malignant gliomas express elevated levels of membrane-associated MMPs and TIMPs, which may be essential in the regulation of vascularization and invasion [149]. Also, ADAMTSases such as ADAMTS4 and ADAMTS5 were reported to be upregulated in proliferating human glioblastoma cells, and this upregulation may be contributing to the invasiveness of glioma [101]. Others have reported that transcription of MMP-9 gene induced by IL-1 and TNF-α in glioma cells is regulated by protein kinase C-zeta (PKC-zeta) through NF-κB [150]. Moreover, expression of the membrane-type1 (MT-1) MMP was shown to have a functional link to vascular endothelial growth factor (VEGF) production in the enhanced tumorigenicity of glioma [151]. Also, IL-1β and TNF-α were shown to upregulate the levels of syndecan-1, a cell surface HSPG, in glioma cells. This upregulation of syndecan-1 expression at the transcriptional level was modulated by NF-κB [82].

In addition to growth factors, the cytokines also can influence both secretion of the ECM proteins and migration of cells, which is a deciding factor in the final outcome of tissue remodeling. For example, we have reported that IFN-α/β, while being inhibitory to glioma cell proliferation and invasion, also inhibits secretion of CS proteoglycan and an active form of MMP-2 [152–154]. On the other hand, others have shown that EGF and proinflammatory cytokines such as IL-1β, TNF-α, and IFN-γ can specifically regulate the expression of components of the plasminogen activation system in astrocytoma cells and therefore may influence the turnover of ECM and migration of cells within the brain [155]. In addition, it was found that in glioma cells the expression of naturally occurring, constitutively active variant EGFRvIII affects glioma cell motility by increase of cellular protrusions/ lamellipodia [30]. In another study the expression of SF/HGF, and TGF-α and its EGFR receptor were shown to have particularly strong motogenic effects on glioma cells [156]. Taken together, understanding the role of ECM components and their interaction with proteases, growth factors, and cytokines is essential for the understanding of glioma invasion.

6. CONCLUSIONS AND FUTURE DIRECTIONS

In conclusion, ECM components in the CNS which serve as a reservoir for growth factors and cytokines play a significant role in maintaining a favorable microenvironment allowing proper development and function of the cells within the CNS. In particular, the immediate availability of growth factors and cytokines during cytogenesis of CNS stem cells, as well as the expression of cytokines and growth factors by the maturing CNS, assures proper CNS development. In addition, results of recent studies point to the role of the stem cell activated by the proinflammatory cytokines in repair mechanisms of the injured CNS following their migration into the damaged CNS areas. Under pathological CNS conditions resulting in inflammation, and the proteolytic degradation of the ECM, the cross-talk between ECM components, growth factors, and cytokines significantly contributes to the disease progression. The role of the ECM-cytokine-growth factor cross-talk may be specifically relevant when considering a role of the NG2 progenitor CNS cell population in the development of CNS neoplasia. In addition, the role of a specific ECM component, such as the laminin family, in the progression of several CNS diseases, has recently generated a considerable interest reflected in a growing body of literature. While the search continues for a specific CNS target molecule(s) associated with a particular CNS pathology, a better understanding of molecules within the laminin family and their role in the CNS inflammatory process may lead to a development of new therapeutic approaches. The results of future research evaluating interactions between ECM components, cytokines, and growth factors in the brain should provide a body of information leading to many answers in the identification of more effective therapeutic target molecules for treatment of CNS pathologies.

REFERENCES

1. Sykova, E. Plasticity of extracellular space. In The Neuronal Environment. Walz W, Ed.; Humana Press Inc., 2002; pp. 57–81.
2. Garwood J, Heck N, Rigato F, Faissner A. The extracellular matrix in neural development, plasticity, and regeneration. In The Neuronal Environment. Walz W, Ed.; Humana Press Inc., 2002; pp. 109–58.
3. Schonherr E, Hausser HJ. Extracellular matrix and cytokines: a functional unit. Dev Immunol 2000;7:89–101.

4. Schwartz MA, Ginsberg MH. Networks and altered crosstalk: integrin signalling spreads. Nat Cell Biol 2002;E:65–68.
5. Schuppan D, Ruhl M. Matrix in signal transduction and growth factor modulation. Braz J Med Biol Res 1994;27:2125–41.
6. BlumckeI, Eggli P, Celio MR. Relationship between astrocytic processes and "perineuronal nets" in rat neocortex. Glia 15:131–40.
7. Brightman MW, Kaya M. Permeable endothelium and the interstitial space of brain. Cell Mol Neurobiol 2000;20:111–30.
8. Brightman MW. The brain's interstitial clefts and their glial walls. J Neurocytol 2002;31:596–603.
9. Mercier F, Kitasako JT, Hatton GL. Fractones and their basal laminae in the hypothalamus. J Comp Neurol 2003;455:324–40.
10. Doetsch F. A niche for adult neural stem cells. Curr Opin Genet Dev 2003;13:543–50.
11. Shen Q, Goderie SK, Jin L, Karanth N, Sun Y, Abramova N, Vincent P, Pumiglia K, Temple S. Endothelial cells stimulate self-renewal and expand neurogenesis of neural stem cells. Science 2004;304:1338–40.
12. Kearns SM, Laywell ED, Kukekov VK, Steindler DA. Extracellular matrix effects on neurosphere cell motility. Exp Neurol 2003;182:240–44.
13. Reynolds B, Weiss S. Generation of neurons and astrocytes from isolated cells of the adult mammalian central nervous system. Science 1992;255:1707–10.
14. Doetsch F, Garcia-Verdugo JM, Alvarez-Buylla A. Cellular composition and three-dimentional organization of the subventricular germinal zone in the adult mammalian brain. J Neurosci 1997;17:5046–61.
15. Campos LS, Decker L, Taylor V, Skarnes W. Notch, epidermal growth factor receptor, and beta 1-integrin pathways are coordinated in neural cells. J Biol Chem 2006;281: 5300–09.
16. Stumm RK, Zhou C, Ara T, Lazarini F, Dubois-Dalcq M, Nagasawa T, Hollt V, Schulz S. CXCR4 regulates interneuron migration in the developing neocortex. J Neurosci 2003;23:5123–30.
17. Krathwohl MD, Kaiser JL. Chemokines promote quiescence and survival of human neural progenitor cells. Stem Cells 2004;22:109–18.
18. Sim FJ, Lang JK, Waldau B, Roy NS, Schwartz TE, Pilcher WH, Chandross KJ, Natesan S, Merrill JE, Goldman SA. Complementary patterns of gene expression by human oligodendrocyte progenitors and their environment predict determinants of progenitor maintenance and differentiation. Ann Neurol 2006;59:763–79.
19. Novak U, and Kaye AH. Extracellular matrix and the brain: components and function. J Clin Neurosci 2000;7:280–90.
20. Steindler DA. Glial boundaries in the developing nervous system. Ann Rev Neurosci 1993;16:445–70.
21. Garcion E, Halilagic A, Faissner A, French-Constant C. Generation of an environmental niche for neural stem cell development by the extracellular matrix molecule tenascin C. Development 2004;131:3423–32.
22. Garwood J, Garcion E, Dobbertin A, Heck N, Calco V, ffrench-Constant C, Faissner A. The extracellular matrix glycoprotein Tenascin-C is expressed by oligodendrocyte precursor cells and required for the regulation of maturation rate, survival and responsiveness to platelet-derived growth factor. Eur J Neurosci 2004;20:2524–40.
23. Scadden DT. The stem cell niche as an entity of action. Nature 2006;441:1075–79.

24. Saghatelyan A, de Chevigny A, Schachner M, Lledo PM. Tenascin-R mediates activity-dependent recruitment of neuroblasts in the adult mouse forebrain. Nat Neurosci 2004; 7:347–56.
25. Chekenya M, and Pilkington GJ. NG2 precursor cells in neoplasia: functional, histogenesis and therapeutic implications for malignant brain tumors. J Neurocytol 2002;31:507–21.
26. Heese O, Disko A, Zirkel D, Westphal M, Lamszus K. Neural stem cell migration toward gliomas. Neuro-oncol 2005;7:476–84.
27. Ben-Hur T, Ben-Yosef Y, Mizrachi-Kol R, Ben-Menachem O, Miller A. Cytokine-mediated modulation of MMPs and TIMPs in multipotential neural precursor cells. J Neuroimmunol 2006;175:12–18.
28. Craig CG, Tropepe V, Morshead CM, Reynolds, BA, Weiss S, Van der Kooy, D. In vivo growth factor expansion of endogenous subependymal neural precursor cell populations in the adult mouse brain. J Neurosci 1996;16:2649–58.
29. Whittemore SR, Morassutti DJ, Walters WM, Liu RH, Magnuson DS. Mitogen and substrate differentially affect the lineage restriction of adult rat subventricular zone neural precursor cell populations. Exp Cell Res 1999;252:75–95.
30. Pedersen MW, Tkach V, Pedersen N, Berezin V, Poulsen HS. Expression of a naturally occurring constitutively active variant of the epidermal growth factor receptor in mouse fibroblasts increases motility. Int J Cancer 2004;108:643–53.
31. Al-Hajj M, Wicha MS, Benito-Hernandez A, Morrison SJ, and Clarke MF. Prospective identification of tumorigenic breast cancer cells. Proc Natl Acad Sci USA 2003;100: 3983–88.
32. Drewes LR. What is the blood-brain barrier? A molecular perspective. Cerebral vascular biology. Adv Exp Med Biol 1999;474:111–22.
33. Vlodavsky I, Folkman J, Sullivan R, Fridman R, Ishai-Michaeli R, Sasse J, Klagsbrun M. Endothelial cell-derived basic fibroblast growth factor: synthesis and deposition into sub-endothelial extracellular matrix. Proc Natl Acad Sci USA 1987;84:2292–96.
34. Schilling L, Wahl M. Mediators of cerebral edema. Adv Exp Med Biol 1999;474:123–41.
35. Hirano A, Kawanami T, Liena JF. Electron microscopy of the blood-brain barrier in disease. Microsc Res Tech 1994;27:543–56.
36. Sixt M, Engelhardt B, Pausch F, Hallmann R, Wendler O, Sorokin LM. Endothelial cell laminin isoforms, laminins 8 and 10, play decisive roles in T cell recruitment across the blood-brain barrier in experimental autoimmune encephalomyelitis. J Cell Biol 2001; 153:933–46.
37. Taipale J, Keski-Oja J. Growth factors in the extracellular matrix. FASEB J 1997;11:51–59.
38. Folkman J, Klagsbrun M, Sasse J, Wadzinski M, Ingber D, Vlodavsky I. A heparin-binding angiogenic protein-basic fibroblast growth factor- is stored within basement membrane. Am J Pathol 1988;130:393–400.
39. Loo BM, Salmivirta M. Heparin/heparan sulfate domains in binding and signaling of fibroblast growth factor 8b. J Biol Chem 2002;277:32616–23.
40. Spillmann D, Witt D, Lindahl U. Defining the interleukin-8-binding domain of heparan sulfate. J Biol Chem 1998;273:15487–93.
41. Lortat-Jacob H, Turnbull JE, Grimaud JA. Molecular organization of the interferon gamma-binding domain in heparan sulfate. Biochem J 1995;310:497–505.
42. Frevert CW, Kinsella MG, Vathanaprida C, Goodman RB, Baskin DG, Proudfoot A, Wells TNC, Wight TN, Martin TR. Binding of interleukin-8 to heparan sulfate and chondroitin sulfate in lung tissue. Am J Res Cell Mol Biol 2003;28:464–72.

43. Marshall LJ, Ramdin LSP, Brooks T, Dphil PC, Shute JK. Plasminogen activator inhibitor-1 supports IL-8 mediated neutrophil transendothelial migration by inhibition of the consecutive shedding of endothelial IL8/heparan sulfate/syndecan-1 complexes. J Immunol 2003; 171:2057–65.
44. Salek-Ardakani S, Arrand JR, Shaw D, Mackett M. Heparin and heparan sulfate bind interleukin-10 and modulate its activity. Blood 2000;96:1879–88.
45. Wolff EA, Greenfield B, Taub DD, Murphy WJ, Bennett KL, Aruffo A. Generation of artificial proteoglycans containing glycosaminoglycan-modified CD44. Demonstration of the interaction between RANTES and chondroitin sulfate. J Biol Chem 1999;274:2518–24.
46. Bourdon MA, Ruoslahti E. Tenascin mediates cell attachment through an RGD-dependent receptor. J Cell Biol 1989;108:1149–55.
47. Norenberg U, Hubert M, Rathjen FG. Structural and functional characterization of tenascin-R (restrictin), an extracellular matrix glycoprotein of glial cells and neurons. Int J Dev Neurosci 1996;14:217–31.
48. Pesheva P, Gloor S, Probstmeier R. Tenascin-R as a regulator of CNS glial cell function. Prog Brain Res 2001;132:103–14.
49. Joester A, Faissner A. The structure and function of tenascins in the nervous system. Matrix Biol 2001;20:13–22.
50. Yamaguchi Y. Lecticans: organizers of the brain extracellular matrix. Cell Mol Life 2000;57:276–89.
51. Murakami T, Ohtsuka A. Perisynaptic barrier of proteoglycans in the mature brain and spinal cord. Arch Histol Cytol 2003;66:195–207.
52. Bruckner G, Grosche J, Hartlage-Rubsamen M, Schmidt S, Schachner M. Region and lamina-specific distribution of extracellular matrix proteoglycans, hyaluronan and tenascin-R in the mouse hippocampal formation. J Chem Neuroanat 2003;26:37–50
53. Steindler DA, Settles D, Erickson HP, Laywell ED, Yoshiki A, Faissner A, Kusakabe M. Tenascin knockout mice: barrels, boundary molecules, and glial scars. J Neurosci 1995; 15:1971–83.
54. Bruckner G, Grosche J, Schmidt S, Hartig W, Margolis RU, Delpech B, Seidenbecher CI, Czaniera R, Schachner M. Postnatal development of perineuronal nets in wild-type mice and in a mutant deficient in tenascin-R. J Comp Neurol 2000;428:616–29.
55. Moore SA, Saito F, Chen J, Michele DE, Henry MD, Messing A, Cohn RD, Ross-Barta SE, Westra S, Williamson RA, Hoshi T, Campbell KP. Deletion of brain dystroglycan recapitulates aspects of congenital muscular dystrophy. Nature 2002;418:422–25.
56. Freitag S, Schachner M, Morellini F. Behavioral alterations in mice deficient for the extracellular matrix glycoprotein tenascin-R. Behav Brain Res 2003 17;145:189–207.
57. Montag-Sallaz M, Montag D. Severe cognitive and motor coordination deficits in tenascin-R-deficient mice. Genes Brain Behav 2003;2:20–31.
58. Putthoff P, Akyuz N, Kutsche M, Zardi L, Borgmeyer U, Schachner M. Structure of the murine tenascin-R gene and functional characterisation of the promoter. Biochem Biophys Res Commun 2003;308:940–49.
59. Powell SK, Kleinman HK. Neuronal laminins and their cellular receptors. Int J Biochem Cell Biol 1997;29:401–14.
60. Emsley JG, Hagg T. alpha6beta1 integrin directs migration of neuronal precursors in adult mouse forebrain. Exp Neurol 2003;183:273–85.
61. Tsai HH, Miller RH. Glial cell migration directed by axon guidance cues. Trends Neurosci 2002;25:173–75.

62. Barallobre MJ, Del Rio JA, Alcantara S, Borrell V, Aguado F, Ruiz M, Carmona MA, Martin M, Fabre M, Yuste R, Tessier-Lavigne M, Soriano E. Aberrant development of hippocampal circuits and altered neural activity in netrin 1-deficient mice. Development. 2000;127:4797–10.
63. Yebra M, Montgomery AM, Diaferia GR, Kaido T, Silletti S, Perez B, Just ML, Hildbrand S, Hurford R, Florkiewicz E, Tessier-Lavigne M, Cirulli V. Recognition of the neural chemoattractant netrin-1 by integrins alpha6beta4 and alpha3beta1 regulates epithelial cell adhesion and migration. Dev Cell 2003;5:695–707.
64. Teyssier JR, Rousset F, Garcia E, Cornillet P, Laubriet A. Upregulation of the netrin receptor (DCC) gene during activation of B lymphocytes and modulation by interleukins. Biochem Biophys Res Commun 2001;283:1031–36.
65. Yamaguchi Y. Brevican: a major proteoglycan in adult brain. Perspect Dev Neurobiol 1996;3:307–17.
66. Bovolenta P, Fernaud-Espinosa I. Nervous system proteoglycans as modulators of neurite outgrowth. Prog Neurobiol 2000;61:113–32
67. Bandtlow CE, Zimmermann DR. Proteoglycans in the developing brain: new conceptual insights for old proteins. Physiol Rev 2000;80:1267–90.
68. Yamaguchi Y. Lecticans: organizers of the brain extracellular matrix. Cell Mol Life Sci 2000;57:276–89.
69. Nishiyama A. NG2 cells in the brain: a novel glial cell population. Hum Cell 2001;14(1):77–82.
70. Seidenbecher CI, Smalla KH, Fischer N, Gundelfinger ED, Kreutz MR. Brevican isoforms associate with neural membranes. J Neurochem 2002;83:738–46.
71. Viapiano MS, Matthews RT, Hockfield S. A novel membrane-associated glycovariant of BEHAB/brevican is up-regulated during rat brain development and in a rat model of invasive glioma. J Biol Chem 2003;278:33239–47.
72. Luckenbill-Edds L. Laminin and the mechanism of neuronal outgrowth. Brain Res Rev 1997;23:1–27.
73. Leprini A, Gherzi R, Vecchi E, Borsi L, Zardi L, Siri A. Rat tenascin-R gene: structure, chromosome location and transcriptional activity of promoter and exon 1. Cytogenet Cell Genet 1998;83:115–23.
74. Koppe G, Bruckner G, Hartig W, Delpech B, Bigl V. Characterization of proteoglycan-containing perineuronal nets by enzymatic treatments of rat brain sections. Histochem J 1997;29:11–20.
75. Li H, Leung TC, Hoffman S, Balsamo J, Lilien J. Coordinate regulation of cadherin and integrin function by the chondroitin sulfate proteoglycan neurocan. J Cell Biol 2000; 149: 1275–88.
76. Asher RA, Morgenstern DA, Fidler PS, Adcock KH, Oohira A, Braistead JE, Levine JM, Margolis RU, Rogers JH, Fawcett JW. Neurocan is upregulated in injured brain and in cytokine-treated astrocytes. J Neurosci 2000 1;20:2427–38.
77. Grumet M, Friedlander DR, Sakurai T. Functions of brain chondroitin sulfate proteoglycans during developments: interactions with adhesion molecules. Perspect Dev Neurobiol 1996; 3:319–30.
78. Schmalfeldt M, Bandtlow CE, Dours-Zimmermann MT, Winterhalter KH, Zimmermann DR. Brain derived versican V2 is a potent inhibitor of axonal growth. J Cell Sci 2000; 113:807–16.
79. Paulus W, Baur I, Dours-Zimmermann MT, Zimmermann DR. Differential expression of versican isoforms in brain tumors. J Neuropathol Exp Neurol 1996;55:528–33.

80. Weber P, Bartsch U, Rasband MN, Czaniera R, Lang Y, Bluethmann H, Margolis RU, Levinson SR, Shrager P, Montag D, Schachner M. Mice deficient for tenascin-R display alterations of the extracellular matrix and decreased axonal conduction velocities in the CNS. J Neurosci 1999 1;19:4245–62.
81. Rezajooi K, Pavlides M, Winterbottom J, Stallcup WB, Hamlyn PJ, Lieberman AR, Anderson PN. NG2 proteoglycan expression in the peripheral nervous system: upregulation following injury and comparison with CNS lesions. Mol Cell Nurosci 2004;25: 572–84.
82. Watanabe A, Mabuchi T, Satoh E, Furuya K, Zhang L, Maeda S, Naganuma H. Expression of syndecans, a heparan sulfate proteoglycan, in malignant gliomas: participation of nuclear factor-kappaB in upregulation of syndecan-1 expression. J Neuro-oncol 2006;77:25–32.
83. Reizes O, Lincecum J, Wang Z, Goldberg O, Huang L, Kaksonen M, Ahima R, Hinkes MT, Barsh GS, Rauvala H, Bernfield M. Transgenic expression of syndecan-1 uncovers a physiological control of feeding behavior by syndecan-3. Cell 2001;106:105–16.
84. Hsueh YP, Sheng M. Regulated expression and subcellular localization of syndecan heparan sulfate proteoglycans and the syndecan-binding protein CASK/LIN-2 during rat brain development. J Neurosci 1999;19:7415–25.
85. Hoover CL, Hilgenberg LG, Smith MA. The COOH-terminal domain of agrin signals via a synaptic receptor in central nervous system neurons. J Cell Biol 2003;161:923–32.
86. Margolis RK, Rauch U, Maurel P, Margolis RU. Neurocan and phosphacan: two major nervous tissue-specific chondroitin sulfate proteoglycans. Perspect Dev Neurobiol 1996;3:273–90.
87. Tang X, Davies JE, Davies SJ. Changes in distribution, cell associations, and protein expression levels of NG2, neurocan, phosphacan, brevican, versican V2, and tenascin-C during acute to chronic maturation of spinal cord scar tissue. J Neurosci Res 2003;71: 427–44.
88. Senn C, Kutsche M, Saghatelyan A, Bosl MR, Lohler J, Bartsch U, Morellini F, Schachner M. Mice deficient for the HNK-1 sulfotransferase show alterations in synaptic efficacy and spatial learning and memory. Mol Cell Neurosci 2002;20:712–29.
89. Kiernan BW, Garcion E, Ferguson J, Frost EE, Torres EM, Dunnett SB, Saga Y,Aizawa S, Faissner A, Kaur R, Franklin RJ, Ffench-Constant C. Myelination and behaviour of tenascin-C null transgenic mice. Eur J Neurosci 1999;11:3082–92.
90. Garcion E, Faissner A, Ffrench-Constant C. Knockout mice reveal a contribution of the extracellular matrix molecule tenascin-C to neural precursor proliferation and migration. Development 2001;128:2485–96.
91. Brakebusch C, Seidenbecher CI, Asztely F, Rauch U, Matthies H, Meyer H, Krug M, Bockers TM, Zhou X, Kreutz MR, Montag D, Gundelfinger ED, Fassler R. Brevican-deficient mice display impaired hippocampal CA1 long-term potentiation but show no obvious deficits in learning and memory. Mol Cell Biol 2002 7417–27.
92. Lu M, Grove EA, Miller RJ. Abnormal development of the hippocampal dentate gyrus in mice lacking the CXCR4 chemokine receptor. Proc Natl Acad Sci USA 2002;99: 7090–95.
93. Nakagami Y, Abe K, Nishiyama N, Matsuki N. Laminin degradation by plasmin regulates long-term potentiation. J Neurosci 2000;20:2003–10.
94. Bukalo O, Schachner M, Dityatev A. Modification of extracellular matrix by enzymatic removal of chondroitin sulfate and by lack of tenascin-R differentially affects several forms of synaptic plasticity in the hippocampus. Neuroscience 2001;104:359–69.

95. Hamel MG, Mayer J, Gottschall PE. Altered production and proteolytic processing of brevican by transforming growth factor beta in cultured astrocytes. J Neurochem 2005;93:1533–41.
96. Lauri SE, Kaukinen S, Kinnunen T, Ylinen A, Imai S, Kaila K, Taira T, Rauvala H. Reg1ulatory role and molecular interactions of a cell-surface heparan sulfate proteoglycan (N-syndecan) in hippocampal long-term potentiation. J Neurosci 1999;19:1226–35.
97. Kaksonen M, Pavlov I, Voikar V, Lauri SE, Hienola A, Riekki R, Lakso M, Taira T, Rauvala H. Syndecan-3-deficient mice exhibit enhanced LTP and impaired hippocampus-dependent memory. Mol Cell Neurosci 2002;21:158–72.
98. Raulo E, Tumova S, Pavlov I, Pekkanen M, Hienola A, Klankki E, Kalkkinen N, Taira T, Kilpelainen I, Rauvala H. The two thrombospondin type I repeat domains of the heparin-binding growth-associated molecule bind to heparin/heparan sulfate and regulate neurite extension and plasticity in hippocampal neurons. J Biol Chem 2005;280:41576–83.
99. Malemud CJ. Matrix metalloproteinases (MMPs) in health and disease: an overview. Front Biosci 2006;11:1696–701.
100. Lepperd D, Lindberg RL, Kappos L, Leib SL. Matrix metalloproteinases: multifunctional effectors of inflammation in multiple sclerosis and bacterial meningitis. Brain Res Brain Res Rev 2001;36:249–57.
101. Held-Feindt J, Paredes EB, Blomer U, Seidenbecher C, Stark AM, Mehdorm HM, Mentlein R. Matrix-degrading proteases ADAMTS4 and ADAMTS5 (disintegrins and metalloproteinases with thrombospondin motifs 4 and 5) are expressed in human glioblastomas. Int J Cancer 2006;118:55–61.
102. Vitkovic L, Bockaert J, Jacque C. "Inflammatory" cytokines:neuromodulators in normal brain? J Neurochem 2000;74:457–71.
103. Schneider H, Pitossie F, Balschun D, Wagner A, Del Rey A, Besedowsky HO. A neuromodulatory role of interleukin-1beta in the hippocampus. Proc Natl Acad Sci USA 1998;95:7778–83.
104. Schwab JM, Beschorner R, Nguyen TD, Meyermann R, Schluesener HJ. Differential cellular accumulation of connective tissue growth factor defines a subset of reactive astrocytes, invading fibroblasts, and endothelial cells following central nervous system injury in rats and humans. J Neurotrama 2001;18:377–88.
105. Lagord C, Berry M, Logan A. Expression of TGFbeta2 but not TGFbeta1 correlates with the deposition of scar tissue in the lesioned spinal cord. Mol Cell Neurosci 2002;20: 69–92.
106. Smith GM, Strunz C. Growth factor and cytokine regulation of chondroitin sulfate proteoglycans by astrocytes. Glia 2005;52:209–18.
107. Garcia de Yebnes E, Ho A, Damani T, Fillit H, Blum M. Regulation of the heparin sulfate proteoglycan, perlecan, by injury and interleukin-1alpha. J Neurochem 1999;73: 812–20.
108. DiProspero NA, Meiners S, Geller HM. Inflammatory cytokines interact to modulate extracellular matrix and astrocytic support of neurite outgrowth. Exp Neurol 1997; 148:628–39.
109. Moran LB, Duke DC, Turkheimer FE, Banati RB, Graeber MB. Towards a transcriptome definition of microglial cells. Neurogenetics 2004;5:95–108.
110. Milner R, Campbell IL. Cytokines regulate microglial adhesion to laminin and astrocyte extracellular matrix via protein kinase C-dependent activation of the alpha6beta1 integrin. J Neurosci 2002;22:1562–72.

111. Milner R, Campbell IL. The extracellular matrix and cytokines regulate microglial integrin expression and activation. J Immunol 2003;170:3850–58.
112. Gilat D, Cahalon L, Hershkoviz R, Lider O. Interplay of T cells and cytokines in the context of enzymatically modified extracellular matrix. Immunol Today 1996;17:16–20.
113. Ohuchida K, Mizumoto K, Murakami M, Qian LW, Sato N, Nagai E, Matsumoto K, Nakamura T, Tanaka M. Radiation to stromal fibroblasts increases invasiveness of pancreatic cancer cells through tumor-stromal interactions. Cancer Res 2004;64:3215–22.
114. He X, Brenchley, PEC, Jayson GC, Hampson L, Davies J, Hampson IN. Hypoxia increases heparanase-dependent tumor cell invasion, which can be inhibited by anti-heparanase antibodies. Cancer Res 2004;64:3928–33.
115. Middleton J, Patterson AM, Gardner L, Schmutz C, Ashton BA. Leukocyte extravasation: chemokine transport and presentation by the endothelium. Blood 2002;100:3853–60.
116. Gotte M. Syndecans in inflammation. FASEB J 2003;17:575–91.
117. Hart MN, Fabry Z, Love-Homan L, Keiner J, Sadewasser KL, Moore SA. Brain microvasculature, smooth muscle and endothelial cells produce granulocyte macrophage colony-stimulating factor and support colony formation of granulocyte-macrophage-like cells. Am J Path 1992;141:421–27.
118. Wang MJ, Kuo JS, Lee WW, Huang HY, Chen WF, Lin SZ. Translational event mediates differential production of tumor necrosis factor-alpha in hyaluronan-stimulated microglia and macrophages. J Neurochem 2006;97:857–71.
119. Brown KA. Factors modifying the migration of lymphocytes across the blood-brain barrier. Int Immunopharmacol 2001;1:2043–62.
120. Giannelli G, De Marzo A, Scagnolari C, Bergamini C, Fransvea E, Bagnato F, Bellomi F, Millefiorini E, Gasperini C, Antonaci S, Antonelli G. Proteolytic balance in patients with multiple sclerosis during interferon treatment. J Interferon Cytokine Res 2002;22:689–92.
121. Siao CJ, Tsirka SE. Extracellular proteases and neuronal cell death. Cell Mol Biol 2002;48:151–61.
122. Kouwenhoven M, Ozenci V, Gomez A, Yarilin D, Giedraitis V, Press R, Link H. Multiple sclerosis: elevated expression of matrix metalloproteinases in blood monocytes. J Autoimmun 2001;16:463–70.
123. Rosenberg GA, Estrada EY, Dencoff JE. Matrix metalloproteinases and TIMPs are associated with blood-brain barrier opening after reperfusion. Stroke 1998;29:2189–95.
124. Dubois B, Masure S, Hurtenbach U, Paemen L, Heremans H, van den Oord J, Sciot R, Meinhardt T, Hammerling G, Opdenakker G, Arnold B. Resistance of young gelatinase B-deficient mice to experimental autoimmune encephalomyelitis and necrotizing tail lesions. J Clin Invest 1999;104:1507–14.
125. Lepperd D, Ford J, Stabler G, Grygar C, Lienert C, Huber S, Miller KM, Hauser SL, Kappos L. Matrix metalloproteinase-9 (gelatinase B) is selectively elevated in CSF during relapses and stable phases of multiple sclerosis. Brain 1998;121:2327–34.
126. Yushchenko M, Mader M, Elitok E, Bitsch A, Dressel A, Tumani H, Bogumil T, Kitze B, Poser S, Weber F. Interferon-beta-1 decreased matrix metalloproteinase-9 serum levels in primary progressive multiple sclerosis. J Neurol 2003;250:1224–28.
127. Gilli F, Bertolotto A, Sala A, Hoffmann F, Copobianco M, Malucchi S, Glass T, Kappos L, Lindberg RL, Lepperd D. Neutralizing antibodies against IFN-beta in multiple sclerosis: antagonization of IFN-beta mediated suppression of MMPs. Brain 2004;127:259–68.
128. Karabudak R, Kurane A, Guc D, Sengelen M, Canpinar H, Kansu E. Effect of interferon beta-1a on serum matrix metalloproteinase-9 (MMP-9) and tissue inhibitor of matrix

metalloproteinase (TIMP-1) in relapsing remitting multiple sclerosis patients. One year follow-up results. J Neurol 2004;251:279–83.
129. Rosenberg GA, Estrada E, Kelly RO, Kornfeld M. Bacterial collagenase disrupts extra-cellular matrix and opens blood-brain barrier in rat. Neurosci Lett 1993;160:117–19.
130. Leib SL, Clements JM, Lindberg RL, Heimgartner C, Loeffler JM, Pfister LA, Tauber MG, Lepperd D. Inhibition of matrix metalloproteinases and tumor necrosis factor alpha converting enzyme as adjuvant therapy in pneumococcal meningitis. Brain 2001; 124:1734–42.
131. Kaul M, Lipton SA. Mechanism of neuronal injury and death in HIV-1 associated dementia. Curr HIV Res 2006;4:307–18.
132. Satoh K, Hata M, Shimizu T, Yokota H, Akatsu H, Yamamoto T, Kosaka K, Yamada T. Lib, transcriptionally induced in senile plaque-associated astrocytes, promotes glial migration through extracellular matrix. Biochem Biophys Res Commun 2005;335: 631–36.
133. Suryadevara R, Holter S, Borgmann K, Persidsky R, Labenz-Zink C, Persidsky Y, Gendelman HE, Wu L, Ghorpade A. Regulation of tissue inhibitor of metalloproteinase-1 by astrocytes: links to HIV-1 dementia. Glia 2003;44:47–56.
134. Giraudon P, Szymocha R, Buart S, Bernard A, Cartier L, Belin MF, Akaoka H. T lymphocytes activated by persistent viral infection differentially modify the expression of metalloproteinases and their endogenous inhibitors, TIMPs, in human astrocytes: relevance to HTLV-1 induced neurological disease. J Immunol 2000;164:2718–27.
135. Cladera J, Martin I O'Shea P. The fusion domain of HIV gp41 interacts specifically with heparin sulfate on the T-lymphocyte cell surface. EMBO J 2001;20:19–26.
136. Tyagi M, Rusnati M, Presta M, Giacca M. Internalization of HIV-1 Tat requires cell surface heparin sulfate proteoglycans. J Biol Chem 2001;276:3254–61.
137. Hallak LK, Spillmann D, Collins PL, Peeples ME. Glycosaminoglycan sulfation requirements for respiratory syncytial virus infection. J Virol 2001;74:10508–13.
138. Bolteus AJ, Berens ME, Pilkington GJ. Migration and invasion in brain neoplasms. Curr Neurol Neurosci Rep 2001;1:225–32.
139. Bellail AC, Hunter SB, Brat DJ, Tan C, Van Meir EG. Microregional extracellular matrix heterogeneity in brain modulates glioma cell invasion. Int J Biochem Cell Biol 2004;36: 1046–69.
140. Hibino S, Shibuya M, Engbring JA, Mochizuki M, Nomizu M, Kleinman HK. Identification of an active site on the laminin alpha5 chain globular domain that binds to CD44 and inhibits malignancy. Cancer Res 2004;64:4810–16.
141. Hibino S, Shibuya M, Hoffman MP, Engbring JA, Hossain R, Mochizuki M, Kudoh S, Nomizu M, Kleinman HK. Laminin alpha5 chain metastasis- and angiogenesis-inhibiting peptide blocks fibroblast growth factor 2 activity by binding to the heparin sulfate chains of CD44. Cancer Res 2005;65:10494–01.
142. Ljubimova JY, Fugita M, Khazenzon NM, Das A, Pikul BB, Newman D, Sekiguchi K, Sorokin LM, Sasaki T, Black KL. Association between laminin-8 and glial tumor grade, recurrance, and patient survival. Cancer 2004;101:604–12.
143. Zhou Z, Doi M, Wang J, Cao R, Liu B, Chan KM, Kortesmaa J, Sorokin L, Cao Y, Tryggvason K. Deletion of laminin-8 results in increased tumor neovascularization and metastasis in mice. Cancer Res 2004;64:4059–63.
144. Johansson FK, Goransson H, Westermark B. Expression analysis of genes involved in brain tumor progression driven by retroviral insertional mutagenesis in mice. Oncogene 2005;24:3896–905.

145. Galli R, Binda E, Orfanelli U, Cipelletti B, Gritti A, De Vitis S, Fiocco R, Foroni C, Dimeco F, Vescovi A. Isolation and characterization of tumorigenic, stem-like neural precursor from human glioblastoma. Cancer Res 2004;64:7011–21.
146. Lin H-X, Dahlin-Huppe K, Stallcup WB. Interaction of the NG2 proteoglycan with the actin cytoskeleton. J Cell Biochem 1996;63:463–77.
147. Mahesparan R, Read TA, Lund-Johansen M, Skaftnesmo KO, Bjerkvig R, Engebraaten O. Expression of extracellular matrix components in a highly infiltrative *in vivo* glioma model. Acta Neuropathol 2003;105:49–57.
148. Tatenhorst L, Senner V, Puttmann S, Paulus W. Regulators of G-protein signaling 3 and 4 (RGS3, RGS4) are associated with glioma cell motility. J Neuropathol Exp Neurol 2004;63:210–22.
149. Nuttall RK, Pennington CJ, Taplin J, Wheal A, Yong VW, Forsyth PA, Edwards DR. Elevated membrane-type matrix metalloproteinases in gliomas revealed by profiling proteases and inhibitors in human cancer cells. Mol Cancer Res 2003;1:333–45.
150. Esteve, PO, Chicone E, Robledo O, Aodjit F, Descoteaux A, Potworowski EF, St-Pierre Y. Protein kinase C-zeta regulates transcription of the matrix metalloproteinase-9 gene induced by IL-1 and TNF-alpha in glioma cells via NF-kappa B. J Biol Chem 2002; 20:277:35150–55.
151. Deryugina EI, Soroceanu L, Strongin AY. Up-regulation of vascular endothelial growth factor by membrane-type 1 matrix metalloproteinase stimulates human glioma xenograft growth and angiogenesis. Cancer Res 2002;62:580–88.
152. Wiranowska M, Naidu AK. Interferon effect on glycasominoglycans in mouse glioma in vitro. J Neuro-Oncol 1994;18:9–17.
153. Wiranowska M, Tresser N, Saporta S. The effect of interferon and anti-CD44 antibody on mouse glioma invasiveness *in vitro*. Anticancer Res 1998;18:3331–38.
154. Wiranowska M, Rojiani AM, Gottschall PE, Moscinski LC, Johnson J, Saporta S. CD44 expression and MMP-2 secretion by glioma cells: effect of interferon and anti-CD44 antibody. Anticancer Res 2000;20:4301–6.
155. Kasza A, Kowanetz M, Poslednik K, Witek B, Kordula T, Koj A. Epidermal growth factor and pro-inflammatory cytokines regulate the expression of components of plasminogen activation system in U373-MG astrocytoma cells. Cytokine 2001;16:187–90.
156. Brockmann MA, Ulbricht U, Gruner K, Fillbrandt R, Westphal M, Lamszus K. Glioblastoma and cerebral microvascular endothelial cell migration in response to tumor-associated growth factors. J Neurosurg 2003;52:1391–99.

Cytokines and the Brain
Edited by C. Phelps and E. Korneva

Acidic Fibroblast Growth Factor, a Satiety Substance, with Diverse Physiological Significance

YUTAKA OOMURA

Department of Integrated Physiology, Kyushu University, Faculty of Medicine, Fukuoka 812-0054, Japan

ABSTRACT

The concentration of acidic fibroblast growth factor (aFGF), which is found in cerebrospinal fluid (CSF), markedly increases after the start of feeding. Food intake was dose-dependently suppressed by picomole amounts of aFGF introduced into the third cerebral ventricle and facilitated by addition of anti-FGF antibody. The ependymal cells, where aFGF is produced, release aFGF by responding to glucose increase in CSF after feeding. Released aFGF diffused into the brain parenchyma and was taken up by neurons in the hypothalamus and hippocampus. Emotional and spatial learning and memory were significantly more reliable after introducing aFGF into the hippocampus. Marked dose-dependent increases in plasma ACTH and corticosterone were detected from 20 min to 2 h after intracerebroventricular or intravenous administration of aFGF. Plasma epinephrine and norepinephrine also increased markedly for up to 120 min. Concomitant increases occurred in the activity of efferent sympathetic nerves. These findings suggest that aFGF activates the hypothalamic–pituitary–adrenal axis and sympathetic outflow. The effects of aFGF on phagocytosis in peritoneal macrophages from thioglycollate-treated mice were examined using flow cytometry. Phagocytosis of fluorescein isothiocyanate-labeled latex particles was enhanced by aFGF in a dose-dependent manner. When viewed together these results indicate that aFGF is not only one of the most potent substance yet found for the suppression of feeding, but it is also extremely effective in memory facilitation and sympathetic activation as well as biodefence.

1. INCREASE IN FIBROBLAST GROWTH FACTOR IN CEREBROSPINAL FLUID AFTER FEEDING

Acidic and basic fibroblast growth factors (aFGF, bFGF) are potent mitogens and are present throughout the nervous system. A postprandial increase in aFGF and bFGF was detected in rat cerebrospinal fluid (CSF), and a dose-dependent increase was also observed after intraperitoneal (i.p.) injection of various concentrations of glucose (Fig. 1 and *inset*). The concentration of aFGF in CSF, which was 11.0 ± 3.0 pg/mL $(0.73 \pm 0.2$ pmol) in

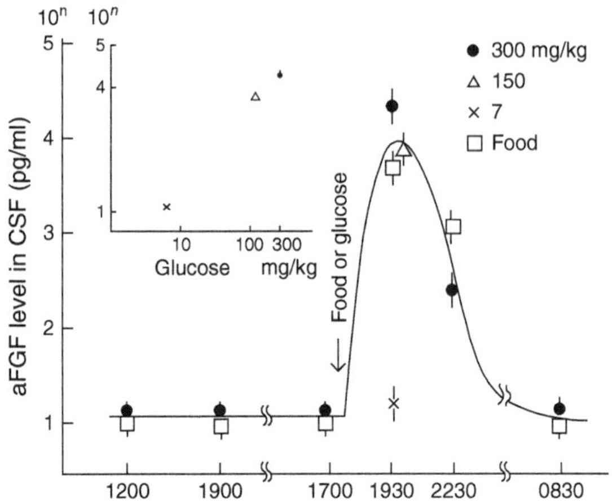

FIGURE 1. aFGF change in CSF before and after feeding or glucose injection. Freely moving rats: □; food intake. i.p. glucose injections: 300, 150, and 7 mg/kg. Each point, mean ± SD of 5–7 determination., Inset: dose-dependent effect of glucose injection. Symbols, same as above; indicates mean values only. Modified from Hanai et al. [1].

pre-feeding state, increased to ~10 ng/mL (0.7 nmol) 2 h after feeding. In rats, intracerebral injection into the third ventricle (III i.c.v.) of aFGF or bFGF (50–200 ng/rat), 1 h before dark period, suppressed food intake in a dose-responsive fashion, whereas central infusion of inactivated FGF was ineffective [1]. The minimum effective dose of aFGF was 3.3 pmol and that of bFGF was 6.6 pmol. Neither aFGF nor bFGF affected water intake. Infusion (III i.c.v.) of either 400 or 800 ng/rat of the first 15 amino acids of aFGF ($aFGF_{1-15}$), 1 h before dark period, dose-dependently decreased food intake for 3 and 8 h, respectively. Whole nighttime food intake (for 8 h), for example, decreased significantly (paired Student's *t*-test, $P < 0.01$) by almost 2 and 3 g from the control intake with the infusion of 400 and 800 ng, respectively. Infusion of up to 800 ng/rat of the last 26 amino acids of aFGF ($aFGF_{114-140}$) did not affect food intake. The $aFGF_{1-15}$ fraction is ~1/16 as potent as the complete aFGF fraction [2]. The $bFGF_{1-24}$ fraction was about equal to the $aFGF_{1-15}$ fraction in effect. However, $aFGF_{1-29}$ had about four times potency compared with that of $aFGF_{1-15}$. All of these fractions of aFGF and bFGF had no mitogenic activity.

2. EFFECTS OF ANTIBODIES TO AFGF AND BFGF ON FOOD INTAKE

The nighttime and total daily food intake were significantly (paired Student's *t*-test, $P < 0.05$) increased by bilateral lateral hypothalamic area (LHA) intraparenchymal injections of anti-aFGF antibody (240 ng/rat) when applied at 1900 h [2]. Daytime food intake was not changed. Similar injections of anti-bFGF antibody also significantly increased night, day, and total daily food intake (paired Student's *t*-test, $P < 0.05$). The increase in food intake, for example, was ~1.2 g for the first 3 h at nighttime and 2 g for the entire dark period (8 h) in both cases. Infusion of 240 ng/rat preimmune immunoglobulin (IG) did not affect food intake. Injections of the antibodies to both the $aFGF_{1-15}$ and the $aFGF_{114-140}$ fractions into the LHA (240 ng/rat) effectively increased food intake. The increase in food intake was ~1.6g for the first 3 h at nighttime. This result, despite the (relative) ineffectiveness of the $aFGF_{114-140}$ fraction, was probably due to the effective blockade of binding on the receptor sites for the entire ($aFGF_{114-140}$) molecule by the anti-$aFGF_{114-140}$ antibody, as well as the anti-$aFGF_{1-15}$ antibody fractions. The net result would appear as a decrease in the effectiveness of the total aFGF population of receptor sites.

3. FGF EFFECTS ON LHA AND VMH NEURONS

3.1. Suppression of rat LHA neurons after application of aFGF

The neuronal electrical activity of glucose-sensitive neurons (GSNs) in the LHA, feeding center, was reversibly suppressed with long latency and duration by electrophoretic application of aFGF using multibarrel glass capillary technique [1]. The latency of the suppression was 7.0 ± 2.3 min and the duration was 10.6 ± 4.1 min. The difference in effects of aFGF on GSNs and non-GSNs was statistically significant (chi-square test, $P < 0.01$). To explain the extremely long latency and duration of the aFGF effect on GSNs, we examined the effects of applying a direct activator of protein kinase C, 1-oleoyl-2-acetyl-glycerol (OAG) on 42 LHA neurons. The effects of OAG on 16 neurons of the 42 were nearly identical to the effects of aFGF, including the long latency and duration [2]. We suggest that the effect of aFGF on GSNs is through its activation of protein kinase C. Neurons in the ventromedial nucleus (VMH), satiety center were not affected by aFGF [1].

3.2. FGF receptor

We investigated the localization of FGF receptor-1 in rat brain by immunohistochemistry using a polyclonal antibody against an acidic peptide sequence of chicken FGF receptor-1. Positive neurons were distributed widely in various brain regions, but were particularly abundant in such regions as the LHA, substantia nigra, locus coeruleus, and raphe nuclei. We could not find the FGF receptor-1 in the VMH neurons [3].

4. SOURCE OF aFGF IN THE BRAIN

Staining the brain with anti-aFGF antibody usually revealed aFGF in ependymal cells in the third cerebral ventricle walls [4]. However, 2 h after a meal or a 300 mg/kg glucose i.p. injection, the ependymal cells were clear and aFGF appeared in neurons in the LHA, dorsomedial septum, hippocampus, amygdala, and other central regions, but not in the VMH [2,4]. When cDNA of the 30–100 bases from the middle part of aFGF were synthesized and hybridized, only ependymal cells were stained [2,4]. Thus, this procedure revealed mRNA in the ependymal cells only, indicating that aFGF is produced in those cells.

5. GLUCOSE SENSOR OF A NEW TYPE IN THE BRAIN

In order to better understand the migration of aFGF, glucose (0.2 mg/10 μL) was applied directly into the third cerebral ventricle (III i.c.v.), and the level of aFGF was determined by

Hydra bioassay [1]. After a meal, glucose in CSF doubles from 2 to 4 mmol/L. Before the glucose injection, aFGF level in CSF was 0.7 pmol, within 15 min this increased to 2.6 nmol; it peaked at ~7.5 nmol at 45–60 min; decreased to ~0.7 nmol at 2 h and to 4.6 pmol at 3 h. The same amount of mannitol or Krebs Ringer solution injected into the III i.c.v. had no effect. The minimum effective dose of 3.3 pmol to suppress feeding is, thus, well within physiological limits. This action of glucose on ependymal cells to release aFGF is comparable to the action of glucose on pancreatic β cells to bring about the release of insulin. Production and release of aFGF is thus an endocrine phenomenon in the brain that is parallel to a visceral phenomenon.

6. aFGF AND MEMORY FACILITATION

The neuronal uptake of aFGF in the hippocamupus 2 h after a meal or glucose i.p. injection led us to investigate the effects of endogenous and exogenous aFGF on learning and memory [5].

6.1. Passive avoidance task

Four groups of mice were injected i.p. with 300 mg/kg glucose, 1, 2, 3, or 5 h before being shocked when they entered a dark box (acquisition). After 24 h the same trials were repeated (retention, affective memory). Behavior of the 1-h, 3-h, and 5-h groups were not significantly different from that of the controls (mice injected with Krebs Ringer solution). The latency of the 2 h group to enter the dark box, 61.3 ± 7.6 s (mean ± SEM), was significantly longer than that of the controls, 20.5 ± 3.5 s (Student's *t*-test, $P < 0.01$, n = 8) and that of the 1, 3, and 5 h groups, 28.2 ± 8.0, 22.9 ± 7.3, and 19.3 ± 4.5 s, respectively [5]. In similar trials, anti-aFGF antibody was administered i.c.v. (1 μg/mouse, bilaterally) and glucose was injected 30 min later. Using on the first trials as controls, all comparisons were made 2 h after glucose administration. In these trials, the latency was 26.6 ± 7.1 s (n = 10), which tended to be longer than the acquisition time (8.2 ± 0.7 s), but was significantly shorter than that of the glucose alone group time of 59.1 ± 8.7 s (n = 10). Infusion of the same amount of preimmune IgG bilaterally did not affect the retention (61.3 ± 12.8 s, n = 10).

When aFGF was infused ICV continuously (20 μg/mL, 0.5 μL/h, bilateral) into rats with implanted osmotic minipumps for 1 week, the passive avoidance latency continued to increase during all of the infusion time. The rate of this increase in latency was significantly greater than the rate during infusion with saline (Student's *t*-test, $P < 0.05$, n = 10).

6.2. Morris water maze task

The water maze pool was a 100-cm diameter, 20-cm deep plastic water tank. A small platform was fixed 1-cm below the surface of the water made opaque by floating styrene foam granules on the surface, so that the platform was invisible to the mouse. A trial was started by placing a mouse into the water at one of four locations. Within each block of four trials, each mouse started at one of the four starting positions, but the sequence of the positions was changed randomly. In each trial, the latency to escape onto the platform was recorded. If the mouse did not find the platform within 120 s, the trial was terminated and a maximum score of 120 s was recorded. All tests on a particular mouse were done on alternate days in 1 week. Each mouse received one block of four trials in 1 day with intertrial intervals of 10 min. There were 1-day intervals between blocks and four blocks were tested in 7 elapsed days. Food was deprived from 7 to 10 h before the trials until to 3 h after, in each testing day.

The mean latency of eight groups (n = 10 per group) to climb on the platform in the pool in the Morris water maze task were measured. Each point (one block) is the mean of four times (10 mice × 4 times each). The first block of performance was significantly more rapid after i.p. glucose administration 2 and 3 h before each first trial, 17.5 ± 2.8 and 16.2 ± 1.7 s respectively, than after saline administration at the same intervals, 34.2 ± 3.9 and 26.5 ± 3.7 s, respectively (Student's *t*-test, at 2 h $P < 0.01$; at 3 h, $P < 0.05$). On the other hand, mice that received glucose injections 1 and 5 h before training, had latencies similar (23.8 ± 2.7 and 20.0 ± 1.9 s) to those of the saline control groups (29.5 ± 3.3 and 25.8 ± 2.7 s, respectively). There were no differences between any groups in the second and fourth blocks. These results indicate that the glucose application on 2- and 3-h groups facilitated spatial memory with time-dependent induction. Either anti-aFGF antibody (n = 10) or preimmune IgG (n = 8) was applied to mice 30 min before glucose injection. The latency of the antibody group to climb on the platform, 35.5 ± 5.3 s, was significantly longer than that of the control IgG group, 17.6 ± 2.2 s, in the first block (Student's *t*-test, $P < 0.05$). This indicates that facilitation of spatial memory by glucose was due to aFGF in the central nervous system.

6.3. Mechanism of aFGF facilitation of learning and memory

6.3.1. Long-term potentiation

Effects of aFGF (0.5–25 ng/mL) on synaptic transmission were investigated in rat hippocampal slices [6]. Stimulation was applied to the Schaffer collateral/commissural afferents and evoked synaptic potentials were recorded at the apical dendtites in CA1 pyramidal cell layer. As shown in Fig. 2, when brief tetanic stimulation (7 pulses at 100 Hz) was applied 30 min after the perfusion of aFGF, evoked synaptic potentials increased in magnitude after the tetanus (short-term potentiation, just after stimulation for 5 min) and facilitated the generation of long-term potentiation (30 min after stimulation). These effects of aFGF were dose-dependent. The facilitation of short-term and long-term potentiation was not evident when aFGF was applied immediately or 10 min after the tetanus. Similar facilitations were obtained by $aFGF_{1–15}$ (101–400 ng/mL). The carboxyl terminal fragment $aFGF_{114–140}$ had no effect. The results suggest that aFGF modulates synaptic efficacy through FGF receptor 1 [3] and can facilitate learning and memory through mechanisms related to the generation of long-term potentiation.

7. PERSPECTIVES

After food intake, centrally released aFGF reaches first the LHA and suppresses food intake through inhibiting the neuronal activity of the GSNs. Second, aFGF reaches the hippocampus and facilitates learning and memory. Therefore, feeding is important, because it not only maintains body energy homeostasis but also leads to preparation of readiness for memory facilitation.

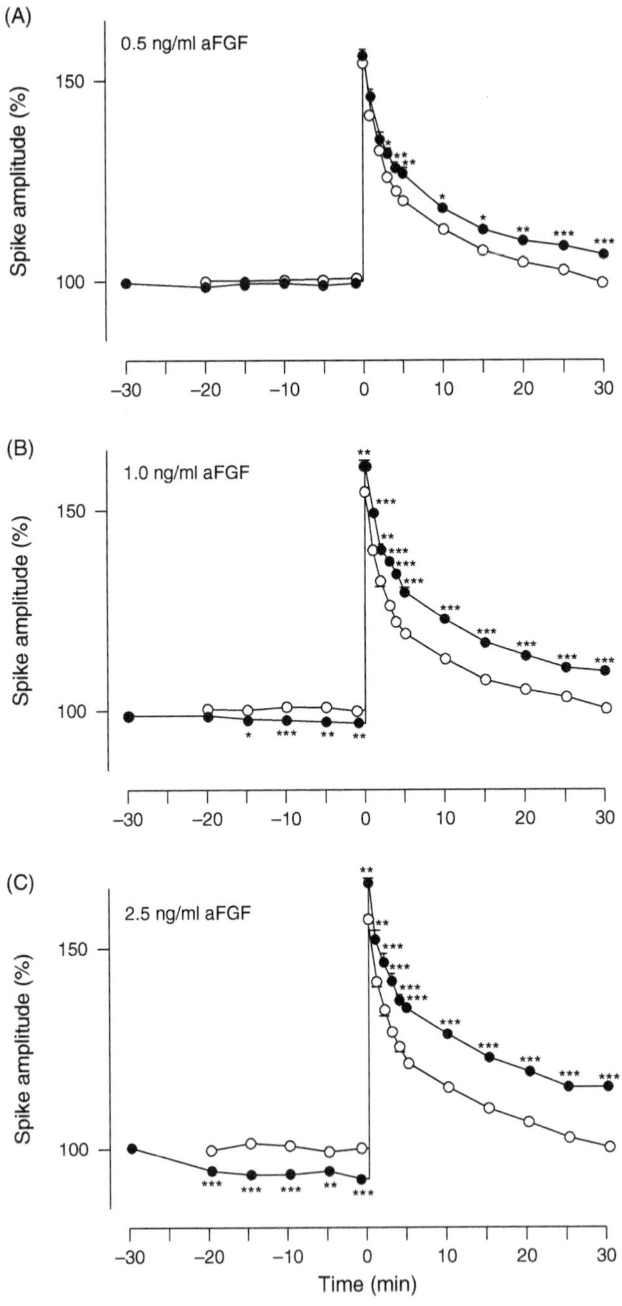

Figure 2. Effect of aFGF on long-term potentiation induced by 7 pules at 100 Hz tetanus. Ordinate, evoked synaptic potential amplitude expressed as a percent change of mean basal evoked synaptic potential amplitude before tetanic stimulation defined as 100 per cent. After the control potentiation (open circles) induced by the first tetanus was observed for 30 min in Kreb Ringer solution, aFGF was perfused for 30 min. The second tetanic stimulation was applied 30 min after beginning of aFGF perfusion (A, B, and C). Effect of 0.5 (n = 9), 1.0 (n = 9), and 2.5 (n = 9) ng/mL aFGF on potentiation (filled circles) induced by the second tetanic stimulation, respectively, $*P < 0.05$; $**P < 0.01$; $***P < 0.001$ [6].

8. ACTIVATION OF AFGF ON HYPOTHALAMIC–PITUITARY–ADRENOCORTICAL AXIS

The peptide aFGF released from the ependymal cells during feeding accesses the parvocellular neurons in the paraventricular nucleus (PVN) and activates hypothalamic–pituitary–adrenal axis [7].

8.1. Effects of aFGF administered i.c.v. on plasma corticosterone and ACTH [8]

Significant increases in plasma corticosterone concentration were evoked in a dose-dependent manner in responese to aFGF administered i.c.v (at 1 and 10 ng; Fig. 3A). The corticosterone levels reached a maximal level at 60 min and remained elevated for up to a further 120 min after the administration. The integrated corticosterone responses (for the 180 min after the administration) also showed a dose-dependent increase (Fig 3. *inset*).

When 10 ng aFGF was administered via the III i.c.v. route, plasma ACTH concentration was already increased significantly at 5 min after the injection; it continued to increase and peaked at 15 min after the injection as shown in Fig 3B. Thereafter, it gradually decreased, but remained elevated until 150 min after the injection and returned to the basal level at 180 min. At this time (180 min), the plasma corticosterone was still at an elevated plateau level after its peak at 60 min (see Fig. 3A).

The responses of corticosterone and ACTH to aFGF administered through the i.c.v. route were significantly greater than those evoked through the intravenous (i.v.) route, the integrated responses in the former experiments being greater than those in the latter by almost twofold at 1 ng and 1.4-fold at 10 ng ($P < 0.05$ in each case).

The pretreatment with anti-CRF antibody in the cerebroventricle significantly attenuated the increase in corticosterone evoked by i.c.v. administration of aFGF but had almost no effect on the response to its i.v. administration. These findings suggest that endogenous aFGF or exogenous aFGF (given i.c.v.) directly stimulates corticotrophin releasing factor (CRF) secretion from parvocellular neurons and that this cortictrophin releasing factor (CRF) then activates the pituitary–adrenal axis.

The aFGF concentration in the CSF may be within the physiological range after an i.c.v. administration of 10 ng (0.7×10^{-12} mol) aFGF. In fact, if we assume that the total volume of the rat CSF is ~300 μL, the concentration of aFGF in the CSF would be ~2 pmol/mL. This would seem to lie within the normal range, because the aFGF cencentration in rat CSF increases from ~0.7 pmol/mL to 0.7 nmol/mL at 15 min, 7.5 nmol/mL at 45 min, and 4.6 pmol/mL at 3 h after 4 mM glucose introduction into the cerebral ventricle [9] and after food intake [1]. The figure of 4 mM glucose corresponds to the glucose level in the CSF after food intake or intraperitoneal injection of 300 mg/kg glucose [1].

8.2. Effect of i.c.v. administration of aFGF on plasma epinephrine and norepinephrine

Administration of either 1 or 10 ng/rat aFGF III i.c.v. induced dose-dependent increases in the plasma levels of both epinephrine (Epi) and norepinephrine (NE) [10]. A significant increase to 250 pg/mL by 1 ng and 600 pg/mL in Epi from the basal 150 pg/mL after 10 ng was first detected at 60 min, and the response seemed to be peaking at 150 min (600 pg/mL, after 1 ng; 1700 pg/mL after 10 ng). At 180 min after the aFGF administration (when sampling ended), plasma epinephrine levels were showing no sign of returning to the basal level. Similar changes

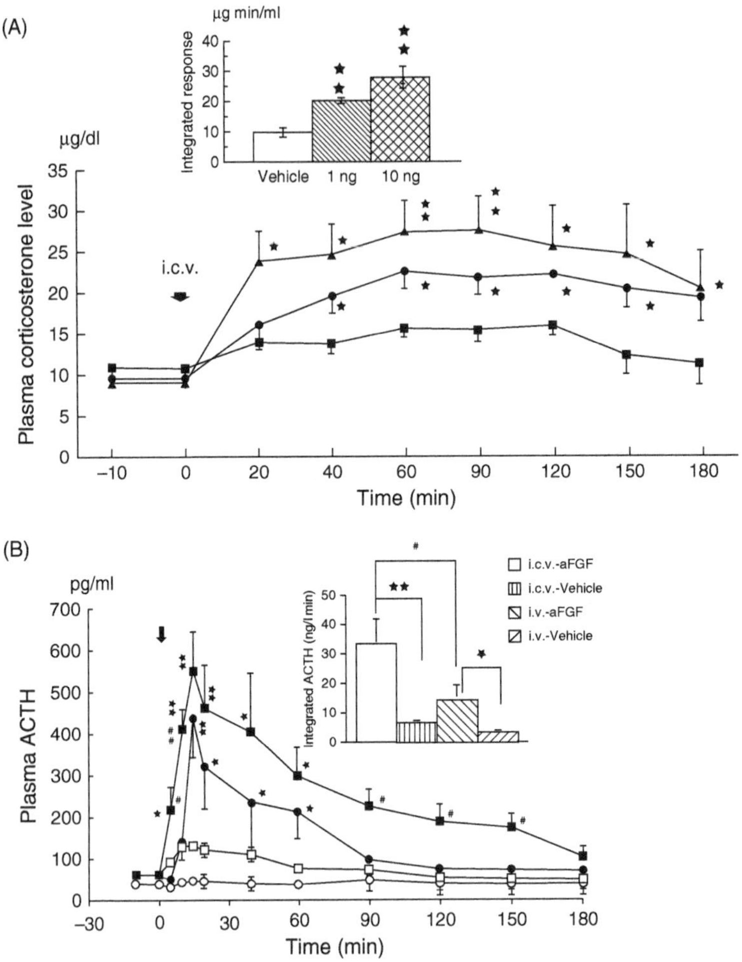

Figure 3. (A) Effect of III i.c.v. aFGF administration of on plasma corticosterone concentration. Vehicle (■, n = 6) or aFGF at 1 (●, n = 7) or 10 ng/rat (▲, n = 5). Inset: corticosterone responses integrated over the 180 min administration of aFGF; stat. Values are means ± SE. $*P < 0.05$, $**P < 0.01$ versus vehicle controls. Note that abscissa is nonlinear. (B) Effect of III i.c.v. or i.v. administration of aFGF on plasma ACTH concentration 10 ng/rat aFGF or vehicle (at arrow). Symbols indicate i.c.v. aFGF (■, n = 6) and its vehicle (□, n = 5), and i.v. aFGF (●, n = 7) and its vehicle (○, n = 6). Inset: ACTH responses integrated over the 180 min after i.c.v. or i.v. administration of aFGF; n, number of animals. Values are means ± SE. $*P < 0.05$, $**P < 0.01$ versus vehicle controls. $\#P < 0.05$, $\#\#P < 0.01$ versus intravenous aFGF [8].

were observed in the case of plasma NE levels. Injection of heat-treated aFGF (10 ng) did not induce increases in plasma catecholamines. Similar increases in Epi and NE also occured after i.v. injection of aFGF, although in absolute terms, aFGF through this route resulted in almost one-tenth of the plasma catecholamine increase when compared with that which occurred when aFGF was administered via the i.c.v. route.

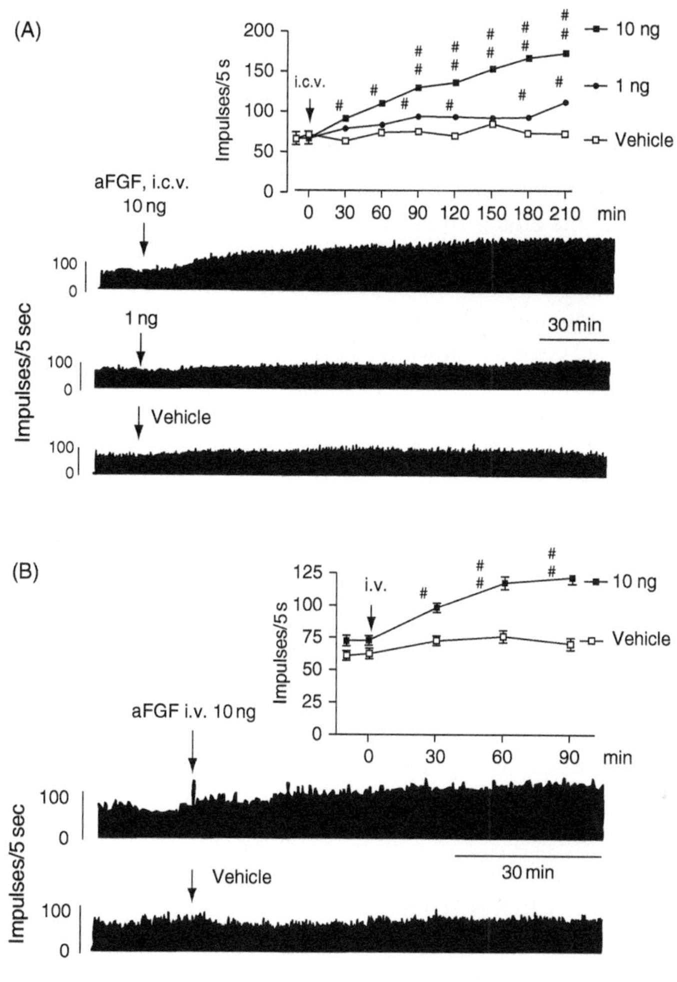

Figure 4. Effect of III i.c.v. or i.v. administration of aFGF on the efferent firing rate in a sympathetic branch of the left adrenal nerve. (A) Efferent discharge rate in response to vehicle (*bottom*) or aFGF at 1 (*middle*) or 10 ng i.c.v. (top). Inset: mean discharge rates (±SE) for vehicle (n = 3) and for 1 (n = 10), and 10 ng i.c.v. (n = 10) aFGF. (B) Efferent discharge rate in response to vehicle (*bottom*) or aFGF at 10 ng i.v. (top). Inset: mean discharge rates (±SE) for vehicle (n = 3) and 10 ng i.v. aFGF (n = 10), #$P < 0.05$. ##$P < 0.01$ versus Vehicle control at same time point [10].

8.3. Effect of aFGF on sympathetic efferent discharge to the adrenal gland, spleen, and brown adipose tissue

In response to 10 ng of aFGF administered i.c.v., the sympathetic efferent activity in the adrenal nerve showed a clear increase (in terms of the multi unit discharge rate) starting at ~15 min and reaching significance at or before 30 min (Fig. 4A). The mean discharge rate showed a dose-dependent increase, and the increases induced by 1 or 10 ng/rat aFGF were both significantly greater than that induced by vehicle (Fig. 4A, *inset*). When 10 ng/rat aFGF was administered i.v., it again markedly facilitated the discharge (Fig. 4B). By comparison with the basal level, the mean discharge rate at 90 min after the 10 ng/rat i.c.v. dose was increased by ~86% versus

67% (i.v.) (see Fig. 4A and B, *insets*) These similar activations were observed in sympathetic efferent discharge to the spleen and brown adipose tissue (BAT) by aFGF administration through the two routes.

8.4. Perspectives

These data suggest that the effects of aFGF on behavior may be mediated, at least in part, by endogenously increased CRF and glucocorticoids. The release of aFGF after feeding and/or peripheral glucose administration facilitates learning and memory in rats and mice as mentioned before. Corticosterone can also both modulate the long-term potentiation [11] in CA1 neurons and promote learning and cognitive processes [12]. In fact, recent clinical data suggest that CRF deficiencies can be detected in the brain of patients with neurodegenerative dementia [13]. The parallel activation of the sympathetic outflow to the adrenal and BAT may show that aFGF has a multifunctional role in the control of neuronal and/or endocrinological activity.

The i.v. administration of aFGF(100 ng/rat) also induces a long-lasting hyperthermia of $> 1°C$ in awake and freely moving rats (our unpublished observation). These results suggest that the long-lasting fever induced by exogenous aFGF given via the i.v. or i.c.v. route is caused, at least in part, by activation of sympathetic nerves innervating the adrenal medulla and BAT through the release of CRF in the brain.

A further area of interest is the link between aFGF and the immune system. Splenic natural killer cell cytotoxicity is reduced by activation of the splenic sympathetic outflow. Furthermore, it has become clear that catecholamines, as well as glucocorticoids, should be viewed as physiological inhibitors of inflammatory responses and as immunosuppressive mediators [14]. These data suggest that aFGF would be expected to affect the immune system.

9. EFFECTS OF AFGF AND ITS FRAGMENTS ON PHAGOCYTOSIS IN MOUSE PERITONEAL MACROPHAGES

9.1. Macrophages and innate and acquired immunity

Macrophages are distributed as resident cells throughout tissues of the normal animal and express altered endocytic and biosynthetic properties after inflammatory recruitment and immune activation. The phagocytic and digestive capacities of macrophages are initial representative responses to activate the immune system. FGF receptor (FGF R-1, 2) has been observed in monocytes which also contain FGF. The effects of aFGF and its fragments on macrophage phagocytosis, including attachment and ingestion, were analyzed using flow cytometry in this study [15].

9.2. The effect of aFGF on phagocytosis

As is shown in Fig. 5, aFGF enhanced the phagocytosis of latex particles in a dose-dependent manner. The threshold concentration of aFGF is lower than 10^{-9} M. The enhancement of phagocytosis at 5×10^{-8} M was $202 \pm 18\%$ (mean $\pm$ SD, $n = 4$) for percentage of phagocytic cells, defined as the % of macrophages that ingested one or more particles (PP) and $249 \pm 35\%$ (mean $\pm$ SD, $n = 4$) for the phagocytic index, defined as the average number of particles ingested per macrophage (PI). This can be compared to the unstimulated control phagocytosis.

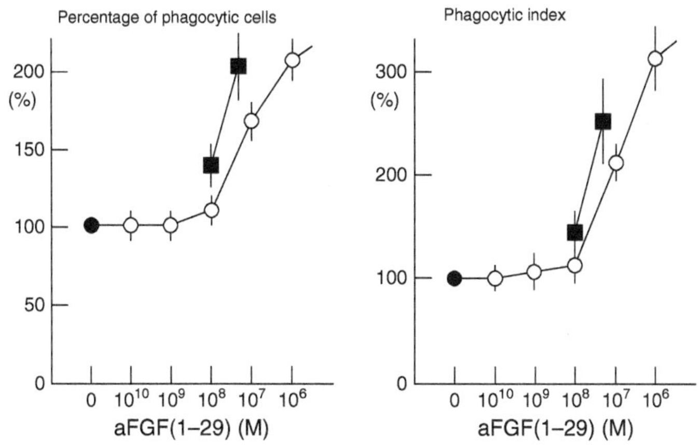

Figure 5. Concentration-response curves showing the enhancement of phagocytosis in peritoneal macrophages by aFGF (1–29) and full length of aFGF. Circles are data of aFGF (1–29), mean ± SD (n = 6 for all concentrations 10^{-8} M, n = 5). Filled squares are data of full length aFGF(1–140) at concentrations 10^{-8} and 5×10^{-8} M, (mean ±SD, n = 4). The left PP and the right PI graphs are represented as the percentage of control phagocytosis (100%) in the absence of aFGF [15].

Treatment with bFGF (2.4×10^{-8} M) also enhanced phagocytosis 136.8 ± 9.5% (mean ± SE, n = 4) for PP and 135.4 ± 11.6% (mean ± SE, n = 4) for PI, respectively, as compared to the control phagocytosis (data not shown).

9.3. Enhancement of phagocytosis by aFGF fragments

Treatment with aFGF (1–15) had little effect on phagocytosis at a concentration of 10^{-7} M. However, treatment with aFGF (1–20) clearly enhanced phagocytic activity. Treatment with the longer fragment aFGF (1–29) demonstrated a significant enhancement, 1.4 times more compared to aFGF (1–20). By eliminating residues 1–8, the enhancement by aFGF (9–29) was slightly reduced [1.2 times more than aFGF (1–20)]. A 26-residue carboxyl terminal fragment of aFGF, aFGF (114–140), had little effect on phagocytosis. Among the fragments used, aFGF (1–29) was most effective in inducing phagocytic enhancement. As shown in Fig. 5, aFGF (1–29) enhanced phagocytosis of latex particles in a dose-dependent manner up to 10^{-6} M. The threshold concentration of aFGF (1–29) may be close to 10^{-9} M. The enhancement of phagocytosis at 5×10^{-8} M was 12.5 ± 5.5% (mean ± SD, n = 6) for PP and 16.7 ± 6.5% (mean ± SD, n = 6) for PI, compared with unstimulated control phagocytosis.

10. CONCLUSIONS

When viewed together (Sections 8 and 9), our data support the contention that aFGF is an immunoregulatory factor and a stimulator for the immune system. However, splenic natural killer cell cytotoxicity is reduced by activation of the splenic sympathetic outflow, and the antigen and antibody reaction is reduced by corticosterone.

Figure 6 summarizes the diverse, overall physiological significance of aFGF and its fragments. After food intake, aFGF is released from the ependymal cells in the third cerebroventricle which respond to the increase in glucose concentration from 2 mM to 4 mM in the CSF.

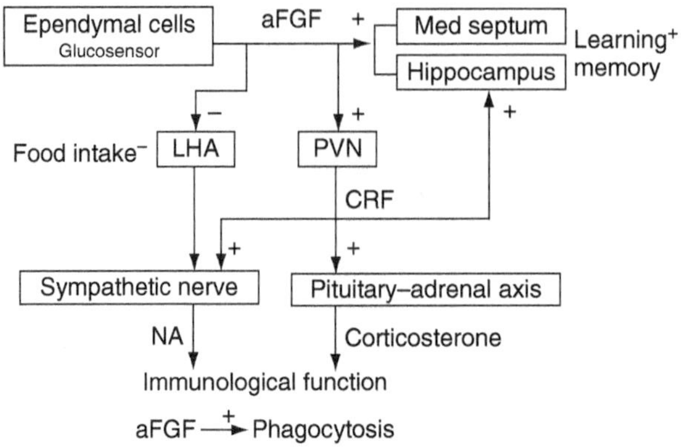

Figure 6. Physiological roles of aFGF in relation to food intake. Ependymal cells located in the cerebroventricle release aFGF responding to an increase in glucose in CSF during food intake. The growth factor subsequently reaches the glucose-sensitive neurons in the lateral hypothalamic area (LHA, the classical feeding center) and inhibits neuronal activity. In addition, aFGF may also reach the hippocampus, where it can facilitate learning and memory. Finally, aFGF reaches the parvocellular neurons in the paraventricular nucleus (PVN) and can activate these neurons, bringing about CRF release. These neurons and CRF activate efferent sympathetic outflow and pituitary–adrenal axis, respectively. Splenic sympathetic activation and released corticosterone modulate immune function. Addition of aFGF in vitro stimulates phagocytosis by macrophages.

Released aFGF diffuses into the brain parenchyma and first reaches the LHA and suppresses food intake through inhibiting the activity of GSNs which induce feeding. This inhibition is antagonized by applications of anti-aFGF or anti-FGF receptor-1 antibody. Second, diffusing aFGF reaches the hippocampus and facilitates synaptic plasticity and improves learning and memory performance. Third, aFGF reaches the parvocellular neurons in the PVN and facilitates CRF release and also activates not only hypothalamo–pituitary–adrenal axis but also splenic sympathetic outflow. Corticosterone and splenic sympathetic activation attenuate immune function. The latter activities could function to reduce autoimmune activity. On the other hand, increasing the amount of aFGF added to culture medium enhances phagocytosis by macrophages and thus biodefence activity. When taken together, this evidences supports the contention that feeding not only maintains the body energy homeostasis but also prepares the brain for other important functions.

REFERENCES

1. Hanai K, Oomura Y, Kai Y, Nishikawa K, Morita H, Plata-Salamon CR. Central action of acidic fibroblast growth factor in feeding regulation. Am J Physiol 1989;256:R217–23.
2. Oomura Y, Sasaki K, Hanai K. Chemical and neuronal regulation of flood intake. In Progress in Obesity Research 1990. Oomura Y, Tarui S, Baba S, Inoue S, Ed.; London: John Libbey, 1991; pp. 3–12.
3. Mastuo A, Tooyama I, Isobe S, Oomura Y, Akiguchi I, Hanai K, Kimura J, Kimura H. Immunohistochemical localization in the rat brain of an epitope corresponding to the fibroblast growth factor receptor-1. Neuroscience 1994;60:49–66.

4. Tooyama I, Hara Y, Yoshihara O, Oomura Y, Sasaki K, Muto T, Suzuki K, Hanai K, Kimura H. Production of antisera to acidic fibroblast growth factor and their application to immunohistochemical study in the rat brain. Neuroscience 1991;40:769–79.
5. Li A-J, Oomura Y, Sasaki K, Suzuki K, Tooyama I, Hanai K, Kimura H, Hori T. A single pre-training glucose injection induces memory facilitation in rodents performing various tasks: contribution of acidic fibroblast growth factor. Neuroscience 1998;85:785–94.
6. Sasaki K, Oomura Y, Figrov A, Yagi H. Acidic fibroblast growth factor facilitates generation of long-term potentiation in rat hippocampal slices. Brain Res Bull 1994;33:505–11.
7. Sasaki K, Oomura Y, Urashima T, Shiokawa A, Tsukada A, Kawarada A, Yanaihara N. Effect of acidic fibroblast growth factor on neuronal activity of the parvocellular part in rat paraventricular nucleus. Neurobiology 1995;3:329–38.
8. Matsumoto I, Oomura Y, Niijima A, Sasaki K, Aikawa T. Acidic fibroblast growth factor activates hypothalamo-pituitary-adreocortical axis in rats. Am J Physiol 1998;274:R503–9.
9. Oomura Y, Sasaki K, Suzuki T, Muto T, Li A-J, Ogita Z, Hanai K, Tooyama I, Kimura H, Yanaihara N. New brain glucosensor and its physiological significance. Am J Clin Nutr 1992;55:278S–82S.
10. Matsumoto I, Niijima A, Oomura Y, Sasaki K, Tsuchiya K, & Aikawa T. Acidic fibroblast growth factor activates adrenomedullary secretion and sympathetic outflow in rats. Am J Physiol 1998;275:R1003–12.
11. Rey M, Carlier E, Talmi M, Soumireu-Mourat B. Corticosterone effects on long-term potentiation in mouse hippocampal slices. Neuroendocrinology 1994;60:36–41.
12. Behan DP, Heinrichs SC, Troncoso JC, Liu XJ, Kawas CH, Ling N, De Souza EB. Displacement of corticotropin releasing factor from its binding protein as a possible treatment for Alzheimer's disease. Nature 1995;378:264–87.
13. De Souza EB, Whitehouse PJ, Kuhar MJ, Price DL, Vale WW. Reciprocal changes in corticotropin-releaseing factor (CRF)-like immunoreactivity and CRF receptors in cerebral cortex of Alzheimer's disease. Nature 1986;319:593–95.
14. Van der Poll T, Lowry SF. Epinephrine inhibits endotoxin-induced IL-1β production: roles of tumor necrosis factor-α and IL-10. Am J Physiol 1997;273:R1885–90.
15. Ichinose M, Sawada M, Sasaki K, Oomura Y. Enhancement of phagocytosis in mouse peritoneal macrophages by fragments of acidic fibroblast growth factor (aFGF). Int J Immunopharmacol 1998;20:193–204.

Cytokines and the Brain
Edited by C. Phelps and E. Korneva

Cytokines and Sleep

JAMES M. KRUEGER, DAVID M. RECTOR, and LYNN CHURCHILL

Program in Neuroscience, Washington State University, Pullman, WA 99164-6520, USA

ABSTRACT

Sleep, like other physiological processes, is regulated in part by humoral substances; cytokines play a major role in this endeavor. The scientific literature dealing with the humoral regulation of sleep began almost 100 years ago when Ishimori [1] showed that the transfer of cerebrospinal fluid from sleep-deprived dogs enhanced sleep in normal recipients. Within just a few years, a French group reported similar findings [2]. Within the past 40 years, several groups have replicated these findings in various species and experimental models [3–9]. Today we recognize that multiple substances are involved with sleep regulation. These molecules, called sleep regulatory substances (SRSs), range from low molecular weight substances with short half lives, such as nitric oxide and adenosine, to peptides such as growth hormone-releasing hormone (GHRH) and proteins including the cytokines interleukin-1β (IL1β) and tumor necrosis factor-α (TNFα). This review will focus on the roles that cytokines play in sleep regulation in health and disease.

1. INTRODUCTION

Sleep regulatory substances (SRSs) were characterized by various experimental approaches. Sleep-promoting factors were isolated from the cerebrospinal fluid or brain, for example, uridine [6,7] and Factor S [8,9]. The sleep-promoting activity of substances known to be linked to sleep was determined, for example, GHRH [10,11], interleukin-1β (IL1β) [12], tumor necrosis factor (TNF) [13]. Genetic mutants with sleep pathologies were studied, for example, orexin-narcolepsy [14,15], and endogenous ligands associated with pharmacological agents were identified, for example, adenosine [16,17]. All of these approaches and new methods, such as genome wide searches, are limited because sleep cannot be isolated as an independent variable. Every physiological process changes with sleep. It is, for example, difficult to know whether changes in the levels of a substance associated with sleep loss are a direct consequence of sleep deprivation or are secondary to other changes, for example, increased body temperature, metabolism, glucocorticoids, etc. As a consequence, sleep researchers have developed lists of criteria that a candidate SRS should meet, before it can be reasonably be proposed that it is involved in sleep regulation (Table 1). Both IL1β and TNFα have met all these criteria and in fact are two of the best characterized SRSs. For instance, TNFα is the only substance for which there is a literature demonstrating that its plasma levels vary in health and disease with sleep propensity in humans [reviewed 18].

Table 1. Criteria for sleep regulatory substances[a]

1.	The SRS should enhance one or more sleep phenotypes, e.g., duration of NREMS
2.	If the SRS is inhibited, spontaneous sleep should be reduced
3.	Levels of the SRS in brain should vary with sleep propensity
4.	The SRS should act on sleep regulatory circuits
5.	The SRS should vary with pathologies that alter sleep, e.g., infections

[a] Derived from Jouvet [219], Inoue [3], Borbely [220], and Krueger and Obal [221].

2. SLEEP

1.1. Sleep states

Sleep is defined electrophysiologically and behaviorally. Electrophysiologically, sleep is usually divided into two states, nonrapid eye movement sleep (NREMS) and rapid eye movement sleep (REMS). Within the sleep literature both states are usually subdivided further. Deep NREMS (also called stage 4 or delta sleep) is associated with high-amplitude electroencephalogram (EEG) delta (0.5–4 Hz) waves, relaxed muscle tone, and many characteristic physiological changes such as increased growth hormone release, a controlled decrease in brain temperature, and reduced brain metabolism. In contrast, REMS is characterized by a low-voltage rapid EEG, muscle atonia, variable physiological measures such as blood pressure and respiration rate and an increase in brain temperature and metabolism. Behaviorally, sleep is characterized by prolonged quiescence, reduced responsiveness, rapid reversibility, and rebound after sleep loss. Functional magnetic imaging, positron emission tomography, and EEG tomography studies indicate that during NREMS and REMS there is differential activation/deactivation of various areas of brain [19–26]. Exactly which areas show changes in blood flow/metabolism depends in part on prior activity of the specific brain areas during wakefulness [27].

Both NREMS and REMS have characteristic phenotypes that are often characterized in sleep studies. The duration of state, sleep cycle length (defined as the time from the onset of one REMS episode until the onset of the next REMS episode) belong here. Further phenotypes are a circadian distribution of sleep episodes, characteristic frequencies in the EEG, such as theta activity during REMS, delta waves during NREMS and EEG delta wave power. The latter measure is posited to reflect the intensity of NREMS. Each of these sleep phenotypes likely has its own SRSs and neural circuitry involved in its regulation but there is likely much overlap because individual SRSs can affect more than one of these parameters.

Sleep is an unusual process in that its function has not been experimentally verified. During sleep one does not eat, drink, socialize, or reproduce and one is more subject to predation; thus what ever sleep does for the brain it must be of sufficient importance to overcome these high evolutionary costs. Furthermore, the field of sleep research has yet to define exactly what it is that sleeps. This has the potential to confuse the discussion of sleep regulation and sleep disorders. For example, traditionally sleep was considered a whole animal phenomenon, the animal was either awake or asleep. However, marine mammals can sleep one cortical hemisphere at a time [28,29]. Furthermore, localized brain regions in humans and animals exhibit sleep-like characteristics that are use-dependent [27,30–34]. The minimal component of brain capable of oscillating between functional states was recently characterized [35] and is discussed in Section 5. A major generalized finding within the sleep literature is that regardless of what part of the brain is lesioned, whether deliberately in experimental animals or as a consequence of stroke or injury in patients, if the patient or animal survives, it sleeps. This suggests that sleep is very robust, self-organizing, an intrinsic property of neural tissue and that no specific area of brain is necessary for sleep.

2.2. Sleep homeostasis

Sleep homeostasis is a defining characteristic of sleep [36]. The brain keeps track of past sleep–wake activity, often over a period of days. If there is sleep loss, there is subsequent sleep rebound during the next sleep period. The sleep rebound is characterized by both increased time in sleep and increased sleep intensity as evidenced, for example, by enhanced EEG delta wave activity [37]. Although sleep rebound after sleep loss is influenced by circadian time, it is somewhat independent of it. The mechanisms of sleep homeostasis may involve the production and release of SRSs and their subsequent actions on neurons and glia. In fact, injection of certain SRSs, including TNFα or IL1β, elicits responses characteristic of sleep after sleep loss, that is, greater duration and intensity (discussed in Sections 2.1 and 2.2). Furthermore, inhibition of either IL1β or TNFα attenuates sleep deprivation-induced sleep rebound [reviewed 18], thereby implicating these cytokines in the sleep homeostatic mechanism.

2.3. Sleep brain anatomy

Historically, von Economo [38] concluded from histological examination of brain lesions in encephalitis lethargica patients that the anterior hypothalamus actively regulates sleep while the posterior hypothalamus regulates wakefulness. Prior to his work, sleep was considered a passive process resulting from the withdrawal of afferent sensory stimulation. In the intervening years, much evidence has supported von Economo's hypothesis, and knowledge of the neural circuitry involved in sleep regulation has been greatly refined and extended (Fig. 1). These developments are reviewed at length elsewhere [39–44]. Important brain regions involved in NREMS include the hypothalamic ventrolateral preoptic area and the median preoptic nucleus. For REMS the laterodorsal tegmental nucleus and the pedunculopontine tegmental nucleus are critical. There are also several arousal systems that project widely through the cerebral cortex including raphe-serotonergic, locus coeruleus-noradrenergic, posterior hypothalamic-histaminergic, basal forebrain-cholinergic and GABAergic, and lateral hypothalamic-orexinergic neurons [reviewed 43]. For the purposes of this review we present evidence for the actions of IL1β and TNFα in the hypothalamus and in certain arousal systems (Sections 3.1 and 3.2). We also present a newer view of brain organization of sleep that posits that cytokines act directly on neuronal assemblies to alter their functional states and that neuronal assembly functional state status is communicated to the classic sleep regulatory circuits mentioned above (Section 6). Neuronal assemblies are collections of highly interconnected neurons and are thought to be a fundamental processing unit of the awake brain [45]. Cortical columns are good examples of neuronal assemblies, for example, the barrels of the somatosensory cortex.

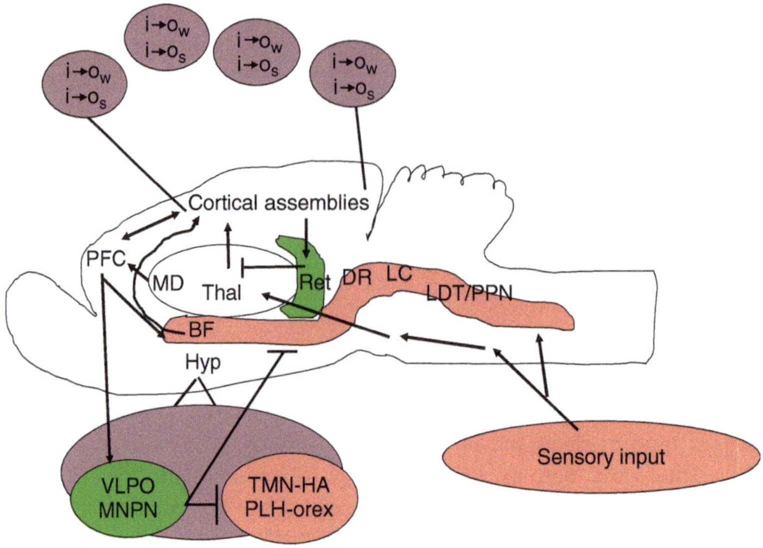

Figure 1. Summary of networks involved in sleep regulation. Cytokines such as IL1β and TNFα act on several of these areas to promote whole animal sleep and a functional sleep-like state in cortical columns (see Text). Cytokines build up in concentration in response to neuronal activity in cortical assemblies stimulated by sensory inputs that project through the brainstem and thalamus into the cortex or by a diffuse subcortical activating system. (Olfactory inputs are an exception in that they do not progress through thalamic inputs to the olfactory cortex and visual inputs do not progress through the brainstem to the thalamus.) When sufficient localized cortical assemblies build up their cytokine levels and release them (purple areas), then several routes for influencing global sleep are possible. The cortical neurons project to the GABAergic neurons in the reticular (Ret) thalamus (thal) which hyperpolarize dorsal thalamic neurons as suggested by Steriade's model [192]. The cortical neurons interconnect with other cortical regions or the mediodorsal thalamus (MD) projects to the prefrontal cortex (PFC) which in turn project to the GABAergic neurons in the anterior hypothalamus (Hyp), such as the ventrolateral preoptic area (VLPO) or median preoptic nucleus (MnPN). The GABAergic neurons in the anterior hypothalamus in turn inhibit the wake-active regions (pink) such as histaminergic neurons in the tuberomammillary nucleus (TMN, orexin/hypocretinergic neurons in the posterior lateral hypothalamus (PLH), cholinergic or GABAergic neurons in the basal forebrain (BF), serotonergic neurons in the dorsal raphe (DR), noradrenergic neurons in the locus coeruleus (LC), and cholinergic neurons in the laterodorsal tegmental nucleus/ pedunculopontine tegmental nucleus (LDT/PPN) as suggested by Saper [41]. Another route might be through the extracellular space around the neurons and glial cells into the cerebrospinal fluid, where the circumventricular organs might influence neuronal activation in the central autonomic nervous system.

3. CYTOKINES AND SLEEP

3.1. TNFα in sleep regulation

The somnogenic properties of TNFα were first described in 1987 [13]. TNFα given intracerebroventricularly (*i.c.v.)*, intravenously (*i.v.*), or intraperitoneally (*i.p.*) enhances duration of NREMS (Table 1, criterion 1). For instance, mice spend about 90 min extra in NREMS during the first 9 h after receiving 3 μg TNFα *i.p.* [46]. NREMS after TNFα treatment is associated with supra-normal EEG slow waves thereby suggesting that it induces a deeper NREMS intensity [13]. TNFα promotes NREMS in all species thus far tested: rabbits [13], mice [46], rats [47], and sheep [48]. TNFα has little effect on REMS if low NREMS-promoting doses are used; however, higher doses can inhibit REMS. Sleep following TNFα treatment appears to be physiological in the sense that sleep architecture remains normal although more time is spent in NREMS, sleep remains easily reversible, postures remain normal, and animals remain responsive to handling. Changes in sleep-coupled autonomic functions, such as the decreases in brain temperature upon entry into NREMS, also persist after TNFα treatment [reviewed 18].

Inhibition of TNFα reduces spontaneous NREMS (Table 1, criterion 2). Thus, treatment with anti-TNFα antibodies [49], the full-length soluble TNF receptor (sTNFR) [50] or sTNFR fragments containing the TNF recognition site [51], all reduce spontaneous NREMS in rabbits and rats. Furthermore, pretreatment of animals with TNF inhibitors prior to sleep deprivation reduces the expected sleep rebound that normally occurs after sleep loss [51]. Substances that inhibit TNFα action or production also inhibit spontaneous sleep, for example, IL4, 10, and 13 (Table 2, Fig. 2). In addition, these substances also inhibit the production of certain other cytokines, for example, IL1β; therefore, their action on sleep may not be specifically related to TNFα. However, they form part of the negative-feedback loops that help to regulate these nuclear factor-κB (NFκB)-sensitive cytokines (Fig. 2). Furthermore, inhibition of TNFα also blocks the increases in NREMS associated with an acute mild increase in ambient temperature [52].

Table 2. Cytokines that affect sleep

Cytokine/growth factor[a]	Effect on NREMS[b]	Reference
Interleukin-1 alpha	↑	[222]
Interleukin-2	↑	[223,224]
Interleukin-6	↑→↓	[181,182,225,226]
Interleukin-8	↑	[227]
Interleukin-15	↑	[224]
Interleukin-18	↑	[228]
Epidermal growth factor	↑	[229]
Acidic fibroblast growth factor	↑	[230–232]
Erythropoietin	↑	[233]
Nerve growth factor	↑	[213,234–238]
Brain derived neurotrophic factor	↑	[239–242]
Glia-derived neurotrophic factor	↑	[243]
Neurotrophin 3	↑	[244]
Neurotrophin 4	↑	[244]
Interferon alpha	↑→↓	[223,245–249]
Interferon gamma	↑	[250]
Tumor necrosis factor-β	↑	[251]
Granulocyte-macrophage colony-stimulating factor	↑	[252]
Interleukin-1 receptor antagonist	↓	[158]
Interleukin-4	↓	[160]
Interleukin-10	↓	[159,253]
Interleukin-13	↓	[254]
Transforming growth factor-β	↓	[254]
Granulocyte colony-stimulating factor	↓	[255]
Insulin-like growth factor	Small dose:↓; high dose:↑	[256,257]
Soluble TNF receptor	↓	[50]
Soluble IL1 receptor	↓	[159]

[a] IL1β and TNFα are omitted from this list, they are reviewed extensively elsewhere and herein [18].
[b] ↑ indicates increase; ↓ decrease; and → no change in duration of NREMS.

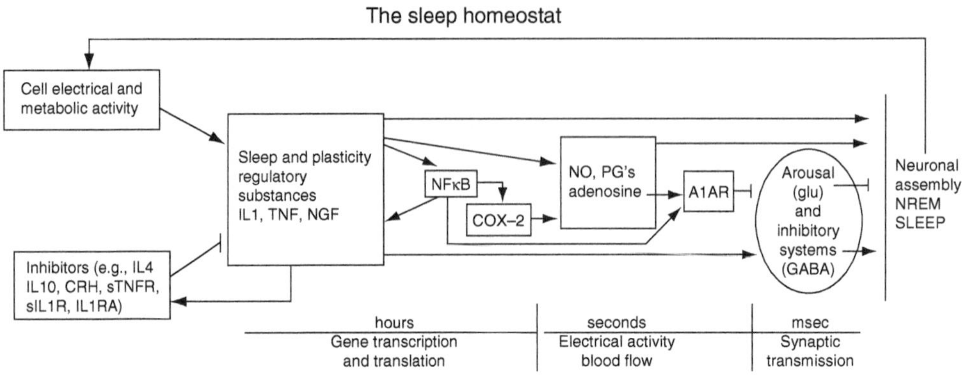

Figure 2. Molecular networks are involved in sleep regulation. Sleep regulatory substances including cytokines such as interleukin-1β (IL1β), tumor necrosis factor-α (TNFα), and nerve growth factor (NGF) are produced in response to cellular activity, determined in part by wakefulness activity. They are regulated in part through feedback mechanisms that involve nuclear factor-κB (NFκB) (positive feedback) and cytokine/hormonal negative-feedback inhibitors. The production and actions of the cytokines involve gene transcription and translation and occur over time periods of hours. As such they likely constitute the sleep homeostat in that they offer a mechanism by which the brain can keep track of past sleep/wake activity. Their direct actions on sleep involve substances that are more labile with half lives in seconds such as nitric oxide (NO) and adenosine. These substances collectively affect NREM sleep. Inhibition of one step does not completely block sleep, since parallel sleep-promoting pathways exist. These redundant pathways provide stability to sleep regulation. Our knowledge of the biochemical events involved in sleep regulation is more extensive than that illustrated. The molecular network shown possesses many of the characteristics of biological networks and engineered systems (this topic is reviewed in several lead articles in *Science* 301:5641, 2003) [258,259]. Thus, the network is modular in that several proteins (cytokines) are working in "overlapping co-regulated groups" in this pathway. Second, the molecular network is robust in that removal of one of the components does not result in complete sleep loss. Third, the network operates as a recurring circuit element with multiple feedback loops affecting other pathways to the extent that similar networks involving many of the same substances and component network parts are used to regulate body temperature, inflammatory responses, the microcirculation, memory, food intake, etc. and these systems, to a limited degree, co-regulate. Specificity for any one physiological process, such as sleep results from multiple interacting molecular and cellular circuits, each possessing different, but similar to each other, reactivity [221]. IL-1RA, IL1 receptor antagonist; sIL1R, soluble IL1 receptor; anti-IL1; CRH, corticotrophin releasing hormone; PGD_2, prostaglandin D_2; sTNFR, soluble TNF receptor; A1AR, adenosine A1 receptor; COX-2, cyclooxygenase-2 glu, glutamic acid; GABA, gamma amino-butyric acid.

Mice lacking the 55 kD TNF receptor (TNFR) do not exhibit NREMS responses if given TNFα, thereby implicating this receptor in TNFα-enhanced sleep [46]. These mice also have less NREMS and REMS than corresponding control strains. In another study the reductions in REMS we described in TNFR-deficient mice were confirmed although that study did not show changes in NREMS [53]. However, in that study inappropriate controls were used, and there was no demonstration that the mice were in fact deficient in the TNFR. Preliminary data from our laboratory using mice lacking both TNFR also indicated that both NREMS and REMS are reduced compared to controls (Bohnet et al., unpublished).

Hypothalamic levels of TNFα bioactivity [54] and TNFα mRNA [55] vary diurnally and are influenced by sleep deprivation (Table 1, criterion 3). The highest levels in rats occur at the beginning of the light period; rats sleep more during the light period. The amplitude of the day–night changes in TNF bioactivity is about 10-fold and in TNFα mRNA about twofold, this likely reflects the predominant post-transcriptional regulation of TNFα. After sleep loss, hypothalamic TNFα mRNA increases [56,57]. Sleep deprivation also increases the expression in brain of the 55 kD TNFR mRNA [57]. TNFα serum levels increase in mice after sleep loss, but not after stress [58]. In healthy humans, blood levels of TNFα correlate with EEG delta wave activity [59]. After sleep deprivation, circulating levels of TNFα [60] and the 55 kD sTNFR, but not the 75 kD sTNFR, are enhanced [61,62]. The 55 kD sTNFR is a normal component of cerebrospinal fluid [63].

In pathologies that are associated with sleepiness, higher plasma levels of TNF occur (Table 1, criterion 5). Thus, sleep apnea patients exhibit elevated TNFα plasma levels [64–70]. AIDS patients have disturbed TNFα and sleep rhythms [59]. Plasma TNFα is higher in chronic fatigue patients [71] chronic insomnia patients [72], myocardial infarct patients [73], excessive daytime sleepiness patients [74], and in pre-eclampsia patients [75]. Postdialysis fatigue is associated with increased TNFα levels [76,77] and cancer patients receiving TNFα report fatigue [78]. Alcoholics have TNFα-associated sleep disturbances [79]. TNFα may also be related to narcolepsy [80–84]. Furthermore, the G-308A TNFα polymorphic variant is linked with metabolic syndrome [85] and sleep apnea [86]. Rheumatoid arthritic subjects receiving the 75 kD sTNFR report reduced fatigue [87], and sleep apnea patients treated with the sTNFR have reduced sleepiness [88]. If obstructive sleep apnea patients are treated surgically, their elevated TNFα plasma levels return to normal [89]. Systemic TNFα, like IL1β and certain other cytokines, likely signals the brain through multiple mechanisms [90]. One mechanism involves vagal afferents because vagotomy attenuates *i.p.*-TNFα-induced NREMS responses [91]. The effects of systemic bacterial products such as endotoxin may also involve TNFα [92]. For instance, endotoxin doses that induce transient increases in sleep in humans also elicit concomitant increases in circulating TNFα [93]. In addition, the sTNFR fragment attenuates bacterial cell wall muramyl dipeptide-enhanced NREMS in rabbits [94].

The sites of action of TNFα-induced NREMS include the preoptic area of the anterior hypothalamus and the locus coeruleus; both areas are involved in sleep regulation (Fig. 1) (Table 1, criterion 4). Microinjection of TNFα into the preoptic area enhances NREMS in rats [95]. In contrast, microinjection of a sTNFR fragment into this area inhibits spontaneous NREMS [95]. TNFα microinjected into the locus coeruleus, after a brief period of excitation, also induces prolonged increases in sleep and EEG synchronization [96]. Anti-TNFα antibodies antagonized these effects. Microinfusion TNFα into the subarachnoid space just beneath the basal forebrain promotes NREMS and reduces REMS in rats [97].

TNFα may also directly act on cortical neuronal assemblies to alter cortical column state. Unilateral application of TNFα onto the surface of the somatosensory cortex induces ipsilateral state-dependent increases in EEG delta wave power (Fig. 3) [98]. Conversely, the sTNFR injected unilaterally after sleep deprivation reduces EEG delta power during NREMS on the injected side but not on the opposite side [98]. The changes in the TNFα-altered EEG power are associated with enhancements of Fos- and IL1-immunoreactivity in the somatosensory cortex and reticular thalamus [99]. These data suggest that TNFα can act locally to induce EEG delta waves characteristic of sleep and that one consequence of such synchronization is to inform the reticular thalamus of this state change (see Section 6).

TNFα is expressed by microglia, astrocytes, and neurons and has various biological actions in the central nervous system, including a role in mediating both brain damage and neuroprotection. Whether TNFα is protective or damaging may depend upon the receptor type present, the 55 kD TNFR or the 75 kD TNFR [100–102] as well as the stimulus context and the presence or absence of substances that modify TNFα activity [103,104]. TNFα participates in mediating several whole organism processes including fever [105,106] and food intake [107]. TNFα also plays a role in brain development [108], learning [109–111] and seems to participate in neuronal connectivity [112–114 and see Section 3].

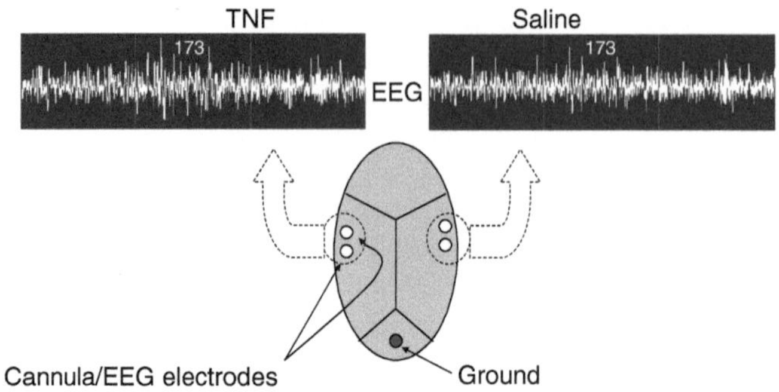

Figure 3. TNFα induces state-dependent enhancements of EEG delta waves on the side injected after unilateral application to the surface of the somatosensory cortex. EEG signals similar to the ones shown were analyzed by fast Fourier transformation for periods of NREMS and REMS for several hours after application of TNFα to one side of the brain. During NREMS, but not during REMS or waking, low doses of TNFα enhanced EEG delta waves. After higher doses of TNFα, EEG delta wave activity increased during NREMS and to a lesser extent during REMS and waking. Similar effects are observed after various lengths of sleep deprivation; after short periods of sleep loss, EEG effects are mostly confined to the immediate periods of NREMS where as after longer periods of deprivation, EEG slow wave intrude into REMS and waking episodes. Data from [98].

2.2. Interleukin-1β in sleep regulation

IL1β was the first cytokine characterized for its sleep-promoting activity [12]. i.c.v., i.p., *or i.v.* injection of IL1β enhances NREMS [reviewed 115] (Table 1, criterion 1). For instance, *i*.c.v. administration of 600 femtomoles of IL1β induces about 2 h of extra NREMS during the first 12 h post-injection in rabbits [116]. Thus far, IL1β has induced excess NREMS in every species tested including rats, mice, rabbits, monkeys, and cats [12,116–121]. Furthermore, humans undergoing IL1β therapy report excessive sleepiness [122]. In cats [118] and rats [123] lower doses of IL1β enhance NREMS while higher doses inhibit NREMS; in rats these effects depend upon the time of day IL1β is administered. For example, a dose of 10 ng of human recombinant IL1β given to rats at the onset of dark hours (the normal wake period for a rat) promotes NREMS while the same dose given at light onset inhibits NREMS. IL1β also enhances EEG delta wave activity during NREMS and is thus thought to induce a more intense NREMS [12]. The effects of IL1β on EEG delta wave activity are dependent upon the route of administration. After *i*.c.v. *or i.v.* IL1β, EEG slow waves are enhanced. In contrast, *i.p.* administration of IL1β to mice or rats reduces EEG delta wave power during NREMS although under these conditions and dose of IL1β used, NREMS was enhanced [124]. The reasons for these differential effects of IL1β on EEG delta activity remain unknown. Regardless, there is ample evidence from several laboratories that injection of low doses of IL1β promote NREMS.

The low NREMS-promoting doses of IL1β have little effect on REMS in rats and rabbits. As the NREMS-promoting doses are increased, duration of REMS is inhibited [123]. Higher doses of IL1β, as mentioned above, inhibit NREMS and also inhibit REMS. The sleep occurring after the low somnogenic doses of IL1β appears normal in the sense that the animals are easily aroused, postures are normal, and normal sleep cycles are observed. However, after higher IL1β doses, sleep becomes fragmented, sleep postures become abnormal, and animals are less responsive to activating stimuli. Similar effects are seen after high doses of endotoxin or during severe infections.

Spontaneous NREMS is reduced if IL1β is inhibited (Table 1, criterion 2). In rabbits, anti-rabbit IL1β [117] and in rats, anti-rat IL1β [125] inhibit duration of spontaneous NREMS. Similarly, a fragment of the soluble IL1 receptor (sIL1R) or the IL1 receptor antagonist (IL1RA) reduces spontaneous NREMS [126] (Fig. 2). Furthermore, substances that inhibit IL1β production, for instance, corticotrophin-releasing hormone (CRH), alpha melanocyte-stimulating hormone, transforming growth factor-β, IL4, 10, and 13, inhibit NREMS [reviewed 18,115] (Fig. 2). These substances also affect other cytokines and hormones; thus, the specificity of these effects to IL1β actions on sleep is unknown. In addition, the sleep rebound that normally occurs after sleep loss is attenuated if animals are pretreated with either a sIL1R fragment or with an anti-IL1 antibody [117,125,127]. Finally, substances such as muramyl dipeptide, a bacterial cell wall product, induce IL1β production and promote NREMS. Inhibitors of IL1β attenuate these muramyl dipeptide-induced responses [128].

Mice lacking the type I IL1 receptor fail to exhibit NREMS responses if administered IL1β, although they can mount NREMS responses if given TNFα [120]. The IL1 type I receptor mutant mice also have less spontaneous NREMS than corresponding controls; this effect is most evident during night-time hours. These data collectively suggest that the type I IL1 receptor is involved in IL1β-enhanced NREMS and in spontaneous NREMS.

The IL1-family of molecules including its receptors, receptor antagonists, and associated proteins are expressed in normal brain [reviewed 108,129,130]. IL1β is produced by neurons, glia, and endothelial cells [115,129,131], and IL1 receptors are found in various cell types including neurons [132]. Production of IL1β, like TNFα and nerve growth factor (NGF), is increased with NFκB activation. In turn, IL1β as well as TNFα and NGF activate NFκB and thereby form a positive-feedback loop (Fig. 2). There are also many negative-feedback loops to dampen IL1β and other cytokine expression and activity (Fig. 2). As mentioned above, these molecules include CRH, IL4, IL10, IL13, and glucocorticoids; all of these substances inhibit sleep (Table 2).

Brain expression of IL1β protein and IL1β mRNA varies with sleep propensity (Table 1, criterion 3). In cats, for example, cerebrospinal fluid levels of IL1-bioactivity vary with the sleep–wake cycle [133]. In rats, hypothalamic levels of IL1β [134] and IL1β mRNA [130] are highest at the beginning of daylight hours, the time when NREMS duration is maximal. Sleep loss enhances hypothalamic IL1β mRNA levels [130,135]. Hippocampal and cortical levels of IL1β mRNA and cortical levels of IL1β-immunoreactivity [Churchill et al., unpublished] also vary with sleep propensity (see Section 4).

Blood levels of IL1β may also vary with sleep, but this literature is not as clear as that relating plasma TNFα levels to sleep propensity, probably due to the difficulty of detecting IL1β in plasma. Peak levels of IL1β occur at sleep onset in human plasma [136]. IL1 blood levels also are enhanced during sleep deprivation [137,138]. How circulating IL1β may affect sleep seems to involve several independent mechanisms including transport of IL1β from blood into brain, permeation of IL1β into the brain through areas such as the median eminence, induction of lipid-soluble small molecules such as prostaglandins, and signaling through the vagus nerve. This latter mechanism has received some direct support in regard to sleep. Vagotomy blocks the NREMS-promoting activity of *i.p.* IL1β [124]. IL1β given *i.p.* enhances hypothalamic IL1β mRNA levels, and this effect is also blocked by vagotomy [139]. Also, IL1β given *i.p.* enhances IL1β mRNA and TNFα mRNA levels in brain regions within the central autonomic system as well as the cortex [140]. Furthermore, excessive food intake enhances both NREMS and hypothalamic levels of IL1β mRNA, and vagotomy blocks these effects as well [141,142]. Regardless, the role of circulating IL1β in physiological sleep remains unknown although there is, as reviewed herein, substantial evidence that central IL1β is involved in every day sleep regulation.

There seem to be many central nervous system sites of action for IL1β NREMS promotion (Table 1; criterion 4). Microinjection of IL1β into the dorsal raphe [143] or into the locus coeruleus [96] enhances NREMS. On the other hand, microinjection of IL1β into the paraventricular nucleus of the hypothalamus promotes wakefulness [144]. In an extensive study of the somnogenic sites of action for IL1β [145], several ventricular and subarachnoid sites were responsive to IL1β. The most potent sites were those close to the anterior hypothalamus. In other studies, IL1β was shown to excite sleep-active neurons and inhibit wake–active neurons in the anterior hypothalamus [146,147]. Within the fever literature there is extensive evidence that anterior hypothalamic neurons are receptive to IL1β [148]. However, if IL1β-induced fevers are blocked, IL1β-enhanced NREMS persists suggesting that independent neural networks are involved in these two responses [12]. The role that cortical IL1β may play in cortical column state determination is discussed in Section 6.

There are several downstream events involved in IL1β-enhanced NREMS (Fig. 2). For instance, if rats are pretreated with an anti-GHRH antibody prior to IL1β administration, the expected IL1β-induced sleep responses are blocked [149]. Such data suggest that GHRH is a downstream event in the IL1β-sleep pathway [150]. However, there is a subpopulation of hypothalamic GABAergic neurons that are receptive to both IL1β and GHRH; stimulation of either receptor enhances intracellular calcium levels [132]. This result could indicate that the somnogenic actions of either GHRH [reviewed 151] or of IL1β could be mediated through the same cells, perhaps those hypothalamic neurons that are sleep or wake active.

IL1β affects a host of other molecules that in turn have the capacity to affect sleep. The list includes prostaglandins [122], NO [152], and adenosine [153] (Fig. 2). For instance, inhibition of nitric oxide synthase blocks IL1β-enhanced NREMS responses [154]. IL1β also interacts with multiple neurotransmitter systems including GABA [155], serotonin [156] and acetylcholine [157]; any or all of these actions could be related to IL1β NREMS-promoting activity.

2.3. Other cytokines in sleep regulation

The regulation of cytokines in the brain is complex and not very well understood. Nevertheless, some cytokine-associated substances, such as the IL1RA, the sTNFR, and the sIL1R seem to act as endogenous antagonists, and indeed these substances inhibit spontaneous sleep (Fig. 2) [51,126,127,158]. Anti-somnogenic cytokines act, in part, by inhibiting production of pro-somnogenic cytokines [159,160]. For example, IL10 inhibits IL1β and TNFα production [161–163] and also inhibits type I and II IL1 receptor gene expression [164]. Furthermore, exogenous IL10 inhibits production or release of other somnogenic substances implicated in sleep regulation, including NGF [162] and NO [165,166], and it increases the production of sleep-inhibitory substances such as CRH [167] and the IL1RA [168]. IL4 inhibits IL1β [169] and TNFα [170] production, and it increases the production of the IL1RA [171,172] and release of the sTNFR [173]. Furthermore, IL4 inhibits production or release of other substances implicated in sleep regulation, for example, NO [174]. Both IL4 and IL10 inhibit sleep [159,160] and are in brain [175].

NFκB and c-Fos (AP-1) are transcription factors that are activated by IL1β, TNFα, and NGF [reviewed 18] (Fig. 2) NFκB activation promotes the production of several other substances in addition to IL1β and TNFα implicated in NREMS regulation including the A1 adenosine receptor (A1AR), cyclooxygenase-2, the GHRH receptor (Taishi et al., unpublished) as well as several of the pro-somnogenic cytokines [reviewed 18]. NFκB is activated within the cortex and hypothalamus during sleep deprivation [176,177]. Adenosine also elicits NFκB nuclear translocation in basal forebrain slices [178] and that action is mediated by the A1AR. A cell soluble peptide inhibitor of NFκB nuclear translocation inhibits NREMS [179].

Microbial toxins, bacterial or viral, manifest their toxicity in part through induction of pro-inflammatory cytokines (Table 2) [18]. In contrast, the anti-inflammatory cytokines IL4, IL10, IL13, TGFβ1 and insulin-like growth factor-1 as well as the IL1 and TNF soluble receptors, as mentioned above, inhibit NREMS [180]. IL6 was originally thought to be non-somnogenic, but studies with homologous IL6 in rats [181] or mice [182] reveal that IL6 enhances NREMS (Table 2). However, mice lacking IL6 appear to have normal spontaneous sleep although their sleep responses to microbial products are altered [182]. Some cytokines do not alter NREMS, for example, basic fibroblast growth factor (FGF) and interferon-β (IFNβ). Many, if not all, of the substances of Table 2 can be made in the central nervous system, their receptors are found in the brain and many alter firing rates of hypothalamic neurons. These facts collectively suggest that a cytokine cascade operates to regulate sleep in health and disease [183]. However, only IL1β and TNFα have been studied extensively for their involvement in physiological sleep regulation.

4. ACTIVITY-DEPENDENT EXPRESSION OF CYTOKINE SRSs

Activity-dependent expression of NGF and BDNF by neurons is well characterized [reviewed 184–186]. Cellular electrical activity alters the synthesis and actions of these regulatory molecules, and in turn, they directly alter electrical properties of cells receptive to them and alter the expression of many molecules necessary for synaptic efficacy and plasticity. These mechanisms are involved in Hebbian synaptic regulation and collectively form the basis for the neurotrophin hypothesis [185]. The syntheses of TNFα [187,188] and IL1β [189] are also enhanced by cellular activity and are sleep dependent [190]. Preliminary data from our laboratory suggest that within the cerebral cortex neurons express both TNFα and IL1β. The number of TNFα-immunoreactive neurons, identified by colocalization of neuronal nuclear marker NeuN, increases in somatosensory cortical columns if afferent activity is enhanced prior to sacrifice by twitching a whisker compared to columns that do not receive input from unstimulated whiskers [191]. Furthermore, during the active night-time hours, both IL1β and TNFα expressions are enhanced in the cortex and their expressions correlate with Fos expression in cortical layers II and III (Churchill, unpublished). Fos expression is often used as a marker for neuronal activity because its nuclear expression is enhanced about 1–2 h after activity within neurons. Such data strongly suggest that cytokine expression in neurons is activity-dependent. We posit that such activity-driven production of IL1β and TNFα play an important role in functional states of cortical columns (see Section 6).

Although the actions of IL1β and TNFα are not studied within the context of Hebbian mechanisms, there are data suggesting TNFα could influence neuronal connectivity through its actions on AMPA receptors. TNFα promotes AMPA receptor expression and increases cytosolic Ca^{++} levels [113]. These actions of TNFα seem to be physiological because an inhibitor of TNFα inhibits AMPA-induced postsynaptic potentials [112] and AMPA-induced changes in cytosolic Ca^{++} [113]. AMPA receptors play a role in EEG synchronization [192] and synaptic plasticity [reviewed 193]. More recently, TNF's involvement in synaptic scaling, a mechanism involved in the stabilization of synaptic networks, was described [114].

5. BRAIN ORGANIZATION OF SLEEP

By the early 1990s it was apparent that multiple pleiotropic SRSs such as IL1β and TNFα participate in sleep regulation. A conceptually simple model of how multiple pleiotropic SRSs and pleiotropic neurons could interact to produce specific sleep responses was developed [194]. Although that was of some explanatory value, the approach did not go beyond the models that other regulatory fields, for example, feeding, had already developed in response to similar problems of substance-activity specificity. Nevertheless, the firm conclusion was reached that SRSs affect each other's production and work in molecular networks involving multiple cells to orchestrate sleep regulation. A major question remained unresolved: What was the mechanism of SRSs production? For homeostatically regulated physiological processes there is a link between their functions and their physiological mechanisms and regulated variables [36]. We were thus led to the link between cell activity and neuronal connectivity and production of SRSs and posited that sleep served to help stabilize synaptic networks in the face of constant activity-induced changes [195]. Hebbian plasticity is a process that strengthens active synapses and weakens inactive synapses [reviewed 196–198]. However, our theory differed from the dominant paradigm of sleep regulation in several important ways by emphasizing that sleep was a fundamental property of neuronal assemblies (also called neuronal groups) and dependent on prior activity of the neuronal assembly, not prior wakefulness. Kavanau [199] reached similar conclusions and proposed a dynamic stabilization theory for sleep function that included a role for intrinsic electrical activity in the synaptic stabilization process.

The central idea of both the Krueger-Obal and the Kavanau's theories was the recognition that the use-dependent-driven changes in synaptic efficacy and connectivity would lead to dysfunction unless there were some process to stabilize synaptic networks that are constantly being modified by activity. In subsequent work this process was called synaptic scaling by others [197]. Synaptic scaling serves to regulate Hebbian plasticity; an increase in network activity causes a slow compensatory decrease in excitatory synaptic efficacy whereas, a decrease in network activity enhances excitatory synaptic strength [198]. The stabilization mechanisms proposed by Krueger and Obal (SRS-induced changes in localized electrical properties of neuronal assemblies) [195] or those proposed by Kavanau (intrinsic spontaneous electrical activity) [199] are scaling mechanisms and in fact, TNFα has now been directly implicated in the synaptic scaling process [114]. Recent similar theories have also invoked synaptic scaling although different semantics, for example, synaptic "homeostasis" versus "stabilization" were used [200].

Our approach in the early 1990s was greatly influenced by three findings. First, the multiple lesion studies led us, and others [195,201], to conclude that sleep is an intrinsic property of any surviving group of neurons (see above). Second, we were influenced by work done in the 1940s and 1950s [202,203], which indicated that cortical islands, isolated from thalamic inputs, wax and wane through states of EEG synchronization and desynchronization. This suggested that perhaps sleep could be a property of small groups of cortical cells. More recently, Amzica and Steriade also concluded that a slow component of the EEG is cortically generated [204]. Such findings reverse one's thoughts from "top down" regulation (i.e., sleep being imposed upon the brain from a sleep regulatory network) to one of "bottom up" regulation (i.e., sleep being a property of neuronal assemblies so that whole organism sleep results from the coordination of neuronal assembly sleep by the sleep regulatory networks). We were also influenced by clinical reports suggesting that the brain could be awake and asleep simultaneously [205]. Finally, we were influenced greatly by the work of Mukhametov with dolphins showing that these marine mammals never have high amplitude EEG slow wave NREMS simultaneously in both cerebral

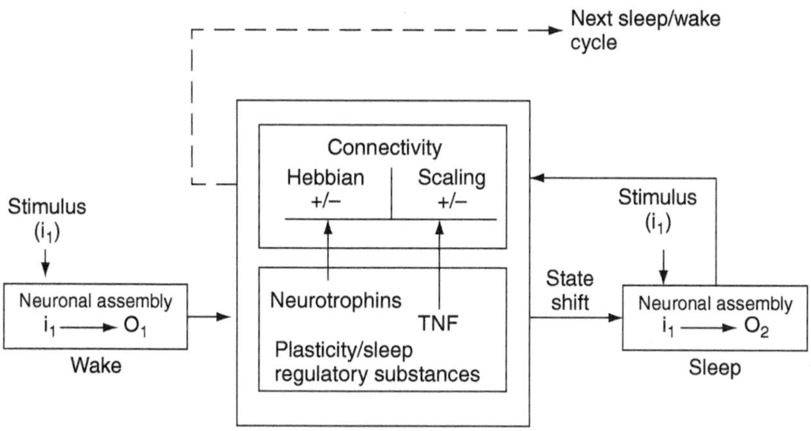

Figure 4. Oscillations of neuronal assemblies between wake-like and sleep-like states. Neuronal activity (left stimulus) acts upon an assembly to induce an output O_1. This activity results in the enhanced production and release of cytokines such as TNFα and neurotrophins; they, in turn, act to change synaptic efficacy and connectivity. As a consequence, after the next input stimulus i_1 (right) there is a different output (O_2); this is a functional state shift because the same input results in a different output. These state shifts of cortical columns have been demonstrated experimentally [35]. As a consequence of the new $i_1 \rightarrow O_2$ relationships, a slightly different set of synapses are activated and that activity also affects cytokine production and release and subsequent reactivity of the assembly. Thus, the next cycle will have different $i_1 \rightarrow O_3$ relationships; the exact assembly synaptic network is never the same yet this mechanism preserves synaptic networks not used during normal waking and simultaneously incorporates new learned patterns [195,206].

hemispheres [28] and that sleep rebound can be uni-hemispheric if only one side of the brain is deprived of sleep [29]. This was a clear demonstration that sleep was a property of something less than the whole brain. Our original version of brain organization of sleep was published in 1993 [195] and in the subsequent 13 years it has been refined [206–208].

The fundamental mechanistic concepts of our theory are relatively straightforward and are extensively supported by the literature (Fig. 4). Briefly, as synapses and circuits are used there is an activity-dependent production of SRSs (evidenced reviewed in Section 4 for TNFα and IL1β). This mechanism is how the brain keeps track of prior activity (see Fig. 2) and is thus intimately involved in sleep homeostasis. If, for example, either IL1β or TNFα is inhibited, sleep rebound after sleep deprivation is greatly attenuated [reviewed 18]. The activity-dependent SRSs act locally within the neuronal assembly that produced them to affect the electrical properties of nearby neurons/glia such that a given input into the neuronal assembly results in a different output. This mechanism allows the brain to target sleep or sleep intensity to areas on the basis of prior activity. Within a neuronal group, the SRS-induced altered input–output relationships can, by definition (and now experimentally see Section 6), be considered a functional state shift. If sufficient numbers of neuronal assemblies are in the sleep-like functional state then whole animal sleep ensues. (This is the only part of our theory for which there is not direct evidence). However, as neuronal assemblies are loosely connected semiautonomous units, synchrony of state between them is not only possible, but in mathematical models, likely [209,210]. This cohesion of neuronal assembly state could be accomplished within the cortex, for example, by the circuits that map columns onto each other and via the arousal systems that map into various layers of the cortex. Experimentally, cohesion of state occurs in both sleep and wakefulness; thus, the probability of a neuronal assembly being in the sleep-like functional state is much higher if the animal is asleep [35]. Finally, the circuits involved in coordinating functional state changes of neuronal assemblies into organism sleep are those sleep regulatory circuits previously identified (Fig. 1). These circuits allow the appropriate species-specific timing and niche adaptation of sleep.

6. ROLE OF CYTOKINES IN EXPERIMENTAL EVIDENCE FOR THE THEORY

There are many ramifications of this theory. Some of the important ones for which there is experimental evidence are

1. Sleep is a fundamental property of neuronal assemblies.
2. Sleep intensity of one part of the brain can be more intense than other parts.
3. SRS levels, including cytokines, are dependent on prior neural activity and sleep history.
4. Changes in SRS levels locally within the cortex will activate neural pathways, such as the corticothalamic projection to the reticular thalamus [211].
5. Cytokines induce the functional sleep like state in cortical columns.

Direct demonstration that cortical columns oscillate between at least two states was determined by probing columns with afferent stimulation and measuring the subsequent amplitudes of surface evoked potentials [35]. One of the states corresponds to whole animal sleep. The probability of entering this functional sleep state is dependent upon its past activity and its past state status. Thus excessive stimulation of the afferent input to a cortical column increases the probability of it being in a functional sleep state (Rector et al., unpublished). The longer a column is in one state the higher the probability it will be found in the other state a few minutes later. Coordination (cohesion) of state between columns is higher during waking than during sleep. Finally, cortical column state determines behavior. If rats are trained to lick in response to stimulation of a single whisker, the error rate is higher if the stimulated whisker's cortical column is in the sleep-like state than if it is in the wake state [212]. Collectively such data suggest that sleep is initiated at the neuronal assembly level and is a fundamental property of neuronal assemblies. This view of brain organization has profound implications for sleep function [195].

TNF-α is expressed to a greater degree in somatosensory cortical neurons after enhancement of afferent input (discussed in Section 4). The number of pyramidal neurons in the somatosensory cortex expressing NGF is also afferent input- and sleep-dependent [213]. Enhanced TNFα, NGF, and IL1β release and their actions within neuronal assemblies are posited to be biochemical mechanisms of sleep homeostasis (Fig. 2) [195]. Localized injection of TNFα onto somatosensory columns induces a functional sleep-like state in the affected columns [214]. Unilateral application of TNFα [98] or IL1β [215] to the surface of the somatosensory cortex induces state-dependent enhancements of EEG delta wave activity ipsilaterally, suggesting that regional sleep intensity is enhanced. Similar state-dependent enhancements of EEG slow waves are also observed regionally after disproportionate stimulation of localized areas of the cortex, whether this is accomplished by afferent stimulation [30,32,34,216], spontaneously [33], or by learning paradigms [31]. Such regional changes are likely related to the activity-dependent changes in blood flow/metabolism described above [27,217]. Furthermore, application of either TNFα or IL1β to the somatosensory cortex activates reticular thalamic neurons as evidenced by enhanced Fos expression [99,218]. Such data suggests that information dealing with cortical column state status is communicated to known sleep regulatory circuits. Collectively, the evidence reviewed in this paragraph strongly implicate cytokines in local regulation of state.

7. CONCLUSIONS

There is strong evidence that the cytokines IL1β and TNFα are involved in physiological sleep regulation and in the sleep responses to pathologies. Other cytokines are also likely involved in these processes but there is insufficient evidence to firmly tie them to sleep regulation. IL1β and TNFα are produced in the brain in response to cellular activity and act locally to affect neuronal assembly state as well as on sleep regulatory circuits to promote whole animal sleep. Cytokines act in concert with other molecules such as adenosine and nitric oxide to regulate state. It seems likely that cytokines are part of the humoral sleep regulatory mechanisms first described in the ancient Greek literature and more recently in the modern scientific literature.

ACKNOWLEDGMENTS

This work was supported by grants from the NIH (Grant numbers NS25378, NS31453, NS27250) to James Krueger and a Beckman Young Investigator Award, The Murdock Foundation, a Sleep Research Society J.C. Gillin Young Investigator Award and by NIH (MH6026 and MH71830) to David Rector.

REFERENCES

1. Ishimori K. True cause of sleep – A hypnogenic substance as evidenced in the brain of sleep-deprived animals. Tokyo Igakkai Zasshi 1909;23:429.
2. Legendre R, Piéron H. Recherches sur le besoin de sommeil consécutif á une veille prlongée. *Z Allg Physiol* 1913;14:235–62.
3. Inoué S. Biology of sleep substances. Boca Raton, FL: CRC Press, Inc., 1989.
4. Monnier M, Hösli L. Dialysis of sleep and waking factor in blood of the rabbit. *Science* 1964;146:796–98.
5. Borbély AA, Tobler I. Endogenous sleep-promoting substances and sleep regulation. *Physiol Rev* 1989;69:605–70.
6. Uchizono K, Higashi A, Iriki M, Nagasaki H, Ishikawa M, Komoda Y, Inoué S, Honda K. Sleep-promoting fractions obtained from the brain-stem of sleep-deprived rats. In Integrative Control Functions of the Brain Vol. I. Ito M, Kubota K, Tsukahara N, Yagi K, Eds; Amsterdam,New York,Oxford: Elsevier/North-Holland Biomedical Press, 1978; pp. 392–96.
7. Komoda Y, Ishikawa M, Nagasaki H, Iriki M, Honda K, Inoué S, Higashi A, Uchizono K. Uridine, a sleep-promoting substance from brainstem of sleep-deprived rats. Biomed Res 1983;4:223–27.
8. Miller TB, Goodrich CA, Pappenheimer JR. Sleep-promoting effects of cerebrospinal fluid from sleep-deprived goats. Proc Natl Acad Sci USA 1967;58:513–17.
9. Fencl V, Koski G, Pappenheimer JR. Factors in cerebrospinal fluid from goats that affect sleep and activity in rats. J Physiol 1971;216:565–89.
10. Ehlers C, Reed TK, Henriksen SJ. Effects of corticotropin-releasing factor and growth hormone-releasing factor on sleep and activity in rats. Neuroendocrinology 1986;42: 467–74.
11. Obal F Jr. Effects of peptides (DSIP, DSIP analogues, VIP, GRF and CCK) on sleep in the rat. Clin Neuropharmacol 1986;9:459–61.

12. Krueger JM, Walter J, Dinarello CA, Wolff SM, Chedid L. Sleep-promoting effects of endogenous pyrogen (interleukin-1). Am J Physiol 1984;246:R994–99.
13. Shoham S, Davenne D, Cady AB, Dinarello CA, Krueger JM. Recombinant tumor necrosis factor and interleukin 1 enhance slow-wave sleep. Am J Physiol 1987;253: R142–49.
14. Lin L, Faraco J, Li R, Kadotani H, Rogers W, Lin X, Qiu X, de Jong PJ, Nishino S, Mignot E. The sleep disorder canine narcolepsy is caused by a mutation in the hypocretin (orexin) receptor 2 gene. Cell 1999;98:365–76.
15. Chemelli RM, Willie JT, Sinton CM, Elmquist JK, Scammell T, Lee C, Richardson JA, Williams SC, Xiong Y, Kisanuki Y, Fitch TE, Nakazato M, Hammer RE, Saper CB, Yanagisawa M. Narcolepsy in orexin knockout mice: Molecular genetics of sleep regulation. Cell 1999;98:437–51.
16. Ticho SR, Radulovacki M. Role of adenosine in sleep and temperature regulation in the preoptic area of rats. Pharmacol Biochem Behav 1991;40:33–40.
17. Benington JH, Kodali SK, Heller HC. Stimulation of A1 adenosine receptors mimics the electroencephalographic effects of sleep deprivation. Brain Res 1995;692:79–85.
18. Obal F Jr, Krueger JM. Biochemical regulation of sleep. Front Biosci 2003;8:520–50.
19. Finelli LA, Achermann P, Borbely AA. Individual "fingerprints" in human sleep EEG topography. Neuropsychopharmacology 2001;25:S57–62.
20. Maquet P. A role for sleep in the processing of memory traces. Contribution of functional neuroimaging in humans. Bull Mem Acad R Med Belg 2004;159:167–70.
21. Braun AR, Balkin TJ, Wesenten NJ, Carson RE, Varga M, Baldwin P, Selbie S, Belenky G, Herscovitch P. Regional cerebral blood flow throughout the sleep-wake cycle. An $H_2(15)O$ PET study. Brain 1997;120:1173–97.
22. Maquet P. Functional neuroimaging of normal human sleep by positron emission tomography. J Sleep Res 2000;9:207–31.
23. Roth C, Achermann P, Borbely AA. Frequency and state specific hemispheric asymmetries in the human sleep EEG. Neurosci Lett 1999;271:139–42.
24. Schwierin B, Achermann P, Deboer T, Oleksenko A, Borbely AA, Tobler I. Regional differences in the dynamics of the cortical EEG in the rat after sleep deprivation. Clin Neurophysiol 1999;110:869–75.
25. Nofzinger EA, Buysse DJ, Germain A, Price JC, Meltzer CC, Miewald JM, Kupfer DJ. Alterations in regional cerebral glucose metabolism across waking and non-rapid eye movement sleep in depression. Arch Gen Psychiatry 2005;62:387–96.
26. DeGennaro L, Ferrara M, Curcio G, Cristiani R, Bertini M. Cortical EEG topography of REM onset: The posterior dominance of middle and high frequencies. Clin Neurophysiol 2002;113:561–70.
27. Maquet P. The role of sleep in learning and memory. Science 2001;294:1048–52.
28. Mukhametov LM. Sleep in marine mammals. Expt Brain Res 1984;8:227–38.
29. Oleksenko AI, Mukhametov LM, Polyakova IG, Supin AY, Kovalzon VM. Unihemispheric sleep deprivation in bottlenose dolphins. J Sleep Res 1992;1:40–44.
30. Kattler H, Dijk DJ, Borbely AA. Effect of unilateral somatosensory stimulation prior to sleep on the sleep EEG in humans. J Sleep Res 1994;3:159–64.
31. Huber R, Ghilardi MF, Massimini M, Tononi G. Local sleep and learning. Nature 2004;430:78–81.
32. Ferrara M, DeGennaro L, Curcio G, Cristiani R, Bertini M. Interhemispheric asymmetry of human sleep EEG in response to selective slow-wave sleep deprivation. Behav Neurosci 2002;116:976–81.

33. Yasuda T, Yasuda K, Brown R, Krueger JM. State dependent effects of the light/dark cycle on the somatosensory and visual cortex EEG in rats. Am J Physiol 2005;289:R1083–89.
34. Vyazovskiy V, Borbely AA, Tobler I. Unilateral vibrissae stimulation during waking induces interhemispheric EEG asymmetry during subsequent sleep in the rat. J Sleep Res 2000;9:367–71.
35. Rector DM, Topchiy IA, Carter KM, Rojas MJ. Local functional state differences between rat cortical columns. Brain Res 2005;1047:45–55.
36. Benington JH. Sleep homeostasis and the function of sleep. Sleep 2000;23:959–66.
37. Pappenheimer JR, Koski G, Fencl V, Karnovsky ML, Krueger JM. Extraction of sleep-promoting Factor S from cerebrospinal fluid and from brains of sleep-deprived animals. J Neurophysiol 1975;38:1299–311.
38. Von Economo C. Sleep as a problem of localization. J Nerv Ment Dis 1930;71:249.
39. Steriade M. The corticothalamic system in sleep. Front Biosci 2003;8:d878–99.
40. McGinty D, Szymusiak R. Hypothalamic regulation of sleep and arousal. Front Biosci 2003;8:d1074–83.
41. Saper CB, Scammell TE, Lu J. Hypothalamic regulation of sleep and circadian rhythms. Nature 2005;437:1257–63.
42. Lu J, Sherman D, Devor M, Saper CB. A putative flip-flop switch for control of REM sleep. Nature 2006;441:589–94.
43. Jones B. Arousal systems. Front Biosci 2003;8:s438–51.
44. Kilduff TS, Peyron C. The hypocretin/orexin ligand-receptor system: implications for sleep and sleep disorders. Trends Neurosci 2000;23:359–65.
45. Koch C. The Quest for Consciousness. Englewood, CO: Roberts and Company, Publishers; 2004.
46. Fang J, Wang Y Krueger JM. Mice lacking the TNF 55 kD receptor fail to sleep more after TNF alpha treatment. J Neurosci 1997;17:5949–55.
47. Nistico G, De Sarro G, Rotiroti D. Behavioral and electrocortical spectrum power changes of interleukins and tumor necrosis factor after microinjection into different areas of the brain. In Sleep, Hormones, and Immunological System. Smirne S, Francesch M, Ferini-Stambi L, Zucconi M, Eds; Milan: Mason, 1992; pp. 11–22.
48. Dickstein JB, Moldofsky H, Lue FA, Hay JB. Intracerebroventricular injection of TNF-alpha promotes sleep and is recovered in cervical lymph. Am J Physiol 1999;276:R1018–22.
49. Takahashi S, Kapas L, Fang J, Krueger JM. An anti-tumor necrosis factor antibody suppresses sleep in rats and rabbits. Brain Res 1995;690:241–44.
50. Takahashi S, Tooley DD, Kapas L, Fang J, Seyer JM, Krueger JM. Inhibition of tumor necrosis factor in the brain suppresses rabbit sleep. Pflügers Arch 1995;431:155–60.
51. Takahashi S, Kapas L, Seyer JM, Wang Y, Krueger JM. Inhibition of tumor necrosis factor attenuates physiological sleep in rabbits. Neuroreport 1996;7:642–46.
52. Takahashi S, Krueger JM. Inhibition of tumor necrosis factor prevents warming-induced sleep responses in rabbits. Am J Physiol 1997;272:R1325–29.
53. Deboer T, Fontana A, Tobler I. Tumor necrosis factor (TNF) ligand and TNF receptor deficiency affects sleep and the sleep EEG. J Neurophysiol 2002;88:839–46.
54. Floyd RA, Krueger JM. Diurnal variations of TNF alpha in the rat brain. Neuroreport 1997;8:915–18.
55. Bredow S, Taishi P, Guha-Thakurta N, Obal Jr F, Krueger JM. Diurnal variations of tumor necrosis factor-alpha mRNA and alpha-tubulin mRNA in rat brain. J Neuroimmunomodulation 1997;4:84–90.

56. Veasey SC, Mackiewicz M, Fenik P, Ro M, Oglivie MD, Pack AL. IL1 beta knockout mice lack the TNF alpha response to sleep deprivation but have normal sleep and sleep recovery. Soc Neurosci Abst 1997;23:792.
57. Taishi P, Gardi J, Chen Z, Fang J, Krueger JM. Sleep deprivation increases the expression of TNF alpha mRNA and TNF 55kD receptor mRNA in rat brain. The Physiologist 1999;42:A4.
58. Hu J, Chen Z, Gorczynski CP, Gorczynski LY, Kai Y, Lee L, Manuel J, Gorczynski RM. Sleep-deprived mice show altered cytokine production manifest by perturbations in serum IL-1ra, TNFα, and IL-6 levels. Brain Behav Immun 2003;17:498–504.
59. Darko DF, Miller JC, Gallen C, White J, Koziol J, Brown SJ, Hayduk R, Atkinson JH, Assmus J, Munnel DT, Naitoh P, McCutchan JA, Mitler MM. Sleep electroencephalogram delta-frequency amplitude, night plasma levels of tumor necrosis factor alpha, and human immunodeficiency virus infection. Proc Natl Acad Sci USA 1995;92:12080–84.
60. Yamasu K, Shimada Y, Sakaizumi M, Soma G, Mizuno D. Activation of the systemic production of tumor necrosis factor after exposure to acute stress. Eur Cytokine Netw 1992;3:391–98.
61. Shearer WT, Reuben JM, Mullington JM, Price NJ, Lee BN, Smith EO, Szuba MP, Van Dongen HP, Dinges DF. Soluble TNF-alpha receptor 1 and IL-6 plasma levels in human subjected to the sleep deprivation model of spaceflight. J Allergy Clin Immunol 2001;107:165–70.
62. Haack M, Pollmacher T, Mullington JM. Diurnal and sleep-wake dependent variations of soluble TNF- and IL-2 receptors in health volunteers. Brain Behav Immun 2004; 18:361–67.
63. Puccioni-Sohler M, Rieckmann P, Kitze B, Lange P, Albrecht M, Flegenhauer K. A soluble form of tumor necrosis factor receptor in cerebrospinal fluid and serum of HTVLV-1-associated myelopathy and other neurological diseases. Neurology 1995;242: 239–42.
64. Vgontzas AN, Papanicolaou DA, Bixler EO, Kales A, Tyson K, Chrousos GP. Elevation of plasma cytokines in disorders of excessive daytime sleepiness: role of sleep disturbance and obesity. J Clin Endocrinol Metab 1997;82:1313–16.
65. Liu H, Liu J, Xiong S, Shen G, Zhang Z, Xu Y. The change in interleukin-6 and tumor necrosis factor in patients with obstructive sleep apnea syndrome. J Tongji Med Univ 2000;20:200–2.
66. Entzian P, Linnemann K, Schlaak M, Zabel P. Obstructive sleep apnea syndrome and circadian rhythms of hormones and cytokines. Am J Respir Crit Care Med 1996;153: 1080–86.
67. Alberti A, Sarchielli P, Gallinella E, Floridi A, Floridi A, Mazzotta G, Gallai V. Plasma cytokine levels in patients with obstructive sleep apnea syndrome: A preliminary study. J Sleep Res 2003;12:305–11.
68. Imagawa S, Yamaguchi Y, Ogawa K, Obara N, Suzuki N, Yamamoto M, Nagasawa T. Interleukin-6 and tumor necrosis factor-alpha in patients with obstructive sleep apnea-hypopnea syndrome. Respiration 2004;71:24–29.
69. Ciftci TU, Kokturk O, Bukan N, Bilgihan A. The relationship between serum cytokine levels with obesity and obstructive sleep apnea syndrome. Cytokine 2004;28:87–91.
70. Minoguchi K, Tazaki T, Yokoe T, Minoguchi H, Watanabe Y, Yamamoto M, Adachi M. Elevated production of tumor necrosis factor-alpha by monocytes in patients with obstructive sleep apnea syndrome. Chest 2004;126:1473–79.

71. Moss RB, Mercandetti A, Vojdani A. TNF-alpha and chronic fatigue syndrome. J Clin Immunol 1999;19:314–16.
72. Vgontzas AN, Zoumakis M, Papanicolaou DA, Bixler EO, Prolo P, Lin HM, Vela-Bueno A, Kales A, Chrousos GP. Chronic insomnia is associated with a shift of interleukin-6 and tumor necrosis factor secretion from nighttime to daytime. Metabolism 2002;51:887–92.
73. Francis J, Chu Y, Johnson AK, Weiss RM, Felder RB. Acute myocardial infarction induces hypothalamic cytokine synthesis. Am J Physiol 2004;286:H2264–271.
74. Vgontzas AN, Bixler EO, Chrousos GP. Metabolic disturbances in obesity versus sleep apnoea: the importance of visceral obesity and insulin resistance. J Intern Med 2003; 254:32–44.
75. Edwards N, Blyton DM, Kesby GJ, Wilcox I, Sullivan CE. Pre-eclampsia is associated with marked alterations in sleep architecture. Sleep 2000;23:619–25.
76. Dreisbach AW, Hendrickson T, Beezhold D, Riesenberg LA, Sklar AH. Elevated levels of tumor necrosis factor alpha in postdialysis fatigue. Int J Artif Organs 1998;21:83–86.
77. Sklar AH, Beezhold D, Newman N, Hendrickson T, Dreisbach AW. Postdialysis fatigue: lack of effect of a biocompatible membrane. Am J Kidney Dis 1998;31:1007–10.
78. Eskander ED, Harvey HA, Givant E, Lipton A. Phase I study combining tumor necrosis factor with interferon-alpha and interleukin-2. Am J Clin Oncol 1997;20:511–14.
79. Irwin M, Rinetti G, Redwine L, Motivala S, Dang J, Ehlers C. Nocturnal proinflammatory cytokine-associated sleep disturbances in abstinent African American alcoholics. Brain Behav Immun 2004;18:349–60.
80. Okun ML, Giese S, Lin L, Einen M, Mignot E, Coussons-Read ME. Exploring the cytokine and endocrine involvement in narcolepsy. Brain Behav Immun 2004;18:326–32.
81. Wieczorek S, Gencik M, Rujescu D, Tonn P, Giegling I, Epplen JT, Dahmen N. TNFα promoter polymorphisms and narcolepsy. Tissue Antigens 2003;61:437–42.
82. Hohjoh H, Nakayama T, Ohashi J, Miyagawa T, Tanaka H, Akaza T, Honda Y, Juji T, Tokunaga K. Significant association of a single nucleotide polymorphism in the tumor necrosis factor-alpha (TNF-alpha) gene promoter with human narcolepsy. Tissue Antigens 1999;54:138–45.
83. Kato T, Honda M, Kuwata S, Juji T, Fukuda M, Honda Y, Kato N. A search for a mutation in the tumour necrosis factor-alpha gene in narcolepsy. Psychiatry Clin Neurosci 1999;53:421–23.
84. Kato T, Honda M, Kuwata S, Juji T, Kunugi H, Nanko S, Fukuda M, Honda Y. Novel polymorphism in the promoter region of the tumor necrosis factor alpha gene: no association with narcolepsy. Am J Med Genet 1999;88:301–4.
85. Sookoian SC, Gonzalez C, Pirola CJ. Meta-analysis on the G-308A tumor necrosis factor alpha gene variant and phenotypes associated with the metabolic syndrome. Obes Res 2005;13:2122–31.
86. Riha RL, Brander P, Vennelle M, McArdle N, Kerr SM, Anderson NH, Douglas NJ. Tumour necrosis factor-alpha (-308) gene polymorphism in obstructive sleep apnoea-hypopnoea syndrome. Eur Respir J 2005;26:673–78.
87. Franklin CM. Clinical experience with soluble TNF p75 receptor in rheumatoid arthritis. Semin Arthritis Rheum 1999;29:171–81.
88. Vgontzas AN, Zoumakis E, Lin HM, Bixler EO, Trakada G, Chrousos GP. Marked decrease in sleepiness in patients with sleep apnea by etanercept, a tumor necrosis factor-α antagonist. J Clin Endocrinol Metab 2004;89:4409–13.

89. Kataoka T, Enomoto F, Kim R, Yokoi H, Fujimori M, Sakai Y, Ando I, Ichikawa G, Ikeda K. The effect of surgical treatment of obstructive sleep apnea syndrome on the plasma TNF-alpha levels. Tohoku J Exp Med 2004;204:267–72.
90. Banks WA. Blood-brain barrier transport of cytokines: A mechanism for neuropathology. Curr Pharm Des 2005;11:973–84.
91. Kubota T, Fang J, Guan Z, Brown RA, Krueger JM. Vagotomy attenuates tumor necrosis factor-alpha-induced sleep and EEG delta-activity in rats. Am J Physiol 2001;280: R1213–20.
92. Mullington J, Korth C, Hermann DM, Orth A, Galanos C, Holsboer F, Pollmacher T. Dose-dependent effects of endotoxin on human sleep. Am J Physiol 2000;278:R947–55.
93. Haack M, Schuld A, Kraus T, Pollmächer T. Effects of sleep on endotoxin-induced host responses in healthy men. Psychosom Med 2001;63:568–78.
94. Takahashi S, Kapas L, Krueger JM. A tumor necrosis factor (TNF) receptor fragment attenuates TNF-alpha- and muramyl dipeptide-induced sleep and fever in rabbits. J Sleep Res 1996;5:106–14.
95. Kubota T, Li N, Guan Z, Brown RA, Krueger JM. Intrapreoptic microinjection of TNF-alpha enhances non-REMS in rats. Brain Res 2002;932:37–44.
96. De Sarro G, Gareri P, Sinopoli VA, David E, Rotiroti D. Comparative, behavioural and electrocortical effects of tumor necrosis factor-alpha and interleukin-1 microinjected into the locus coeruleus of rat. Life Sci 1997;60:555–64.
97. Terao A, Matsumura H, Yoneda H, Saito M. Enhancement of slow-wave sleep by tumor necrosis factor-alpha is mediated by cyclooxygenase-2 in rats. Neuroreport 1998;9:3791–96.
98. Yoshida H, Peterfi Z, Garcia-Garcia F, Kirkpatrick R, Yasuda T, Krueger JM. State-specific asymmetries in EEG slow wave activity induced by local application of TNF alpha. Brain Res 2004;1009:129–36.
99. Churchill L, Yasuda K, Yasuda T, Blindheim KA, Falter M, Garcia-Garcia F, Krueger JM. Unilateral cortical application of tumor necrosis factor alpha induces asymmetry in Fos- and interleukin-1 beta-immunoreactive cells within the corticothalamic projection. Brain Res 2005;1055:15–24.
100. Peschon JJ, Torance DS, Stocking KL, Glaccum MB, Otten C, Willis CR, Charrier K, Morrissey PJ, Ware CB, Mohler KM. TNF receptor-deficient mice reveal divergent roles for p55 and p75 in several models of inflammation. J Immunol 1998;160:943–52.
101. Fontaine V, Mohand-Said S, Hanoteau S, Fuchs C, Pfizenmaier K, Eisel U. Neurodegenerative and neuroprotective effects of tumor necrosis factor (TNF) in retinal ischemia: opposite roles of TNF receptor 1 and TNF receptor 2. J Neurosci 2002;22: RC216(1–7).
102. Yang L, Lindholm K, Konishi Y, Li R, Shen Y. Target depletion of distinct tumor necrosis factor receptor subtypes reveals hippocampal neuron death and survival through different signal transduction pathways. J Neurosci 2002;22:3025–32.
103. Schubert P, Ogata T, Marchini C, Ferroni S, Rudolphi K. Protective mechanisms of adenosine in neurons and glial cells. Ann NY Acad Sci 1997;825:1–10.
104. Carlson NG, Bacchi A, Rogers SW, Gahring LC. Nicotine blocks TNF-alpha mediated neuroprotection to NMDA by an alpha-bungarotoxin-sensitive pathway. J Neurobiol 1998;35:29–36.
105. Dinarello CA. Infection, fever, and exogenous and endogenous pyrogens: some concepts have changed. J Endotoxin Res 2004;10:201–22.

106. Saper CB, Breder CD. Endogenous pyrogens in the CNS: role in febrile response. Prog Brain Res 1992;93:419–28.
107. Plata-Salaman CR. Central nervous system mechanisms contributing to the cachexia-anorexia syndrome. Nutrition 2000;16:1009–12.
108. Merrill JE. Tumor necrosis factor alpha, interleukin 1 and related cytokines in brain development: normal and pathological. Dev Neurosci 1992;14:1–10.
109. Gerber J, Bottcher T, Hahn M, Siemer A, Bunkowski S, Nau R. Increased mortality and spatial memory deficits in TNF-alpha-deficient mice in ceftriaxone-treated experimental pneumococcal meningitis. Neurobiol Dis 2004;16:133–38.
110. Das UN. Can memory be improved? A discussion on the role of ras, GABA, acetylcholine, NO, insulin, TNF-alpha, and long-chain polyunsaturated fatty acids in memory formation and consolidation. Brain Dev 2003;25:251–61.
111. Albensi BC, Mattson MP. Evidence for the involvement of TNF and NF-kappaB in hippocampal synaptic plasticity. Synapse 2000;35:151–59.
112. Beattie EC, Stellwagen D, Morishita W, Bresnahan JC, Ha BK, Von Zastrow M, Beattie MS, Malenka RC. Control of synaptic strength by glial TNF alpha. Science 2002; 295:2282–85.
113. De A, Krueger JM, Simasko SM. Tumor necrosis factor alpha increases cytosolic calcium response AMPA and KCl in primary cultures of rat hippocampal neurons. Brain Res 2003;981:133–42.
114. Stellwagen D, Malenka RC. Synaptic scaling mediated by glial TNF-alpha. Nature 2006;440:1054–59.
115. Krueger JM, Obal F Jr, Fang J, Kubota T, Taishi P. The role of cytokines in physiological sleep regulation. Ann NY Acad Sci 2001;933:211–21.
116. Takahashi S, Kapas L, Fang J, Krueger JM. Somnogenic relationships between tumor necrosis factor and interleukin- 1. Am J Physiol 1999;276:R1132–40.
117. Opp MR, Krueger JM. Interleukin-1 is involved in responses to sleep deprivation in the rabbit. Brain Res 1994;639:57–65.
118. Susic V, Totic S. 'Recovery' function of sleep: effects of purified human interleukin-1 on the sleep and febrile response of cats. Metab Brain Dis 1989;4:73–80.
119. Friedman EM, Boinski S, Coe CL. Interleukin-1 induces sleep-like behavior and alters call structure in juvenile rhesus macaques. Am J Primatol 1995;35:145–53.
120. Fang J, Wang Y, Krueger JM. The effects of interleukin-1 beta on sleep are mediated by the type I receptor. Am J Physiol 1998;274:R655–60.
121. Tobler I, Borbély AA, Schwyzer M, Fontana A. Interleukin-1 derived from astrocytes enhances slow wave activity in sleep EEG of the rat. Eur J Pharmacol 1984; 104:191–92.
122. Dinarello CA. Interleukin-1 and interleukin-1 antagonism. Blood 1991;77:1627–52.
123. Opp MR, Obal F Jr, Krueger JM. Interleukin 1 alters rat sleep: temporal and dose-related effects. Am J Physiol 1991;260:R52–58.
124. Hansen MK, Krueger JM. Subdiaphragmatic vagotomy blocks the sleep and fever-promoting effects of interleukin-1 beta. Am J Physiol 1997;273:R1246–53.
125. Opp MR, Krueger JM. Anti-interleukin-1 beta reduces sleep and sleep rebound after sleep deprivation in rats. Am J Physiol 1994;266:R688–95.
126. Takahashi S, Kapas L, Fang J, Seyer JM, Wang Y, Krueger JM. An interleukin-1 receptor fragment inhibits spontaneous sleep and muramyl dipeptide-induced sleep in rabbits. Am J Physiol 1996;271:R101–8.

127. Takahashi S, Kapas L, Hansen M, Zhang J, Seyer JM, Krueger JM. An interleukin-1 (IL-1) soluble receptor fragment inhibits IL-1 beta-induced sleep and non-rapid-eye-movement-sleep rebound after sleep deprivation in rabbits. Sleep Res 1995;24A:457.
128. Imeri L, Opp MR, Krueger JM. An IL-1 receptor and an IL-1 receptor antagonist attenuate muramyl dipeptide- and IL-1-induced sleep and fever. Am J Physiol 1993; 265:R907–13.
129. Vitokovic L, Bockaert J, Jacque C. "Inflammatory" cytokines: neuromodulators in normal brain? J Neurochem 2000;74:457–71.
130. Taishi P, Bredow S, Guha-Thakurta N, Obal F Jr, Krueger JM. Diurnal variations of interleukin-1 beta mRNA and beta-actin mRNA in rat brain. J Neuroimmunol 1997;75:69–74.
131. Breder CD, Dinarello CA, Saper CB. Interleukin-1 immunoreactive innervation of the human hypothalamus. Science 1988;240:321–24.
132. De A, Churchill L, Obal F Jr, Simasko SM, Krueger JM. GHRH and IL1beta increase cytoplasmic Ca^{2+} levels in cultured hypothalamic GABAergic neurons. Brain Res 2002;949:209–12.
133. Lue FA, Bail M, Jephthah-Ocholo J, Carayanniotis K, Gorczynski R, Moldofsky H. Sleep and cerebrospinal fluid interleukin-1 like activity in the cat. Int J Neurosci 1998;42:179–83.
134. Nguyen KT, Deak T, Owens SM, Kohno T, Fleshner M, Watkins LR, Maier SF. Exposure to acute stress induces brain interleukin-1 beta protein in the rat. J Neurosci 1998;18:2239–46.
135. Mackiewicz M, Sollars PJ, Ogilvie MD, Pack AI. Modulation of IL-1beta gene expression in the rat CNS during sleep deprivation. Neuroreport 1996;7:529–33.
136. Moldofsky H, Lue FA, Eisen J, Keystone E, Gorczynski RM. The relationship of interleukin-1 and immune functions to sleep in humans. Psychosom Med 1986;48:309–18.
137. Hohagen F, Timmer J, Weyerbrock A, Fritsch-Montero R, Ganter U, Krieger S, Berger M, Bauer J. Cytokine production during sleep and wakefulness and its relationship to cortisol in healthy humans. Neuropsychobiology 1993;28:9–16.
138. Uthgenannt D, Schoolmann D, Pietrowsky R, Fehm H-L, Born J. Effects of sleep on the production of cytokines in humans. Psychosom Med 1995;57:97–104.
139. Hansen MK, Taishi P, Chen Z, Krueger JM. Vagotomy blocks the induction of interleukin-1 beta mRNA in the brain of rats in response to systemic interleukin-1 beta. J Neurosci 1998;18:2247–53.
140. Churchill L, Taishi P, Wang M, Brandt J, Cearley C, Krueger JM. Brain distribution of cytokine mRNA induced by systemic administration of interleukin-1 or tumor necrosis factor α. Brain Res 2006;1120:64–73.
141. Hansen M, Kapas L, Fang J, Krueger JM. Cafeteria diet-induced sleep is blocked by subdiaphragmatic vagotomy. Am J Physiol 1998;274:R168–74.
142. Hansen MK, Taishi P, Chen Z, Krueger JM. Cafeteria-feeding induces interleukin-1 beta mRNA expression in rat liver and brain. Am J Physiol 1998;43:R1734–39.
143. Manfridi A, Brambilla D, Bianchi S, Mariotti M, Opp MR, Imeri L. Interleukin-1 beta enhances non-rapid eye movement sleep when microinjected into the dorsal raphe nucleus and inhibits serotonergic neurons *in vitro*. Eur J Neurosci 2003; 18:1041–49.
144. Slisli Y, De Beaurepaire R. Interleukin-1 beta and calcitonin, but not corticotropin-releasing factor, alter sleep cycles when microinjected into the rat hypothalamic lateral paraventricular area. Neurosci Lett 1999;265:29–32.

145. Terao A, Matsumura H, Saito M. Interleukin-1 induces slow-wave sleep at the prostaglandin D2-sensitive sleep-promoting zone in the rat brain. J Neurosci 1998;18:6599–607.
146. Alam MN, McGinty D, Bashir T, Kumar S, Imeri L, Opp MR, Szymusiak R. Interleukin-1 beta modulates state-dependent discharge activity of preoptic area and basal forebrain neurons: Role in sleep regulation. Eur J Neurosci 2004;20:207–16.
147. Baker FC, Shah S, Stewart D, Angara C, Gong H, Szymusiak R, Opp MR, McGinty D. Interleukin 1 beta enhances non-rapid eye movement sleep and increases c-Fos protein expression in the medial preoptic nucleus of the hypothalamus. Am J Physiol Regul Integr Comp Physiol 2005;288:R998–1005.
148. Shibata M. Hypothalamic neuronal responses to cytokines. Yale J Biol Med 1990; 63:147–56.
149. Obal F Jr, Fang J, Payne LC, Krueger JM. Growth hormone-releasing hormone mediates the sleep-promoting activity of interleukin-1 in rats. Neuroendocrinology 1995;61:559–65.
150. Taishi P, De A, Alt J, Gardi J, Obal F Jr, Krueger JM. Interleukin-1beta stimulates growth hormone-releasing hormone receptor mRNA expression in the rat hypothalamus *in vitro* and *in vivo*. J Neuroendocrinol 2004;16:113–18.
151. Obal Jr F, Krueger JM. GHRH and sleep. Sleep Med Rev 2004;8:367–77.
152. McCann SM, Kimura M, Karanth S, Yu WH, Rettori V. Role of nitric oxide in the neuroendocrine response to cytokines. Front Horm Res 2002;29:117–29.
153. Luk WP, Zhang Y, White TD, Lue FA, Wu C, Jiang CG, Zhang L, Moldofsky H. Adenosine: a mediator of interleukin-1 beta-induced hippocampal synaptic inhibition. J Neurosci 1999;19:4238–44.
154. Kapás L, Shibata M, Kimura M, Krueger JM. Inhibition of nitric oxide synthesis suppresses sleep in rabbits. Am J Physiol 1994;266:R151–57.
155. Miller LG, Galpern WG, Dunlap K, Dinarello CA, Turner TJ. Interleukin-1 augments gamma-aminobutyric acid A receptor function in brain. Mol Pharmacol 1991;39:105–8.
156. Imeri L, Mancia M, Opp MR. Blockade of 5-hydroxytryptamine (serotonin)-2 receptors alters interleukin-1 induced changes in rat sleep. Neuroscience 1999;92:745–49.
157. Tringali G, Mirtella A, Mancuso C, Guerriero G, Preziosi P, Navarra P. The release of immunoreactive interleukin-1 beta from rat hypothalamic explants is modulated by neurotransmitters and corticotropin-releasing hormone. Pharmacol Res 1997;36:269–73.
158. Opp MR, Postlethwaite AE, Seyer JM, Krueger JM. Interleukin1 receptor antagonist blocks somnogenic and pyrogenic responses to an interleukin 1 fragment. Proc Natl Acad Sci USA 1992;89:3726–30.
159. Opp MR, Smith EM, Hughes Jr TK. Interleukin-10 (cytokine synthesis inhibitory factor) acts in the central nervous system of rats to reduce sleep. J Neuroimmunol 1995;60:165–68.
160. Kushikata T, Fang J, Wang Y, Krueger JM. Interleukin-4 inhibits spontaneous sleep in rabbits. Am J Physiol 1998;275:R1185–91.
161. Fiorentino DF, Zlotnik A, Mosmann TR, Howard M, O'Garra A. IL-10 inhibits cytokine production by activated macrophages. J Immunol 1991;147:3815–22.
162. Kanaan SA, Poole S, Saade NE, Jabbur S, Safieh-Garabedian B. Interleukin-10 reduces the endotoxin-induced hyperalgesia in mice. J Neuroimmunol 1998;86:142–50.
163. Thomassen MJ, Divis LT, Fisher CJ. Regulation of human alveolar macrophage inflammatory cytokine production by interleukin-10. Clin Immunol Immunopathol 1996;80:321–24.
164. Dickensheets HL, Donnelly RP. IFN-gamma and IL-10 inhibit induction of IL-1 receptor type I and type II gene expression by IL-4 and IL-13 in human monocytes. J Immunol 1997;159:6226–33.

165. Dugas N, Palacios-Calender M, Dugas B, Riveros-Moreno V, Delfraissy JF, Kolb JP, Moncada S. Regulation by endogenous interleukin-10 of the expression of nitric oxide synthase induced after ligation of CD23 in human macrophages. Cytokine 1998; 10:680–89.
166. Laffranchi R, Spinas GA. Interleukin 10 inhibits insulin release from and nitric oxide production in rat pancreatic islets. Eur J Endocrinol 1996;135:374–78.
167. Stefano GB, Prevot V, Beauvillain JC, Hughes TK. Interleukin-10 stimulation of corticotrophin releasing factor median eminence in rats; evidence for dependence upon nitric oxide production. Neurosci Lett 1998;256:167–70.
168. Joyce DA, Steer JH, Kloda A. Dexamethazone antagonizes IL-4 and IL-10 induced release of IL-1RA by monocytes but augments IL4, IL10, TGF-beta-induced suppression of TNF-alpha release. J Interferon Cytokine Res 1996;16:511–17.
169. Mori N, Shirakawa F, Murakami S, Oda S, Eto S. Interleukin-4 inhibits the production of interleukin-1 by adult T-cell leukemia cells. Eur J Haematol 1995;55:121–25.
170. Lee JD, Rhoades K, Economou JS. Interleukin-4 inhibits the expression of tumor necrosis factors alpha and beta, interleukins-1 beta and 6 and interferon-gamma. Immunol Cell Biol 1995;73:57–61.
171. Hart PH, Ahern MJ, Smith MD, Finlay-Jones JJ. Comparison of the suppressive effects of interleukin-10 and interleukin 4 on synovial fluid macrophages and blood monocytes from patients with inflammatory arthritis. Immunology 1995;84:536–42.
172. Sone S, Orino E, Mizuno K, Yano S, Nishioka Y, Haku T, Nii A, Ogura T. Production of IL1 and its receptor antagonist is regulated differently by IFN-gamma and IL4 in human monocytes and alveolar macrophages. Eur Respir J 1994;7:657–63.
173. Joyce DA, Gibbons DP, Green P, Steer JH, Feldman M, Brennan FM. Two inhibitors of pro-inflammatory cytokine release, interleukin-10 and interleukin-4, have contrasting effects on release of soluble p75 tumor necrosis factor receptor by cultured monocytes. Eur J Immunol 1994;24:2699–705.
174. Kasai K, Hattori Y, Nakanishi N, Manaka K, Banba N, Motohashi S, Shimoda S. Regulation of inducible nitric oxide production by cytokines in human thyrocytes in culture. Endocrinology 1995;136:4261–70.
175. Ledeboer A, Breve JJ, Wierinckx A, van der Jagt S, Bristow AF, Leysen JE, Tilders FJ, Van Dam AM. Expression and regulation of interleukin-10 and interleukin-10 receptor in rat astroglial and microglial cells. Eur J Neurosci 2002;16:1175–85.
176. Chen, Z, Gardi J, Kushikata T, Fang J, Krueger JM. Nuclear factor kappa B-like activity increases in murine cerebral cortex after sleep deprivation. Am J Physiol 1999;276: R1812–18.
177. Brandt JA, Churchill L, Rehman A, Ellis G, Mémet S, Israël A, Krueger JM. Sleep-deprivation increases activation of nuclear factor kappa B in lateral hypothalamic cells. Brain Res 2004;1004:91–97.
178. Basheer R, Rainnie DG, Porkka-Heiskanen T, Ramesh V, McCarley RW. Adenosine, prolonged wakefulness, and A1-activated NF-κB DNA binding in the basal forebrain of the rat. Neuroscience 2001;104:731–39.
179. Kubota T, Kushikata T, Fang J, Krueger JM. Nuclear factor kappa B (NFκB) inhibitor peptide inhibits spontaneous and interleukin-1β-induced sleep. Am J Physiol 2000;279: R404–13.
180. Krueger JM, Majde JA. Sleep and the immune response. In Encyclopedia of Sleep Medicine. Lee-Chong TL, Ed.; John Wiley & Sons, Inc, Hoboken, NJ 2006; pp. 767–72.

181. Hogan D, Morrow JD, Smith EM, Opp MR. Interleukin-6 alters sleep of rats. J Neuroimmunol 2003;137:59–66.
182. Morrow JD, Opp MR. Diurnal variation of lipopolysaccharide-induced alterations in sleep and body temperature of interleukin-6-deficient mice. Brain Behav Immun 2005;19:40–51.
183. Krueger JM, Majde JA. Host defense. In Principles and Practice of Sleep Medicine. Kryger MH, Roth T, Dement WC, Eds; Philadelphia: Elsevier Saunders, 2005; pp. 256–65.
184. Vaillant AR, Zanassi P, Walsh GS, Aumont A, Alonso A, Miller FD. Signaling mechanisms underlying reversible, activity-dependent dendrite formation. Neuron 2002;34:985–98.
185. Schinder AF, Poo M-M. The neurotrophin hypothesis for synaptic plasticity. Trends Neurosci 2000;23:639–45.
186. Zhang LI, Poo M-M: Electrical activity and development of neural circuits. Nature Neuroscience 2001;4:1207–14.
187. De Bock F, Dornand J, Rondouin G. Release of TNF alpha in the rat hippocampus following epileptic seizures and excitotoxic neuronal damage. Neuroreport 1996;7:1125–29.
188. De A, Krueger JM, Simasko SM. Glutamate induces expression and release of tumor necrosis factor alpha in cultured hypothalamic cells. Brain Res 2005:1053:54–61.
189. Schneider H, Pitossi F, Balschun D, Wagner A, del Rey A, Besedovsky HO. A neuromodulatory role of interleukin-1 beta in the hippocampus. Proc Natl Acad Sci USA 1998;95:7778–83.
190. Yi PL, Tsai CH, Lin JG, Lee CC, Chang FC. Kindling stimuli delivered at different times in the sleep-wake cycle. Sleep 2004;27:203–12.
191. Fix C, Churchill L, Hall S, Krueger JM. The number of tumor necrosis factor α-immunoreactive cells increases in layer IV of the barrel field in response to whisker deflection in rats. Sleep 2006;29:A11.
192. Bazhenov M, Timofeev I, Steriade M, Sejnowski TJ. Model of thalamocortical slow-wave sleep oscillations and transitions to activated states. J Neurosci 2002;22:8691–04.
193. Malinow R, Malenka RC. AMPA receptor trafficking and synaptic plasticity. Ann Rev Neurosci 2002;25:103–26.
194. Krueger JM, Obal F Jr. Sleep regulatory substances. In Monographs in Clinical Neuroscience. Schwartz WJ, Ed.; Basel: Karger, 1997; pp. 175–94.
195. Krueger JM, Obal Jr F. A neuronal group theory of sleep function. J Sleep Res 1993;2:63–69.
196. Turrigiano GG, Nelson SB. Hebb and homeostasis in neuronal plasticity. Curr Opin Neurobiol 2000;10:358–64.
197. Turrigiano GG, Leslie KR, Desai NS, Rutherford LC, Nelson SB. Activity-dependent scaling of quantal amplitude in neocortical neurons. Nature 1998;391:892–96.
198. Abbott LF, Nelson SB. Synaptic plasticity: taming the beast. Nat Neurosci 2000;3:1178–83.
199. Kavanau JL. Sleep and dynamic stabilization of neural circuitry: a review and synthesis. Behav Brain Res 1994;63:111–26.
200. Tononi G, Cirelli C. Sleep function and synaptic homeostasis. Sleep Med Rev 2006; 10:49–62.
201. Kavanau JL. Memory sleep and dynamic stabilization of neural circuitry: evolutionary perspectives. Neurosci Biobehav Rev 1996;20:289–311.
202. Kristiansen K, Courtois G. Rhythmic activity from isolated cerebral cortex EEG. Clin Neurophysiol 1949;1:265–72.
203. Henry CE, Scoville WB. Suppression-burst activity from isolated cerebral cortex in man. Electroencephalogr Clin Neurophysiol 1952;4:1–22.

204. Amzica F, Steriade M. Neuronal and glial membrane potentials during sleep and paroxysmal oscillations in the neocortex. J Neurosci 2000;20:6648–65.
205. Mahowald MW, Schenck CH. Dissociated states of wakefulness and sleep. Neurology 1992;42:44–51.
206. Krueger JM, Obal F Jr. Sleep function. Frontiers in Biosci 2003;8:511–19.
207. Krueger JM, Obal F Jr, Kapas L, Fang J. Brain organization and sleep function. Behav Brain Res 1995;69:177–85.
208. Krueger JM, Obal F Jr, Fang J. Why we sleep. Sleep Med Rev 1999;3:119–29.
209. Earl MG, Strogatz SH. Synchronization in oscillator networks with delayed coupling: a stability criterion. Phys Rev E Stat Nonlin Soft Matter Phys 2003;67:036204.
210. Strogatz SH, Stewart I. Coupled oscillators and biological synchronization. Sci Am 1993;269:102–9.
211. Guillery RW, Harting JK. Structure and connections of the thalamic reticular nucleus: advancing views over half a century. J Comp Neurol 2003;463:360–71.
212. Walker AJ, Topchiy I, Kouptsov K, Rector DM. ERP differences during conditioned lick response in the rat. Sleep 2005;28:A15.
213. Brandt J, Churchill L, Guan Z, Fang J, Chen L, Krueger JM. Sleep deprivation but not a whisker trim increases nerve growth factor within barrel cortical neurons. Brain Res 2001;898:105–12.
214. Churchill L, Rector D, Yasuda K, Rojas MJ, Schactler S, Fix C, Hall S, Yasuda T, Krueger JM. Tumor necrosis factor α increases surface evoked potentials in the barrel field by whisker deflection during sleep in rats. Sleep 2006;29:A12–13.
215. Yasuda T, Yoshida H, Garcia-Garcia F, Kay D, Krueger JM. Interleukin-1β has a role in cerebral cortical state dependent electroencephalographic slow-wave activity. Sleep 2005;28:177–84.
216. Vyazovskiy V, Borbely AA, Tobler I. Interhemispheric sleep EEG asymmetry in the rat is enhanced by sleep deprivation. J Neurophysiol 2002;88:2280–86.
217. Orban P, Rauchs G, Balteau E, Degueldre C, Luxen A, Maquet P, Peigneux P. Sleep after spatial learning promotes covert reorganization of brain activity. Proc Natl Acad Sci USA 2006;103:7124–29.
218. Yasuda K, Churchill L, Yasuda T, Blindheim K, Falter M, Krueger JM. Unilateral cortical application of interleukin-1β (IL1β) induces asymmetry in Fos- and IL1β-immunoreactivity: Implications for sleep regulation. Brain Res 2007;1131:44–59.
219. Jouvet M. Neuromediateurs et facteurs hypnogenes. Rev Neurol (Paris) 1984;140:389–400.
220. Borbely AA, Tobler I. The search for an endogenous "sleep substance." Trends Pharmacol Sci 1980;1:356–58.
221. Krueger JM, Obál F Jr. Sleep factors. In: Saunders NA, Sullivan CE, Eds; New York: Marcel Dekker, Inc. 1994; pp. 79–112.
222. Obal F Jr, Opp M, Cady AB, Johannsen L, Postlethwaite AE, Poppleton HM, Seyer JM, Krueger JM. Interleukin-1alpha and an interleukin-1beta fragment are somnogenic. Am J Physiol 1990;259:R439–46.
223. De Sarro GP, Masuda Y, Ascioti C, Audino MG, Nistico G. Behavioral and ECoG spectrum changes by intracerebral infusion of interferons and interleukin 2 in rats are antagonized by naloxone. Neuropharmacology 1990;29:167–79.
224. Kubota T, Brown RA, Fang J, Krueger JM. Interleukin-15 and interleukin-2 enhance non-REMS in rabbits. Am J Physiol 2001;281:R1004–12.
225. Krueger JM. Interleukin-6 is pyrogenic but not somnogenic. Physiol Behav 1989;48:1069–72.

226. Späth-Schwalbe E, Hansen K, Schmidt F, Schrezenmeier H, Marshall L, Burger K, Fehm HL, Born J. Acute effects of recombinant human interleukin-6 on endocrine and central nervous sleep functions in healthy men. J Clin Endocrinol Metab 1998;83:1573–79.
227. Garcia-Garcia F, Yoshida H, Krueger JM. Interleukin-8 promotes non-rapid eye movement sleep in rabbits and rats. J Sleep Res 2004;13:55–61.
228. Kubota T, Fang J, Brown RA, Krueger JM. Interleukin-18 promotes sleep in rabbits and rats. Am J Physiol 2001;281:R828–38.
229. Kushikata T, Fang J, Chen Z, Wang Y, Krueger JM: Epidermal growth factor enhances spontaneous sleep in rabbits. Am J Physiol 1998;275:R509–14.
230. De Saint Hilaire Z, Nicolaidis S. Enhancement of slow wave sleep parallel to the satiating effect of acidic fibroblast growth factor in rats. Brain Res Bull 1992;29:525–28.
231. Knefati M, Somogyi C, Kapas L, Bourcier T, Krueger JM. Acidic fibroblast growth factor (FGF) but not basic FGF induces sleep and fever in rabbits. Am J Physiol 1995;269:R87-R91
232. Galan JM, Cuevas B, Dujovny N, Gimenez-Gallego G. Cuevas P. Sleep promoting effects of intravenously administered acidic fibroblast growth factor. Neurol Res 1996;18:567–69.
233. Garcia-Garcia F, Krueger JM. Intracerebroventricular injection of erythropoietin enhances sleep in the rat. Brain Res Bull 2003;61:541–46.
234. Takahashi S, Krueger JM. Nerve growth factor enhances sleep in rabbits. Neurosci Lett 1999;264:149–52.
235. Yamuy J, Morales FR, Chase MH. Induction of rapid eye movement sleep by the microinjection of nerve growth factor into the pontine reticular formation of the cat. Neuroscience 1995;66:9–13.
236. Yamuy J, Sampogna S, Chase MH. Neurotrophin-receptor immunoreactive neurons in mesopontine regions involved in the control of behavioral states. Brain Res 2000;866:1–14.
237. Tononi G, .Pompeiano M, Cirelli C. The locus coeruleus and immediate-early genes in spontaneous and forced wakefulness. Brain Res Bull 1994;35:589–96.
238. Kapas L, Obal Jr F, Book AA, Schweitzer JB, Wiley RG, Krueger JM. The effects of immunolesions of nerve growth factor-receptive neurons by 192 IgG-saporin on sleep. Brain Res 1996;712:53–59.
239. Kushikata T, Fang J, Krueger JM. Brain-derived neurotrophic factor enhances spontaneous sleep in rats and rabbits. Am J Physiol 1999;276:R1334–38.
240. Cirelli C, Tononi G. Differential expression of plasticity-related genes in waking and sleep and their regulation by the noradrenergic system. J Neurosci 2000;20: 9187–94.
241. Taishi P, Sanchez C, Wang Y, Fang F, Harding JW, Krueger JM. Conditions that affect sleep alter the expression of molecules associated with synaptic plasticity. Am J Physiol 2001;281:R839–45.
242. Cirelli C, Tononi G. Gene expression in the brain across the sleep-waking cycle. Brain Res 2000;885:303–21.
243. Kushikata T, Kubota T, Fang J, Krueger JM. Glial cell line-derived neurotrophic factor promotes sleep in rats and rabbits. Am J Physiol 2001;280:R1001–6.
244. Kushikata T, Kubota T, Fang J, Krueger JM. Neurotrophin 3 and 4 enhance spontaneous sleep in rabbits. Neurosci Lett 2003;346:161–64.
245. Birmanns B, Saphier D, Abramsky O. Alpha-interferon modifies cortical EEG activity: dose-dependence and antagonism by naloxone. J Neurol Sci 1990;100:22–26.

246. Krueger JM, Dinarello CA, Shoham S, Davenne D, Walter J, Kubillus S. Interferon alpha-2 enhances slow-wave sleep in rabbits. Int J Immunopharmacol 1987;9:23–30.
247. Kimura M, Majde JA, Toth LA, Opp MR, Krueger JM. Somnogenic effects of rabbit and recombinant human interferons in rabbits. Am J Physiol 1994;267:R53–61.
248. Toth LA. Strain differences in the somnogenic effects of interferon inducers in mice. J Interferon Cytokine Res 1996;16:1065–72.
249. Späth-Schwalbe E, Lange T, Perras B, Fehm HL, Born J. Interferon-alpha acutely impairs sleep in healthy humans. Cytokine 2000;12:518–21.
250. Kubota T, Majde JA, Brown RA, Krueger JM. Tumor necrosis factor receptor fragment attenuates interferon-gamma- induced non-REMS in rabbits. J Neuroimmunol 2001;119: 192–98.
251. Kapas L, Krueger JM. Tumor necrosis factor-beta induces sleep, fever, and anorexia. Am J Physiol 1992;263:R703–7.
252. Kimura M, Kodama T, Aguila MC, Zhang S-Q, Inoué S. Granulocyte-macrophage colony-stimulating factor modulates rapid eye movement (REM) sleep and non-REMS in rats. J Neurosci 2000;20:5544–51.
253. Kushikata T, Fang J, Krueger JM. Interleukin-10 inhibits spontaneous sleep in rabbits. J Interferon Cytokine Res 1999;19:1025–30.
254. Kubota T, Fang J, Kushikata T, Krueger JM. Interleukin-13 and transforming growth factor-beta1 inhibit spontaneous sleep in rabbits. Am J Physiol 2000;279:R786–92.
255. Schuld A, Mullington J, Hermann D, Hinze-Selch D, Fenzel T, Holsboer F, Pollmacher T. Effects of granulocyte colony-stimulating factor on night sleep in humans. Am J Physiol 1999;276:R1149–55.
256. Obal F Jr, Kapas L, Gardi J, Taishi P, Bodosi B, Krueger JM. Insulin-like growth factor-1 (IGF-1)-induced inhibition of growth hormone secretion is associated with sleep suppression. Brain Res 1999;818:267–74.
257. Obal F Jr, Kapas L, Bodosi B, Krueger JM. Changes in sleep in response to intracerebral injection of insulin-like growth factor-1 (IGF-1) in the rat. Sleep Res Online 1998;1: 87–91.
258. Alon U. Biological networks: The tinkerer as an engineer. Science 2003;301:1866–67.
259. Laughlin SB, Sejnowski TJ. Communication in neuronal networks. Science 2003;301:1870–74.

C. CHEMOKINES IN THE BRAIN

Cytokines and the Brain
Edited by C. Phelps and E. Korneva

Chemokines, their Receptors and Significance in Brain Function

TULLIO FLORIO and GENNARO SCHETTINI

Section of Pharmacology, Department of Oncology, Biology and Genetic, University of Genoa, Genoa, Italy
Pharmacology, School of Medicine, University of Genova, Viale Benedetlo xv2, 16100 Genova, Italy

ABSTRACT

Chemo-attractant cytokines, now known as chemokines, comprise the largest and most diverse subset of cytokines identified to date. Chemokines are characterized by their capacity to bring about the directional migration and activation of leucocytes and other somatic cell types during inflammation; cell-mediated immune reactions; to regulate cell adhesion, angiogenesis, leukocyte trafficking and homing, as well as lymphopoiesis and hematopoiesis. Chemokines are produced by a wide variety of leukocytes and other cell types in response to inflammatory agents, antigens, and endogenous cytokines. Studies of the *in vivo* effects of neutralizing antibodies and homologous deletions of chemokine genes reveal that chemokines play a central role in host defense against infectious organisms. Chemochines are relevant in normal central nervous system (CNS) physiology and development, as well as in the pathogenesis of diverse conditions such as tumor metastasis, riperfusion injuries and stroke. Chemokine structure, expression in CNS, and systematic classification schemes are presented according to (1) their constitutive expression and (2) their inducibility in response to inflammatory stimuli. The classification of chemokine receptor expression in specific CNS cells and their principal intracellular signal transduction pathways are also included, in addition to their various physiological roles. Consideration is given to the fascinating hypothesis that chemokines may influence neural and glial cell migration and proliferation during CNS development. The immuno-competence of the CNS, including inflammatory cell recruitment into the CNS as a function of chemokines in the evolution of pathological and host defense processes, is explored for head trauma, brain ischemia and trauma, AIDS dementia, Alzheimer's disease, and brain tumor progression.

1. INTRODUCTION

About 20 years ago a new chemotactic factor for neutrophils initially named interleukin-8 (IL-8), now called CXCL8, was identified. CXCL8/IL8 was the first member of a new class of biological mediators, the chemokines. These factors were originally believed to be exclusively related to the immune system and derived from the larger family of cytokines [1]. Since then, the chemokine family has grown and more than 50 chemokines and at least 20 specific receptors have been discovered [2]. Chemokines are widely conserved during evolution with high amino acid similarities and diffuse expression among numerous animal species such as mammals, birds, and fish. Many different cellular types produce chemokines, and their actions cover a large area of functions. These proteins are structurally and functionally related and exert their biological activity by binding to cell surface receptors. Since their discovery, it has become gradually evident that chemokines play a fundamental role not only in immune system function but also in development, homeostasis, angiogenesis and angiostatic processes, tumor, and metastasis progression. They are also important in the central nervous system (CNS) development, physiology, and pathology [3–5].

2. STRUCTURE AND CLASSIFICATION OF CHEMOKINES AND CHEMOKINE RECEPTORS

2.1. Chemokines

Discovery of the first chemokine molecule was followed by an overwhelmingly rapid discovery of new members of this fascinating family of biological mediators, identification of their receptors, in addition to identifying interactions occurring between members of the family. The definition of some peculiarities of members of the chemokine family was followed by the discovery of their pleiotropism: particularly their pleiotropism, promiscuity, and redundancy. Chemokines are classified according to their chemical structure [6]. They are small (8–14 kDa), mainly basic molecules, that, on the basis of the number and spacing of conserved cysteines, are subdivided into four groups named CXC, CC, XC, and CX3C, following the systematic nomenclature, α, β, γ and δ chemokines, respectively, using the Greek letters (Fig. 1). While in the CXC or α-chemokine family the cysteines are separated by a single amino acid, in the CC, or β family, the first two cysteines are adjacent. The XC or γ-chemokines are represented by two chemokines, called XCL1/lymphotactin-α and XCL2/lymphotactin-β, which contain only two of the four conserved cysteines found in the other subfamilies. Finally, the CX3C or δ-chemokines include only one member, called CX3CL1/fractalkine (and its murine homologue neurotactin), in which the first two cysteines are separated by three amino acids; this molecule is peculiar because it exists in both soluble and membrane-bound forms [7]. Indeed, CX3CL1/fractalkine primary transcript is a transmembrane glycoprotein that can be released after proteolysis by members of the ADAM family of disintegrins [8]. Recently, a similar mechanism was identified also for CXCL16 [9] (Table 1).

The CXC chemokine group can be further subclassified in ELR+ or ELR−, according to the presence of a conserved tripeptide motif glutamic acid-leucine-arginine (ELR) at the N-terminal of the protein, before the CXC domain. The ERL motif is relevant because presence of the motif has been correlated with the chemokine's effects on neutrophil chemotaxis and induction of neo-angiogenesis [10].

In general, chemokine subfamilies show similar, often overlapping, specificity. Chemokines belonging to the CC family are chemoattractant for monocytes, basophils, eosinophils, and T lymphocytes, but have little or no action on neutrophils; chemokines belonging to CXC-ELR− family attract lymphocytes and monocytes, but are not very effective on neutrophils, while the CXC-ELR+ chemokines act predominantly on neutrophils. In addition to the above classification scheme, a more recent chemokine classification plan distinguishes between "inflammatory" (or inducible) chemokines and "homeostatic" (or constitutive) chemokines [11] (Table 2).

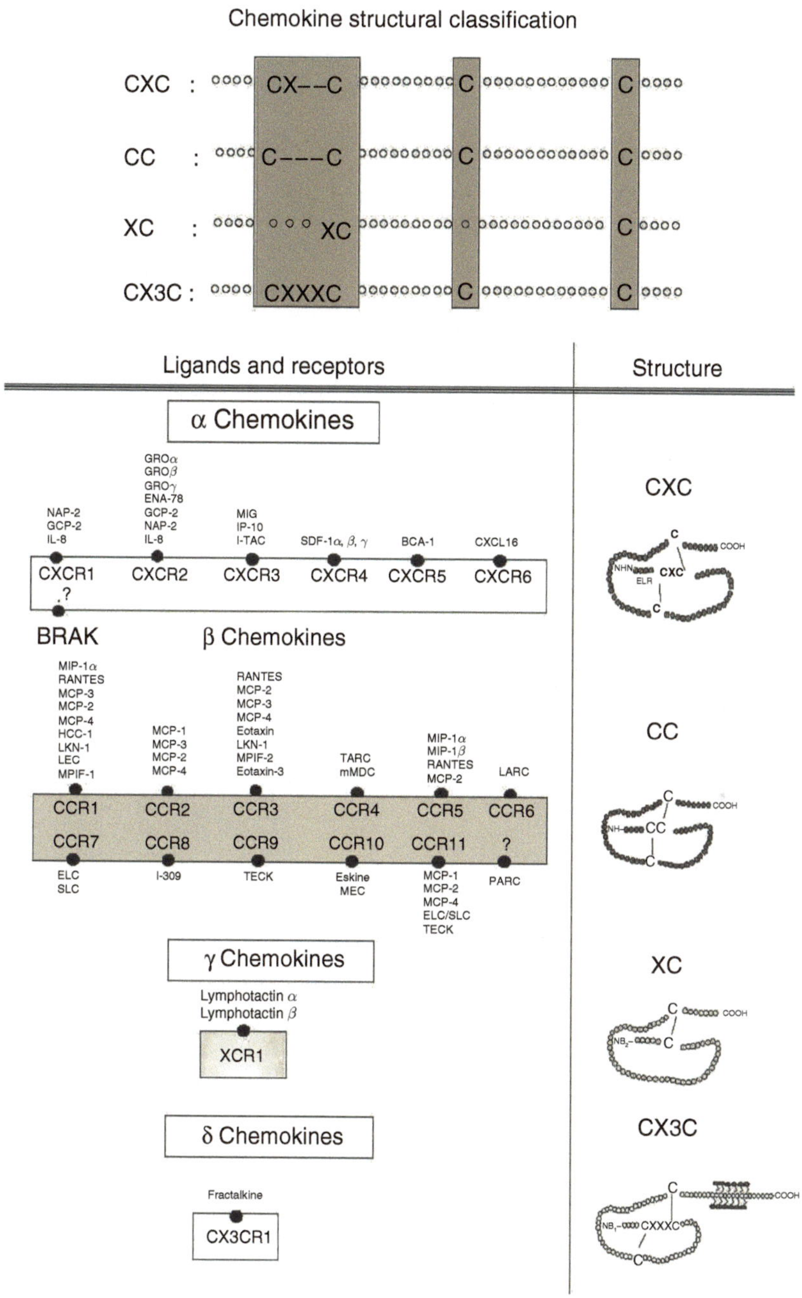

Figure 1. Upper panel: Schematic representation of the structural characteristics of the chemokine subfamilies identifying the position of the conserved cysteines (C) in each group (X: any amino acid other than cysteine; °: other amino acids; −: gaps within the alignaments). Lower panel: Classification of chemokines based on their structural characteristics (stylized beside). The list of chemokines signaling through each receptor is indicates. "?" indicates that receptors for those ligands are still unknown.

Table 1. Classification of chemokines

Systematic name	Old name	Systematic name	Old name
SUBFAMILY	**CXC**	CCL3	Macrophage inflammatory protein-1α (MIP-1α)
CXCL1	Growth-related oncogene α (Gro α)	CCL4	Macrophage inflammatory protein-1β (MIP-1β)
CXCL2	Growth-related oncogene β (Gro β)	CCL5	Regulated on activation normal T cell expressed and secreted (RANTES)
CXCL3	Growth-related oncogene γ (Gro γ)	CCL7	Monocyte chemoattractant protein-3 (MCP-3)
CXCL4	Platelet factor-4 (PF-4)	CCL8	Monocyte chemoattractant protein-2 (MCP-2)
CXCL5	Epitelial cell-derived neutrophil-activating factor 78 (ENA-78)	CCL11	Eotaxin
CXCL6	Granulocyte chemoattractant protein (GCP-2)	CCL13	Monocyte chemoattractant protein-4 (MCP-4)
CXCL7	Neutrophil-activating protein (NAP-2)	CCL14	Hemofiltrate CC chemokine (HCC-1)
CXCL8	Interleukin-8 (IL-8)	CCL15	Leukotactin-1 (Lkn-1)
CXCL9	Monokine induced by γ-interferon (Mig)	CCL16	Liver-expressed chemokine (LEC)
CXCL10	γ-interferon-inducible protein-10 (IP-10)	CCL17	Thymus- and activation-related chemokine (TARC)
CXCL11	Interferon-inducible T cell α-chemoattractant (I-TAC)	CCL18	Pulmonary- and activation-regulated chemokine (PARC)
CXCL12	Stromal cell-derived factor-1 (SDF-1)	CCL19	Epstein–Barr virus-induced receptor ligand chemokine (ELC)
CXCL13	B cell-activating chemokine-1 (BCA-1)	CCL20	Liver- and activation-related chemokine (LARC)
CXCL14	Breast and kidney chemokine (BRAK)	CCL21	6Ckine; secondary lymphoid tissue chemokine (SLC)
CXCL15		CCL22	Macrophage-derived chemokine (MDC)
CXCL16		CCL23	Myeloid progenitor inhibitory factor-1 (MPIF-1)
SUBFAMILY	**XC**	CCL24	Myeloid progenitor inhibitory factor-2 (MPIF-2)
XCL1	Lymphotactin-α	CCL25	Thymus lymphoma cell-stimulating factor (TECK)
XCL2	Lymphotactin-β	CCL26	Eotaxin-3
SUBFAMILY	**CX3C**	CCL27	ESkine
CX3CL1	Fractalkine	CCL28	Mucosae-associated epithelial chemokine (MEC)
SUBFAMILY	**CC**		
CCL1	I-309		
	Thymus-derived chemotactic agent (TCA3)		
CCL2	Monocyte chemoattractant protein-1 (MCP-1)		

2.2. Chemokine receptors

As is the case of most of the first messenger molecules, chemokines also exert their biological activity through the interaction with cell membrane receptors belonging to the superfamily of seven-transmembrane domain receptors that signal through coupled heterotrimeric G proteins. Often multiple chemokines can bind the same receptor, and a single chemokine can bind several receptors, but the chemokine–chemokine receptor interactions are almost always restricted within a single subclass. This relationship is complex and has been referred to as being

Table 2. Classification of chemokines according to the their constitutive expression (homeostatic), their inducibility in response to extracellular stimuli (inflammatory) and the molecules that according to the cell type analyzed can display both behavior

Homeostatic chemokines	
CXCL12	CCL14
CXCL13	CCL15
	CCL16
	CCL18
	CCL19
	CCL25
	CCL27
CX3CL1	
Inflammatory chemokines	
CXCL1	CCL1
CXCL2	CCL2
CXCL3	CCL3
CXCL5	CCL4
CXCL6	CCL5
CXCL7	CCL7
CXCL8	CCL8
CXCL9	CCL11
CXCL10	CCL13
CXCL11	CCL23
CXCL14	CCL24
	CCL26
Inflammatory and homeostatic chemokines	
	CCL17
	CCL20
	CCL21
	CCL22
XCL1	
XCL2	

promiscuous, although monogamous chemokine–receptor relationships also exist, such as CXCR4/SDF-1, CXCR5/BCA, CCR6/LARC, CCR9/TECK, and CX3CR1/fractalkine [3].

Chemokine receptor sizes range from 340 to 370 amino acids in length, sharing from 25 to 80% amino acid homology. They contain a cysteine in each of the four extracellular loops, a DRYLAIVHA (or a similar) sequence in the second intracellular loop and a short acidic N-terminal segment, which was recently proven to be involved in interaction with G proteins [12]. Numerous serines and threonines are present in the C terminal tail, which are phosphorylated upon ligand–receptor interaction.

The classification of chemokine receptors is based on the chemokine group to which their ligand(s) belong. To date, six CXC receptors have been identified, named from CXCR1 to CXCR6, and 11 CC receptors (from CCR1 to CCR11), along with a single receptor for fractalkine and one for lymphotactin α/β, called CX3CR1 and XCR1, respectively. Chemokines also interact with Duffy and D6, two 7TMD nonsignaling molecules that for this functional characteristic are not included in the receptor superfamily. Moreover, there are two chemokines: CXCL14/BRAK belonging to the CXC subfamily and CCL18/PARC, of the CC family, for which a specific receptor has not yet been unanimously identified. On the other hand, it has been reported that

CXCL12/SDF-1 was able to bind the orphan chemokine receptor RDC1 in T lymphocytes [13] that was initially renamed CXCR7. However, subsequent studies reported that CXCL12/SDF-1 does not bind RDC1 in all cell types [14], and thus, further data will be required in order to definitely assess this issue. Interestingly, it has also been reported that CCR2, CCR5, CXCR1, CXCR2, and CXCR4 can homo-or heterodimerize as consequence of the ligand binding, a process that was proposed to be required for the signal transduction of these receptors [15–18].

3. SIGNAL TRANSDUCTION

Despite the structural similarity and the apparent redundancy after binding by the different ligands, the chemokine receptors can bring about diverse intracellular responses as an expression of the pleiotropic biological effects brought about by these proteins (Fig. 2). The effects of receptor stimulation by different ligands can result in the activation of a wide range of signal transduction pathways. This is one of the most intriguing aspects of this receptor biology: how the same receptor can elicit different responses upon stimulation by diverse ligands or, vice

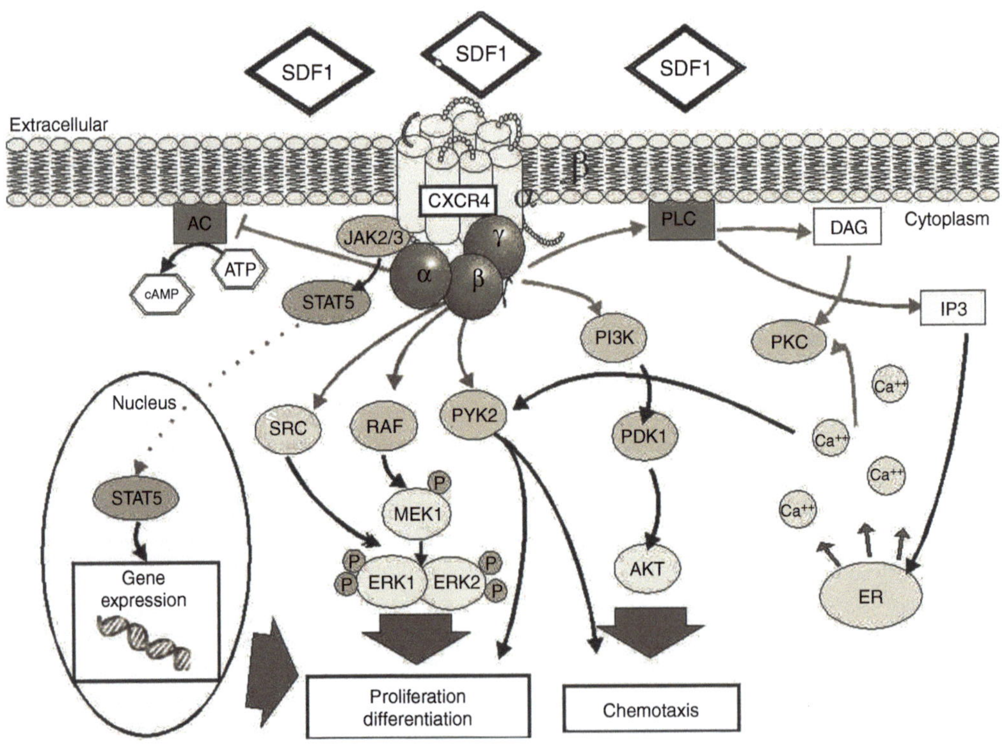

Figure 2. Schematic representation of the main signal transduction pathways activated by CXCL12/SDF-1 through CXCR4. The pathways preferentially leading to either proliferation or differentiation are indicated. AC, adenylyl cyclase; JAK, janus kinase; STAT, signal transducer and activator of transcription; α, ß, γ, subunits of G protein; PLC, phospholipase C; ER, endoplasmic reticulum; IP3 inositol trisphosphate; PYK2, protein tyrosine kinase 2; PI-3K, phosphoinositide 3-kinase; PDK1, PI3K-dependent kinase; MEK, MAP kinase; ERK1 and ERK2, extracellular signal-regulated kinase 1 and 2; Akt, protein kinase B.

versa, the same ligand can elicit multiple responses by activating different receptors. Is the "diverse reactivity" of the receptor moiety for each ligand due to different bridges or attraction(s) among amino acid residues? Are these (likely) varied ligand–receptor interactions responsible for the activation of specific signalling? Is the same receptor, expressed on different cells, exerting similar function upon binding of the same specific ligand? Is the same receptor, expressed on the same cell exerting different functions upon binding with different ligands? Although these questions have been answered in part, they are far from receiving a full explanation; moreover, it is likely that the redundancy and pleiotropism may occur as hypothesized.

After chemokine binding occurs, all chemokine receptors activate heterotrimeric G proteins to transduce the signal intracellularly (Fig. 2). Most of the G proteins involved belong to the Gi/Go, pertussis toxin sensitive, subfamily, leading to the activation of classical second messenger systems such as AC-cAMP, PLC/IP3-Ca^{2+}/DAG-PKC as well as PI-3K-AKT, ERK1/2, focal adhesion kinases (FAK, Pyk2, etc). Within the G protein complex (Gi/Go), the $\beta\gamma$ subunit seems to play a major role, rather than the α subunit. For instance, the stimulation of cell migration that can be very important in patterning during brain development and repair appears to require the functional coupling of the receptor to $G\alpha_{i,}$ as migration is completely inhibited by treatment with PTX [19]. This is the case even though the $G\alpha_i$ itself may not be necessary for cell migration [20]. It is indeed the $\beta\gamma$ subunits of the G protein that carries out the bulk of the signal for chemokine receptors. Moreover, it has been reported that only the $\beta\gamma$ subunits released from G_i-coupled receptors, but not those released by the G_s- or G_q-coupled receptors, can mediate cell migration [19]. Mutational studies, using CXCR4 as a model, showed that the third intracellular loop is responsible for the G protein binding and ERK1/2 activation. However, the chemotactic activity of the receptor requires not only the intracellular signaling through this portion of the receptor but also the second intracellular loop and the carboxyl-terminus, suggesting that this important biological response requires a complex array of signals [21]. The second messenger cascade triggered after chemokine receptor stimulation, by releasing the $\beta\gamma$ subunits, provokes the rapid activation of phospholipase C, which leads to IP3 generation and transient elevations of cytosolic [Ca^{2+}]. This stimulation pathway is often used to test the responsiveness of chemokine receptors to their different ligands, being one of the most characteristic intracellular signal elicited by chemokines [22,23]. As expected, chemokine-mediated activation of phospholipase C not only results in IP3 production but also leads to the formation of diacylglycerol and subsequent activation of protein kinase C [24,25].

Beside the increase in intracellular [Ca^{2+}] induced by IP3, chemokine receptor activation also induces a regulation of Ca^{2+} or other ion fluxes, through channels or different pathways [26,27]. For instance, in microglial cells, the activation of CCR3 by CCL5/RANTES or CCL11/eotaxin induced a rapid calcium influx from the extracellular environment that was sensitive to the dihydropyridine derivative nifedipine [27]. This observation suggests a modulation of L-type voltage sensitive calcium channels, although the lack of voltage-dependent ion fluxes in microglial cells indicates that this effect is mediated by a different class of yet unidentified nifedipine-sensitive channels [27]. CCL4/MIP-1β and CXCL12/SDF-1 stimulation of the CCR5 and CXCR4 receptors, respectively, induces the activation of both Ca^{2+}-activated potassium channels and chloride channels in primary cultures of human macrophages [26]. The activation of Ca^{2+}-activated potassium channels was also identified as a component of the proliferative responses of GH3 cells to CXCL12/SDF-1 [28].

Other important players in chemokine signaling are the PI-3K and the subsequent activation of Akt, the MAPK cascade, and in particular the pathway involving ERK1/2 activation, as well as the phosphorylation of FAK and PTK 2 [29–32]. Chemokines directly stimulate PI-3Kγ through the G-protein βγ subunits, with consequent PIP3 formation and Akt activation. This pathway is involved, but not exclusively, in chemotaxis [23,33], as shown by a marked attenuation, but not complete inhibition of chemotaxis *in vitro,* in cells derived from PI-3Kγ animals [34]. Chemokines also stimulate the MAPK pathway through the activation of the classical components including Ras, Raf-1, and MAPK/ERK kinase without involvement of G proteins [29,35]. However, although MAPK activation by chemokines is well documented, beside the classical pathway, alternative linking signaling was not definitively established [29,33]. A role for PI-3K in ERK1/2 activation was proposed in light of the reduced activation of this MAP kinase observed after treatment with specific PI-3K inhibitors, such as wortmannin or LY294002, an observation confirmed in mice lacking PI-3Kγ subunit [32,33,36]. Moreover, beside ERK1/2, other components of the MAP kinase family, such as p38 and JNK, were activated by different chemokine receptors [37–40].

Phosphorylation is an extremely important process, which controls numerous cellular functions, whose regulation depends not only on kinases but also on other important players such as phosphatases. A role for PTPs in the regulation of signals generated by chemokines has emerged. It has been reported that SHP1, a PTP expressed also in normal and injured CNS cells, and SHIP act as negative regulators of chemokine signaling, while the ubiquitously expressed PTP, SHP2, appears to enhance chemokine signaling [41–44]. α and β chemokine receptor signaling also involves JAK/STAT pathway. Activation of different STAT transcription factors in T cells through CCR2 in response to stimulation with CCL2/MCP-1 and through CCR5 in response to CCL5/RANTES has been reported, while CXCL12/SDF-1 can activate, through CXCR4, the JAK/STAT pathway in a PTX-independent manner [45,46].

As previously mentioned, chemokines induce receptor homo-heterodimerization which results in the activation of the JAK/STAT pathway [46,47]. This receptor heterodimerization may increase the cell sensitivity to chemokine stimulation [48]. However, the activation of this pathway by chemokines may not be a general principle, and its occurrence could depend both on the receptor and on the cell type involved. Nuclear translocation of STAT in response to RANTES has been observed in both T cells and astrocytes [49]. CXCR4 signaling through STAT proteins was not found in astrocytes but only in hematopoietic progenitors and T cells [50], and it was demonstrated to be dispensable for the CXCL12/SDF-1-induced chemotaxis [51]. Moreover, it has been reported that in an ovarian cancer cell line, CXCR1/2 stimulation by CXCL8/IL-8 [52], as well as CXCR4 activation by CXCL12/SDF-1 [53,54], induced epidermal growth factor receptor phosphorylation, indicating that a "cross-talk" between chemokine and growth factor signaling pathways can exist. Moreover, it has been proposed that such transactivation involves the G-protein-dependent activation of the cytosolic tyrosine kinases of the Src family [53,54]. Such cross-talk has been reported for other GPCRs whose signal transduction is linked to tyrosine kinase receptors, such as the EGF receptor and PDGF receptor [55].

CXCL12/SDF-1 stimulation of T cells enhanced the tyrosine phosphorylation of SHP2 that is constitutively associated with the CXCR4 receptor [42]. Conversely, CXCL12/SDF-1-induced chemotaxis, actin polymerization, and ERK1/2 activation were all enhanced in macrophages, T and B cells derived from SHP1-deficient mice. In addition, chemotaxis, Ca^{2+} influx, and actin polymerization were increased in response to CXCL12/SDF-1 in both immature and mature hematopoietic cells derived from SHIP-deficient mice [56], suggesting a negative role for these phosphatases in CXCR4 intracellular signalling. Finally, the β-chemokine receptor CCR5 also signals through the tyrosine phosphatases, SHP1 and SHP2 [41].

4. EXPRESSION OF CHEMOKINE AND THEIR RECEPTORS IN THE BRAIN: CELLULAR LOCALIZATION AND PHYSIOLOGICAL ROLES

Despite the large bulk of evidence clearly defining the role of chemokines in immune surveillance and in inflammatory cell recruitment in different diseases, including those that develop in the CNS, little is known about their physiological roles in the CNS. Since the end of last century, considerable progress has been made in the description of chemokine expression in the brain. Both *in vitro* and *in vivo* studies have characterized the presence of a large number of chemokines and chemokine receptors in different regions of the CNS and their specific cellular localization (Table 3). The cell types expressing and producing chemokines and chemokine receptors in the CNS are represented by astrocytes, microglia, oligodendrocytes, neurons, and endothelial cells (for review see [57]). Their expression has been described in normal brain, in particular during development and in several brain pathological conditions, such as inflammatory and neuro-degenerative diseases (MS, EAE, AD, ADC, brain injury, and tumors) [58]. A few studies have proposed a classic chemotactic activity for neural cells [59], but the involvement of chemokines in directing CNS cell migration during development remains to be elucidated and confirmed. The latter represents an important perspective for the discovery of new therapeutic approaches in degenerative diseases. As far as the α-chemokines are concerned, immunohistochemical analysis of brain tissues for the expression of CXCL8/IL-8 receptors revealed that CXCR1 and CXCR2 are expressed in subsets of projecting neurons in diverse regions of the brain and spinal cord [60]. CXCL8/IL-8 itself is expressed by activated and neoplastic astrocytes and enhances the survival

Table 3. Chemokine receptor expression in the CNS and their principal intracellular signalling

Receptor	Brain cell types	Signal transduction
CCR1	Neurons, astrocytes, microglia, endothelia	↑ $[Ca^{++}]$, ERK1/2, FAK, STAT1
CCR2	Neurons, astrocytes, microglia	↑ $[Ca^{++}]$, ERK1/2, PKC, PI3K/AKT, ROS
CCR3	Neurons, astrocytes, microglia, endothelia	↑ $[Ca^{++}]$ (VSCC), ERK1/2, p38
CCR4	Neurons, endothelia	↑ $[Ca^{++}]$, ERK1/2, AKT
CCR5	Neurons, astrocytes, microglia, endothelia	↑ $[Ca^{++}]$, ERK1/2, PI3K/AKT, p38, SHP1/2, Syk, BK_{Ca}, STAT3
CCR8	Brain (general)	↑ ERK1/2
CCR9	Neurons	↑ ERK1/2, FAK
CCR10	Neurons, astrocytes	
CCR11	Astrocytes, microglia	
CXCR1	Neurons, oligodendrocytes	↑ $[Ca^{++}]$, ERK1/2
CXCR2	Neurons, astrocytes, microglia, oligodendrocytes	↑ $[Ca^{++}]$, ERK1/2, PI3K/AKT
CXCR3	Neurons, astrocytes	↑ ERK1/2, PI3K/AKT, src
CXCR4	Neurons, astrocytes, microglia, endothelia	↑ $[Ca^{++}]$, ERK1/2, PI3K/AKT, FAK, Pyk2, src, BK_{Ca}, STAT5 ↓cAMP
CXCR5	Neurons	↑ ERK1/2, FAK
CXCR6	Astrocytes	↑ ERK1/2, FAK
CX3CR1	Neurons, astrocytes, microglia	↑ $[Ca^{++}]$, ERK1/2, AKT, SAPK/JNK

of hippocampal neurons *in vitro* [60]. The other ligand for CXCR1–2 receptors, CXCL1/GROα, is produced by subsets of astrocytes and neurons in a developmentally regulated pattern and is a potent promoter of oligodendrocyte precursor proliferation in response to their major mitogen PDGF, but has no effect on more mature oligodendrocyte precursors. Also in the postnatal spinal cord its expression is consistent with the pattern of emergence of oligodendrocytes, confirming a developmentally regulated activity of this chemokine [61]. Finally, CXCL1/GROα overexpression in oligodendrocytes induces neutrophil invasion and astrogliosis [62]. In addition, CXCL8/IL-8 and CXCL1/GROα [63] and CXCL12/SDF-1 [64] appear also to play a neuromodulatory role on cerebellar synaptic activity.

A growing body of evidence suggests that chemokines participate in the regulation of neuronal signaling in various ways. The first suggestions came from the observation that transgenic mice expressing high levels of KC (the murine homolog of CXCL1/GROα) frequently developed a progressive neurological dysfunction, characterized by ataxia, postural instability, and rigidity. No significant damage to neurons, myelin or axons occured in these mice, thus the syndrome was explainable by hypothesizing a direct receptor-mediated effect(s) [62]. Indeed, in mouse Purkinje neurons CXCL1/GROα as well as CXCL8/IL-8, through the activation of CXCR2, are able not only to generate Ca^{2+} transients but also to enhance the synaptic activity by increasing neurotransmitter release and to suppress the induction of long term depression [63]. CXCR3–5 is another chemokine receptor expressed in the CNS. CXCR3 is constitutively expressed in a subpopulation of neurons in various cortical and subcortical regions and CXCL10/IP-10, one of its ligands, was detected in astrocytes derived from healthy and AD brains [65]. CXCL10/IP10, whose overexpression has been found in neurodegenerative diseases such as AD, MS, and ADC, has been shown to induce apoptosis in fetal neurons through an increase in $[Ca^{2+}]i$, upon binding to CXCR3 receptor. The increased $[Ca^{2+}]i$, in turn, causes the release of cytochrome C from mitochondria and the activation of the caspase cascade: first the activation of the initiator caspase 9 that activates the effector caspase 3, leading to apoptosis [66].

CXCR4 expression has been detected at high levels on vascular endothelial cells, microglia, astrocytes, and neurons from both central and peripheral nervous systems [67–70]. In specific neuronal subpopulation, which includes cerebellar Purkinje cells, hippocampal hilar neurons, cerebral cortical neurons, high level of CXCR4 receptor RNA expression was detected [71]. CXCL12/SDF-1, the only known ligand for CXCR4, has been detected in astrocytes, microglia, cortical neurons, and cerebellar granule cells and is released in basal conditions by cerebellar granule cells and after LPS stimulation by cultured astrocytes [68–70]. CXCL12/SDF-1 also induces an enhancement in the spontaneous synaptic activity and a slow inward current in Purkinje neurons; these effects were reduced by ionotropic glutamate receptor antagonists, but not by tetrodotoxin, and thus, were most likely the consequence of extrasynaptic glutamate released by surrounding cells [64]. The observation that CXCL12/SDF-1 stimulated the release of glutamate from astrocytes, and thus, was able to influence the synaptic activity independently of the presence of its receptor on neurons confirmed these hypotheses. It was proposed that this chemokine was involved in the communication between glia and neurons and in particular in the regulation of synaptic transmission mediated by glial activation [72]. This pathway of intercommunication between glia and neurons, mediated by glutamate, was greatly enhanced by microglia as well as by the HIV-1-coating protein gp120, thus representing a possible mechanism of damage in neurotoxicity [72].

The expression of the human CXCR5 and of its murine homolog showed a similar pattern in neurons, with a high level in the granule and Purkinje cell layers of the cerebellum [73,74]. Differently from the CXC family, β-chemokines are weakly expressed in normal brains but their production appears to be increased in inflamed and activated brain tissues. Indeed, their expression in the brain can be both constitutive and inducible (see Table 2). Among the CCR family, CCR1, CCR2, CCR3, CCR4, CCR5, CCR9/10, and CCR11 expression has been detected in the brain both in physiological and pathological conditions. CCL5/RANTES differentially promoted proliferation or survival in human fetal forebrain astrocytes, depending on the age of the fetuses from which the cells were isolated. This suggested that this chemokine is possibly involved in the regulation of the expansion of astrocytes in the early phases of development, and then it promotes differentiation and maintains the astrocyte population [75]. Finally, both CCL2/MCP-1 and CCL3/MIP-1α induce proliferation in glial cultures [75].

CCR3 and CCR5 have been detected in normal and inflamed human CNS tissues associated with both glial and neuronal cells [76–79]. Their expression was found in subpopulations of large hippocampal and neocortical pyramidal neurons, human and macaque fetal neurons, as well as on microglia and astrocyte cells in both normal and encephalitic brain [76,78,80]. CCL5/RANTES was also reported to be produced by primary astrocytes, human astrocytoma cells, and it is upregulated in HIV-1 infected astrocytes [81,82]. In normal brains, CCR3 staining is reported in pyramidal neurons of the entorhinal cortex, frontal cortex, and large neurons in the dentate nucleus of the cerebellum. CCL3/MIP-1α, CCL4/MIP-1β, and CCL2/MCP-1 are slightly expressed in cultured microglia in unstimulated conditions. In response to LPS or cytokines, such as TNF-α and IL-1β, both CCL3/MIP-1α, CCL4/MIP-1β secretion was increased. CCL2/MCP-1 release was stimulated by LPS, while IFN-γ did not significantly induce the expression of this chemokine [83]. CCL2/MCP-1, CCL6/MCP-2, and CCL7/MCP-3 expression is detected in CNS tissues of patients with MS, in contrast with normal control brains [84]. The local injection of the neurotoxin kainic acid increased the *in vivo* production of CCL2/MCP-1 in astrocytes and macrophages of the adult rat brain [85]. CCR2 expression was shown in axonal and dendritic processes of human fetal neurons and NT2.N cells. Both astrocytes and NT2.N cells express and release also the CCR2 ligand CCL2/MCP-1, with the possible occurrence of a paracrine/autocrine loop of activation. These results are consistent with a report showing CCL2/MCP-1 immunoreactivity in developing neurons in the cerebellum and pons nuclei of human fetal brain [86]. By reverse transcriptase–polymerase chain reaction (RT–PCR) analysis, CCR1, CCR4, CCR5, and CCR9/10 mRNA was identified in primary culture of hippocampal neurons, with CCR1 also detected in cultured astrocyte populations [87,88] and TER1/CCR8 identified in brain-derived cells [89]. CCR11 mRNA was detected in cultured neonatal mouse astrocytes, while a lower expression level was identified in microglia [88].

CX3CL1/fractalkine occurs in both as membrane-anchored and in soluble forms. It is the only chemokine expressed in the CNS in higher amounts than in the immune system and peripheral tissues. This protein is encoded as a transmembrane molecule that displays adhesion properties and can be also cleaved from the cell membrane to produce a soluble form that creates a chemotactic gradient for receptive cells [7]. Both anchored and soluble forms bind CX3R1, so far, the only identified receptor for CX3CL1/fractalkine [57,88]. Neurons are the principal source of CX3CL1/fractalkine in the brain. CX3CL1/fractalkine has been detected in several discrete regions of adult rat brain, including cortex, hippocampus, caudate-putamen, thalamus, and olfactory bulb, while it is absent in cerebellum, brain stem, and white matter regions, including the corpus callosum and the fimbria/fornix. CX3CL1/fractalkine expression has also been described in microglia, while in astrocytes it is expressed at low constitutive levels, but it can be upregulated by TNF-α and IL-1β [90]. Microglial cells express the CX3CR1 receptor at high levels also and migrate in response to fractalkine stimulation. The CX3CR1 is expressed at a much lower level on astrocytes, and, in these cells, it does not mediate chemotaxis [90].

Evidence of functional expression of CX3CR1 in primary culture of hippocampal neurons has also been reported [91].

Chemokines also play an inhibitory role on neuronal transmission. Several CXC and CC chemokines and soluble CX3CL1/fractalkine reduced calcium oscillations in synaptically coupled hippocampal neurons *in vitro,* by decreasing the glutamate release from the presynaptic neuron [87], while CXCL8/IL-8 reduced calcium currents in cholinergic neurons through the G-protein-mediated inhibition of L- and N-type calcium channels [92]. Thus, some chemokines can be considered as neuromodulators; their mechanism of action and their cellular targets can significantly vary, because they act at both pre- and post-synaptic level and on surrounding glia, mostly through the regulation of neurotransmitter release, but also through direct modulation of ion channel activity.

5. CHEMOKINES IN BRAIN DEVELOPMENT

5.1. Neuronal and glial migration

A fascinating hypothesis is that chemokines may influence the migration of the multipotent progenitor cells during brain development. The migratory characteristics of numerous cell types appear to be influenced by the action of soluble chemoattractant factors present in their immediate environment and chemokines may be good candidates for these roles, including neuropoiesis, based on their effects on differentiation of neuronal subsets [93]. The neurotropic properties of CCL5/RANTES, CCL3/MIP-1α, and CCL2/MCP-1 were examined on mouse embryonic dorsal root ganglia cells. CCL5/RANTES elicited migration and differentiation into a nociceptive phenotype of a subset of these cells *in vitro*. Its temporal and spatial expression was consistent with a neuropoietic effector molecule. Conversely, in response to CCL2/MCP-1 the migration of dorsal root ganglia cells was much less robust, while CCL3/MIP1α was ineffective. CXCL12/SDF-1 has also been described to trigger migration both in a human neuronal cell line and in rat E15 neuronal progenitors, *in vitro* [59]. The migration induced by CXCL12/SDF-1 was inhibited by PTX and AMD3100 (a CXCR4 antagonist) and by the inhibition of the MAP kinase pathway with PD98059. In $CXCR4^{-/-}$ and $SDF\text{-}1^{-/-}$ animals, fetal cerebellar development was markedly affected, presenting an aberrant laminar structure, with premature migration of the external granular layer cells into the internal granular layer at E17, an event that normally occurs after birth. Secondarily, the cerebellum was grossly malformed, lacking foliation [94,95]. Nonetheless, other parts of cerebellum appeared to be normally developed in these mice, including the cerebellar radial glial cells (the Bergmann glia), which have an important role in the migration of external granular layer cells [94,95]. Thus, it appears that this abnormal inward migration of external granule cells was due to the absence of CXCR4-mediated signalling. At birth, CXCL12/SDF-1 is expressed in cerebellum and olfactory bulbs, but its content decreases within 2 weeks, in contrast with other brain regions (cortex, thalamus, and hippocampus) where its expression, low at birth, progressively increases during the first two post-natal weeks. The spatio-temporal distribution of CXCL12/SDF-1 transcripts correlates with granule cell migration across the molecular layer in the cerebellum [96]. The migration into the cortical marginal zone of the Cajal-Retzius neurons, which is a transient neuronal population that coordinates the migration and layer arrangement of all the cortical neurons [97], and the migration of neural stem cells into the cerebellum [98] are controlled by the meningeal secretion of CXCL12/SDF-1. The correct migration of the cortical interneurons was regulated by the activation of CXCR4 [99]. CXCR4 null mice were also used to study the development of cerebellar cortical inhibitory interneurons and pontine neurons that form synaptical connections with granule cells. Although the precursors of the cortical inhibitory interneurons tend to follow their normal developmental pattern, they do not correctly reach the superficial external granular layer, due to the abnormal scattered foci of proliferating granule cells that impede their dispersion. In CXCR4 null mice the transit of pontine neurons from the rhombic lip through the anterior extramural stream to the basilar pons is disrupted. Missing CXCR4, pontine neurons do not reach the pons but dipserse deeply into the brainstem, causing hypoplasia of pontine nuclei [100]. Astrocyte migration was observed *in vitro* after CCL2/MCP-1, CCL1/TCA3, and CCL3/MIP-1α treatment [88]. Murine astrocytes migrate in response to subnanomolar concentrations of CCL3/MIP-1α but are insensitive to CCL4/MIP-1β and CXCL12/SDF-1 [101,102].

5.2. Regulation of synaptic activity

Chemokines are now recognized as important modulators of synaptic activity in the brain. In the cerebellum CXCL8/IL-8 and CXCL1/GRO-α modulate fast synaptic transmission and long-term synaptic plasticity [63], and CXCR4 activation was shown to play a role in synaptic transmission, as well [64]. The activation of chemokine receptors can inhibit excitatory transmission between hippocampal neurons, and in particular, CCL22/MDC and soluble CX3CL1/fractalkine blocked the spontaneous glutamatergic excitatory postsynaptic currents in these neurons. Intriguingly, CX3CL1/fractalkine, but not CCL22/MDC, inhibited voltage-sensitive calcium currents [87]. These observations, however, are not sufficient to fully explain the mechanisms responsible for the inhibition of glutamate release.

5.3. Proliferation

During development, gliosis, and tumor progression, proliferation represents another important critical event. (ELR-)CXC chemokines have been shown to have potent mitogenic effects on resident cells in several tissues, including CNS cells [10]. Indeed, CXCL1/GROα is a potent promoter of oligodendrocyte precursor proliferation. The proliferative response of immature spinal cord oligodendrocyte precursors to their major mitogen PDGF is dramatically enhanced by CXCL1/GROα and elevated levels of this chemokine correlate with elevated oligodendrocyte progenitor proliferation in *Jimp* mutant mice [61]. Also CXCL12/SDF-1 induced a dose-dependent proliferation of cultured type I rat astrocytes *in vitro*, an effect occurring through the activation of MAP kinases and PI-3K [32]. The expression of CXL12/SDF-1, in the rat brain, correlates with that of CXCR4, suggesting paracrine and/or autocrine signaling [32,96].

5.4. Studies in genetically modified animals

Experiments using genetically modified animals have been useful for studying physiological and pathological chemokine functions, although chemokines redundancy and functional overlap may jeopardize interpretation of the experimental results. Although most chemokine knockouts (KOs) are fertile, viable and exhibit mendelian inheritance of the nullizygous trait, without gross developmental alterations of immune functions, genetic disruption of CXCR4 or CXCL12/SDF-1 in mice gives, so far, the only known nonviable phenotype among the chemokine and chemokine receptor KOs [94,95]. In these experimental models, the animals die perinatally, displaying complex and lethal developmental alterations in the immune system, ventricular septal defects, defective gastric vasculogenesis, and defective fetal cerebellar development. This phenotype indicates a specific role for CXCL12/SDF-1 and CXCR4 in proper neuronal patterning. Indeed, CXCR4 mRNA is expressed in proliferative areas of the brain including the cerebellum, cerebral cortex, hippocampus and spinal cord during development [95] and accordingly with these observations we found that CXCL12/SDF-1 induces astrocyte proliferation [32]. Neural and oligodendrocyte precursors derived from CXCR4-deficient mice displayed impaired survival and migration activity, implying that CXCL12/SDF-1 signaling regulates survival and outward chemotactic migration of these precursor cells during embryonic and postnatal CNS development [103].

Transgenic mice overexpressing KS (the murine homologue of CXCL1/GRO-α) develop a neurological syndrome that includes loss of the landing reflex and abnormalities of gait, signs seen in cerebellar mutant mice, and is generally associated with altered cerebellar function [62]. Extensive microgliosis and astrogliosis, without signs of neutrophil activation or an effect on CNS myelination, were observed at the neuropathological examination of these animals [62]. Conversely, mice deficient for the CXCL8/IL-8 receptor homolog, the sole murine receptor on neutrophils that functionally binds this chemokine, and for the murine CXCL1/GROα, and CXCL2/GROβ genes do not show any apparent neurological abnormalities [104,105]. Transgenic mice overexpressing CCL2/MCP-1 in the CNS, under the control of myelin basic protein promoter, produced a pronounced mononuclear cell infiltrate in the brain that was detected in myelin-rich areas concentrating in close proximity to blood vessels. In these mice CCL2/MCP-1 is synthesized and secreted by oligodendrocytes accumulated in the brain vessels where it recruited monocytes and macrophages [106]. In a different transgenic mouse line that constitutively expressed low levels of CCL2/MCP-1 in the CNS under control of the astrocyte-specific glial fibrillary acidic protein promoter, the proteolipid protein-induced experimental autoimmune encephalomyelitis caused a significantly milder clinical disease than in littermate controls. This suggests that sustained, tissue-specific expression of CCL2/MCP-1 *in vivo* downregulates the Th1 autoimmune response, culminating in milder clinical disease [107].

Although expressed in defined structures of the cerebellum [73], the phenotype of gene-targeted mice lacking the B-lymphocyte chemoattractant receptor 1, a murine homolog of CXCR5, showed anomalies in B-cell migration and localization of these cells within specific anatomic compartments, without revealing of any alterations in the CNS [108]. Using three different *in vivo* models, it was shown that CX3CR1 deficiency dysregulates microglial responses, resulting in neurotoxicity. For example peripheral lipopolysaccharide injection in CX3CR1$^{-/-}$ mice caused cell-autonomous microglial neurotoxicity. In experimental models of Parkinson disease-related toxicity and of amyotrophic lateral sclerosis, CX3CR1$^{-/-}$ mice showed more extensive neuronal cell loss than CX3CR1$^{+/+}$ control animals. Conversely the activation CX3CR1 signaling in control mice protected against microglial neurotoxicity, whereas CX3CR1 antagonists increased neuronal vulnerability [109].

6. CHEMOKINE AND CHEMOKINE RECEPTOR INVOLVEMENT IN BRAIN PATHOLOGIES

6.1. Role in brain inflammation and neurodegenerative diseases

CNS inflammatory reactions are shaped, in part, by the unique anatomic character conferred by the blood–brain barrier and the absence of fully developed lymphatic antigen-presentation sites in the CNS. Although the CNS has long been considered to be an "immunologically-privileged site," primarily because of observations involving delayed rejection of xenografts and tumor allografts, immunopathological states such as EAE and MS document unequivocally that the CNS is indeed an immunocompetent organ. Inflammatory cell recruitment into the CNS is a critical step in the evolution of pathological and host-defense processes such as head trauma, stroke, viral encephalitis, and MS.

The best-characterized biological activity of chemokines is chemotaxis, but chemokines also reversibly activate leukocyte integrins, consistent with the role in modulating leukocyte-endothelial interaction, defining the cellular composition of inflammatory infiltrates. In some cases, chemokine expression follows initial leukocyte entry in the CNS, suggesting that chemokines amplify, rather than initiate, inflammatory cell infiltration. However, the expression of chemokine genes can precede CNS leukocyte infiltration, as shown for the expression of the CXCL10/IP-10 gene after lymphocytic choriomeningitis virus infection [110]. Moreover, transgenic mice and microinjection studies provide further confirmation that localized production of chemokines can function to promote the recruitment of selected leukocytes to the CNS [62,106].

6.2. Multiple sclerosis

Multiple sclerosis is an inflammatory demyelinating disease associated with an autoimmune response directed against myelin proteins within the CNS with loss of neurological functions. MS lesions are characterized by a chronic inflammation, with a progressive immune-mediated destruction of the myelin sheath, glial cell activation, loss of axons, and a recruitment of immune cells into the CNS, mainly represented by T cells and monocytes. The inflammatory process and the infiltration of the immune cells into the CNS are mediated in response to chemotactic signals such as chemokines. Chemokine expression in MS has been observed both in CSF and by *in situ* and immunohistochemical analysis in MS lesions. The α-chemokines CXCL10/IP-10 and CXCL9/Mig seem to participate in the MS pathogenesis, being detected in the CSF during MS attacks. Expression of the corresponding chemokine receptors CXCR3 was also found in lymphocytes in the perivascular inflammatory infiltrate, and in MS lesions in lymphocytes, macrophages, and microglia [111–113].

The presence of β-chemokines CCL2/MCP-1, CCL8/MCP-2, and CCL7/MCP-3, CCL5/RANTES, CCL3/MIP-1α, and CCL4/MIP-1β was described in acute and chronic-active MS brain lesions [83]. CCL3/MIP-1α expression was predominantly associated with glial cells, CCL4/MIP-1β with macrophages/microglia, CCL5/RANTES with perivascular leukocytes and CCL2/MCP-1 with macrophages and astrocytes in chronic active MS lesions [111,113]. The presence of the corresponding receptors CCR2, CCR3, and CCR5 on foamy macrophages and activated microglia in the lesions, strengthen the relevance of the chemokines in MS pathology. CCR3 and CCR5 are expressed on reactive astrocytes while CCR2 and CCR5 are also present on infiltrating T cells [111,114].

To study the role of chemokines in MS, EAE is considered an excellent animal model. EAE is induced by immunization with antigens derived from myelin, such as PLP and MOG. Multifocal perivascular CNS inflammatory infiltrates primarily consisting of T cells, both Ag-specific and Ag-nonspecific, and monocytes, with little or no polymorphonuclear cells, are found in this model [115]. The expression of CCL5/RANTES, CCL3/MIP-1α, CCL4/MIP-1β, CCL2/MCP-1, CXCL10/IP-10 mRNA, and proteins have been correlated with the inflammatory lesions [116]; CCL3/MIP-1α is involved in the control of mononuclear cell accumulation during acute EAE, while CCL2/MCP-1 regulates mononuclear cell infiltration during relapsing EAE [117,118].

Studies in KO mice, deficient in CCL3/MIP-1α, CCL2/MCP-1, and CCR1, CCR2, and CCR5 gene expression [119–121] show that CCR2-deficient mice are resistant to EAE, fail to develop mononuclear infiltrates and display a decreased Ag-specific proinflammatory response in the secondary lymphoid organs [119,120]. Similarly, mice not expressing CCL2/MCP-1 are resistant to MOG-induced EAE [122], while CCL3/MIP-1α- and CCR5-deficient mice are not resistant to the MOG-induced disease, although CCR1-deficient mice had less severe clinical signs [121]. Using neutralizing antibodies, the importance of CXCL10/IP-10 in MS pathogenesis has been further strengthened. In two different mouse models the brain infection with mouse hepatitis virus brain infection results in an acute encephalomyelitis followed by a chronic demyelination disease with clinical and histological similarities with MS and the EAE model [123]. In such studies antibodies against CXCL10/IP-10 decreased the clinical and histological disease incidence and the accumulation of inflammatory mononuclear cells during the pathogenesis of EAE and also in the mouse hepatitis virus model [123].

6.3. Brain ischemia and trauma

Stroke and trauma elicit robust inflammation in the brain. Ischemic brain injury secondary to arterial occlusion is characterized by acute local inflammation and reperfusion. Transient ischemia often causes greater tissue damage than persistent ischemia, which is characterized by leukocyte infiltration into the damaged brain. The enhanced expression of chemokines, such as CXCL1/GROα, CXCL10/IP-10, CCL2/MCP-1, CXCL8/IL-8, CCL7/MCP3, CCL3/MIP-1α, CCL4/MIP-1β, and of the chemokine receptor CXCR3 has been observed in experimental cerebral ischemia, generated by occlusion of common carotid artery and middle cerebral artery [124–127]. The use of anti-CXCL8/IL8 neutralizing antibodies significantly reduced brain edema and infarct size, suggesting that chemokines could be considered as novel potential therapeutic targets for stroke and neurotrauma [128].

As there is evidence that several chemokines have a pivotal role in atherosclerosis, they may have also an indirect role in the pathogenesis of stroke. It has been reported that CCL2/MCP-1-deficient mice have less arterial lipid deposition in hypercholesterolemia models such as low density lipoprotein receptor deficiency or apoB overexpression [129]. Also CXCL8/IL-8, CXCL12/SDF-1, CXCL10/IP-10, CCL1/I309, CXCR2 have all been associated with atherosclerotic lesions in animal models [36,130]. Similarly, $CCR2^{-}/^{-}$ mice showed a reduction of disease in a model of apoE deficiency [131]. The traumatic brain injuries are characterized by migration of mononuclear inflammatory cells to the CNS parenchyma and activation of astrocytes and are studied mostly using a rodent brain stab wound animal model. Following brain damage, induced by stab injury, nitrocellulose stab or implants in adult mice, an increase in CCL2/MCP-1 expression, produced by reactive astrocytes, is observed. [132]. The expression of CCL3/MIP-1α, CCL4/MIP-1β, CCL5/RANTES, and CXCL10/IP-10 was also elevated in the brain after cortical injury and endotoxin addition, while CCL2/MCP-1 is the only chemokine produced after sterile injury [133]. Prior to any sign of retrograde neuronal degeneration, temporal and regional distribution of CCL2/MCP-1 expression, following a visual cortex lesion, demonstrated the presence of CCL2/MCP-1 in the thalamus [134]. In experimentally induced brain injury a rapid overexpression of CXCL1/GROα and CXCL2/MIP-2 occurred, followed by a homologous neuronal down-regulation only of CXCR2 but not of CXCR1 receptors. This discrepancy was proposed to render neurons more vulnerable to the lesional stimuli. However, the higher concentrations of the ligands caused a recruitment of a CXCR2 overexpressing microglia that could be involved in both neuronal repair process or in exacerbating the damage [135].

Peripheral nerve axotomy induces the increase of CX3CL1/fractalkine and CX3CR1 in facial motor nucleus with parallel cellular and morphological changes in microglia [136]. However, studies in KO mice excluded the hypothesized role of CX3CL1/fractalkine as messenger between injured neurons and microglia, because the absence of CX3CR1 did not result in impaired activation, proliferation, differentiation, and recruitment of microglial cells [137]. Stimuli including induction of EAE or experimental cerebral ischemia *in vivo* did not affect brain expression of CX3CL1/fractalkine mRNA, although *in vitro* this chemokine was released from cell membranes after excitotoxic stimulus [138].

6.4. AIDS dementia complex

The involvement of chemokines in AIDS pathology emerged when CCL5/RANTES, CCL3/MIP-1α, CCL4/MIP-1β, and CXCL12/SDF-1 were found to act as suppressive factors of HIV-1 infection [139–141]. CCR5 and CCR3 were then recognized as key co-receptors for M tropic viruses while CXCR4 is the co-receptor for T-tropic viruses. These receptors are required for viral entry, with CD4 interactions with the HIV-1 coat protein gp120 altering its conformation, allowing co-receptor binding and subsequent fusion with the membrane and HIV-1 entry into the cells [139–141]. CCR2, CCR8, CXCR6, and CX3CR1 receptors can have similar HIV co-receptor functions, but their role in HIV infection is less prominent, likely due to lower affinity of the viral envelop proteins for these receptors [142]. Approximately 25–50% of HIV-infected children and adults develop a progressive encephalopathy, named ADC, characterized by cognitive, motor and sensory impairment. As in the CNS, HIV-1 only infects microglia and macrophages and, nonproductively, astrocytes and oligodendrocytes, neuronal death in ADC patients is not the consequence of a viral infection of neurons. Thus, the neuronal damage must be caused by an indirect mechanism mediated by cells that have been infected or indirectly activated. Pro-inflammatory cytokines TNFα, interferon α and β, IL-1β, and IL-6 that may be released from activated microglia and astrocytes, and several other factors, including arachidonic acid metabolites, free radicals, NO, platelet activating factor, all known to be a potential cause of neuronal death, are increased in the brains of ADC patients [143]. As demonstrated by both *in vitro* and *in vivo* studies, some viral proteins shed by infected cells, such as the viral envelope glycoprotein gp120 and the regulatory protein Tat, seem to be the best candidates as initiators of the cascade process that leads to neurotoxicity [144,145]. Toggas et al. [146] showed that transgenic mice constitutively expressing gp120 in astrocytes have brain damage similar to that observed in ADC patients, with gp120 expression sufficient to induce pathological effects on neurons, astrocytes, and microglia. The mechanism of gp120 induced neuronal damage is still unknown. The hypothesis that the neurotoxic process could arise from the gp120-chemokine receptor interaction is supported by the ability of gp120 to bind and, more importantly, to generate an intracellular signal through chemokine receptors, even independently of CD4 [69,87,102,147]. Gp120 could modulate the activation of chemokine receptors or, alternatively, interfere with the physiological function of chemokines in the brain. CXCR4 receptor is expressed in the CNS by neurons, astrocytes, microglia, and endothelial cells [68,69,101]. This receptor has been considered the best candidate as mediator of gp120 toxicity, although it is still under discussion as to whether gp120 exerts its toxicity directly by binding CXCR4 on neurons, or indirectly by binding the receptor expressed on glial cells and thus triggering neurotoxin release. Interestingly, different chemokines, including CCL22/MDC, CCL5/RANTES, CXCL12/SDF-1, and soluble CXC3CL1/fractalkine, block $gp120_{IIIB}$ (T-tropic strain)-induced apoptosis in hippocampal neurons. This neuroprotective effect can be ascribed to the ability of chemokines to interfere with gp120 binding, thus recovering their trophic function by competing at their binding sites, at least for CXCL12/SDF-1α [87]. At odds with this report other authors found that CCL5/RANTES and CCL4/MIP-1β prevented apoptosis induced by $gp120_{SF2}$ (another T-tropic strain) in mixed cerebrocortical cultures, while CXCL12/SDF-1 not only did not protect from gp120, but induced neurotoxicity on its own [143]. The over-stimulation of CXCR4 by gp120 and progeny virions, as well as by its natural ligand CXCL12/SDF-1, can induce apoptosis in human and rat neurons *in vitro* and *in vivo* [148,149]. This apparently ambiguous role of CXCL12/SDF-1 might be explained in light of the discovery that activation of CXCR4 on astrocytes induces a rapid and dose-dependent release of glutamate through autocrine/paracrine secretion of TNFα and prostaglandin synthesis [72]. In glial-neuronal cultures exposed to CXCL12/SDF-1 or gp120 a perturbation of this pathway could lead to the release of toxic amounts of glutamate. Indeed, the induction of glutamate release by gp120, as well as gp120-induced neuronal death, were both strongly enhanced by the presence of activated microglia, and inhibited by antagonists of CXCR4 receptor.

Chemokines promote the recruitment of monocytes and lymphocytes that facilitate HIV entry and spread within the brain and these effects may also contribute to the development of ADC. The expression of CCL2/MCP-1, CCL3/4/MIP-1α/β, CCL5/RANTES, CX3CL1/Fraktalkine, CXCL10/IP-10, and CXCL11/I-TAC has been found to be altered in the CNS of ADC patients [150–154]. The receptors CCR1, CCR3, CCR5, CXCR3, CXCR4 have been found upregulated in brain tissues from ADC cases as compared to nondemented HIV+ or uninfected controls [77,152]. An increased expression of CCR3 and CCR5 has been detected also in the CNS of macaques infected with the SIV as in the brains of children with AIDS [77,78,155]. Similarly, an elevated expression of the β-chemokines CCL3/MIP-1a, CCL4/MIP-1β, CCL7/MCP-3, and CCL5/RANTES was seen in the brains of macaque monkeys with SIV encephalitis [156], as in the brain tissues of patients with ADC. CCL2/MCP-1 was also detected in brain tissues of AIDS patients without encephalitis and, in non-HIV controls, in microglia within the cortical and subcortical gray matter.

6.5. Alzheimer's disease

Alzheimer's disease, the most common form of dementia, is characterized by the formation of extracellular Aβ peptide plaque deposits that are surrounded by activated astrocytes and microglia, the presence of abnormal neurons containing tangles of tau protein and the progressive loss of selected neuronal populations. Aβ peptides have been shown to be potent activators of microglia and macrophages. However, in the CNS of AD patients, inflammation and leukocyte infiltration clearly occurs in vulnerable regions of the brain, although abnormal or excessive migration of inflammatory cells is not so relevant as in other neurological disorders [157]. Insoluble Aβ peptides deposits, neurofibrillary tangles, and damaged neurons can behave as stimuli for inflammation in the AD brain, thus inflammation mediators, such as chemokines and their receptors, can be found upregulated in resident CNS cells and may contribute to plaque-associated inflammation and neurodegeneration. Both α- and β-chemokines have been reported to be expressed in the AD brain.

In addition, the expression of CXCR2, CXCR3, CCR2, and CCR5 receptors and their ligands has been observed in AD tissues. CXCR2 is expressed in hippocampal and cortical neurons and also in dystrophic neurites associated with a subset of senile plaques [79]. CCL2/MCP-1 has been localized to mature senile plaques and reactive microglia, but was not found in immature senile plaques. CCR3 and CCR5 were present on microglia of both control and AD brains, with an increased expression in reactive microglia in AD. CCL3/MIP-1α was constitutively expressed, at a low level, by neurons and microglia, whereas MIP-1β was predominantly expressed by a subpopulation of reactive astrocytes. CCR3 and CCR5 reactive microglial cells and CCL4/MIP-1β reactive astrocytes were found to be associated with amyloid deposits [79]. CXCL10/IP-10 is expressed in a subpopulation of astrocytes in the normal brain and is markedly elevated in the AD brain, while its receptor, CXCR3, is expressed by neurons and neuronal processes in various cortical and subcortical regions [65]. CXCL10/IP-10 and CXCL9/Mig activated the ERK1/2 pathway in mouse cortical neurons *in vitro* [65].

Whether Aβ peptides contribute to inflammation and leukocyte infiltration in the AD brain by stimulating chemokine production is still an open question. In the presence of Aβ peptide astrocytes and oligodendrocytes produced and released CCL2/MCP-1 and CCL5/RANTES while in human monocytes the expression of CXCL8/IL-8, CCL2/MCP-1, CCL3/MIP-1α, and CCL4/MIP-1β was enhanced [157,158]. Thus, the chemokines produced by the plaque-surrounding cells could play a role in the recruitment and accumulation of astrocytes and microglia in amyloid plaques and in the associated inflammation.

6.6. Brain tumors

Chemokines regulate tumor progression through the control of the angiogenic/angiostatic process, cell migration, and metastasis, influence of tumor cell proliferation and finally the infiltration of immune cells into the tumor mass. CXCL8/IL-8, one of the most studied chemokines, has been shown to contribute to human cancer progression acting as a mitogenic, motogenic, and angiogenic factor, particularly being involved in vascular endothelial cell proliferation and tumor neovascularization [159]. CXCL8/IL-8 is expressed in several human cancer tissues and cell lines and can act as an autocrine growth factor in different tumors [159]. In gliomas its expression is increased after anoxia, and the CXCL8/IL-8 expression as antisense RNA inhibited the growth of human glioma cells, *in vitro* [160]. CXCL8/IL-8 was identified, by RT–PCR and histologically, within tumor cells in human pituitary adenomas [161].

Surprisingly, CXC chemokines can be either angiogenic or angiostatic factors. Chemokines containing an ELR motif prior to the first cysteines, including CXCL1/GROα, CXCL5/ENA-78, CXCL7/NAP-2, and CXCL8/IL-8, show angiogenic activity through interaction with CXCR2, while the ELR-negative CXC chemokines CXCL9/MIG-1 and CXCL10/IP-10 exert an angiostatic activity. CXCL12/SDF-1 has been described to induce vascularization, despite the lack of the ELR motif [4,162], and also CCL2/MCP-1, a member of the CC chemokine family, can act as a direct mediator of angiogenesis [163]. Chemokines may have therapeutic effects through their angiostatic action or by boosting the immune response against the tumor through their ability to induce migration of T, NK cells, dendritic cells, and macrophages. CXCL10/IP-10 and CXCL9/MIG have shown an anti-tumor action generated by the increase of

T-cell recruitment and the inhibition of angiogenesis. These chemokines are induced by γ-interferon and are believed to be responsible, at least in part, for the anti-tumor effects of IL-12 through a γ-interferon dependent mechanism [164]. In addition, CCL7/MCP-3, CCL3/MIP-1α, CCL2/MCP-1, and CCL1/TCA-3 have also been shown to have anti-tumor activity [5].

Meningiomas, astrocytomas, and glioblastomas have all been shown to express CCL2/MCP-1, a chemokine isolated in the supernatants of a cultured glioma cell line whose inhibition by antibodies was able to block monocyte chemotaxis induced by tumor fluids of glioblastomas and astrocytomas [165]. Moreover, in human glioblastomas, the expression of all the CXC receptors and their ligands was reported in a high percentage of the cases analyzed [166]. Ectopic expression of p53 induced the expression of CX3CL1/fractalkine, a chemokine that favors chemotaxis of monocytes and cytotoxic T cells and that is largely expressed in the CNS, suggesting the involvment of this chemokine in the immune control to prevent cells from undergoing malignant trasformation [167].

CXCL12/SDF-1 and its receptor CXCR4 have been described as being expressed in brain tumors such as neuroblastomas and glioblastomas, and their level of expression in human gliobastoma tissues directly correlates with tumor malignancy grade [166,168–171]. CXCR4 expression was upregulated in human glioblastoma cell lines and experimental approaches aimed to inhibit its function blocked cell proliferation [172]. CXCR4 expression can be enhanced by cytokines like TNF-α and IL-1β [170], whose levels are elevated in astroglioma tumors and in some cases could be upregulated by CXCL12/SDF-1 itself [173]. CXCL12/SDF-1 was also produced and released after LPS stimulation by cultured astrocytes and directly stimulates astrocyte proliferation, suggesting that like CXCL8/IL-8, CXCL12/SDF-1 could also act as a growth factor. The concomitant expression of CXCR4 and CXCL12/SDF-1 should lead to autocrine and paracrine regulation of cell growth [32,69] that was indeed demonstrated in both glioblastoma multiforme established cell lines and in the primary cultures of postsurgical specimens of human glioblastoma [166,174]. In astroglioma cells, CXCR4 stimulation induces expression of both CXCL10/IP-10 and CXCL8/IL-8 and enhances the constitutive expression of CCL2/MCP-1 through the ERK1/2 pathway [170]. Besides their role in cell proliferation and angiogenesis, chemokines and their receptors also seem to be involved in the process of tumor cell migration, invasion, and metastasis [175]. A critical role for chemokines in organ-specific metastatic destination of tumor cells has been suggested. Indeed CXCL12/SDF-1 and CCL21/6Ckine are overexpressed in organs that are preferentially colonized by breast cancer metastasis and the relative receptors were expressed on the tumor cells [44,176]. Furthermore, the blockade of CXCR4 with a neutralizing antibody *in vivo* impaired metastases to regional lymph nodes and lung [176]. Although the metastatic process is probably more complex and other molecules may be involved, these results indicate that chemokines and theirs receptors have a pivotal role in the ability of tumor cells to invade other tissues. Based on *in vitro* studies, a role for CXCL12/SDF-1 in the metastatic process has also been proposed in neuroblastoma and ovarian cancer [169,177].

REFERENCES

1. Walz A, Peveri P, Aschauer H, Baggiolini M Purification and amino acid sequencing of NAF, a novel neutrophil-activating factor produced by monocytes. Biochem Biophys Res Commun 1987;149:755–61.
2. Laing KJ, Secombes CJ. Chemokines. Dev Comp Immunol 2004;28:443–60.

3. Bajetto A, Bonavia R, Barbero S, Florio T, Schettini G. Chemokines and their receptors in the central nervous system. Front Neuroendocrinol 2001;22:147–84.
4. Belperio JA, Keane MP, Arenberg DA, Addison CL, Ehlert JE, Burdick MD, Strieter RM. CXC chemokines in angiogenesis. J Leukoc Biol 2000;68:1–8.
5. Rossi D, Zlotnik A. The biology of chemokines and their receptors. Annu Rev Immunol 2000;18:217–42.
6. Rollins BJ. Chemokines. Blood 1997;90:909–28.
7. Bazan JF, Bacon KB, Hardiman G, Wang W, Soo K, Rossi D, Greaves DR, Zlotnik A, Schall TJ. A new class of membrane-bound chemokine with a CX3C motif. Nature 1997; 385:640–44.
8. Umehara H, Bloom ET, Okazaki T, Nagano Y, Yoshie O, Imai T. Fractalkine in vascular biology: from basic research to clinical disease. Arterioscler Thromb Vasc Biol 2004;24:34–40.
9. Gough PJ, Garton KJ, Wille PT, Rychlewski M, Dempsey PJ, Raines EW. A disintegrin and metalloproteinase 10-mediated cleavage and shedding regulates the cell surface expression of CXC chemokine ligand 16. J Immunol 2004;172:3678–85.
10. Strieter RM, Burdick MD, Mestas J, Gomperts B, Keane MP, Belperio JA. Cancer CXC chemokine networks and tumour angiogenesis. Eur J Cancer 2006;42:768–78.
11. Moser B, Loetscher P. Lymphocyte traffic control by chemokines. Nat Immunol 2001;2:123–28.
12. Lagane B, Ballet S, Planchenault T, Balabanian K, Le Poul E, Blanpain C, Percherancier Y, Staropoli I, Vassart G, Oppermann M, Parmentier M, Bachelerie F. Mutation of the DRY motif reveals different structural requirements for the CC chemokine receptor 5-mediated signaling and receptor endocytosis. Mol Pharmacol 2005;67:1966–76.
13. Balabanian K, Lagane B, Infantino S, Chow KY, Harriague J, Moepps B, Arenzana-Seisdedos F, Thelen M, Bachelerie F. The chemokine SDF-1/CXCL12 binds to and signals through the orphan receptor RDC1 in T lymphocytes. J Biol Chem 2005;280:35760–66.
14. Infantino S, Moepps B, Thelen M. Expression and regulation of the orphan receptor RDC1 and its putative ligand in human dendritic and B cells. J Immunol 2006; 176:2197–207.
15. Rodriguez-Frade JM, Vila-Coro AJ, de Ana AM, Albar JP, Martinez AC, Mellado M. The chemokine monocyte chemoattractant protein-1 induces functional responses through dimerization of its receptor CCR2. Proc Natl Acad Sci USA 1999;96:3628–33.
16. Mellado M, Rodriguez-Frade JM, Vila-Coro AJ, Fernandez S, Martin de Ana A, Jones DR, Toran JL, Martinez AC. Chemokine receptor homo- or heterodimerization activates distinct signaling pathways. Embo J 2001;20:2497–507.
17. Wilson S, Wilkinson G, Milligan G. The CXCR1 and CXCR2 receptors form constitutive homo- and heterodimers selectively and with equal apparent affinities. J Biol Chem 2005;280:28663–74.
18. Percherancier Y, Berchiche Y A, Slight I, Volkmer-Engert R, Tamamura H, Fujii N, Bouvier M and Heveker N Bioluminescence resonance energy transfer reveals ligand-induced conformational changes in CXCR4 homo- and heterodimers. J Biol Chem 2005; 280:9895–903.
19. Neptune ER and Bourne HR Receptors induce chemotaxis by releasing the betagamma subunit of Gi, not by activating Gq or Gs. Proc Natl Acad Sci USA 1997;94:14489–94.
20. Neptune ER, Iiri T and Bourne HR Galphai is not required for chemotaxis mediated by Gi-coupled receptors. J Biol Chem 1999;274:2824–28.

21. Roland J, Murphy BJ, Ahr B, Robert-Hebmann V, Delauzun V, Nye KE, Devaux C, Biard-Piechaczyk M. Role of the intracellular domains of CXCR4 in SDF-1-mediated signaling. Blood 2003;101:399–406.
22. Bajetto A, Bonavia R, Barbero S, Piccioli P, Costa A, Florio T, Schettini G. Glial and neuronal cells express functional chemokine receptor CXCR4 and its natural ligand stromal cell-derived factor 1. J Neurochem 1999;73:2348–57.
23. Thelen M. Dancing to the tune of chemokines. Nat Immunol 2001;2:129–34.
24. Bokoch GM. Chemoattractant signaling and leukocyte activation. Blood 1995;86: 1649–60.
25. Premack BA, Schall TJ. Chemokine receptors: gateways to inflammation and infection. Nat Med 1996;2:1174–78.
26. Liu QH, Williams DA, McManus C, Baribaud F, Doms RW, Schols D, De Clercq E, Kotlikoff MI, Collman RG, Freedman BD. HIV-1 gp120 and chemokines activate ion channels in primary macrophages through CCR5 and CXCR4 stimulation. Proc Natl Acad Sci USA 2000;97:4832–37.
27. Hegg CC, Hu S, Peterson PK, Thayer SA. Beta-chemokines and human immunodeficiency virus type-1 proteins evoke intracellular calcium increases in human microglia. Neuroscience 2000;98:191–99.
28. Florio T, Casagrande S, Diana F, Bajetto A, Porcile C, Zona G, Thellung S, Arena S, Pattarozzi A, Corsaro A, Spaziante R, Robello M, Schettini G. The chemokine SDF1{alpha} induces proliferation and growth hormone release in GH4C1 rat pituitary adenoma cell line, through multiple intracellular signals. Mol Pharmacol 2006;69:539–46.
29. Ganju RK, Brubaker SA, Meyer J, Dutt P, Yang Y, Qin S, Newman W, Groopman JE. The alpha-chemokine, stromal cell-derived factor-1alpha, binds to the transmembrane G-protein-coupled CXCR-4 receptor and activates multiple signal transduction pathways. J Biol Chem 1998;273:23169–75.
30. Ganju RK, Dutt P, Wu L, Newman W, Avraham H, Avraham S, Groopman JE. Beta-chemokine receptor CCR5 signals via the novel tyrosine kinase RAFTK. Blood 1998;91:791–97.
31. Tilton B, Ho L, Oberlin E, Loetscher P, Baleux F, Clark-Lewis I, Thelen M. Signal transduction by CXC chemokine receptor 4. Stromal cell-derived factor 1 stimulates prolonged protein kinase B and extracellular signal-regulated kinase 2 activation in T lymphocytes. J Exp Med 2000;192:313–24.
32. Bajetto A, Barbero S, Bonavia R, Piccioli P, Pirani P, Florio T, Schettini G. Stromal cell-derived factor-1alpha induces astrocyte proliferation through the activation of extracellular signal-regulated kinases 1/2 pathway. J Neurochem 2001;77:1226–36.
33. Sasaki T, Irie-Sasaki J, Jones RG, Oliveira-dos-Santos AJ, Stanford WL, Bolon B, Wakeham A, Itie A, Bouchard D, Kozieradzki I, Joza N, Mak TW, Ohashi PS, Suzuki A, Penninger JM. Function of PI3Kgamma in thymocyte development, T cell activation, and neutrophil migration. Science 2000;287:1040–46.
34. Hirsch E, Katanaev VL, Garlanda C, Azzolino O, Pirola L, Silengo L, Sozzani S, Mantovani A, Altruda F, Wymann MP. Central role for G protein-coupled phosphoinositide 3-kinase gamma in inflammation. Science 2000;287:1049–53.
35. Bonacchi A, Romagnani P, Romanelli RG, Efsen E, Annunziato F, Lasagni L, Francalanci M, Serio M, Laffi G, Pinzani M, Gentilini P, Marra F. Signal transduction by the chemokine receptor CXCR3: activation of Ras/ERK, Src, and phosphatidylinositol 3-kinase/Akt controls cell migration and proliferation in human vascular pericytes. J Biol Chem 2001;276:9945–54.

36. Abi-Younes S, Sauty A, Mach F, Sukhova GK, Libby P, Luster AD. The stromal cell-derived factor-1 chemokine is a potent platelet agonist highly expressed in atherosclerotic plaques. Circ Res 2000;86:131–38.
37. Popik W, Pitha PM. Early activation of mitogen-activated protein kinase kinase, extracellular signal-regulated kinase, p38 mitogen-activated protein kinase, and c-Jun N-terminal kinase in response to binding of simian immunodeficiency virus to Jurkat T cells expressing CCR5 receptor. Virology 1998;252:210–17.
38. Sun Y, Cheng Z, Ma L, Pei G. Beta-arrestin2 is critically involved in CXCR4-mediated chemotaxis, and this is mediated by its enhancement of p38 MAPK activation. J Biol Chem 2002;277:49212–19.
39. Vlahakis S R, Villasis-Keever A, Gomez T, Vanegas M, Vlahakis N, Paya CV. G protein-coupled chemokine receptors induce both survival and apoptotic signaling pathways. J Immunol 2002;169:5546–54.
40. Riol-Blanco L, Sanchez-Sanchez N, Torres A, Tejedor A, Narumiya S, Corbi AL, Sanchez-Mateos P, Rodriguez-Fernandez JL. The chemokine receptor CCR7 activates in dendritic cells two signaling modules that independently regulate chemotaxis and migratory speed. J Immunol 2005;174:4070–80.
41. Ganju RK, Brubaker SA, Chernock RD, Avraham S, Groopman JE. Beta-chemokine receptor CCR5 signals through SHP1, SHP2, and Syk. J Biol Chem 2000;275: 17263–68.
42. Chernock RD, Cherla RP, Ganju RK. SHP2 and cbl participate in alpha-chemokine receptor CXCR4-mediated signaling pathways. Blood 2001;97:608–15.
43. Wain CM, Westwick J, Ward SG. Heterologous regulation of chemokine receptor signaling by the lipid phosphatase SHIP in lymphocytes. Cell Signal 2005;17:1194–202.
44. Fernandis AZ, Prasad A, Band H, Klosel R, Ganju RK. Regulation of CXCR4-mediated chemotaxis and chemoinvasion of breast cancer cells. Oncogene 2004;23:157–67.
45. Mellado M, Rodriguez-Frade JM, Aragay A, del Real G, Martin AM, Vila-Coro AJ, Serrano A, Mayor F Jr, Martinez AC. The chemokine monocyte chemotactic protein 1 triggers Janus kinase 2 activation and tyrosine phosphorylation of the CCR2B receptor. J Immunol 1998;161:805–13.
46. Vila-Coro AJ, Rodriguez-Frade JM, Martin De Ana A, Moreno-Ortiz MC, Martinez AC, Mellado M. The chemokine SDF-1alpha triggers CXCR4 receptor dimerization and activates the JAK/STAT pathway. Faseb J 1999;13:1699–10.
47. Vila-Coro AJ, Mellado M, Martin de Ana A, Lucas P, del Real G, Martinez AC, Rodriguez-Frade JM. HIV-1 infection through the CCR5 receptor is blocked by receptor dimerization. Proc Natl Acad Sci USA 2000;97:3388–93.
48. Mellado M, Vila-Coro AJ, Martinez C, Rodriguez-Frade JM. Receptor dimerization: a key step in chemokine signaling. Cell Mol Biol 2001;47:575–82.
49. Bakhiet M, Tjernlund A, Mousa A, Gad A, Stromblad S, Kuziel WA, Seiger A, Andersson J. RANTES promotes growth and survival of human first-trimester forebrain astrocytes. Nat Cell Biol 2001;3:150–57.
50. Zhang XF, Wang JF, Matczak E, Proper JA, Groopman JE. Janus kinase 2 is involved in stromal cell-derived factor-1alpha-induced tyrosine phosphorylation of focal adhesion proteins and migration of hematopoietic progenitor cells. Blood 2001;97:3342–48.
51. Moriguchi M, Hissong BD, Gadina M, Yamaoka K, Tiffany HL, Murphy PM, Candotti F, O'Shea JJ. CXCL12 signaling is independent of Jak2 and Jak3. J Biol Chem 2005;280: 17408–14.

52. Venkatakrishnan G, Salgia R, Groopman JE. Chemokine receptors CXCR-1/2 activate mitogen-activated protein kinase via the epidermal growth factor receptor in ovarian cancer cells. J Biol Chem 2000;275:6868–75.
53. Porcile C, Bajetto A, Barbieri F, Barbero S, Bonavia R, Biglieri M, Pirani P, Florio T, Schettini G. Stromal cell-derived factor-1alpha (SDF-1alpha/CXCL12) stimulates ovarian cancer cell growth through the EGF receptor transactivation. Exp Cell Res 2005;308:241–53.
54. Cabioglu N, Summy J, Miller C, Parikh NU, Sahin AA, Tuzlali S, Pumiglia K, Gallick GE, Price JE. CXCL-12/stromal cell-derived factor-1alpha transactivates HER2-neu in breast cancer cells by a novel pathway involving Src kinase activation. Cancer Res 2005;65:6493–97.
55. Drube S, Stirnweiss J, Valkova C, Liebmann C. Ligand-independent and EGF receptor-supported transactivation: lessons from beta2-adrenergic receptor signalling. Cell Signal 2006;18:1633–46.
56. Kim CH, Hangoc G, Cooper S, Helgason CD, Yew S, Humphries RK, Krystal G, Broxmeyer HE. Altered responsiveness to chemokines due to targeted disruption of SHIP. J Clin Invest 1999;104:1751–59.
57. Bajetto A, Bonavia R, Barbero S, Schettini G. Characterization of chemokines and their receptors in the central nervous system: physiopathological implications. J Neurochem 2002;82:1311–29.
58. Biber K, de Jong E K, van Weering HR, Boddeke HW. Chemokines and their receptors in central nervous system disease. Curr Drug Targets 2006;7:29–46.
59. Lazarini F, Tham TN, Casanova P, Arenzana-Seisdedos F, Dubois-Dalcq M. Role of the alpha-chemokine stromal cell-derived factor (SDF-1) in the developing and mature central nervous system. Glia 2003;42:139–48.
60. Horuk R, Martin AW, Wang Z, Schweitzer L, Gerassimides A, Guo H, Lu Z, Hesselgesser J, Perez HD, Kim J, Parker J, Hadley TJ, Peiper SC. Expression of chemokine receptors by subsets of neurons in the central nervous system. J Immunol 1997;158:2882–90.
61. Robinson S, Tani M, Strieter RM, Ransohoff RM, Miller RH. The chemokine growth-regulated oncogene-alpha promotes spinal cord oligodendrocyte precursor proliferation. J Neurosci 1998;18:10457–63.
62. Tani M, Fuentes ME, Peterson JW, Trapp BD, Durham SK, Loy JK, Bravo R, Ransohoff RM, Lira SA. Neutrophil infiltration, glial reaction, and neurological disease in transgenic mice expressing the chemokine N51/KC in oligodendrocytes. J Clin Invest 1996;98:529–39.
63. Giovannelli A, Limatola C, Ragozzino D, Mileo AM, Ruggieri A, Ciotti MT, Mercanti D, Santoni A, Eusebi F. CXC chemokines interleukin-8 (IL-8) and growth-related gene product alpha (GROalpha) modulate Purkinje neuron activity in mouse cerebellum. J Neuroimmunol 1998;92:122–32.
64. Limatola C, Giovannelli A, Maggi L, Ragozzino D, Castellani L, Ciotti MT, Vacca F, Mercanti D, Santoni A, Eusebi F. SDF-1alpha-mediated modulation of synaptic transmission in rat cerebellum. Eur J Neurosci 2000;12:2497–504.
65. Xia MQ, Bacskai BJ, Knowles RB, Qin SX, Hyman BT. Expression of the chemokine receptor CXCR3 on neurons and the elevated expression of its ligand IP-10 in reactive astrocytes: in vitro ERK1/2 activation and role in Alzheimer's disease. J Neuroimmunol 2000;108:227–35.
66. Sui Y, Stehno-Bittel L, Li S, Loganathan R, Dhillon NK, Pinson D, Nath A, Kolson D, Narayan O, Buch S. CXCL10-induced cell death in neurons: role of calcium dysregulation. Eur J Neurosci 2006;23:957–64.

67. Gupta SK, Lysko PG, Pillarisetti K, Ohlstein E, Stadel JM. Chemokine receptors in human endothelial cells. Functional expression of CXCR4 and its transcriptional regulation by inflammatory cytokines. J Biol Chem 1998;273:4282–87.
68. Ohtani Y, Minami M, Kawaguchi N, Nishiyori A, Yamamoto J, Takami S, Satoh M. Expression of stromal cell-derived factor-1 and CXCR4 chemokine receptor mRNAs in cultured rat glial and neuronal cells. Neurosci Lett 1998;249:163–66.
69. Bajetto A, Bonavia R, Barbero S, Piccioli P, Costa A, Florio T, Schettini G. Glial and neuronal cells express functional chemokine receptor CXCR4 and its natural ligand stromal cell-derived factor 1. J Neurochem 1999;73:2348–57.
70. Bajetto A, Bonavia R, Barbero S, Florio T, Schettini G. Chemokines and their receptors in the central nervous system. Front Neuroendocrinol 2001;22:147–84.
71. Wong M L, Xin WW, Duman RS. Rat LCR1: cloning and cellular distribution of a putative chemokine receptor in brain. Mol Psychiatry 1996;1:133–40.
72. Bezzi P, Domercq M, Brambilla L, Galli R, Schols D, De Clercq E, Vescovi A, Bagetta G, Kollias G, Meldolesi J, Volterra A. CXCR4-activated astrocyte glutamate release via TNFalpha: amplification by microglia triggers neurotoxicity. Nat Neurosci 2001;4:702–10.
73. Kaiser E, Forster R, Wolf I, Ebensperger C, Kuehl WM, Lipp M The G protein-coupled receptor BLR1 is involved in murine B cell differentiation and is also expressed in neuronal tissues. Eur J Immunol 1993;23:2532–39.
74. Flynn G, Maru S, Loughlin J, Romero IA, Male D. Regulation of chemokine receptor expression in human microglia and astrocytes. J Neuroimmunol 2003;136:84–93.
75. Rezaie P, Trillo-Pazos G, Everall IP, Male DK. Expression of beta-chemokines and chemokine receptors in human fetal astrocyte and microglial co-cultures: potential role of chemokines in the developing CNS. Glia 2002;37:64–75.
76. He J, Chen Y, Farzan M, Choe H, Ohagen A, Gartner S, Busciglio J, Yang X, Hofmann W, Newman W, Mackay C R, Sodroski J, Gabuzda D. CCR3 and CCR5 are co-receptors for HIV-1 infection of microglia. Nature 1997;385:645–49.
77. Vallat AV, De Girolami U, He J, Mhashilkar A, Marasco W, Shi B, Gray F, Bell J, Keohane C, Smith TW, Gabuzda D. Localization of HIV-1 co-receptors CCR5 and CXCR4 in the brain of children with AIDS. Am J Pathol 1998;152:167–78.
78. Klein RS, Williams KC, Alvarez-Hernandez X, Westmoreland S, Force T, Lackner AA, Luster AD. Chemokine receptor expression and signaling in macaque and human fetal neurons and astrocytes: implications for the neuropathogenesis of AIDS. J Immunol 1999;163:1636–46.
79. Xia MQ, Hyman BT. Chemokines/chemokine receptors in the central nervous system and Alzheimer's disease. J Neurovirol 1999;5:32–41.
80. Westmoreland SV, Rottman JB, Williams KC, Lackner AA, Sasseville VG. Chemokine receptor expression on resident and inflammatory cells in the brain of macaques with simian immunodeficiency virus encephalitis. Am J Pathol 1998;152:659–65.
81. Sanders VJ, Pittman CA, White MG, Wang G, Wiley CA, Achim CL. Chemokines and receptors in HIV encephalitis. Aids 1998;12:1021–26.
82. Cota M, Kleinschmidt A, Ceccherini-Silberstein F, Aloisi F, Mengozzi M, Mantovani A, Brack-Werner R, Poli G. Upregulated expression of interleukin-8, RANTES and chemokine receptors in human astrocytic cells infected with HIV-1. J Neurovirol 2000; 6:75–83.
83. McManus CM, Brosnan CF, Berman JW. Cytokine induction of MIP-1 alpha and MIP-1 beta in human fetal microglia. J Immunol 1998;160:1449–55.

84. McManus C, Berman JW, Brett FM, Staunton H, Farrell M, Brosnan CF. MCP-1, MCP-2 and MCP-3 expression in multiple sclerosis lesions: an immunohistochemical and in situ hybridization study. J Neuroimmunol 1998;86:20–29.
85. Calvo CF, Yoshimura T, Gelman M, Mallat M. Production of monocyte chemotactic protein-1 by rat brain macrophages. Eur J Neurosci 1996;8:1725–34.
86. Meng SZ, Oka A, Takashima S. Developmental expression of monocyte chemoattractant protein-1 in the human cerebellum and brainstem. Brain Dev 1999;21:30–35.
87. Meucci O, Fatatis A, Simen AA, Bushell TJ, Gray PW, Miller RJ. Chemokines regulate hippocampal neuronal signaling and gp120 neurotoxicity. Proc Natl Acad Sci USA 1998;95:14500–5.
88. Dorf ME, Berman MA, Tanabe S, Heesen M, Luo Y. Astrocytes express functional chemokine receptors. J Neuroimmunol 2000;111:109–21.
89. Jinno A, Shimizu N, Soda Y, Haraguchi Y, Kitamura T, Hoshino H. Identification of the chemokine receptor TER1/CCR8 expressed in brain-derived cells and T cells as a new coreceptor for HIV-1 infection. Biochem Biophys Res Commun 1998;243:497–502.
90. Maciejewski-Lenoir D, Chen S, Feng L, Maki R, Bacon KB. Characterization of fractalkine in rat brain cells: migratory and activation signals for CX3CR-1-expressing microglia. J Immunol 1999;163:1628–35.
91. Meucci O, Fatatis A, Simen AA, Miller RJ. Expression of CX3CR1 chemokine receptors on neurons and their role in neuronal survival. Proc Natl Acad Sci USA 2000;97:8075–80.
92. Puma C, Danik M, Quirion R, Ramon F, Williams S. The chemokine interleukin-8 acutely reduces Ca(2+) currents in identified cholinergic septal neurons expressing CXCR1 and CXCR2 receptor mRNAs. J Neurochem 2001;78:960–71.
93. Tran PB, Miller RJ. Chemokine receptors: signposts to brain development and disease. Nat Rev Neurosci 2003;4:444–455.
94. Ma Q, Jones D, Borghesani PR, Segal RA, Nagasawa T, Kishimoto T, Bronson RT, Springer TA. Impaired B-lymphopoiesis, myelopoiesis, and derailed cerebellar neuron migration in CXCR4- and SDF-1-deficient mice. Proc Natl Acad Sci USA 1998;95:9448–53.
95. Zou YR, Kottmann AH, Kuroda M, Taniuchi I, Littman DR. Function of the chemokine receptor CXCR4 in haematopoiesis and in cerebellar development. Nature 1998;393: 595–99.
96. Tham TN, Lazarini F, Franceschini IA, Lachapelle F, Amara A, Dubois-Dalcq M. Developmental pattern of expression of the alpha chemokine stromal cell-derived factor 1 in the rat central nervous system. Eur J Neurosci 2001;13:845–56.
97. Borrell V, Marin O. Meninges control tangential migration of hem-derived Cajal-Retzius cells via CXCL12/CXCR4 signaling. Nat Neurosci 2006;9:1284–93.
98. Reiss K, Mentlein R, Sievers J, Hartmann D. Stromal cell-derived factor 1 is secreted by meningeal cells and acts as chemotactic factor on neuronal stem cells of the cerebellar external granular layer. Neuroscience 2002;115:295–305.
99. Stumm RK, Zhou C, Ara T, Lazarini F, Dubois-Dalcq M, Nagasawa T, Hollt V, Schulz S. CXCR4 regulates interneuron migration in the developing neocortex. J Neurosci 2003;23:5123–30.
100. Vilz TO, Moepps B, Engele J, Molly S, Littman DR, Schilling K. The SDF-1/CXCR4 pathway and the development of the cerebellar system. Eur J Neurosci 2005;22:1831–39.
101. Tanabe S, Heesen M, Berman MA, Fischer MB, Yoshizawa I, Luo Y, Dorf ME. Murine astrocytes express a functional chemokine receptor. J Neurosci 1997;17:6522–28.
102. Lazarini F, Casanova P, Tham TN, De Clercq E, Arenzana-Seisdedos F, Baleux F, Dubois-Dalcq M. Differential signalling of the chemokine receptor CXCR4 by stromal

cell-derived factor 1 and the HIV glycoprotein in rat neurons and astrocytes. Eur J Neurosci 2000;12:117–25.
103. Dziembowska M, Tham TN, Lau P, Vitry S, Lazarini F, Dubois-Dalcq M. A role for CXCR4 signaling in survival and migration of neural and oligodendrocyte precursors. Glia 2005;50:258–69.
104. Boisvert WA, Santiago R, Curtiss LK, Terkeltaub RA. A leukocyte homologue of the IL-8 receptor CXCR-2 mediates the accumulation of macrophages in atherosclerotic lesions of LDL receptor-deficient mice. J Clin Invest 1998;101:353–63.
105. Broxmeyer HE, Cooper S, Cacalano G, Hague NL, Bailish E, Moore MW. Involvement of Interleukin (IL) 8 receptor in negative regulation of myeloid progenitor cells in vivo: evidence from mice lacking the murine IL-8 receptor homologue. J Exp Med 1996;184:1825–32.
106. Fuentes ME, Durham SK, Swerdel MR, Lewin AC, Barton DS, Megill JR, Bravo R, Lira SA. Controlled recruitment of monocytes and macrophages to specific organs through transgenic expression of monocyte chemoattractant protein-1. J Immunol 1995;155:5769–76.
107. Elhofy A, Wang J, Tani M, Fife BT, Kennedy KJ, Bennett J, Huang D, Ransohoff RM, Karpus WJ. Transgenic expression of CCL2 in the central nervous system prevents experimental autoimmune encephalomyelitis. J Leukoc Biol 2005;77:229–37.
108. Forster R, Mattis AE, Kremmer E, Wolf E, Brem G, Lipp M. A putative chemokine receptor, BLR1, directs B cell migration to defined lymphoid organs and specific anatomic compartments of the spleen. Cell 1996;87:1037–47.
109. Cardona AE, Pioro EP, Sasse ME, Kostenko V, Cardona SM, Dijkstra IM, Huang D, Kidd G, Dombrowski S, Dutta R, Lee JC, Cook DN, Jung S, Lira SA, Littman DR, Ransohoff RM. Control of microglial neurotoxicity by the fractalkine receptor. Nat Neurosci 2006;9:917–24.
110. Asensio VC, Kincaid C, Campbell IL. Chemokines and the inflammatory response to viral infection in the central nervous system with a focus on lymphocytic choriomeningitis virus. J Neurovirol 1999;5:65–75.
111. Balashov KE, Rottman JB, Weiner HL, Hancock WW. CCR5(+) and CXCR3(+) T cells are increased in multiple sclerosis and their ligands MIP-1alpha and IP-10 are expressed in demyelinating brain lesions. Proc Natl Acad Sci USA 1999;96: 6873–78.
112. Simpson JE, Newcombe J, Cuzner ML, Woodroofe MN. Expression of the interferon-gamma-inducible chemokines IP-10 and Mig and their receptor, CXCR3, in multiple sclerosis lesions. Neuropathol Appl Neurobiol 2000;26:133–42.
113. Simpson JE, Newcombe J, Cuzner ML, Woodroofe MN. Expression of monocyte chemoattractant protein-1 and other beta-chemokines by resident glia and inflammatory cells in multiple sclerosis lesions. J Neuroimmunol 1998;84:238–49.
114. Simpson J, Rezaie P, Newcombe J, Cuzner ML, Male D, Woodroofe MN. Expression of the beta-chemokine receptors CCR2, CCR3 and CCR5 in multiple sclerosis central nervous system tissue. J Neuroimmunol 2000;108:192–200.
115. Zamvil S, Nelson P, Trotter J, Mitchell D, Knobler R, Fritz R, Steinman L. T-cell clones specific for myelin basic protein induce chronic relapsing paralysis and demyelination. Nature 1985;317:355–58.
116. Jiang Y, Salafranca MN, Adhikari S, Xia Y, Feng L, Sonntag MK, deFiebre CM, Pennell NA, Streit WJ, Harrison JK. Chemokine receptor expression in cultured glia and rat experimental allergic encephalomyelitis. J Neuroimmunol 1998;86:1–12.

117. Kennedy KJ, Strieter RM, Kunkel SL, Lukacs NW, Karpus WJ. Acute and relapsing experimental autoimmune encephalomyelitis are regulated by differential expression of the CC chemokines macrophage inflammatory protein-1alpha and monocyte chemotactic protein-1. J Neuroimmunol 1998;92:98–108.
118. Karpus WJ, Ransohoff RM. Chemokine regulation of experimental autoimmune encephalomyelitis: temporal and spatial expression patterns govern disease pathogenesis. J Immunol 1998;161:2667–71.
119. Fife BT, Huffnagle GB, Kuziel WA, Karpus WJ. CC chemokine receptor 2 is critical for induction of experimental autoimmune encephalomyelitis. J Exp Med 2000;192:899–905.
120. Izikson L, Klein RS, Charo IF, Weiner HL, Luster AD. Resistance to experimental autoimmune encephalomyelitis in mice lacking the CC chemokine receptor (CCR)2. J Exp Med 2000;192:1075–80.
121. Rottman JB, Slavin AJ, Silva R, Weiner HL, Gerard CG, Hancock WW. Leukocyte recruitment during onset of experimental allergic encephalomyelitis is CCR1 dependent. Eur J Immunol 2000;30:2372–77.
122. Huang DR, Wang J, Kivisakk P, Rollins BJ, Ransohoff RM. Absence of monocyte chemoattractant protein 1 in mice leads to decreased local macrophage recruitment and antigen-specific T helper cell type 1 immune response in experimental autoimmune encephalomyelitis. J Exp Med 2001;193:713–26.
123. Fife BT, Paniagua MC, Lukacs NW, Kunkel SL, Karpus WJ. Selective CC chemokine receptor expression by central nervous system-infiltrating encephalitogenic T cells during experimental autoimmune encephalomyelitis. J Neurosci Res 2001;66:705–14.
124. Che X, Ye W, Panga L, Wu DC, Yang GY. Monocyte chemoattractant protein-1 expressed in neurons and astrocytes during focal ischemia in mice. Brain Res 2001;902: 171–77.
125. Wang X, Li X, Schmidt DB, Foley JJ, Barone FC, Ames RS, Sarau HM. Identification and molecular characterization of rat CXCR3: receptor expression and interferon-inducible protein-10 binding are increased in focal stroke. Mol Pharmacol 2000;57:1190–98.
126. Wang X, Ellison JA, Siren AL, Lysko PG, Yue TL, Barone FC, Shatzman A, Feuerstein GZ. Prolonged expression of interferon-inducible protein-10 in ischemic cortex after permanent occlusion of the middle cerebral artery in rat. J Neurochem 1998;71: 1194–204.
127. Wang X, Yue TL, Barone FC, Feuerstein GZ. Monocyte chemoattractant protein-1 messenger RNA expression in rat ischemic cortex. Stroke 1995;26:661–65;
128. Mukaida N, Matsumoto T, Yokoi K, Harada A, Matsushima K. Inhibition of neutrophil-mediated acute inflammation injury by an antibody against interleukin-8 (IL-8). Inflamm Res 1998;47(Suppl 3):S151–57.
129. Gosling J, Slaymaker S, Gu L, Tseng S, Zlot CH, Young SG, Rollins BJ, Charo IF. MCP-1 deficiency reduces susceptibility to atherosclerosis in mice that overexpress human apolipoprotein B. J Clin Invest 1999;103:773–78.
130. Boisvert WA, Curtiss LK, Terkeltaub RA. Interleukin-8 and its receptor CXCR2 in atherosclerosis. Immunol Res 2000;21:129–37.
131. Dawson TC, Kuziel WA, Osahar TA, Maeda N. Absence of CC chemokine receptor-2 reduces atherosclerosis in apolipoprotein E-deficient mice. Atherosclerosis 1999; 143:205–11.
132. Glabinski AR, Balasingam V, Tani M, Kunkel SL, Strieter RM, Yong VW, Ransohoff RM. Chemokine monocyte chemoattractant protein-1 is expressed by astrocytes after mechanical injury to the brain. J Immunol 1996;156:4363–68.

133. Hausmann EH, Berman NE, Wang YY, Meara JB, Wood GW, Klein RM. Selective chemokine mRNA expression following brain injury. Brain Res 1998;788:49–59.
134. Muessel MJ, Berman NE, Klein RM. Early and specific expression of monocyte chemoattractant protein-1 in the thalamus induced by cortical injury. Brain Res 2000;870:211–21.
135. Valles A, Grijpink-Ongering L, de Bree FM, Tuinstra T, Ronken E. Differential regulation of the CXCR2 chemokine network in rat brain trauma: implications for neuroimmune interactions and neuronal survival. Neurobiol Dis 2006;22:312–22.
136. Harrison JK, Jiang Y, Chen S, Xia Y, Maciejewski D, McNamara RK, Streit WJ, Salafranca MN, Adhikari S, Thompson DA, Botti P, Bacon KB, Feng L. Role for neuronally derived fractalkine in mediating interactions between neurons and CX3CR1-expressing microglia. Proc Natl Acad Sci USA 1998;95:10896–901.
137. Jung S, Aliberti J, Graemmel P, Sunshine MJ, Kreutzberg GW, Sher A, Littman DR. Analysis of fractalkine receptor CX(3)CR1 function by targeted deletion and green fluorescent protein reporter gene insertion. Mol Cell Biol 2000;20:4106–14.
138. Chapman GA, Moores K, Harrison D, Campbell CA, Stewart BR, Strijbos PJ. Fractalkine cleavage from neuronal membranes represents an acute event in the inflammatory response to excitotoxic brain damage. J Neurosci 2000;20:RC87.
139. Bleul CC, Farzan M, Choe H, Parolin C, Clark-Lewis I, Sodroski J, Springer TA. The lymphocyte chemoattractant SDF-1 is a ligand for LESTR/fusin and blocks HIV-1 entry. Nature 1996;382:829–33.
140. Oberlin E, Amara A, Bachelerie F, Bessia C, Virelizier JL, Arenzana-Seisdedos F, Schwartz O, Heard JM, Clark-Lewis I, Legler DF, Loetscher M, Baggiolini M, Moser B. The CXC chemokine SDF-1 is the ligand for LESTR/fusin and prevents infection by T-cell-line-adapted HIV-1. Nature 1996;382:833–35.
141. Garzino-Demo A, DeVico AL, Cocchi F, Gallo RC. Beta-chemokines and protection from HIV type 1 disease. AIDS Res Hum Retroviruses 1998;14(Suppl 2):S177–84.
142. Murphy PM. Viral exploitation and subversion of the immune system through chemokine mimicry. Nat Immunol 2001;2:116–22.
143. Kaul M, Lipton SA. Mechanisms of neuronal injury and death in HIV-1 associated dementia. Curr HIV Res 2006;4:307–18.
144. Scorziello A, Florio T, Bajetto A, Thellung S, Schettini G. TGF-beta1 prevents gp120-induced impairment of Ca2+ homeostasis and rescues cortical neurons from apoptotic death. J Neurosci Res 1997;49:600–7.
145. Bonavia R, Bajetto A, Barbero S, Albini A, Noonan DM, Schettini G. HIV-1 Tat causes apoptotic death and calcium homeostasis alterations in rat neurons. Biochem Biophys Res Commun 2001;288:301–8.
146. Toggas SM, Masliah E, Rockenstein EM, Rall GF, Abraham CR, Mucke L. Central nervous system damage produced by expression of the HIV-1 coat protein gp120 in transgenic mice. Nature 1994;367:188–93.
147. Scorziello A, Florio T, Bajetto A, Schettini G. Intracellular signalling mediating HIV-1 gp120 neurotoxicity. Cell Signal 1998;10:75–84.
148. Corasaniti MT, Piccirilli S, Paoletti A, Nistico R, Stringaro A, Malorni W, Finazzi-Agro A, Bagetta G. Evidence that the HIV-1 coat protein gp120 causes neuronal apoptosis in the neocortex of rat via a mechanism involving CXCR4 chemokine receptor. Neurosci Lett 2001;312:67–70.
149. Zheng J, Thylin MR, Ghorpade A, Xiong H, Persidsky Y, Cotter R, Niemann D, Che M, Zeng YC, Gelbard HA, Shepard RB, Swartz JM, Gendelman HE. Intracellular CXCR4

signaling, neuronal apoptosis and neuropathogenic mechanisms of HIV-1-associated dementia. J Neuroimmunol 1999;98:185–200.
150. Schmidtmayerova H, Nottet H S, Nuovo G, Raabe T, Flanagan CR, Dubrovsky L, Gendelman HE, Cerami A, Bukrinsky M, Sherry B. Human immunodeficiency virus type 1 infection alters chemokine beta peptide expression in human monocytes: implications for recruitment of leukocytes into brain and lymph nodes. Proc Natl Acad Sci USA 1996;93:700–4.
151. Kelder W, McArthur JC, Nance-Sproson T, McClernon D, Griffin DE. Beta-chemokines MCP-1 and RANTES are selectively increased in cerebrospinal fluid of patients with human immunodeficiency virus-associated dementia. Ann Neurol 1998; 44:831–35.
152. Poluektova L, Moran T, Zelivyanskaya M, Swindells S, Gendelman HE, Persidsky Y. The regulation of alpha chemokines during HIV-1 infection and leukocyte activation: relevance for HIV-1-associated dementia. J Neuroimmunol 2001;120:112–28.
153. Tong N, Perry SW, Zhang Q, James HJ, Guo H, Brooks A, Bal H, Kinnear SA, Fine S, Epstein LG, Dairaghi D, Schall TJ, Gendelman HE, Dewhurst S, Sharer LR, Gelbard HA. Neuronal fractalkine expression in HIV-1 encephalitis: roles for macrophage recruitment and neuroprotection in the central nervous system. J Immunol 2000;164:1333–39.
154. Cinque P, Bestetti A, Marenzi R, Sala S, Gisslen M, Hagberg L, Price RW. Cerebrospinal fluid interferon-gamma-inducible protein 10 (IP-10, CXCL10) in HIV-1 infection. J Neuroimmunol 2005;168:154–63.
155. McManus CM, Weidenheim K, Woodman SE, Nunez J, Hesselgesser J, Nath A, Berman JW. Chemokine and chemokine-receptor expression in human glial elements: induction by the HIV protein, Tat, and chemokine autoregulation. Am J Pathol 2000;156:1441–53.
156. Sasseville VG, Smith MM, Mackay CR, Pauley DR, Mansfield KG, Ringler DJ, Lackner AA. Chemokine expression in simian immunodeficiency virus-induced AIDS encephalitis. Am J Pathol 1996;149:1459–67.
157. Akiyama H, Barger S, Barnum S, Bradt B, Bauer J, Cole GM, Cooper NR, Eikelenboom P, Emmerling M, Fiebich BL, Finch CE, Frautschy S, Griffin W S, Hampel H, Hull M, Landreth G, Lue L, Mrak R, Mackenzie IR, McGeer PL, O'Banion MK, Pachter J, Pasinetti G, Plata-Salaman C, Rogers J, Rydel R, Shen Y, Streit W, Strohmeyer R, Tooyoma I, Van Muiswinkel FL, Veerhuis R, Walker D, Webster S, Wegrzyniak B, Wenk G, Wyss-Coray T. Inflammation and Alzheimer's disease. Neurobiol Aging 2000;21:383–21.
158. Prat E, Baron P, Meda L, Scarpini E, Galimberti D, Ardolino G, Catania A, Scarlato G. The human astrocytoma cell line U373MG produces monocyte chemotactic protein (MCP)-1 upon stimulation with beta-amyloid protein. Neurosci Lett 2000;283:177–80.
159. Xie K. Interleukin-8 and human cancer biology. Cytokine Growth Factor Rev 2001;12:375–91.
160. Desbaillets I, Diserens AC, de Tribolet N, Hamou MF, Van Meir EG. Regulation of interleukin-8 expression by reduced oxygen pressure in human glioblastoma. Oncogene 1999;18:1447–56.
161. Suliman ME, Royds JA, Baxter L, Timperley WR, Cullen DR, Jones TH. IL-8 mRNA expression by in situ hybridisation in human pituitary adenomas. Eur J Endocrinol 1999;140:155–58.
162. Salcedo R, Oppenheim JJ. Role of chemokines in angiogenesis: CXCL12/SDF-1 and CXCR4 interaction, a key regulator of endothelial cell responses. Microcirculation 2003;10:359–70.

163. Salcedo R, Ponce ML, Young HA, Wasserman K, Ward JM, Kleinman HK, Oppenheim JJ, Murphy WJ. Human endothelial cells express CCR2 and respond to MCP-1: direct role of MCP-1 in angiogenesis and tumor progression. Blood 2000;96:34–40.
164. Kanegane C, Sgadari C, Kanegane H, Teruya-Feldstein J, Yao L, Gupta G, Farber JM, Liao F, Liu L, Tosato G. Contribution of the CXC chemokines IP-10 and Mig to the antitumor effects of IL-12. J Leukoc Biol 1998;64:384–92.
165. Leung SY, Wong MP, Chung LP, Chan AS, Yuen ST. Monocyte chemoattractant protein-1 expression and macrophage infiltration in gliomas. Acta Neuropathol (Berl) 1997;93:518–27.
166. Bajetto A, Barbieri F, Dorcaratto A, Barbero S, Daga A, Porcile C, Ravetti JL, Zona G, Spaziante R, Corte G, Schettini G, Florio T. Expression of CXC chemokine receptors 1–5 and their ligands in human glioma tissues: role of CXCR4 and SDF1 in glioma cell proliferation and migration. Neurochem Int 2006;49:423–32.
167. Shiraishi K, Fukuda S, Mori T, Matsuda K, Yamaguchi T, Tanikawa C, Ogawa M, Nakamura Y, Arakawa H. Identification of fractalkine, a CX3C-type chemokine, as a direct target of p53. Cancer Res 2000;60:3722–26.
168. Rempel SA, Dudas S, Ge S, Gutierrez JA. Identification and localization of the cytokine SDF1 and its receptor, CXC chemokine receptor 4, to regions of necrosis and angiogenesis in human glioblastoma. Clin Cancer Res 2000;6:102–11.
169. Geminder H, Sagi-Assif O, Goldberg L, Meshel T, Rechavi G, Witz IP, Ben-Baruch A. A possible role for CXCR4 and its ligand, the CXC chemokine stromal cell-derived factor-1, in the development of bone marrow metastases in neuroblastoma. J Immunol 2001;167: 4747–57.
170. Oh JW, Drabik K, Kutsch O, Choi C, Tousson A, Benveniste EN. CXC chemokine receptor 4 expression and function in human astroglioma cells. J Immunol 2001;166:2695–704.
171. Zhou Y, Larsen PH, Hao C, Yong VW. CXCR4 is a major chemokine receptor on glioma cells and mediates their survival. J Biol Chem 2002;277:49481–87.
172. Sehgal A, Keener C, Boynton AL, Warrick J, Murphy GP. CXCR-4, a chemokine receptor, is overexpressed in and required for proliferation of glioblastoma tumor cells. J Surg Oncol 1998;69:99–104.
173. Han Y, He T, Huang DR, Pardo CA, Ransohoff RM. TNF-alpha mediates SDF-1 alpha-induced NF-kappa B activation and cytotoxic effects in primary astrocytes. J Clin Invest 2001;108:425–35.
174. Barbero S, Bonavia R, Bajetto A, Porcile C, Pirani P, Ravetti JL, Zona GL, Spaziante R, Florio T, Schettini G. Stromal cell-derived factor 1alpha stimulates human glioblastoma cell growth through the activation of both extracellular signal-regulated kinases 1/2 and Akt. Cancer Res 2003;63:1969–74.
175. Zhang J, Sarkar S, Yong VW. The chemokine stromal cell derived factor-1 (CXCL12) promotes glioma invasiveness through MT2-matrix metalloproteinase. Carcinogenesis 2005;26:2069–77.
176. Muller A, Homey B, Soto H, Ge N, Catron D, Buchanan ME, McClanahan T, Murphy E, Yuan W, Wagner SN, Barrera JL, Mohar A, Verastegui E, Zlotnik A. Involvement of chemokine receptors in breast cancer metastasis. Nature 2001;410:50–56.
177. Scotton CJ, Wilson JL, Scott K, Stamp G, Wilbanks GD, Fricker S, Bridger G, Balkwill FR. Multiple actions of the chemokine CXCL12 on epithelial tumor cells in human ovarian cancer. Cancer Res 2002;62:5930–38.

III. IMMUNE RESPONSE IN THE BRAIN

Cytokines and the Brain
Edited by C. Phelps and E. Korneva

Immune and Inflammatory Responses in the Central Nervous System: Modulation by Astrocytes

MILENA PENKOWA[1], JUAN HIDALGO[2], and MICHAEL ASCHNER[3]

[1]*Department of Medical Anatomy, The Panum Institute, University of Copenhagen, DK-2200, Copenhagen, Denmark;*
[2]*Institute of Neurosciences and Department of Cellular Biology, Physiology and Immunology, Animal Physiology Unit, Faculty of Sciences, Autonomous University of Barcelona, Bellaterra, Barcelona, Spain 08193;*
[3]*Department of Pediatrics, Vanderbilt University Medical Center, Nashville, TN 37232, USA*

ABSTRACT

Beyond their long-recognized support functions, astrocytes are active partners of neurons in processing information, synaptic integration, and production of trophic factors, just to name a few. Both microglia and astrocytes produce and secrete a number of cytokines, modulating and integrating the communication between hematogenous cells and resident cells of the central nervous system (CNS). This review will address (1) the functions of astrocytes in the normal brain and (2) their role in surveying noxious stimuli within the brain, with particular emphasis on astrocytic responses to damage or disease, a process referred to as reactive astrogliosis/ astrocytosis. In addition, the review will discuss (3) the role of astrocytes as an abundant cellular source for immunoregulatory (cytokines) factors, and their fundamental roles in the type and extent of CNS immune and inflammatory responses. (4) Recent experimental evidence on the role of astroglia in the etiology of neurological diseases will be highlighted, along with (5) the role of oxidative stressors generated within astrocytes in this process.

1. INTRODUCTION

A tacit concept unique to the nervous system is that its functions are overwhelmingly due to the properties of its electrically excitable cells, the neurons. However, there is an even more numerous class of nonexcitable cells in the nervous system, collectively referred to as the neuroglia. These comprise the astroglia or astrocytes, the oligodendroglia or oligodendrocytes, and the microglia. Elucidation of the true functional nature of the neuroglia depended on advances in histological staining by Golgi and Ramony Cajal around 1870 and 1890, respectively. The term neuroglia had come from an essentially erroneous concept by Virchow, which he put forward in 1850, namely that neurons were embedded in a connective tissue to which he gave the name neuroglia, or nerve glue. Although erroneous, this term has persisted as the preferred generic term for these cells, or in its shortened form of "glia."

Astrocytes are ubiquitous and are the predominant glial cell in the brain, where they have many functions, both during health and disease. Until recently, astrocytes were considered to be supportive cells, which mainly formed a histoanatomical part of the blood-brain barrier (BBB). However, new data have emerged showing that astrocytes have essential physiological roles in monitoring and modulating the neuronal microenvironment and in information processing or signaling in the brain. Astrocytes produce trophic factors and cytokines, regulate neurotransmitter and ion concentrations, and remove toxins and debris from the extracellular fluid (ECF), thus maintaining an extracellular milieu that is optimally suited for neuronal function. As established over the last few decades, astrocytic functional impairments, as well as physiological reactions of astrocytes to injury, have the potential to induce and/or exacerbate neuronal dysfunction.

This review considers contemporary evidence provoking reformulation of concepts of the interdependence of immune responses in the central nervous system (CNS), as well as the role of astrocytes in the etiology of CNS autoimmune disorders. There is also a brief overview of astrocytic functions, which features the array of astrocytic functions that serve either to protect or to amplify cell injury, thus also potentially contributing to neurodegenerative disorders.

2. ASTROCYTES IN THE NORMAL BRAIN

Astrocytes are comprised of two glial cell types: stellar fibrillary astroglia and protoplasmatic astroglia. The stellar-fibrillary astrocytes have elongated thin cell processes and are mainly seen in the white matter [1]. The protoplasmatic astrocytes reveal several short and ramified cell processes and they are mainly seen in the gray matter [2,3]. Moreover, in the molecular layer of the cerebellum, an intermediate astroglial phenotype, the Bergmann glia, exists [4]. In addition, tanycytes that are localized in the periependymal area of the central canal and the third ventricle are usually considered as a specialized and distinct type of astroglia [5].

The modern experimental approach to astrocytes is generally considered to have started with the pioneering electrophysiological studies of Kuffler and his colleagues in the 1960s on astrocyte-like glial cells in primitive animals, such as the leech and in lower vertebrates [6]. Kuffler's work was most instrumental in disposing of the last remnant of the old speculative approach, which had led electron microscopists and others to propose that astrocytes formed the extracellular space of the brain, since they were seen by electron microscopy as "enlarged watery structures" with a seemingly absent extracellular space. The work of Kuffler and his associates clearly showed that for the glia in the nervous system of leech and the optic nerve of amphibians, and by implication for other glial cells, this concept was incorrect. The amphibian glial cells were found to have a normal high intracellular K^+ and, in fact, to be characterized by a membrane potential, which was essentially the same as the Nernst potential for K^+ (e.g., -80 to -90 mV intracellular negative). Later work on "electrically silent" cells in the mammalian CNS identified putative glial cells, which had essentially the same characteristic, that is, a nonexcitable cell with large negative membrane potentials and apparently only sensitive to changes in $[K^+]o$. This data led to one of the earliest functions proposed for glial cells, namely the control of extracellular K^+ by glial cells, particularly astrocytes for taking up K^+ released by active neurons [6]. Moreover, the finding of an essentially pure K^+-dependent membrane potential implied that the cell membranes were impermeant to sodium and possibly chloride.

In addition to the role of astrocytes in K^+ buffering (see above), they also have diverse housekeeping functions. Astroglial cells influence the formation and maintenance of the BBB, which limits the entry of potentially harmful molecules into the brain, whereas the entry of essential nutrients into the CNS is facilitated [7] by the presence of numerous transporters within capillary endothelia. Astrocytes are also responsible for the induction of tight junctions between the endothelial cells of the BBB. These junctions (zonula occludens) are characterized by a high electrical resistance (around 2000 ohm/cm2), indicative of a low conductance to even small ions. The importance of astrocytes in the maintenance of the BBB is demonstrated by the failure of barrier repair and the appearance of vasogenic edema following astrocyte ablation [8]. Communication between astrocytes and endothelial cells via nitric oxide diffusion is a likely contributing factor in controlling the blood flow [9]. In addition, astrocytes provide metabolites to neurons [4], and neurotransmitter-mediated metabolic coupling between astrocytes and neurons may be associated with lactate release from astrocytes for utilization as an energy substrate in neurons [10,11]. Moreover, astrocytes help to preserve the stability of the extracellular environment by regulating extracellular pH and removing neurotransmitters, excess ions, ammonia, free radicals, and water from the brain microenvironment [10,12,13]. One of the best-characterized examples of this type of astrocytic system is the uptake of the excitatory transmitter glutamate from the synaptic cleft by astrocytes. This task is facilitated by specific glutamate transporter subtypes, such as GLAST and GLT1 [13]. It is estimated that approximately 80% of synaptic glutamate is removed by astrocytes. Consequently, astrocytes have a major role in the prevention of glutamate-induced neurotoxicity, and in addition provide the neurons with glutamine, the precursor for glutamate synthesis. In addition, astroglia preserve neuronal homeostasis by providing neurotrophic and protective factors [4,14], contributing to neuronal function and survival. Neurons and astrocytes also establish contact via electrical coupling through gap junctions, both in vitro and in vivo, suggesting the possible existence of rapid regulatory cross talk networks between neurons and astrocytes during synaptic transmission. Finally, astrocytes contribute prominently to neuronal development, providing cues and signals that are necessary for optimal neuroblast migration or neurite outgrowth and guidance [15,16].

3. REACTIVE ASTROGLIOSIS

The large numbers and the branching-process morphology of astrocytes make them ideally suited to survey the CNS for noxious stimuli. Astrocytes react profoundly to brain damage or disease by increasing both their number and size, a process referred to as reactive astrogliosis/astrocytosis. Reactive astrogliosis represents a remarkably homotypic response of astrocytes to all types of injuries of the CNS. In this capacity, the astrocytes function as microsensors of the injured microenvironment regardless of their location in the CNS. However, evidence has accumulated that there may be a heterogeneous response of the astrocyte, depending upon the nature of the injury (i.e., anisomorphic gliosis – response in an open injury – versus isomorphic gliosis) and their distance to the lesion site [14,17].

The morphological changes characteristics of reactive astrocytes reflect a significant increase in their metabolism and protein synthesis in order to compensate for increased (patho)physiological demands. In addition, the cell hypertrophy of reactive astrocytes is reflective of their increased content in intracellular filaments and cytoskeletal proteins [14], which stabilizes the brain tissue around the injury; the cell cytoplasm becomes packed with glial fibrillary acidic protein (GFAP)-positive intermediate filaments, and GFAP immunostaining is a commonly used marker for reactive astrocytes.

During pathological conditions, reactive astrocytes have numerous and diverse functions. A major role of these cells is to initiate and regulate CNS inflammatory and immune response. Reactive astroglia express and secrete important neuroimmunological factors, such as adhesion molecules, major histocompatibility complex (MHC) class I and II, costimulatory molecules, cytokines, growth factors, acute-phase reactants, signaling molecules, eicosanoids, calcium-binding proteins, enzymes, recognition and transport factors, receptor molecules, and antioxidant proteins; it is also likely that reactive astrocytes participate in antigen presentation in the CNS [14,18–22]. Reactive astrocytes also modulate the CNS microenvironment by producing and secreting molecules of the extracellular matrix [12,18]. In addition, reactive astrogliosis may regulate the elimination of toxic compounds produced by either tissue damage, injured cells and degenerating neurons, or the influx of substances resulting from a breakdown of the BBB. Moreover, reactive astrocytes will eventually replace microglia from the surface of neurons, after which the astroglia enwrap damaged neurons with thin sheet-like lamellar cell processes [23]. These flat lamellar processes adhere to each other, creating a well-organized multilayered stack of astroglial lamellae encircling the neurons, a process generally referred to as synaptic stripping because of the attendant loss of synapses. The astrocytes retract their lamellar cell processes prior to neuroregeneration, and the neuronal surfaces eventually reestablish synapses [24].

Following a traumatic or direct injury, an especially pronounced reactive astrocytosis process is noted, where astrocytes contribute to the formation of a glial scar [18,25,26]. The glial scar consists of a dense network of reactive astrocytes, which display "chunky" interdigitating cell processes, as well as mesodermal cells such as fibroblasts, endothelial and hematogenous cells, and microglia/macrophages, in addition to an associated extracellular tissue matrix, the latter showing a dramatic change in its composition [14,18,27].

Central nervous system neurons do not effectively regenerate following injury, and the glial scar is thought to constitute a major barrier to this process. Nonetheless, a few axons may bypass the lesion scar by growing through adjacent connective tissues or vasculature or bridges of surviving parenchymal tissues [28]. The glial scar impediment to neurodegeneration does not appear to be intrinsic to neurons, but rather to the microenvironment of the CNS, where astrocytes appear to play a prominent role in its formation [18,29]. The reasons for the inhibitory nature of the astrocytic scar remain unclear, but likely involve the production of growth-inhibitory factors as well as the formation of a frank physical barrier [27].

Despite this long-held consideration of the glial scar in general, and astrocytes in particular as an impediment to neuronal regrowth and sprouting following injury, a number of recent studies suggest that astrocytes themselves are not necessarily inhibitory and that the formation of a glial barrier around the injured necrotic area could be beneficial, in that it separates and isolates the still unharmed brain tissue and thereby protects this tissue from secondary damage; furthermore, some data have clearly indicated that reactive astrocytes may in fact promote neuronal regeneration and sprouting [14,30,31]. In agreement with this, attempts that prevent reactive astrogliosis have not been successful in promoting neuron regrowth, as well as brain repair [12,14,18,25,26]. Furthermore, factors that increase reactive astrocytes during certain conditions appear to stimulate brain tissue repair and wound healing [12,25,33–35]. Moreover, recent results suggest that formation of an astroglial scar does not represent a barrier to regrowing axons, and that it is instead the lesion-induced basal membrane, which is the primary factor in the neurite scar impermeability [27].

4. ASTROCYTES AND CYTOKINES

New insight into astrocytic and microglial cell function has necessitated the revision of the long-standing view that the CNS is insulated from the effects of the immune system. As early as the late 1980s, the results of in vitro studies led to postulates that astrocytes might function as antigen-presenting cells (APCs), namely those cells with the ability to present antigens to lymphocytes [36].

4.1. Immunoregulatory molecules regulate central nervous system inflammation and reactive astrocytosis

Thus, reactive astroglia can secrete and respond to a wide variety of cytokines involved in immune responses [18,19,21,22,32,37,38]. Astrocytes are a cellular source for both pro- and anti-inflammatory cytokines in the CNS such as interleukin-1α,β (IL-1α,β), IL-6, IL-10, IL-12, tumor necrosis factor (TNF), colony-stimulating factors (CSFs: M-CSF, GM-CSF, and G-CSF), interferon (IFN)-α,β, IFN-γ, chemokines (RANTES, monocyte chemoattractant protein-1, IL-8, IFN-γ-inducible protein-10), and transforming growth factor-β (TGF-β) [13,14, 21,22,26,39,40]. As an abundant cellular source for these immunoregulatory factors, astrocytes have fundamental roles in the type and extent of CNS immune and inflammatory responses. Accordingly, in the CNS, astrocytes are important for activating other immune-responsive cells, such as monocytes/macrophages/microglia and lymphocytes [4,14,19].

Increased expression of these proinflammatory factors is routinely noted during neuropathological processes [19,21,38]. The inflammatory reaction is important for coping with CNS damage, as illustrated by means of mice with deficiency of proinflammatory cytokines such as IL-1, IL-6, TNF-α, and M-CSF [4,24]. For example, when mice with genetic IL-6 deficiency are exposed to CNS damage, the inflammatory responses of astroglia, macrophages/microglia, and lymphocytes are significantly hampered, as reflected in impaired defense reactions, and consequent increased brain tissue damage and cell (particularly neuronal) death [40-45]. However, discrepancies do exist in the literature regarding the role of IL-6 in genetically modified mice [46,47]. Some of the latter may be attributed to the utilization of different paradigms of experimental injury models. It is also likely that such discrepancies are related to the fact that many cytokines and the inflammation itself [48] may represent a process with both positive and negative contributions, as demonstrated by using transgenic mice with astrocyte-targeted expression of IL-6, IL-3, IFN-α, and TNF [49,50]. These mice show chronic neuroinflammation and neurodegeneration, and in the case of the IL-3 and TNF transgenic mice, demyelination and progressive motor disease are also seen [49,50]. In agreement with these in vivo models, it has been demonstrated that IL-6 protects neurons against ischemic damage and excitotoxicity [38-51]; yet, it is also largely responsible for the neuropathology associated with the multiple sclerosis (MS; Fig. 1) animal model, experimental autoimmune encephalomyelitis (EAE), since IL-6 KO mice are completely resistant to EAE [52-55]. Administration of the superantigen staphylococcal enterotoxin B, which acts through a TNF-receptor pathway, overcomes this resistance, but IL-6 is still considered crucial in perpetuating the disease [56]. It has also been shown that TNF mediates damage to myelin and oligodendrocytes [57,58], and that immunotherapeutic inhibition of TNF may be beneficial during demyelinating diseases, such as MS (see Fig. 1) and

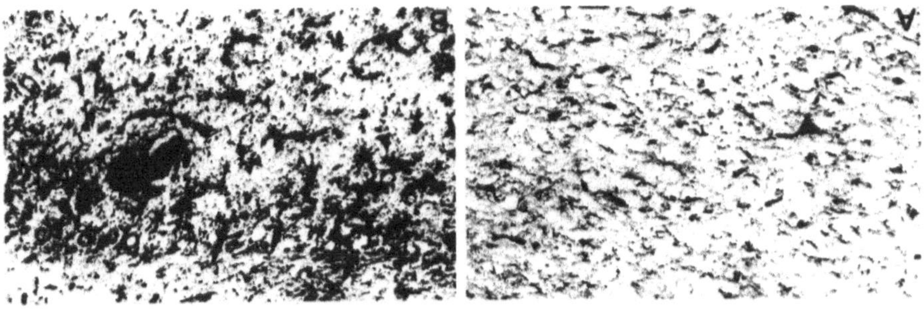

Figure 1. Astrocytes are shown by GFAP immunoreactivity. (A) In sections from healthy control patients, GFAP expression was seen in some scattered astrocytes throughout the CNS. (B) In MS patients with MS lesions, GFAP immunostaining is increased relative to controls and shows reactive astrocytes in the MS lesions in the white matter. Reactive astrogliosis is characterized by hypertrophy and hyperplasia.

EAE [58,59]. Again, results obtained in genetically modified mice have not only convincingly demonstrated an essential role of TNF during EAE, but also provided evidence of a dual role for this cytokine with both anti- and proinflammatory properties in the CNS [60]. In addition, other cytokines, such as IL-2 and IFN-γ, may also inhibit oligodendrocytes [13]. Accordingly, the astrocytic production of proinflammatory cytokines can be a beneficial defense reaction that is associated with the attenuation of CNS damage, but in the case of prolonged and/or large quantitative increase of expression, these cytokines may induce chronic inflammation and neurodegeneration. Therefore, it is crucial that molecules that directly downregulate the inflammatory machinery also be produced and released into the ECF, as to maintain optimal CNS homeostasis. Interestingly, astrocytes are an important source of such anti-inflammatory cytokines and growth factors [14,18,21,26,39,61,62], which likely contribute to the neuroprotective and regenerative functions of astrocytes. In agreement with this, astrocytes are important for brain tissue repair and neuroregeneration during and after pathological conditions of the CNS. Thus, in situations where astrocyte function is compromised, related brain damage and neuronal dysfunction may increase [12,18,25,26,32,37]. Consequently, astroglia have major functions during both the acute phase and later stages of CNS inflammation [4,32].

Astrocytes that are isolated and cultured from Lewis rat brains express much higher levels of constitutive Ia than astrocytes cultured from Brown Norway rats. Hyperinduction of Ia by the susceptible Lewis rats was also shown to be astrocyte-specific, since peritoneal macrophages of susceptible and resistant strains showed the same Ia induction profile. In addition, the structural damage caused by EAE is greatly lessened by antibodies to the class II molecules [32].

Given that the expression of MHC type II antigens is common both to astrocytes and to microglia, the debate regarding which cell type represents "real" APCs continues. Several studies favor that astrocytes are more significant than microglia in this regard. First, astrocytes in the adult CNS far outnumber the microglia and are therefore more immediately accessible for antigen-presenting functions. Second, in vitro studies demonstrate that the amount of Ia expressed on IFN-γ-treated astrocytes correlates with the in vivo susceptibility of the CNS to immune-mediated encephalitis (see above). Furthermore, given recent studies that corroborate the ability of astrocytes to produce and release numerous cytokines in vivo (see above section on astrocytes and cytokines) [13,14,21,26], a substantial body of evidence favors their potential involvement in antigen presentation.

5. ASTROCYTES AND OXIDATIVE STRESS

Central nervous system inflammation and immune responses normally result in increased levels of free radicals or reactive oxygen species (ROS) [21,63–65]. Reactive oxygen species are highly toxic in the CNS, and when production of ROS is increased to a degree that overcomes the neutralizing effects of endogenous antioxidants, a process that is generally referred to as oxidative stress occurs. Reactive oxygen species cause neurotoxicity, neurodegeneration, and cell death. Oxidative stress is a hallmark of neurodegenerative diseases, such as Parkinson's disease, Alzheimer's disease and amyotrophic lateral sclerosis, and other neurological disorders, such as ischemia and trauma [66,67]. Cells can detoxify ROS and thereby inhibit oxidative stress by means of the endogenous antioxidant defense system and/or diet-dependent factors that includes the following: antioxidant factors such as glutathione (GSH), cysteine, ascorbate, α-tocopherol, ubiquinol, carotenoids and flavonoids, and nonenzymatic and enzymatic proteins such as metallothionein (MT), Cu, Zn-superoxide dismutase (Cu, Zn-SOD), Mn-SOD, catalase, xanthine oxidase, glutathione reductase, and glutathione peroxidase [67,68]. The role of astrocytes in maintaining optimal levels of GSH is best exemplified by in vitro studies where the maintenance of adequate intracellular GSH levels in neurons for extended periods of time occurs only when they are cocultured with astrocytes. In addition to releasing cysteine for neuronal GSH biosynthesis, astrocytes also release the dipeptide cysteine-glycine (CysGly). Glutathione from astrocytes serves as a substrate for the astrocytic ectoenzyme γ-glutamyltranspeptidase (γ-GT) [69]. The product CysGly, thus formed, serves as a precursor for the synthesis of neuronal GSH either through the direct uptake of this dipeptide or through cysteine and glycine in the ECF.

Nevertheless, the CNS contains relatively low levels of antioxidants despite the fact that ROS production is relatively high [67]. Interestingly, among cells of the CNS, astrocytes are the main cell source of antioxidant defense systems [40,69,70] and accordingly, astrocytes are essential for protecting other neural cell types against toxicity from ROS. Hence, oligodendroglia contain less than one-half the content of reduced GSH compared to astrocytes. Also, astrocytes are the principal CNS cell source of MT [34,43,61], which is an extraordinarily efficient free radical scavenger and antioxidant [61,68,71–78]. Moreover, MT protects against ROS with much higher molar efficiency (almost 800-fold) relative to GSH, and MT can quench ROS with an affinity more than 300-fold higher than GSH [78]; furthermore, MT can functionally substitute for Cu, Zn-SOD in the cellular defense against oxidative stress [79]. The oxidative metal release from MT appears to be controlled by zinc-thiol/disulfide exchange, hence GSH redox status may directly or indirectly affect zinc metabolism and MT homeostasis [80]. Thus, the antioxidative effects of MT are extremely potent, and in agreement with this, reduced MT levels during neuropathology are followed by significantly increased ROS formation and oxidative stress, as well as increased neurodegeneration and cell death [25,26,40,59,74,81]. In support of these findings, increases in endogenous MT expression in transgenic mice and treatment of animals with exogenous MT during pathological conditions of the CNS are followed by significantly reduced tissue damage, oxidative stress, and cell death [33,34,43,59,82]. Furthermore, treating rats and mice with MT during CNS pathology significantly decreases the mortality and clinical symptoms of the disease [34,43]. Accordingly, MT treatment could be a promising new therapy for neuropathological conditions.

ACKNOWLEDGMENTS

This is review was supported in part by Public Health Service Grant NIEHS 07331 and 10563 (MA), the Ministerio de Ciencia y Tecnología and Feder (SAF2002-01268) and Direcció General de Recerca (2001SGR 00203) (JH).

REFERENCES

1. Bignami A, Eng LF, Dahl D, Uyeda CT. Localization of the glial fibrillary acidic protein in astrocytes by immunofluorescence. Brain Res 1972;43:429–35.
2. Cammer W, Tansey FA. Carbonic anhydrase immunostaining in astrocytes in the rat cerebral cortex. J Neurochem 1988;50:319–22.
3. Didier M, Harandi M, Aguera M, Bancel B, Tardy M, Fages C, Calas A, Stagaard M, Mollgard K, Belin MF. Differential immunocytochemical staining for glial fibrillary acidic (GFA) protein, S-100 protein and glutamine synthetase in the rat subcommissural organ, nonspecialized ventricular ependyma and adjacent neuropil. Cell Tissue Res 1986; 245:343–51.
4. Raivich G, Bohatschek M, Kloss CU, Werner A, Jones LL, Kreutzberg GW. Neuroglial activation repertoire in the injured brain: Graded response, molecular mechanisms and cues to physiological function. Brain Res Rev 1999;30:77–105.
5. Flament-Durand J, Brion JP. Tanycytes: Morphology and functions: A review. Int Rev Cytol 1985;96:121–55.
6. Kuffler SW, Nicholls JG, Orkand RK. Physiological properties of glial cells in the CNS system of amphibia. J Neurophysiol 1966;29:768–87.
7. Aschner M. Neuron-astrocyte interactions: Implications for cellular energetics and antioxidant levels. Neurotoxicology 2000;21:1101–7.
8. Bush TG, Puvanachandra N, Horner CH, Polito A, Ostenfeld T, Svendsen CN, Mucke L, Johnson MH, Sofroniew MV. Leukocyte infiltration, neuronal degeneration, and neurite outgrowth after ablation of scar-forming, reactive astrocytes in adult transgenic mice. Neuron 1999;23:297–308.
9. Mi H, Haeberle H, Barres BA. Induction of astrocyte differentiation by endothelial cells. J Neurosci 2001;21:1538–47.
10. Tsacopoulos M, Magistretti PJ. Metabolic coupling between glia and neurons. J Neurosci 1996;16:877–85.
11. Waagepetersen HS, Bakken IJ, Larsson OM, Sonnewald U, Schousboe A. Comparison of lactate and glucose metabolism in cultured neocortical neurons and astrocytes using 13C-NMR spectroscopy. Dev Neurosci 1998;20:310–20.
12. Aschner M, Allen JW, Kimelberg HK, LoPachin RM, Streit WJ. Glial cells in neurotoxicity development. Annu Rev Pharmacol Toxicol 1999;39:151–73.
13. Benveniste EN. Cytokines: Influence on glial cell gene expression and function. Chem Immunol 1997;69:31–75.
14. Ridet JL, Malhotra SK, Privat A, Gage FH. Reactive astrocytes: Cellular and molecular cues to biological function. Trends Neurosci 1997;20:570–7.
15. Nakashima K, Wiese S, Yanagisawa M, Arakawa H, Kimura N, Hisatsune T, Yoshida K, Kishimoto T, Sendtner M, Taga T. Developmental requirement of gp130 signaling in neuronal survival and astrocyte differentiation. J Neurosci 1999;19:5429–34.

16. Liu Y, Wu Y, Lee JC, Xue H, Pevny LH, Kaprielian Z, Rao MS. Oligodendrocyte and astrocyte development in rodents: An in situ and immunohistological analysis during embryonic development. Glia 2002;40:25–43.
17. Fernaud-Espinosa I, Nieto-Sampedro M, Bovolenta P. Differential activation of microglia and astrocytes in aniso- and isomorphic gliotic tissue. Glia 1993;8:277–91.
18. Eddleston M, Mucke L. Molecular profile of reactive astrocytes – Implications for their role in neurologic disease. Neuroscience 1993;54:15–36.
19. Hopkins SJ, Rothwell NJ. Cytokines and the nervous system I: Expression and recognition. Trends Neurosci 1995;18:83–8.
20. Murphy S (Ed.). Astrocytes: Pharmacology and Function. New York: Academic, 1995.
21. Xiao BG, Link H. Immune regulation within the central nervous system. J Neurol Sci 1998;157:1–12.
22. Dong Y, Benveniste EN. Immune function of astrocytes. Glia 2001;36:180–90.
23. Graeber MB, Kreutzberg GW. Delayed astrocyte reaction following facial nerve axotomy. J Neurocytol 1988;17:209–20.
24. Raivich G, Jones LL, Werner A, Bluthmann H, Doetschmann T, Kreutzberg GW. Molecular signals for glial activation: Pro- and anti-inflammatory cytokines in the injured brain. Acta Neurochir Suppl (Vienna) 1999;73:21–30.
25. Penkowa M, Carrasco J, Giralt M, Moos T, Hidalgo J. CNS wound healing is severely depressed in metallothionein-I and II-deficient mice. J Neurosci 1999;19:2535–45.
26. Penkowa M, Carrasco J, Giralt M, Molinero A, Hernández J, Campbell IL, Hidalgo J. Altered central nervous system cytokine-growth factor expression profiles and angiogenesis in metallothionein-I+II deficient mice. J Cereb Blood Flow Metab 2000;20: 1174–89.
27. Stichel CC, Muller HW. The CNS lesion scar: New vistas on an old regeneration barrier. Cell Tissue Res 1998;294:1–9.
28. Clemente CD. Structural regeneration in the mammalian central nervous system and the role of neuroglia and connective tissue. In Regeneration in the Nervous System. Windle WF, Ed.; Springfield, IL: Charles C. Thomas, 1955; pp. 147–61.
29. Costa S, Planchenault T, Charriere-Bertrand C, Mouchel Y, Fages C, Juliano S, Lefrancois T, Barlovatz-Meimon G, Tardy M. Astroglial permissivity for neuritic outgrowth in neuron-astrocyte cocultures depends on regulation of laminin bioavailability. Glia 2002; 37:105–13.
30. Hatten ME, Liem RK, Shelanski ML, Mason CA. Astroglia in CNS injury. Glia 1991; 4:233–43.
31. Kawaja MD, Gage FH. Reactive astrocytes are substrates for the growth of adult CNS axons in the presence of elevated levels of nerve growth factor. Neuron 1991;7:1019–30.
32. Aschner M. Astrocytes as mediators of immune and inflammatory responses in the CNS. Neurotoxicology 1998;19:269–81.
33. Giralt M, Penkowa M, Lago N, Molinero A, Hidalgo J. Metallothionein-1+2 protect the CNS after a focal brain injury. Exp Neurol 2001;173:114–28.
34. Penkowa M, Hidalgo J. Metallothionein I+II expression and their role in experimental autoimmune encephalomyelitis. Glia 2000;2:247–63.
35. Swartz KR, Liu F, Sewell D, Schochet T, Campbell I, Sandor M, Fabry Z. Interleukin-6 promotes post-traumatic healing in the central nervous system. Brain Res 2001;896: 86–95.
36. Erb P, Kennedy M, Hagmann I, Wassmer P, Huegli G, Fierz W, Fontana A. Accessory cells and the activation and expression of different T cell functions. In Regulation of Immune

Gene Expression. Feldmann M and McMichael A, Eds; Clifton, NJ: The Humana Press, 1986; pp. 187.
37. Aschner M. Astrocytic functions and physiological reactions to injury: The potential to induce and/or exacerbate neuronal dysfunction – A forum position paper. Neurotoxicology 1998;19:7–17.
38. Van Wagoner NJ, Benveniste EN. Interleukin-6 expression and regulation in astrocytes. J Neuroimmunol 1999;100:124–39.
39. Ghirnikar RS, Lee YL, Eng LF. Inflammation in traumatic brain injury: Role of cytokines and chemokines. Neurochem Res 1998;23:329–40.
40. Penkowa M, Giralt M, Carrasco J, Hadberg H, Hidalgo J. Impaired inflammatory response and increased oxidative stress and neurodegeneration after brain injury in interleukin-6 deficient mice. Glia 2000;32:271–85.
41. Klein MA, Moller JC, Jones LL, Bluethmann H, Kreutzberg GW, Raivich G. Impaired neuroglial activation in interleukin-6 deficient mice. Glia 1997;19:227–33.
42. Penkowa M, Hidalgo J. IL-6 deficiency leads to reduced metallothionein I+II expression and increased oxidative stress in the brain stem after 6-aminonicotinamide treatment. Exp Neurol 2000;163:72–84.
43. Penkowa M, Molinero A, Carrasco J, Hidalgo J. IL-6 deficiency reduces the inflammatory response and increases oxidative stress and neurodegeneration after kainic acid-induced seizures. Neuroscience 2001;102:805–18.
44. Galiano M, Liu ZQ, Kalla R, Bohatschek M, Koppius A, Gschwendtner A, Xu S, Werner A, Kloss CU, Jones LL, Bluethmann H, Raivich G. Interleukin-6 (IL6) and cellular response to facial nerve injury: Effects on lymphocyte recruitment, early microglial activation and axonal outgrowth in IL6-deficient mice. Eur J Neurosci 2001;14:327–41.
45. Pavelko KD, Howe CL, Drescher KM, Gamez JD, Johnson AJ, Wei T, Ransohoff RM, Rodriguez M. Interleukin-6 protects anterior horn neurons from lethal virus-induced injury. J Neurosci 2003;23:481–92.
46. Clark WM, Rinker LG, Lessov NS, Lowery SL, Cipolla MJ. Efficacy of antioxidant therapies in transient focal ischemia in mice. Stroke 2001;32:1000–4.
47. Morganti-Kossmann MC, Rancan M, Stahel PF, Kossmann T. Inflammatory response in acute traumatic brain injury: A double-edged sword. Curr Opin Crit Care 2002;8:101–5.
48. Wyss-Coray T, Mucke L. Inflammation in neurodegenerative disease – A double-edged sword. Neuron 2002;35:419–32.
49. Campbell IL. Structural and functional impact of the transgenic expression of cytokines in the CNS. Ann N Y Acad Sci 1998;840:83–96.
50. Campbell IL. Transgenic mice and cytokine actions in the brain: Bridging the gap between structural and functional neuropathology. Brain Res Rev1998;26:327–36.
51. Gruol DL, Nelson TE. Physiological and pathological roles of interleukin-6 in the central nervous system. Mol Neurobiol 1997;15:307–39.
52. Mendel I, Katz A, Kozak N, Ben-Nun A, Revel M. Interleukin-6 functions in autoimmune encephalomyelitis: A study in gene-targeted mice. Eur J Immunol 1998;28:1727–37.
53. Samoilova EB, Horton JL, Hilliard B, Liu TS, Chen Y. IL-6-deficient mice are resistant to experimental autoimmune encephalomyelitis: Roles of IL-6 in the activation and differentiation of autoreactive T cells. J Immunol 1998;161:6480–6.
54. Eugster HP, Frei K, Kopf M, Lassmann H, Fontana A. IL-6-deficient mice resist myelin oligodendrocyte glycoprotein-induced autoimmune encephalomyelitis. Eur J Immunol 1998;28:2178–87.

55. Okuda Y, Sakoda S, Bernard CC, Fujimura H, Saeki Y, Kishimoto T, Yanagihara T. IL-6-deficient mice are resistant to the induction of experimental autoimmune encephalomyelitis provoked by myelin oligodendrocyte glycoprotein. Int Immunol 1998;10:703–8.
56. Eugster HP, Frei K, Winkler F, Koedel U, Pfister W, Lassmann H, Fontana A. Superantigen overcomes resistance of IL-6-deficient mice towards MOG-induced EAE by a TNFR1 controlled pathway. Eur J Immunol 2001;31:2302–12.
57. Probert L, Akassoglou K, Pasparakis M, Kontogeorgos G, Kollias G. Spontaneous inflammatory demyelinating disease in transgenic mice showing central nervous system-specific expression of tumor necrosis factor alpha. Proc Natl Acad Sci USA 1995;92:11294–8.
58. Selmaj KW, Raine CS. Tumor necrosis factor mediates myelin and oligodendrocyte damage in vitro. Ann Neurol 1988;23:339–46.
59. Penkowa M, Hidalgo J. Metallothionein treatment reduces proinflammatory cytokines IL-6 and TNF-α and apoptotic cell death during experimental autoimmune encephalomyelitis (EAE). Exp Neurol 2001;170:1–14.
60. Probert L, Eugster HP, Akassoglou K, Bauer J, Frei K, Lassmann H, Fontana A. TNFR1 signalling is critical for the development of demyelination and the limitation of T-cell responses during immune-mediated CNS disease. Brain 2000;123:2005–19.
61. Aschner M. Astrocyte metallothioneins (MTs) and their neuroprotective role. Ann N Y Acad Sci 1997;825:334–47.
62. Yoshida K, Toya S. Neurotrophic activity in cytokine-activated astrocytes. Keio J Med 1997;46:55–60.
63. Floyd RA. Neuroinflammatory processes are important in neurodegenerative diseases: An hypothesis to explain the increased formation of reactive oxygen and nitrogen species as major factors involved in neurodegenerative disease development. Free Radical Biol Med 1999;26:1346–55.
64. Floyd RA. Antioxidants, oxidative stress, and degenerative neurological disorders. Proc Soc Exp Biol Med 1999;222:236–45.
65. Floyd RA, Hensley K, Jaffery F, Maidt L, Robinson K, Pye Q, Stewart C. Increased oxidative stress brought on by pro-inflammatory cytokines in neurodegenerative processes and the protective role of nitrone-based free radical traps. Life Sci 1999;65:1893–9.
66. Lipton P. Ischemic cell death in brain neurons. Physiol Rev 1999;79:1431–568.
67. Halliwell B. Role of free radicals in the neurodegenerative diseases: Therapeutic implications for antioxidant treatment. Drugs Aging 2001;18:685–716.
68. Viarengo A, Burlando B, Ceratto N, Panfoli I. Antioxidant role of metallothioneins: A comparative overview. Cell Mol Biol 2000;46:407–17.
69. Dringen R. Metabolism and functions of glutathione in brain. Prog Neurobiol 2000; 62:649–71.
70. Thorburne SK, Juurlink BH. Low glutathione and high iron govern the susceptibility of oligodendroglial precursors to oxidative stress. J Neurochem 1996;67:1014–22.
71. Aschner M. The functional significance of brain metallothioneins. FASEB J 1996; 10:1129–36.
72. Aschner M. Metallothionein (MT) isoforms in the central nervous system (CNS): Regional and cell-specific distribution and potential functions as an antioxidant. Neurotoxicology 1998;19:653–60.
73. Giralt M, Gasull T, Blanquez A, Hidalgo J. Effect of endotoxin on rat serum, lung and liver lipid peroxidation and on tissue metallothionein levels. Rev Esp Fisiol 1993;49: 73–8.

74. Hidalgo J, Aschner M, Zatta P, Vasak M. Roles of the metallothionein family of proteins in the central nervous system. Brain Res Bull 2001;55:133–45.
75. Hidalgo J, Penkowa M, Giralt M, Carrasco J, Molinero A. Metallothionein expression and oxidative stress in the brain. Meth Enzymol 2002;348:238–49.
76. Molinero A, Carrasco J, Hernandez J, Hidalgo J. Effect of nitric oxide synthesis inhibition on mouse liver and brain metallothionein expression. Neurochem Int 1999;33:559–66.
77. Schwarz MA, Lazo JS, Yalowich JC, Reynolds I, Kagan VE, Tyurin V, Kim YM, Watkins SC, Pitt BR. Cytoplasmic metallothionein overexpression protects NIH 3T3 cells from tert-butyl hydroperoxide toxicity. J Biol Chem 1994;269:15238–43.
78. Thornalley PJ, Vasak M. Possible role for metallothionein in protection against radiation-induced oxidative stress. Kinetics and mechanism of its reaction with superoxide and hydroxyl radicals. Biochim Biophys Acta 1985;827:36–44.
79. Tamai KT, Gralla EB, Ellerby LM, Valentine JS, Thiele DJ. Yeast and mammalian metallothioneins functionally substitute for yeast copper-zinc superoxide dismutase. Proc Natl Acad Sci USA 1993;90:8013–7.
80. Maret W. Oxidative metal release from metallothionein via zinc-thiol/disulfide interchange. Proc Natl Acad Sci USA 1994;91:237–41.
81. Carrasco J, Penkowa M, Hadberg H, Molinero A, Hidalgo J. Enhanced seizures and hippocampal neurodegeneration following kainic acid induced seizures in metallothionein-I+II deficient mice. Eur J Neurosci 2000;12:2311–22.
82. Molinero A, Penkowa M, Hernández J, Camats J, Giralt M, Lago N, Carrasco J, Campbell IL, Hidalgo J. Metallothionein-I overexpression decreases brain pathology in transgenic mice with astrocyte-targeted expression of interleukin 6. J Neuropathol Exp Neurol 2003; 62:315–28.

Edited by C. Phelps and E. Korneva

Immune Response in the Brain: Glial Response and Cytokine Production

AKIO SUZUMURA

Department of Neuroimmunology, Research Institute of Environmental Medicine, Nagoya University, Furo-cho, Chikusa, Nagoya 464-8601, Japan

ABSTRACT

Although the brain has been considered as an immunologically privileged site, the evidence to date suggests that this is no longer the case. Cytokines such as interferon (IFN)-γ, tumor necrosis factor (TNF)-α, and interleukin (IL)-3 induce class I major histocompatibility complex (MHC) antigen expression on neural cells. IFN-γ, the most potent inducer of MHC antigen, also induces class II MHC antigen expression on microglia and astrocytes, which enable them to function as antigen-presenting cells. Thus, in some pathological conditions, invading T cells can interact with neural cells to induce central nervous system (CNS) damage. Glial cells have also been shown to produce various cytokines and chemokines. Almost all cytokines and chemokines known to occur in the immune system are also produced in the CNS. In this chapter, the glial responses contributing to neuroimmune interactions are reviewed, with a focus on production and functions of cytokines in the CNS.

1. INTRODUCTION

The brain has long been considered as an immunologically privileged site based on a large body of evidence: the lack of major histocompatibility complex (MHC) antigen expression on neural cells; the lack of lymphoid drainage in the central nervous system (CNS); and the presence of the blood–brain barrier (BBB), which blocks the invasion of immune cells or high molecular substances including antibodies into the brain. However, as has been shown by research published in the 1980s, some cytokines or viral infections induce MHC antigen expression on both neuronal and glial cells. Interferon-γ (IFN-γ), the most potent inducer of MHC antigen, also induces class II MHC antigen expression on microglia and some populations of astrocytes, which enable them to function as antigen-presenting cells (APCs). In order to effectively present antigens to T cells, APCs have to express other costimulatory molecules. Microglia and astrocytes have been shown to express these costimulatory molecules, and this expression is also enhanced by exposure to IFN-γ. Thus, if activated T cells enter the CNS, either microglia or some populations of astrocytes are able to present CNS antigens to expand a T-cell clone specific for a particular CNS antigen. In fact, it has also been shown that activated T cells can enter the brain through an intact BBB. Consequently, in certain pathological conditions, glial cells may alter their functions to actively interact with immune cells. In most cases these glial changes are mediated by cytokines.

Another remarkable glial response in pathological conditions is the production of cytokines. In the late 1980s, many laboratories, including ours, have demonstrated the production of cytokines by glial cells. Almost all cytokines known to occur in the immune system were produced in the CNS. Thus, the brain should no longer be considered as an immunologically privileged site. In this chapter, I will review the glial responses in neuroimmune interactions in the CNS with a focus on the production and functions of cytokines.

2. GLIAL RESPONSE TO INTERACT WITH IMMUNE CELLS

2.1. Induction of major histocompatibility complex antigens

In normal or unstimulated conditions *in vivo* and *in vitro*, neuronal and glial cells do not usually express class I or class II MHC antigens on their surface, whereas microglia only weakly express class I MHC antigens *in vitro*. Consequently, in the normal brain, neural cells cannot interact with their own immune cells in a specific manner. However, it has been shown that both neuronal and glial cells can be induced to express class I MHC antigens in response to lymphokines [1,2]. As a result of this induction, the cytotoxic T cells acquire the capacity to lyse the CNS cells in a MHC-restricted manner. Although IFN-γ is a principal factor for the induction of MHC antigen expression, tumor necrosis factor-α (TNF-α) can also induce class I MHC antigen expression on astrocytes, but not on oligodendrocytes [3].

Lymphokines, especially IFN-γ, also induce the expression of class II MHC antigens on astrocytes [4] and microglia *in vitro* [5]. This expression is associated with the induction of mRNA for class II MHC antigens. Induction of class II MHC antigens is also observed *in vivo* in certain pathological conditions. In the brains of *experimental allergic encephalomyelitis (EAE)*, microglia near the infiltrating T cells are reported to be class II MHC antigen-positive [6–8], suggesting that a T cell-derived cytokine, most probably IFN-γ, can induce class II MHC antigen expression *in vivo* as well.

After axotomy there are increased numbers, relative to controls, of microglia in and around facial nerve nuclei. Moreover, these cells are reportedly class II MHC antigen-positive [9]. Since the BBB is not damaged in this experimental condition and since there is no definitive evidence that neural cells produce INF-γ in the CNS, it is unlikely that IFN-γ is responsible for the induction of class II MHC antigen expression in this model. Another candidate for the induction of class II MHC antigens in microglia is interleukin (IL)-3. We have shown that IL-3 induces, in a dose-dependent manner, surface expression and mRNA for class II MHC antigens in microglia, which is completely inhibited by anti-IL-3 antibody [10]. Although we do not detect IL-3 or IL-3 mRNA in either microglia or astrocytes in the mouse cellular system under study, it has been reported that rat microglia produce IL-3 *in vitro* [11], and that IL-3 mRNA is detected in some populations of astrocytes and neurons by *in situ* hybridization [12]. Therefore, it is possible that IL-3 derived from degenerating neurons, reactive astrocytes, or microglia *in vivo* may themselves induce class II MHC antigens on microglia in certain pathological conditions.

In contrast to IL-3, granulocyte–macrophage colony-stimulating factor (GM-CSF) down-regulates IFN-γ-induced class II MHC antigen expression in microglia. The suppression occurs in a dose-dependent manner and is neutralized by anti-GM-CSF antibody [10]. As we have shown previously, GM-CSF is produced by astrocytes [13] and induces the proliferation of microglia *in vitro* [14,15]. It is possible that astrocytes downregulate immunoregulatory functions of microglia. However, so far, there is no evidence that GM-CSF contributes to the modulation of microglia proliferation and suppression of their Ia antigen expression in either physiological or pathological conditions *in vivo*.

All the macrophage deactivating cytokines, or inhibitory cytokines, such as IL-10, IL-4, and transforming growth factor-β (TGF-β) downregulate the INF-γ-induced class II MHC antigen expression in microglia [16–18]. As we and other groups have shown, astrocytes and microglia produce IL-10 [18] and TGF-β [19,20], but neither cells produce IL-4, although both cell types express IL-4 receptors [17,21]. Thus, microglia may downregulate their own immunoregulatory functions in an autocrine fashion, or the astrocyte may suppress the functions of microglia in a paracrine manner. It is also possible that invading T helper cells, especially T helper 2 (Th2), may downregulate class II MHC antigen expression in microglia by these inhibitory cytokines.

Induction of MHC antigen expression on glial cells occurs without breakdown of BBB, without invasion of immune cells. We, and another group, have shown that infection with neurotropic corona virus induces class I MHC antigen expression on oligodendrocytes and astrocytes [22,23] and class II MHC antigen on astrocytes [23]. These inductions permit glia to interact with invading immune cells to produce CNS pathology.

3. CYTOKINE PRODUCTION AND EXPRESSION OF RECEPTORS FOR CYTOKINES IN GLIA CELLS.

Microglia produce various cytokines, such as IL-1, IL-1 receptor antagonist (IL-1ra), IL-5, IL-6, IL-10, IL-12, IL-15, IL-18, IL-23, IL-27, TGF-β, TNF-α, IFN-γ, as shown in Table 1, and chemokines [18,19,24–33]. Rat microglia reportedly produce IL-3 in culture [11]. Only a trace amount of IL-1, but not the other cytokines, is detectable in the supernatant of unstimulated microglial culture. However, lipopolysaccharides (LPS), and/or IFN-γ in some cases, induce cytokine production. Since microglia express receptors for most of the cytokines produced (see Table 1), these components may function as an autocrine regulator. They also express receptors for cytokines, which are produced by other cells, but not by themselves, such as IL-2 or GM-CSF. Thus, the latter may function as paracrine mediators (for functions of these cytokines on microglia, refer to our previous review [34]). Unstimulated microglia do not express the IL-2 receptors (IL-2R); however, LPS treatment will induce IL-2R expression on these cells. Moreover, IL-2 can also induce the proliferation of LPS-stimulated microglia [35]. Although IL-2 treatment has been shown to induce the proliferation of oligodendrocytes as well [36], we could not confirm these effects [2]. Microglia also express the receptor for IL-4, the cytokine produced by T cells, but not in the CNS. Thus, IL-4 may be a paracrine mediator exerting its effects only in cases of an inflammatory process occurring in the CNS, but not in the normal brain. We have shown the production of IL-5 and the upregulation by IFN-γ in murine microglia by means of RT-PCR for mRNA expression and the bioassay to assess IL-5 activity [28]. However, since we have not detected IL-5 receptors on neural cells, the functions of IL-5 in the CNS remain to be elucidated. In contrast to murine microglia, Lee et al. [37] failed to detect IL-5 mRNA expression in human microglia as assessed by RT-PCR, while they detected mRNA for the IL-5 receptor.

Astrocytes produce cytokines very similar to those produced by microglia (Table 2). However, microglia, rather than astrocytes, seem to be a principal source of most critical cytokines,

Table 1. Cytokine production and receptor expression in microglia

	Production	Receptor expression
IL-1	Yes	Yes
IL-1ra	Yes	Yes
IL-2	ND	Yes**
IL-3	No (yes)*	Yes
IL-4	No	Yes
IL-5	Yes	No
IL-6	Yes	Yes
IL-7	ND	Yes
IL-10	Yes	Yes
IL-12	Yes	Yes
IL-13	No	Yes
IL-15	Yes	Yes
IL-18	Yes	Yes
IL-23	Yes	ND
IL-27	Yes	ND
TNFα	Yes	Yes
IFNγ	Yes	Yes
TGFβ	Yes	Yes
M-CSF	Yes	Yes
GM-CSF	No	Yes
Chemokines	IL-8, IP-10, MIP-1α,β, MCP-1, RANTES, fractalkine	CCR2, CCR3, CCR5, CXCR4, CX3CR1

ND, not determined; ( ), most probably yes; *, reportedly yes, although we could not confirm in our mouse system; **, inducible.

Table 2. Cytokine production and receptor expression in astrocytes

	Production	Receptor expression
IL-1	Yes	Yes
IL-2	No	No
IL-3	(Yes) no	ND
IL-4	No	Yes
IL-5	Yes	Yes*
IL-6	Yes	Yes
IL-10	Yes	(Yes)
IL-12	No	No
TNFα	Yes	(Yes)
IFNγ	No	(Yes)
TGFβ	Yes	ND
M-CSF	Yes	ND
G-CSF	Yes	ND
GM-CSF	Yes	ND
Chemokines	IL-8, MCP-1, MIP-1α,β, RANTES, fractalkine	CCR1, CCR2, CCR5, CXCR3, CXCR4, CX3CR1

ND, not determined; ( ), most probably yes; *, reportedly yes, although we could not confirm in our mouse system.

such as TNF-α, IL-1, and IL-12 as discussed later [26,29], in both pathological conditions and in culture systems. Astrocytes sometimes have suppressive effects on microglia or microglia-derived cytokines. For, example, in contrast to microglia after stimulation with LPS and IFN-γ, astrocytes produce IL-12 p40, but not p35 [29]. When the astrocyte-derived p40 forms a homodimer, it may suppress the functional heterodimer IL-12 p70 produced by microglia. In addition, the production of IL-12 p70 by activated microglia was inhibited by coculture with astrocytes [38]. Thus, it is possible that astrocytes suppress microglial cytokine production and/or the effects of produced cytokines. The suppression of IFN-γ-induced MHC class II expression on microglia by astrocyte-derived GM-CSF is another example of this type of interaction [10].

As discussed above, IFN-γ activates various functions of glial cells including the induction of cytokines. The production of IFN-γ was thought to be restricted to lymphoid cells. However, it has recently been shown that human fetal forebrain cells can be induced to express IFN-γ mRNA and produce IFN-γ protein when stimulated with *trypanosome lymphocyte-triggering factor* (TLTF) [39]. The authors claimed that astrocytes were the major producer of IFN-γ in response to TLTF. We have shown recently that microglia produce IFN-γ in response to IL-12 and/or IL-18 [32]. Thus, microglia may be another source of INF-γ production in the CNS. It has been shown that APCs such as macrophages and dendritic cells also produce IFN-γ in response to IL-12 [40].

4. CYTOKINE NETWORKS IN THE CENTRAL NERVOUS SYSTEM

4.1. Immunoregulatory cytokines, which affect the functions of antigen-presenting cells

The immune response is initiated when a protein antigen is presented to T cells by APCs within lymphoid organs. The APC processes antigen, either foreign or self, by internalizing and digesting it into peptide fragments. Subsequently, processed peptide fragments are expressed on the surface of APCs as a MHC–peptide complex. When the MHC–peptide complex interacts with T-cell receptors (TCRs), T-cell activation occurs. Class II MHC molecules present antigen to CD4-positive T cells, while class I MHC molecules present antigens to CD8-positive T cells. Binding of the MHC–peptide complex to the TCRs is critical, but not sufficient, for the activation of T cells. There should be several costimulatory molecules that interact with the ligands on T cells in order for sufficient activation to occur. These costimulatory molecules on APC are B7.1, B7.2, leukocyte function-associated molecule 3 (LFA-3), intercellular adhesion molecule-1 (ICAM-1), ICAM-2, and ICAM-3. The molecules bind to ligands on T cells to form ligand pairs such as B7.1–CD28, B7.2–CTLA4, LFA-3–CD2, ICAM-1, 2, or 3–LFA-1. Interaction of T cells and APCs occurs in a MHC-restricted manner. The T cells recognize a foreign antigen only when the antigen is complexed with self-MHC molecules on APCs. Therefore, the cells expressing class II MHC and costimulatory molecules constitutively are considered to be professional APCs. These include macrophages, B cells, dendritic cells, and Langerhans cells. Nonprofessional APCs differ from the professional APCs by expressing little or no MHC class II molecules constitutively, and by not having a complete set of costimulatory molecules. Candidates for nonfunctional APCs in the CNS are microglia, astrocytes, and endothelial cells [4,41–45]. They usually do not express class II MHC antigen constitutively, although some populations of microglia reportedly may express class II MHC antigens constitutively [46]. These cells induced the expression of class II MHC molecules after treatment with certain inflammatory cytokines, especially IFN-γ [4,5,41,44], in addition to expressing some of the costimulatory molecules as well [47–49]. There is published evidence that endothelial cells [50], astrocytes [41], and pericytes [51] can process and present protein antigens to primed CD4-positive T cells *in vitro*, but the specific role of these cells as APCs *in vivo* is still unclear. Astrocytes do not usually express class II MHC antigens *in vivo*, even in the presence of inflammatory cells [42]. Since microglia have functional characteristics very similar to macrophages and can be induced to express class II MHC antigens as discussed above, *microglia* are the most possible candidates for APCs in the CNS. The expression of costimulatory molecules, such as B7, ICAM, LFA3, in microglia, but only a few in astrocytes, further supports this hypothesis. Menendez Iglesias et al. [49] detected B7-2, but not B7-1, in murine microglia only after stimulation with LPS and IFN-γ. Satoh et al. [48] have shown that human microglia, but not astrocytes, express both B7-1 and B7-2, suggesting that microglia is a much more suitable candidate for local APCs in the CNS. In fact, microglia when stimulated with IFN-γ reportedly presented antigen to ovalbumin-specific or myelin basic protein (MBP)-specific T cells *in vitro* [44,45]. In a carefully executed study, Hickey and Kimura [43] have shown that microglia function as APCs in pathological conditions *in vivo*. They used bone marrow chimeras of EAE-susceptible and -resistant animals, and found that EAE lesions developed only when the perivascular microglia were replaced with those of an EAE-susceptible strain, suggesting that antigen presentation by perivascular microglia is critical for the development of EAE lesions.

Professional APCs such as dendritic cells and macrophages produce IL-12 and IL-18. Both cytokines have been shown to be key cytokines in the development of autoimmune processes, regulating differentiation of naïve T cells into Th1. In order to exert its activity, IL-12 has to form a heterodimer of P35 and P40; the homodimer of P40 suppresses the functional heterodimer. Immature IL-18 is cleaved by caspase-1 to become a functionally mature IL-18 that induces the differentiation of Th1 and the cytotoxic activity of NK and T cells. It has been reported that both microglia and astrocytes produce IL-12 upon stimulation with LPS [38], while we detected functional IL-12 p70 production only in microglia, but not in astrocytes, after stimulation with LPS and IFN-γ [30]. Since soluble TNF receptors reportedly suppress IL-12 production by human microglia [52], TNF signal may also be involved in IL-12 production. Microglia and astrocytes also express IL-18 mRNA after stimulation with LPS [32,53]. LPS-stimulated microglia have enough IL-18 bioactivity to induce INF-γ production by thymocytes and splenocytes in synergism with IL-12. This suggests that microglia express caspase-1 as well. In fact, caspase-1 mRNA expression is elevated in microglia in multiple sclerosis (MS) plaques [54] where IL-18 is also reported to be elevated [55]. Interestingly, there is a group of microglia that produce only IL-12 P40, but not IL-12 P35, resulting in the failure to produce functional IL-12 p70 heterodimers [29]. The population did not produce IL-18 even after LPS stimulation (unpublished observation). Therefore, microglia may have subpopulations, which regulate the differentiation of T cells in a different manner.

4.2. The roles of microglia in the central nervous system cytokine network

Both microglia and astrocytes produce the same cytokines, such as IL-1, IL-6, TNF-α, and TGF-β. However, there are several differences in the response to stimulation in these two cell types. For example, microglia produce TNF-α in response to lower doses of LPS than are required for astrocytes and more rapidly than astrocytes as well. IL-6 production is induced by TNF-α in astrocytes, but not in microglia [27]. Similarly, GM-CSF produced by astrocytes induces IL-6 production in microglia, but not in astrocytes [56]. These observations indicate that microglia and astrocytes may mutually regulate their individual cytokine production. Since

Table 3. Effects of inhibitory cytokines on microglial functions

	IL-4	IL-10	TGF-β
Proliferation	↑	→	↓
Enzyme activity	↑↑	↓	↓
IFN-γ-induced Ia expression	↓	↓	↓
LPS-induced cytokine production	→	↓	↓
GM-CSF-induced IL-6 production	↓	↓	↓
Cytokine receptor expression	→(↑)[a]	↓	→

↑, Upregulate; →, no effect; ↓, downregulate.
[a] IL-4 upregulates IL-4 receptor, but does not affect the expression of other receptors on microglia.

microglia are activated in the earlier phase than are astrocytes under various pathological conditions, microglia may initiate the cascade of cytokine actions in the CNS cytokine network. Inhibitory signals are also included in the network (Table 3). TGF-β, produced by astrocytes and microglia, suppresses all the functions of microglia. It suppresses M- and GM-CSF-induced proliferation of microglia, LPS-induced activation of enzymatic activity in microglia, IFN-γ-induced class II MHC antigen expression and cytokine production by microglia. TGF-β along with IL-4 and IL-10 is known to be a *macrophage-deactivating* factor. Therefore, these cytokines may function as negative regulators in the CNS cytokine network by suppressing cytokine production and activation of microglia. In fact, it has been found that these inhibitory cytokines exert their influence on microglia differently. TGF-β functions as if it is a total inhibitory factor [16]. IL-10 suppresses cytokine production and IFN-γ-induced class II MHC antigen expression in microglia, but does not suppress the proliferation or the activation of lysosomal enzymes in microglia [18]. IL-4 also suppresses IFN-γ-induced class II MHC antigen expression in microglia [17]. However, unlike other inhibitory cytokines, IL-4 induces the proliferation of microglia in either unstimulated or M-, or GM-CSF-stimulated conditions. IL-4 does not suppress LPS-induced cytokine production, though it suppresses GM-CSF-induced IL-6 production by microglia [56]. We also found that IL-10, but neither TGF-β nor IL-4, suppressed the expression of cytokine receptors [57]. Thus, it would appear that all these three inhibitory cytokines regulate the functions of microglia in a distinct manner, and that IL-10 may be the most potent inhibitor for the functions of cytokines on microglia because it suppresses both cytokine production and receptor expression.

5. CYTOKINES IN THE CENTRAL NERVOUS SYSTEM PATHOLOGIES

5.1. Demyelination

Several lines of evidences suggest that TNF-α plays a critical role in the pathogenesis of *inflammatory demyelination*, either directly or indirectly via induction of other cytokines, nitric oxide (NO), or free radicals (Fig. 1). Increased cerebrospinal fluid levels of TNF-α have been demonstrated in patients with MS [58]. TNF-α-positive microglia and astrocytes have been identified, especially in new active plaques. *In vitro* studies have demonstrated that TNF-α kills oligodendrocytes, myelin-forming cells in the CNS [59,60], and that microglia are the principal

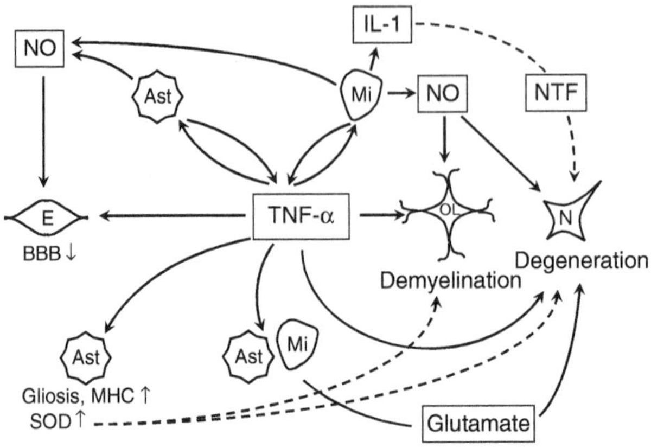

Figure 1. The roles of TNF-α in the development of CNS pathology. Ast, astrocytes; Mi, microglia; OL, oligodendrocytes; N, neuron; E, endothelial cells. NO, nitric oxide; NTF, neurotropic factors; BBB, blood–brain barrier. ——— destructive; - - - - - - protective.

effectors for oligodendrocyte killing [61]. It has also been shown that anti-TNF-α antibody suppresses the development of EAE, an animal model of MS [62,63]. Demyelination has been demonstrated to be much more severe in transgenic mice producing TNF-α in the CNS [64]. In addition, TNF-α induces inflammatory cytokines or chemokines in endothelial cells and impairs the tight junctions of the BBB [65]. Up until now, several substances that suppress TNF-α production have been used for the treatment of EAE and MS. Most of them, such as phosphodiesterase inhibitors, *N*-acetyl-L-cysteine, have been shown to effectively suppress the development of EAE and MS [66–68], further supporting the hypothesis that TNF-α is critical for the development of inflammatory demyelination. However, experimental demyelination could also be induced in TNF-α knockout mice, though EAE was delayed in the onset and inflammatory leukocytes failed to move normally into the CNS parenchyma [69]. More recently, TNF-α has been identified as a factor that promotes *remyelination* [70]. Thus, although TNF-α is an important cytokine, it may not be the sufficient effector molecule for inflammation and demyelination. It is also possible that TNF-α may exert different effects on inflammatory demyelination, depending on whether the TNF signaling through type 1 TNF receptor (TNFR1) or TNFR2 is dominant.

5.2. Gliosis

Gliosis is a rather common pathological finding observed as a glial scar following inflammation, demyelination, ischemia, and neuronal degeneration. It consists of astrocyte proliferation, hypertrophy, and increased synthesis of glial fibrillary acidic protein (GFAP), a phenotypic marker for astrocytes. Evidence to date suggests critical roles for cytokines in the development of *astrocytic gliosis*. Fontana et al. [71] first demonstrated that factors from activated lymphocytes stimulated astrocyte proliferation and designated the factor(s) as glial cell-stimulating factor (GSF). Merrill et al. [36] also demonstrated increased proliferation of astrocytes after treatment with lymphokines. Using enriched cultures of astrocytes and recombinant cytokines, Selmaj et al. [72] showed that TNF-α is a primary factor to bring about the proliferation of rat astrocytes. However, Giulian et al. [73,74] have shown that IL-1 derived from microglia [24] is the principle factor to induce astrocyte proliferation in gliosis. In contrast, Yong et al. [75] claimed that the primary factor that induced the proliferation of human astrocytes was IFN-γ and not IL-1 or TNF-α. These differences in experimental results may be attributed to either species differences or redundancy of functions for these cytokines. Alternatively, it is possible that other factors induced by either IL-1, TNF-α, or IFN-γ may also play a role in the proliferation of astrocytes. In view of these diverse results, it can be concluded that cytokines contribute to the pathogenesis of gliosis. However, precise identification of individual cytokine contributions to the overall process will require additional experimental inquiries.

5.3. Neuronal degeneration

TNF-α has also been implicated as an effector for neuronal degeneration [76–78]. TNF-α exerts its cytotoxicity directly via TNFR1. Alternatively, it also induces NO or free radicals to form the toxic peroxinitrite. It has been shown that β-amyloid stimulates microglia to produce factors toxic to neurons. It is possible that neuronal apoptosis induced by β-amyloid is also mediated by glia-derived TNF-α [79]. Combs et al. [80] concluded that the most critical factor in β-amyloid-induced, microglia-mediated neuronal apoptosis might be NO, because neurotoxicity was decreased by the selective inhibitors against inducible nitric oxide synthase. Apoptosis of motor neurons and dorsal root ganglion neurons by peripherin aggregates is also reportedly mediated by TNF-α [81]. TNF-α also exerts its neurotoxicity by activating astrocytes to release glutamate [82]. Recently, we have shown that the most neurotoxic factor from activated microglia is glutamate [83]. TNF-α dose not exert direct neurotoxicity, but induces neurotoxicity via glutamate production by microglia. Glutamate disturbs the mitochondrial respiratory chain to cause energy depletion in neurons, which results in neuronal damage toward cell death [84]. In contrast, IL-1, but not TNF-α, may be involved in neurotoxicity during some variants of viral encephalitis [85]. The protein Fas associated with death domain (FADD) is an adaptor protein of the TNF receptor family death pathway. A number of FADD-positive dopaminergic neurons in the substantia nigra pars compacta have been shown to be significantly decreased in patients with Parkinson's disease (PD), as compared to levels in normal subjects [86]. This decrease correlated with the known selective vulnerability of nigral dopaminergic neurons in PD. On the basis of the latter, the authors concluded that the TNF–FADD pathway contributed to the susceptibility of dopaminergic neurons in PD to the effects of TNF-mediated apoptosis [86].

Interestingly, cytokines described above as toxic also have protective roles for neurons against oxidative stress. TNF-α and IL-1 have been shown to increase the level of manganese superoxide dismutase (Mn-SOD) in astrocytes, in a dose- and time-dependent manner [87]. Since SOD functions as protective against oxidative stress, and since the increased Mn-SOD activity has been demonstrated in the substantia nigra of parkinsonian patients [88], these cytokines may function to protect degenerating neurons, via induction of SOD. IL-1 reportedly increases the production of nerve growth factors by astrocytes [89]. Therefore, a balance between toxic and protective factors induced by cytokines may determine neuronal damage (see Fig. 1).

5.4. Other pathological conditions in the central nervous system

Microglia undergo various morphological changes to become either ramified, amoeboid, or rod-shaped. We have shown that all of these morphological changes could be reproduced *in vitro* with various cytokines [14,15]. Microglia also form a unique phenotype of multinucleated giant cells (MNGC), which are observed in AIDS encephalopathy, tuberculosis, etc. Lee et al. [90] have demonstrated that treatment with IL-3, IL-4, IFN-γ, and GM-CSF induces MNGC in rat microglia, while addition of IL-1, IL-6, or TNF-α failed to form MNGC. In mouse experiments using microglia, there was no single cytokine that induced MNGC in culture. However, when stimulated with IL-4 or IL-13 in the presence of GM-CSF or M-CSF, MNGC formation occurred in the cultures of mouse microglia [91]. The different results between these studies may be attributable to species differences. Nevertheless, the results of these studies indicate that introduction of cytokines, most probably those that are T cell-derived, can induce MNGC formation without infectious agents.

REFERENCES

1. Wong GHW, Bartlett PF, Clark-Lewis I, Battye F Schrader JW. Inducible expression of H-2 and Ia antigens on brain cells. Nature 1984;310:688–91.
2. Suzumura A, Silberberg DH. Expression of H-2 antigen on oligodendrocytes is induced by soluble factors from concanavalin A activated T cells. Brain Res 1985;36:71–175.
3. Lavi E, Suzumura A, Murasko DM, Murray EM, Silberberg DH, Weiss SR. Tumor necrosis factor induces expression of MHC antigens on mouse astrocytes. J Neuroimmunol 1988;18:245–53.
4. Fierz W, Endler B, Reske K, Wekerke H, Fontana A. Astrocytes as antigen-presenting cells. I. Induction of Ia antigen expression on astrocytes by T cells via immune interferon and its effect on antigen presentation. J Immunol 1985;134:3785–93.
5. Suzumura A, Mezitis SGE, Gonatus N, Silberberg DH. MHC antigen expression on bulk isolated macrophage–microglia from newborn mouse brain: Induction of Ia antigen expression by gamma-interferon. J Neuroimmunol 1987;15:263–78.
6. Matsumoto Y, Hra N, Tanaka R, Fujiwara M. Immunohistochemical analysis of the rat central nervous system during experimental allergic encephalomyelitis, with special reference to Ia positive cells with dendritic morphology. J Immunol 1986;36:3668–76.
7. Butter C, O'Neil JK, Baker D, Gschmeissner SE, Turk JL. An immunoelectron microscopical study of class II major hsitocompatibility complex during chronic relapsing experimental allergic encephalomyelitis in Biozzi AB/H mice. J Neuroimmunol 1991;33:37–42.
8. Gehrmann J, Gold R, Linington C, Lannes-Vieira J, Wekerle H, Kreutzberg GW. Microglial involvement in autoimmune inflammation of the central and peripheral nervous system. Glia 1993;7:50–9.
9. Streit WJ, Graeber MB, Kreutzberg GW. Expression of Ia antigen on perivascular and microglial cells after sublethal and lethal motoneuron injury. Exp Neurol 1989;105:115–26.
10. Imamura K, Suzumura A, Sawada M, Mabuchi C, Marunouchi T. Induction of MHC class II antigen expression on murine microglia by interleukin-3. J Neuroimmunol 1994;55:119–25.
11. Gebicke-Haerter PJ, Appel K, Taylor GD, Schobert A, Rich IN, Northoff H, Berger M. Rat microglial interleukin-3. J Neuroimmunol 1994;50:203–14.
12. Farrar WL, Vinocour M, Hill JM. In situ hybridization histochemistry localization of interleukin-3 mRNA in mouse brain. Blood 1989;73:137–40.
13. Ohno K, Suzumura A, Sawada M, Marunouchi T. Production of granulocyte/macrophage colony stimulating factor by cultured astrocytes. Biochem Biophys Res Commun 1990;169:719–24.

14. Suzumura A, Sawada M, Yamamoto H, Marunouchi T. Effects of colony stimulating factors on isolated microglia. J Neuroimmunol 1990;30:111–20.
15. Suzumura A, Marunouchi T, Yamamoto H. Morphological transformation of microglia in vitro. Brain Res 1991;545:301–6.
16. Suzumura A, Sawada M, Yamamoto H, Marunouchi T. Transforming growth factor beta suppresses activation and proliferation of microglia in vitro. J Immunol 1993;151:2150–8.
17. Suzumura A, Sawada M, Ito Y, Marunouchi T. IL-4 induces proliferation and activation of microglia but suppressed their induction of class II major histocompatibility complex antigen expression. J Neuroimmunol 1994;53:209–18.
18. Mizuno T, Sawada M, Marunouchi T, Suzumura A. Production of interleukin-10 by mouse glial cells in culture. Biochem Biophys Res Commun 1994;205:1907–15.
19. Wahl SM, Allen JB, McCarney-Fransis N, Morgnti-Kossmann MC, Kossmann T, Ellingsworth L, Mai UEH, Mergenhagen SE, Orenstein JM. Macrophage-, and astrocyte-derived transforming growth factor beta as a mediator of central nervous system dysfunction in acquired immunodeficiency syndrome. J Exp Med 1991;173:981–91.
20. Constam DB, Philipp J, Malipiero UV, ten Dijke P, Schachner M, Fontana A. Differential expression of transforming growth factor beta1,2, and 3 by glioblastoma cells, astrocytes, and microglia. J Immunol 1992;148:1404–10.
21. Sawada M, Ito Y, Suzumura A, Marunouchi T. Expression of cytokine receptors in cultured neuronal and glial cells. Neurosci Lett 1993;160:131–4.
22. Suzumura A, Lavi E, Weiss SR, Silberberg DH. Corona virus infection induces H-2 antigen expression on oligodendrocytes and astrocytes. Science 1986;232:991–3.
23. Massa PT, Dorries R, ter Meulen V. Viral particles induce Ia antigen expression on astrocytes. Nature 1986;320:543–6.
24. Giulian D, Baker TJ, Shin L-CN, Lackman LB. Interleukin-1 of the central nervous system is produced by ameboid microglia. J Exp Med 1986;164:594–604.
25. Frei K, Malipiero UV, Leist TP, Zinkernagel RM, Schwab ME, Fontana A. On the cellular source and function of interleukin-6 produced in the central nervous system in viral diseases. Eur J Immunol 1989;19:689–94.
26. Sawada M, Kondo N, Suzumura A, Marunouchi T. Production of tumor necrosis factor alpha by microglia and astrocytes in culture. Brain Res 1989;491:394–7.
27. Sawada M, Suzumura A, Marunouchi T. TNFα induces IL-6 production by astrocytes but not by microglia. Brain Res 1992;583:296–9.
28. Sawada M, Suzumura A, Ito Y, Marunouchi T. Production of interleukin-5 by mouse astrocytes and microglia in culture. Neurosci Lett 1993;155:175–8.
29. Suzumura A, Sawada M, Takayanagi T. Production of interleukin-12 and the expression of its receptors by murine microglia. Brain Res 1998;787:139–42.
30. Liu JS, Amaral TD, Brosnan CF, Lee SC. IFNs are critical regulators of IL-1 receptor antagonist and IL-1 expression in human microglia. J Immunol 1998;161:1989–96.
31. Prinz M, Hanisch UK. Murine microglial cells produce and respond to interleukin-18. J Neurochem 1999;72:2215–8.
32. Kawanokuchi J, Mizuno T, Takeuchi H, Kato H, Wang J, Mitsuma N, Suzumura A. Production of interferon-γ by microglia. Mult Scler 2006;12:558–64.
33. Sonobe Y, Yawata I, Kawanokuchi J, Takeuchi H, Mizuno T, Suzumura A. Production of IL-27 and IL-12 family cytokines by microglia and their subpopulations. Brain Res 2005;1040:202–7.
34. Sawada M, Suzumura A, Marunouchi T. Cytokine network in the central nervous system and its roles in growth and differentiation of glial and neuronal cells. Int J Dev Neurosci 1995;13:253–64.

35. Sawada M, Suzumura A, Marunouchi T. Induction of functional IL-2 receptor in mouse microglia. J Neurochem 1995;64:1973–9.
36. Merrill JE, Kutsunai S, Mohlstrom C, Hofman F, Groopman J, Golde DW. Proliferation of astroglia and oligodendroglia in response to human T cell-derived factors. Science 1984;224:1428–30.
37. Lee YB, Nagai A, Kim SU. Cytokines, chemokines, and cytokine receptors in human microglia. J Neurosci Res 2002;69:94–103.
38. Aloisi F, Penna G, Cerase J, Menendez Iglesias B, Adrini L. IL-12 production by central nervous system microglia is inhibited by astrocytes. J Immunol 1997;159:1602–12.
39. Bakhiet M, Hamadien M, Tjernlund A, Mousal A Seiger A. African trypanosomes activate human fetal brain cells to proliferation and IFN-γ production. Neuroreport 2002;13:53–6.
40. Frucht DM, Fukao T, Bogdan C, Schindler H, O'Shea JJ, Koyasu, S. IFN-α production by antigen presenting cells: Mechanisms emerge. Trends Immunol 2001;22:556–60.
41. Fontana A, Fierz W, Wekerle H. Astrocytes present myelin basic protein to encephalitogenic T-cell lines. Nature 1984;307:273–6.
42. Hickey WF, Osborn JP, Kirby WM. Expression of Ia molecules by astrocytes during acute experimental allergic encephalomyelitis in the Lewis rat. Cell Immunol 1985;91:528–35.
43. Hickey WF, Kimura H. Perivascular microglial cells of the CNS are bone marrow-derived and present antigen in vivo. Science 1988;239:290–2.
44. Frei K, Siepl C, Groscurth P, Bodmer S, Schwerdel C, Fontana A. Antigen presentation and tumor cytotoxicity by interferon-gamma-treated microglial cells. Eur J Immunol 1987;12:237–43.
45. Matsumoto Y, Ohmori K, Fujiwara M. Immune regulation by brain cells in the central nervous system; microglia but not astrocytes present myelin basic protein to encephalitogenic T-cells under *in vivo*-mimicking conditions. Immunology 1992;76:209–16.
46. Ford A, Goodsall AL, Hickey WF, Sedgwick JD. Normal adult ramified microglia separated from other central nervous system macrophages by flow cytometric sorting. J Immunol 1995;154:4309–21.
47. De Simone R, Giampaolo A, Giometto B, Gallo P, Levi G, Peschle C, Aloisi F. The costimulatory molecule B7 is expressed in human microglia in culture and multiple sclerosis acute lesions. J Neuropathol Exp Neurol 1995;54:175–87.
48. Satoh J, Lee YB, Kim SU. T-cell costimulatory molecules B71 (CD80) and B7-2 (CD86) are expressed in human microglia but not in astrocytes in culture. Brain Res 1995;704:92–6.
49. Menendez Iglesias B, Cerase J, Ceracchini C, Levi G, Aloisi F. Analysis of B7-1 and B7-2 costimulatory ligands in cultured mouse microglia: Upregulation by interferon-gamma and lipopolysaccharide and downregulation by interleukin-10, prostaglandin E2 and cyclic AMP-elevating agents. J Neuroimmunol 1997;72:83–93.
50. Sedgwick JD, Hughes CC, Male DK, MacPhee IA, ter Meulen V. Antigen-specific damage to brain vascular endothelial cells mediated by encephalitogenic and nonencephalitogenic CD4 T cell lines in vitro. J Immunol 1990;145:2474–81.
51. Fabry Z, Waldschmidt MM, Moore SA, Hart MN. Antigen presentation by brain microvessel smooth muscle and endothelium. J Neuroimmunol 1990;28:63–71.
52. Becher B, Dodelet V, Fedorowicz V, Antel JP. Soluble tumor necrosis factor receptor inhibits interleukin 12 production by stimulated human adult microglial cells *in vitro*. J Clin Invest 1996;98:1539–43.
53. Conti B, Park LC, Calingasan NY, Kim Y, Kim H, Bae Y, Gibson GE, Joh TH. Cultures of astrocytes and microglia express interleukin 18. Mol Brain Res 1999;67:46–52.

54. Ming X, Li W, Maeda Y, Blumberg B, Raval S, Cook S, Dowling PC. Caspase-1 expression in multiple sclerosis plaques and cultured glial cells. J Neurol Sci 2002;197:9–18.
55. Balashov KE, Rottman JB, Weiner HL, Hancock WW. CCR5(+) and CXCR3(+) T cells are increased in multiple sclerosis and their ligands MIP-1alpha and IP-10 are expressed in demyelinating brain lesions. Proc Natl Acad Sci USA 1999;96:6873–8.
56. Suzumura A, Sawada M, Marunouchi T. Selective induction of interleukin-6 in mouse microglia by granulocyte–macrophage colony stimulating factor. Brain Res 1996;717:192–8.
57. Sawada M, Suzumura A, Hosoya H, Marunouchi T, Nagatsu T. IL-10 inhibits both production of cytokine and expression of cytokine receptors in microglia. J Neurochem 1999;72:1466–71.
58. Maimone D, Gregory S, Arnason BG, Reder AT. Cytokine levels in the cerebrospinal fluid and serum of patients with multiple sclerosis. J Neuroimmunol 1991;32:67–74.
59. Selmaj KW, Raine CS. Tumor necrosis factor mediates myelin and oligodendrocyte damage *in vitro*. Ann Neurol 1988;23:339–47.
60. Selmaj KW, Raine CS, Farooq M. Cytokine cytotoxicity against oligodendrocytes. Apoptosis induced by lymphotoxin. J Immunol 1991;147:1522–9.
61. Merrill J., Ignarro LJ, Sherman MP, Melinek J, Lane TE. Microglial cell cytotoxicity of oligodendrocytes is mediated through nitric oxide. J Immunol 1993;151:2132–41.
62. Ruddle NH, Bergman CM, McGrath KM, Lingenheld EG, Grunnet ML, Padula SJ, Clark RB. An antibody to lymphotoxin and tumor necrosis factor prevents transfer of experimental allergic encephalomyelitis. J Exp Med 1990;172:1193–200.
63. Selmaj KW, Raine CS, Cross AH. Anti-tumor necrosis factor therapy abrogates autoimmune demyelination. Ann Neurol 1991;30:694–700.
64. Taupin V, Renno T, Bourbonniere L, Peterson AC, Rodriguez M, Owens T. Increased severity of experimental autoimmune encephalomyelitis, chronic macrophage/microglial reactivity, and demyelination in transgenic mice producing tumor necrosis factor-alpha in the central nervous system. Eur J Immunol 1997;27:905–13.
65. Deli MA, Descamps L, Dehouck MP, Cecchelli R, Joo F, Abraham CS, Torpier G. Exposure of tumor necrosis factor-alpha to luminal membrane of bovine brain capillary endothelial cells cocultured with astrocytes induces a delayed increase of permeability and cytoplasmic stress fiber formation of actin. J Neurosci Res 1995;41:717–26.
66. Rott O, Cash E, Fleischer B. Phosphodiesterase inhibitor pentoxifylline, a selective suppressor of T helper type 1- but not type 2-associated lymphokine production, prevents induction of experimental autoimmune encephalomyelitis in Lewis rats. Eur J Immunol 1993;23:1745–51.
67. Lehmann D, Karussis D, Misrachi-Koll R, Shezen E, Ovadia H, Abramsky O. Oral administration of the oxidant-scavenger *N*-acetyl-L-cysteine inhibits acute experimental autoimmune encephalomyelitis. J Neuroimmunol 1994;50:35–42.
68. Suzumura A, Nakamuro T, Tamaru T, Takayanagi T. Drop in relapse rate of multiple sclerosis patients using combination therapy of three different phosphodiesterase inhibitors. Mult Scler 2000;6:56–8.
69. Riminton S, Korner D, Strickland DH, Lemckert FA, Pollard J, Sedgwick JD. Challenging cytokine redundancy: Inflammatory cell movement and clinical course of experimental allergic encephalomyelitis are normal in lymphotoxin-deficient, but not in tumor necrosis factor-deficient, mice. J Exp Med 1998;187:1517–28.
70. Arnett HA, Mason J, Marino M, Suzuki K, Matsushima GK, Ting JP. TNFα promotes proliferation of oligodendrocyte progenitors and remyelination. Nat Neurosci 2001;4:1116–22.

71. Fontana A, Dubs R, Merchant R, Balsiger S, Grob PJ. Glia cell stimulating factor (GSF): A new lymphokine. Part 1. Cellular sources and partial purification of murine GSF, role of cytoskeleton and protein synthesis in its production. J Neuroimmunol 1982;2:55–71.
72. Selmaj KW, Farooq M, Norton WT, Raine CS, Brosnan CF. Proliferation of astrocytes *in vitro* in response to cytokines. A primary role for tumor necrosis factor. J Immunol 1990;144:129–35.
73. Giulian D, Lackman LB. Interleukin-1 stimulation of astroglial proliferation after brain injury. Science 1985;228:497–9.
74. Gulian D, Young DG, Woodward J, Brown DC, Lachman LB. Interleukin-1 is an astroglial growth factor in the developing brain. J Neurosci 1988;8:709–14.
75. Yong VW, Moumdjian R, Yong FP, Ruijs TC, Freedman MS, Cashman N, Antel JP. γ-Interferon promotes proliferation of adult human astrocytes *in vitro* and reactive gliosis in the adult mouse brain *in vivo*. Proc Natl Acad Sci USA 1991;88:7016–20.
76. Talley AK, Dewhurst S, Perry SW, Dollard SC, Gummuluru S, Fine SM, New D, Epstein LG, Gendelman HE, Gelbard HA. Tumor necrosis factor alpha-induced apoptosis in human neuronal cells: protection by the antioxidant *N*-acetylcysteine and the genes bcl-2 and crmA. Mol Cell Biol 1995;15:2359–66.
77. Shen Y, Li R, Shiosaki K. Inhibition of p75 tumor necrosis factor receptor by antisense oligonucleotides increases hypoxic injury and beta-amyloid toxicity in human neuronal cell line. J Biol Chem 1997;272:3550–3.
78. Downen M, Amaral TD, Hua LL, Zhao ML, Lee SC. Neuronal death in cytokine-activated primary human brain cell culture: Role of tumor necrosis factor-alpha. Glia1999;28:114–27.
79. McGeer PL, McGeer EG. The inflammatory response system of brain: Implications for therapy of Alzheimer and other neurodegenerative diseases. Brain Res Rev 1995;21:195–218.
80. Combs CK, Karlo JC, Kao SC, Landreth GE. β-Amyloid stimulation of microglia and monocytes results in TNF-alpha-dependent expression of inducible nitric oxide synthase and neuronal apoptosis. J Neurosci 2001;21:1179–88.
81. Robertson J, Beaulieu JM, Doroudchi MM, Durham HD, Julien JP, Mushynski WE. Apoptotic death of neurons exhibiting peripherin aggregates is mediated by the proinflammatory cytokine tumor necrosis factor-α. J Cell Biol 2001;155:217–26.
82. Bezzi P, Domercq M, Brambilla L, Galli R, Schols D, De Clercq E, Vescovi A, Bagtta G, Kollias G, Meldolesi J, Volterra A. CXCR4-activated astrocyte glutamate release via TNF-α: Amplification by microglia triggers neurotoxicity. Nat Neurosci 2001;4:702–10.
83. Takeuchi H, Mizuno T, Zhang G, Wang J, Kawanokuchi J, Kuno R, Suzumura A. Neuritic beading induced by activated microglia is an early feature of neuronal dysfunction toward neuronal death by inhibition of mitochondrial respiration and axonal transport. J Biol Chem 2005;280:10444–54.
84. Takeuchi H, Jin S, Wang J, Zhang G, Kawanokuchi J, Kuno R, Sonobe Y, Mizuno T, Suzumura A. Tumor necrosis factor-α induces neurotoxicity via glutamate release from hemichannels of activated microglia in an autocrine manner. J Biol Chem 2006;281:21362–8.
85. Takikita S, Takano T, Narita T, Takikita M, Ohno M, Shimada M. Neuronal apoptosis mediated by IL-1β expression in viral encephalitis caused by a neuroadapted strain of the mumps virus (Kilham Strain) in hamsters. Exp Neurol 2001;172:47–59.
86. Hartmann A, Mouatt-Prigent A, Faucheux BA, Agid Y, Hirisch EC. FADD: A link between TNF family receptors and caspases in Parkinson's disease. Neurology 2002;58:308–10.
87. Mokuno K, Ohtani K, Suzumura A, Kiyosawa K, Hirose Y, Kawai K, Kato K. Induction of manganese superoxide dismutase by cytokines and lipopolysaccharide in cultured mouse astrocytes. J Neurochem 1994;63:612–6.

88. Saggu H, Cooksey J, Dexter D, Wells FR, Lees A, Nenner P, Marsden CB. A selective increase in particulate superoxide dismutase activity in parkinsonian substantia nigra. J Neurochem 1989;53:692–7.
89. Friedman WJ, Thakur S, Seidman L, Rabson AB. Regulation of nerve growth factor mRNA by interleukin-1 in rat hippocampal astrocytes is mediated by NFkB. J Biol Chem 1996;271:31115–20.
90. Lee TT, Martin FC, Merrill JE. Lymphokine induction of rat microglia multinucleated giant cell formation. Glia 1993;8:51–1.
91. Suzumura A, Tamaru T, Yoshikawa M, Takayanagi T. Multinucleated giant cell formation by microglia: Induction by interleukin(IL)-4 and IL-13. Brain Res 1999;849:239–43.

IV. CYTOKINES IN PATHOPHYSIOLOGICAL BRAIN RESPONSES

A. BRAIN–IMMUNE INTERACTION

Cytokines and the Brain
Edited by C. Phelps and E. Korneva

Lymphocytes and Adrenergic Sympathetic Nerves: The Role of Cytokines

YUKIKO KANNAN-HAYASHI, MITSUAKI MORIYAMA, and YOICHI NAKAMURA

Laboratory of Integrative Physiology, Division of Veterinary Sciences, Graduate School of Life and Environmental Sciences, Osaka Prefecture University, 1-1 Gakuen-cho, Naka-ku, Sakai 599-8531, Department of Human Life Sciences, Baika Junior College, 2-19-5, Shukunosho, Ibaraki, Osaka 567-8578, Japan

ABSTRACT

Growing evidence indicates that the sympathetic nervous system (SNS) is closely linked to the immune system. Primary and secondary lymphoid organs receive extensive sympathetic noradrenergic innervation. Under stimulation, norepinephrine (NE) released from the sympathetic nerve terminals in these organs, or circulating catecholamines (CAs) such as epinephrine, affects lymphocyte circulation, proliferation, and cytokine and antibody production through adrenergic receptors (ARs) expressed on lymphocytes and other immune cells. Although the mechanisms of adrenergic regulation of immune cells are very complicated, NE and epinephrine appear to promote humoral immunity rather than cellular immunity by suppressing the helper T (Th)1 response and upregulating T cell-dependent antibody production, through stimulation of the β_2-AR-cyclic AMP (cAMP)-protein kinase (PK) A pathway. The SNS also strongly affects natural killer (NK) cell function, especially during stress. The immune system influences SNS activity by cytokines such as interleukin (IL)-1, IL-6, and tumor necrosis factor-α (TNF-α), which travel to the brain and stimulate the hypothalamic or other regional neurons regulating the sympathetic outflow. IL-1, IL-2, IL-3, IL-6, IL-12, and granulocyte–macrophage colony-stimulating factor (GM-CSF) can stimulate the sympathetic neurite outgrowth, with or without the mediation of nerve growth factor (NGF) production. Lymphocytes are not only regulated by NE from the sympathetic nerve terminals, but also synthesize NE and other CAs, and the lymphocyte-derived CAs may regulate themselves in an auto- and paracrine way.

ABBREVIATIONS

AC, adenylate cyclase; AR, adrenergic receptor; β-ARK, β-AR kinase; BBB, blood–brain barrier; CA, catecholamine; cAMP, cyclic AMP; CGRP, calcitonin gene-related peptide; CNS, central nervous system; Con A, concanavalin A; CRF, corticotropin-releasing factor; DAG, diacylglycerol; DBH, dopamine beta-hydroxylase; EAE, experimental autoimmune encephalomyelitis; Gi proteins, inhibitory G proteins; GM-CSF, granulocyte–macrophage colony-stimulating factor; G proteins, GTP-binding proteins; HPA, hypothalamic–pituitary–adrenal; IFN-γ, interferon-γ;

IL, interleukin; IL-1RI, type 1 IL-1 receptors; IP_3, inositol 1,4,5-triphosphate; KLH, keyhole limpet hemocyanin; LIF, leukemia inhibitory factor; LPS, lipopolysaccharide; MAPK, mitogen-activated protein kinase; MLN, mesenteric lymph nodes; α-MT, α-methyl-*p*-L-tyrosine; NK, natural killer; NE, norepinephrine; NGF, nerve growth factor; NNT-1/BSF-3, novel neurotrophin-1/B cell-stimulating factor-3; NT, neurotrophin; 6-OHDA, 6-hydroxydopamine; PALS, periarterial lymphatic sheath; PHA, phytohemagglutinin; PK, protein kinase; SNS, sympathetic nervous system; SCG, superior cervical ganglia; SCID, severe combined immunodeficient; SRBC, sheep red blood cell; Tc, cytotoxic T; Th, helper T; TH, tyrosine hydroxylase; TNF-α, tumor necrosis factor-α; TNP, trinitrophenyl; VIP, vasoactive intestinal polypeptide.

1. INTRODUCTION

The sympathetic nervous system (SNS) is closely linked to the immune system, and the complex bidirectional interactions between the two systems regulate the cellular activities of each other [1]. This is revealed by the facts that all lymphoid organs are densely innervated with sympathetic neurons, and that their postsynaptic norepinephrine (NE) and circulating catecholamines (CAs) such as epinephrine modulate various immune functions by binding to adrenergic receptors (ARs) expressed on immune cells. Although NE has been generally regarded as being immunosuppressive, the action of NE is now known to be more complicated. In contrast, the immune system appears to influence the SNS activity by releasing cytokines that bind to cytokine receptors expressed on neuronal and/or non-neuronal cells. Cytokines can affect the SNS either by traveling to the hypothalamus or other centers in the central nervous system (CNS) that regulate autonomic outflow, by signaling to the local sympathetic ganglia, or by directly contacting the sympathetic neurons present in lymphoid organs or inflammatory sites. Monocyte-derived proinflammatory cytokine, interleukin (IL)-1, causes secretion of hormones of the hypothalamic–pituitary–adrenal (HPA) axis that results in an increase of NE turnover in lymphoid organs. IL-1 also induces nerve growth factor (NGF) secretion from fibroblasts, glial cells, and other cell types. Accumulating evidence indicates that other cytokines also affect sympathetic nerve growth. However, the role of immune cell-derived cytokines in the activity and function of the SNS during immune and inflammatory responses is still unclear. It has been suggested that immune cells are not only adrenergically stimulated by locally innervated sympathetic neurons, but also synthesize CAs themselves, and the CA synthesis is enhanced during lymphocyte activation. In this section, we will review what is currently known about the ways the SNS and immune systems influence each other, and particularly discuss the role of lymphocyte-derived cytokines in sympathetic nerve growth and lymphocyte-derived CAs in the regulation of immune responses.

2. INNERVATION OF LYMPHOID ORGANS

Fluorescent histochemical analysis of CAs and immunofluorescent labeling of the NE-synthesizing enzymes tyrosine hydroxylase (TH) and dopamine β-hydroxylase (DBH) have shown that all primary and secondary lymphoid organs are richly innervated by sympathetic neurons. The sympathetic innervation of the rodent spleen has been extensively studied because of its high density of sympathetic innervation [2,3]. Rogausch et al. [3] showed that the density of the sympathetic innervation is several times higher in the spleen than in other peripheral organs, and the spleen receives stronger tonic sympathetic input than do mesenteric lymph nodes (MLN).

Noradrenergic postganglionic innervation of the spleen originates mainly from the superior mesenteric/celiac ganglia [4,5]. The nerve fibers enter with the vasculature, travel along the trabeculae and along the branching vasculature, and are distributed mainly in the white pulp along the central artery and its branches. Some of the fibers extend away and pass into the T cell-rich periarterial lymphatic sheath (PALS) to reach the marginal zone that contains macrophages and B cells. Fibers are also in the marginal sinus, which is where lymphocytes enter the spleen. However, the B cell-containing follicles are sparsely innervated and the red pulp contains only scattered fibers. Electron microscopic studies have revealed that the nerve terminals are juxtaposed with the lymphocytes and macrophages in the white pulp [6,7]. These neuroimmune junctions do not contain cell processes. Neuropeptide Y, a 36-amino-acid peptide, colocalizes with NE in the postganglionic sympathetic nerve fibers of the spleen [8]. Nonsympathetic nerves also innervate rat spleen. Vasoactive intestinal polypeptide (VIP)-containing nerves course along large arteries and the central artery in the white pulp, and are also present in the venous/trabecular system and the red pulp [9].

In the thymus, sympathetic nerve fibers are found in the capsular and septal system. These fibers form varicose plexuses in the subcapsular cortex and at the corticomedullary junction but are extremely rare in the medulla [10,11]. The dense plexuses present in the outer cortex, where immature thymocytes reside and develop, and the nerve fibers are adjacent to thymocytes. In the deeper cortex and medulla, they are adjacent to thymic epithelial cells. The thymus is also innervated with parasympathetic nerves because acetylcholinesterase (AchE)-positive nerve fibers enter the gland together with the vasculature [12]. These nerves are mainly in the corticomedullary border and medulla. Sensory nerve fibers that are positive for SP and the calcitonin gene-related peptide (CGRP) are also present in the capsule, interlobular septa and corticomedullary boundary [11,13,14]. They occur in arteries, veins, and the microvasculature, associated with the perivascular and paravascular plexus. Some of these nerve fibers come in close contact with mast cells. VIP-positive fibers are in the capsular/septal system, cortex, and medulla [9].

In the lymph nodes, noradrenergic nerve fascicles enter the nodes with the vasculature in the hilar region and prevail in the perivascular plexus in the medullary region, and enter the cortical and paracortical regions surrounding the germinal centers. Nerve fibers are occasionally seen in T-cell areas [15–17]. Sensory nerves coding for coexisting SP and CGRP and VIP-positive nerves also innervate the lymph nodes [9,16,18].

3. EXPRESSION OF ARS ON LYMPHOCYTES

Norepinephrine and epinephrine stimulate two principal receptors on their target cells, namely the α- and β-ARs. The α-ARs are subdivided into two types (α_1 and α_2), while the β-ARs are subdivided into β_1, β_2 and β_3 subtypes, and each of these subtypes consists of several different additional subtypes. The coupling of the ARs to GTP-binding proteins (G proteins) leads to the activation of a number of different effector enzymes, which produce intracellular second messengers and alterations in the biological activity [19–29]. α_1-ARs couple with the G_α subunits of the G_q class to activate phospholipase C, which increases inositol 1,4,5-triphosphate (IP_3) and diacylglycerol (DAG) levels [19–25]. IP_3 causes the release of Ca^{2+} from the endoplasmic reticulum, which generates a rapid rise in cytosolic Ca^{2+} levels [25]. DAG activates a Ca^{2+}-phospholipid-dependent protein kinase (PK) C that phosphorylates and alters the activity of certain cellular proteins including cyclic AMP (cAMP) response element-binding protein (CREB) [21,23]. In addition to the classic system, the phospholipase A_2–arachidonic acid system and the tyrosine kinase phosphorylation system are also involved [19,20,22]. The β-ARs couple with stimulatory G proteins (Gs proteins) to activate adenylate cyclase (AC). The activated AC then increases intracellular cAMP and PKA activity [26–33]. PKA is activated by being phosphorylated and serves to activate various downstream effector molecules, including transcription factors. The activation of β_2-ARs has also been shown to affect the inhibitory G proteins (Gi proteins) and stimulate mitogen-activated protein kinase (MAPK) [27,29,31,33]. When activated, the α_2-ARs couple with the Gi proteins, which inhibit AC activity and thus block the subsequent formation of cAMP [34–36]. These signal transduction cascades interact with each other as a complicated network.

Adrenergic receptors are unequally distributed among lymphocyte subpopulations. It is generally noted that the high-affinity β_2-ARs are expressed at high levels in natural killer (NK) cells, at intermediate levels in cytotoxic T (Tc) cells and B cells, and at lowest levels in helper T (Th) cells [37–42]. β_2-ARs have been detected in Th1 cells but not in Th2 cells [43,44]. The density of ARs also appears to depend on cell maturity and state of activation. Although thymocytes have significantly fewer β-ARs than the other lymphocyte populations, mature medullary thymocytes are reported to have higher numbers of receptors than splenic T cells [45]. Stimulation of T cells with IL-2 has been shown to upregulate the β-AR expression to induce a cAMP response to β-AR agonists [42,46,47]. However, human peripheral blood T cells stimulated with IL-2 and phytohemagglutinin (PHA) show a biphasic pattern in their β-AR density in which there is an early increase in density, followed by a later decrease in density [48]. Stimulation of the CD3 complex, which is associated with the T-cell receptor (TCR), maintained the β-AR levels on the cell surface of Th1 cell clones, but not Th2 cell clones [43]. Different mitogens have different effects on the β-AR density on fresh and cloned T cells: concanavalin A (Con A) increases the number or β-ARs per cell, whereas PMA/ionophore decreases it [49]. Other studies showed decreased lymphocyte β-AR density and agonist-stimulated cAMP production by mitogens, alloantigens, or cytokines such as IL-1, IL-2, interferon-γ (IFN-γ), and granulocyte–macrophage colony-stimulating factor (GM-CSF) [50–53]. Moreover, patients with rheumatoid arthritis are reported to have fewer β_2-ARs in synovial fluid T cells than peripheral blood T cells and a disease activity-correlated decrease of β_2-ARs in peripheral blood $CD8^+$ T cells [54]. The β-AR kinase (β-ARK) is a member of the growing family of G protein-coupled receptor kinases (GRKs), which is serine-threonine kinase involved in the process of homologous desensitization of G-coupled receptors including β-ARs. PHA was shown to increase the β-ARK expression in T cells, and PKC was suggested to mediate the upregulation of the β-ARK expression [53,55]. The AR activity of lymphocytes is also affected by CAs. Epinephrine increases the α_2 or β_2-AR activity in mononuclear blood cells and T cells, but decreases it in NK cells [39,42,56]. The immune responses and the SNS activity appear to adjust the sympathetic supply to lymphocytes by changing the density and activity of lymphocyte ARs.

4. EFFECT OF ADRENERGIC STIMULATION ON T LYMPHOCYTES

Several lines of evidence from *in vitro* studies suggest that CAs or β_2-AR agonists inhibit the T-cell proliferation that is induced by mitogens (such as PHA and Con A) or by the immobilized anti-CD3 antibody, and this inhibition is usually accompanied by increases in cAMP and PKA levels in the stimulated lymphocytes that are proportional to the degree of proliferation inhibition [57–59]. Suppressive effects of a high dose of an α_2-AR agonist on lymphocyte reactivity were also reported [60,61], while the lower dose of the α_2-AR agonist shows a stimulatory effect on rat thymocyte and splenocyte proliferation accompanied by increased IL-2 production [61]. *In vivo* continuous α- or β-AR infusion suppresses the *in vitro* T-cell mitogen response [62,63]. Furthermore, exercise, stress, or conditioned immunosuppression associated with increased NE and epinephrine levels exerts an inhibitory effect on mitogen-induced T-cell proliferation and IL-2 or IFN-γ production via β_1-AR or β_2-AR stimulation [64–67]. In addition, T-cell activation has been shown to lead to β_2-AR-mediated inhibition of IL-2 production and IL-2R expression [68–70]. Moreover, several studies have shown that NE or β_2-AR agonists reduce IFNγ and IL-12 production and increase IL-4, IL-5, and IL-10 production of T cells, suggesting that priming of naive Th cells with β_2-AR stimulation leads to the inhibition of Th1 and the enhancement of Th2 cell differentiation [71–75]. Sanders et al. [44] demonstrated that clones of murine Th1 cells, but not Th2 cells, express a high level of β_2-ARs and that exposure of Th1 cells to the β_2-AR agonist terbutaline before activation by antigen-presenting B cells inhibited their IFN-γ production. However, Loza et al. [75] recently showed that β_2-AR activation directly modulate CD3-stimulated cytokine production in Th2 cells in human peripheral blood lymphocytes. Heijink et al. [76] reported that the production of cytokine (IFN-γ and IL-4/IL-5) in polarized Th1 and Th2 cells is not affected by the activation of AC/cAMP-linked β_2-ARs. Stimulation of β_2-ARs on naive murine Th cells induced the Th cells to develop into Th1 cells [77]. Stimulation of long-term cultures of allergen-specific T cell lines by β_2-AR agonists enhanced the secretion of a Th1 cytokine (IFN-γ) and reduced the secretion of Th2 cytokines (IL-4 and IL-5) [78]. From these findings, the ability of adrenergic ligands to influence Th cell differentiation appears to depend on the AR density (which is related to cell maturity or state of activation) and the cytokine milieu in which T cells reside during their initial activation and differentiation. Several recent studies have provided further evidence that activated or impaired β_2-ARs influence the immune response in autoimmune/inflammatory disease. β_2-AR stimulation has been suggested to suppress the development of autoimmune myocarditis by inhibiting T-cell activation and by shifting the imbalance in Th1/Th2 cytokine toward Th2 cytokine [79]. In contrast, allergen provocation in asthma patients has been suggested to reduce β_2-AR control by chemokine release and activation of β-ARK, and to enhance Th2-like activity [80,81]. However, β-ARK was shown to be downregulated in $CD4^+$ cells in spleen and MLN of adjuvant arthritic rats [82].

Tc-mediated cytotoxic killing of allogenic tumor cells in a mixed mouse lymphocyte culture was found to be stimulated by a β-AR and inhibited by an α-AR [83]. However, in another study, CAs exerted biphasic effects on Tc-mediated cytotoxic activity against syngenic tumor cells. Low concentrations of NE stimulated the *in vitro* cytotoxic killing of the MOPC-815 plasmacytoma by splenic Tc cells from mice rejecting a large MOPC-315 tumor as a consequence of L-phenylalanine mustard treatment, and higher concentrations of NE inhibited the anti-MOPC315 cytotoxic activity [84]. The inhibitory effect of higher doses of NE is mediated by the inhibition of tumor necrosis factor-α (TNF-α) production, and involves β-AR signaling [85]. Thus, NE seems to exert opposite effects on Tc functions in a manner that is dependent on not only the AR subtypes but also the NE concentrations and/or the nature of the target cells.

5. EFFECT OF ADRENERGIC STIMULATION ON B LYMPHOCYTES

Norepinephrine and β_2-AR effect on B cells have been markedly studied by Sanders and colleagues *in vitro* and *in vivo*, and their studies suggest B-cell activation by β_2-AR stimulation. Sanders and Powell-Oliver [86] found that stimulation of β_2-ARs on B cells enhances

IgM production induced by TH-dependent antigens. β_2-AR stimulation also enhances the level of B cell receptor-induced costimulatory molecule B7-2 (CD86) expression in murine splenic B cells by increasing B7-2 mRNA stability, via protein tyrosine kinase-, PKA-, PKC-, and MAPK-dependent mechanisms [87]. In addition, β_2-AR stimulation upregulates the CD86-induced IgG1 production in CD40 and IL-4-activated B cells by increasing the expression of the transcription factor Oct-2 coactivator OCA-B in a PKA-dependent manner [88,89]. Moreover, β_2-AR stimulation promotes IgE production by increasing CD23 expression in PKA-independent and p38 MAPK-dependent manner [90]. In the studies of other groups, β_2-AR stimulation was found to potentiate IL-4-induced IgE production from lipopolysaccharide (LPS)-activated murine B cells [71]. However, Li et al. [91] showed that NE increases the B-cell proliferation and maturation into IgM-, IgG- and IgA-secreting cells induced by LPS and a membrane proteoglycan from *Klebsiella pneumoniae*, but not by anti-mouse mu-chain antibodies (anti-mu). They found that NE, rather, inhibits the B-cell proliferation induced by anti-mu via β-ARs and cAMP-dependent pathway [91]. The *in vivo* study in which the influence of CAs was examined on the lymphocyte reactivity during chronic mild stress showed the β_2-AR-meditated inhibition of mitogen-induced B-cell proliferation and T cells, concomitant with a significant increase in β_2-AR density [67]. Furthermore, epilepsy-prone mice, in which splenic NE levels are higher than in epilepsy-resistant mice, show the β_2-AR-meditated reduction of anti-sheep red blood cell (SRBC) IgG production when mice are immunized with SRBC [92]. Taken together, these results indicate that CAs regulate B-cell activation mainly via β_2-ARs, depending on the activator, and with a distinct biochemical mechanism.

6. EFFECT OF ADRENERGIC STIMULATION ON NATURAL KILLER CELLS

Catecholamines have been mostly reported to inhibit NK activity mainly via β_2-ARs and cAMP production [70,93–97]. Norepinephrine, epinephrine, and β_2-AR agonists inhibit both NK cell cytotoxicity and antibody-dependent cell cytotoxicity by reducing the expression of CD1 (also called FcγRIII, which is an NK cell receptor responsible and necessary for antibody-dependent cell cytotoxicity and cytokine secretion), IL-2-mediated upregulation of the activation marker CD69, secretion of TNF-α, IFN-γ and GM-CSF, and the production of perforin and granzyme B [95–97]. Brief incubation of NK cells with epinephrine, and to a lesser extent NE, causes them to detach from cultured human endothelial cells [98]. This epinephrine-induced detachment was mediated by β_2-ARs and cAMP [98]. This suggests that CAs, via β_2-ARs, can induce the recruitment of NK cells from the marginating pool to the circulating pool, possibly by altering the adhesive interactions between NK cells and endothelial cells. Supporting this notion are *in vivo* studies showing that CA infusion acutely increases NK cells in the peripheral blood via a spleen-independent β_2-AR mechanism [99–101]. During the remarkable increase of NK cells in the peripheral blood, the blood NK activity is suppressed by β-AR stimulation [100]. Splenic NK activity is also reduced by the stimulation of the splenic sympathetic nerves [102]. Among the different types of lymphocytes, NK cells appear to be most sensitive to stress, which supports the hypothesis that NK cells are regulated by the SNS. Exhaustive exercise by healthy volunteers was found to increase the number and activity of NK cells via β_2-AR mediation [65]. However, another study showed that physical and psychological stress was found to increase the number of NK cells and to decrease their activity by the SNS combined with the β_1- or β_2-AR-mediated mechanism [103–106].

7. EFFECT OF DEPLETION OF NORADRENERGIC INNERVATION ON LYMPHOCYTES

One approach to study the role of the SNS in the immune system is to destroy the sympathetic nerve terminals in lymphoid organs, either chemically or surgically, followed by the assessment of immune reactivity *in vivo* or *in vitro*. The most common sympathectomy method is to chemically eliminate sympathetic nerve terminals by using 6-hydroxydopamine (6-OHDA), which is a "false" transmitter substance that selectively destroys postganglionic adrenergic elements. When 6-OHDA is administered systemically to adults, the sympathectomy is limited to the periphery because 6-OHDA does not cross the blood–brain barrier (BBB). This treatment destroys the nerve terminals but not the cell bodies and thus the nerve fibers can regenerate over time. However, if 6-OHDA is administered to neonates, the cell bodies are destroyed, which results in long-term depletion of the nerve fibers in the periphery. In addition, since the neonatal BBB is not mature, 6-OHDA can access the CNS, which causes the destruction of the central noradrenergic and dopaminergic pathways. After acute treatment with 6-OHDA, NE levels and noradrenergic nerve terminal numbers are significantly decreased in most peripheral organs, and they are reduced by over 90% in spleen and lymph nodes [15,107–109]. Reinnervation occurs by a common pattern with the postnatal development of noradrenergic innervation of the lymphoid organs [108]. Because the effect of CAs on immune responses is variable, as revealed by *in vitro* and *in vivo* studies, the results of 6-OHDA-induced denervation frequently have contradictory effects on immune responses. In DBA/2 mice, 6-OHDA enhanced mitogen-induced proliferation of splenic T cells (but not B cells) by peripheral axotomy [110]. However, 6-OHDA had no such effect in C57BL/6 mice, even though the two strains had similar levels of splenic NE depletion [110]. In addition, 6-OHDA denervation increases keyhole limpet hemocyanin (KLH)-stimulated production of Th1 and Th2 cytokines and increases the production of anti-KLH IgM and IgG by splenocytes [111–113]. However, 6-OHDA decreases Con A-stimulated Th1 cytokine production, but not Con A-stimulated Th2 cytokine production [112]. In mice with increased circulating NE and Th2 responses as a result of thermal injury, 6-OHDA denervation was found to reduce Th2 cytokine production [114]. 6-OHDA was also found to have an organ-specific effect on immune function, that is 6-OHDA treatment of adult mice reduces *in vitro* Con A-induced proliferation of spleen and LN cells, but increases Con A-induced IFN-γ production [115]. 6-Hydroxydopamine treatment of mice does not alter Con A-induced IL-2 production in LN cells, but decreases it in splenocytes [115]. The SNS was found to have bimodal roles in type II collagen-induced arthritis [116]. In mice with collagen-induced arthritis, early sympathectomy, sympathectomy performed 7 days before immunization, increased the production of Th2 cytokines (IL-4 and IL-10) by LN cells and reduced arthritis scores in the mice. In contrast, late sympathectomy, sympathectomy performed on day 56, increased the production of Th1 cytokines (IFN-γ and TNF-α) by LN and splenic cells and resulted in increased arthritis scores. Bimodal effects of sympathectomy were also demonstrated during experimental autoimmune encephalomyelitis (EAE) in wild-type and IL-4 knockout mice [117]. When mice are actively sensitized with a peptide corresponding to residues 35–55 of myelin oligodendrocyte glycoprotein peptide, sympathectomy causes paralysis with higher disease scores in the wild type, but suppresses EAE in IL-4 knockout mice. 6-Hydroxydopamine treatment prior to or following antigen sensitization decreases the antigen-specific Tc/s cell generation in LN [118,119]. Reduction of splenic Tc/s generation was also reported by neonatal injection of 6-OHDA to mice [118]. These data emphasize that SNS regulation of T-cell function is heterogeneous and can vary in a time-, organ-, strain-, and lymphocyte activator (antigen or mitogen)-specific fashion, and even the absence of a single cytokine can severely alter the SNS influence in immune responses.

Sympathetic nervous system regulation of B-cell function is also heterogeneous. LPS-stimulated proliferation and polyclonal IgG secretion of B cells are enhanced in LN, but are reduced in spleen [117]. Kohm and Sanders [120] treated severe combined immunodeficient (SCID) mice with 6-OHDA before reconstitution with a KLH-specific Th2 cell clone (β_2-AR negative) and resting trinitrophenyl (TNP)-specific B cells (β_2-AR positive) isolated from the spleen of unimmunized mice. The serum levels of TNP-specific IgM and IgG resulting from the application of TNP–KLH were significantly reduced in 6-OHDA-treated SCID mice, and the decreased antibody levels were partially recovered by the administration of a β_2-AR-selective agonist. However, neonatal sympathectomy of *lpr/lpr* mice, which develops a lymphoproliferative, autoimmune lupus-like disease, with 6-OHDA further increases the blood concentration of IgM and IgG2a and lymphadenopathy [121].

Sympathectomy of mice with 6-OHDA at birth changes the function of splenic NK cells with development [122], and central and peripheral treatment of 6-OHDA reverses the cold stress-induced suppression of NK cytotoxicity [123], suggesting that the SNS regulates NK function during development and acute stress.

A few reports described the effects of surgical sympathetic denervation on the immune response. Superior cervical ganglionectomy exhibited decreased cellularity and NK activity, but increased LPS-induced cell proliferation and plaque-forming cell response to SRBCs in submaxillary lymph nodes [124,125]. Immobilization-induced suppression of splenic NK cytotoxicity is also attenuated by surgical denervation of the splenic nerve [126]. The physiological roles of the SNS in the immune system were also examined by using DBH-deficient mice, when housed in specific pathogen-free conditions, which cannot produce NE or epinephrine, but they produced dopamine, had normal numbers of blood leukocytes, and normal T- and B-cell development and *in vitro* immune function [127]. However, when challenged *in vivo* by infection with the intracellular pathogens *Listeria monocytogenes* or *Mycobacterium tuberculosis*, DBH-deficient mice are more susceptible to infection, and have impaired T-cell function, including Th1 cytokine production. When immunized with TNP–KLH, the DBH-deficient mice produce less Th1 cytokine-dependent IgG2a anti-TNP antibody. These results suggest that endogenous NE and epinephrine are not required for normal development of the immune system, but that they are important in the modulation of T cell-mediated immunity to infection and immunization, and play central roles in stress-induced NK suppression.

8. EFFECTS OF CYTOKINES ON THE SYMPATHETIC NERVOUS SYSTEM

It is believed that the immune system communicates with the SNS by two routes (Fig. 1). First, the immune system is known to influence the hypothalamic or other CNS centers that regulate the autonomic output. Proinflammatory IL-1β has been suggested to signal the brain, via the circulation or through the afferents of the vagal nerves, and to activate the central corticotropin-releasing factor (CRF)-containing neurons and/or prostaglandin E2 production, then to trigger the activation of the SNS and NE release in the spleen [128–133]. Other immune cell-derived cytokines, such as IL-2 and TNF-α were reported to activate central noradrenergic turnover, mediating sympathetic outflow to lymphoid organs [134,135].

Second, the immune system appears to more directly regulate the activity of the sympathetic nerves in the lymphoid organs and inflammatory sites or in the sympathetic ganglia. Cultured

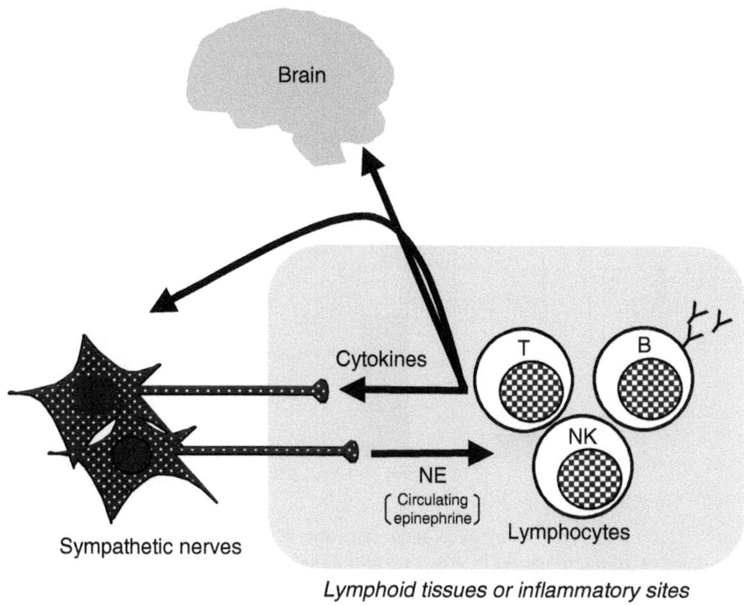

Figure 1. Signaling pathways mediating interactions between the SNS and the immune system. Sympathetic neurons regulate various immune responses via NE and circulating epinephrine in lymphoid organs and inflammatory sites. In contrast, certain cytokines secreted during the immune response act on the SNS by being transported to the brain through the circulation or through the afferents of the vagus nerves. The action of these cytokines on the SNS triggers the activation of sympathetic outflow. Alternatively, they act on the sympathetic nerves directly and locally in lymphoid organs or inflammatory sites, or they may arrive at sympathetic ganglia through the circulation.

sympathetic neurons were demonstrated to express IL-1β, type 1 IL-1 receptors (IL-1RI), IL-1β, and IL-6, or TNF-α stimulation increased the expression of IL-1RI [136–138]. IL-1 stimulation of neurons also enhances their intracellular NF-κB DNA-binding activity [137] and inhibits splenic vasoconstrictor tonus at the postganglionic, prejunctional level [139]. Thus, IL-1 may modulate sympathetic regulation in an autocrine or paracrine manner, and it could be potentiated in inflammatory conditions. Many studies have also shown that IL-1α or IL-1β treatment of glial cells or targets of sympathetic or other NGF-sensitive neurons in both the brain and the periphery induces them to produce NGF, with or without costimulation of TNF-α or platelet-derived growth factor (PDGF)-BB [140–146]. In our *in vitro* study, IL-1β promoted neurite outgrowth from murine sympathetic superior cervical ganglia (SCG), which was completely inhibited by anti-NGF, suggesting that IL-1 mediates NGF production, which in turn induces the growth of sympathetic neurons [147]. We also found the TNF-α induced NGF-dependent neurite outgrowth (Fig. 2; unpublished data).

Sympathetic neurons have also been demonstrated to express gp130 and leukemia inhibitory factor (LIF) receptor, subunits of the receptor that binds cytokines of the IL-6 family including IL-6, IL-11, ciliary neurotrophic factor (CNTF), LIF, oncostatin M (OSM), cardiotrophin-1, and novel neurotrophin-1/B cell-stimulating factor-3 (NNT-1/BSF-3) [148–150]. Although IL-6 and IL-11 induce homodimerization of gp130 to activate the signaling cascade, other members of the IL-6 family induce heterodimerization of gp130 and LIF receptors. IL-6 family members have been reported to have both negative and positive effects on neuronal

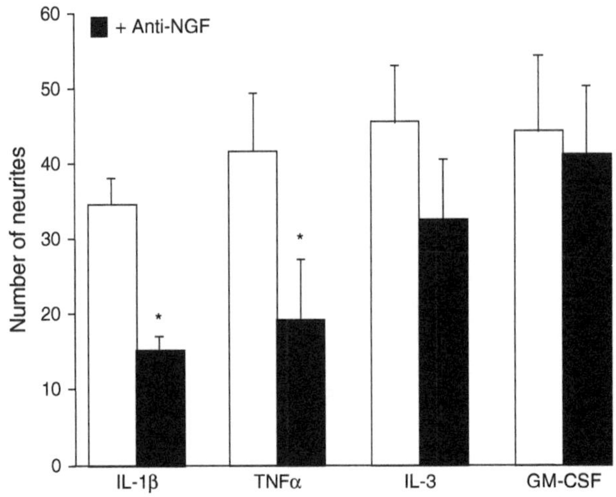

Figure 2. Inhibitory effects of anti-NGF on cytokine-induced neurite outgrowth from SCG. Murine SCG were cultured for 1 day in the presence of 10 ng/ml murine IL-1β, 1 ng/ml murine TNF-α, 10 U/ml murine IL-3, or 1 U/ml murine GM-CSF, with or without 5 μg/ml anti-NGF. Bars represent the mean numbers of neurites (+SE). In this experiment, NGF (10 ng/ml) induced 163 ± 12.3 neurites without anti-NGF, and 8 ± 6.8 neurites with anti-NGF. $*p < 0.01$, compared to anti-NGF (−) ($n = 5$–29).

differentiation and the survival of cultured sympathetic neurons [150–157]. Among them, NNT-1/BSF-3 was found to be mainly produced in lymph nodes and spleen, and to support the survival of chick sympathetic neurons [155]. The results of *in vivo* studies imply that IL-6 and LIF have facilitatory role in sympathetic sprouting in dorsal root ganglia (DRG) induced by nerve injury [158,159].

IL-2 and IL-12 were reported to enhance neurite outgrowth of cultured sympathetic neurons [160,161].

There is recent evidence that T and B cells express NGF and other neurotrophins (NTs) [brain-derived neurotrophic factor (BDNF), NT-3, NT-4/5], and that antigenic or mitogenic activation increases NGF levels in Th2 cells [160–167]. Barouch et al. [165] found that β-AR agonists as well as T- and B-cell mitogens stimulate murine splenocytes to express NGF. Lymphocyte-derived NGF is considered to play dual role as a neurotrophic factor for neighboring peripheral neurons and as a para-autocrine factor for immune cells acting through a high-affinity NGF receptor (TrkA) with or without a low-affinity NGF receptor (p75) [168–175].

To determine whether cytokines have a role in sympathetic neuroimmune interactions within lymphoid organs, we used a coculture model of rodent explants [147,176]. Adult lymphoid tissue explants (about 1 mm^3) were cultured together with neonatal SCG in a Matrigel layer at a distance of 1 mm. After 1 day in culture, many neurites grew toward the thymus and the spleen (Figs 3 and 4). Although this effect was weaker with MLN, activated MLN isolated from mice 5 to 10 days after infection with the nematode *Nippostrongylus brasiliensis* caused significantly increased neurite outgrowth compared to normal MLN (see Fig. 4). An NGF-blocking antibody inhibited most of the neurite outgrowth toward the thymus (84% inhibition) and the spleen (79% inhibition), and to a lesser extent toward MLN (64% inhibition). Interestingly, the

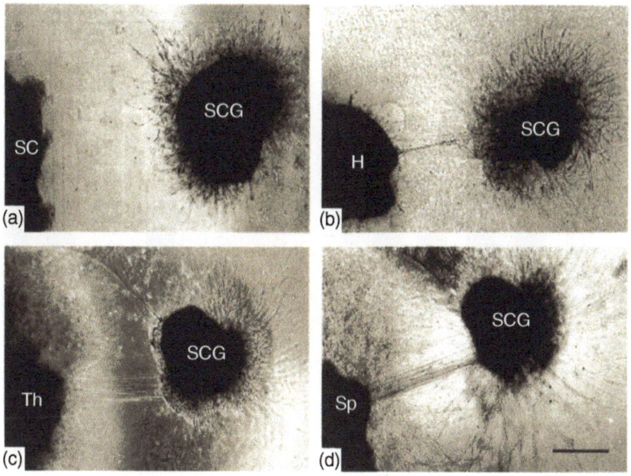

Figure 3. Phase contrast photomicrographs of rat SCG and target tissues after 2 days of coculture in Matrigel. (a) Coculture of SCG and spinal cord (SC; a negative control). Only a few long neuritis and some short fibers were visible. (b) SCG and heart (H; a positive control). Many neurites have extended directly toward the heart explants. (c) SCG and thymus (Th). (d) SCG and spleen (Sp). Neurite outgrowth toward thymus and spleen explants was similar to that seen with heart explants. Scale bar, 500 μm.

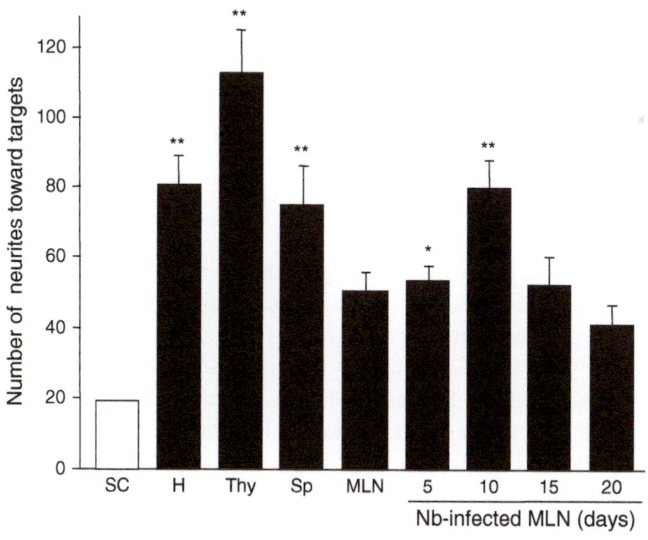

Figure 4. SCG neurite outgrowth induction by murine lymphoid tissues. The mean number of neurites and SE was calculated from 19 to 26 SCG after 1 day of coculture with SC, heart (H), thymus (Thy), spleen (Sp), MLN, or MLN isolated from mice infected 5, 10, 15, or 20 days previously with *Nippostrongylus brasiliensis* (Nb5-20). $*p < 0.05$, $**p < 0.01$, compared to SC.

blocking effect of anti-NGF on Nb-MLN (day 10)-induced neurite outgrowth was much less than the blocking effect on the thymus, spleen, and MLN (49% inhibition). To determine if cytokines are involved in the induction of neurite outgrowth by lymphoid tissues, anti-mouse cytokines were administered to cocultures of SCG and lymphoid tissues. Neurite outgrowth

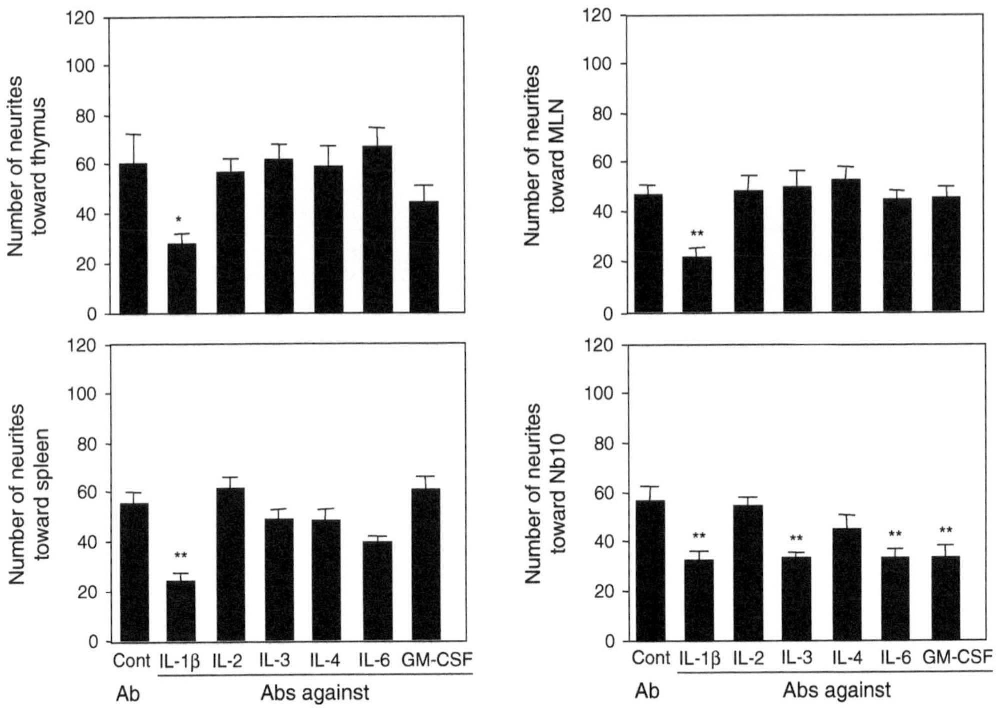

Figure 5. Inhibition of murine lymphoid tissues-induced neurite outgrowth from SCG by anti-cytokine mAbs. Thymus, spleen, MLN and Nb-MLN-10d (Nb10) were cocultured with SCG in the presence of blocking Abs against the indicated cytokines. The mean number of neurite and SE was calculated from 18 to 30 SCG after 1 day of culture. $*p < 0.05$, $**p < 0.01$, compared to control Ab.

toward the thymus, spleen, or MLN was significantly suppressed by anti-IL-1β (50–60% suppression), but not by anti-IL-2, IL-3, IL-4, IL-6, or anti-GM-CSF (Fig. 5). We confirmed that exogenous IL-1β promoted neurite outgrowth from SCG, and the effect was completely inhibited by anti-NGF (see Fig. 2). Thus, IL-1β would seem to stimulate neuritogenesis via NGF production within lymphoid tissues. As IL-1β is preferentially secreted by macrophages and acts as a major T-cell activator, it is likely to stimulate NGF secretion by T cells. In addition to anti-IL-1β, anti-IL-3, anti-IL-6, and anti-GM-CSF significantly inhibited the neurite outgrowth toward Nb-MLN (see Fig. 5). Thus, IL-3, IL-6, and GM-CSF might play important roles in stimulating nerve growth during inflammation. IL-3 and GM-CSF are hematopoietic cytokines that act as neurotrophic factors on central cholinergic neurons *in vitro* and *in vivo* [177]. However, because it is unclear whether they have their effects on sympathetic neurons, we investigated their neurotrophic effects in more detail [178]. Exogenous IL-3 or GM-CSF added to murine SCG caused neurite outgrowth, and this effect was not blocked by anti-NGF antibody (see Fig. 2). Similarly, the addition of physiological doses of IL-3 or GM-CSF to the culture of dissociated SCG neurons increased the number of process-bearing neurons. These effects were completely blocked by their corresponding antibodies but not by anti-NGF, indicating that their action is specific and completely independent of NGF. In addition, IL-3 and, to a lesser extent, GM-CSF protect NGF-differentiated SCG neurons from apoptotic cell death caused by NGF withdrawal. IL-3 and GM-CSF stimulation of the differentiated neurons caused in a rapid elevation of MAPK activity, and PD98059, an inhibitor of MAPKK activity, blocked both the neuritogenic and neuroprotective effects of IL-3 and GM-CSF. Immunocytochemical studies showed that IL-3 and GM-CSF receptors (both the α subunit and the signal-transducing common β subunit) are present on the NGF-differentiated neurons. These data indicate that IL-3 and GM-CSF are able to stimulate sympathetic nerve growth, via their specific cytokine receptors on neurons, which lead to the activation of the MAPK pathway that then mediates the observed neurotrophic effects (Table 1). During inflammation, some of the locally produced cytokines may act directly to promote sympathetic nerve growth. By traveling through the blood, they may also act indirectly by stimulating NGF production in lymphoid organs or inflammatory sites, or sympathetic ganglia (Fig. 6).

Table 1. Neurotrophic action of IL-3 and GM-CSF on cultured murine SCG neurons

Process outgrowth of SCG neurons were promoted by IL-3 and GM-CSF, which were not inhibited by anti-NGF
IL-3 and, to a lesser extent, GM-CSF protected NGF-differentiated neurons from apoptotic cell death caused by NGF withdrawal
IL-3 and GM-CSF stimulation resulted in a rapid elevation of MAPK activity, and PD98059 blocked the neurotrophic actions of both cytokines
IL-3 and GM-CSF receptors were detected on the cultured neurons immunocytochemically

Kannan et al. [178].

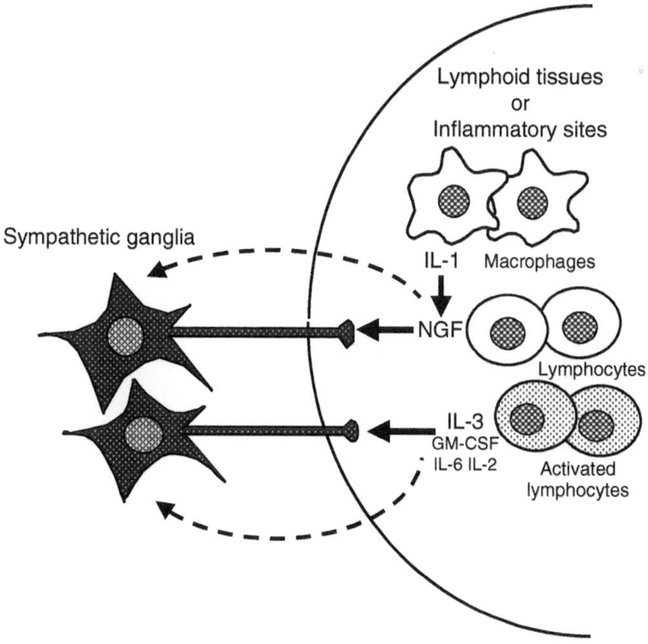

Figure 6. Local regulation of cytokines on the sympathetic nerves. IL-1 seems to constantly stimulate nerve growth via NGF production within lymphoid tissues. Moreover, some cytokines such as IL-2, IL-3, IL-6, and GM-CSF produced by activated lymphocytes may also directly stimulate sympathetic nerve growth. These cytokines may also affect the nerve growth of sympathetic ganglia through the circulation.

In vivo studies using immunodeficient or autoimmunological animals, the spleen of athymic nude mice has increased NE levels and fluorescent catecholamine-containing nerve fibers [179]. Moreover, dense TH^+ immunoreactivity was detected in the perivascular and subcapsular regions and, to a lesser extent, in the underlying cortical regions of thymus in SCID mice [180]. Severe combined immunodeficient mice lack functional T and B cells as a result of a spontaneous autosomal-recessive mutation that is located on chromosome 16 that blocks the expression of functional cell surface antigen receptors [181]. MRL-*lpr/lpr* mice serve as a genetic model of systemic lupus erythematosus, a disease in which the number of $CD4^-/CD8^-$ T cells is increased, and AIDS mice are characterized by the rapid and profound lymphoproliferation and dysregulation of cytokine production accompanied by hypergammaglobulinemia and immunosuppression. Both mice reduced NE levels and nerve destruction in the spleen

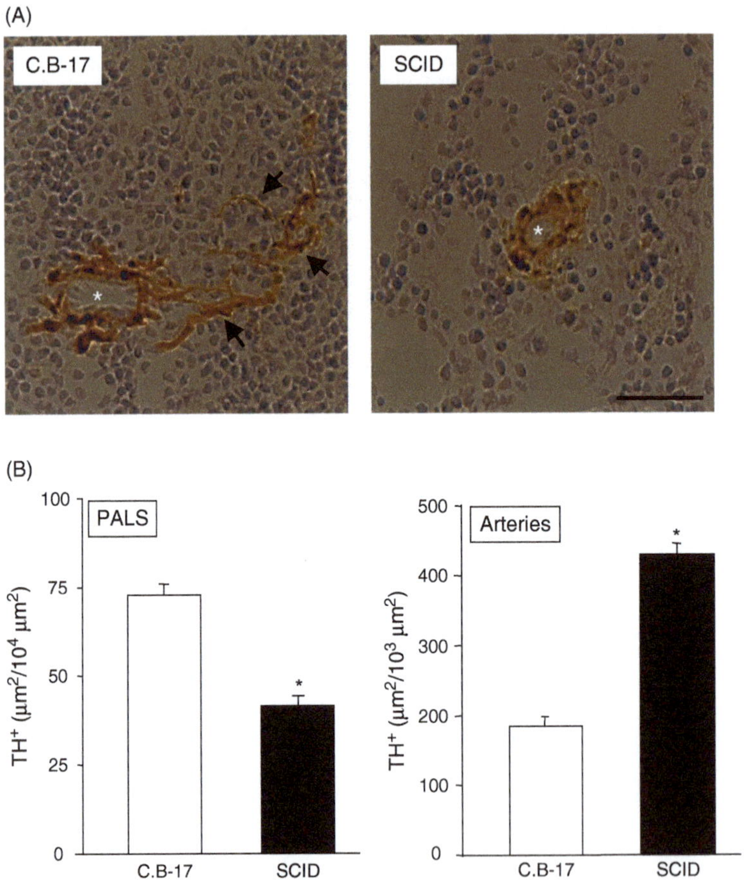

Figure 7. (a) Photomicrographs showing TH^+ fibers in the white pulp of wild-type (C.B-17) and SCID mice. TH^+ fibers were densely aggregated (white asterisks) around central arteries. Some TH^+ fibers were found to extend into the parenchyma of the PALS in C.B-17 mice (arrows), while only very few fibers did so in SCID mice. Bar, 50 μm. (b) The magnitude of the TH^+ areas in C.B-17 (open columns) and SCID (closed columns) mice was estimated using image analysis software. Each value represents the mean ± SE of 23 (C.B-17) and 28 (SCID) photographic images from 6 (C.B-17) and 7 (SCID) mice. $^*p < 0.01$, compared to C.B-17.

[182,183]. Interestingly, noradrenergic nerve density in the spleens of arthritic rats declined in the white pulp and the regions distal to the hilus, but increased in the red pulp and the hilar regions [184]. This indicates that noradrenergic nerves die back and undergo a compensatory sprouting response, and that activated immune cells migrated and localized in the red pulp to signal the noradrenergic nerves during chronic inflammatory stages of arthritis. Serpe et al. [185], using a peripheral nerve lesion model, showed that facial motoneuron loss is exacerbated in SCID mice, but can be reversed by T- and B-cell reconstitution. In addition, Armstrong et al. [186] showed that the neuropeptide pituitary adenylate cyclase-activating polypeptide (PACAP) restored motor neuron axotomy-induced gene induction in SCID mice preinfused with $CD4^+$ T cells. Thus, lymphocytes, particularly TH cells, may play a supporting role in some types of neurons. We examined the innervation and cytokine levels in the spleen of SCID mice, and we found that SCID mice have significantly fewer TH+ fibers in the PALS, while the density of innervation of their splenic central arteries is increased compared to wild-type C.B-17 mice (Fig. 7) [187]. The change in the splenic innervation of SCID mice was eliminated by the injection of C.B-17 splenic T cells, but not B cells. Following T-cell reconstitution, the level of IL-3, but not the level of IL-1β significantly increased in the spleen of SCID mice. Furthermore, the administration of an anti-IL-3 antibody blocked the T cell-induced increase in the innervation of the PALS. We also found that treatment of C.B-17 mice with 6-OHDA caused significant loss of fibers in the spleen one day after treatment. Reinnervation was observed 10 days later, and recovery appeared to be completed by 21 days. The regeneration of fibers in the PALS and vasculature was significantly blocked by anti-IL-3. Our results suggest that T cells play a significant role in promoting the sympathetic nerve growth in the spleen. This effect appears to be mediated, at least in part, by IL-3.

9. CATECHOLAMINE SYNTHESIS BY LYMPHOCYTES

There is increasing evidence that lymphocytes not only receive NE from the sympathetic nerve terminals, but also produce CAs themselves. Bergquist et al. [188] detected dopamine in human lymphocytes taken from the cerebrospinal fluid. Subsequently, Musso et al. [189] detected NE in T cells but not in B cells collected from human peripheral blood. Marino et al. [190] detected dopamine and epinephrine as well as NE in human peripheral blood mononuclear cells. In addition, dopamine and NE were detected in mouse splenocytes and some T- and B-cell hybridomas [191]. CA synthesis in lymphocytes was demonstrated by the detection of TH, an increase of CAs by the addition of tyrosine or L-DOPA (precursors of CA synthesis), and a decrease of CAs by the addition of a TH blocker in the culture medium [192–194]. In addition, lymphocytes stimulated with PHA show enhanced TH gene expression and enzyme activity [192]. As cells in the S+G2/M phase are more likely to be TH^+ than cells in the G0/G1 phase, TH expression correlates with the cell cycle.

As it is not known whether rat lymphocytes synthesize CAs, we examined the CA contents in lymphocytes isolated from the thymus, spleen, and MLN of adult Sprague-Dawley rats (unpublished data). We found that splenic lymphocytes contain much higher levels of NE and epinephrine than thymic and MLN cells. We also found that Con A, KLH or LPS stimulation significantly increased the NE levels in splenic lymphocytes during 24 and 48 h of culture (Fig. 8), and that treatment of the cells with α-methyl-*p*-L-tyrosine (α-MT), an inhibitor of TH, completely inhibited the NE increase. Thus, it appears that

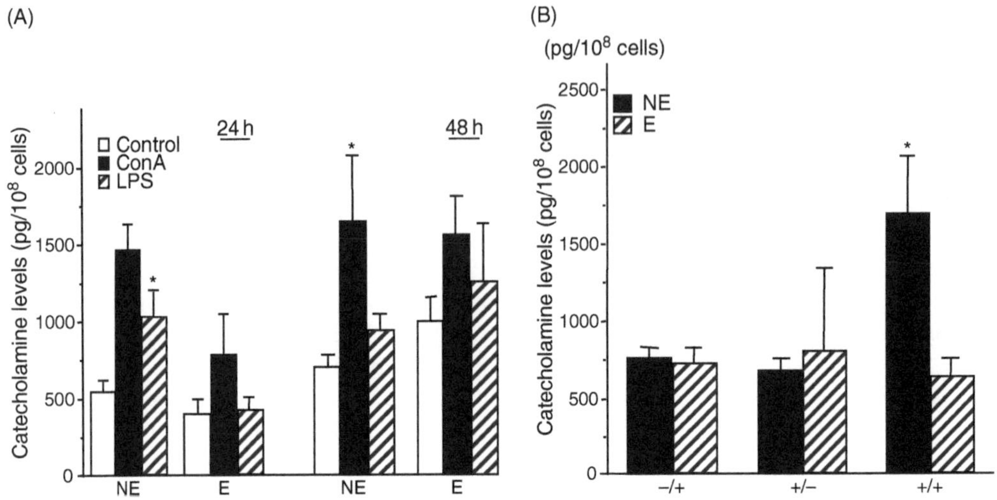

Figure 8. Changes in CA levels in splenic lymphocytes during mitogen and KLH stimulation. Freshly isolated lymphocytes (10^8 cells) contained 10281 ± 50.7 pg NE and 305 ± 111.6 pg epinephrine (E). (A) Splenic lymphocytes were cultured in the presence of Con A (2 μg/ml) or LPS (25 μg/ml) for 2 days. $^*p < 0.05$, $^*p < 0.01$, compared to control. (B) Rats were injected 1 mg KLH and spleen were isolated after 9 days. Lymphocytes were cultured in the presence or absence of KLH (150 mg/ml) for 1 day. −/+: no priming and KLH stimulation in culture; +/−: KLH priming and culture without KLH; +/+: KLH priming and stimulation in culture. $^*p < 0.01$, compared to −/+ and +/−.

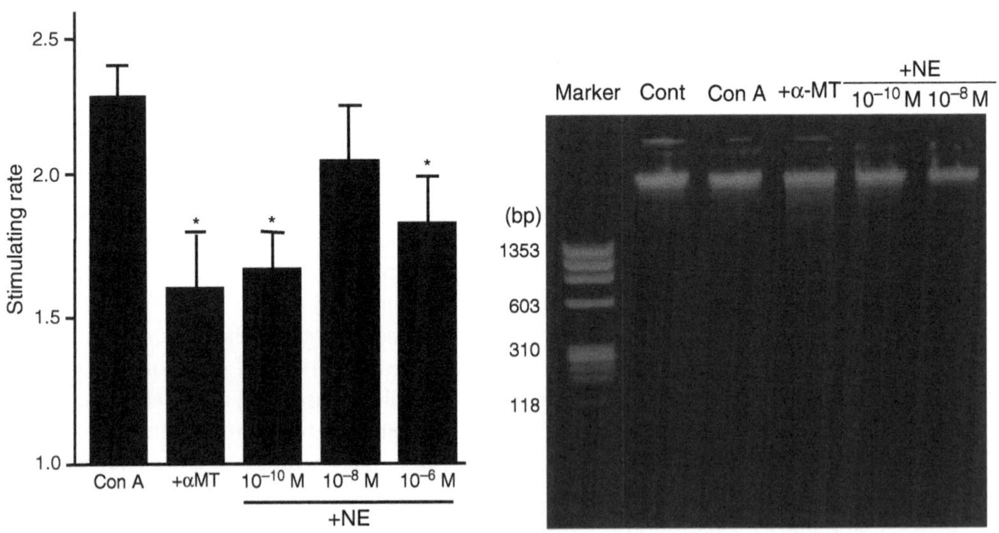

Figure 9. Effect of α-MT and NE on lymphocyte proliferation and apoptosis. Lymphocytes (2×10^5 cells) were cultured with 2 μg/ml Con A and 10^{-3} M α-MT for 1 day. NE was added 3 and 7 h after cell plating. (Left) BrdU was added 2 h before cell harvesting and the proliferation assay was performed by ELISA. The vertical bar shows the increase in the rate of stimulation in relation to the OD value of the control medium. $^*p < 0.05$, compared to Con A ($n = 8$–22). (Right) Agarose gel electrophoresis of lymphocyte DNA was performed after culture in the presence of Con A with or without α-MT and NE for 1 day, as mentioned above.

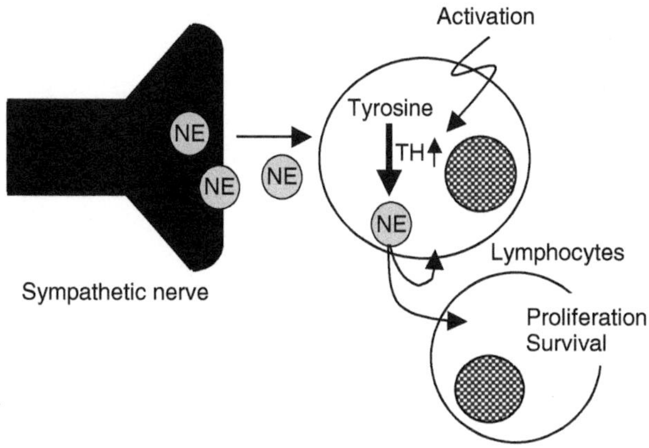

Figure 10. A schematic diagram of NE synthesis and function in lymphocytes. NE synthesis could be accelerated during immune responses and the NE synthesized might positively regulate lymphocyte proliferation and survival (and probably other functions as well) in an autocrine/paracrine fashion.

both T and B cells increase their TH-dependent NE synthesis during mitogen or antigen-specific activation. These observations are consistent with the above-mentioned finding of Reguzzoni et al. [192].

!Since the physiological roles of the CAs synthesized by lymphocytes are unclear, we examined whether α-MT treatment alters the proliferation and apoptosis of lymphocytes. We found that α-MT inhibits the proliferation of Con A-stimulated lymphocytes and increases their rate of apoptosis (Fig. 9). These effects were reversed by the addition of 10^{-9}–10^{-8} M NE (see Fig. 9). α-methyl-p-L-tyrosine treatment did not alter the levels of IL-2 secretion by the lymphocytes. These observations suggest that lymphocyte-derived NE participates in lymphocyte activation by at least enhancing proliferation (in an IL-2-independent manner) and serves to suppress apoptosis at low concentrations by acting in an autocrine/paracrine way (see Fig. 10). Further studies will be needed to better understand the roles played by immune cell-derived CAs.

10. CONCLUSIONS

Lymphoid organs are densely innervated with sympathetic neurons. Postsynaptic NE and circulating epinephrine modulate various immune functions by binding to ARs expressed on immune cells. Histofluorescent and immunohistochemical studies have established that lymphoid organs, particularly the spleen, receive predominant sympathetic noradrenergic and NPY innervation. The nerve fibers enter with the vasculature, travel along the trabeculae and along the branching vasculature, and are distributed mainly in the white pulp along the central artery and its branches. Some of the fibers extend away and pass into the T cell-rich PALS to reach the marginal zone that contains macrophages and B cells. ARs are unequally distributed among lymphocyte subpopulations. The high-affinity β_2-ARs tend to be expressed at high levels on NK cells, at intermediate levels on Tc cells and B cells, and at low levels on Th cells. Th1 cells but not Th2 cells express detectable levels of β_2-ARs. The density of ARs varies widely and appears to depend on the degree of lymphocyte maturation and activation and by the presence or absence of CAs.

In vitro and *in vivo* studies involving chemical and surgical sympathectomy have shown that mechanisms of the SNS regulation of T and B cell function are very complicated. The regulatory mechanisms are heterogenous and can vary in a time-, organ-, strain-, and lymphocyte activator (antigen or mitogen)-specific fashion. In general, NE and epinephrine promote humoral immunity rather than cellular immunity by suppressing the Th1 response and by upregulating T cell-dependent antibody production. The SNS also strongly affects NK cell function, especially during stress. The SNS plays a central role in the stress-induced suppression of NK cytotoxicity. In addition, CAs, via β_2-ARs, acutely increase NK cells in the peripheral blood, by the recruitment from the marginating pool to the circulating pool, possibly by altering the adhesive interactions between NK cells and endothelial cells.

Moreover, immune cells not only influence the CNS centers that regulate the autonomic output, but also regulate more directly the activity of the sympathetic nerves in the lymphoid organs and inflammatory sites or in the sympathetic ganglia. A proinflammatory IL-1β signal to the brain activates the central CRF-containing neurons and/or prostaglandin E2 production, which triggers the activation of the SNS, resulting in NE release in the spleen. IL-1β might locally support the sympathetic innervation of lymphoid organs by inducing immune cells to secrete NGF. T cell-derived cytokines such as IL-2, IL-3, IL-6, and GM-CSF can stimulate sympathetic nerve growth. Moreover, T cells play an important role in sympathetic innervation and reinnervation following sympathectomy in lymphoid tissues. Such innervation appears to be at least partly mediated by IL-3.

Lymphocytes can synthesize CAs by themselves, especially when they are in an activated stage. Lymphocyte-derived CAs may directly regulate immune responses in a similar fashion to other cytokines, namely in an autocrine/paracrine way.

Further studies are needed to fully understand the physiology and pathophysiology of the interaction between the SNS and the immune system. Such studies will be very helpful in devising therapeutic strategies for autoimmune, allergic, autonomic or stress-related diseases.

REFERENCES

1. Elenkov IJ, Wilder RL, Chrousos GP, Vizi ES. The sympathetic nerve – An integrative interface between two supersystems: The brain and the immune system. Pharmacol Rev 2000;52:595–638.
2. Felten DL, Felten SY, Carlson SL, Olschowka JA, Livnat S. Noradrenergic and peptidergic innervation of lymphoid tissue. J Immunol 1985;135:755–65s.
3. Rogausch H, Bock T, Voigt KH, Besedovsky H. The sympathetic control of blood supply is different in the spleen and lymph nodes. Neuroimmunomodulation 2004;11:58–64.
4. Bellinger DL, Felten SY, Lorton D, Felten DL. Origin of noradrenergic innervation of the spleen in rats. Brain Behav Immun 1989;3:291–311.
5. Quinson N, Robbins HL, Clark MJ, Furness JB. Locations and innervation of cell bodies of sympathetic neurons projecting to the gastrointestinal tract in the rat. Arch Histol Cytol 2001;64:281–94.
6. Felten SY, Olschowka J. Noradrenergic sympathetic innervation of the spleen: II. Tyrosine hydroxylase (TH)-positive nerve terminals form synapticlike contacts on lymphocytes in the splenic white pulp. J Neurosci Res 1987;18:37–48.

7. Saito H. Innervation of the guinea pig spleen studied by electron microscopy. Am J Anat 1990;189:213–35.
8. Romano TA, Felten SY, Felten DL, Olschowka JA. Neuropeptide-Y innervation of the rat spleen: Another potential immunomodulatory neuropeptide. Brain Behav Immun 1991;5:116–31.
9. Bellinger DL, Lorton D, Horn L, Brouxhon S, Felten SY, Felten DL. Vasoactive intestinal polypeptide (VIP) innervation of rat spleen, thymus, and lymph nodes. Peptides 1997;18:1139–49.
10. Leposavic G, Micic M, Ugresic N, Bogojevic M, Isakovic K. Components of sympathetic innervation of the rat thymus during late fetal and postnatal development: Histofluorescence and biochemical study. Sympathetic innervation of the rat thymus. Thymus 1992;19:77–87.
11. Kranz A, Kendall MD, von Gaudecker B. Studies on rat and human thymus to demonstrate immunoreactivity of calcitonin gene-related peptide, tyrosine hydroxylase and neuropeptide Y. J Anat 1997;191:441–50.
12. Danko J, Ondrasovic M, Svicky E, Jenca A, Pospieszny N, Ondrasovicova O. Histochemical study of innervation and NADPH-D activity of the thymus. Anat Histol Embryol 2003;32:233–5.
13. Weihe E, Muller S, Fink T, Zentel HJ. Tachykinins, calcitonin gene-related peptide and neuropeptide Y in nerves of the mammalian thymus: Interactions with mast cells in autonomic and sensory neuroimmunomodulation? Neurosci Lett 1989;100:77–82.
14. Lorton D, Bellinger DL, Felten SY, Felten DL. Substance P innervation of the rat thymus. Peptides 1990;11:1269–75.
15. Felten DL, Livnat S, Felten SY, Carlson SL, Bellinger DL, Yeh P. Sympathetic innervation of lymph nodes in mice. Brain Res Bull 1984;13:693—99.
16. Fink T, Weihe E. Multiple neuropeptides in nerves supplying mammalian lymph nodes: Messenger candidates for sensory and autonomic neuroimmunomodulation? Neurosci Lett 1988;90:39–44.
17. Panuncio AL, De La Pena S, Gualco G, Reissenweber N. Adrenergic innervation in reactive human lymph nodes. J Anat 1999;194:143–6.
18. Nilsson G, Alving K, Ahlstedt S, Hokfelt T, Lundberg JM. Peptidergic innervation of rat lymphoid tissue and lung: Relation to mast cells and sensitivity to capsaicin and immunization. Cell Tissue Res 1990;262:125–33.
19. Nishio E, Nakata H, Arimura S, Watanabe Y. alpha-1-Adrenergic receptor stimulation causes arachidonic acid release through pertussis toxin-sensitive GTP-binding protein and JNK activation in rabbit aortic smooth muscle cells. Biochem Biophys Res Commun 1996;219:277–82.
20. Romanelli A, van de Werve G. Activation of mitogen-activated protein kinase in freshly isolated rat hepatocytes by both a calcium- and a protein kinase C-dependent pathway. Metabolism 1997;46:548–55.
21. Taguchi K, Yang M, Goepel M, Michel MC. Comparison of human alpha1-adrenoceptor subtype coupling to protein kinase C activation and related signalling pathways. Naunyn Schmiedebergs Arch Pharmacol 1998;357:100–10.
22. Zhong H, Minneman KP. Differential activation of mitogen-activated protein kinase pathways in PC12 cells by closely related alpha1-adrenergic receptor subtypes. J Neurochem 1999;72:2388–96.
23. Thonberg H, Fredriksson JM, Nedergaard J, Cannon B. A novel pathway for adrenergic stimulation of cAMP-response-element-binding protein (CREB) phosphorylation: Mediation

via alpha1-adrenoceptors and protein kinase C activation. Biochem J 2002;364: 73–9.
24. Chen ZJ, Minneman KP. Recent progress in alpha1-adrenergic receptor research. Acta Pharmacol Sin 2005;26:1281–7.
25. Luo DL, Gao J, Lan XM, Wang G, Wei S, Xiao RP, Han QD. Role of inositol 1,4,5-trisphosphate receptors in alpha1-adrenergic receptor-induced cardiomyocyte hypertrophy. Acta Pharmacol Sin 2006;27:895–900.
26. Green SA, Holt BD, Liggett SB. Beta 1- and beta 2-adrenergic receptors display subtype-selective coupling to Gs. Mol Pharmacol 1992;41:889–93.
27. Crespo P, Cachero TG, Xu N, Gutkind JS. Dual effect of beta-adrenergic receptors on mitogen-activated protein kinase. Evidence for a beta gamma-dependent activation and a G alpha s-cAMP-mediated inhibition. J Biol Chem 1995;270:25259–65.
28. Pelzer S, You Y, Shuba YM, Pelzer DJ. Beta-adrenoceptor-coupled Gs protein facilitates the activation of cAMP-dependent cardiac Cl^- current. Am J Physiol 1997;273:H2539–48.
29. Soeder KJ, Snedden SK, Cao W, Della Rocca GJ, Daniel KW, Luttrell LM, Collins S. The beta3-adrenergic receptor activates mitogen-activated protein kinase in adipocytes through a Gi-dependent mechanism. J Biol Chem 1999;274:12017–22.
30. Can A, Sayar K, Friedman E, Ambrosio C, Erdemli E, Gurdal H. The effects of agonist stimulation and beta(2)-adrenergic receptor level on cellular distribution of Gs(alpha) protein. Cell Signal 2000;12:303–9.
31. Zamah AM, Delahunty M, Luttrell LM, Lefkowitz RJ. Protein kinase A-mediated phosphorylation of the beta 2-adrenergic receptor regulates its coupling to Gs and Gi. Demonstration in a reconstituted system. J Biol Chem 2002;277:31249–56.
32. Rapacciuolo A, Suvarna S, Barki-Harrington L, Luttrell LM, Cong M, Lefkowitz RJ, Rockman HA. Protein kinase A and G protein-coupled receptor kinase phosphorylation mediates beta-1 adrenergic receptor endocytosis through different pathways. J Biol Chem 2003;278:35403–11.
33. Santos IN, Sumitame M, Caceres VM, Moreira MF, Krieger MH, Spadari-Bratfisch RC. Evidence for two atypical conformations of beta-adrenoceptors and their interaction with Gi proteins. Eur J Pharmacol 2005;513:109–18.
34. Remaury A, Larrouy D, Daviaud D, Rouot B, Paris H. Coupling of the alpha 2-adrenergic receptor to the inhibitory G-protein Gi and adenylate cyclase in HT29 cells. Biochem J 1993;292:283–88.
35. Dorn GW 2nd, Oswald KJ, McCluskey TS, Kuhel DG, Liggett SB. Alpha 2A-adrenergic receptor stimulated calcium release is transduced by Gi-associated G (beta gamma)-mediated activation of phospholipase C. Biochemistry 1997;36:6415–23.
36. Takesono A, Zahner J, Blumer KJ, Nagao T, Kurose H. Negative regulation of alpha2-adrenergic receptor-mediated Gi signalling by a novel pathway. Biochem J 1999;343:77–85.
37. Paietta E, Schwarzmeier JD. Differences in beta-adrenergic receptor density and adenylate cyclase activity between normal and leukaemic leukocytes. Eur J Clin Invest 1983;13:339–46.
38. Landmann RM, Burgisser E, Wesp M, Buhler FR. Beta-adrenergic receptors are different in subpopulations of human circulating lymphocytes. J Recept Res 1984;4:37–50.
39. Van Tits LJ, Michel MC, Grosse-Wilde H, Happel M, Eigler FW, Soliman A, Brodde OE. Catecholamines increase lymphocyte beta 2-adrenergic receptors via a beta 2-adrenergic, spleen-dependent process. Am J Physiol 1990;258:E191–202.

40. Knudsen JH, Kjaersgaard E, Christensen NJ. Individual lymphocyte subset composition determines cAMP response to isoproterenol in mononuclear cell preparations from peripheral blood. Scand J Clin Lab Invest 1995;55:9–14.
41. Anstead MI, Hunt TA, Carlson SL, Burki NK. Variability of peripheral blood lymphocyte beta-2-adrenergic receptor density in humans. Am J Respir Crit Care Med 1998;157:990–2.
42. Wahle M, Stachetzki U, Krause A, Pierer M, Hantzschel H, Baerwald CG. Regulation of beta2-adrenergic receptors on CD4 and CD8 positive lymphocytes by cytokines in vitro. Cytokine 2001;16:205–9.
43. Ramer-Quinn DS, Baker RA, Sanders VM. Activated T helper 1 and T helper 2 cells differentially express the beta-2-adrenergic receptor: A mechanism for selective modulation of T helper 1 cell cytokine production. J Immunol 1997;159:4857–67.
44. Sanders VM, Baker RA, Ramer-Quinn DS, Kasprowicz DJ, Fuchs BA, Street NE. Differential expression of the beta2-adrenergic receptor by Th1 and Th2 clones: Implications for cytokine production and B cell help. J Immunol 1997;158:4200–10.
45. Fuchs BA, Albright JW, Albright JF. Beta-adrenergic receptors on murine lymphocytes: Density varies with cell maturity and lymphocyte subtype and is decreased after antigen administration. Cell Immunol 1988;114:231–45.
46. Dailey MO, Schreurs J, Schulman H. Hormone receptors on cloned T lymphocytes. Increased responsiveness to histamine, prostaglandins, and beta-adrenergic agents as a late stage event in T cell activation. J Immunol 1988;140:2931–6.
47. Karaszewski JW, Reder AT, Anlar B, Kim WC, Arnason BG. Increased lymphocyte beta-adrenergic receptor density in progressive multiple sclerosis is specific for the CD8+, CD28− suppressor cell. Ann Neurol 1991;30:42–7.
48. Korichneva IL, Tkachuk VA. Alterations in beta-adrenoceptor density on T-lymphocytes upon activation with interleukin-2 and phytohaemagglutinin. Biomed Sci 1990; 1:84–8.
49. Radojcic T, Baird S, Darko D, Smith D, Bulloch K. Changes in beta-adrenergic receptor distribution on immunocytes during differentiation: An analysis of T cells and macrophages. J Neurosci Res 1991;30:328–35.
50. Van Oosterhout AJ, Van Heuven-Nolsen D, De Boer SF, Thijssen JH, Nijkamp FP. Endotoxin-induced reduction of beta-adrenergic binding sites on splenic lymphocytes in vivo and in vitro: Its modulation by anterior hypothalamic lesions. Life Sci 1989; 44:57–65.
51. Van Oosterhout AJ, Stam WB, Vanderschueren RG, Nijkamp FP. Effects of cytokines on beta-adrenoceptor function of human peripheral blood mononuclear cells and guinea pig trachea. J Allergy Clin Immunol 1992;90:340–8.
52. Cazaux CA, Sterin-Borda L, Gorelik G, Cremaschi GA. Down-regulation of beta-adrenergic receptors induced by mitogen activation of intracellular signaling events in lymphocytes. FEBS Lett 1995;364:120–4.
53. Genaro AM, Cremaschi GA, Gorelik G, Sterin-Borda L, Borda ES. Downregulation of beta adrenergic receptor expression on B cells by activation of early signals in alloantigen-induced immune response. Neuroimmunomodulation 2000;8:114–21.
54. Baerwald CG, Laufenberg M, Specht T, von Wichert P, Burmester GR, Krause A. Impaired sympathetic influence on the immune response in patients with rheumatoid arthritis due to lymphocyte subset-specific modulation of beta 2-adrenergic receptors. Br J Rheumatol 1997;36:1262–9.

55. De Blasi A, Parruti G, Sallese M. Regulation of G protein-coupled receptor kinase subtypes in activated T lymphocytes. Selective increase of beta-adrenergic receptor kinase 1 and 2. J Clin Invest 1995;95:203–10.
56. Jetschmann JU, Benschop RJ, Jacobs R, Kemper A, Oberbeck R, Schmidt RE, Schedlowski M. Expression and in-vivo modulation of alpha- and beta-adrenoceptors on human natural killer (CD16+) cells. J Neuroimmunol 1997;74:159–64.
57. Carlson SL, Trauth K, Brooks WH, Roszman TL. Enhancement of beta-adrenergic-induced cAMP accumulation in activated T-cells. J Cell Physiol 1994;161:39–48.
58. Selliah N, Bartik MM, Carlson SL, Brooks WH, Roszman TL. cAMP accumulation in T-cells inhibits anti-CD3 monoclonal antibody-induced actin polymerization. J Neuroimmunol 1995;56:107–12.
59. Sempere T, Urbina M, Lima L. 5-HT1A and beta-adrenergic receptors regulate proliferation of rat blood lymphocytes. Neuroimmunomodulation 2004;11:307–15.
60. Felsner P, Hofer D, Rinner I, Porta S, Korsatko W, Schauenstein K. Adrenergic suppression of peripheral blood T cell reactivity in the rat is due to activation of peripheral alpha 2-receptors. J Neuroimmunol 1995;57:27–34.
61. Colic M, Cupic V, Pavicic L, Vucevic D, Varagic VM. Xylazine, an alpha 2-adrenergic agonist, modulates proliferation of rat thymocytes in vivo and in vitro. Meth Find Exp Clin Pharmacol 2000;22:557–62.
62. Felsner P, Hofer D, Rinner I, Mangge H, Gruber M, Korsatko W, Schauenstein K. Continuous in vivo treatment with catecholamines suppresses in vitro reactivity of rat peripheral blood T-lymphocytes via alpha-mediated mechanisms. J Neuroimmunol 1992;37:47–57.
63. Murray DR, Polizzi SM, Harris T, Wilson N, Michel MC, Maisel AS. Prolonged isoproterenol treatment alters immunoregulatory cell traffic and function in the rat. Brain Behav Immun 1993;7:47–62.
64. Luecken LJ, Lysle DT. Evidence for the involvement of beta-adrenergic receptors in conditioned immunomodulation. J Neuroimmunol 1992;38:209–19.
65. Murray DR, Irwin M, Rearden CA, Ziegler M, Motulsky H, Maisel AS. Sympathetic and immune interactions during dynamic exercise. Mediation via a beta 2-adrenergic-dependent mechanism. Circulation 1992;86:203–13.
66. Exton MS, Gierse C, Meier B, Mosen M, Xie Y, Frede S, Goebel MU, Limmroth V, Schedlowski M. Behaviorally conditioned immunosuppression in the rat is regulated via noradrenaline and beta-adrenoceptors. J Neuroimmunol 2002;131:21–30.
67. Edgar VA, Silberman DM, Cremaschi GA, Zieher LM, Genaro AM. Altered lymphocyte catecholamine reactivity in mice subjected to chronic mild stress. Biochem Pharmacol 2003;65:15–23.
68. Feldman RD, Hunninghake GW, McArdle WL. Beta-adrenergic-receptor-mediated suppression of interleukin 2 receptors in human lymphocytes. J Immunol 1987; 139:3355–9.
69. Ramer-Quinn DS, Swanson MA, Lee WT, Sanders VM. Cytokine production by naive and primary effector CD4+ T cells exposed to norepinephrine. Brain Behav Immun 2000; 14:239–55.
70. Peng YP, Qiu YH, Jiang JL, Wang JJ. Effect of catecholamines on IL-2 production and NK cytotoxicity of rats in vitro. Acta Pharmacol Sin 2004;25:1354–60.
71. Coqueret O, Petit-Frere C, Lagente V, Moumen M, Mencia-Huerta JM, Braquet P. Modulation of IgE production in the mouse by beta 2-adrenoceptor agonist. Int Arch Allergy Immunol 1994;105:171–6.

72. Panina-Bordignon P, Mazzeo D, Lucia PD, D'Ambrosio D, Lang R, Fabbri L, Self C, Sinigaglia F. Beta2-agonists prevent Th1 development by selective inhibition of interleukin 12. J Clin Invest 1997;100:1513–9.
73. Borger P, Hoekstra Y, Esselink MT, Postma DS, Zaagsma J, Vellenga E, Kauffman HF. Beta-adrenoceptor-mediated inhibition of IFN-gamma, IL-3, and GM-CSF mRNA accumulation in activated human T lymphocytes is solely mediated by the beta2-adrenoceptor subtype. Am J Respir Cell Mol Biol 1998;19:400–7.
74. Agarwal SK, Marshall GF Jr. Beta-adrenergic modulation of human type-1/type-2 cytokine balance. J Allergy Clin Immunol 2000;105:91–8.
75. Loza MJ, Foster S, Peters SP, Penn RB. Beta-agonists modulate T-cell functions via direct actions on type 1 and type 2 cells. Blood 2006;107:2052–60.
76. Heijink IH, Vellenga E, Borger P, Postma DS, Monchy JG, Kauffman HF. Polarized Th1 and Th2 cells are less responsive to negative feedback by receptors coupled to the AC/cAMP system compared to freshly isolated T cells. Br J Pharmacol 2003;138: 1441–50.
77. Swanson MA, Lee WT, Sanders VM. IFN-gamma production by Th1 cells generated from naive CD4+ T cells exposed to norepinephrine. J Immunol 2001;166:232–40.
78. Holen E, Elsayed S. Effects of beta2 adrenoceptor agonists on T-cell subpopulations. APMIS. 1998;106:849–57.
79. Nishii M, Inomata T, Niwano H, Takehana H, Takeuchi I, Nakano H, Shinagawa H, Naruke T, Koitabashi T, Nakahata J, Izumi T. Beta2-Adrenergic agonists suppress rat autoimmune myocarditis: Potential role of beta2-adrenergic stimulants as new therapeutic agents for myocarditis. Circulation 2006;114:936–44.
80. Borger P, Jonker GJ, Vellenga E, Postma DS, De Monchy JG, Kauffman HF. Allergen challenge primes for IL-5 mRNA production and abrogates beta-adrenergic function in peripheral blood T lymphocytes from asthmatics. Clin Exp Allergy 1999;29:933–40.
81. Heijink IH, Vellenga E, Oostendorp J, de Monchy JG, Postma DS, Kauffman HF. Exposure to TARC alters beta2-adrenergic receptor signaling in human peripheral blood T lymphocytes. Am J Physiol Lung Cell Mol Physiol 2005;289:L53–9.
82. Lombardi MS, Kavelaars A, Cobelens PM, Schmidt RE, Schedlowski M, Heijnen CJ. Adjuvant arthritis induces down-regulation of G protein-coupled receptor kinases in the immune system. J Immunol 2001;166:1635–40.
83. Hatfield SM, Petersen BH, DiMicco JA. Beta adrenoceptor modulation of the generation of murine cytotoxic T lymphocytes in vitro. J Pharmacol Exp Ther 1986;239:460–6.
84. Cook-Mills JM, Mokyr MB, Cohen RL, Perlman RL, Chambers DA. Neurotransmitter suppression of the in vitro generation of a cytotoxic T lymphocyte response against the syngeneic MOPC-315 plasmacytoma. Cancer Immunol Immunother 1995;40:79–87.
85. Kalinichenko VV, Mokyr MB, Graf LH Jr, Cohen RL, Chambers DA. Norepinephrine-mediated inhibition of antitumor cytotoxic T lymphocyte generation involves a beta-adrenergic receptor mechanism and decreased TNF-alpha gene expression. Immunology 1999;163:2492–9.
86. Sanders VM, Powell-Oliver FE. Beta 2-adrenoceptor stimulation increases the number of antigen-specific precursor B lymphocytes that differentiate into IgM-secreting cells without affecting burst size. J Immunol 1992;148:1822–8.
87. Kohm AP, Mozaffarian A, Sanders VM. B cell receptor- and beta 2-adrenergic receptor induced regulation of B7-2 (CD86) expression in B cells. J Immunol 2002;168:6314–22.

88. Kasprowicz DJ, Kohm AP, Berton MT, Chruscinski AJ, Sharpe A, Sanders VM. Stimulation of the B cell receptor, CD86 (B7-2), and the beta 2-adrenergic receptor intrinsically modulates the level of IgG1 and IgE produced per B cell. J Immunol 2000;165:680–90.
89. Podojil JR, Kin NW, Sanders VM. CD86 and beta2-adrenergic receptor signaling pathways, respectively, increase Oct-2 and OCA-B expression and binding to the 3′-IgH enhancer in B cells. J Biol Chem 2004;279:23394–404.
90. Pongratz G, McAlees JW, Conrad DH, Erbe RS, Haas KM, Sanders VM. The level of IgE produced by a B cell is regulated by norepinephrine in a p38 MAPK- and CD23-dependent manner. J Immunol 2006;177:2926–38.
91. Li YS, Kouassi E, Revillard JP. Differential regulation of mouse B-cell activation by beta-adrenoceptor stimulation depending on type of mitogens. Immunology 1990; 69:367–72.
92. Green-Johnson JM, Zalcman S, Vriend CY, Dolina S, Nance DM, Greenberg AH. Role of norepinephrine in suppressed IgG production in epilepsy-prone mice. Life Sci 1996;59:1121–32.
93. Whalen MM, Bankhurst AD. Effects of beta-adrenergic receptor activation, cholera toxin and forskolin on human natural killer cell function. Biochem J 1990; 272:327–31.
94. Takamoto T, Hori Y, Koga Y, Toshima H, Hara A, Yokoyama MM. Norepinephrine inhibits human natural killer cell activity in vitro. Int J Neurosci 1991; 58:127–131.
95. Gan X, Zhang L, Solomon GF, Bonavida B. Mechanism of norepinephrine-mediated inhibition of human NK cytotoxic functions: Inhibition of cytokine secretion, target binding, and programming for cytotoxicity. Brain Behav Immun 2002;16:227–46.
96. Shimamiya T, Wakabayashi S, Terada N. Effects of adrenaline and cortisone on the early activation of lymphocytes. Biol Sci Space 2003;17:221–2.
97. Dokur M, Boyadjieva N, Sarkar DK. Catecholaminergic control of NK cell cytolytic activity regulatory factors in the spleen. J Neuroimmunol 2004;151:148–57.
98. Benschop RJ, Schedlowski M, Wienecke H, Jacobs R, Schmidt RE. Adrenergic control of natural killer cell circulation and adhesion. Brain Behav Immun 1997;11:321–32.
99. Schedlowski M, Hosch W, Oberbeck R, Benschop RJ, Jacobs R, Raab HR, Schmidt RE. Catecholamines modulate human NK cell circulation and function via spleen-independent beta 2-adrenergic mechanisms. J Immunol 1996;156:93–9.
100. Shakhar G, Ben-Eliyahu S. In vivo beta-adrenergic stimulation suppresses natural killer activity and compromises resistance to tumor metastasis in rats. J Immunol 1998; 160: 3251–8.
101. Oberbeck R, Schmitz D, Wilsenack K, Schuler M, Pehle B, Schedlowski M, Exton MS. Adrenergic modulation of survival and cellular immune functions during polymicrobial sepsis. Neuroimmunomodulation 2004;11:214–23.
102. Katafuchi T, Take S, Hori T. Roles of sympathetic nervous system in the suppression of cytotoxicity of splenic natural killer cells in the rat. J Physiol 1993;465:343–57.
103. Landmann RM, Muller FB, Perini C, Wesp M, Erne P, Buhler FR. Changes of immunoregulatory cells induced by psychological and physical stress: Relationship to plasma catecholamines. Clin Exp Immunol 1984;58:127–35.
104. Klokker M, Secher NH, Madsen P, Pedersen M, Pedersen BK. Adrenergic beta 1- and beta 1 + 2-receptor blockade suppress the natural killer cell response to head-up tilt in humans. J Appl Physiol 1997;83:1492–8.
105. Ben-Eliyahu S, Shakhar G, Page GG, Stefanski V, Shakhar K. Suppression of NK cell activity and of resistance to metastasis by stress: A role for adrenal catecholamines and beta-adrenoceptors. Neuroimmunomodulation 2000;8:154–64.

106. Engler H, Dawils L, Hoves S, Kurth S, Stevenson JR, Schauenstein K, Stefanski V. Effects of social stress on blood leukocyte distribution: The role of alpha- and beta-adrenergic mechanisms. J Neuroimmunol 2004;156:153–62.
107. Sudo A. Decrease in adrenaline content of various organs of the rat after 6-hydroxydopamine. Eur J Pharmacol 1985;114:79–83.
108. Lorton D, Hewitt D, Bellinger DL, Felten SY, Felten DL. Noradrenergic reinnervation of the rat spleen following chemical sympathectomy with 6-hydroxydopamine: Pattern and time course of reinnervation. Brain Behav Immun 1990;4:198–222.
109. Kawamura M, Schwartz JP, Nomura T, Kopin IJ, Goldstein DS, Huynh TT, Hooper DR, Harvey-White J, Eisenhofer G. Differential effects of chemical sympathectomy on expression and activity of tyrosine hydroxylase and levels of catecholamines and DOPA in peripheral tissues of rats. Neurochem Res 1999;24:25–32.
110. Lyte M, Ernst S, Driemeyer J, Baissa B. Strain-specific enhancement of splenic T cell mitogenesis and macrophage phagocytosis following peripheral axotomy. J Neuroimmunol 1991;31:1–8.
111. Kruszewska B, Felten SY, Moynihan JA. Alterations in cytokine and antibody production following chemical sympathectomy in two strains of mice. J Immunol 1995;155:4613–20.
112. Callahan TA, Moynihan JA. Contrasting pattern of cytokines in antigen- versus mitogen-stimulated splenocyte cultures from chemically denervated mice. Brain Behav Immun 2002;16:764–73.
113. Bellinger DL, Stevens SY, Thyaga Rajan S, Lorton D, Madden KS. Aging and sympathetic modulation of immune function in Fischer 344 rats: Effects of chemical sympathectomy on primary antibody response. J Neuroimmunol 2005;165:21–32.
114. Takahashi H, Tsuda Y, Kobayashi M, Herndon DN, Suzuki F. Increased norepinephrine production associated with burn injuries results in CCL2 production and type 2T cell generation. Burns 2004;30:317–21.
115. Madden KS, Moynihan JA, Brenner GJ, Felten SY, Felten DL, Livnat S. Sympathetic nervous system modulation of the immune system. III. Alterations in T and B cell proliferation and differentiation in vitro following chemical sympathectomy. J Neuroimmunol 1994;49:77–87.
116. Harle P, Mobius D, Carr DJ, Scholmerich J, Straub RH. An opposing time-dependent immune-modulating effect of the sympathetic nervous system conferred by altering the cytokine profile in the local lymph nodes and spleen of mice with type II collagen-induced arthritis. Arthritis Rheum 2005;52:1305–13.
117. Pal E, Yamamura T, Tabira T. Autonomic regulation of experimental autoimmune encephalomyelitis in IL-4 knockout mice. J Neuroimmunol 1999;100:149–55.
118. Madden KS, Felten SY, Felten DL, Sundaresan PR, Livnat S. Sympathetic neural modulation of the immune system. I. Depression of T cell immunity in vivo and vitro following chemical sympathectomy. Brain Behav Immun 1989;3:372–89.
119. Li X, Taylor S, Zegarelli B, Shen S, O'Rourke J, Cone RE. The induction of splenic suppressor T cells through an immune-privileged site requires an intact sympathetic nervous system. J Neuroimmunol 2004;153:40–9.
120. Kohm AP, Sanders VM. Suppression of antigen-specific Th2 cell-dependent IgM and IgG1 production following norepinephrine depletion in vivo. Immunology 1999;162:5299–308.
121. del Rey A, Roggero E, Kabiersch A, Schafer M, Besedovsky HO. The role of noradrenergic nerves in the development of the lymphoproliferative disease in Fas-deficient, lpr/lpr mice. J Immunol 2006;176:7079–86.

122. Reder A, Checinski M, Chelmicka-Schorr E. The effect of chemical sympathectomy on natural killer cells in mice. Brain Behav Immun 1989;3:110–8.
123. Jiang XH, Guo SY, Xu S, Yin QZ, Ohshita Y, Naitoh M, Horibe Y, Hisamitsu T. Sympathetic nervous system mediates cold stress-induced suppression of natural killer cytotoxicity in rats. Neurosci Lett 2004;357:1–4.
124. Alito AE, Romeo HE, Baler R, Chuluyan HE, Braun M, Cardinali DP. Autonomic nervous system regulation of murine immune responses as assessed by local surgical sympathetic and parasympathetic denervation. Acta Physiol Pharmacol Latinoam 1987; 37:305–19.
125. Arce A, Castillon P, Cardinali DP, Esquifino AI. Effect of local autonomic denervation on in vitro responsiveness of lymphocytes from rat submaxillary lymph nodes. J Auton Nerv Syst 1997;62:155–62.
126. Shimizu N, Kaizuka Y, Hori T, Nakane H. Immobilization increases norepinephrine release and reduces NK cytotoxicity in spleen of conscious rat. Am J Physiol 1996; 271:R537–44.
127. Alaniz RC, Thomas SA, Perez-Melgosa M, Mueller K, Farr AG, Palmiter RD, Wilson CB. Dopamine beta-hydroxylase deficiency impairs cellular immunity. Proc Natl Acad Sci USA 1999;96:2274–8.
128. Shimizu N, Hori T, Nakane H. An interleukin-1 beta-induced noradrenaline release in the spleen is mediated by brain corticotropin-releasing factor: An in vivo microdialysis study in conscious rats. Brain Behav Immun 1994;8:14–23.
129. Kannan H, Tanaka Y, Kunitake T, Ueta Y, Hayashida Y, Yamashita H. Activation of sympathetic outflow by recombinant human interleukin-1 beta in conscious rats. Am J Physiol 1996;270:R479–85.
130. Ek M, Kurosawa M, Lundeberg T, Ericsson A. Activation of vagal afferents after intravenous injection of interleukin-1beta: Role of endogenous prostaglandins. J Neurosci 1998;18:9471–9.
131. Kenney MJ, Blecha F, Wang Y, McMurphy R, Fels RJ. Sympathoexcitation to intravenous interleukin-1beta is dependent on forebrain neural circuits. Am J Physiol Heart Circ Physiol 2002;283:H501–5.
132. Safieh-Garabedian B, Poole S, Haddad JJ, Massaad CA, Jbbur SJ, Saade NE. The role of the sympathetic efferents in endotoxin-induced localized inflammatory hyperalgesia and cytokine upregulation. Neuropharmacology 2002;42:864–72.
133. Wieczorek M, Swiergiel AH, Pournajafi-Nazarloo H, Dunn AJ. Physiological and behavioral responses to interleukin-1beta and LPS in vagotomized mice. Physiol Behav 2005;85:500–11.
134. Zalcman S, Green-Johnson JM, Murray L, Wan W, Nance DM, Greenberg AH. Interleukin-2-induced enhancement of an antigen-specific IgM plaque-forming cell response is mediated by the sympathetic nervous system. J Pharmacol Exp Ther 1994; 271:977–82.
135. Tolchard S, Hare AS, Nutt DJ, Clarke G. TNF alpha mimics the endocrine but not the thermoregulatory responses of bacterial lipopolysaccharide (LPS): Correlation with FOS-expression in the brain. Neuropharmacology 1996;35:243–8.
136. Freidin M, Bennett MV, Kessler JA. Cultured sympathetic neurons synthesize and release the cytokine interleukin 1 beta. Proc Natl Acad Sci USA 1992;89:10440–3.
137. Bai Y, Hart RP. Cultured sympathetic neurons express functional interleukin-1 receptors. J Neuroimmunol 1998;91:43–54.

138. Morris JL, Jobling P, Shimizu T, Gibbins IL. Interleukin-1 receptor immunoreactivity in sympathetic vascular and non-vascular neurons in guinea-pig coeliac ganglion. Neurosci Lett 2002;333:54–8.
139. Rogausch H, del Rey A, Kabiersch A, Reschke W, Ortel J, Besedovsky H. Endotoxin impedes vasoconstriction in the spleen: Role of endogenous interleukin-1 and sympathetic innervation. Am J Physiol 1997;72:R2048–54.
140. Lindholm D, Heumann R, Meyer M, Thoenen H. Interleukin-1 regulates synthesis of nerve growth factor in non-neuronal cells of rat sciatic nerve. Nature 1987;330:658–9.
141. Alheim K, Andersson C, Tingsborg S, Ziolkowska M, Schultzberg M, Bartfai T. Interleukin 1 expression is inducible by nerve growth factor in PC12 pheochromocytoma cells. Proc Natl Acad Sci USA 1991;88:9302–6.
142. Hattori A, Iwasaki S, Murase K, Tsujimoto M, Sato M, Hayashi K, Kohno M. Tumor necrosis factor is markedly synergistic with interleukin 1 and interferon-gamma in stimulating the production of nerve growth factor in fibroblasts. FEBS Lett 1994;340:177–80.
143. Pluss KM, Pfeilschifter J, Muhl H, Huwiler A, Boeckh C, Otten U. Modulatory role of platelet-derived growth factor on cytokine-induced nerve growth factor synthesis in rat glomerular mesangial cells. Biochem J 1995;312:707–11.
144. Manni L, Aloe L. Role of IL-1 beta and TNF-alpha in the regulation of NGF in experimentally induced arthritis in mice. Rheumatol Int 1998;18:97–102.
145. Wu VW, Nishiyama N, Schwartz JP. A culture model of reactive astrocytes: Increased nerve growth factor synthesis and reexpression of cytokine responsiveness. J Neurochem 1998;71:749–56.
146. Juric DM, Carman-Krzan M. Interleukin-1 beta, but not IL-1 alpha, mediates nerve growth factor secretion from rat astrocytes via type I IL-1 receptor. Int J Dev Neurosci 2001;19:675–83.
147. Kannan Y, Bienenstock J, Ohta M, Stanisz AM, Stead RH. Nerve growth factor and cytokines mediate lymphoid tissue-induced neurite outgrowth from mouse superior cervical ganglia in vitro. J Immunol 1996;157:313–20.
148. Clarke GA, White CA, Moss DJ. Substrate-bound GP130/F11 will promote neurite outgrowth: Evidence for a cell surface receptor. Eur J Cell Biol 1993;61:108–15.
149. Wong V, Pearsall D, Arriaga R, Ip NY, Stahl N, Lindsay RM. Binding characteristics of ciliary neurotrophic factor to sympathetic neurons and neuronal cell lines. J Biol Chem 1995;270:313–8.
150. Ng YP, He W, Ip NY. Leukemia inhibitory factor receptor signaling negatively modulates nerve growth factor-induced neurite outgrowth in PC12 cells and sympathetic neurons. J Biol Chem 2003;278:38731–9.
151. Fann MJ, Patterson PH. Neuropoietic cytokines and activin A differentially regulate the phenotype of cultured sympathetic neurons. Proc Natl Acad Sci USA 1994;91:43–7.
152. Cheng JG, Pennica D, Patterson PH. Cardiotrophin-1 induces the same neuropeptides in sympathetic neurons as do neuropoietic cytokines. J Neurochem 1997;69: 2278–84.
153. Marz P, Cheng JG, Gadient RA, Patterson PH, Stoyan T, Otten U, Rose-John S. Sympathetic neurons can produce and respond to interleukin 6. Proc Natl Acad Sci USA 1998;95:3251–6.
154. Guo X, Chandrasekaran V, Lein P, Kaplan PL, Higgins D. Leukemia inhibitory factor and ciliary neurotrophic factor cause dendritic retraction in cultured rat sympathetic neurons. J Neurosci 1999;19:2113–21.

155. Senaldi G, Varnum BC, Sarmiento U, Starnes C, Lile J, Scully S, Guo J, Elliott G, McNinch J, Shaklee CL, Freeman D, Manu F, Simonet WS, Boone T, Chang MS. Novel neurotrophin-1/B cell-stimulating factor-3: A cytokine of the IL-6 family. Proc Natl Acad Sci USA 1999;96:11458–63.
156. Dziennis S, Habecker BA. Cytokine suppression of dopamine-beta-hydroxylase by extracellular signal-regulated kinase-dependent and -independent pathways. J Biol Chem 2003;278:15897–904.
157. Kaur N, Kim IJ, Higgins D, Halvorsen SW. Induction of an interferon-gamma Stat3 response in nerve cells by pre-treatment with gp130 cytokines. J Neurochem 2003;87:437–47.
158. Ramer MS, Murphy PG, Richardson PM, Bisby MA. Spinal nerve lesion-induced mechanoallodynia and adrenergic sprouting in sensory ganglia are attenuated in interleukin-6 knockout mice. Pain 1998;78:115–21.
159. Thompson SW, Majithia AA. Leukemia inhibitory factor induces sympathetic sprouting in intact dorsal root ganglia in the adult rat in vivo. J Physiol 1998;506:809–16.
160. Haugen PK, Letourneau PC. Interleukin-2 enhances chick and rat sympathetic, but not sensory, neurite outgrowth. J Neurosci Res 1990;25:443–52.
161. Lin H, Hikawa N, Takenaka T, Ishikawa Y. Interleukin-12 promotes neurite outgrowth in mouse sympathetic superior cervical ganglion neurons. Neurosci Lett 2000;278:129–32.
162. Ehrhard PB, Erb P, Graumann U, Otten U. Expression of nerve growth factor and nerve growth factor receptor tyrosine kinase Trk in activated CD4-positive T-cell clones. Proc Natl Acad Sci USA 1993;90:10984–88.
163. Santambrogio L, Benedetti M, Chao MV, Muzaffar R, Kulig K, Gabellini N, Hochwald G. Nerve growth factor production by lymphocytes. J Immunol 1994;153:4488–95.
164. Lambiase A, Bracci-Laudiero L, Bonini S, Bonini S, Starace G, D'Elios MM, De Carli M, Aloe L. Human CD4+ T cell clones produce and release nerve growth factor and express high-affinity nerve growth factor receptors. J Allergy Clin Immunol 1997;100:408–14.
165. Barouch R, Appel E, Kazimirsky G, Braun A, Renz H, Brodie C. Differential regulation of neurotrophin expression by mitogens and neurotransmitters in mouse lymphocytes. J Neuroimmunol 2000;103:112–21.
166. Moalem G, Gdalyahu A, Shani Y, Otten U, Lazarovici P, Cohen IR, Schwartz M. Production of neurotrophins by activated T cells: Implications for neuroprotective autoimmunity. J Autoimmun 2000;15:331–45.
167. Edling AE, Nanavati T, Johnson JM, Tuohy VK. Human and murine lymphocyte neurotrophin expression is confined to B cells. J Neurosci Res 2004;77:709–17.
168. Thorpe LW, Perez-Polo JR. The influence of nerve growth factor on the in vitro proliferative response of rat spleen lymphocytes. J Neurosci Res 1987;18:134–9.
169. Otten U, Ehrhard P, Peck R. Nerve growth factor induces growth and differentiation of human B lymphocytes. Proc Natl Acad Sci USA 1989;86:10059–63.
170. Brodie C, Gelfand EW. Functional nerve growth factor receptors on human B lymphocytes. Interaction with IL-2. J Immunol 1992;148:3492–7.
171. Aloe L, Bracci-Laudiero L, Alleva E, Lambiase A, Micera A, Tirassa P. Emotional stress induced by parachute jumping enhances blood nerve growth factor levels and the distribution of nerve growth factor receptors in lymphocytes. Proc Natl Acad Sci USA 1994; 91:10440–4.
172. Brodie C, Oshiba A, Renz H, Bradley K, Gelfand EW. Nerve growth-factor and anti-CD40 provide opposite signals for the production of IgE in interleukin-4-treated lymphocytes. Eur J Immunol 1996;26:171–8.

173. Melamed I, Kelleher CA, Franklin RA, Brodie C, Hempstead B, Kaplan D, Gelfand EW. Nerve growth factor signal transduction in human B lymphocytes is mediated by gp140trk. Eur J Immunol 1996;26:1985–92.
174. D'Onofrio M, de Grazia U, Morrone S, Cuomo L, Spinsanti P, Frati L, Gulino A, Ragona G. Expression of neurotrophin receptors in normal and malignant B lymphocytes. Eur Cytokine Netw 2000;11:283–91.
175. Rosini P, De Chiara G, Bonini P, Lucibello M, Marcocci ME, Garaci E, Cozzolino F, Torcia M. Nerve growth factor-dependent survival of CESS B cell line is mediated by increased expression and decreased degradation of MAPK phosphatase 1. J Biol Chem 2004;279:14016–23.
176. Kannan Y, Stead RH, Goldsmith CH, Bienenstock J. Lymphoid tissues induce NGF-dependent and NGF-independent neurite outgrowth from rat superior cervical ganglia explants in culture. J Neurosci Res 1994;37:374–83.
177. Konishi Y, Chui DH, Hirose H, Kunishita T, Tabira T. Trophic effect of erythropoietin and other hematopoietic factors on central cholinergic neurons in vitro and in vivo. Brain Res 1993;609:29–35.
178. Kannan Y, Moriyama M, Sugano T, Yamate J, Kuwamura M, Kagaya A, Kiso Y. Neurotrophic action of interleukin 3 and granulocyte–macrophage colony-stimulating factor on murine sympathetic neurons. Neuroimmunomodulation 2000;8:132–41.
179. Besedovsky HO, del Rey A, Sorkin E, Burri R, Honegger CG, Schlumpf M, Lichtensteiger W. T lymphocytes affect the development of sympathetic innervation of mouse spleen. Brain Behav Immun 1987;1:185–93.
180. Mitchell B, Kendall M, Adam E, Schumacher U. Innervation of the thymus in normal and bone marrow reconstituted severe combined immunodeficient (SCID) mice. J Neuroimmunol 1997;75:19–27.
181. Bosma GC, Custer R P, Bosma MJ. A severe combined immunodeficiency mutation in the mouse. Nature 1983;301:527–30.
182. Breneman SM, Moynihan JA, Grota LJ, Felten DL, Felten SY. Splenic norepinephrine is decreased in MRL-lpr/lpr mice. Brain Behav Immun 1993;7:135–43.
183. Kelley SP, Moynihan JA, Stevens SY, Grota LJ, Felten DL. Sympathetic nerve destruction in spleen in murine AIDS. Brain Behav Immun 2003;17:94–109.
184. Lorton D, Lubahn C, Lindquist CA, Schaller J, Washington C, Bellinger DL. Changes in the density and distribution of sympathetic nerves in spleens from Lewis rats with adjuvant-induced arthritis suggest that an injury and sprouting response occurs. J Comp Neurol 2005;489:260–73.
185. Serpe, CJ, Tetzlaff JE, Coers S, Sanders VM, Jones KJ. Functional recovery after facial nerve crush is delayed in severe combined immunodeficient mice. Brain Behav Immun 2002;16:808–12.
186. Armstrong BD, Abad C, Chhith S, Rodriguez W, Cheung-Lau G, Trinh V, Waschek JA. Restoration of axotomy-induced PACAP gene induction in SCID mice with CD4+ T-lymphocytes. Neuroreport 2004;15:2647–50.
187. Kannan-Hayashi Y, Okamura K, Hattori S, Kuwamura M, Higuchi E, Terayama H, Moriyama M, Mukamoto M, Okada M, Ohsugi Y, Nakamura Y. Neuritogenic effects of T cell-derived IL-3 on mouse splenic sympathetic neurons *in vivo*. J Immunol *in press*.
188. Bergquist J, Tarkowski A, Ekman R, Ewing A. Discovery of endogenous catecholamines in lymphocytes and evidence for catecholamine regulation of lymphocyte function via an autocrine loop. Proc Natl Acad Sci USA 1994;91:12912–6.

189. Musso NR, Brenci S, Setti M, Indiveri F, Lotti G. Catecholamine content and in vitro catecholamine synthesis in peripheral human lymphocytes. J Clin Endocrinol Metab 1996;81:3553–7.
190. Marino F, Cosentino M, Bombelli R, Ferrari M, Lecchini S, Frigo G. Endogenous catecholamine synthesis, metabolism storage, and uptake in human peripheral blood mononuclear cells. Exp Hematol 1999;27:489–95.
191. Josefsson E, Bergquist J, Ekman R, Tarkowski A. Catecholamines are synthesized by mouse lymphocytes and regulate function of these cells by induction of apoptosis. Immunology 1996;88:140–6.
192. Reguzzoni M, Cosentino M, Rasini E, Marino F, Ferrari M, Bombelli R, Congiu T, Protasoni M, Quacci D, Lecchini S, Raspanti M, Frigo G. Ultrastructural localization of tyrosine hydroxylase in human peripheral blood mononuclear cells: Effect of stimulation with phytohaemagglutinin. Cell Tissue Res 2002;310:297–04.
193. Ferrari M, Cosentino M, Marino F, Bombelli R, Rasini E, Lecchini S, Frigo G. Dopaminergic D1-like receptor-dependent inhibition of tyrosine hydroxylase mRNA expression and catecholamine production in human lymphocytes. Biochem Pharmacol 2004;67:865–73.
194. Qiu YH, Cheng C, Dai L, Peng YP. Effect of endogenous catecholamines in lymphocytes on lymphocyte function. J Neuroimmunol 2005;167:45–52.

Cytokines and the Brain
Edited by C. Phelps and E. Korneva

Cytokines in Neural Signaling to the Brain

LISA E. GOEHLER

Department of Psychology, University of Virginia, Charlottesville, VA, USA

ABSTRACT

Cytokines signal the brain via multiple pathways, which include both direct action at brain barrier regions, including circumventricular organs, vascular endothelium, choroid plexus, and meninges, and via peripheral nerves. Immune-sensitive neural signals are carried by neurons associated with general visceral and somatic sensory cranial and spinal nerves. Although general viscerosensory nerves (glossopharyngeal and vagus) seem to function to activate symptoms of sickness, particularly sickness behavior, cytokine activation of trigeminal and spinal nerves enhances pain transmission. Peripheral nerves and sensory ganglia contain cytokine-expressing immune cells that respond to infection or inflammation. Cytokine receptors have been localized in primary sensory neurons of the vagus and dorsal root ganglia, as well as in chemosensory cells in specialized structures, the carotid bodies and vagal paraganglia of the glossopharyngeal and vagal nerves. Functional studies implicate peripheral nerves most notably in the induction of sickness behaviors, primarily social withdrawal and pain-related behavior. Cytokine signaling via peripheral nerves may play a role in affective changes associated with systemic infection as well as in chronic inflammatory conditions such as inflammatory bowel syndrome.

1. INTRODUCTION

Peripheral immune activation leads to the elaboration of immune-derived mediators that collectively serve to coordinate host–defense responses, including those mediated by the brain. Among theses mediators, proinflammatory cytokines, primarily interleukin-1β (IL-1β), IL-6, and tumor necrosis factor (TNF)-α, appear to play a particularly important role in the induction of a constellation of brain-mediated responses that include fever, changes in sleep and autonomic function, and the induction of sickness behavior. Recent evidence has identified multiple pathways by which these cytokines signal the brain. These mechanisms can be generally categorized as either humoral signals detected at brain–barrier tissues or locally generated signals that activate peripheral nerves, including both cranial and spinal nerves. The purpose of this review is to explore the role of cytokines in these neural pathways for immune-to-brain communication: the mechanism by which they respond to immune signals, and in turn influence central nervous system (CNS) functioning.

2. CRANIAL NERVE VISCEROSENSORY PATHWAYS: NEURAL SURVEILLANCE OF PATHOGEN INTERFACES

Two general viscerosensory cranial nerves have been implicated in cytokine-to-brain signaling: the glossopharyngeal (which also carries special visceral gustatory signals) and the vagus. These two nerves together innervate most of the alimentary canal, as well as many other important visceral tissues including lung and lymph nodes. These tissues are notable as major points of entry for diverse pathogens.

2.1. The glossopharyngeal nerve

The glossopharyngeal nerve (the ninth cranial nerve) innervates the posterior two-thirds of the tongue as well as other posterior oral structures. Specialized immune structures including the tonsils are located in this region, thus the glossopharyngeal nerve is well positioned for a role in immunosensory surveillance. In support of this idea, application of either lipopolysaccharide (LPS) or IL-1 into the soft palate (receptive field of the glossopharyngeal nerve) induces a fever that can be blocked by the prior section of the glossopharyngeal nerve [1]. Sectioning the glossopharyngeal nerve was ineffective in blocking fevers induced by systemic (intraperitoneal) injections of LPS or IL-1, supporting the idea that this nerve signals immune activation locally within the oral cavity.

In addition to innervating the oral cavity, sensory fibers of the glossopharyngeal nerve innervate the carotid bodies. The carotid bodies are located at the carotid bifurcation and consist of a very large collection of chemosensory glomus cells, which are sensitive to blood gasses and likely other chemical stimuli in the general circulation [2]. The carotid bodies have recently been demonstrated to express IL-1 receptor type 1 immunoreactivity [3], indicating that in addition to monitoring stimuli relevant to respiratory reflexes, these structures may well participate in signaling systemic immune-related signals.

2.2. The vagus nerve

Like the glossopharyngeal nerve, the vagus nerve (the tenth cranial nerve) is well positioned to interact with pathogen products and cytokines. *Vagus* means wanderer; this nerve innervates nearly every internal structure, from the larynx to the colon. Internal tissues commonly in contact with pathogens, notably the lungs, gastrointestinal tract, and liver, are richly supplied with vagal afferents potentially capable of signaling immune activation in these tissues [4]. The cell bodies of vagal sensory neurons occupy two ganglia, the nodose (or inferior vagal) and the jugular (or superior vagal) ganglia, which lie just outside the caudal cranium. These two ganglia form a complex with the petrosal ganglion, which contains the cell bodies of sensory neurons contributing to the glossopharyngeal nerve. The central projections of these pseudounipolar neurons terminate in the dorsal vagal complex of the caudal brainstem (see below). In this way, the vagus nerve is positioned to detect cytokines generated in response to local infection or inflammation in tissues commonly in contact with pathogens, and rapidly signal the brain. The vagus may thus serve as an early warning or a sentinel system functioning to initiate prompt responses to infection.

2.2.1. Distribution of vagal sensory fibers

If vagal sensory nerve fibers function as sentinels for the early activation of brain-mediated host–defense responses, then one would expect these fibers to innervate lymphoid tissues, such as spleen and lymph nodes. Although the spleen is located in the abdomen and might be expected to receive vagal innervation, it clearly does not [5,6]. This may be related to the fact that the spleen operates as a filter for circulating immune and pathogen components, and thus is not quite on the first line of defense. Rather, pathogens typically arrive at lymph nodes first before the general circulation. Lymph nodes thus provide a site of early immune activation, as these are the major locations in which antigen-presenting cells interface with the T cells that serve to coordinate immune responses. The lymphatic system comprises an interconnecting network of conducting vessels that carry immune cells and antigens, including microorganisms, from lymph node to node, progressively to the heart. Lymph nodes are innervated by both sympathetic and sensory neuropeptide-containing nerve fibers [7–9]. Much of the lymphatic system, notably the pelvic, mesenteric, deep cervical, and mediastinal ducts and nodes, lie within the range of vagal afferent peripheral terminal fields as well. Vagal sensory neurons likely innervate these lymph nodes, based on the findings that injections of the retrograde tracer Fluorogold into cervical and pelvic lymph nodes labeled neurons in the nodose and the jugular ganglia [10]. These observations are consistent with a role of vagal afferents in monitoring early stage activation in immune-related tissues.

Prior to entry into the lymphatic system, pathogens must cross the epithelial barrier tissue in the lung and gut. Abundant immune-type tissues and cells are found throughout the gastrointestinal tract, which is not surprising as this is a barrier site for infectious agents. Specialized immune tissues, including lymphoid nodules (which are organized somewhat like lymph nodes) and Peyer's patches of the small intestine, reside directly beneath the epithelium. In addition, macrophages and dendritic cells line the epithelium and overlie the Peyer's patches [11]. Berthoud and Neuhuber [4] reported that anterograde tracing of vagal sensory neurons revealed vagal nerves that innervate the submucosal and epithelial regions of the intestine and are closely associated with a cell type described as possessing several long dendrite-like processes. This description is similar to that for immune cells within the nerve itself, which are positive for major histocompatibility complex (MHC)-II immunoreactivity [12]. In addition, vagal sensory fibers were found in close association with mast cells [4,12]. These findings indicate that vagal sensory neurons occupy a position in which they might be sensitive to cytokines produced by immune cells responding to local infection. In this way, vagal sensory fibers could rapidly respond to pathogens in the gastrointestinal system.

The intrinsic (enteric) neurons that reside throughout the extent of the gastrointestinal tract and control secretion and motility are sensitive to immune activation within the gut [13,14]. Indeed, enteric responses directed toward expelling pathogens provide a critical initial host–defense mechanism. Enteric neurons express receptors for IL-1 and TNF, and the activation of these receptors facilitates enteric neuronal excitability [14]. Vagal sensory neurons monitor enteric ganglia [4]. Thus, cytokine responses may provide a mechanism, via vagal sensory fibers, by which cytokines may indirectly signal the brain.

In addition to activating vagal afferents directly, cytokines may activate vagal immunosensitive pathways via the chemoreceptive cells located in the vagal paraganglia, and/or similar vagally innervated structures, the neuroepithelial bodies, which are found in lung airways [15]. The vagal paraganglia are collections of glomus cells interspersed throughout the vagus nerve, which are innervated by vagal sensory neurons [4,16]. These glomus cells cluster around blood and lymph vessels, suggesting that these cells are likely monitoring substances circulating in body fluids. Glomus cells of the vagal paraganglia, like those of the carotid bodies, express IL-1 receptors [17,18]. Immune cells expressing LPS-induced IL-1 immunoreactivity codistribute with these glomus cells expressing IL-1 receptors (Fig. 1) [12], providing an alternative arrangement whereby vagal sensory nerves may monitor immune-related stimuli circulating in either blood or lymph.

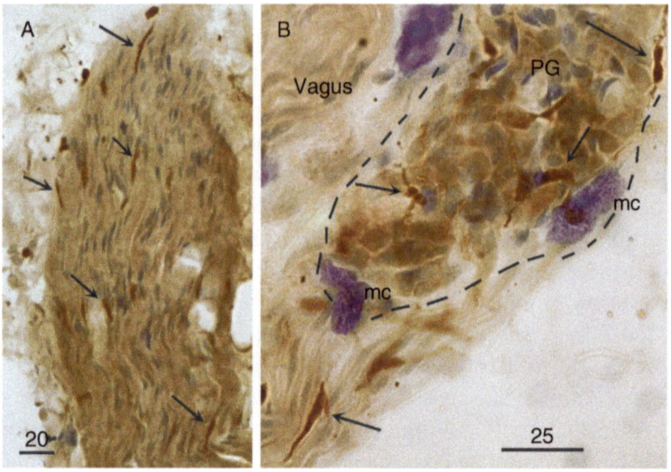

Figure 1. Immune cells in the abdominal vagus nerve and associated paraganglia express IL-1β immunoreactivity (brown staining) following intraperitoneal injection of lipopolysaccharide (LPS). These immune cells are elongated in shape within the nerve (arrows in A). Within the paraganglia (PG, demarcated with dashed lines), immune cells intersperse themselves among the chemosensory glomus cells (lightly stained) and show LPS-induced IL-1β immunoreactivity (thin arrows). The tissue was counterstained with cresyl violet, revealing blue cell nuclei and purple granular staining of mast cells (mc). Scale bars are depicted in micrometers.

2.2.2. *Immune cells and the vagus*

The vagus nerve, like other nerves, is enriched with several types of immune cells (see Fig. 1) [12]. Immunohistochemical studies have shown that most of these cells are myeloid cells of the monocyte, macrophage, and dendritic cell family, based on morphological features and markers such as constitutive expression of MHC-II, a protein that enables antigen presentation to T cells, a critical step in the induction of systemic immune responses. Mast cells also occupy the vagus nerve (see Fig. 1) [12]. These cells are potential sources of cytokines and other proinflammatory mediators including histamine and substance P. Numerous dendritic-like cells are interspersed among vagal nerve fibers, and within the paraganglia, their processes encircle adjacent chemosensory (glomus) cells (see Fig. 1B). These dendritic-like cells are also found among the cell bodies within the vagal ganglia (Fig. 2). In addition, the connective tissue surrounding the nerve contains collections of myeloid cells, mostly ED-1 (CD68) and complement receptor-3-positive macrophages, as well as mast cells and possible lymphoid cells. When treated with intraperitoneal LPS, dendritic-like cells, as well as some macrophages, express IL-1 immunoreactivity (see Fig. 1) [12]. The role of these immune cells is unknown, but their sensitivity to LPS suggests that they may serve a sentinel function, alerting the brain, via cytokine expression in the vagus nerve, regarding the presence of infectious microorganisms.

2.2.3. *Experimental evidence for the vagus as an immunosensory interface*

Experimental evidence supporting a functional role for the vagus in immunosensory signaling initially relied on studies that involved cutting the vagus nerve in the abdomen, below the diaphragm (subdiaphragmatic vagotomy). This is a partial lesion, leaving thoracic structures,

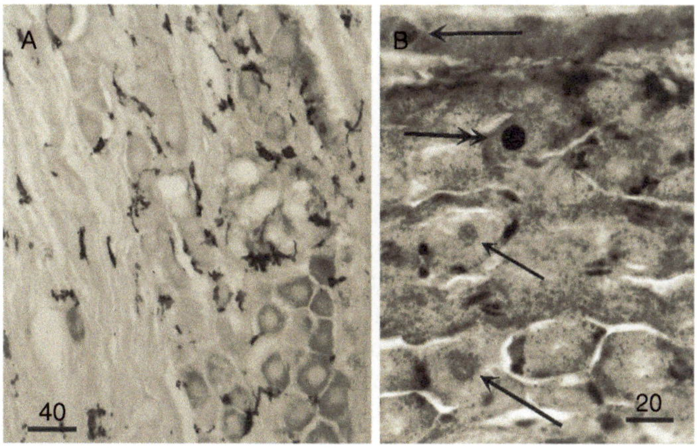

Figure 2. (A) Immune cells in the vagus nerve and vagal (nodose) ganglia express major histocompatibility complex-II (MHC-II; darkly stained cells with irregular processes), which may serve as a source for IL-1β or other cytokines signaling vagal sensory neurons within the ganglia. (B) Vagal sensory neurons in these ganglia express c-Fos immunoreactivity (dark reaction product in the cell nucleus) following intraperitoneal injection of IL-1β indicative of functional activation. Double-headed arrow shows an intensely stained sensory neuron, whereas single arrowheads depict lighter stained nuclei. Scale bars are depicted in micrometers.

notably lung and lymph nodes with intact vagal innervation, but sectioning the vagus above these structures is not compatible with life. Consequently, results from vagotomy studies need to be interpreted with care. When animals recovered, they were challenged with different immune stimulants, and the effects of the surgery on illness responses were observed. The results from these studies showed that vagotomy can block or attenuate a wide range of illness responses, including hyperalgesia, fever, hypersomnelence, hypothalamic–pituitary–adrenal (HPA) activation, conditioned taste aversion, and social withdrawal [19–35]. In general, the effects of vagotomy are most pronounced when the immune stimulus is presented to peritoneal cavity and when the dose of stimulant is low [36,37]. These findings suggest that the vagus nerve may contribute to the signaling of immune activity locally in visceral tissues and that higher doses of immune stimulants such as cytokines recruit additional immunosensory pathways associated with the brain, for example, brain barrier tissues. Additionally, or alternatively, vagal sensory nerves left intact (innervating thoracic structures including lung and lymph nodes) may contribute to cytokine signaling following subdiaphragmatic vagotomy.

Although the results from vagotomy studies support some role for this nerve in immunosensory signaling, the conclusions from these studies are complicated by the fact that the vagus carries both sensory and motor nerve fibers. Thus, cutting the vagus may inhibit illness responses not because it interrupts sensory signaling but because it produces side effects or impairs immune functioning as a result of interrupting parasympathetic outflow. These alternative explanations for the effects of vagotomy have, for the most part, been ruled out. Vagotomized animals develop fevers identical to controls when the thermogenic stimulus is not associated with cytokine treatment [38,39]. In addition, vagotomy does not impair either cytokine expression following LPS treatment or the entry of cytokines or LPS into the systemic circulation [40]. In fact, vagotomy blocks fever responses to low doses of intraperitoneally injected IL-1, even when the injected IL-1 reaches the systemic circulation [24]. Taken together, findings from vagotomy studies support the idea that vagotomy effects follow from the interruption of vagal sensory transmission of cytokine signals.

If in fact, vagal sensory neurons carry cytokine signals, peripheral administration of cytokines should evoke evidence of activation in the primary sensory neurons. Indeed, IL-1 induces c-fos mRNA [41] and c-Fos protein (see Fig. 2) [42] in vagal sensory neurons, and increases electrically recorded neural firing in hepatic vagal sensory fibers [43]. In addition, peripheral injections of LPS induce c-Fos immunoreactivity in vagal sensory neurons, as does staphylococcal enterotoxin B (SEB), a product of gram-positive bacteria [44,45]. The activation of sickness responses by SEB is a T-cell mediated process that likely takes place in lymph nodes and leads to the relatively rapid (1–2 h) induction of the proinflammatory cytokines IL-2 and TNF [46,47]. Further, vagal sensory neurons respond to live bacterial infections in the gut [48,49]. The finding that several types of bacterial stimuli induce activation in vagal sensory neurons suggests that this neural pathway may carry cytokine signals induced by a variety of pathogens.

2.3. Cytokine receptors in the vagus

The expression of c-Fos protein in vagal sensory neurons following peripheral cytokine treatment implies that these neurons express cytokine receptors. Receptors for both IL-1 and TNF have been demonstrated on neurons in the vagal ganglia, as well as in other cells of these structures. In addition to neurons, sensory ganglia contain satellite cells, which may be analogous to CNS glia, as well as macrophages/dendritic-like immune cells and endothelial cells associated with the vasculature that are potential sources of either cytokines or cytokine receptors. Indeed, Ek et al. [41] demonstrated mRNA expression for type 1 IL-1 receptors in the primary sensory neurons and satellite cells in the vagal ganglia. A functional role for IL-1 was demonstrated by Mascarucci et al. [50] that systemic injections of IL-1 or LPS provoke glutamate release by central terminals of vagal neurons (see below). Thus, vagal sensory nerves may signal the presence of local tissue cytokines in the lung and gut epithelium, as well as cytokines in the blood perfusing the nerve or ganglia. The vagal response to LPS may have followed from circulating IL-l generated by macrophages in response to the LPS injection, or by cells in the vagal ganglia, which express the LPS receptor Toll-like receptor (TLR)-4 [51] as well as TLR-9 [52], which responds to bacterial DNA. Taken together, findings from receptor localization and expression studies support the role of vagal sensory neurons in signaling peripheral immune activation.

Based on the sensitivity of TNF signals to subdiaphragmatic vagotomy [22,25], it might be expected that vagal sensory nerves express TNF receptors as well. Although Emch et al. [53] reported TNF receptor 1 (TNFR1; p55) immunoreactivity on the central projection of vagal sensory nerves within the brainstem, as well as on the cell bodies within the vagal sensory ganglia, TNFR1 was absent on the peripherally projecting fibers. Interestingly, activation of TNFR1 receptors enhances glutamate release from vagal terminals [54], suggesting a role for TNF primarily as a neuromodulator, enhancing immune-related signaling to the brain. Tumor necrosis factor is yet to be described in immune cells associated with the vagus, although it has been described in lymph nodes following treatment with SEB [47]. However, it is quite likely that the macrophage type cells closely associated with the nerve and paraganglia express TNF, as they do in culture and in other tissues. In addition, levels of TNF rapidly rise in the general circulation following peripheral administration of LPS [40], providing a potential source of circulating cytokine possibly relevant to receptors expressed in the ganglia and the brainstem [54].

3. CYTOKINES IN SPINAL AND TRIGEMINAL NERVES AND PAIN MODULATION

Although cytokine signaling to cranial viscerosensory nerves induces the familiar constellation of brain-mediated illness responses, cytokines produced in damaged or inflamed trigeminal or spinal nerves modulate spinal mechanisms of pain transmission. Modulation of pain transmission may influence these stress responses and affective states, and potentially influencing ongoing sickness responses. Like the vagus, the spinal nerves contain perineurial immune cells, and MHC-II-positive dendritic-like cells are interspersed among the nerve fibers [12]. These cells express TNF and IL-1 during inflammatory neuritis [55]. The release of TNF, in particular, has been shown to dramatically facilitate pain transmission in the spinal cord [56,57]. Exogenous (epineurial) treatment with TNF produces behavioral allodynia [57], an enhanced pain state whereby normally nonpainful stimuli become painful. Blocking the actions of TNF in models of neuritis prevent enhanced pain sensitivity (hyperalgesia) [57]. Similar findings obtain using IL-1, whereby exogenous IL-1 enhances pain states [58], which can be blocked by treatment with the IL-1 receptor antagonist [58]. Indeed, responses to neuropathic pain were dramatically reduced in animals lacking IL-1 R1 [59], providing strong evidence for a role of IL-1 in inflammatory pain.

In addition to innervating somatic tissues, C fibers of the spinal dorsal root ganglia (DRG) innervate visceral organs. These fibers respond to inflammatory conditions within the gut and seem to play a role in visceral pain and hypersensitivity [60,61]. However, they do not seem to play a role in the induction of sickness responses, suggesting a functional dichotomy or specialization of roles in inflammation between vagal and spinal visceral nerves. Thus, whereas spinal visceral nerves are primarily proinflammatory, activation of cholinergic vagal motor nerves seems to be anti-inflammatory [62].

3.1. Cytokine receptors in spinal and trigeminal ganglia

Cells associated with the DRG express IL-1, TNF, and IL-6 in models of inflammatory or neuropathic pain [54,63–70], as well as in their receptors. Sensory neurons in the DRG express TNFR1 [54,65], and TNF activates second messenger systems (p38 mitogen-activated kinase [56], protein kinase A [71]) in these sensory neurons. Blocking the actions of TNF prevented the activation of p38 [56], and blocking protein kinase A blocks the effects of TNF on sensory neuron excitability [71]. Endogenous TNF is produced by immune cells in the ganglia [65], and taken together, these findings strongly support a role for endogenous, locally generated TNF in the pathogenesis of neuropathic pain. It is likely that IL-1 and IL-6 contribute as well.

3.2. Peripheral actions of cytokine-responsive trigeminal and dorsal root ganglia neurons

The interaction of cytokines and spinal nerve sensory neurons is apparently bidirectional. Intraplantar (hindpaw) administration of capsaicin induced hyperalgesia and the expression of cytokines (IL-1, IL-6, TNF) in the hindpaw skin [72]. However, the expression of cytokines was absent in animals previously treated with capsaicin (to lesion capsaicin-sensitive sensory neurons in the sciatic nerve) [70]. This finding indicates that peripheral nerves, in addition to responding to cytokine signals, can influence the expression of these same cytokines, as well as the release of proinflammatory peptides, such as calcitonin gene-related peptide (CGRP) [70]. This raises the possibility that pathological pain states may be complicated by a positive feedback loop in which cytokines can activate pain-transmitting neurons, which can in turn upregulate the expression of the cytokines.

4. CENTRAL PROJECTIONS OF CYTOKINE-RESPONSIVE NERVES: INTERFACE WITH BRAIN REGIONS SUBSERVING SICKNESS RESPONSES

Cytokines produce physiological and behavioral illness responses by activating brain neurocircuitry that mediates these responses. The identity of this neurocircuitry has been probed using the expression of the activation marker c-Fos following treatment with cytokines or LPS [73,74]. These studies have shown that immune-responsive brain nuclei are overwhelmingly associated with autonomic functions, including the hypothalamus, amygdala, visceral thalamus, periaquiductal gray, and cingulate and infralimbic cortex [74]. Although the neurocircuitry mediating some autonomic functions, including fever [74,75], and the activation of the HPA axis [76] have been described in at least rough detail, neurocircuitry driving other responses, notably behavioral ones, are as yet unclear.

4.1. Cranial nerve viscerosensory projections

Sensory fibers associated with the vagal and glossopharyngeal ganglia collect signals from the tissues that they innervate and convey this information to the brainstem dorsal vagal complex: the nucleus of the solitary tract (nTS) and the area postrema (a circumventricular organ). These nuclei coordinate local, protective reflexes, such as emesis and gastric retention. In addition, they relay a wide variety of viscerosensory signals to forebrain regions concerned with the integration of visceral information with ongoing behavior and other sensory inputs. Notably, brain regions by driven ascending pathways emanating from the dorsal vagal complex [10,77] overlap significantly with those shown to respond to peripheral immune stimulation [73,74]. This arrangement is consistent with the idea that one pathway by which cytokines signal the brain to activate illness responses is via peripheral nerves that in turn drive the dorsal vagal complex and its projections to higher brain regions.

Several lines of evidence support the idea that viscerosensory (vagus and/or glossopharyngeal) nerves convey cytokine-derived information to the dorsal vagal complex. As mentioned earlier, following intraperitoneal administration of IL-1, the vagal/glossopharyngeal sensory neurotransmitter glutamate [78] is released into these structures [50], consistent with the idea that cytokine exposure leads to functional activation of vagal sensory neurons. Both of these brainstem structures, the nTS and the area postrema, express activation markers (e.g., c-Fos) following peripheral administration of immune stimulants including IL-1. Although these findings support the idea that cytokines signal the brain via vagal sensory fibers, it is important to note that activation of dorsal vagal complex could also occur via humoral routes [79,80]. The area postrema is a circumventricular organ in which the blood–brain barrier is weak (hence it is sensitive to circulating signals unavailable to the brain parenchyma), and it contains immune cells that respond to peripherally administered LPS by expressing IL-1 immunoreactivity [80,81]. Some of these IL-1-positive immune cells make direct contact with neurons in the area postrema [81], and lesion of the area postrema attenuates HPA axis activation in response to peripheral IL-1 injection [79], consistent with a role in immunosensory signaling. In addition, it has been suggested that the nTS may respond to cytokine signals directly [53]. Thus, the dorsal vagal complex seems to function as a crossroad for converging immune-related signaling.

4.2. Cytokine-induced potentiation of pain states: trigeminal and spinal mechanisms

Somatic and visceral sensory nerves derived from spinal and trigeminal ganglia collect information from internal tissues including blood vessels and (in the case of spinal fibers) internal organs, as well as from skin, muscle, joint, bone, etc. Those responding to inflammation are A and C fibers that serve to induce and facilitate pain transmission within the spinal cord and the brainstem. The role of cytokines, notably TNF, seems primarily to enhance pain signal transmission, at least in part via enhancement of sensory neurotransmitter release into the CNS [54].

Cytokines, including IL-1 produced by glia in the spinal cord, play a critical role in the modulation of pain states [59]. The induction of spinal cytokines seems to follow from the release of the chemokine fractalkine from primary sensory neurons, which seem to serve as a pain-related signal to spinal cord glia, which is critical for the induction of enhanced pain states, such as allodynia [82]. Prostaglandins likely play a role as well [83]. Pain-related signals from the spinal dorsal horn and trigeminal nuclei of the brainstem are propagated via ascending neural projections to brain regions that integrate information associated with neuroendocrine, physiological and emotional responses to challenges, including host defense (as above).

5. PERSPECTIVES

The immune system is frequently described as "diffuse sensory system" [84]. This concept is based on the fact that peripherally generated immune mediators lead to host defense, or "sickness" responses that are mediated by the brain, via mechanisms similar to those in responses to psychological stressors or homeostatic challenges [85]. This sensory system is comprised of multiple (humoral, neural) pathways that each may be relatively important for specific aspects of nervous system responses to infection.

As noted above, the fact that viscerosensory nerves innervate tissues, notably tonsils, gastro-intestinal subepithelium, and lymph nodes, that are in contact with pathogens early on during infection is consistent with a major role for these nerves in signaling cytokines generated early on as well. Consistent with this idea, vagotomy has impaired brain responses to a live gastro-intestinal bacterial infection in rats [86]. Because the vagus is an autonomic nerve, one would assume that vagal sensory nerves would carry cytokine signals relevant to autonomic functions or ingestive behavior, such as fever, HPA axis, feeding, rather than affective aspects of sickness. However, evidence from vagotomy studies provides support for the case that whereas vagal sensory fibers appear to contribute something to most sickness responses, they also appear to be important for affective responses. For instance, although vagotomy can be ineffective in blocking anorexia resulting from the peripheral administration of immune stimulants [87], it consistently blocks social withdrawal [19,20,88], a component of behavioral depression, even when it fails to block fever [88]. Although this idea at first seems counterintuitive, it is consonant with other recent findings regarding the function of the vagus nerve. Animal and human studies implicate the vagus in the modulation of affective states [89]. In addition, the vagus may also carry feedback signals regarding peripheral responses, such as circulating epinephrine, to behavioral arousal that have been shown to facilitate memory [90]. Thus, several lines of enquiry implicate the vagus, especially the sensory component, in the regulation or the modulation of cognitive and affective functions. This may be a hallmark of neurally mediated cytokine pathways. The major sensory modality influenced by cytokine signaling in the spinal DRG seems to be pain, particularly its enhancement. In addition to influencing local reflexes, pain (like sickness) has profound influences on behavior, cognition, and affect.

From a clinical perspective, studies implicating the vagus in the modulation of mood may shed light on potential mechanisms by which infection and inflammation produce both the perception of "sickness" and the accompanying fatigue. Inflammation in the gastrointestinal tract (irritable bowel syndrome, colitis) is associated with increased expression of proinflammatory cytokines, as well as affective symptoms, notably anxiety and depression [91,92]. These symptoms are specifically associated with illness, as opposed to being a function of personality [91]. Although it is likely that affective symptoms follow from the cognitive response to a painful and stressful disorder, it is also possible that such symptoms may be enhanced by cytokines induced during inflammation that serve to induce components of sickness behavior. Interestingly, subclinical infection with live bacteria (*Camplyobacter jejuni* or *Citrobacter rodentium*) induces anxiety-like behavior in mice and c-Fos protein in vagal sensory neurons, in the absence of circulating cytokines [48,49,93]. These findings suggest that vagal sensory nerves innervating the gastrointestinal tract may serve as a conduit by which such infections influence behavior (Fig. 3).

In summary, cytokines are generated in peripheral nerves in response to inflammation and peripheral immune activation. Neurons in cranial and spinal sensory ganglia express cytokine receptors and are capable of responding to locally generated cytokines, and likely circulating cytokines as well. These cytokines serve to initiate brain-mediated sickness responses, including the enhancement of pain. Although cytokine signaling via nerves, notably the vagus, contributes to a wide variety of sickness responses, it may be relatively more important for modulating changes in affective states, and possibly cognition or arousal, that accompany inflammatory and infectious conditions.

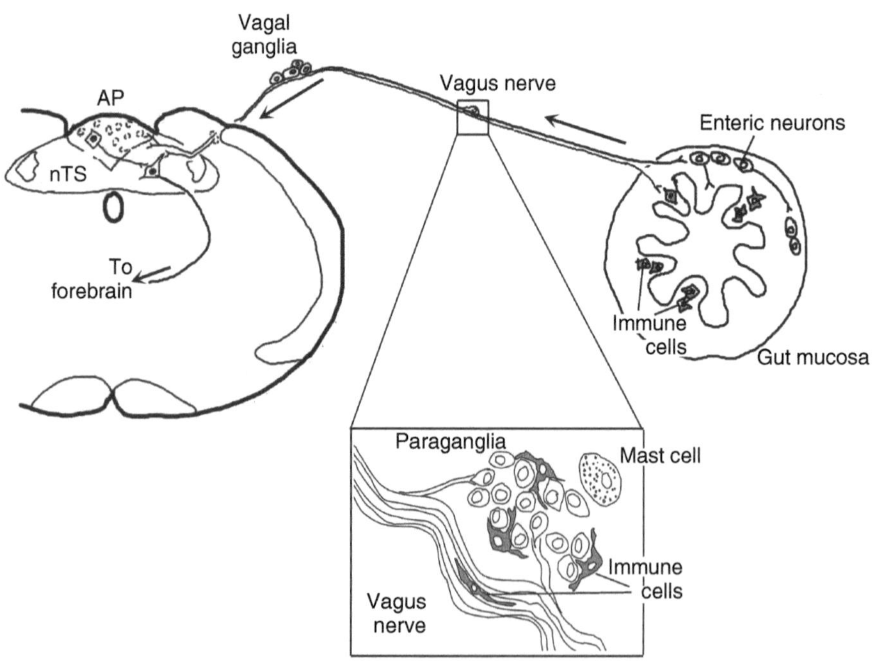

Figure 3. A diagram of potential pathways by which cytokines generated during infection or inflammation in the gut may signal vagal paraganglia and sensory neurons, which subsequently relay immune-related signals to neurons in the area postrema (AP) and the nucleus of the solitary tract (nTS) that project to forebrain regions involved in mediating sickness responses.

ACKNOWLEDGMENTS

The author would like to thank Dr. Ronald Gaykema for critical review of the manuscript. This work was supported by NIH grants MH 55283, MH 64648, MH 68834, and MH 50431.

REFERENCES

1. Romeo H, Tio DL, Rahman SU, Chiappelli F, Taylor AN. The glossopharyngeal nerve as a novel pathway in immune-to-brain communication: Relevance to neuroimmune surveillance of the oral cavity. J Neuroimmunol 2001;115:91–100.
2. Matsuura S. Chemoreceptor properties of glomus tissue found in the carotid region of the cat. J. Physiol Lond 1973;235:57–73.
3. Wang X, Wang B-R, Duan X-L, Zhang P, Ding Y-Q, Jia Y, Jiao X-Y, Ju G. Strong expression of interleukin-1 receptor type 1 in the rat carotid body. J Histochem Cytochem 2002;50:1677–84.
4. Berthoud H-R, Neuhuber W.L Functional and chemical anatomy of the afferent vagal system. Autonom Neurosci Basic Clin 2000;85:1–17.
5. Cano G, Sved AF, Rinaman L, Rabin BS, Card JP. Characterization of the central nervous system innervation of the rat spleen using viral transneuronal tracing. J Comp Neurol 2001;439:1–18.
6. Nance DM, Burns, J. Innervation of the spleen in the rat: evidence for absence of afferent innervation. Brain Behav Immun 1989;3:281–90.
7. Felton DL, Livnat S, Felton SY, Carlson SL, Bellinger DL, Yeh P. Sympathetic innervation of lymph nodes in mice. Brain Res Bull 1984;13:693–6.
8. Fink T, Weihe E. Multiple neuropeptides in nerves supplying mammalian lymph nodes: Messenger candidates for sensory and autonomic neuroimmunomodulation. Neurosci Lett 1988;19:39–44.
9. Popper P, Mantyh CR, Vigna SR, Maggio JE, Mantyh PW. The localization of sensory nerve fibers and receptor binding sites for sensory neuropeptides in canine lymph nodes. Peptides 1988;9:257–67.
10. Goehler LE, Gaykema RPA, Hansen MK, Maier SF, Watkins LR. Vagally mediated fever: a visceral chemosensory modality. Auton Neurosci 2000;85:49–59.
11. Nagura H, Ohtani H, Masuda T, Kimura M, Nakamura S. HLA-DR expression on M cells overlaying Peyer's patches is a common feature of human small intestine. Acta Pathol Jpn 1991;41:818–23.
12. Goehler LE, Gaykema RPA, Nguyen KT, Lee JL, Tilders FJH, Maier SF, Watkins LR. Interleukin-1β in immune cells of the abdominal vagus nerve: An immune to nervous system link? J Neurosci 1999;17:2799–806.
13. Sharkey KA, Kroese, ABA. Consequences of intestinal inflammation on the enteric nervous system: Neuronal activation induced by inflammatory mediators. Anat Rec 2001:262:79–90.
14. Sharkey KA, Mawe GM. Neuroimmune and epithelial interactions in intestinal inflammation. Curr Opin Pharmacol 2002;2:669–77.
15. Adriaensen D, Timmermans J-P, Brouns I, Berthoud H-R, Neuhuber WL, Scheuermann DW. Pulmonary intraepithelial vagal nodose afferent nerve terminals are confined to neuroepithelial bodies: An anterograde tracing and confocal microscopy study in adult rats. Cell Tissue Res 1998;293:395–405.

16. Berthoud H-R, Kressel M, Neuhuber WL. Vagal afferent innervation of the rat abdominal paraganglia as revealed by anterograde DiI-tracing and confocal microscopy. Acta Anat 1995;152:127–32.
17. Goehler LE, Relton JK, Dripps D, Keichle R, Tartaglia N, Maier SF, Watkins LR. Vagal paraganglia bind biotinylated interleukin-1 receptor antagonist: A possible mechanism for immune-to-brain communication. Brain Res Bull 1997;43:357–64.
18. Wang X, Wang BR, Duan WL, Liu HL, Ju G. The expression of IL-1 receptor type 1 in nodose ganglion and vagal paraganglia in the rat. Chin J Neurosci 2000;16:90–3.
19. Bret-Dibat JL, Bluthe R-M, Kent S, Kelley KW, Dantzer R. Lipopolysaccharide and interleukin-1 depress food-motivated behavior in mice by a vagal-mediated mechanism. Brain Behav Immun 1995;9:242–6.
20. Bluthe R-M, Walter V, Parnet P, Laye S, Lestage J, Verrier D, Poole S, Stenning BE, Kelley KW, Dantzer R. Lipopolysaccharide induces sickness behavior in rats by a vagal mediated mechanism. C R Acad Sci III 1994;317:499–503.
21. Fleshner M, Goehler LE, Hermann J, Relton JK, Maier SF, Watkins LR. Interleukin 1β induced corticosterone elevation and hypothalamic NE depletion is vagally mediated. Brain Res Bull 1995;37:605–10.
22. Fleshner M, Goehler LE, Schwartz BA, McGorry M, Martin D, Watkins LR, Maier SF. Thermogenic and corticosterone responses to intravenous cytokines (IL-1β and TNF-α) are attenuated by subdiaphragmatic vagotomy. J Neuroimmunol 1998;86:134–41.
23. Gaykema RPA, Dijkstra I, Tilders FJH. Subdiaphragmatic vagotomy suppresses endotoxin-induced activation of the hypothalamic corticotropin-releasing hormones neurons and ACTH secretion. Endocrinology 1995;136:4717–20.
24. Gaykema RPA, Goehler LE, Hansen MK, Maier SF, Watkins LR. Subdiaphragmatic vagotomy blocks interleukin-1β-induced fever but does not reduce interleukin-1β levels in the circulation. Auton Neurosci Basic Clin 2000;85:72–7.
25. Goehler LE, Busch CR, Tartaglia N, Relton J, Sisk D, Maier SF, Watkins LR. Blockade of cytokine induced conditioned taste aversion by subdiaphramatic vagotomy: Further evidence for vagal mediation of immune–brain interactions. Neurosci Lett 1995; 185:163–6.
26. Hansen MK, Krueger JM. Subdiaphragmatic vagotomy blocks the sleep- and fever-promoting effects of interleukin-1β. Am J Physiol 1997;273:R1246–53.
27. Kapcala LP, He JR, Gao Y, Pieper JO, DeTolla LJ. Subdiaphragmatic vagotomy inhibits intra-abdominal interleukin-1β stimulation of adrenocorticotropin secretion. Brain Res 1996;728:247–54.
28. Opp MR, Toth LA. Somnogenic and pyrogenic effects of interleukin-1 beta and lipopolysaccharide in intact and vagotomized rats. Life Sci 1998;62:923–36.
29. Romanovsky AA, Simons CT, Szekely M, Kulchitsky VA. The vagus nerve in the thermoregulatory response to systemic inflammation. Am J Physiol 1997;273:R407–13.
30. Sehic E, Blatteis CM. Blockade of lipopolysaccharide-induced fever by subdiaphragmatic vagotomy in guinea pigs. Brain Res 1996;726:160–6.
31. Wan W, Wetmore L, Sorensen CM, Greenberg AH, Nance DM. Neural and biochemical mediators of endotoxin and stress-induced c-fos expression in the rat brain. Brain Res Bull 1994;34:7–14.
32. Watkins LR, Wiertelak EP, Goehler L, Mooney-Heiberger K, Martinez J, Furness L, Smith KP, Maier SF. Neurocircuitry of illness-induced hyperalgesia. Brain Res 1994; 639:283–99.

33. Watkins LR, Weirtelak EP, Goehler LE, Smith KP, Martin D, Maier SF. Characterization of cytokine induced hyperalgesia. Brain Res 1994;654:15–26.
34. Watkins LR, Goehler LE, Relton JK, Tartaglia N, Silbert L, Martin D, Maier SF. Blockade of interleukin-1-induced fever by subdiaphragmatic vagotomy: Evidence for vagal mediation of immune–brain communication. Neurosci Lett 1995;183:27–31.
35. Watkins LR, Goehler LE, Relton J, Brewer MT, Maier SF. Immune-to-brain communication: Systemic tumour necrosis factor-alpha (TNF-alpha) produces behavioral hyperalgesia via vagal afferents. Brain Res 1995;692:244–50.
36. Bluthe R-M, Michaud B, Kelly KW, Dantzer, R. Vagotomy blocks behavioral effects of interleukin-1 injected via the intraperitoneal route but not by other systemic routes. NeuroReport 1996;7:2823–7.
37. Hansen MK, O'Conner KA, Goehler LE, Watkins LR, Maier SF. The role of the vagus nerve in interleukin-1β-induced fever is dependent on dose. Am J Physiol 2001;280:R929–34.
38. Milligan E, McGorry MM, Fleshner M, Gaykema RPA, Goehler LE, Watkins LR, Maier SF. Subdiaphragmatic vagotomy does not prevent fever following intracerebroventricular prostaglandin: Further evidence for the importance of vagal afferents in immune-to-brain communication. Brain Res 1997;766:240–3.
39. Sugimoto N, Simons CT, Romanovsky AA. Vagotomy does not affect thermal responsiveness to intrabrain prostaglandin E2 and cholecystokinin octapeptide. Brain Res 1999;844:157–63.
40. Hansen MK, Nguyen KT, Fleshner M, Goehler LE, Gaykema RPA, Maier SF, Watkins LR. Effects of vagotomy on circulating levels of endotoxin, pro-inflammatory cytokines, and corticosterone following intraperitoneal lipoplysaccharide. Am J Physiol 2000;278: R331–6.
41. Ek M, Kurosawa M, Lundeberg T, Ericsson A. Activation of vagal afferents after intravenous injection of interleukin-1β: Role of endogenous prostaglandins. J Neurosci 1998;18:9471–9.
42. Goehler LE, Gaykema RPA, Hammack SE, Maier SF, Watkins LR. Interleukin-1 induces c-Fos immunoreactivity in primary afferent neurons of the vagus nerve. Brain Res 1998;804:306–10.
43. Niijima A. The afferent discharges from sensors for interleukin-1β in the hepatoportal system in the anesthetized rat. J Auton Nerv Syst 1996;61:287–91.
44. Gaykema RPA, Goehler LE, Tilders FJH, Bol JGM, McGorry MM, Maier SF, Watkins LR. Bacterial endotoxin induces Fos immunoreactivity in primary afferent neurons of the vagus nerve. Neuroimmunomodulation 1998;5:234–40.
45. Gaykema RPA, Goehler LE, Armstrong CB, Khorsand J, Maier SF, Watkins, LR. Differential FOS expression in rat brain induced by lipopolysaccharide and staphylococcal enterotoxin B. Neuroimmunomodulation 1999;6:220.
46. Shurin G, Shanks N, Nelson L, Hoffman G, Huang L, Kusnecov AW. Hypothalamic–pituitary–adrenal activation by the bacterial superantigen staphylococcal enterotoxin B: Role of macrophages and T cells. Neuroendocrinology 1997;65:18–28.
47. Litton MJ, Sander S, Murphy, O'Garra, Abrams JS. Early expression of cytokines in lymph nodes after treatment in vivo with staphylococcus enterotoxin B. J Immunol Meth 1994;175:47–58.
48. Goehler LE, Gaykema RPA, Opitz N, Reddaway R, Badr NA, Lyte M. Activation in vagal afferents and central autonomic pathways: Early responses to intestinal infection with *Campylobacter jejuni*. Brain Behav Immun 2005:19:334–44.

49. Lyte M, Wang L, Opitz N, Gaykema RPA, Goehler LE. Anxiety-like behavior during initial stage of infection with agent of colonic hyperplasia *Citrobacter rodentium*. Physiol Behav 2006;89:350–7.
50. Mascarucci P, Perego C, Terrazzino S, DeSimoni MG. Glutamate release in the nucleus tractus solitarius induced by peripheral lipopolysaccharide and interleukin-1β. Neuroscience 1998;86:1285–90.
51. Hosoi T, Okuma Y, Matsuda T, Nomura Y. Novel pathway for LPS-induced afferent vagus nerve activation: Possible role of nodose ganglion. Autonom Neurosci Basic Clin 2005:120:104–7.
52. Sako K, Okuma Y, Hosoi T, Nomura Y. STAT3 activation and c-FOS expression in the brain following peripheral administration of bacterial DNA. J Neuroimmunol 2005;158:40–9.
53. Emch GS, Hermann GE, Rogers RC. TNF-alpha induces c-Fos generation in the nucleus of the solitary tract that is blocked by NBQX and MK801. Am J Physiol 2001;281:R1394–400.
54. Hermann GE, Holmes GM, Rogers RC. TNF(alpha) modulation of visceral and spinal sensory processing. Curr Pharm Des 2005;11:1391–409.
55. Gazda LS, Milligan ED, Hansen MK, Twining CM, Poulos NM, Chacur M, O'Conner KA, Armstrong C, Maier SF, Watkins LR, Myers RR. Sciatic inflammatory neuritis (SIN): Behavioral allodynis in parallel with peri-sciatic proinflammatory cytokine and superoxide production. J Peripher Nerv Syst 2001;6:111–29.
56. Scafers M, Svensson C, Sommer C, Sorkin LS. Tumor necrosis factor-α induces mechanical allodynia after spinal nerve ligation by activation of p38 MAPK in primary sensory neurons. J Neurosci 2003;23:2517–21.
57. Sorkin LS, Doom CM. Epineurial application of TNF elicits an acute mechanical hyperalgesia in the awake rat. J Peripher Nerv Syst 2000;5:96–100.
58. Clark AK, D'Aquisto F, Gentry C, Marchand F, McMahon SB, Malcangio M. Rapid co-release of interleukin-1beta and caspase 1 in spinal cord inflammation. J Neurochem 2006;99:868–80.
59. Wolf G, Gabay E, Tal M, Yirmiya R, Shavit Y. Genetic impairment of interleukin-1 signaling attenuates neuropathic pain, autotomy, and spontaneous ectopic neuronal activity, following nerve injury in mice. Pain 2006;120:315–24.
60. Kirkup AJ, Brunsden AM, Grundy D. Receptors and transmission in the brain–gut axis: Potential for novel therapies: I. Receptors on visceral afferents. Am J Physiol Gastrointest Liver Physiol 2001;280:G787–94.
61. Holzer P. Efferent-like roles of afferent neurons in the gut: Blow flow regulation and tissue protection. Auton Neurosci Basic Clin 2006:125:70–5.
62. Pavlov VA, Tracey KJ. The cholinergic anti-inflammatory pathway. Brain Behav Immun 2005;19:493–9.
63. Cunningham ET Jr, Stalder AK, Sanna PP, Liu SS, Bloom FE, Howes EL Jr., Campbell IL, Margolis TP. Distribution of tumor necrosis factor receptor messenger RNA in normal and herpes simplex virus infected trigeminal ganglia in the mouse. Brain Res 1997;30:99–106.
64. Lee HL, Lee KM, Son SJ, Hwang SH, Cho HJ. Temporal expression of cytokines and their receptors mRNA in a neuropathic pain model. Neuroreport 2004;15:2807–11.
65. Li,Y, Ji A, Weihe E, Schafer K-H. Cell-specific expression and lipopolysaccharide-induced regulation of tumor necrosis factor a (TNFα) and TNF receptors in rat dorsal root ganglia. J Neurosci 2004;24:9623–31.
66. Li M, Shi J, Tang J-R, Chen D, Ai B, Chen J, Wang L-N, Cao F-Y, Li L-L, Lin C-Y, Guan X-M. Effects of complete Freund's adjuvant on immunohistochemical distribution of IL-1β and IL-1R I in neurons and glia cells of dorsal root ganglion. Acta Pharmacol Sin 2005;26:192–8.

67. Ozaktay AC, Kallakuri S, Takebayashi T, Cavanaugh JM, Asik I, Deleo JA, Weinstein JN. Effects of interleukin-1 beta, interleukin-6, and tumor necrosis factor on sensitivity of dorsal root ganglion and peripheral receptive fields in rats. Eur Spine J 2006;15:1529–37.
68. Liu L, Yang TM, Liedtke W, Simon SA. Chronic IL-1 signaling potentiates voltage-dependent sodium currents in trigeminal nociceptive neurons. J Neurophysiol 2006;95:1478–90.
69. Cunha TM, Verri WA Jr, Silva JS, Poole S, Cunha FQ, Ferreira SH. A cascade of cytokines mediates mechanical inflammatory hypernociception in mice. Proc Natl Acad Sci USA 2005;102:1755–60.
70. Hou L, Li W, Wang X. Mechanism of interleukin-1b-induced calcitonin gene-related peptide production from dorsal root ganglion neurons of neonatal rats. J Neurosci Res 2003;73:188–97.
71. Zhang J-M, Li H, Liu B, Brull, SJ. Acute topical application of tumor necrosis factor α evokes protein kinase A-dependent responses in rat sensory neurons. J Neurophysiol 2002;88:1387–92.
72. Saade NE, Massaad CA, Ochoa-Chaar CI, Jabbur SJ, Safieh-Garabedian, Atweh SF. Upregulation of proinflammatory cytokines and nerve growth factor by intraplantar injection of capsaicin in rats. J Physiol 2002;343:241–52.
73. Brady LS, Lynn AB, Herkenham M, Gottesfels Z. Systemic interleukin-1 induces early and late patterns of c-fos mRNA expression in brain. J Neurosci 1994;14:4951–64.
74. Elmquist JK, Scammell TE, Saper CB. Mechanisms of CNS response to systemic immune challenge: The febrile response. Trends Neurosci 1997;20:565–70.
75. Zhang Y-H, Lu J, Elmquist JK, Saper CB. Lipopolysaccharide activates specific populations of hypothalamic and brainstem neurons that project to the spinal cord. J Neurosci 2000;20:6578–86.
76. Ericsson A, Arias C, Sawchenko PE. Evidence for an intramedullary prostaglandin-dependent mechanism in the activation of stress-related neuroendocrine circuitry by intravenous interleukin-1. J Neurosci 1997;17:7166–79.
77. Ricardo JA, Koh ET. Anatomical evidence of direct projections from the nucleus of the solitary tract to the hypothalamus, amygdala, and other forebrain structures in the rat. Brain Res 1978;153:1–26.
78. Schaffar N, Roa H, Kessler JP, Jean A. Immunohistochemical detection of glutamate in rat vagal sensory neurons. Brain Res 1997;778:302–8.
79. Lee HY, Whiteside MB, Herkenham M. Area postrema removal abolishes stimulatory effects of intravenous interleukin-1β on hypothalamic–pituitary–adrenal axis activity and c-fos mRNA in the hypothalamic paraventricular nucleus. Brain Res Bull 1998;46:495–503.
80. Konsman JP, Kelley K, Dantzer R. Temporal and spatial relationship between lipopolysaccharide-induced expression of Fos, interleukin-1β and inducible nitric oxide synthase in rat brain. Neuroscience 1999;89:535–48.
81. Goehler LE, Erisir A, Gaykema RPA. Neural–immune interface in the area postrema. Neuroscience 2006;140:1415–34.
82. Milligan E, Zapata V, Schoeniger D, Chacur M, Green P, Poole S, Martin D, Maier SF, Watkins LR. An initial investigation of spinal mechanisms underlying pain enhancement induced by fractalkine, a neuronally released chemokine. Eur J Neurosci 2005;22:2775–82.
83. Samad TA, Moore KA, Sapirstein A, Billet S, Allchorne A, Poole S, Bonventre JV, Woolf CJ. Interleukin-1beta-mediated induction of Cox-2 in the CNS contributes to inflammatory pain hypersensitivity. Nature 2001;410:471–5.

84. Besedovsky HO, del Rey A. Immune-neuroendocrine circuits: Integrative role of cytokines. Front Neuroendocrinol 1992;13:61–94.
85. Larson SJ, Dunn AJ. Behavioral effects of cytokines. Brain Behav Immun 2001;15:371–87.
86. Wang X, Wang B-R, Zhang X-J, Xu Z, Ding,Y-Q, Ju G. Evidences for vagus nerve in maintenance of immune balance and transmission of immune information from gut to brain in STM-infected rats. World J Gastroenterol 2002;8:540–5.
87. Porter MH, Hrupka BJ, Langhans W, Schwartz GJ. Vagal and splanchnic afferents are not necessary for the anorexia produced by peripheral IL-1β, LPS and MDP. Am J Physiol 1998;275:R384–9.
88. Konsman JP, Luheshi GN, Bluthe R-M., Dantzer R. The vagus nerve mediates behavioural depression, but not fever, in response to peripheral immune signals; a functional anatomical analysis. Eur J Neurosci 2000;12:4434–46.
89. Zagon A. Does the vagus nerve mediate the sixth sense? Trends Neurosci 2001;24:671–3.
90. Clark KB, Smith DC, Hassert DC, Browning RA, Noritoku DK, Jensen RA. Post-training electrical stimulation of vagal afferents with concomitant vagal efferent inactivation enhances memory storage processes in the rat. Neurobiol Learn Mem 1998;70:364–73.
91. Simrin M, Axelsson J, Gillberg R, Abrahamsson H, Svedlund J, Bjornsson ES. Quality of life in inflammatory bowel disease in remission: The impact of IBS-like symptoms and associated psychological factors. Am J Gastroenterol 2002;97:389–96.
92. Addolorato G, Capristo E, Stafanini GF, Gasbarrini, G. Inflammatory bowel disease: A study of the association between anxiety and depression, physical morbidity, and nutritional status. Scand J Gastroenterol 1997;32:1013–21.
93. Lyte M, Varcoe JJ, Bailey MT. Anxiogenic effect of subclinical bacterial infection in mice in the absence of overt immune activation. Physiol Behav 1998;65:63–9.

Cytokines and the Brain
Edited by C. Phelps and E. Korneva

Interleukin-2 Gene Expression in Central Nervous System Cells after Stress and Antigen Application

ELENA A. KORNEVA and TATIANA B. KAZAKOVA

State Organization "Institute for Experimental Medicine of Russian Academy of Medical Sciences", Department of General Pathology and Pathophysiology, St. Petersburg, Russia

ABSTRACT

In this article we summarize the data concerning the expression of immediate early gene (IEG), c-fos, in brain cells, which encodes for one of the interleukin-2 (IL-2) cytokine transcriptional factors. The main molecular features and tissue-specific differences of IL-2 and IL-2 receptors in the brain are also discussed. Various forms of stress stimulated the expression of c-fos and IL-2 genes in central nervous system (CNS) neurons has been shown after different stressor stimuli. Stress activated IEG expression in cells of definite hypothalamic structures.

Antigen injection led to the activation of c-fos and IL-2 gene expression specifically in the cells of definite hypothalamic nuclei and in other areas of rat brain. A temporospatial pattern of activation of hypothalamic structures was found consistently in response to exposure to antigen, which was different from that induced by other stressors.

Short synthetic immunomodulating peptides influenced IEG expression in immune and nervous system cells. Definite physical factors, like Extremely high-frequency electromagnetic millimeter waves (EHF) skin irradiation, modulated (mostly decreased) the stress-induced stimulation of IEG genes in hypothalamic neurons. We discuss the possible role of the JAK-STAT and Ras-MAPK signal transduction pathways in IL-2 gene expression in lymphocytes and in nerve cells.

1. INTRODUCTION

Cytokines represent a large group of mediators with a wide spectrum of biological activities, which include the regulation of immune system, the development of immune responses, the dynamic hematopoiesis [1]. Most cytokines are produced not only by immune cells but also by nerve cells [2,3] and participate in the interaction of the nervous and immune systems [3,4]. Clearly, cytokines are signal molecules that function as messengers in both the nervous and immune systems.

Cytokine gene expression in brain cells in health and in pathological conditions (multiple sclerosis, encephalitis, Alzheimer's disease, AIDS, and others) has been described earlier [5–9].

Interleukins, interferon-α (INF-α), tumor necrosis factor-α (TNF-α), transforming growth factor-β (TGF-β), and their receptors are thought to be produced by central nervous system (CNS) cells [3,8,10]. These cytokines appear to play a physiological role in neuronal and glial cells.

Interleukin-2 (IL-2) is one of the cytokines expressed in brain cells [3]. The functional role of IL-2 in the CNS deserves intense attention. The interactions of IL-2 with various elements of the signal transduction pathways remain obscure.

In this chapter we attempt to answer a number of questions regarding the expression of IL-2 gene and IL-2 receptors in brain cells after exposure to various stimuli.

2. INTERLEUKIN-2 AS AN ENDOGENOUS CYTOKINE OF CENTRAL NERVOUS SYSTEM

The expression of IL-2 gene and IL-2 receptors is found in brain cells [3,11,12]. IL-2 was first detected by immunohistochemical and radioimmunological methods in the neurons, astroglia, and microglia of healthy and sick animals [12–14]. The greatest number of IL-2 synthesizing cells was found in the hippocampus, cerebral cortex, hypothalamus, and cerebellum [10,11]. It was shown that after trauma and during aging, the amount of IL-2 in brain cells increased [10,15], while in immune cells, the expression of IL-2 decreased with age [16,17].

IL-2 mRNA isolated from embryonic cells of the human brain had identical sequence with IL-2 mRNA obtained from lymphocytes [18]. But in the brain cells, other forms of IL-2 mRNA and IL-2 protein with molecular weights (MWs) different from those synthesized in T lymphocytes were also found [14]. The MW of IL-2 mRNA synthesized in T lymphocytes corresponds to 1.5–2 kb. In cells of the embryonic and mature human brain of mouse and rat, there are two forms of IL-2 mRNA with MW equal to 1.5–2 kb and with MW equal to 5 kb. The presence of IL-2-like protein, a dimeric form, was also noted in CNS cells. For example, in fish the MW of IL-2 in brain cells is 28 kDa, while MW of lymphocyte-derived IL-2 is14 kDa. In rats the MW of IL-2 in the brain cells and in T lymphocytes corresponds to 23 and 17 kDa, respectively [12,19–21].

The question of IL-2 penetration into brain tissue through the blood–brain barrier is controversial. There are positive and negative observations [22–24]. It was observed that IL-2 entry into the brain is low because of the absence of blood-to-brain transporters, circulating factors, and of CNS-to-blood efflux system [25]. It seems that the lack of these elements may protect the brain from circulating IL-2. In contrast, the synthesis of endogenous IL-2 by CNS cells has been established.

The presence of IL-2 receptors on the membranes of CNS cells was demonstrated by radioligand binding, by immunochemical methods, and by the determination of mRNA [2]. Proteins of three subunits of IL-2 receptors (α, β, γ) were found in brain cells. Their structure is identical to those of immune cells. In contrast, the MW of IL-2 receptor-like proteins in the brain can be different from those expressed by T cells [11].

Effects of IL-2 were studied mainly in brain cell cultures and after the administration of exogenous rIL-2 to animals [21,26]. The IL-2 content of the nerve cells was analyzed in pathological conditions, such as cancer, Alzheimer's disease, trauma [27–30]. IL-2 was shown to promote the survival of neurons [31] to increase the release of hypothalamic and pituitary hormones, and other mediators, including luteinizing hormone (LH), its releasing factor, corticotropin releasing factor (CRF), growth hormone-releasing hormone, somatostatin, arginine–vasopressin, met-enkephalin, β-endorphin, and acetylcholine [10,32–35]. IL-2 accelerates the metabolism of norepinephrine and dopamine in the hypothalamus [36], the synthesis of mRNA of propiomelanocortin [37] releases adrenocorticotropin hormone (ACTH), LH, and thyroid-stimulating hormone from the pituitary gland [38], and exerts an analgesic effect [39,40]. In rat skin a time-related expression of IL-2 was observed during wound healing [41].

The understanding of the physiological effects of IL-2 in the CNS and its influence on the expression of the neuronal cell genome is of considerable significance. This is important for basic science, and also for clinical sciences. IL-2 has been used already for therapeutic aims (e.g., for treatment of cancer or of various immunodeficiency states). Such therapy may lead to neurological, neuroendocrine, or emotional side effects [42–44]. There is evidence for a link between IL-2 and depressive illness, which indicates that this cytokine may contribute to the development of affective disorders [45].

3. INTERLEUKIN-2 GENE EXPRESSION IN BRAIN CELLS OF RATS EXPOSED TO STRESS

Gene expression in neurons reveals the cells and structures that react to a certain external stimulus and of those participating in the development of specific responses. Genomics is a new science in modern biology, which deals with the analysis of genetic processes that underlie cell physiology and performs comparative assessments of genome expression under various conditions. The method of DNA microarray (cDNA microchip technology) [46,47] and the method of a serial analysis of gene expression [48,49] are the tools of this science. These techniques allow the detection of gene expression in response to defined stimuli [50,51]. Another method, which is widely used, is the hybridization of mRNA to cDNA. This is carried out after the isolation of total RNA from cells. A labeled cDNA probe is used for hybridization [52]. mRNA is an intermediate messenger for the synthesis of functional proteins. mRNA transcription does not always lead to the production of proteins. So, a combined analysis of the synthesis of both mRNA and its corresponding protein is necessary.

The c-fos protein, which is the product of the c-fos proto-oncogene, is a well-known marker of neuronal cell activation [53–57]. c-Fos protein forms complexes with protein transcription factors (e.g., AP-1, NF-AT) of a number of inducible genes, including the IL-2 gene [17]. At the same time, a number of investigators found that IL-2 β and γ receptor subunits have a critical role in the regulation of c-fos gene expression through the SRE element of its promoter area [58–60]. A regulating role of IL-2 and IL-6 in the expression of proto-oncogene, c-fos, was found in cultures of human adenoma cells [61,62].

The expression of c-fos mRNA in brain cells begins within several minutes after exposure to a stimulus and reaches its maximum in 30–60 min. Maximum synthesis of c-fos protein in nerve cells was observed in 1 and 3 h after exposure to a stimulus and in 4–6 h a decrease in the protein contents of nerve cells occurs [53,56–58].

IL-2 mRNA expression in T lymphocytes of mice [63,64] begins within 2 h after the application of a stimulus, with maximum synthesis occurring in 8–10 hours [65]. The immediate early gene (IEG) response in cells of different brain areas after various stimuli allows for the characterization of the individual neurons and brain structures that respond to certain stimuli [66–70]. This way a "functional map" of the brain can be created.

Stress causes responses not only in neurons, but also in immunocompetent cells. In this respect the hypothalamus is important because it controls vegetative functions and processes [71–73], which include immune function. For example, the electrical stimulation of the lateral hypothalamic area (LHA) results in increased activity of natural killer cells [74]. In contrast, ventromedial hypothalamic nucleus (VMH) stimulation inhibits the activity of natural killer cells [73]. It should be noted that different areas of the hypothalamus are activated by various stress conditions [68,70].

Various stressful conditions are known to activate c-fos gene transcription and c-fos protein synthesis in different areas of the brain. Acute stress (e.g., swimming and restraint stress) induces the synthesis of c-fos mRNA in the hypothalamic paraventricular nucleus (PVH) in 30 min after exposure. c-Fos appears in neurons synthesizing and secreting CRF, arginine–vasopressin, and other substances that are involved in the regulation of ACTH and glucocorticoid hormone secretion [26,33–35]. In addition, stress-induced synthesis of c-fos protein takes place in nerve cells of the cerebral cortex, cerebellum, hippocampus, corpus striatum, and definable part of the brain stem [35,75–82].

One of the stimuli-inducing expression of c-fos gene is psychoemotional stress, which activates the cells of PVN, locus coeruleus (LC) [77], and limbicoreticular brain structures [78]. c-Fos mRNA is synthesized in response to different kinds of stress in hypothalamic paraventricular nucleus (PVN), supraoptic (SO) and dorsomedial hypothalamic (DMH) nuclei, LC, and central amygdaloid nucleus (ACe) complex [79,80].

All forms of stress lead to c-fos mRNA and c-fos protein synthesis in various brain areas. According to Kovács [81], the areas that show c-fos gene expression in response to a given stressful stimulus are commonly grouped into three categories that are as follows:

(1) Areas that show increased activation after handling
(2) Areas that are involved in conveying stressor-specific information to the stress-related pathway
(3) Regions that mediate stereotypic neuroendocrine, autonomic, and behavioral responses to stress [79].

The effects of various kinds of stress (e.g., pain, surgical intervention, trauma, infections) were intensively studied on c-fos gene expression in neuronal cells [53,54,76]. However, little information was available on the effects of stress on IL-2 gene expression in brain cells, and on its possible correlation with the expression of the c-fos gene.

3.1. Expression of c-fos and interleukin-2 genes in brain cells of rats exposed to rotational stress

The c-fos family proteins are expressed in the cells of the nervous system in response to stress of different nature: both mild stress that has a stimulatory effect on the immune system and severe stress that suppresses the activity of immunocompetent cells [83–85]. Rotation stress stimulates the activity of the immune system, the synthesis of c-fos mRNA, c-fos-like protein, and IL-2 mRNA in splenocytes of CBA mice [68,71,86]. Two hours after exposure, it increases the number of c-fos mRNA-containing cells in hypothalamic and thalamic structures and in the sensomotoric area of the cerebral cortex. In intact animals, single cells containing c-fos mRNA are found in the above-mentioned structures [68].

In 2 h after the rotation stress, the number of cells containing IL-2 mRNA is increased in hypothalamic structures. The most intensive expression of both genes (e.g., c-fos and IL-2) after stress application occurs in LHA and cerebral cortex. Labeled cells are also found in the thalamus. Expression of c-fos mRNA after rotation stress is more pronounced than that of IL-2 mRNA [68].

3.2. Expression of c-fos and interleukin-2 genes in brain cells of rats exposed to pain or combined stimulation

Different kinds of painful stimulation – mechanical, electrical and combined – lead to expression of c-fos and IL-2 genes. In 2 hours after a needle prick, c-fos mRNA and c-fos-like protein-positive cells are found in the hippocampus, PVN, suprachiasmatic (SCH) hypothalamic nuclei, LHA, thalamus, nucleus caudate (CdN), ACe, and sensomotor areas of the cortex [68].

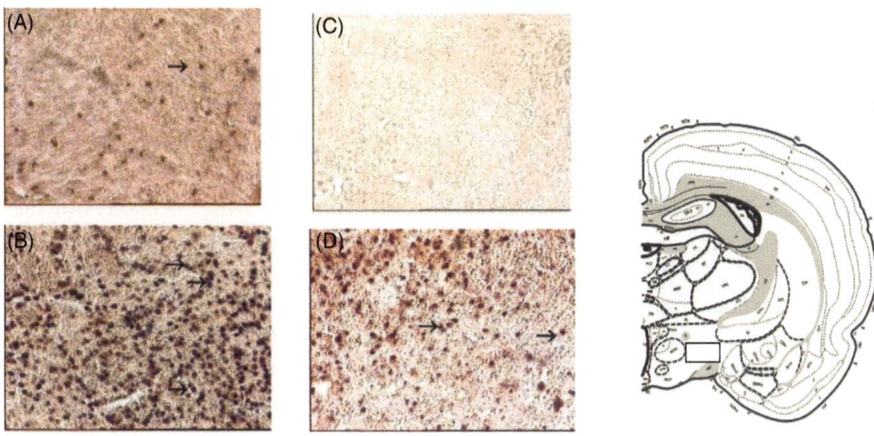

Figure 1. Photomicrographs of rat brain sections of the LHA with c-fos mRNA and IL-2 mRNA expressing cells. c-fos mRNA (A) and IL-2 mRNA (C) expressing cells in intact rats × 10 (*in situ* hybridization of mRNA–digoxin–labeled cDNA). c-fos mRNA (B) and IL-2 mRNA (D) expressing cells in 2 h after injection of 10 μl 5% mustard in vegetable oil into the rat's gastrocnemius muscle of the left hind limb. The arrows show the position of the labeled cells. Scheme of the brain (level 26) according to Swanson's maps of the rat brain (E) [89]. □ The analyzed LHA structure.

In 2 hours after the injection of 5% mustard in vegetable oil, the expression of c-fos mRNA and c-fos-like protein was found in the cells of SO, DMH, VMH, arcuate hypothalamic nuclei (ARH), LHA, thalamus, CdN, ACe and sensomotor areas of the cortex. IL-2-positive cells were detected in the same structures of the brain where c-fos gene expression was noted (Fig. 1). This indicates that stress affects brain cells and modulates c-fos and IL-2 gene expression. The c-fos mRNA-positive cells may be found even after a 30-min stimulation, but the IL-2 mRNA synthesis that follows later is less intensive [68].

Noxious mechanical stimulation (NMS) of a foot causes an increased number of c-fos-positive cells in LHA, DMH, VMH, and anterior hypothalamic area (AHA) by 115, 101, 199, and 157%, respectively, as compared to control animals [87,88]. Interestingly, physical factors, like EHF skin irradiation, modulate (mostly decrease) stress-induced stimulation of IEG gene expression in hypothalamic neurons [86].

4. EXPRESSION OF C-FOS AND INTERLEUKIN-2 GENES IN BRAIN CELLS AFTER INJECTION OF ANTIGEN

4.1. Effects of lipopolysaccharide injection on c-fos gene expression in brain cells

Lipopolysaccharide (LPS) stimulates cytokine expression in the immune system and also causes a series of reactions in the CNS; it activates the hypothalamic–pituitary–adrenal axis (HPA) and the thermoregulatory system [57,58,90–92]. Lipopolysaccharide modulates the expression of genes and the secretion of neuropeptides in cells of various brain structures [93]. Thus, after 3, 6, 9, and 12 h of LPS injection i/p, it stimulated the expression of immediate early c-fos genes and the production of nerve growth factor (NGFI-B) and the neuropeptides, CRF, oxitocin, and vasopressin. The expression of c-fos gene and NGFI-B was shown by immunohistochemistry and *in situ* hybridization in sections of the paraventricular nucleus [94]. Using these techniques, the space–time pattern of c-fos-, CRF-, oxitocin-, and vasopressin-synthesizing neurons has been determined.

The i/p or intracerebral injections of LPS stimulate c-fos gene synthesis in spinal cord cells also [94,95]. In brain cells, LPS induces mRNA synthesis of pro- and anti-inflammatory cytokines, such as the components of IL-1β family (ligand, signal receptor, receptor-accessory proteins, receptor-antagonist), TNF-α, TGF-β, glycoprotein 130 (a component of IL-6 signaling receptors), as well as neuropeptide Y, Y5 receptor, leptin, and pro-opiomelanocortin [95–99].

The transcriptional modulation of IEG expression induced by LPS correlates with the expression of cytokines (IL-1β, TNF-α, TGF-β, and others) in certain structures of the brain (in PVN, medial preoptic area, organum vasculosum of the lamina terminalis, SO, magnocellular division of the PVN, the ARH/median eminence, the LC, the nucleus of the solitary tract, and the area postrema). Less-intensive expression of genes was noted in the DMH, ACe, the ventral part of the tuberomammillary nucleus, the laterodorsal tegmental nucleus, the external lateral part of the parabracial nucleus, the dorsal division of the ambiguous nucleus, and the lateral reticular nucleus. Lipopolysaccharide-induced alterations in the metabolism of cytokines in brain cells (i.e., superexpression) may lead to neurological and neuropsychiatric disorders [95–100].

The intensity of c-fos protein synthesis after LPS injection was found to be dependent on the animal's age and the dosage of LPS [57]. The number of c-fos-positive neurons in the preoptic area, PVH, and organum vasculosum laminae terminalis of rats increases only 12 days after birth. The authors consider that in the period from the 1st to 9th day of the postnatal development, the number of c-fos immunoreactive cells determined in the brain of young rats does not reflect the specificity of their structural localization. In 12 days after birth, the application of LPS leads to differentiated stimulation of c-fos-positive neurons in the structures of the hypothalamus, thalamus, amygdala, hippocampus, and other areas of the brain. In rats, 500 μg/kg of LPS is most effective for the stimulation of c-fos-positive cells in these structures of the brain. The dynamics of c-fos protein expression in CNS cells and the level of cytokines in peripheral blood depend on the method of LPS administration (e.g., central or peripheral) [100,101].

The intracerebroventricular injection of LPS causes the expression of proinflammatory cytokine genes in the brain cells and reduces the level of cytokines (IL-6, TNF-α) in blood. In contrast, the i/v application of LPS increases the blood level of these cytokines [99–101]. Continuous administration of LPS stimulates the expression of IL-1 and IL-2 in the brain.

IL-2 stimulates cholinergic neurons and activates neural nitric oxide synthase (NOS) [102]. The released NO diffuses into CRF neurons and CRF is released. IL-2 stimulates ACTH secretion. In contrast, IL-1α, induced by LPS in neurons, blocks NO-induced secretion of Luteinizing hormone releasing hormone (LHRH) from LHRH-synthesizing neurons and also inhibits LH secretion. IL-1α inhibits growth hormone secretion stimulated by NO and costimulates the secretion of somatostatin [102].

4.2. Expression of c-fos and interleukin-2 genes in brain cells of rats after injection of tetanus toxoid

Injection of LPS results in the development of fever and other components of an acute phase of inflammation. The injection of tetanus toxoid, which is a weak antigen, fails to result in the development of pronounced nonspecific reactions. Purified nonsorbed anatoxin causes an immune response of low intensity after primary immunization, reaching a serum titer of 1:4.

The intravenous injection of antigen leads to dynamic gene expression in brain cells, including the induction of c-fos mRNA, c-fos-like protein, and IL-2 mRNA synthesis [103–105]. Three types of structures may be distinguished on the basis of reaction dynamics. The first type is characterized by maximum activation of gene expression in 2 hours after the

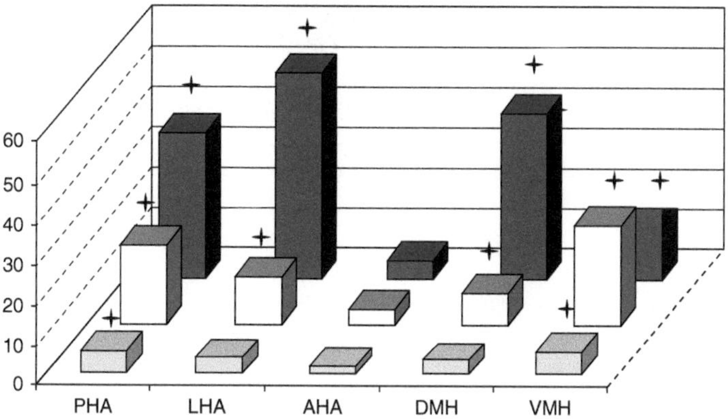

Figure 2. Activation of c-fos mRNA, c-fos-like protein, and IL-2 mRNA synthesis in the rat's hypothalamic cells after injection of antigen. ■, the number of c-fos mRNA-expressing cells; □, the number of c-fos-positive cells, ▮, the number of IL-2 mRNA-expressing cells. Y axis, the number of labeled cells; X axis, hypothalamic structures. Tetanus toxoid was injected i/v 200 μg/kg of the body mass in 250 μl. $^*p < 0.05$ as compared to the number of cells found after injection of the apyrogenic saline.

injection of tetanus toxoid. c-Fos mRNA and c-fos-like protein are expressed early in the posterior hypothalamic area (PHA), LHA, DMH, and VMH, and IL-2 mRNA in the PHA, DMH, and VMH.

The second group includes the structures where mRNA is activated within 6 hours after antigen injection. c-Fos mRNA and c-fos-like protein are expressed after 6 hours in the anterior hypothalamic area (AHA) and PVH, and IL-2 mRNA is expressed in the PVH in such fashion.

The third group includes the hypothalamic structures, which do not show appreciable changes in c-fos mRNA, c-fos-like protein, and IL-2 mRNA-containing cells in response to the injection of tetanus toxoid. The ARH and SO hypothalamic nuclei belong to this group [104–106] (Fig. 2).

It should be noted that 16 h after the injection of tetanus toxoid or apyrogenic saline, the number of cells containing c-fos mRNA, c-fos-like protein, and IL-2 mRNA returns to the baseline levels of intact control rats.

The intensity of synthesis of IL-2 mRNA is considerably lower than that of c-fos mRNA and c-fos-like protein. The expression of c-fos gene occurs earlier than that of IL-2 gene. The synthesized c-fos protein is a transactivation factor not only for IL-2 gene but also for other inducible genes. In addition, the above-mentioned difference might be associated with the following factors:

(1) c-fos protein stimulates the expression of other genes including its own gene [82].
(2) The c-fos gene product in the cytoplasm is insufficient for binding with c-Jun protein, which would form complexes with AP-1 and NF-AT and bind to IL-2 gene sequences.
(3) Among the detectable c-fos-like proteins, a significant proportion belongs to other, non-c-fos representatives of the Fos family.
(4) The AP-1 complex is formed in small quantity due to the weak expression of c-Jun protein.
(5) In rats IL-2 gene is not expressed in all neurons and in all hypothalamic structures in response to stress [103–106]. Stress (e.g., exposure to an NMS) enhances the expression of c-fos and IL-2 genes in cells of different hypothalamic structures including SO and ARH, while the injection of the antigen (tetanus toxoid) fails to cause gene expression in the cells of SO and ARH [97,106]. There is a very intense c-fos gene expression after NMS in SO and ARH (Fig. 3). Thus, the expression of immediate early response genes in brain neurons depends on the nature of the applied stimulus.

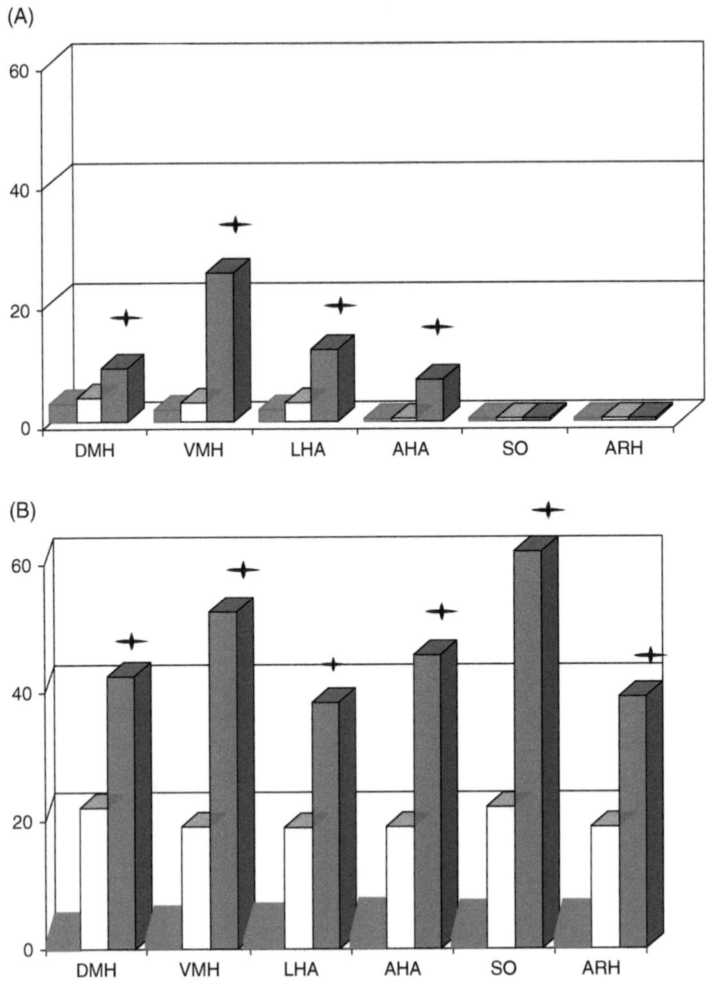

Figure 3. Expression of c-fos-like protein in cells of different hypothalamic structures of rats in 2 h after injection of tetanus toxoid (A) or mechanical noxious stimulation (B). Groups of animals: ■, intact; □, control (after injection of saline); ■, after i/v injection of tetanus anatoxin 200 μg/kg in 250 μl. ■, intact; □, control (after injection of saline); ■, after mechanical noxious stimulation (left hind foot of the rat was squeezed between two surfaces of a surgical intestine forceps 10 times for 10 s each during 10 minutes). Y axis, the number of c-fos-positive cells. *$p < 0.05$ as compared to the number of c-fos-positive cells found in the control group of rats.

5. EXPRESSION OF C-FOS AND INTERLEUKIN-2 GENES BY BRAIN CELLS OF RATS AFTER THE APPLICATION OF AN IMMUNOMODULATING PEPTIDES

Unfavorable conditions, diseases, and ageing are known to affect the function of the immune, endocrine, and nervous systems. During the last decade, there was an intensive search for new biologically active compounds that are capable of correcting the function of cells, which are impaired by destabilizing factors [107]. Biologically active endogenous peptides or their structural analogs have relevance to this problem, which may act on both the immune and nervous systems. Some peptides of the thymus, epiphysis, myelopeptides, opioid peptides influence the intensity of immune response, the phagocytic and antigen-presenting function of macrophages as well as the intensity of synthesis of IL-1, IL-2, IL-6, and TNF [107–111]. In recent years, a number of biologically active structural analogs of peptides from the thymus, epiphysis, cerebral cortex, and other cellular structures and organs have been synthesized [112,113].

Experimental and clinical investigations showed that the application of some short peptides promoted the restoration of impaired functions of the immune and neuroendocrine systems. For example, an analog of the thymus peptides, vilon (Lys-Glu), of the epiphysis peptides, epitalon (Ala-Glu-Asp-Gly), and of the cerebral cortex peptides, cortagen (Ala-Glu-Asp-Pro), are immunomodulators [112–119]; they inhibit apoptosis [120] and cortagen stimulates the regeneration of nerve fibers [121]. Vilon and epitalon were shown to be geroprotectors [122]. The long-term application of these peptides to animals prolongs their life span [123,124]. Long and short peptides were proven *in vitro* to penetrate the membrane of spleen lymphocytes, fibroblasts, neurons of the spinal cord and brain, which were isolated from rat embryos [125–127].

Scientists in the California Technological Institute in Pasadena discovered that short peptides activate the ubiquitin system that controls the degradation of proteins [128]. It was shown that dipeptides modify the activity of the ubiquitin-dependent pathway. They cause the allosteric modulation of the E3 enzyme, which mediates the degradation of the Ptr2 transporter repressor peptide. Dipeptides with destabilizing N-end residues reinforce E3 activity that results in reinforced derepression of Ptr2 expression and promotes increased import of different peptides into the cell.

The peptides vilon (Lys-Glu) and epitalon (Ala-Glu-Asp-Gly) were obtained on the basis of the amino acid analysis of complex preparations of the thymus or the epiphysis [116–118]. These peptides influence the expression of c-fos and IL-2 genes in cells of different structures of the brain [114].

Twenty-four hours after a single intramuscular injection of vilon, the number of c-fos-positive cells in the PVH was increased. Epitalon had a stimulating effect on the synthesis of c-fos-like proteins in cells of AHN, SO, PVH, and ARH.

Daily intramuscular injections of epitalon for 5 days induced IL-2 mRNA synthesis in cells of the PVH, VMH, DMH, LHA, AHA, and medial amygolalar nucleus (MEA) [129]. Vilon induced the synthesis of IL-2 mRNA only in single cells (2–3) of the investigated structures. In intact animals, single IL-2 mRNA-positive cells were noted in the analyzed hypothalamic structures, and also in animals after the injection of saline (0–1) (Fig. 4). The maximum level of neuronal cells expressing IL-2 mRNA after intramuscular epitalon injection was 60% in LHA and 50% in DMH cells that were present in the structure under test (0.1 mm^2). In the VMH, the number of IL-2 mRNA-positive cells was 41% and in PVH and AHA 34 and 15%, respectively. Thus, in fact, only epitalon injections stimulated IL-2 mRNA synthesis in the cells of certain hypothalamic structures.

Intranasal epitalon application induced IL-2 mRNA synthesis in DMH and AHN cells (1.4 and 1.6%, comparatively) [129]. Vilon after intranasal application induced IL-2 mRNA expression only in LHA cells (3.2%). These data indicate that epitalon and vilon stimulate IL-2 mRNA expression in different areas of the hypothalamus. The mode of epitalon or vilon administration and other experimental conditions influence brain cell responses to these agents [130].

Vilon and epitalon application under low stressful conditions (e.g., "handling") leads to a decreased quantity of IL-2 protein-positive cells. Twenty four hours after intramuscular epitalon

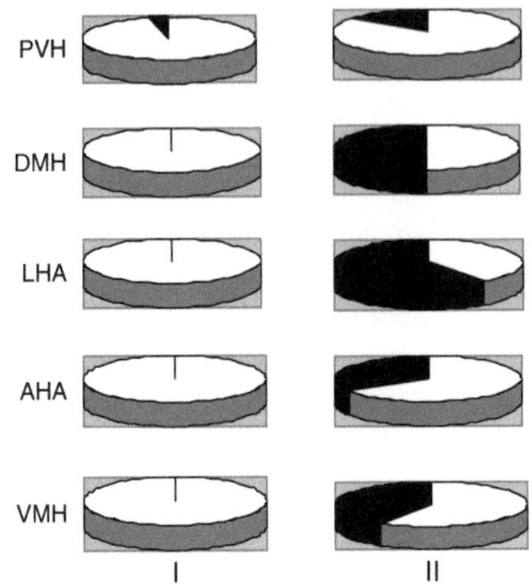

Figure 4. Scheme of the distribution of IL-2 mRNA-positive cells in different structures of the rat hypothalamus after injection of immunomodulating peptides. Vilon (I) or epitalon (II) are injected i/m once per day for 5 days (10 μg/μl/kg). The quantity of the histologically stained cells on the square area (0.1 mm^2) of the structure under test were taken as 100%.

injection, IL-2-positive cells decreased in AHN, PVH, and SO. Two hours after intranasal application of vilon decreases were shown in the PVH, and application of epitalon decreased in the PVH and SO. If the animals are allowed to adapt to the experimental conditions for 5 days, it will lead to a decrease of IL-2 protein containing cells in the PVH after vilon, and in PVH and LHA after epitalon intranasal applications. One may suggest that these peptides inhibit IL-2 protein synthesis, but it is also possible that the balance has been altered between IL-2 synthesis and consumption.

The combined analysis of two proteins (c-fos – as a marker of neuron activation and transcription factor for IL-2 gene and for other inducible genes and IL-2 protein) in intact animals that were exposed to various stimuli revealed a decreased number of IL-2 protein positive cells in the hypothalamic structures. This analysis also revealed differences in the pattern of changes of c-fos and IL-2 protein positive cells under stressful conditions or after adaptation to stress [131].

6. THE POSSIBLE PATHWAYS OF TRANSMISSION OF INTERLEUKIN-2 SIGNALS INTO CENTRAL NERVOUS SYSTEM CELLS

The initial stage of T-lymphocyte stimulation by IL-2 is known to be the binding of IL-2 with the IL-2 receptor (IL-2-R), the activation of tyrosine kinases, and the phosphorylation of the β and γ chains of IL-2-R [132–134]. The phosphorylated tyrosine residues of IL-2-R bind the signal and adapter proteins. The bound proteins include STAT trans-signal proteins [STAT 1, STAT 3 and STAT 5 (A and B) isoforms], which are phosphorylated by Jak kinases, form homo or heterodimer complexes, and are transposed into the nucleus where they recognize the specific sequences of a number of inducible genes (e.g., TNF-α, TNF-β, INF-γ, c-fos, IL-2) [135,136].

However, the phosphorylation of the IL-2-R β subunit is not necessary for binding the signal proteins in all the cases. After the activation of lymphocytes by IL-2, the phosphorylation of IL-2-R β chain was not a necessary condition for the binding of STAT 1 and STAT 3, unlike STAT 5 A and B isoforms. The binding took place due to the interaction of the protein with the acidic subdomain of the IL-2 receptor [137].

The IL-2 receptor complex binds proteins containing Src-homolog 2 (SH2) domains and adapter proteins containing phosphotyrosine-binding domains. It activates three main signal pathways:

(1) Erk-Ras pathway leading to the activation of MAP kinase cascade and cellular proliferation [138–140]
(2) The JAK-STAT pathway that regulates gene expression [141]
(3) The phosphoinositol pathway that is associated with the activation of phosphoinositol 3-kinase, which is important for driving the cellular cycle, for the prevention of apoptosis, and for the regulation of the cytoskeleton rearrangement [142–144]

Cytokine administration activates the JAK-STAT pathway, which regulates inducible gene expression and activates immune competent cells, astrocytes, glial and neuronal cells of the brain [145–147]. The cytokine-like ciliary neurotrophic factor and INF-γ activate STAT1 and STAT3 proteins in neuroblastoma cells [148]. The interaction of the JAK-STAT and Ras-MAPK signaling pathways for the expression of IL-6 and c-fos genes was shown in human and rat neuronal cell cultures [149,150]. Schumann et al. [151] proposed that there are two independent pathways of cytokine signal transmission in the CNS: JAK/STAT and Erk2/MAP. This was shown in human and rat neuroblastoma cells and in hippocampal neurons treated with IL-6. It is likely that in cultured brain cells, STAT3 proteins are involved in the regulation of proliferation, differentiation, and apoptosis [152]. The presence of STAT-binding sites was found in promoters of genes coding substance P, vasoactive intestinal peptide (VIP), c-fos, somatostatin, cholecystokinin, enkephalin, and choline acetyltransferase. These genes are associated with cholinergic processes of neuromodulation [153,154]. The JAK-STAT pathway signals astrocytes, glia cells, and neurons in animals exposed to different kinds of stress, mechanical injury, ischemia, and to various chemical agents [155–157]. This pathway also signals cultured cells after treatment with various drugs [145,146].

One may conclude that the JAK/STAT and Erk2/MAP signal transduction pathways mediate signal transmission by IL-2 and by some other cytokines in brain cells.

7. CONCLUSION

It is known at present that IL-2 mRNA and IL-2 protein are expressed by cells of the nervous system, and that IL-2 mRNA expression takes place in certain brain structures after the application of various stimuli [68,104]. The space–time pattern of c-fos and IL-2 mRNA expression in brain cells varies according to the stimulus used for activation. Antigen stimulates IL-2 mRNA in cells of a number of hypothalamic structures. It is likely that IL-2 participates in the reaction of the brain to antigen and in the process of nervous–immune system interaction. There are differences in the space–time pattern of IL-2 gene expression by the brain cells in response to the antigenic and to nonantigenic stimuli. The immunomodulating peptide, epitalon, modulates brain cell activation at the level of c-fos and IL-2 gene expression [114]. There are reasons to think that the regulatory effect of epitalon is exerted not only in immune cells but also in brain cells.

These data are consistent with the hypothesis that the immunoregulatory and neuroregulatory effects of cytokines are based on the participation of multiple forms of cytokines [158]. It was shown that neuroleptics, when penetrate the brain, have an inhibitory effect on the release of the proinflammatory cytokines, IL-1β and IL-2 [159].

It is suggested that the JAK-STAT and Ras-MAPK signal transduction pathways are involved in IL-2 signal transduction in brain cells.

ACKNOWLEDGMENTS

This work was supported by Richard Fox foundation, USA. It was also supported in part by the Russian Fund for Basic Research (grants No. 03-04-49241 and No. 06-04-49265)

REFERENCES

1. Balkwill FR, Burke F. The cytokine network. Immunol Today 1989;10:299–303.
2. Sawada M, Itoh Y, Suzumura A, Marunouchi T. Expression of cytokine receptors in cultured neuronal and glial cells. Neurosci Lett 1993;160:131–4.
3. Picrre BASt, Merrill E, Dopp JM. Effects of cytokines on CNS cells: Glia: In Cytokines and the CNS. Ransohoff R and Benveniste EN, Eds; Boca Raton, New York, London, Tokyo: CRC Press, 1996; pp. 151–68.
4. Licinio J, Wong Ma-Li. Cytokines in the brain: From localization and function to clinical implications. In New Foundation of Biology, Neuroimmune Biology. Berezi I and Gorczynski RM, Eds; Amsterdam, London, New York, Oxford, Paris, Shannon, Tokyo: Elsevier, 2001; pp. 1–494.
5. Hofman FM, Hinton DR, Johnson K, Merrill JE. Tumor necrosis factor identified in multiple sclerosis brain. J Exp Med 1989;170:607–12.
6. Dickson W, Lee SC, Bronsnan CF, Sinicropi S, Vlassara H, Yen Shu-Hui C. Neuroimmunology of aging and Alzheimer's disease with emphasis on cytokines. In Cytokines and the CNS. Ransohoff R and Benveniste EN, Eds; Boca Raton, New York, London, Tokyo: CRC Press, 1996; pp. 239–67.
7. Dickson W, Lee SC, Mattiace LA, Yen SH, Bronsnan CF. Microglia and cytokines in neurological diseases, with special reference to AIDS and Alzheimer's disease. Glia 1993;7:75–83.
8. Campbell IL, Chiang Chi-Sh. Cytokine involvement in central nervous system disease. Implications from transgenic mice. Ann N-Y Acad Sci 1995;301–12.
9. Ransohoff R, Hamilton T, Tani M, Stoler MH, Shick HE, Major JA, Estes ML, Thomas DM, Tuohy VK. Astrocyte expression of mRNA encoding cytokines IL-10 and JE/MCP-1 in experimental autoimmune encephalomyelitis. FASEB J 1993;7:592–8.
10. Morris ChS, Esiri MM. The expression of cytokines and their receptors in normal and mildly reactive human brain. J Neuroimmunol 1998;92:85–97.
11. Hanisch U-K, Quirion R. Interleukin-2 as a neuroregulatory cytokine. Brain Res 1996;21:246–84.
12. Araujo DM, Lapchak PA, Collier B, Quirion R. Localization of interleukin-2 immunoreactivity and interlukin-2 receptors in the rat brain: Interaction with cholinergic system. Brain Res 1989;498:257–66.
13. Lapchak PA, Araujo DM, Quirion R, Beaudet A. Immunoautoradiographic localization of interleukin-2-like immunoreactivity and interlukin-2 receptors (Tac antigen-like immunoreactivity) in the rat brain. Neuroscience 1991;44:173–84.
14. Otero GC, Merrill JE. Molecular cloning of IL-2 α, IL-2β, and IL-2 Rγ cDNAs from human oligodendroglioma cell line: Presence of IL-2R mRNAs in the human central nervous system. Glia 1995;14:295–302.

15. Nieto-Sampedro M, Chandi KG. Interleukin-2-like activity in injured rat brain. Neurochem Res 1987;12:723–7.
16. Whisler R, Chen M, Beiqing L, Carle K. Impaired induction of c-fos/c-jun genes and of transcriptional regulatory proteins binding distinct c-fos/c-jun promoter elements in activated human T cells during aging cell. Immunology 1997;175:42–50.
17. Pahlavani MA, Harris MD, Richardson A. The increase in the induction of IL-2 expression with caloric restriction is correlated to changes in the transcription factor NF-AT. Cell Immunol 1997;180:10–9.
18. Eizenberg O, Faber-Elman A, Lotan M, Schwartz M. Interleukin-2 transcripts in human and rodent brains: Possible expression by astrocytes. J Neurochem 1995;64:1928–36.
19. Eitan S, Zisling R, Cohen A, Belkin M, Hirschberg DL, Lotan M, Schwartz M. Identification of an interleukin-2-like substance as a factor cytotoxic to oligodendrocytes and associated with central system regeneration. Proc Natl Acad Sci USA 1992;89:–5442–6.
20. Sei Y, Vitkoviç LL, Yokoyama MM. Cytokines in the central nervous system: Regulatory roles in neuronal function, cell death and repair. Neuroimmunomodulation 1995;2:121–33.
21. Jiang Ch-L, Lu Ch-L. Interleukin-2 and its effects in the central nervous system. Biol Signals Recept 1998;7:148–56.
22. Saris SC, Patronas NJ, Rosenberg SA, Alexander JT, Frank J, Schwarzentruber DJ, Rubin JT, Barba D, Oldfield EH. The effect of intravenous interleukin-2 on brain water content. J Neurosurg 1989;71:169–74.
23. Martin R, Schwulera U, Menke G, Rudolph W, Buch K, Fasold H, Lissner R, Thrun A, Krauseneck P, Bogdahn U. Interleukin-2 and blood brain barrier in cats pharmacokinetics and tolerance following intrathecal and intravenous administration. Eur Cyt Netw 1992;3:399–406.
24. Banks WA, Kastin AJ. Blood to brain transport of interleukin links the immune and central nervous systems. Life Sci 1991;48:117–21.
25. Banks WA, Nichoff ML, Zalcman SS. Permeability of the mouse blood–brain barrier to murine interleukins-2: Predominance of a saturable efflux system. Brain Behav Immun 2004;18:434–42.
26. Hanisch U-K, Seto D, Quirion R. Modulation of hippocampal acetylcholine release: A potent central action of interleukin-2. J Neurosci 1993;13:3368–74.
27. Liang SM, Liang CM, Chiuch CC. Visualization of interleukin-2-like molecules in MPP (+)-lesioned rat brain. Biochem Biophys Res Commun 1989;165:1312–8.
28. Whittington R, Faulds D. Interleukin-2. A review of its pharmacological properties and therapeutic use in patients with cancer. Drugs 1993;46:446–514.
29. Araujo DM, Lapchak PA. Induction of immune system mediators in the hippocampal formation in Alzheimer's and Parkinson's diseases: Selective effects on specific interleukins and interleukin receptors. Neuroscience 1994;61:745–54.
30. Dantzer R. Timing and specificity of the cognitive changes induced by interleukin-2 and interferon-α treatments in cancer patients. Somat Med 2001;63:376–86.
31. Awatsuji H, Furukawa Y, Nakajiama M, Furukawa S, Hayashi K. Interleukin-2 as a neurotrophic factor for supporting the survival of neurons cultured from various regions of fetal rat brain. J Neurosci Res 1993;35:305–11.
32. Jones TH, Kennedy RL. Cytokines and hypothalamic–pituitary function. Cytokine 1993;5:531–8.
33. Rubber J, Bloom FE. IL-2 induces vasopressin release from the hypothalamus and the amygdala: Role of nitric oxide-mediated signalling. J Neurosci 1994;14:6187–95.
34. Karanth Sh, Lyson K, Aguila MC, McCann SM. Effects of luteinizing-hormone-releasing hormone, α-melanocyte-stimulating hormone, naloxone, dexamethasone and

indomethacin on interleukin-2-induced corticotropin-releasing factor release. Neuroimmunomodulation 1995;2:166–73.
35. Karanth Sh, Lyson K, McCann SM. Effects of cholinergic agonists and antagonists on interleukin-2-induced corticotropin-releasing factor release from the mediobasal hypothalamus. Neuroimmunomodulation 1999;6:168–74.
36. Zalcman S, Green-Johnson JM, Murray L, Nance DM, Dyck D, Anisman H, Greenberg AH. Cytokine-specific central monoamine alterations induced by interleukin-1, -2 and -6. Brain Res 1994;643:40–9.
37. Harbuz MS, Staphanou A, Knight RA, Clover-Gonzalez AJ, Lightman SL. Action of interleukin-2 and interleukin-4 on CRF mRNA and POMC mRNA in the anterior pituitary. Brain Res 1992;6:214–22.
38. Karanth S, MacCann SM. Anterior pituitary hormone control by interleukin 2. Proc Natl Acad Sci USA 1991;88:2691–965.
39. Jiang CL, Xu D, Zhao YF, Song CY, Zheng ZC, Lu CL, Liu XY. Relationship between structure and analgesic function of human interleukin-2. Sci Chin 1996;26:8–12.
40. Wang Y, Li L, Liu JF, Wang ZY, Jiang CL, Liu XY. The analgesic domain of IL-2. Biochem Biophys Res Commun 1997;230:542–5.
41. Wang HJ, Ding YC. Analysis of ELISA on the time-related expression of IL-2 and TNF-alfa during the healing process of wound in the rat skin (in Chinese). Fa Yi Xue Za Zhi 2003;19:10–12.
42. Meyers CA, Valentine AD. Neurological and psychiatric adverse effects of immunological therapy. CNS Drugs 1995;3:56–60.
43. Lacosta S, Merali Z, Anisman H. Central monoamine activity following acute and repeated systemic interleukin-2 administration. Neuroimmunomodulation 2000;8:83–90.
44. Capuron L, Ravand A, Dantzer R. Early symptoms in cancer patients receiving interleukins 2 and or interferon alfa-2b therapy. J Clin Oncol 2000;18:2143–51.
45. Sudom K, Turrin NP, Hayley S, Anisman H. Influence of chronic interleukins-2 infusion and stressors on sickness behaviors and neurochemical change in mice. Neuroimmunomodulation 2004;11:341–50.
46. Schena M. Genome analysis with gene expression microarrays. BioEssays 1996;18:427–31.
47. Lockhart DJ, Winzeler LA. Genomics, gene expression and DNA arrays. Nature 2000;405:827–36.
48. Velculescu VE, Zhang L, Vogelstein B, Kinzler KW. Serial analysis of gene expression. Science 1995;270:484–7.
49. van Hal NLW, Vorst O, van Houwelingen AMML, Kok EJ, Peijnenburg Ad, Aharoni A, van Tunen AJ, Keijer J. The application of DNA microarrays in gene expression analysis. J Biotech 2000;78:271–80.
50. Anisimov SV, Lakatta EG, Boheler KR. Discovering altered genomic expression patterns in heart transcriptome determination by serial analysis of gene expression. Eur J Heart Fail 2001;3:271–81.
51. Brazma A, Vilo J. Gene expression data analysis. FEBS Lett 2000;480:17–24.
52. Boehringer Mannheim. Nonradioactive in situ hybridization. Application manual GmbH. Biochemia 1992;1–75.
53. Bullitt E. Expression of c-fos-like protein as a marker for neuronal activity following noxious stimulation in the rat. J Comp Neurol 1990;296;517–30.
54. Bullitt E, Lee ChL, Right AR, Willcockson H. The effect of stimulus duration on noxious-stimulus induced c-fos expression in the rodent spinal cord. Brain Res 1992;580:172–9.

55. Hunt SP, Pini A, Evan G. Induction of c-fos-like protein in spinal cord neurons following sensory stimulation. Nature 1987;328:632–4.
56. Morgan JI, Curran T. Role of ion flux in the control of c-fos expression. Nature 1986;322:552–5.
57. Oladehin A, Blatteis CM. Lipopolysaccharide-induced Fos expression in hypothalamic nuclei of neonatal rats. Neuroimmunomodulation 1995;2:282–9.
58. Yokoyama Ch., Sasaki K. Regional expression of c-Fos-like immunoreactivity in rat cerebral cortex after stress; restraint and intraperitoneal lipopolysaccharide. Brain Res 1999;816:267–75.
59. Hatakeyama M, Kawahara A, Mori H, Shibuya H. C-fos gene induction by interleukin 2: Identification of the critical cytoplasmic regions within the interleukin 2 receptor β chain. Proc Natl Acad Sci USA 1992;89:2022–6.
60. Kono MY, Minami Y, Taniguchi T. The IL-2 receptor complex and signal transduction: Role of the beta-chain. Semin Immunol 1993;5:299–307.
61. Kawahara A, Minami Y, Miyzaki T, James NI, Taniguchi T. Critical role of the interleukin 2 (IL-2) receptor γ-chain-associated Jak3 in the IL-2-inducd c-fos and c-myc, but not bcl-2, gene induction. Proc Natl Acad Sci USA 1995;89:8724–8.
62. Pereda MP, Goldberg V, Chervin A, Carrizo G, Molina A, Andrada J, Sauer J, Renner U, Stalla GK, Arzt E. Interleukin-2 (IL-2) and IL-6 regulate c-fos protooncogene expression in human pituitary adenoma explants. Mol Cell Endocrinol 1996;124:33–42.
63. Mouzaki A, Zubler R, Doucet A, Rungger D. Trans-active factors controlling the IL-2 gene in adult human T-cell subsets. Mediat Inflamm 1992;1:33–7.
64. Golovko OI, Grishina TV, Novikova NS, Nosov MA, Mulberg AA, Korneva EA, Kazakova TB. Regulation of interleukin-2 gene expression by nuclear factors from spleen and brain cells in T lymphocyte cells in normal and immunosuppressive conditions (in Russian). Neurochemia 1996;13:195–205.
65. Mouzaki A, Weil R, Muster L, Rungger D. Silencing and trans-activation of the mouse IL-2 gene in Xenopus oocytes by proteins from resting and mitogen-induced primary T-lymphocytes. EMBO J 1991;10:1399–406.
66. Mũller R, Bravo R, Burckhardt J, Curran T. Induction of c-fos gene and protein by growth factors precedes activation of c-myc. Nature 1984;312:716–20.
67. Guthrie KM, Gall CM. Odors increase Fos in olfactory bulb neurons including dopaminergic cells. Neuroreport 1995;6:2145–9.
68. Barabanova SV, Golovko OI, Novikova NS, Nosov MA, Korneva EA, Kazakova TB. Effects of stress on c-fos and interleukin-2 gene expression in the cells of the nervous and immune systems (in Russian). Nurochemia 1998;15:380–7.
69. Hata R, Mies G, Wiessner C, Hossmann K-A. Differential expression of c-fos and hsp72 mRNA in focal cerebral ischemia of mice. NeuroReport 1998;9:27–32.
70. Kazakova TB, Barabanova SV, Novikova NS, Nosov MA, Rogers VJ, Korneva EA. Induction of c-fos and interleukin-2 gene expression in the central nervous system following stressor stimuli. Int J Pathol 2000;7:53–61.
71. Korneva EA, Khai LM. Effects of destruction of some hypothalamic areas on the immunogenesis process (in Russian). Russ J Physiol 1963;49:42–8.
72. Korneva EA, Klimenko VM, Shkhinek EK. Neurohumoral support of the immune homeostasis (in Russian). Leningrad:Nauka 1978;248.
73. Korneva EA, Grigoryev VA, Klimenko VM, Stolyarov YD. Electro-physiological phenomena of the brain in the immune reactions. Electro-physiological immune reactions (in Russian). Leningrad:Nauka 1989;150.

74. Wenner M, Kawamura N, Ishikawa T. Reward linked to increased natural killer cell activity in rats. Neuroimmunomodulation 2000;7:1–5.
75. Cullinan WE, Herman JP, Battaglia DF, Akil H, Watson SJ. Pattern and time course of immediate early gene expression in rat brain following acute stress. Neuroscience 1995;64:477–505.
76. Honkaniemi J, Kainu T, Ceccatelli S, Rechardt L, Hokfelt T, Pelto-Huikko M. Fos and jun rat central amygdaloid nucleus after stress. Mol Neurosci 1992;3:849–52.
77. Sudakov KV, Umruchin PE, Koplik EV, Anokhin KV. C-fos gene expression and emotional stress in rats: The blocking role of peptides which invoke delta-sleep. Russ J Physiol 2000;86:617–24.
78. Coccatelli S, Villar MJ, Goldstein MM, Hokfelt T. Expression of c-Fos immunoreactivity in transmitter-characterized neurons after stress. Proc Natl Acad Sci 1989;86:9569–73.
79. Imaki T, Shibaski T, Demura H. Regulation of gene expression in the central nervous system by stress: Molecular pathways of stress responses. Endocrinol J 1995;42:121–30.
80. Imaki T, Shibasaki T, Wang XQ, Demura H. Intracerebroventricular administration of corticotrophin-releasing factor antagonist attenuates c-fos mRNA expression in the paraventricular nucleus after stress. Neuroendocrinology 1995;61:445–51.
81. Kovács KJ. C-Fos as a transcription factor: A stressful (re) view from a functional map. Neurochem Int 1998;33:287–97.
82. Herdegen T, Leah JD. Inducible and constitutive transcription factors in the mammalian nervous system: Control of gene expression by Jun, Fos and Krox, and CREB/ATF proteins. Brain Res Rev 1998;28:370–490.
83. Korneva EA, Rybakina EG, Fomicheva EE, Kozinetz IA, Shkhinek EK. Altered interleukin-1 production in mice exposed to rotation stress. Int J Tissue React 1992;14:219–24.
84. Levi-Montalcini R, Aloe L, Alleva E. Prog Neuroendocr Immunol 1990;3:1–10.
85. Rybakina EG, Martinova NA, Kalinin AG, Kozinetz IA, Shanin SN, Fomicheva EE. Altered LAF production in rats exposed to cold stress. Int J Tissue React 1997;19:135–9.
86. Korneva EA, Kazakova TB. Present approaches to stress influence analysis on the metabolic processes in the nervous and immune system cells (in Russian). Med Immunol 1999;1:17–22.
87. Novikova NS, Kazakova TB, Rogers VJ, Korneva EA. C-fos gene expression induced in cells in specific hypothalamic structures by noxious mechanical stimulation and it's modification by exposure of the skin to extremely high frequency irradiation. Neuroendocrinol Lett 2002;23:315–20.
88. Novikova NS, Kazakova TB, Rogers VJ, Korneva EA. C-fos gene expression in the rat spinal cord and brain cells after stress application under different kinds of halathane anaesthesia (in Russian). Russ J Physiol 2002;88:1378–87.
89. Swanson LW. Brain Maps Computer Graphics Files. Amsterdam, The Netherlands: Elsevier Sci B.V., 1992.
90. Sager SM, Price KJ, Kasting NW, Sharp FR. Anatomical patterns of c-Fos immunostaining in rat brain following systemic endotoxin administration. Brain Res Bull 1995;36:381–92.
91. Gaykema RP, Goehler LE, Tilders FJ, Bol JG, McGorry M, Fleshner M, Maier SF, Watkins LR. Bacterial endotoxin induces Fos immunoreactivity in primary afferent neurons of the vagus nerve. Neuroimmunomodulation 1998;5:234–40.
92. Oladehin A, Barriga-Briceno JA, Blatteis CM. Lipopolysaccharide (LPS)-induced Fos expression in the brain of febrile rats. In Temperature Regulation: Recent Physiological and Pharmacological Advances. Milton AS, Ed.; Basel: Birkhaüser, 1994; pp. 81–95.

93. Rivest S, Laflamme N. Neuronal activity and neuropeptide gene transcription in the brains of immune-challenged rats. J Neuroendocrinol 1995;7:501–25.
94. Plata-Salaman CR, Ilyin SE, Gayte D, Flynn MC. Gram-negative and gram-positive bactirial products induce differential cytokine profiles in the brain: Analysis using an integrative molecular-behavioral in vivo model. Int J Mol Med 1998;1:387–99.
95. Turrin NP, Gayle D, Ilyin SE, Flynn MC, Langhans W, Schwartz GJ, Plata-Salama'n CR. Pro-inflammatory and anti-inflammatory cytokine mRNA induction in the periphery and brain following intraperitoneal administration of bacterial lipopolysaccharide. Brain Res Bull 2001;54:443–53.
96. Plata-Salama'n CR. Turrin NP. Cytokine interactions and cytokine balance in the brain: Relevance to neurology and psychiatry. Mol Psychiatry 1999;4:303–6.
97. Konsman JP, Kelley K, Dantzer R. Temporal and spatial relationships between lipopolysaccharide-induced expression of Fos, interleukin-1 beta and inducible nitric oxide synthase in rat brain. Neuroscience 1999;89:535–48.
98. Jackson ATK. Chemical specificity of endotoxin-induced c-fos expressing neurons in the rat hypothalamus. M.Sc. Thesis, Dept Physiology Univ Manitoba, April 1999.
99. Chen G, McCuskey RS, Reichlin S. Blood interleukin-6 and tumor necrosis factor-α elevation after intracerebroventricular injection of *Escherichia coli* endotoxin in the rat is determined by two opposing factors: Peripheral induction by LPS transferred from brain to blood and inhibition of peripheral response by a brain-mediated mechanism. Neuroimmunomodulation 2000;8:59–69.
100. Nance DM, Greenberg AH, Jackson ATK. Central catecholamine involvement in the hypothalamic induction of c-fos after endotoxin treatment. Soc Neurosci Nov 1995;CA 45.
101. Wan W, Janz L, Vriend CY, Sorensen CM, Greenberg AH, Nance DM. Differential induction of c-fos immunoreactivity in hypothalamus and brain stem nuclei following central and peripheral administration of endotoxin. Brain Res Bull 1993;32:581–7.
102. McCann SM, Kimura M, Karanth S, Yu WH, Mastronardi CA. The mechanism of action of cytokines to control the release of hypothalamic and pituitary hormones in infection. Ann NY Acad Sci 2000;917:4–18.
103. Korneva EA, Kazakova TB, Nosov MA. C-fos mRNA and c-Fos-like protein expression in the hypothalamic cells after injection of antigen (in Russian). Allergol Immunol 2000;1:37–44.
104. Korneva EA, Barabanova SV, Golovko OI, Nosov MA, Novikova NS, Kazakova TB. C-fos and IL-2 gene expression in rat brain cells and splenic lymphocytes after nonantigenic and antigenic stimuli. Ann NY Acad Sci 2000;917:197–209.
105. Nosov MA, Barabanova SV, Glushikhina MS, Kazakova TB, Korneva EA. Antigen-induced activation of hypothalamic cells by c-fos gene expression (in Russian). Russ J Physiol 2001;87:331–40.
106. Nosov MA, Barabanova SV, Glushikhina MS, Kazakova TB, Korneva EA. Antigen-induced activation of hypothalamic cells (assessed by expression of the c-fos gene). Neurosci Behav Physiol 2002;32:523–8.
107. Aprikyan VS, Galojan KA. Influence of the hypothalamic polypeptides on the cell mediated immune response (in Russian). Neurochemia 1998;15:189–94.
108. Devoino L, Cheido M, Idova G, Morazova N, Chipens G. Low molecular weight peptides in neuroimmuno-modulation: Central effect and interaction with monoamine systems. In

Neuropeptides and Immunopeptides. O'Doriso M and Panerai A, Eds; New York: New-York Academy of Science, 1990; pp. 449–51.
109. Goncharova V, Romanjuk A, Chalisova N, Akoev G, Chumasov E, Sotnikov O. Bovine brain factor promotes neurite extension from chick embryo dorsal root ganglia in vitro. Neurosci Lett 1987;82:1–5.
110. Tooyama I, Sasaki K, Oomura Y, Li A-J, Kimura H. Effect of acidic fibroblast growth factor on basal forebrain cholinergic neurons in senescence-accelerated mice. Exp Gerontol 1997;32:171–9.
111. Matsumoto I, Oomura Y, Niijima A, Sasaki K, Aikawa T. Acidic fibroblast growth factor activates hypothalamic–pituitary–adrenocortical axis in rats. Am Physiol Soc 1998:503–509.
112. Khavinson VKh, Serii SV, Malinin VV. Medium possess immunomodulating activity. Pat. No. 2080120, Russ 1997.
113. Morozov VV, Khavinson VKh. Natural and synthetic peptides as therapeutics for immune dysfunction. Int J Immunopharmacol 1997;9/10:–501–3.
114. Kazakova TB, Barabanova SV, Nosov MA, Parkchomenko EP, Gluschichina MS, Korneva EA. Interleukin-2 gene expression in murine lymphocytes and CNS after structural analogs of thymic and epiphysis peptides application. Biogerontology from molecules to human, Abstract. Adv Gerontol 2000;5:54.
115. Kazakova TB, Burov SV, Golovko OI, Grishina TV, Novikova NS, Mulberg AA, Semko TV, Korneva EA. Biological activity of LHRH structural analog in the immune response regulation in T lymphocytes (in Russian). Bull Exp Biol Med 1996;122:334–7.
116. Khavinson VKh, Morozov VG, Malinin VV, Kazakova TB, Korneva EA. Influence of peptide Lys-Glu on expression of interleukin-2 gene in lymphocytes (in Russian). Bull Exp Biol Med 2000;130:330–2.
117. Khavinson VKh, Kvetnoy IM, Aschmarin IP. Peptidergic regulation of homeostasis (in Russian). Uspechi sovr Biol 2002;122:190–203.
118. Kazakova TB, Barabanova SV, Khavinson VKh, Glushikhina MS, Parkhomenko EP, Malinin VV, Korneva EA. In vitro effect of short peptides on expression of interleukin-2 gene in splenocytes (in Russian). Bull Exp Biol Med 2002;133:614–6.
119. Khavinson VKh, Popuchiev VV, Kvetnoy IM, Yuzhakov VV, Kotlova LN. The regulating influence of epitalon on the stomach endocrine cells in pinealectomized rats (in Russian). Bull Exp Biol Med 2000;130:651–3.
120. Khavinson VKh, Kvetnoy IM. Peptide bioregulators inhibit apoptosis (in Russian). Bull Exp Biol Med 2000;130:657–9.
121. Turchaninova LN, Kolosova LI, Malinin VV, Moiseeva AB, Nozdrachev AD, Khavinson VKh. Tetrapeptide cortagen influence on the sciatic nerve regeneration (in Russian). Bull Exp Biol Med 2000;130:656–9.
122. Khavinson VKh, Goncharova N, Lapin B. Synthetic tetrapeptide epitalon restores disturbed neuroendocrine regulation in senescent monkeys. Neuroendocrinol Lett 2001;22:251–4.
123. Khavinson VKh. Increased life duration with peptide bioregulators (in Russian). Clin Gerontol 2000;6:107–109.
124. Khavinson VKh, Anisimov VN, Zavarzina NU. Vilon influence on the biological age factors and murine life duration (in Russian). Bull Exp Biol Med 2000;130:88–91.
125. Rakowicz-Szulczyn'ska EM, Horst A. Incorporation of antigen 125 J IgG into particular cell compartments: Binding by chromatin. Cell Biochem Funct 1985;3:61–9.

126. Le Roux I, Joliot AH, Bloch-Gallego E, Prochiantz A. Neurotrophic activity of the Antennapedia homeodomain depends on its specific DNA-binding properties. Proc Natl Acad Sci USA 1993;90:9120–4.
127. Levine JE, Prystowsky MB. Polypeptide growth factors in the nucleus: A review of function and translocation. Neuroimmunomodulation 1995;2:290–5.
128. Turner GC, Du F, Varshavsky A. Peptides accelerate their uptake by activating a ubiquitin-dependent proteolytic pathway. Nature 2000;403:579–82.
129. Kazakova TB, Barabanova SV, Novikova NS, Glushikhina MS, Khavinson VKh, Malinin VV, Korneva EA. Synthesis of IL-2 mRNA in cells of rat hypothalamic structures after injection of short peptides (in Russian). Bull Exp Biol Med 2005;139:718–20. Bull Exp Biol Med 2005;139:688–91.
130. Barabanova SV, Artjuhina ZE, Kazakova TB, Khavinson VKh, Malinin VV, Korneva EA. The interleukin-2 changes in the rat hypothalamic structures after peptide applications under low stressful conditions (in Russian). Bull Exp Biol Med 2006;141:371–4.
131. Saltzman EM, Thom RR, Casnellie JE. Activation of a tyrosine protein kinase is an early event in the stimulation of T lymphocytes by interleukin-2. J Biol Chem 1988;263:6956–9.
132. Gesbert F, Garbay C, Bertoglio J. Interleukin-2 stimulation induced tyrosine phosphorylation of p120-Cbl and CrkL and formation of multimolecular signaling complexes in T lymphocytes and natural killer cells. J Biol Chem 1998;273:3986–93.
133. Asao H, Kumaki S, Takeshita T, Nakamura M, Sugamura K. IL-2-dependent in vivo and in vitro tyrosine phosphorylation of IL-2 receptor γ chain. FEBS Lett 1992;304: 141–5.
134. Weil D, Dautry F. Induction of tumor necrosis factor-alpha and -beta and interferon-gamma mRNA by interleukin-2 in murine lymphocytes lines. Oncogene Res 1988;3:409–14.
135. Lu L, Zhu J, Zheng Z, Yan M, Xu W, Sun L, Theze J, Liu X. Jak-Stat pathway is involved in the induction of TNF-beta gene during stimulation byinterleukin-2. Eur J Immunol 1998;28:805–10.
136. Imada K, Leonard WJ. The Jak-STAT pathway. Mol Immunol 2000;37:1–11.
137. Delespine-Carmagnat M, Bouvlier G, Bertoglio J. Association of STAT1, STAT3 and STAT5 proteins with the IL-2 receptor involves different subdomains of the IL-2 receptor β chain. Eur J Immunol 2000;30:59–68.
138. O'Shea JJ, Gadina M, Schreiber RD. Cytokine signalling in 2002: New surprises in the Jak/Stat pathway. Cell 2002;109:S121
139. Graves JD, Dawnward J, Izquierdo-Pastor M, Rayter S, Warne PH, Cantrell DA. The growth factor IL-2 activates p21 ras proteins in normal human T lymphocytes. J Immunol 1992;148:2417–22.
140. Perkins GR, Marvel J, Collins MK. Interleukin-2 activates extracellular signal-regulated protein kinase 2. J Exp Med 1993;178:1429–34.
141. Xu W, Yan M, Lu L, Sun L, Theze J, Zheng Z, Liu X. The p38 MAPK pathway is involved in the IL-2 induction of TNF-β gene via the EBS element. Biochem Biophys Res Commun 2001;289:979–86.
142. Ahmed NN, Grimes HL, Bellacosa A, Chan TQ, Tsichlis PN. Transduction of interleukin-2 antiapoptotic and proliferative signals via Akt protein kinase. Proc Natl Acad Sci USA 1997;94:3627–32.
143. Brennan P, Babbage JW, Burgering BM, Groner B, Reif K, Cantrell DA. Phosphatidylinositol 3-kinase couples the interleukin-2 receptor to the cell cycle regulator E2F. Immunity 1997;7:679–89.

144. Arrieumerlou C, Donnadieu E, Brennan P, Keryer G, Bismuth G, Cantrell DA, Trautmann A. Involvement of phosphoinositide 3-kinase and Rac in membrane ruffling induced by IL-2 in T cells. Eur J Immunol 1998;28:1877–85.
145. Halfer H, Postert C, Friedrich M, Ringelstein EB, Stogbauer F. Activation of the Jak-Stat- and MAPK-pathways by oncostatin M is not sufficient. Brain Res Mol Brain Res 2000;80:198–206.
146. Paillaud E, Costa S, Fages C, Plassat JL, Rochette-Egly C, Monville C, Tardy M. Retinoic acid increases proliferation rate of GL-15 glioma cells, involving activation of STAT-3 transcription factor. J Neurosci Res 2002;67:670–9.
147. Bonni A, Sun Y, Nadal-Vicens M, Bhatt A, Frank DA, Rozovsky I, Stahl N, Yancopoulos GD, Greenberg ME. Regulation of gliogenesis in the central nervous system by JAK-STAT signaling pathway. Science 1997;278:477–83.
148. Kaur N, Wohihueter AL, Halvorsen SW. Activation and inactivation of signal transducers and activators of transcription by ciliary neurotrophic factor in neuroblastoma cells. Cell Signal 2002;14:419–29.
149. Kuroda H, Sugimoto T, Horii Y, Sawada T. Signaling pathway of ciliary neurotrophic factor in neuroblastoma cell lines. Med Pediatr Oncol 2001;36:118–21.
150. Plata-Salaman CR, Turin NP. Cytokine interactions and cytokine balance in the brain: Relevance to neurology and psychiatry. Mol Psychiatry 1999;4:303–6.
151. Schumann G, Huell M, Machein U, Hocke G, Fiebich B. Interleukin-6 activates signal transducer and activator of transcription and mitogen-activated protein kinase signal transduction pathways and induces de novo protein synthesis in human neuronal cells. J Neurochem 1999;73:2009–17.
152. Cattaneo E, Magrassi L, De-Fraja C, Conti L, Di Gennaro I, Butti G, Govoni S. Variations in the levels of the JAK/STAT and ShcA proteins in human brain tumors. Anticancer Res 1998;18:2381–7.
153. Rajotte D, Sadowski HB, Haman A, Gopalbhai K, Meloche S, Liu L, Krystal G, Hoang T. Contribution of both STAT and SRF/TCF to c-fos promoter activation by granulocyte–macrophage colony-stimulating factor. Blood 1996;–88:2906–16.
154. Servidei T, Aoki Y, Lewis SE, Symes A, Fink JS, Reeves SA. Coordinate regulation of Stat signalling and c-fos expression by tyrosine phosphatase SHP-2. J Biol Chem 1998;273:6233–41.
155. Peterson WM, Wang Q, Tzekova R, Wiegand SJ. Ciliary neurotrophic factor and stress stimuli activate the Jak-STAT pathway in retinal neurons and glia. J Neurosci 2000;20:4081–90.
156. Suzuki Sh, Tanaka K, Nogawa Sh, Dembo T, Kosakai A, Fukuuchi Y. Phosphorylation of signal transducer and activator of transcription-3 (Stat3) after focal cerebral ischemia in rats. Exp Neurol 2001;170:63–71.
157. Planas AM, Soriano MA, Berruezo M, Justicia C, Estrada A, Pitarch S, Ferrer I. Induction of Stat3, signal transducer and transcription factor, in reactive microglia following transient focal cerebral ischemia. Eur J Neurosci 1996;8:2612–8.
158. Jiang Ch-L, Xu D, Lu Ch-L, Wang Y-X, You Z-D. Interleukin-2: Structural and biological relatedness to opioid peptides. Neuroimmunomodulation 2000;8:20–4.
159. Kowalski J, Labuzck K, Herman ZS. Flupentixol and trifluperidol reduce interleukin-1 beta and interleukin-2 release by rat mixed glial and microglial cell cultures. J Pharmacol 2004;5:563–70.

Cytokines and the Brain
Edited by C. Phelps and E. Korneva

Sex Hormones and Cytokines in Central Nervous System Pathology and Repair

ANDRÉS GOTTFRIED-BLACKMORE, GIST F. CROFT, and KAREN BULLOCH

The Rockefeller University, Lab. of Neuroendocrinology, Box 165,1230 York Ave, New York, NY 10065, USA

ABSTRACT

The adult central nervous system (CNS) is far more capable of repair and regeneration than previously believed. Research in the field of neuroendocrine immunology indicates that vital repair mechanisms are dependent on cytokines derived from the initial inflammatory or innate immune response, which in turn are regulated by steroid hormones. It is now becoming clear that the ability of the CNS to restore itself is dependent upon regional mechanisms that are, in part, modulated by sex hormones through the activation of specific nuclear and membrane receptors located in microglia, astrocytes, neurons, neural stem cells, and other cell types. These receptors in turn stimulate multiple pathways that affect the functional balance of these cellular constituents during the brain's injury/healing response. Adding to the complexity of these interactions is the fact that sex hormones, in particular estrogens, can be regionally synthesized in response to brain trauma, and their presence in the local environment may be more germane to inflammation and healing than circulating hormone levels. The details and ramifications of our knowledge about sex hormones in the CNS injury/healing response are the subjects of the following article.

1. INTRODUCTION

Long-held theories of central nervous system (CNS) function in the adult seriously underestimate the ability of this tissue to undergo plasticity, repair, and regeneration. Yet a new body of evidence in the field of neuroimmunology increasingly demonstrates that the brain's capacity to repair and restore itself is embedded in regional mechanisms that are under hormonal regulation. The initiation of vital repair mechanisms in response to brain injury is dependent on cytokines derived from the initial inflammatory innate immune response [1].

Following injury, endogenous production of cytokines precedes and may modulate the induction of neurotrophic factors. For example, the proinflammatory cytokine IL-1β is implicated in the induction of nerve growth factor (NGF) [2] and is required for ciliary neurotrophic factor (CNTF) and insulin-like growth factor-1 (IGF-1) expression in astrocytes [3,4]. Moreover, IL-1β-deficient mice present impaired remyelination associated with delayed oligodendrocyte progenitor differentiation [4], suggesting the requirement of this cytokine.

In contrast, tumor necrosis factor (TNF)-α has been implicated in the negative regulation of neurotrophins. Venters et al. [5,6] have shown that this cytokine potently inhibits IGF-1-dependent survival of neurons and promotes microglia-mediated neurodegeneration. However, there is indirect evidence from the TNF-α knockout mice that TNF-α may be involved in reparative and/or developmental mechanisms. These mice show significant delays in remyelination and a reduction in pools of oligodendrocyte progenitors [7]. Collectively, these data suggest that tipping the balance between neurodegeneration and neuroprotection could depend on the temporal expression and the sites of action of these pleiotropic factors, as seen in embryonic development and suggested in the neuropathologies of chronic inflammation.

While the reparative mechanisms and the extent of neuroplasticity in the adult are not completely understood, it is becoming evident that there are multiple factors and cellular components under cytokine regulation. The temporal and regional modulation of growth factors is critical to not only the regeneration of new axons and synapses in the CNS, but also the proliferation and differentiation of newly born neurons (neurogenesis) in the adult nervous system.

Sex steroids have multiple effects in the CNS including neuroprotection [8,9]. Exploring the effects of sex steroids on the magnitude and severity of the immune response, the expression of immune mediator receptors, and the impact on cytokine release by CNS immunocytes will be key in understanding their involvement in repair and neuroprotection following injury. Thus it becomes increasingly important for basic and clinical research to understand and incorporate the mechanisms by which sex steroids modulate the nervous system's response to injury. In this article we review the literature and analyze the interdependence of the immune response and sex hormonal status with particular emphasis on estrogens.

2. EFFECTS OF STERIOD HORMONES IN CENTRAL NERVOUS SYSTEM INFLAMMATION

2.1. Clinical correlates and neuroprotective studies

In general, women show a more robust immune response than males; for example, females have 10 times higher propensity to autoimmune disorders, more potent antigen-presenting cells, higher circulating immunoglobulin levels, and stronger antigen-specific humoral immune responses [rev. by [10]]. Likewise, there are clear documented differences between the sexes in susceptibility to and recovery from many types of neurodegenerative diseases. Premenopausal females, in both humans and rodents, are less susceptible to acute brain injury such as cerebral ischemia, neurotrauma, and certain neurotoxic agents. There is also evidence from clinical trials and animal studies that the sex hormone 17 beta estradiol (E2β) is associated with a decreased incidence, delayed onset, and delayed progression of acute and chronic brain disorders ranging from stroke and schizophrenia to Alzheimer's disease, multiple sclerosis, and Parkinson's disease [11–15]. Estrogens, however, are not the only gonadal steroids that can influence the progression and outcome of neuropathology.

Recent reports indicate that, besides estrogens, androgens as well as progesterone (P) and dihydroepiandrosterone (DHEA) have potent neuroprotective actions. Animal studies show that androgens protect against several types of injury, such as brain trauma, and enhance recovery after stroke and spinal cord injury [16–18]. The effects of other androgens, such as testosterone (T), on neuroprotection are not as clear, although T deficiency has been associated with increased symptoms of Parkinson's disease [19,20]. The sex differences in susceptibility and ability to recover from neurodegenerative diseases raise important questions about how and why hormones affect the injury/healing response. The fact that women in their reproductive years are more prone to autoimmunity, coupled with studies that demonstrate amelioration of some diseases/injuries by E2β or P, in both genders, suggest that the differences are due to production of sex hormones rather than to a genetic program.

2.2. Glial steroid receptor expression

While the protective effects of E2β and other gonadal steroids on neurons have been clearly demonstrated in CNS damage models including ischemia, glutamate excitotoxicity, oxidative stress, amyloid B toxicity, and neuronal apoptosis [9,16,21–35], the mechanisms by which they affect CNS inflammation and cytokine production are less well understood. To address this issue, several studies have focused on microglia and astrocytes, given that these cells are pivotal in orchestrating the inflammatory response in the CNS [10,36–41]. Evidence to date suggest that these immunocytes of the CNS, like their peripheral cousins, the macrophages, express sex hormone receptors and are targets for direct steroid modulation of cytokines and growth factors in the brain.

Estrogen receptors (ERα, ERβ) and androgen receptors (AR) are expressed in the normal brain and upregulated after injury [42–45]. *In vivo* studies in rats indicate that ERα and ERβ gene expression is modulated differentially after injury. Following ischemia, there is a significant increase (6–8-fold) in ERα expression and a moderate decrease (50%) in ERβ expression in the ipsilateral cortex. E2β treatment prevents the loss of ERβ gene expression by injury and induces the expression of Bcl-2, an antiapoptotic gene, in the ipsilateral region [46,47]. In addition to the modulation of nuclear ERα and ERβ following ischemic injury, a novel estrogen receptor, ER-X, which mediates rapid nongenomic effects of 17 alpha-estradiol (E2α) and E2β in the picomolar range, is also upregulated [48,49]. Other ERs associated with the plasma membrane that modulate various signaling pathways have been described [50], making E2β a truly pleiotropic factor.

The intra- and extracellular locations of these receptors, coupled with their sensitivity and temporal responsiveness to estrogens, underscore the multiple levels of complexity at play in the CNS injury response. Although many of the studies above do not identify the particular cell types expressing hormone receptors, investigators have addressed this issue. A comprehensive review of steroid receptor expression in glial cells has been done by Garcia-Ovejero et al. [51]. Although expressed at low levels in the normal brain, reactive astrocytes and microglia transiently upregulate ERs and ARs in both excitotoxic/chemical injury to the rat hippocampus and a stab wound to the parietal cortex and hippocampus [44]. Likewise, progesterone receptor (PR) expression is upregulated by CNS injury in neurons and glia, in male and female rats [52]. Other studies also document glial expression of ERs and ARs in other species (mice and lower primates) [9,43,53].

In a recent study, Sierra et al. [54,55] isolated microglia from adult mice and showed by RT-PCR that these cells express mRNA for glucocorticoid (GR) and mineralocorticoid (MR) receptors, as well as low levels of ERα mRNA, but transcripts for ERβ, ARs, or PRs were not detected. The activation of microglia by lipopolysaccharide (LPS) injections ip downregulated the expression of GR, MR, and ERα in 24 h after stimulation. Activation of microglia with LPS at this time point did not induce the expression of ARs, PRs, or ERβ.

In vitro, microglia cell lines, which express partially activated phenotypes [56], are reported to express mRNA and proteins for both ERα and ERβ [57,58]. Yet, there are few reports with primary cultures of microglia or *in vivo* microglia corroborating that this cell type expresses ERs [54,55]. Given that transformation may induce spurious gene expression, results obtained from cell lines should be confirmed in primary cultures and preferentially *in vivo*. Certainly, both neurons and glial cells express steroid hormone receptors and become increasingly responsive to the regional actions of these steroids in the injured CNS.

3. REGULATION OF INFLAMMATORY MEDIATORS BY SEX STEROIDS

3.1. Cytokine regulation by E2β and other hormones

The mechanisms by which E2β affects cytokines within the CNS are not as well documented as E2β modulation of cytokines in peripheral, immune-related pathologies such as cardiovascular disease, [59,60], osteoporosis [61,62], and autoimmunity [63]. However, studies of hormonal effects on cytokine synthesis in rodent models have been informative.

Several reports indicate that E2β, as well as P, are capable of reducing the secretion of IL-1β and TNF-α from stimulated peripheral blood monocytes [64–66]. Interestingly, E2β also has been noted to have a dose-dependent, biphasic effect on a number of cell responses including cytokine secretion [67]. E2β at physiological doses (10^{-9}–10^{-8} M) for the estrogen receptor ERα and ERβ increases IL-1β synthesis/secretion, while pharmacological doses (10^{-6}–10^{-5} M) attenuate it in peripheral macrophages [63]. Lymphocytes from stimulated female mice produce more IFN-γ than those from males. Furthermore, *in vitro* physiological doses of E2β increase the expression of this cytokine in spleen cells as well as the transcription from its promoter in transfected cell lines [68], suggesting that activated ERs may activate other transcription factors directly and/or signaling pathways that regulate gene expression. Such potentiating effects of E2β on cytokine expression may explain the higher incidence of immune pathologies among women. In contrast, E2β, acting through ERα, is reported to be required for mounting appropriate immune microglia responses and regulating cytokines that mediate the switch from innate to adaptive immunity following a CNS viral challenge [69].

Many proinflammatory cytokines are induced and/or signal through nuclear factor kappa-B (NFκB), and it is now recognized that activated ERs can interact with this transcription factor and alter its transcriptional capabilities [70–72], yet not necessarily its activation [57,73]. Estrogen receptor antagonism of NFκB seems to be cell type-specific, suggesting that for ERs, there may be cell type-specific cofactors that are important modulators of the ER/NFκB interaction [74]. Interestingly, cytokine-activated NFκB can block ER-mediated transcription in osteoblasts and breast cancer cells [75], underpinning the complex interactions these transcription factors display.

E2β blocks several inflammatory mediators produced by reactive microglia *in vitro*. Lipopolysaccharide, phorbol ester, or IFN-γ induce microglial increases in inducible nitric oxide synthase (iNOS) and matrix metalloproteinase-9 expression, superoxide, nitric oxide (NO), and prostaglandin-E2 release, as well as phagocytic activity and NFκB activation. E2β pretreatment blocks these events, with the exception of NFκB activation [57,58,76]. In the neural HIV inflammatory model, microglia are activated by the viral Tat regulatory protein leading to superoxide and NO production, phagocytosis, and TNF-α release. *In vitro*, HIV microglial activation is suppressed by E2β pretreatment through interference with Tat-mediated MAPK activation [77].

Cytokines and surface receptors critical for adaptive immunity, such as IL-10, TNF-α, IFN-γ, major histocompatibility complex (MHC) Class I, CD40, and CD86, are also regulated by E2β in N9 microglia cells [78]. β-Amyloid protein induction of NO production in microglia is modulated by steroids too. E2β shows a biphasic effect depending on the dose and milieu, in which physiological doses of E2β stimulate NO secretion and toxicity [79] whereas high doses are inhibitory [57].

In the mouse CNS, E2β treatment decreases the recruitment of total inflammatory cells as well as TNF-α-positive macrophages and T cells at the onset of experimentally induced autoimmune encephalitis [80]. Microglia, in contrast, show only a moderate attenuation of the peak TNF-α expression in response to E2β [80]. This emphasizes the notion that E2β can differentially affect brain inflammation by influencing efflux and afflux of peripheral immunocytes, for example, lymphocytes, as well as CNS resident immunocytes, astrocytes, and microglia.

In contrast to the aforementioned studies, Woodfork et al. (18) reported that estrogen has no effect on LPS-activated J774A.1 macrophages (a cell line expressing ERα) in the secretion of NO, TNF-α, IL-6, and monocyte chemoattractant protein-1 (MCP-1). Other studies have also failed to demonstrate anti-inflammatory actions of E2β [54,81,82]. However, tumor-producing cell lines may not reflect normal, *in vivo*, cell function or may loose their sensitivity to estrogen modulation when attaining immortality.

Some of the most consistent works on the anti-inflammatory effects of E2β on microglia come from Vegeto et al. [83,84] showing that *in vivo* E2β administration prior to LPS injection can decrease microglia proliferation as well as expression of inflammatory mediators such as MCP-1 and TNF-α. Additionally, in a transgenic mouse model of chronic inflammation, this group demonstrates that hormone ablation increases microglia reactivity at β-amyloid deposits, and that E2β administration decreases microglia reactivity and expression of scavenger receptor-A [83]. Yet the expression of ERs in brain microglia was not confirmed, leaving the possibility that E2β anti-inflammatory effects may be mediated through other cells mediating inflammation such as astrocytes and endothelial cells.

Besides interfering with gene expression, E2β and P have also been shown to modulate microglia cytokine function. Drew and Chavis [76] have reported that *in vitro*, estriol, E2β, and P treatment can block the microglia LPS-induced TNF-α production and suppress the release of NO induced by TNF-α or IFN-γ. Steroid hormones also modulate inflammation at other levels. For example, P reduces injury-induced lipid peroxidation and blood–brain barrier leakage through its antioxidative properties and possibly by blocking proinflammatory cytokines like TNF-α [52].

4. STEROID HORMONES AND CENTRAL NERVOUS SYSTEM REPAIR

4.1. Sex hormone effects on central nervous system repair mechanisms

Neuroprotection afforded by sex steroids not only comes from promoting neural survival and blocking inflammation but can also be attributed to the brain's repair mechanisms. Although women have higher incidences of multiple sclerosis (MS), men show a more severe disease course suggesting protective effects of estrogens in the CNS. Oligodendrocytes and Schwann cells, two important cell types involved in nerve regeneration, express ERs, PRs, and ARs [rev. by [51]]. Oligodendrocyte proliferation and differentiation is enhanced by P and E2β [85,86], as well as oligodendrocyte myelin formation and myelin basic protein expression [87,88].

Androgens and estrogens have been proposed to enhance brain repair, possibly through steroid actions on surrounding astrocytes [13,89]. Glial fibrillary acidic protein (GFAP) expression reflects astrocyte hypertrophy/proliferation and has been a marker associated with nerve injury. In the hamster facial motor neuron damage model, T treatment at the time of axotomy reduces the loss of synaptic contacts to the cell body, decreases astrocyte GFAP mRNA and protein, and enhances regeneration [90]. Garcia-Estrada et al. [91] have shown that DHEA, T, E2β, and P decrease gliosis and the proliferation of GFAP-positive reactive astrocytes following penetrating injury. These findings are important because gliosis and scar formation interfere with repair and plasticity. Similar effects have been seen in other damage models such as striatal neurotoxicity [92]. It is interesting to note that neural injury seems to increase ER expression in astrocytes and AR expression in microglia [44,51], indicating that androgen effects on astrogliosis may be mediated through microglia. Another possible explanation is that androgens are converted to estrogens within the CNS [44,93].

4.2. Neural stem cells

Although it is clear that the mammalian CNS does not undergo global remodeling or regeneration, evidence is now accumulating that new neurons are born in the adult brain as part of normal allostasis and in response to CNS acute or long-term trauma (allostatic load) [52,94–96]. Many studies demonstrate that the brain has several active sites of neurogenesis throughout life: the subventricular zone (SVZ), the rostral migratory stream and olfactory bulb, and the subgranular layer (SGL) and dentate gyrus of the hippocampus [97–101].

Multiple forms of injury, such as ischemia, seizures, and traumatic brain injury, induce proliferation of progenitors and immature neurons in the SVZ and the dentate gyrus [102]. Adult rat CNS lesion studies show increases in cell number and proliferation of progenitors in the SVZ and dentate gyrus neurogenic zones [103,104]. These findings are of particular relevance to the development of new therapies for the treatment of brain damage, because generation of new neurons in the adult animal has been correlated with improved performance in cognitive learning and memory tasks [105–107].

Recently, a novel route of cell replacement in normal and injured brain has been proposed. There is now convincing evidence that bone marrow stromal stem cells (MSCs), which reside in adult bone marrow along with hematopoietic stem cells (HSCs), are able to escape the traditionally held notions of lineage restriction and differentiate into neural phenotypes *in vitro* and *in vivo*. These MSC enter the brain parenchyma, adopt CNS resident cell-type characteristics, and functionally integrate into the neuronal circuitry [108–110]. Mezey et al. [111–113] and Brazelton et al. [114] demonstrated that whole bone marrow transplants could repopulate the recipient's brain with cells displaying morphology of glia and neurons, and expressing neuronal markers of mature neurons, for example, pyramidal cells in the cortex and granule cells in the hippocampus.

In rat models of CNS ischemic injury, MSC repopulation of damaged brain regions correlates with functional improvements [115]. Yet, it remains to be seen whether these new cells act by integrating as replacements for resident neurons or as support cells, which produce growth factors and stimulate repair of resident cell types [108,115]. The relative accessibility of MSCs for autologous transplant makes these recent findings extremely interesting and will no doubt generate considerable effort for future research in neuroimmunology.

4.2.1. Neurogenesis, inflammation, and hormonal regulation

In vivo and *in vitro* studies show that adult neurogenesis and plasticity are responsive to environmental factors. Increases in neurogenesis have been observed following ischemia, stress, dietary restrictions, kainic acid neurotoxicity, and other forms of CNS damage [116].

As discussed above, cytokine expression is an integral feature of the brain's response to injury. While the *in vivo* effects of cytokines on neurostem cell (NSC) and neurogenesis are not thoroughly understood, it is clear that there is a robust interaction between cytokines in normal and injured brain, and in the processes of adult neurogenesis. A study done by Monje et al. [117] demonstrates this by showing that neuroinflammation inhibits neurogenesis and that inflammatory blockade with indomethacin restores this process. Transgenic mice that constitutively express IL-6 in astrocytes display dramatically decreased adult neurogenesis in the dentate gyrus [118]. Yet at lower physiological levels, cytokines may be required or beneficial. During development, the IL-6 cytokine family acts synergistically with bone morphogenetic proteins (BMPs) on progenitor cells to elaborate astrocytic differentiation [119]. Additionally, adult NSC can be differentiated *in vitro* through IL-1β and IL-6 produced from astrocytes found in neurogenic zones of the brain [120]. We, as well as others, have shown IL-6 expression in a putative progenitor cell line that is under E2β control [121].

Another cytokine, IL-1β, has been shown to restrict the proliferation of oligodendrocyte precursors and promote their differentiation [122]. An *in vivo* role for IL-1β in NSC and/or neurogenesis regulation is further suggested by the rich distribution of mRNA for IL-1 receptor in the granule cell layer of the dentate gyrus, a major region of adult neurogenesis [123]. Regulation of this cytokine is also modulated by E2β in microglia and macrophages [63,64,66].

In vivo, the behavior of NSC is determined by the integration of a multitude of extrinsic and intrinsic signals such as Delta-Notch, BMPs and Sonic Hedgehog, Wnt, growth factors, hormones, and cell–cell interactions [124]. Of particular importance is that the fate of NSC during development, allostasis, and allostatic load [96] is influenced by sex hormones, of which E2β is one of the best characterized.

E2β affects distinct elements of nervous system development [125,126]. Cells in neurogenic zones express ERα and ERβ [127,128], and the regulation of the expression of the receptors has been shown to correlate with distinct stages of differentiation of cerebellar neurons and glia. For example, in the cerebellum on postnatal day 4, the expression of ERα increases while there is a significant decrease in ERβ expression [129]. In related *in vitro* studies with a cerebellar neural precursor cell line, EtC.1 [130,131] cells were shown to express functional ERs and E2β was shown to affect gene/protein expression [121].

Sex hormones also affect adult neurogenesis. T modulates the survival of adult neurons derived from NSC [132], and E2β transiently increases the number of newborn neurons in the dentate gyrus of the adult female rat [133]. These findings are in concert with those showing that estrogen enhances synaptic density and plasticity in the adult hippocampus [134].

In vitro experiments with NSCs derived from embryonic and adult rats show functional expression of ERα and ERβ [127]. The authors demonstrate that E2β increases embryonic NSC proliferation but not adult NSC proliferation. Furthermore, E2β significantly increases the neuron/astrocyte ratio of embryonic NSC but does not affect that of adult NSC. These effects were inhibited by the ER antagonist ICI-182-780 [127]. In other *in vitro* experiments, E2β significantly increases neuronal differentiation of adult and embryonic-derived NSCs, and preferentially directs these neurons toward a specific functional phenotype [135].

Neurotrophic factors are important regulators of NSC in development and injury. Insulin-like growth factor-1 regulates survival of neurons during development [136], and IGF-I and/or insulin are necessary for the maintenance of the stem cell *in vitro* [137]. Insulin-like growth factor receptor and its ligands are also present in the cells of adult neurogenic zones [137]. Furthermore, this molecule increases adult hippocampal neurogenesis [138]. Sex hormones such as E2β have been shown to increase the expression of IGF-1 after trauma [139]. E2β also

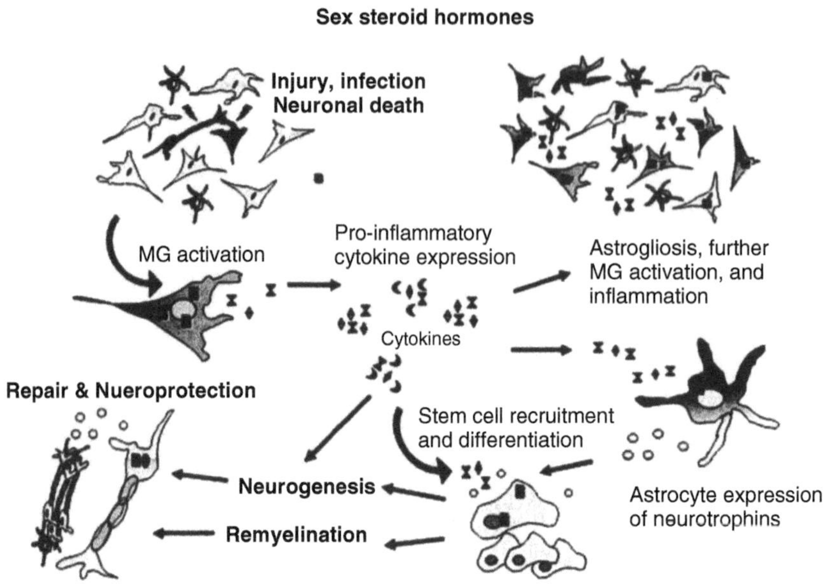

Figure 1. Sex steroid hormones influence the key cell types and molecules involved in the brain's injury and repair response prornoting cellular plasticity and protection.

regulates IGF concentration and retention in tanycytes [140,141], and is involved, along with IGF-1, in regulating synaptic plasticity in this region through astrocytes and possibly tanycytes [89]. Moreover, E2β can activate the IGF-1 signaling pathway [142,143] and promote neuroprotection [144], suggesting a truly neurotrophic role for this hormone in the brain.

Following brain injury, FGF-2 also promotes neurogenesis [145]. Exogenously supplemented epidermic growth factor (EGF) or fibroblast growth factor (FGF)-2 induces replacement of a subset of hippocampal CA1 pyramidal neurons with new neurons following ischemia. These neurons establish appropriate functional synaptic connections and ameliorate injury-induced deficits in learning and memory [146]. Although no studies to date have shown a direct effect of E2β on FGF-2 or EGF, other researches have shown that E2β upregulates related members of this family [147], suggesting that a modulator effect is possible.

Thus sex hormones can significantly affect proliferation and differentiation of NSC directly or through the stimulation of other cells to secrete growth and differentiation factors both during normal adult regeneration and as regional regulators of the CNS injury response.

5. NEUROSTEROIDOGENESIS

5.1. Synthesis of central nervous system steroid hormones

Androgens and estrogens have long been recognized to be synthesized in the gonads and adrenals in response to circulating hypothalamic factors, that is, follicle-stimulating hormone (FSH) and luteinizing hormone (LH), as part of reproductive function. However, during periods of infection and inflammation, gonadal function can be altered. This alteration is mediated by cytokine regulation of the hypothalamic–pituitary gonadal axis. For example, IL-1β modulates GnRH secretion leading to the subsequent alteration of the gonadotropin surge and inhibition of the proestrus LH surge [148]. Several other major cytokines produced by microglia, like IL-2, IL-6, TNF-α, and IFN-γ, affect the release of anterior pituitary hormones, by acting on the hypothalamus and/or directly on the pituitary gland itself [149]. What is particularly germane to this issue is whether the sex hormones that affect the brain's response to injury are circulating or are synthesized within the CNS or both.

A strong body of evidence now shows that steroid hormones are synthesized and have definitive metabolic roles within the CNS as well as the vasculature, adipose tissue, and bone [150]. They can act in paracrine or intracrine fashion and are synthesized *de novo* from cholesterol or from circulating C19 substrates. The biosynthesis of E2β is catalyzed by a heme enzyme, aromatase 450 (P450arom), a member of the cytochrome P450 superfamily [150]. The gene encoding the P450arom, CYP19, is the sole member of the CYP19 family.

Although E2β and other sex hormones are produced primarily by the gonads in premenopausal women, the major source of E2β in men and in postmenopausal women is production by extragonadal P450arom. This view is further supported by the finding that ER-mediated transcription in nonreproductive tissues can occur independently of circulating hormone levels [151]. In the brain, p450arom as well as the various enzymes that catalyze the conversion and synthesis of steroid hormones from cholesterol is widely expressed [152–154]. Recently our laboratory has shown that a cerebellar neural precursor cell line and a microglia cell line express functional p450arom and have the capacity to synthesize both androgens and estrogens [121,155,156]. Recent work in our laboratory has shown that microglia express various functional enzymes that metabolize androgens into active hormones [157].

5.2. Regulation of central nervous system steroidogenesis

P450arom transcription is regulated by tissue-specific activation of promoters located in the 5' untranslated regions of an alternative exon, for example, promoter I.1 in placenta, promoter I.4 in adipose tissue, and promoter I.F in brain [158]. After transcription, alternative splicings yield mature mRNAs that are different only in their 5'-UTR but are identical in amino acid sequence in the gene product, regardless of tissues. This transcriptional control maintains P450arom expression well regulated in different tissues.

While promoter I.F is regulated in the CNS [150], transcription can occur from other promoters such as promoter I.4 (more appropriately considered the mesenchymal promoter) [159]. Regulation of this promoter in the CNS provides a potential mechanism by which E2β may serve as a fundamental component of the injury response because promoter I.4 mediates transcription initiated by three factors: class I cytokines, for example, (IL-6, IL-11, and TNF-α), glucocorticoids, and the transcription factor, Sp-1 [160], all key factors during inflammation. Specifically, P450arom transcription occurs when (1) cytokines bind to their receptors activating the JAK1/STAT3 pathway leading to phosphorylated STAT3 binding to the IFN-gamma activation site (GAS) element of promoter 1.4; (2) glucocorticoids bind to their receptors which in turn bind to the glucocorticoid response element (GRE); and (3) Sp-1 binds to its site [161].

Consistent with these findings, Garcia-Segura et al. [93] has reported P450arom upregulation in the CNS following damage, particularly in astrocytes. If, as mentioned previously, the hypothalamic–pituitary–adrenal (HPA)–gonadal axis is disrupted during CNS trauma, p450arom activation within the local milieu of the CNS may be the primary source for E2β in males and in pre- and postmenopausal females. Indeed, neuronal damage is increased if p450arom function is impaired during a CNS challenge [162]. Supporting these results, p450arom knockout female mice display greater neuronal damage in an ischemia model compared to ovariectomized WT females, suggesting that local, nongonadal E2β formation could have therapeutic implications in brain damage [163]. It is evident that localized E2β synthesis, through P450arom, plays a vital role in the response of the brain to injury.

6. CONCLUSION

The effects of sex hormones and other humoral factors on CNS injury and repair occur through the activation of specific nuclear and membrane receptors located in microglia, astrocytes, neurons, NSC, and other cells. These receptors, in turn, stimulate multiple pathways that result in the functional balance of these cellular constituents during the brain's injury/healing response. Recent evidence reveals multiple isoforms of ER genes differentially distributed within a diverse range of brain regions and expressed during different phases of development. Deepening the complexity is the possibility that E2β can act at multiple binding sites with different affinities in the same cell type, and that the expression of ERs may be modulated by the hormone itself.

E2β and other sex hormone effects must also be considered in terms of circulating serum concentration as well as biosynthesis within the CNS. The source and target for this hormone may well differ, during different phases of the life cycle of men and women. Thus it is imperative that these factors are considered when developing strategies in drug design for estrogen-related illness. We are just becoming aware of the extent, diversity, and importance of sex hormone signaling and modulation in normal and damaged CNS function. This awareness will inevitably lead to a greater understanding of the fundamental mechanisms of neuroendocrine–immune interaction and the response of the CNS to injury and disease.

REFERENCES

1. Nguyen MD, Julien JP, Rivest S. Innate immunity: the missing link in neuroprotection and neurodegeneration? Nat Rev Neurosci 2002;3(3):216–27.
2. DeKosky ST, Styren SD, O'Malley ME, Goss JR, Kochanek P, Marion D, Evans CH, Robbins PD. Interleukin-1 receptor antagonist suppresses neurotrophin response in injured rat brain. Ann Neurol 1996;39(1):123–7.
3. Herx LM, Rivest S, Yong VW. Central nervous system-initiated inflammation and neurotrophism in trauma: IL-1 beta is required for the production of ciliary neurotrophic factor. J Immunol 2000;165(4):2232–9.
4. Mason JL, et al. Interleukin-1beta promotes repair of the CNS. J Neurosci 2001; 21(18):7046–52.
5. Venters HD, Dantzer R, Kelley KW. A new concept in neurodegeneration: TNFalpha is a silencer of survival signals. Trends Neurosci 2000;23(4):175–80.
6. Venters HD, et al. A new mechanism of neurodegeneration: A proinflammatory cytokine inhibits receptor signaling by a survival peptide. Proc Natl Acad Sci USA 1999;96(17): 9879–84.
7. Arnett HA, et al. TNF alpha promotes proliferation of oligodendrocyte progenitors and remyelination. Nat Neurosci 2001;4(11):1116–22.
8. McEwen BS. Invited review: Estrogens effects on the brain: Multiple sites and molecular mechanisms. J Appl Physiol 2001;91(6):2785–801.
9. Behl C. Sex hormones, neuroprotection and cognition. Prog Brain Res 2002;138:135–42.

10. Mor G, et al. Estrogen and microglia: A regulatory system that affects the brain. J Neurobiol 1999;40(4):484–96.
11. Roof RL, Hall ED. Gender differences in acute CNS trauma and stroke: Neuroprotective effects of estrogen and progesterone. J Neurotrauma 2000;17(5):367–88.
12. Behl C. Oestrogen as a neuroprotective hormone. Nat Rev Neurosci 2002; 3(6):433–42.
13. Chowen JA, et al. Sex steroids and the brain: Lessons from animal studies. J Pediatr Endocrinol Metab 2000;13(8):1045–66.
14. Bisagno V, Bowman R, Luine V. Functional aspects of estrogen neuroprotection. Endocrine 2003;21(1):33–41.
15. Lee SJ, McEwen BS. Neurotrophic and neuroprotective actions of estrogens and their therapeutic implications. Annu Rev Pharmacol Toxicol 2001;41:569–91.
16. Stein DG. Brain damage, sex hormones and recovery: A new role for progesterone and estrogen? Trends Neurosci 2001;24(7):386–91.
17. Cardounel A, Regelson W, Kalimi M. Dehydroepiandrosterone protects hippocampal neurons against neurotoxin-induced cell death: Mechanism of action. Proc Soc Exp Biol Med 1999;222(2):145–9.
18. Kalimi M, et al. Anti-glucocorticoid effects of dehydroepiandrosterone (DHEA). Mol Cell Biochem 1994;131(2):99–104.
19. Okun MS, McDonald WM, DeLong MR. Refractory nonmotor symptoms in male patients with Parkinson disease due to testosterone deficiency: A common unrecognized comorbidity. Arch Neurol 2002;59(5):807–11.
20. Okun MS, et al. Beneficial effects of testosterone replacement for the nonmotor symptoms of Parkinson disease. Arch Neurol 2002;59(11):1750–3.
21. Morale MC, et al. Estrogen, neuroinflammation and neuroprotection in Parkinson's disease: Glia dictates resistance versus vulnerability to neurodegeneration. Neuroscience 2006;138(3):869–78.
22. McCullough LD, Hurn PD. Estrogen and ischemic neuroprotection: An integrated view. Trends Endocrinol Metab 2003;14(5):228–35.
23. Corasaniti MT, et al. 17beta-Estradiol reduces neuronal apoptosis induced by HIV-1 gp120 in the neocortex of rat. Neurotoxicology 2005;26(5):893–903.
24. Guerra B, et al. Plasma membrane oestrogen receptor mediates neuroprotection against beta-amyloid toxicity through activation of Raf-1/MEK/ERK cascade in septal-derived cholinergic SN56 cells. J Neurochem 2004;91(1):99–109.
25. Wu TW, et al. 17Beta-Estradiol induced Ca^{2+} influx via L-type calcium channels activates the Src/ERK/cyclic-AMP response element binding protein signal pathway and BCL-2 expression in rat hippocampal neurons: A potential initiation mechanism for estrogen-induced neuroprotection. Neuroscience 2005;135(1):59–72.
26. Cordey M, et al. Estrogen activates protein kinase C in neurons: Role in neuroprotection. J Neurochem 2003;84(6):1340–8.
27. Belcredito S, et al. Estrogen neuroprotection: The involvement of the Bcl-2 binding protein BNIP2. Brain Res Brain Res Rev 2001;37(1–3):335–42.
28. Wang Q, et al. Estrogen provides neuroprotection in transient forebrain ischemia through perfusion-independent mechanisms in rats. Stroke 1999;30(3):630–7.
29. Garcia-Segura LM, Azcoitia I, DonCarlos LL. Neuroprotection by estradiol. Prog Neurobiol 2001;63(1):29–60.
30. Platania P, et al. 17beta-Estradiol rescues spinal motoneurons from AMPA-induced toxicity: A role for glial cells. Neurobiol Dis 2005;20(2):461–70.

31. Heyer A, et al. In vitro gender differences in neuronal survival on hypoxia and in17beta-estradiol-mediated neuroprotection. J Cereb Blood Flow Metab 2005; 25(4):427–30.
32. Wen Y, et al. Estrogen attenuates nuclear factor-kappa B activation induced by transient cerebral ischemia. Brain Res 2004;1008(2):147–54.
33. Sawada H, Shimohama S. Estrogens and Parkinson disease: Novel approach for neuroprotection. Endocrine 2003;21(1):77–9.
34. Rau SW, et al. Estradiol attenuates programmed cell death after stroke-like injury. J Neurosci 2003;23(36):11420–6.
35. Rau SW, et al. Estradiol differentially regulates c-Fos after focal cerebral ischemia. J Neurosci 2003;23(33):10487–94.
36. Hagberg H, Mallard C. Effect of inflammation on central nervous system development and vulnerability. Curr Opin Neurol 2005;18(2):117–23.
37. Panickar KS, Norenberg MD. Astrocytes in cerebral ischemic injury: Morphological and general considerations. Glia 2005;50(4):287–98.
38. Jankowsky JL, Patterson PH. The role of cytokines and growth factors in seizures and their sequelae. Prog Neurobiol 2001;63(2):125–49.
39. Ladeby R, et al. Microglial cell population dynamics in the injured adult central nervous system. Brain Res Brain Res Rev 2005;48(2):196–206.
40. Raivich G, et al. Neuroglial activation repertoire in the injured brain: Graded response, molecular mechanisms and cues to physiological function. Brain Res Brain Res Rev 1999;30(1):77–105.
41. Raivich G, et al. Molecular signals for glial activation: Pro- and anti-inflammatory cytokines in the injured brain. Acta Neurochir Suppl 1999;73:21–30.
42. Cooke B, et al. Sexual differentiation of the vertebrate brain: Principles and mechanisms. Front Neuroendocrinol 1998;19(4):323–62.
43. Gahr M, Metzdorf R, Aschenbrenner S. The ontogeny of the canary HVC revealed by the expression of androgen and oestrogen receptors. Neuroreport 1996;8(1):311–5.
44. Garcia-Ovejero D, et al. *Glial expression of estrogen and androgen receptors after rat brain injury.* J Comp Neurol 2002;450(3):256–71.
45. Kruijver FP, et al. Sex differences in androgen receptors of the human mamillary bodies are related to endocrine status rather than to sexual orientation or transsexuality. J Clin Endocrinol Metab 2001;86(2):818–27.
46. Dubal DB, et al. Differential modulation of estrogen receptors (ERs) in ischemic brain injury: A role for ERalpha in estradiol-mediated protection against delayed cell death. Endocrinology 2006;147(6):3076–84.
47. Wise PM, et al. Estradiol is a neuroprotective factor in in vivo and in vitro models of brain injury. J Neurocytol 2000;29(5–6):401–10.
48. Toran-Allerand CD. Estrogen and the brain: Beyond ER-alpha, ER-beta, and 17beta-estradiol. Ann N Y Acad Sci 2005;1052:136–44.
49. Toran-Allerand CD, et al. ER-X: A novel, plasma membrane-associated, putative estrogen receptor that is regulated during development and after ischemic brain injury. J Neurosci 2002;22(19):8391–401.
50. Bryant DN, et al. Multiple pathways transmit neuroprotective effects of gonadal steroids. Endocrine 2006;29(2):199–207.
51. Garcia-Ovejero D, et al. Glia–neuron crosstalk in the neuroprotective mechanisms of sex steroid hormones. Brain Res Brain Res Rev 2005;48(2):273–86.

52. Bulloch K. Regional Neural Regulation of Immunity: Anatomy and Function. In: Handbook of Physiology. Coping with the Environment: Neural and Endocrine Mechanisms (MC Ewen BS, ed), pp. 353–80. New York: Oxford University Press (2001).
53. Finley SK, Kritzer MF. Immunoreactivity for intracellular androgen receptors in identified subpopulations of neurons, astrocytes and oligodendrocytes in primate prefrontal cortex. J Neurobiol 1999;40(4):446–57.
54. Sierra A, Gottfried-Blackmore A, Milner TA, McEwen BS, Bulloch K. Steroid Hormone Receptor Expression and Function in Microglia GLIA 2008. In press.
55. Sierra A, Gottfried-Blackmore A, McEwen BS, Bulloch K. Steroid receptors in microglia. In: Society for Neuroscience 2005 [Poster # 974.1], 2005, Washington, DC, USA.
56. Nakamura Y, Si QS, Kataoka K. Lipopolysaccharide-induced microglial activation in culture: Temporal profiles of morphological change and release of cytokines and nitric oxide. Neurosci Res 1999;35(2):95–100.
57. Bruce-Keller AJ, et al. Antiinflammatory effects of estrogen on microglial activation. Endocrinology 2000;141(10):3646–56.
58. Vegeto E, et al. *Estrogen prevents the lipopolysaccharide-induced inflammatory response in microglia.* J Neurosci 2001;21(6):1809–18.
59. Hodgin JB, Maeda N. Minireview: Estrogen and mouse models of atherosclerosis. Endocrinology 2002;143(12):4495–501.
60. Wagner JD, Kaplan JR, Burkman RT. Reproductive hormones and cardiovascular disease mechanism of action and clinical implications. Obstet Gynecol Clin North Am 2002; 29(3):475–93.
61. Turner RT, Riggs BL, Spelsberg TC. Skeletal effects of estrogen. Endocr Rev 1994; 15(3): 275–300.
62. Watts NB. Therapies to improve bone mineral density and reduce the risk of fracture: Clinical trial results. J Reprod Med 2002;47(1 Suppl):82–92.
63. Cutolo M, et al. Estrogens, the immune response and autoimmunity. Clin Exp Rheumatol 1995;13(2):217–26.
64. Morishita M, Miyagi M, Iwamoto Y. Effects of sex hormones on production of interleukin-1 by human peripheral monocytes. J Periodontol 1999;70(7):757–60.
65. Rogers A, Clowes JA, Pereda CA, Eastell R. Different effects of raloxifene and estrogen on interleukin-1beta and interleukin-1 receptor antagonist production using in vitro and ex vivo studies. Bone 2007;40(1):105–10.
66. Rogers A, Eastell R. The effect of 17beta-estradiol on production of cytokines in cultures of peripheral blood. Bone 2001;29(1):30–4.
67. Calabrese EJ. Estrogen and related compounds: Biphasic dose responses. Crit Rev Toxicol 2001;31(4–5):503–15.
68. Fox HS, Bond BL, Parslow TG. Estrogen regulates the IFN-gamma promoter. J Immunol 1991;146(12):4362–7.
69. Soucy G, et al. Estradiol is required for a proper immune response to bacterial and viral pathogens in the female brain. J Immunol 2005;174(10):6391–8.
70. Liu H, Liu K, Bodenner DL. Estrogen receptor inhibits interleukin-6 gene expression by disruption of nuclear factor kappaB transactivation. Cytokine 2005;31(4):251–7.
71. Sharma RV, Gurjar MV, Bhalla RC. Selected contribution: Estrogen receptor-alpha gene transfer inhibits proliferation and NF-kappaB activation in VSM cells from female rats. J Appl Physiol 2001;91(5):2400–6; discussion 2389–90.

72. McKay LI, Cidlowski JA. Cross-talk between nuclear factor-kappa B and the steroid hormone receptors: Mechanisms of mutual antagonism. Mol Endocrinol 1998;12(1): 45–56.
73. Galea E, et al. Estrogen inhibits NF kappa B-dependent inflammation in brain endothelium without interfering with I kappa B degradation. Neuroreport 2002;13(11):1469–72.
74. McKay LI, Cidlowski JA. Molecular control of immune/inflammatory responses: Interactions between nuclear factor-kappa B and steroid receptor-signaling pathways. Endocr Rev 1999;20(4):435–59.
75. Bodine PV, Harris HA, Komm BS. Suppression of ligand-dependent estrogen receptor activity by bone-resorbing cytokines in human osteoblasts. Endocrinology 1999;140(6):2439–51.
76. Drew PD, Chavis JA. Female sex steroids: Effects upon microglial cell activation. J Neuroimmunol 2000;111(1–2):77–85.
77. Bruce-Keller AJ, et al. Pro-inflammatory and pro-oxidant properties of the HIV protein Tat in a microglial cell line: Attenuation by 17 beta-estradiol. J Neurochem 2001; 78(6):1315–24.
78. Dimayuga FO, et al. Estrogen and brain inflammation: Effects on microglial expression of MHC, costimulatory molecules and cytokines. J Neuroimmunol 2005;161(1–2):123–36.
79. Harris-White ME, et al. Estrogen (E2) and glucocorticoid (Gc) effects on microglia and A beta clearance in vitro and in vivo. Neurochem Int 2001;39(5–6):435–48.
80. Ito A, et al. Estrogen inhibits systemic T cell expression of TNF-alpha and recruitment of TNF-alpha(+) T cells and macrophages into the CNS of mice developing experimental encephalomyelitis. Clin Immunol 2002;102(3):275–82.
81. Dovio A, et al. Inhibitory effect of physiological concentrations of cortisol but not estradiol on interleukin (IL)-6 production by human osteoblast-like cell lines with different constitutive IL-6 expression. Cytokine 2001;15(1):47–52.
82. Suuronen T, et al. Anti-inflammatory effect of selective estrogen receptor modulators (SERMs) in microglial cells. Inflamm Res 2005;54(5):194–203.
83. Vegeto E, et al. The endogenous estrogen status regulates microglia reactivity in animal models of neuroinflammation. Endocrinology 2006;147(5):2263–72.
84. Vegeto E, et al. Estrogen receptor-alpha mediates the brain antiinflammatory activity of estradiol. Proc Natl Acad Sci USA 2003;100(16):9614–9.
85. Marin-Husstege M, et al. Oligodendrocyte progenitor proliferation and maturation is differentially regulated by male and female sex steroid hormones. Dev Neurosci 2004; 26(2–4):245–54.
86. Ghoumari AM, Baulieu EE, Schumacher M. Progesterone increases oligodendroglial cell proliferation in rat cerebellar slice cultures. Neuroscience 2005;135(1):47–58.
87. Ghoumari AM, et al. Progesterone and its metabolites increase myelin basic protein expression in organotypic slice cultures of rat cerebellum. J Neurochem 2003;86(4): 848–59.
88. Jung-Testas I, et al. The neurosteroid progesterone increases the expression of myelin proteins (MBP and CNPase) in rat oligodendrocytes in primary culture. Cell Mol Neurobiol 1996;16(3):439–43.
89. Garcia-Segura LM, et al. Role of astroglia in estrogen regulation of synaptic plasticity and brain repair. J Neurobiol 1999;40(4):574–84.
90. Jones KJ, et al. Androgenic regulation of the central glia response following nerve damage. J Neurobiol 1999;40(4):560–73.

91. Garcia-Estrada J, et al. Dehydroepiandrosterone, pregnenolone and sex steroids down-regulate reactive astroglia in the male rat brain after a penetrating brain injury. Int J Dev Neurosci 1999;17(2):145–51.
92. Tomas-Camardiel M, et al. Comparative study of the neuroprotective effect of dehydroepiandrosterone and 17beta-estradiol against 1-methyl-4-phenylpyridium toxicity on rat striatum. Neuroscience 2002;109(3):569–84.
93. Garcia-Segura LM, et al. Aromatase expression by astrocytes after brain injury: Implications for local estrogen formation in brain repair. Neuroscience 1999; 89(2):567–78.
94. Peterson DA. Stem cells in brain plasticity and repair. Curr Opin Pharmacol 2002; 2(1): 34–42.
95. Gross CG. Neurogenesis in the adult brain: Death of a dogma. Nat Rev Neurosci 2000; 1(1):67–73.
96. Bulloch K, McEwen BS. Regulation of the injury–immune response in the central nervous system: Allostasis and allostatic load in immunity. In: Hormones, Brain and Behavior, Vol. 1, Ch. 15. Pfaff D, Ed.; New York: Academic Press/Elsevier Science, 2002; pp. 773–801.
97. Altman J, Das GD. Autoradiographic and histological evidence of postnatal hippocampal neurogenesis in rats. J Comp Neurol 1965;124(3):319–35.
98. Gould E, et al. Neurogenesis in the neocortex of adult primates. Science 1999;286(5439): 548–52.
99. Doetsch F, et al. Subventricular zone astrocytes are neural stem cells in the adult mammalian brain. Cell 1999;97(6):703–16.
100. Gage FH. Mammalian neural stem cells. Science 2000;287(5457):1433–8.
101. Alvarez-Buylla A, Seri B, Doetsch F. Identification of neural stem cells in the adult vertebrate brain. Brain Res Bull 2002;57(6):751–8.
102. Parent JM, Valentin VV, Lowenstein DH. Prolonged seizures increase proliferating neuroblasts in the adult rat subventricular zone-olfactory bulb pathway. J Neurosci 2002;22(8):3174–88.
103. Szele FG, Chesselet MF. Cortical lesions induce an increase in cell number and PSA-NCAM expression in the subventricular zone of adult rats. J Comp Neurol 1996; 368(3):439–54.
104. Gould E, Tanapat P. Lesion-induced proliferation of neuronal progenitors in the dentate gyrus of the adult rat. Neuroscience 1997;80(2):427–36.
105. Kempermann G, Kuhn HG, Gage FH. Experience-induced neurogenesis in the senescent dentate gyrus. J Neurosci 1998;18(9):3206–12.
106. Gould E, et al. Learning enhances adult neurogenesis in the hippocampal formation. Nat Neurosci 1999;2(3):260–5.
107. Shors TJ, et al. Neurogenesis in the adult is involved in the formation of trace memories. Nature 2001;410(6826):372–6.
108. Stewart R, Przyborski S. Non-neural adult stem cells: Tools for brain repair? Bioessays 2002;24(8):708–13.
109. Woodbury D, Reynolds K, Black IB. Adult bone marrow stromal stem cells express germline, ectodermal, endodermal, and mesodermal genes prior to neurogenesis. J Neurosci Res 2002;69(6):908–17.
110. Woodbury D, et al. Adult rat and human bone marrow stromal cells differentiate into neurons. J Neurosci Res 2000;61(4):364–70.

111. Mezey E, et al. Turning blood into brain: Cells bearing neuronal antigens generated in vivo from bone marrow. Science 2000;290(5497):1779–82.
112. Mezey E, et al. Transplanted bone marrow generates new neurons in human brains. Proc Natl Acad Sci USA 2003;100(3):1364–9.
113. Crain BJ, Tran SD, Mezey E. Transplanted human bone marrow cells generate new brain cells. J Neurol Sci 2005;233(1–2):121–3.
114. Brazelton TR, et al. From marrow to brain: Expression of neuronal phenotypes in adult mice. Science 2000;290(5497):1775–9.
115. Zhao LR, et al. Human bone marrow stem cells exhibit neural phenotypes and ameliorate neurological deficits after grafting into the ischemic brain of rats. Exp Neurol 2002; 174(1):11–20.
116. Grassi Zucconi, G., Giuditta A. Is it only neurogenesis? Rev Neurosci 2002;13(4): 375–82.
117. Monje ML, Toda H, Palmer TD. Inflammatory blockade restores adult hippocampal neurogenesis. Science 2003;302(5651):1760–5.
118. Vallieres L, et al. Reduced hippocampal neurogenesis in adult transgenic mice with chronic astrocytic production of interleukin-6. J Neurosci 2002;22(2):486–92.
119. Taga T, Fukuda S. Role of IL-6 in the neural stem cell differentiation. Clin Rev Allergy Immunol 2005;28(3):249–56.
120. Barkho BZ, et al. Identification of astrocyte-expressed factors that modulate neural stem/ progenitor cell differentiation. Stem Cells Dev 2006;15(3):407–21.
121. Gottfried-Blackmore A, Croft G, Clark J, McEwen BS, Jellinck PH, Bulloch K. Characterization of a cerebellar granule progenitor cell line E(t)C. 1 and its responsiveness to 17-beta-estradiol. Brain Res. 2007 Dec 19;1186:29–40. Epub 2007 Sep 8.
122. Vela JM, et al. Interleukin-1 regulates proliferation and differentiation of oligodendrocyte progenitor cells. Mol Cell Neurosci 2002;**20**(3):489–502.
123. Cunningham ET Jr, et al. In situ histochemical localization of type I interleukin-1 receptor messenger RNA in the central nervous system, pituitary, and adrenal gland of the mouse. J Neurosci 1992;12(3):1101–14.
124. Gaiano N, Fishell G. The role of notch in promoting glial and neural stem cell fates. Annu Rev Neurosci 2002;25:471–90.
125. Gorski RA. Sexual dimorphisms of the brain. J Anim Sci 1985;61 Suppl 3:38–61.
126. Simerly RB. Wired for reproduction: Organization and development of sexually dimorphic circuits in the mammalian forebrain. Annu Rev Neurosci 2002;25:507–36.
127. Brannvall K, Korhonen L, Lindholm D. Estrogen-receptor-dependent regulation of neural stem cell proliferation and differentiation. Mol Cell Neurosci 2002; 21(3):512–20.
128. Fried G, et al. Estrogen receptor beta is expressed in human embryonic brain cells and is regulated by 17beta-estradiol. Eur J Neurosci 2004;20(9):2345–54.
129. Belcher SM. Regulated expression of estrogen receptor alpha and beta mRNA in granule cells during development of the rat cerebellum. Brain Res Dev Brain Res 1999;115(1): 57–69.
130. Bulloch K, Stallcup WB, Cohn M. The derivation and characterization of neuronal cell lines from rat and mouse brain. Brain Res 1977;135(1):25–36.
131. Bulloch K, Stallcup WB, Cohn M. A new method for the establishment of neuronal cell lines from the mouse brain. Life Sci 1978;22(6):495–504.
132. Rasika S, Alvarez-Buylla A, Nottebohm F. BDNF mediates the effects of testosterone on the survival of new neurons in an adult brain. Neuron 1999;22(1):53–62.

133. Tanapat P, et al. Estrogen stimulates a transient increase in the number of new neurons in the dentate gyrus of the adult female rat. J Neurosci 1999;19(14):5792–801.
134. McEwen B, et al. Tracking the estrogen receptor in neurons: Implications for estrogen-induced synapse formation. Proc Natl Acad Sci USA 2001;98(13):7093–100.
135. Kishi Y, et al. Estrogen promotes differentiation and survival of dopaminergic neurons derived from human neural stem cells. J Neurosci Res 2005;79(3):279–86.
136. Dudek H, et al. Regulation of neuronal survival by the serine–threonine protein kinase Akt. Science 1997;275(5300):661–5.
137. Arsenijevic Y, et al. Insulin-like growth factor-I is necessary for neural stem cell proliferation and demonstrates distinct actions of epidermal growth factor and fibroblast growth factor-2. J Neurosci 2001;21(18):7194–202.
138. Anderson MF, et al. Insulin-like growth factor-I and neurogenesis in the adult mammalian brain. Brain Res Dev Brain Res 2002;134(1–2):115–22.
139. Toran-Allerand CD, Singh M, Setalo G Jr. Novel mechanisms of estrogen action in the brain: New players in an old story. Front Neuroendocrinol 1999;20(2):97–121.
140. Duenas M, et al. Gonadal hormone regulation of insulin-like growth factor-I-like immunoreactivity in hypothalamic astroglia of developing and adult rats. Neuroendocrinology 1994;59(6):528–38.
141. Duenas M, et al. Interaction of insulin-like growth factor-I and estradiol signaling pathways on hypothalamic neuronal differentiation. Neuroscience 1996;74(2):531–9.
142. Kahlert S, et al. Estrogen receptor alpha rapidly activates the IGF-1 receptor pathway. J Biol Chem 2000;275(24):18447–53.
143. Garcia-Segura LM, et al. Interaction of the signalling pathways of insulin-like growth factor-I and sex steroids in the neuroendocrine hypothalamus. Horm Res 1996;46(4–5): 160–4.
144. Quesada A, Micevych PE. Estrogen interacts with the IGF-1 system to protect nigrostriatal dopamine and maintain motoric behavior after 6-hydroxdopamine lesions. J Neurosci Res 2004;75(1):107–16.
145. Yoshimura S, et al. FGF-2 regulates neurogenesis and degeneration in the dentate gyrus after traumatic brain injury in mice. J Clin Invest 2003;112(8):1202–10.
146. Nakatomi H, et al. Regeneration of hippocampal pyramidal neurons after ischemic brain injury by recruitment of endogenous neural progenitors. Cell 2002;110(4):429–41.
147. Dhandapani KM, et al. Astrocyte-derived transforming growth factor-{beta} mediates the neuroprotective effects of 17{beta}-estradiol: Involvement of nonclassical genomic signaling pathways. Endocrinology 2005;146(6):2749–59.
148. Kalra PS, et al. Endogenous opioid peptides mediate the interleukin-1-induced inhibition of the release of luteinizing hormone (LH)-releasing hormone and LH. Endocrinology 1990;127(5):2381–6.
149. Jones TH, Kennedy RL. Cytokines and hypothalamic–pituitary function. Cytokine 1993; 5(6):531–8.
150. Simpson ER, et al. Aromatase – A brief overview. Annu Rev Physiol 2002;64:93–127.
151. Ciana P, et al. In vivo imaging of transcriptionally active estrogen receptors. Nat Med 2003;9(1):82–6.
152. Tsutsui K, et al. Novel brain function: Biosynthesis and actions of neurosteroids in neurons. Neurosci Res 2000;36(4):261–73.
153. Robel P, Baulieu EE. Neurosteroids: Biosynthesis and function. Crit Rev Neurobiol 1995;9(4):383–94.

154. Stoffel-Wagner B. Neurosteroid metabolism in the human brain. Eur J Endocrinol 2001; 145(6):669–79.
155. Jellinck PH, et al. Metabolism of dehydroepiandrosterone by rodent brain cell lines: Relationship between 7-hydroxylation and aromatization. J Steroid Biochem Mol Biol 2005;93(1):81–6.
156. Jellinck PH, et al. Dehydroepiandrosterone (DHEA) metabolism in the brain: Identification by liquid chromatography/mass spectrometry of the delta-4-isomer of DHEA and related steroids formed from androstenedione by mouse BV2 microglia. J Steroid Biochem Mol Biol 2006;98(1):41–7.
157. Gottfried-Blackmore A, Sierra A, Jellinck PH, McEwen BS, Bulloch K. Brain Microglia Express Steroid-Converting Enzymes in the Mouse. J Steroid Biochem Mol Biol. 2008. In press.
158. Simpson ER, et al. Cytochromes P450 11: Expression of the CYP19 (aromatase) gene: An unusual case of alternative promoter usage. FASEB J 1997;11(1):29–36.
159. Sasano H, et al. Aromatase in the human central nervous system. Clin Endocrinol (Oxf) 1998;48(3):325–9.
160. Zhao Y, et al. Tumor necrosis factor-alpha stimulates aromatase gene expression in human adipose stromal cells through use of an activating protein-1 binding site upstream of promoter 1.4. Mol Endocrinol 1996;10(11):1350–7.
161. Zhao Y, et al. Aromatase P450 gene expression in human adipose tissue. Role of a Jak/STAT pathway in regulation of the adipose-specific promoter. J Biol Chem 1995; 270(27):16449–57.
162. Sierra A, Azcoitia I, Garcia-Segura L. Endogenous estrogen formation is neuroprotective in model of cerebellar ataxia. Endocrine 2003;21(1):43–51.
163. McCullough LD, et al. Aromatase cytochrome P450 and extragonadal estrogen play a role in ischemic neuroprotection. J Neurosci 2003;23(25):8701–5.

Cytokines and the Brain
Edited by C. Phelps and E. Korneva

Involvement of Brain Cytokines in Stress-induced Immunosuppression

TOSHIHIKO KATAFUCHI

Department of Integrative Physiology, Graduate School of Medical Sciences, Kyushu University, Fukuoka 812-8582, Japan

ABSTRACT

In the brain–immune system interactions, cytokines produced in the periphery and the brain during inflammatory and noninflammatory stress play an important role as signal molecules. The central administration of interleukin-1β (IL-1β) and interferon-α (IFN-α) is shown to suppress the splenic natural killer (NK) cell activity in rats, which is mediated by, at least in part, the sympathetic innervation to the spleen. The central IL-1β and IFN-α increase the splenic sympathetic nerve activity, and an electrical stimulation of the nerve results in a suppression of splenic NK cell activity through a β-adrenergic receptor-mediated process. Furthermore, the findings that (1) immobilization (IMB) stress produced an elevation of extracellular concentration of noradrenaline in the spleen, (2) the IMB-induced reduction of splenic NK activity was partially blocked by splenic denervation, (3) pretreatment with central injection of neutralizing anti-IL-1β antibody attenuated the IMB-induced NK suppression, and (4) hypothalamic IL-1β and IFN-α mRNA were increased after 1 h IMB suggested that IL-1β and IFN-α produced in the brain may be key substances mediating the IMB stress-induced immunosuppression.

1. INTRODUCTION

Several lines of evidence have indicated that the central nervous system (CNS) and the immune system communicate with each other. In this bidirectional communication between the two systems, cytokines that are synthesized in the peripheral immune system signal the brain, thereby producing a variety of autonomic, endocrine, and behavioral responses [1,2]. In addition, cytokines are newly induced in the brain by immunological challenges. Furthermore, not only the inflammatory but also noninflammatory stress such as immobilization (IMB) also induces cytokine expression in the brain.

In this paper, we review briefly (1) cytokine expression in the brain during inflammatory stress immunologically induced by systemic administration of immune activators; (2) the mechanisms of brain cytokine-induced immunosuppression, a possible negative feed back loop in brain–immune interactions; and (3) an involvement of brain cytokines in stress-induced immunosuppression.

2. CYTOKINE EXPRESSION IN THE BRAIN DURING INFLAMMATORY STRESS

It is well known that the brain produces cytokines and possesses their receptors in both glial and neurons. Cytokines are usually not found in the brain and if present they are at very low levels in the normal state. Production of most cytokines in the brain is induced not only during pathological and inflammatory events in the organ, such as injury, ischemia, infection, and degeneration [3], but also during peripheral inflammation. Among inflammatory stresses, immunologically induced stresses produced by the administration of immune activators are considered to be the best stimuli for inducing cytokines in the brain. For example, systemic injection of lipopolysaccharide (LPS), which is used most commonly as a model for bacterial infection, induces interleukin-1β (IL-1β) [4–6], tumor necrosis factor-α (TNF-α), and IL-6 [7,8] in the cerebrospinal fluid and in various brain regions at the protein and gene levels. We have recently observed that an intraperitoneal (i.p.) injection of polyriboinosinic:polyribocytidylic acid (poly I:C), a double-stranded synthetic RNA that mimics a viral infection, induces an enhanced expression of interferon-α (IFN-α) in the cortex and hypothalamus, thereby producing central fatigue [9,10].

Multiple routes that may explain how the circulating LPS and/or cytokines signal the brain have been suggested. For example, systemic injection of LPS can induce fever, one of the central effects of immune activators, through actions of LPS itself or peripherally produced cytokines induced by multiple routes, including the following: (1) endothelial cells of the cerebral microvessel [11]; (2) cells in circumventricular organs (CVO) such as the organum vasculosum of the lamina terminalis and the area postrema, which lack a functional blood–brain barrier [12,13]; (3) visceral vagal afferent nerves [14]; and (4) cytokine-specific transporters [15]. Although it is not fully understood how peripheral signals can induce the production of cytokines in the brain, it is possible that these pathways may play a role in the mechanisms of cytokine expression in the brain. It has been reported that double-stranded RNA including poly I:C is recognized by toll-like receptors (TLR)-3, which are present in the periphery and in brain cells, and activate nuclear factor-κB to produce IFNs [16], while LPS-induced signal transduction is mediated by TLR-4 [17]. Thus, it is possible that the peripheral LPS- and poly I:C-induced expression of cytokines in the brain may share common pathways of signal transduction.

3. IMMUNOSUPPRESSION INDUCED BY BRAIN CYTOKINES

Cytokines produced in both the peripheral immune system and the brain evoke a variety of acute phase responses, such as fever, anorexia, slow wave sleep, activation of the sympathetic nervous system (SNS), and the hypothalamic–pituitary–adrenal (HPA) axis. These responses are considered to be part of the biological defense system and have adaptive values for the infected host [1,2]. In addition, it has been also reported that central administrations of IL-1β [18] and IFN-α [19], but not IL-2, IL-6, or TNF-α [20], suppress peripheral cellular immunity through the activation of the SNS and/or the HPA axis. It may be disputable whether or not the central cytokine-induced immunosuppression is also included in the adaptive responses. However, it has been proposed, for example, that the activation of the HPA axis raises plasma levels of glucocorticoids, which, in turn, inhibits a broad spectrum of immunological and inflammatory responses to limit overreaction of the biological defense system, thereby forming a negative feedback [1]. Therefore, central IL-1β- and IFN-α-induced immunomodulation may also be regarded as a part of the negative feedback loop. Furthermore, the activation of the HPA axis and the SNS may not solely suppress the cellular immunity, but alter the T helper (TH)1/TH2 balance to the TH2 dominance through the actions of glucocorticoids and catecholamines [21].

3.1. Interleukin-β-induced immunosuppression

Intracerebroventricular (icv) injection of IL-1β in rats has been shown to suppress various immune responses such as natural killer (NK) cell activity, mitogenic response to phytohemagglutinin, and IL-2 production of lymphocytes isolated from blood and spleen [18,22], specific antibody production [23], and secretion of IL-1 from splenic macrophages [24]. Immunosuppression, induced by icv IL-1 administration, was mediated by the activation of both the HPA axis and the SNS, since both chemical [18] or surgical [24] sympathectomy and adrenalectomy [22] only partially blocked the immunosuppression. The reduction of immune responses was blocked by α-melanocyte-stimulating hormone (α-MSH), which is known to inhibit the action of IL-1 [25]. The activation of the splenic sympathetic nerve following icv injection of IL-1β was directly confirmed by the recording of the efferent nerve activity in anesthetized rats [25]. The IL-1β-induced increase in nerve activity was blocked by icv pretreatment with α-MSH as well as by IL-1 receptor antagonist. Since the electrical stimulation of the splenic sympathetic nerve suppresses the splenic NK activity through a β-adrenergic receptor-mediated process [26], the enhancement of the splenic nerve activity following icv injection of IL-1β at least in part, plays a role in the IL-1β-induced immunosuppression.

3.2. Interferon-α-induced immunosuppression

Similar to IL-1β, an icv injection of IFN-α also suppresses splenic NK activity in mice and rats [19,27]. However, the IFN-α-induced reduction of NK activity was completely blocked by the denervation of the splenic sympathetic nerve, but not by adrenalectomy. Therefore, it is concluded that the reduction of NK activity induced by central IFN-α is exclusively mediated by the sympathetic nerve, while the IL-1-induced suppression is dependent on both the SNS and the HPA axis. Microinjection studies have revealed that a site of action for IFN-α in the brain is the medial preoptic area (MPO) in the hypothalamus. The microinjection of IFN-α into the lateral preoptic area (LPO), the lateral hypothalamus (LHA), the ventromedial hypothalamus (VMH), and the paraventricular nucleus (PVN) caused no changes in NK activity [28]. As was expected, both icv and microinjection of IFN-α into the MPO, but not PVN, increased the splenic sympathetic nerve activity in the rat [26,29].

It has been reported that IFN-α exerts its effects through opioid receptors in the brain, since IFN-α inhibits the binding of dihydromorphine to mouse brain homogenates [30] and of naloxone to rat brain membranes [31]. In fact, the suppression of NK cytotoxicity and the increase in the splenic sympathetic nerve activity induced by icv IFN-α were abolished by pretreatment with opioid antagonists [17,24].

3.3. Involvement of corticotropin-releasing factor and prostaglandins

Both central IL-1β- and IFN-α-induced immunosuppression is completely blocked by pretreatment with an anti-corticotropin-releasing factor (CRF) antibody [18] and a CRF antagonist (α-helical CRF_{9-41}) [19], respectively. In addition, the enhancement of the splenic sympathetic activity induced by icv IL-1β was also blocked by α-helical CRF_{9-41} [25]. An icv injection of CRF decreased the splenic and blood NK cell activity [32,19] and antibody production in rats

[23]. The CRF-induced reduction of NK cell activity was completely blocked by icv pretreatment with a ganglionic blocker, which also blocked an increase in plasma noradrenaline (NA), whereas the same dose of the blocker did not affect the increase in plasma levels of adrenocorticotropic hormone (ACTH) and corticosterone [32]. Therefore, it is likely that the SNS, but not the HPA axis, plays a significant role in the CRF-induced immunosuppression. Consistently, splenic sympathetic nerve activity [33] and NA release from the splenic sympathetic nerve terminals [34] increased following icv infusion of CRF in rats.

It has been demonstrated that in rats, prostaglandin E_2 (PGE_2), which is a principal mediator of many proinflammatory cytokines, decreases proliferative responses of splenic lymphocytes by its action within the brain [35]. Since this is accompanied by an increase in plasma ACTH and corticosterone levels, it is likely that the activation of the HPA axis plays a role in this response. However, the splenic sympathetic nerve activity also increases after icv infusion of PGE_2 [33] through its action on EP1 receptor subtype for PGE_2 [36]. Furthermore, an icv injection of PGE_2, but not that of PGD_2 or PGF_{2a}, increased the NA turnover in the spleen [37]. These data indicate that the SNS is also involved in the PGE_2-induced immunosuppression. It has been shown that IL-1β-induced ACTH release [38] and fever [39] are abolished by the inhibition of a key enzyme of prostanoids syntheses (cyclooxygenase, COX). Similarly the

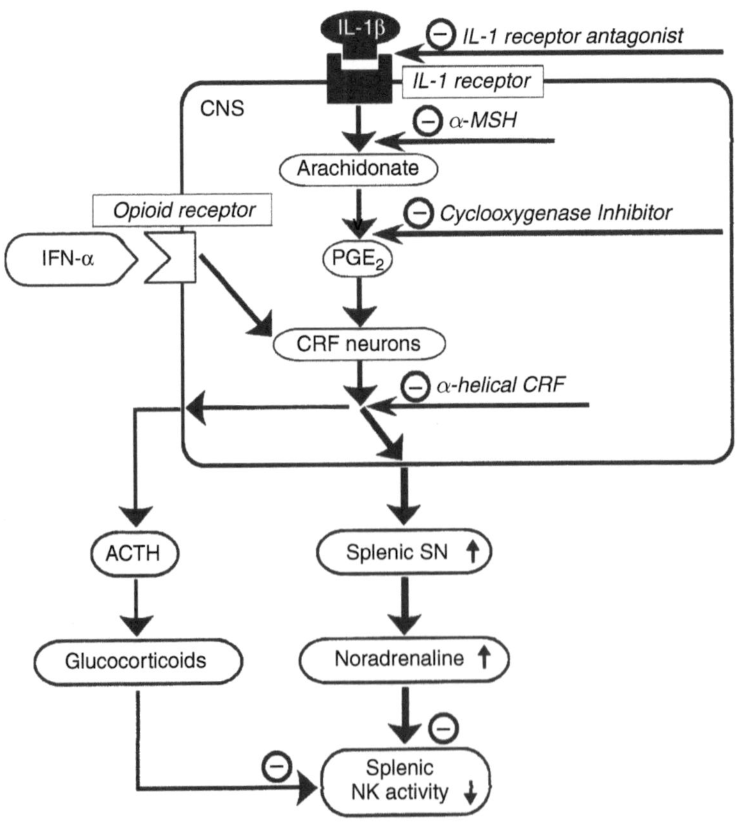

Figure 1. Mechanisms of IL-1β- and IFN-α-induced suppression of splenic NK cell activity. For details, refer to the text.

enhancement of the splenic sympathetic activity [25] and the NA turnover in the spleen induced by central IL-1β [37] are also blocked by COX inhibitors, suggesting an involvement of prostanoid synthesis in the brain in the IL-1β-induced reduction of cellular immunity.

As mentioned above, the icv IL-1β-induced enhancement of the splenic sympathetic activity was blocked by a CRF antagonist, α-helical CRF_{9-41}, and a COX inhibitor, sodium salicylate [25]. An increase in the splenic nerve activity by icv PGE_2 was blocked by icv pretreatment with α-helical CRF_{9-41}, whereas the CRF-induced enhancement of the nerve activity was not affected by sodium salicylate at a dose 100 times more than that required for suppressing the IL-1β-induced activation of the nerve [33]. Therefore, it is likely that an activation of prostanoid-dependent process is followed by that of the CRF system in the brain, resulting in the enhanced splenic nerve activity. A similar sequential relationship between prostanoid-dependent processes and CRF mechanisms has been suggested in the IL-1-induced activation of the HPA axis [40]. Figure 1 illustrates the mechanisms of IL-1β- and IFN-α-induced suppression of the splenic NK cell activity through the SNS and the HPA axis.

4. ROLE OF BRAIN CYTOKINES IN THE IMMOBILIZATION-INDUCED IMMUNOSUPPRESSION

It is well known that noninflammatory stress such as IMB and foot shock induces suppression of immune functions. Several studies have been undertaken to investigate the mechanism of this stress-induced immunosuppression, since it is considered to be a good model for the interaction between the brain and the immune system.

4.1. Involvement of sympathetic nerve in stress-induced immunosuppression

It has been shown that foot shock stress-induced suppression of proliferative responses of splenocytes was blocked by surgical denervation of the splenic sympathetic nerve [41], suggesting that the sympathetic nerve may be a significant mediator of the immunosuppression. Shimizu et al. [42] have shown that IMB for 90 min resulted in a reduction of the splenic NK cell activity in rats, and the suppression was attenuated by splenic denervation. They also demonstrated that extracellular concentration of NA in the spleen measured by *in vivo* microdialysis in conscious rats markedly increased during IMB. Since the increase in the splenic NA levels was almost completely abolished by the splenic denervation, it was apparently derived from the splenic sympathetic nerve terminals, but not from adrenal medulla.

Irwin et al. [43] have demonstrated that foot shock stress-induced reduction of splenic NK cell activity was mediated by brain CRF independently to the activation of the HPA axis. As described already, an icv injection of CRF increases the electrical activity of the splenic sympathetic nerves [25,33], and the electrical stimulation of the splenic sympathetic nerve resulted in the suppression of splenic NK cell activity through a β-adrenergic receptor-mediated process [26]. These findings, taken together, suggest that the stress-induced suppression of NK cell activity is mediated, at least partly, by the activation of the splenic sympathetic nerve.

4.2. Significant role of brain cytokines in the stress-induced immunosuppression

In addition to the inflammatory stress, noninflammatory stress such as IMB has also shown to induce cytokine expression in the brain. In agreement with a previous finding by Northern blot

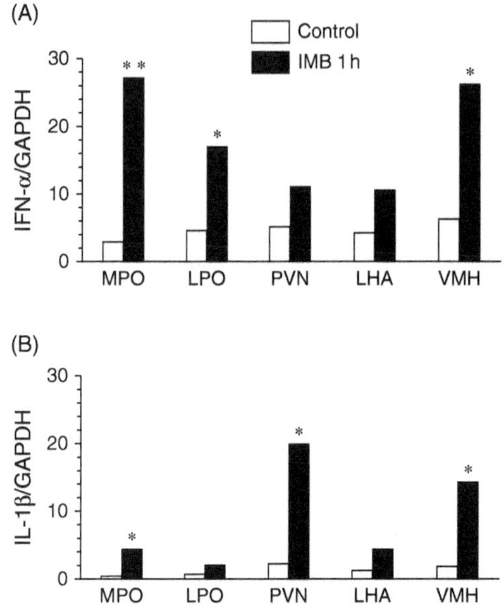

Figure 2. Changes in IFN-α and IL-1β mRNA in the hypothalamus. (A) IFN-α and (B) IL-1β mRNA, respectively. Amounts of mRNAs were quantitatively measured using a real-time capillary PCR method. Each value of cytokine mRNA was divided by the amount of a house keeping gene, glyceraldehyde 3-phosphate dehydrogenase (GAPDH), mRNA in the same sample for standardization. Abbreviations are defined in the text.

analysis [44], we have found that the amount of mRNA for IFN-α and IL-1β is increased after IMB stress for 1 h in the rat hypothalamus as assessed quantitatively by real-time capillary RT-PCR. The amount of IFN-α mRNA increased in the MPO, LPO, and VMH, while IL-1β mRNA increased in the MPO, PVN, and VMH (Fig. 2). As described above, both IL-1β and IFN-α in the brain can induce suppression of the peripheral immunity, which is dependent on central CRF, and is mediated, at least in part, by activation of the SNS. These findings raise a possibility that these brain cytokines are involved in stress-induced immunosuppression. Supporting this hypothesis, the suppression of splenic NK cell activity produced by IMB in mice was blocked by pretreatment with anti-IL-1β neutralizing antibody administered icv 10 min before IMB, while the pretreatment with nonspecific IgG was without effect [45]. In addition, the reduction of lymphocyte proliferation induced by foot shock stress was also attenuated by icv pretreatment with anti-IL-1 antibody [46].

Immediate early gene c-*fos* is commonly used as a marker for functionally activated neurons. We have found that after 90 min of IMB, Fos protein encoded by c-*fos* was found most heavily concentrated in the parvocellular region of the PVN and moderately in the LHA, VMH, and MPO (our unpublished data). When the PVN, which is known to have direct connections with the sympathetic preganglionic neurons in the spinal cord, was chemically activated by microinjection of an excitatory amino acid, glutamate, into the PVN, the splenic sympathetic nerve activity was enhanced [29]. Based on the immunohistochemical and electrophysiological findings, it is thus suggested that activation of the PVN shown by the Fos expression contributed to the IMB-induced suppression of the splenic NK cell activity through the sympathetic nerves. Furthermore, synthesis of PGE_2 was involved in the IMB-induced Fos expression in the PVN, since the icv pretreatment with a COX inhibitor, diclofenac, attenuated the IMB-induced Fos expression (our unpublished data).

It is possible that the MPO is also involved in the NK suppression, since the MPO is the only effective site in the hypothalamus where implanted IFN-α brings about a reduction in NK activity [28], and IFN-α mRNA is also markedly increased after IMB as described previously. In contrast to the PVN, both chemical and electrical stimulation of the MPO decreased the splenic sympathetic nerve activity [29]. It has been reported that a majority of MPO neurons decrease their firing rate by the direct applications of IFN-α and IL-1β [47,48], probably through an action of PGE_2 as a second mediator [49]. Since the MPO sends an inhibitory projection to the PVN [50], the suppression of the MPO neurons may cause a disinhibition of the PVN neurons, which, in turn, induces an activation of the PVN, thereby resulting in the enhancement of the nerve activity. In accordance with this finding, electrical lesion of the MPO increased the splenic sympathetic nerve activity, resulting in the suppression of splenic NK cell activity, which was completely blocked by the denervation [29].

5. CONCLUSION

The CNS has its own network system consisting of the afferent sensory, autonomic, and humoral pathways to induce changes in the neuronal activity, and the efferent route to regulate the autonomic, endocrine, and homeostatic systems including higher brain functions such as learning and memory and emotion. In contrast, the immune system also has its own regulating system. However, it is now evident that information from immune activators and circulating cytokines is

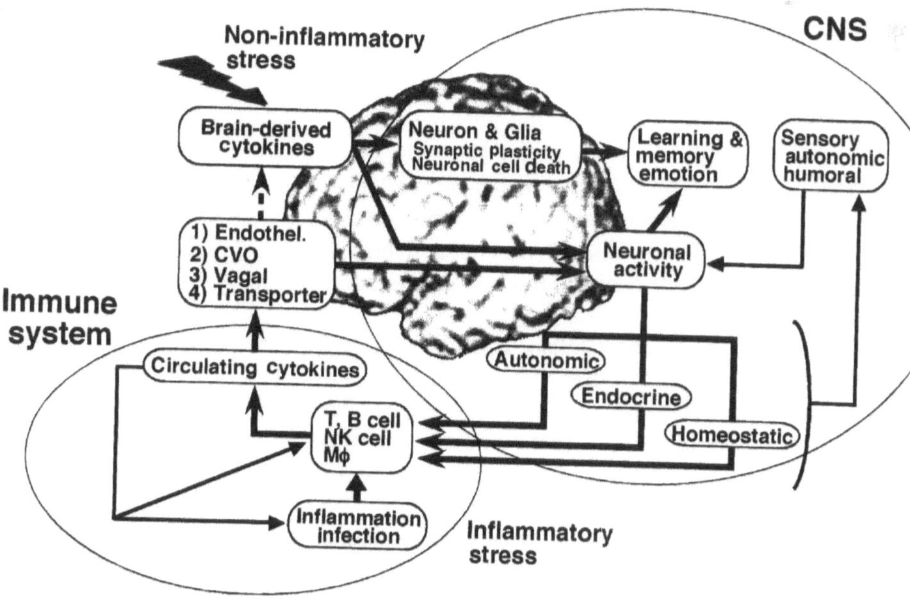

Figure 3. Illustration demonstrating the brain–immune system interaction. Endothel., endothelial cells of the brain microvessel; CVO, circumventricular organ; Vagal, the afferent vagal nerve; Transporter, specific cytokine transporter. For details, refer to the text.

transmitted to the brain through at least four pathways, as described in the text. Cytokines are also induced in the brain by noninflammatory stress such as IMB. These cytokines act on neurons and glial cells, thereby affecting synaptic plasticity and neuronal cell death, as well as influencing peripheral immune functions through the autonomic, endocrine, and homeostatic systems (Fig. 3) . It is concluded from the available evidence that the neuronal network of the hypothalamic-SNS is one of the important communication channels that mediate the central modulation of cellular immunity, and this network is deeply involved in the IMB-induced suppression of NK cell activity.

ACKNOWLEDGMENTS

This work was supported by Grant-in Aid for Scientific Research (19603004 (C) to T.K.) and Special Coordination Funds of the Ministry of Education, Culture, Sports, Science and Technology, the Japanese Government.

REFERENCES

1. Hori T, Katafuchi T, Oka T. Central cytokines: Effects on peripheral immunity, inflammation and nociception. In: Psychoneuroimmunology, Vol. 1. Ader R, Felten DL, and Cohen N, Eds; San Diego, CA: Academic Press, 2001; pp. 517–45.
2. Rothwell NJ, Hopkins SJ. Cytokines and the nervous system. II. Actions and mechanisms of actions. Trends Neurosci 1995;18:130–6.
3. Hopkins SJ, Rothwell NJ. Cytokines and the nervous system. I. Expression and recognition. Trends Neurosci 1995;18:83–8.
4. Ban E, Haour F, Lenstra R. Brain interleukin-1 gene expression induced by peripheral lipopolysaccharide administration. Cytokine 1992;4:48–54.
5. Hillhouse EW, Mosley K. Peripheral endotoxin induces hypothalamic immunoreactive interleukin-1β in the rat. Br J Pharmacol 1993;109:289–90.
6. Quan N, Sundar SK, Weiss JM. Induction of interleukin-1 in various brain regions after peripheral and central injections of lipopolysaccharide. J Neuroimmunol 1994;49:125–34.
7. Coceani F, Lees J, Mancilla J, Berlizario J, Dinarello CA. Interleukin-6 and tumor necrosis factor in cerebrospinal fluid: Changes during pyrogen fever. Brain Res 1993;612:165–71.
8. Klir JJ, Roth J, Szelenyi Z, McClellan JL, Kluger, MJ. Role of hypothalamic interleukin-6 and tumor necrosis factor-α in LPS fever in rat. Am J Physiol 1993;265:R512–7.
9. Katafuchi T, Kondo T, Yasaka K, Kubo K, Take S, Yoshimura M. Prolonged effects of polyriboinosinic:polyribocytidylic acid on spontaneous running wheel activity and brain interferon-α mRNA in rats: A model for immunologically induced fatigue. Neuroscience 2003;120:837–45.
10. Katafuchi T, Kondo T, Take S, Yoshimura M. Involvement of enhanced expression of brain interferon-α and serotonin transporter in poly I:C-induced fatigue in rats. Eur J Neurosci 2005;22:2817–26.
11. Matsumura K, Cao K, Ozaki M, Morii H, Nakadate K, Watanabe Y. Brain endothelial cells express cyclooxygenase-2 during lipopolysaccharide-induced fever: Light and electron microscopic immunocytochemical studies. J Neurosci 1998;18:6279–6289.

12. Blatteis CM, Sehic E. Prostaglandin E2: A putative fever mediator. In: Fever: Basic Mechanisms and Management. Mackowiak PA, Ed.; Philadelphia, PA: Lippincott Raven Publishers, 1997; pp. 117–45.
13. Ota K, Katafuchi T, Takaki A, Hori T. AV3V neurons that send axons to hypothalamic nuclei respond to the systemic injection of IL-1β. Am J Physiol 1997;272:R532–40.
14. Watkins LR, Maier SF, Geohler LE. Cytokine-to-brain communication: A review and analysis of alternative mechanisms. Life Sci 1995;57:1011–26.
15. Banks WA, Kastin AJ, Broadwell RD. Passage of cytokines across the blood–brain barrier. Neuroimmunomodulation 1995;2:241–8.
16. Alexopoulou L, Holt AC, Medzhitov R, Flavell RA. Recognition of double-stranded RNA and activation of NF-κB by Toll-like receptor 3. Nature 2001;413:732–8.
17. Chow JC, Young DW, Golenbock DT, Christ WJ, Gusovsky F. Toll-like receptor-4 mediates lipopolysaccharide-induced signal transduction. J Biol Chem 1999;274:10689–92.
18. Sundar SK, Cierpial MA, Kilts C, Ritchie JC, Weiss JM. Brain IL-1-induced immunosuppression occurs through activation of both pituitary–adrenal axis and sympathetic nervous system by corticotropin-releasing factor. J Neurosci 1990;10:3701–6.
19. Take S, Mori T, Katafuchi T, Hori T. Central interferon-α inhibits natural killer cytotoxicity through sympathetic innervation. Am J Physiol 1993;265: R453–9.
20. Connor TJ, Song C, Leonard BE, Merali Z, Anisman H. An assessment of the effects of central interleukin-1β, -2, -6, and tumor necrosis factor-α administration on some behavioural, neurochemical, endocrine and immune parameters in the rat. Neuroscience 1998;84:923–33.
21. Elenkov IJ, Papanicolaou DA, Wilder RL, Chrousos GP. Modulatory effects of glucocorticoids and catecholamines on human interleukin-12 and interleukin-10 production: Clinical implications. Proc Assoc Am Physicians 1996;108:374–81.
22. Sundar SK, Becker KJ, Cierpial MA, Carpenter MD, Rankin LA, Fleener SL, Ritchie JC, Simson PE, Weiss JM. Intracerebroventricular infusion of interleukin 1 rapidly decreases peripheral cellular immune responses. Proc Natl Acad Sci USA 1989;86:6398–402.
23. Irwin M. Brain corticotropin-releasing hormone- and interleukin-1β-induced suppression of specific antibody production. Endocrinology 1993;133:1352–60.
24. Brown R, Li Z, Vriend CY, Nirula R, Janz L, Falk J, Nance DM, Dyck DG, Greenberg AH. Suppression of splenic macrophage interleukin-1 secretion following intracerebroventricular injection of interleukin-1β: Evidence for pituitary–adrenal and sympathetic control. Cell Immunol 1991;132:84–93.
25. Ichijo T, Katafuchi T, Hori T. Central administration of interleukin-1β enhances splenic sympathetic nerve activity in rats. Brain Res Bull 1994;34:547–53.
26. Katafuchi T, Take S, Hori T. Roles of sympathetic nervous system in the suppression of cytotoxicity of splenic natural killer cells in the rat. J Physiol 1993;465:343–57.
27. Take S, Mori T, Katafuchi T, Kaizuka Y, Hori T. Central interferon-α suppresses the cytotoxic activity of natural killer cells in the mouse spleen. Ann N Y Acad Sci 1992; 650:46–50.
28. Take S, Uchimura D, Kanemitsu Y, Katafuchi T, Hori T. Interferon-α acts at the preoptic hypothalamus to reduce natural killer cytotoxicity in rats. Am J Physiol 1995;268:R1406–10.
29. Katafuchi T, Ichijo T, Take S, Hori T. Hypothalamic modulation of natural killer cell activity in rats. J Physiol 1993;471:209–21.
30. Blalock JE, Smith EM. Human leukocytic interferon (HuIFN-α): Potent endorphin-like opioid activity. Biochem Biophys Res Commun 1981;101:472–8.

31. Menzies RA, Patel R, Hall NRS, O'Grady MP, Rier SE. Human recombinant interferon alpha inhibits naloxone binding to rat brain membranes. Life Sci 1992;50:PL227–32.
32. Irwin M, Hauger R, Brown M, Britton KT. CRF activates autonomic nervous system and reduces natural killer cytotoxicity. Am J Physiol 1988;255:R744–7.
33. Katafuchi T, Ichijo T, Hori T. Sequential relationship between actions of CRF and PGE_2 in the brain on splenic sympathetic nerve activity in rats. J Auton Nerv Syst 1997;67:200–6.
34. Shimizu N, Hori T, Nakane H. An interleukin-1β-induced noradrenaline release in the spleen is mediated by brain corticotropin-releasing factor: An *in vivo* microdialysis study in conscious rats. Brain Behav Immun 1994;7:14–23.
35. Rassnick S, Zhou D, Rabin BS. Central administration of prostaglandin E_2 suppresses in vitro cellular immune responses. Am J Physiol 1995;269:R92–7.
36. Ando T, Ichijo T, Katafuchi T, Hori T. Intracerebroventricular injection of prostaglandin E_2 increases splenic sympathetic nerve activity in rats. Am J Physiol 1995;269:R662–8.
37. Terao A, Kitamura H, Asano A, Kobayashi M, Saito M. Roles of prostaglandins D_2 and E_2 in interleukin-1-induced activation of norepinephrine turnover in the brain and peripheral organs of rats. J Neurochem 1995;65:2742–7.
38. Katsuura G, Arimura A, Koves K, Gottschall PE. Involvement of organum vasculosum lamina terminalis and preoptic area in interleukin-1β-induced ACTH release. Am J Physiol 1990;258:E163–71.
39. Stitt JT. Prostaglandin E as the neural mediator of the febrile response. Yale J Biol Med 1986;59:137–49.
40. Watanabe T, Morimoto A, Sakata Y, Murakami N. ACTH response induced by interleukin-1 is mediated by CRF secretion stimulated by hypothalamic PGE. Experientia 1990;46: 481–4.
41. Wan W, Vriend CY, Wetmore L, Gartner JG, Greenberg AH, Nance DM. The effects of stress on splenic immune function are mediated by the splenic nerve. Brain Res Bull 1993;30:101–5.
42. Shimizu N, Kaizuka Y, Hori T, Nakane H. Immobilization increases norepinephrine release and reduces NK cytotoxicity in spleen of conscious rat. Am J Physiol 1996;271:R537–44.
43. Irwin M, Vale W, Rivier C. Central corticotropin-releasing factor mediates the suppressive effect of stress on natural killer cytotoxicity. Endocrinology 1990;126:2837–44.
44. Minami M, Kuraishi Y, Yamaguchi T, Nakai S, Hirai Y, Satoh M. Immobilization stress induces interleukin-1β mRNA in the rat hypothalamus. Neurosci Lett 1991;123:254–6.
45. Katafuchi T, Kondo T, Take S, Hori T. Hypothalamic modulation of delayed-type hypersensitivity and splenic natural killer cell activity. In Brain and Biodefence. Oomura Y, Hori T, Eds; Tokyo, Basel: Japan Scientific Societies Press & S. Karger AG, 1998; pp. 3–11.
46. Saperstein,A, Brand H, Audhya T, Nabriski D, Hutchinson B, Rosenzweig S, Hollander CS. Interleukin 1β mediates stress-induced immunosuppression via corticotropin-releasing factor. Endocrinology 1992;130:152–8.
47. Nakashima T, Hori T, Kuriyama K, Matsuda T. Effects of interferon-α on the activity of preoptic thermosensitive neurons in tissue slices. Brain Res 1988;454:361–7.
48. Hori T, Shibata M, Nakashima T, Yamasaki M, Asami A, Asami T, Koga H. Effects of interleukin-1 and arachidonate on the preoptic and anterior hypothalamic neurons. Brain Res Bull 1988;20:75–82.

49. Katafuchi T, Duan S, Take S, Yoshimura M. Modulation of glutamate-induced outward current by prostaglandin E_2 in rat dissociated preoptic neurons. Brain Res 2005;1037:180–6.
50. Saphier D, Feldman S. Effects of stimulation of the preoptic area on hypothalamic paraventricular nucleus unit activity and corticosterone secretion in freely moving rats. Neuroendocrinology 1986;42:167–73.

Cytokines and the Brain
Edited by C. Phelps and E. Korneva

Neuroprotective Effects of Inflammation in the Nervous System

JORGE CORREALE[1,2], MARCELA FIOL[1,2] and ANDRÉS VILLA[3]

[1]Department of Neurology, Raúl Carrea Institute for Neurological Research, FLENI; [2]School of Biological Sciences, Austral University; [3]Department of Neurology, José María Ramos Mejía Hospital, School of Medicine, Buenos Aires University, Buenos Aires, Argentina

ABSTRACT

Inflammation in the nervous system is widely recognized as contributing to a number of neurological conditions. However, the central nervous system (CNS) has also been classically recognized as occupying a privileged site with respect to immune-related phenomena. This dichotomy is widely understood to be a functional manifestation of known CNS and peripheral nervous system (PNS) barriers, in addition to the variable immunocompetence of certain CNS and PNS cells. Moreover, the variable capacities of, for example, CNS cells to produce cytokines and/or induce neurotrophic factors in certain disease states or after trauma are also recognized as contributing. In light of these considerations, this chapter explores the damage-to-benefit ratio for several classes of cells and their secretory products after CNS and PNS lesions, trauma, and diseases. Included are an evaluation of macrophage, microglial, and Schwann cell activation, as well as T-lymphocyte responses during several experimental pathophysiological models and disease states. These cellular activities range from being protective to pathogenic factors. The contributions of autoreactive antibodies (Abs) to disease processes and nervous system repair are also considered for their effects that categorically parallel described cellular responses. However, in experimental model systems involving myelination, overall observations point to Abs synthesized following CNS injuries as being potentially therapeutic, and these examples are reviewed. Moreover, data presented from the studies of cytokines in experimental and natural disease states demonstrate that certain cytokines have the capacity to generate multiple and opposite effects including producing damage, protection, or regeneration. Examples are described involving tumor necrosis factor-α (TNF-α), interleukin (IL)-1β, IL-6, and interferon (IFN)-γ. Ultimately, our understanding of the inflammatory processes that protect versus those that contribute to neuronal damage will be a prerequisite to designing effective new therapies for diseases affecting the nervous system.

1. INTRODUCTION

Inflammation is a key component of host–defense responses closely linked to immune system activation and essential for defense mechanisms limiting invading pathogen proliferation. However, an excessive, prolonged or unregulated inflammatory response can also be highly detrimental. Traditionally, inflammation has been widely recognized as a contributor to neurological diseases, such as multiple sclerosis (MS), Guillain-Barré syndrome (GBS), chronic inflammatory demyelinating neuropathy (CIDP), AIDS dementia, Alzheimer's disease (AD), cerebral ischemia, traumatic brain injury, and more recently Parkinson's disease (PD) and amyotrophic lateral sclerosis (ALS) [1,2].

Classically, the central nervous system (CNS) has always been viewed as the prime example of an "immune-privileged site"; however, recent evidence has challenged this traditional viewpoint. Indeed, the CNS shows a well-organized innate immune reaction during both systemic infection and local injury. Currently, it is clear that distinctive CNS features contribute to generate the variable susceptibility of this compartment to immune-mediated injuries [3]. The most prominent element involved in these mechanisms is the blood–brain barrier (BBB), a physical and metabolic barrier separating the CNS from the systemic circulation, creating a unique and stable environment for neuronal activity. Tight junctions present between endothelial cells of the BBB limit the entry of large molecules, and immune cell recruitment across the BBB is a highly regulated process involving adhesion molecules, chemokines, cytokines, and matrix metalloproteinases [4]. An additional mechanism is the capacity of resident CNS cells, particularly astrocytes and microglial cells, to regulate immune reactivity within the CNS. Specifically, expression of major histocompatibility complex (MHC) Class II and of costimulatory molecules has been demonstrated *in vitro* and *in situ* on rodent and human microglial cells at sites of inflammation or injury [5–7]. Furthermore, *in vitro* functional studies have demonstrated that rodent microglial cells can act as efficient antigen-presenting cells (APCs) [6,8,9]. Although *in vitro* experiments had initially suggested that astrocytes were cells with potential antigen-presenting capacity [10,11], further studies have now demonstrated that astrocytes do not express the crucial costimulatory molecules [6], nor do they support proliferation of naive CD4 T cells after the induction of MHC Class II [12]. Therefore, the incomplete antigen-presenting capacity of astrocytes may suppress or anergize invading T cells, a mechanism through which CNS inflammation and tissue damage is limited [13,14]. Nevertheless, because microglial cells and astrocytes coexist within the CNS, the ultimate analysis of the interactions between both cell populations as APCs *in vivo* remains a subject for further study.

Interactions of the immune system with peripheral nerves also contribute to the development of a number of inflammatory neuropathies. The peripheral nervous system (PNS) preserves the intrinsic capacity to regenerate; however, the main target of immune attack in the PNS are Schwann cells, which can also act as modulators of inflammation through the secretion of different cytokines and inflammatory mediators [15]. Schwann cells are capable of secreting different trophic factors such as nerve growth factor (NGF), ciliary neurotrophic factor (CNTF), and glia-derived neurotrophic factor (GDNF), all of which contribute to axonal outgrowth, sprouting, and neuronal survival [16–18]. Different reports have shown Schwann cell HLA-DR reactivity in human nerve biopsy specimens [19,20]. Moreover, MHC Class II can be induced in rat and human Schwann cell cultures after stimulation with IFN-γ, or by coculture with T lymphoblasts [21–23]. IFN-γ also induces the expression of intracellular adhesion molecule (ICAM)-1 on rat Schwann cells [22,23]. Overall, these findings suggest that under certain stimulatory conditions, Schwann cells can acquire the role of APCs. In fact, *in vitro* experiments have shown that when exposed to IFN-γ or IFN-γ together with tumor necrosis factor (TNF)-α, Schwann cells are able to present P2 protein to specific T cell lines [22]. Nevertheless, whether Schwann cells actually act as APCs *in vivo* is still a matter of debate [24–26].

Inflammation in the nervous system has been studied in the context of autoimmunity and infection for some time. During Wallerian degeneration (WD) following axotomy, a considerable inflammatory response involving hematogenous macrophages, Schwann cells, and inflammatory mediators develops in the distal stump, setting the stage for success or failure of the subsequent regeneration [27–31].

There are also substantial data demonstrating the active involvement of inflammatory processes in acute CNS injuries, for example, stroke as well as brain and spinal cord trauma [32–34]. Inflammation develops in concert with cytokine release from the microglia following injury and has been thought to be a major contributor to secondary tissue damage; different studies have indeed shown that inhibition of inflammatory mediators ameliorates tissue injury after acute insults [35–37]. Likewise, microglia and astrocyte activation, also indicative of inflammation, can occur in chronic degenerative CNS disorders such as AD, PD, and ALS. Furthermore, serum and colony-stimulating factor (CSF) from these patients show elevated innate immune system molecule levels (interleukin (IL)-6, TNF-α, and IL-1β), and inhibition of these mediators can have a neuroprotective effect [38–40].

When appropriately activated and regulated in the majority of tissues, inflammation is beneficial for the host in combating invading pathogens, removing damaged cells, and promoting repair and recovery. The same may be true in the nervous system. Thus, it is likely that the contribution of the immune response to the pathogenesis of several neurological diseases does not necessarily involve harmful phenomena exclusively, but rather may depend on a variety of actions, of both detrimental and beneficial nature. This chapter is a review of the mechanisms underlying the protective role of the immune response in the nervous system.

2. THE PROTECTIVE ROLE OF MACROPHAGES

Experimental studies in adult rodents have demonstrated that after crush injury of CNS fibers, axonal regeneration occurs when the CNS environment is switched to that of the PNS [41,42]. Although initial observations had suggested that astrocytic gliosis was linked to the failure of axonal regeneration in the adult mammalian CNS, further *in vitro* and *in vivo* experiments now provide evidence that astrocytes are essential for growth support, and that macrophages can reverse CNS growth failure [43–45]. In contrast, the PNS can repair itself after injury. Differences observed between CNS and PNS repair after injury correlate with the degree of myelin debris clearance and macrophage invasion in each system [46–48]. Thus, after crush injury to the sciatic nerve in rodents, a large number of macrophages are attracted to degenerating axons and enter peripheral nerves, both before and during the period of maximal Schwann cell proliferation. In contrast, the degenerating optic nerve attracts few macrophages, and the removal of myelin is much slower [46,47]. After CNS injury, microglial cells become phagocytic at the site of the lesion, but only in a limited capacity. Microglial cells fail to clear myelin debris or damaged extracellular matrix [46], or myelin-associated growth inhibitors such as Nogo and Myelin associated glycoprotein (MAG), which have been demonstrated to have strong inhibitory effects on axonal growth [49,50]. Although hematogenous macrophages accumulate intensively at injury sites in both the CNS and the PNS, remote macrophage migration within the CNS does not occur [46,51], further reducing CNS debris removal while increasing injured neuron exposure to axonal regeneration inhibitors. Interestingly, *in vitro* studies have demonstrated that phagocytic activity and nitric oxide production by macrophages and brain-derived microglial cells are enhanced after exposure to sciatic nerve segments, but inhibited by exposure to optic nerve segments [52]. Another difference between CNS and PNS responses to nerve injuries is the slow release of cytokines such as TNF-α from microglia after CNS injuries, in contrast to rapid cytokine release from macrophages of injured peripheral nerves.

Early TNF-α release is essential in priming microglial cells to effectively activate phagocytosis. These findings suggest that axon regeneration failure after injury in the CNS can be attributed to the inability of the environment surrounding the injured tissue to produce adequate inflammatory responses necessary for healing.

Recently, new endeavors have been undertaken to make the CNS environment more permissive toward inducing regeneration after injuries. For example, local transplantation of macrophages stimulated with the PNS tissue into transected optic nerves prevented CNS regrowth failure, suggesting that healing of injured mammalian CNS requires the active participation of the immune system [53,54]. Likewise, transplantation of activated macrophages stimulated *ex vivo* with sciatic nerve segments into transected spinal cord induced axonal regrowth and partial functional recovery [55]. Interestingly, PNS-activated macrophages were more beneficial for axonal growth than were CNS-activated or nonactivated macrophages [53], and the recovery was found to be dependent upon the number of transplanted macrophages [55]. An association between macrophages and remyelination efficiency has been observed in several models of CNS demyelination. For instance, in lysolecithin-induced demyelination, early macrophage depletion results in a significant decrease in oligodendrocyte (OGD) remyelination [56]. Also, if macrophage depletion is delayed until after the second half of the remyelination phase, there is no effect on repair outcome, suggesting that macrophages are required in early stages of CNS remyelination.

One of the main reasons for macrophage presence at injury sites is myelin debris removal. Debris phagocytosis helps remyelinating cells to engage demyelinated axons, and also fosters OGD progenitor maturation and axonal regrowth [57–60]. In addition, macrophages may themselves be a source of growth factors and cytokines directly or indirectly promoting proliferation, differentiation, and migration of cellular lineages through astrocyte activation. For example, insulin-like growth factor (IGF-1) and transforming growth factor-β1 (TGF-β1) are produced by macrophages in both cuprizone and lysolecithin models of demyelination/remyelination. Both growth factors have been associated with OGD progenitor differentiation [61,62]. Likewise, activated macrophages and microglial cells are able to induce dopaminergic sprouting in the injured striatum through GDNF and brain-derived growth factor (BDNF) production [63]. Reactive macrophages and microglial cells also produce cytokines such as IL-1, which induce astrocytosis [64,65]. Of interest, *in vitro* and *in vivo* investigations have demonstrated that reactive astrocytes are able to contribute to axonal growth via the production of cell surface and extracellular matrix molecules [66–68]. IL-1-stimulated astrocytes have been found to produce basic fibroblast growth factor (bFGF) and CNTF in different animal models, both growth factors enhancing dopaminergic cell survival [69–71].

Perhaps the greatest current interest regarding the role of inflammation in chronic CNS disorders lies with AD. One of the pathology hallmarks of this disease is the presence of β-amyloid deposits in brain parenchyma and cerebral vasculature. These deposits, also known as amyloid plaques, together with the presence of neurofibrillary tangles cause progressive synaptic dysfunction and neuronal death, in strong correlation with the progressive cognitive impairment and personality changes of the disease. Many studies have provided evidence that microglial cells are present in amyloid deposits both in human and in animal models of the disease [72,73]. Nevertheless, the role of microglial cells in AD remains a matter of controversy. In vitro experiments have shown that microglia become activated in the presence of β-amyloid, secreting neurotoxic reactive oxygen species [74]. β-Amyloid can also stimulate inducible nitric oxide synthase (iNOS) expression by neuronal cells, inducing subsequent nitric oxide-mediated neuronal cell apoptosis [75]. However, there is no clear evidence for microglial-mediated neurotoxicity *in vivo* [76,77]. In contrast to this view, different studies support the notion that microglial cells have neuroprotective actions in AD. Indeed, activated microglial cells release neurotrophic factors, which have beneficial properties for neurons [33]. In this regard, microglial inhibition causes extensive damage in acute neurotoxicity models [78]. Microglial cells also represent a natural mechanism by which protein aggregates and debris can be removed through phagocytosis [79], and clearance of β-amyloid from the brain following β-amyloid immunization was associated with enhanced microglial cell activity around amyloid deposits [80]. There are multiple mechanisms by which microglial cell activation enhances β-amyloid clearance from the brain. First, by the expression of scavenger receptors, such as integrin-αβ, scavenger receptor-A, and scavenger receptor-BI [81]. Second, by binding soluble β-amyloid to microglial cell receptors as heparin sulfate proteoglycans and insulin receptors, resulting in phagocytosis of β-amyloid [82,83]. Third, through degrading enzymes release, such as metalloproteinases [84]. Collectively, these observations suggest that microglial cells may play a leading role in protecting neurons during the course of AD.

The pathological hallmark of demyelinating autoimmune neuropathies, such as GBS and CIDP, are lymphocyte and macrophage infiltrates. In fact, macrophages actively strip off myelin lamellae from axons, induce vesicular disruption of the myelin sheath, and phagocytose both intact and damaged myelin [85,86]. Besides, in primary axonal forms of GBS, macrophages also contribute to axonal loss prior to secondary myelin breakdown [87]. Likewise, macrophages play a critical role in experimental autoimmune neuritis (EAN), an animal model of peripheral nerve inflammation. In fact, depletion of macrophages in animal models reduces the severity of the disease [88]. In contrast, activation of macrophages by IFN-γ enhances disease severity [89]. Macrophages in the PNS not only participate in tissue destruction but also promote recovery through diverse mechanisms. Macrophages participate in the induction of T-cell apoptosis by secreting mediators such as nitric oxide and TNF-α [90,91]. Macrophages are also able to secrete anti-inflammatory cytokines such as IL-10 and TGF-β [92,93], provoking a time-limited inflammatory response with a pattern of cytokine release similar to that observed after injury-induced inflammatory responses of non-neural tissues, so that the inflammatory response is turned off. An example of the case in point would be how serial determinations of TGF-β1 in GBS patients correlate with disease recovery [94].

Hematogenous macrophages participate actively in myelin clearance through a mechanism involving complement receptor type 3 [95]. Finally, macrophages are also involved in PNS repair. It is a known fact that secretion of growth factors and cytokines induces Schwann cell proliferation and modulation of extracellular matrix components, promoting remyelination and axonal regeneration [96–99].

On the basis of these observations, it seems reasonable to propose that in nervous system damage as well as in lesions in other tissues such as skeletal muscle and skin, activated macrophages and microglial cells are essential for healing at an early stage after injury. Nevertheless, because there are features of macrophage activity that are detrimental to tissue repair, it is also evident that the damage to benefit ratio of macrophage activation or of the transplantation of these immune cells into the nervous system at injury sites remains unclear.

3. THE PROTECTIVE ROLE OF T CELLS

Immunopathological studies of MS lesions and observations in experimental allergic encephalomyelitis (EAE), an animal model for MS, clearly point to a primary role of myelin-specific T cells in the pathogenesis of the disease [100]. However, despite their pathogenic potential, autoreactive T cells against myelin antigens are part of the normal cell repertoire present in healthy subjects [101,102]. Furthermore, a neuroprotective role has been suggested for T cells after CNS injury. In a model of lysolecithin-induced demyelination, T-cell-deficient mice show significantly reduced spontaneous remyelination compared to control animals of matching genetic background, indicating that T cells are necessary for efficient remyelination [103]. Likewise, using different experimental paradigms in rodents, CNS trauma has been shown to elicit a systemic T-cell-mediated neuroprotective response, reducing neuronal loss, which can also be transferred to animals with new injuries through splenocytes activated *ex vivo* with myelin antigens [104,105]. Interestingly, this protective effect is lost in adult animals subjected to thymectomy at birth, and in severe combined immunodeficiency (SCI) mice, which lack mature B and T cells [106]. Furthermore, depletion of $CD4^+CD25^+$ regulatory T cells enhances the spontaneous T-cell-dependent protective response, hence improving postinjury neuronal survival [107]. Likewise, in nude mice replenished with a population of splenocytes lacking $CD4^+CD25^+$ regulatory T cells, significantly more neurons survive after optic nerve injury than in nude mice replenished with a complete splenocyte population [107]. These data support the notion that T-cell-mediated neuroprotection is likely to be a physiological response triggered by CNS injury, directed at self-antigens residing near the damage site. However, such spontaneous response, though beneficial, may ultimately not prove effective enough.

Inflammation also plays an important role in the pathophysiology of ischemic stroke. Polymorphonuclear cells rapidly enter the injured brain tissue, and white blood cells migrate across the BBB. The infarcted zone is infiltrated with polymorphonuclear cells, lymphocytes, and macrophages producing different deleterious metabolites. Following ischemia, endothelial cells, microglial cells, and astrocytes produce IL-1β, TNF-α, and IL-6, which can directly induce neuronal cell death and vessel wall injury [108]. Oral and nasal vaccination with proteins associated with CNS myelin decreases infarct size in animal models of stroke [109,110], and promotes retinal ganglion cell survival after optic nerve crush [111]. Interestingly, adoptive transfer experiments have demonstrated that injury reduction following stroke is mediated by myelin oligodendrocyte glycoprotein (MOG)-induced IL-10-secreting $CD4^+$ T cells [109]. In addition, MOG-treated animals show a dramatic reduction of infiltrating microglia/macrophage cells, which can contribute to secondary infarct expansion after ischemic injury by enhancing nitric oxide synthesis.

Recent research has demonstrated that after partial optic nerve crush injury or spinal cord contusion in rats, injection of syngeneic-activated T cells, specific for CNS antigens such as myelin basic protein (MBP), can significantly reduce secondary neuron degeneration [112,113]. The effect is long lasting and manifests itself both morphologically and functionally. T cells specific for a different self-antigen such as heat shock protein 60, or to a foreign antigen (e.g., ovoalbumin), fail to protect neurons against secondary degeneration, despite accumulation at the lesion site. In an attempt to both reduce the risk of autoimmune diseases such as EAE and to retain the benefit of neuroprotection, animals were vaccinated with T cells specific for MBP-altered peptide ligand or against cryptic MBP epitopes. These weakly pathogenic T cell lines were as effective in preventing secondary degeneration as the original anti-MBP T cell lines [113,114]. The use of T cells specific for certain self-antigen epitopes would therefore seem feasible, in order to induce a beneficial immune response without simultaneously triggering an autoimmune disease process. Neuroprotection was also observed when rats with spinal cord contusion were injected locally or systemically with dendritic cells (DCs) pulsed with MBP-derived or other related peptides [115]. Interestingly, no symptoms of EAE were observed in animals vaccinated with these cells. Moreover, neuroprotection was not observed in the absence of mature T cells either, suggesting that DC-mediated protection is achieved via the induction of a T-cell-dependent immune response, which is antigen-specific.

These findings seem to imply that T-cell response against specific self-antigens might not necessarily be detrimental, and could, under certain circumstances, play a neuroprotective role following nervous system injuries.

4. ANTIBODIES AS MEDIATORS OF CENTRAL NERVOUS SYSTEM REPAIR

It has often been suggested that autoreactive antibodies (Abs) may prove pathogenic in demyelinating diseases like MS [116]. However, several studies have shown that CNS-specific Abs can also enhance tissue repair following injury. In the Theiler murine encephalomyelitis virus (TMEV) model of MS, immunization with spinal cord homogenates (SCH) induced substantial CNS remyelination when compared with control animals [117]. Interestingly, passive transfer of antiserum or immunoglobulins from uninfected animals immunized with SCH was also effective in promoting remyelination. These observations led to the isolation and characterization of two murine monoclonal Abs against SCH (SCH79.08 and SCH94.03), which enhance remyelination in the TMEV model, demonstrating their beneficial role in humoral immune responses against SCH [118,119]. Other monoclonal Abs, several of which are routine markers for the OGD lineage (O1, O4, A2B5, and NHK-1), also promote remyelination in the same animal model [119]. Although these Abs display diverse binding specificity, all bind to antigens expressed on the OGD cell surface at distinct stages of differentiation, suggesting that enhanced remyelination induced by these Abs involves direct myelin-producing cell stimulation, regardless of the developmental stage. Interestingly, although they all belong to the IgMκ subclass, there is no common pattern in germ line Ig gene usage [63,119]. Like murine Abs, systemic administration of serum-derived human IgM monoclonal Abs directed against OGD surface molecules or polyclonal human IgM results in significant remyelination both in the TMEV and in the lysolecithin demyelination models [120,121]. These particular Abs are naturally occurring polyreactive auto-Abs present in the serum of healthy humans and rodents. Recently, *in vivo* studies using magnetic resonance imaging and biotin-streptavidin conjugated with ultrasmall superparamagnetic iron oxide (USPIO) as contrast material have shown that these Abs enter the spinal cord of animals with demyelinating lesions, binding directly to them, a finding consistent with the hypothesis that they work directly on CNS glia to induce remyelination [122]. In contrast, after treatment with polyclonal human IgG, enhancement of remyelination was far less effective in either model, suggesting that this group of immunoglobulins is unable to influence OGD behavior.

Interestingly, sHIgM12 and sHIgM42, two human IgM Abs isolated from individuals with gammopathy, bound to the surface of neurons supporting neurite outgrowth in mixed neuron/glia aggregates, to a degree equal to that induced by laminin [123]. Although the antigen target of these Abs remains to be identified, treatment of granule cells with inhibitors of sphingolipid synthesis and scialic acid cleavage enzymes suggests that mAb-binding sites are carbohydrates associated with lipids, probably neuronal gangliosides [123]. These Abs could potentially play an important role in neuronal survival and axonal repair following nervous system injuries.

Different mechanisms through which myelin-reactive Abs facilitate remyelination have been proposed. First, studies in mixed primary glial cultures demonstrate that OGD-specific Abs induce changes in cytoskeletal structures [124,125], preceded by Ab-induced calcium influx in both astrocytes and OGDs [126,127]. The ability of Abs to induce a transient increase of calcium influx in OGD in culture correlates with their ability to enhance myelin repair *in vivo*, suggesting that the former may play an important role in OGD structure and function regulation [127]. Nevertheless, a definitive link between both phenomena remains to be established. Second, *in vitro* experiments have shown that myelin-reactive Abs and human immunoglobulins may stimulate remyelination by enhancing opsonization and clearance of myelin debris by phagocytic cells, a function mediated by the Fc or constant region of the Abs [128,129]. However, investigations in mice chronically infected with TMEV have demonstrated that different Abs show different responses. While some Ab fragments $F(ab')_2$ retain biological function in *in vivo* remyelination assays, suggesting that myelin repair is independent of Fc function, in other Abs an intact pentameric IgM structure is required for *in vivo* remyelination [130]. Third, remyelination-promoting Abs may exhibit immunomodulatory effects. In this respect, experiments in mouse models of MS have shown that the murine Ab SCH94.03 significantly influences acquired immunity, reducing the number of T cells infiltrating the CNS of SJL/J mice infected with TMEV and suppressing the humoral immune response. Moreover, treatment with SCH94.03 mAb was beneficial in an adaptive transfer model of EAE [131]. In contrast, the human counterpart rHIgM22 mAb operates through mechanisms independent of immunomodulation, most probably by direct action on resident CNS cells [132]. Fourth, treatment with antimyelin Abs is able to rescue premyelinating OGDs from *in vitro* apoptosis induced by either hydrogen peroxide or TNF-α. Interestingly, this rescue is dependent on Ca^{2+} influx through 6-cyano-7nitroquinoxaline-2-3-dione (CNQX)-sensitive channels. In mice chronically infected with TMEV, rescue correlated with the significant reduction of caspases gene expression and caspase-3 activity in the spinal cord and brainstem of Ab-treated animals, suggesting that these Abs may promote remyelination via a protective effect on glial cells [133]. Likewise, recent studies have shown that anti-MAG Abs protect OGDs from glutamate-mediated oxidative stress *in vitro* [134]. Lastly, myelin-reactive Abs may block myelin-associated inhibitors of axon outgrowth, such as MAG and Nogo, allowing axonal regeneration and functional recovery as demonstrated in animal models of spinal cord transsection, stroke, and demyelination [134–136].

Overall, these observations indicate that Abs synthesized following CNS injuries may participate in repair, suggesting that such Abs could be administered exogenously as therapeutic molecules.

5. THE DUAL ROLE OF PROINFLAMMATORY CYTOKINES

The role of soluble mediators such as cytokines, growth factors, neuropeptides, and hormones in the pathogenesis of several disorders of the nervous system has now been demonstrated. Access of cytokines produced in the periphery to the CNS occurs through different mechanisms. First, the BBB can increase its permeability during inflammatory responses. Second, cytokine traffic can occur through regions of the brain where the classical BBB structure is replaced by a blood–CSF barrier, which is more permeable. Finally, cytokines that cross the BBB can use saturable transport systems [137]. Cytokines present in the nervous system originate not only from the immune system but also from resident cells. Results of *in vitro* and *in vivo* studies indicate that both astrocytes and microglial cells are a source for various cytokines, which can in turn cause amplification or suppression of immune responses [138]. Glial cells are also targets for many cytokines in the CNS, and can in turn release neuroactive substances in response to cytokine stimulation. Final outcome regarding the response to CNS injury will ultimately depend on both the profile of cytokines secreted and the particular substances released by glial cells in response to those cytokines. Interestingly, cytokines well known for their promotion of inflammatory responses can also have immunosuppressive functions, assisting in recovery or repair processes (Table 1).

Table 1. Neuroprotective effects of proinflammatory cytokines

Cytokine	Model	Proposed mechanisms
TNF-α	Animal models of demyelination, ischemia, and traumatic brain injury in TNF-α, and TNF-α receptors KO mice	Preferential TNFR2 (p75)activation
	Glutamate-induced cell death in neuronal cultures	Up regulation of Bcl-2 and Bcl-XL through PI3K-dependent NF-κβ activation
	MPTP-induced neurotoxicity	Not described
	Animal models of WD after PNS injury	Induction of NGF production by local fibroblasts
IL-1	Demyelination induced by cuprizone in IL-1-deficient mice	Induction of IGF-1 production by microglia cells and astrocytes
	Traumatic brain injury in rodent models	Induction of CNTF secretion
	AMPA-mediated neuronal death in organotypic hippocampal cultures	Not described
	Cultured explants of sciatic nerve	NGF synthesis by activated macrophages
	Animal models of WD after PNS injury	Induction of NGF production by local fibroblasts. Co-mitogen effect in neonatal rat Schwann cells cultures
IFN-γ	EAE animals treated with mAbs against IFN-γ	Increase in the number of apoptotic CNS-infiltrating cells
	Abrogation of IFN-γ and IFN-γ receptor genes in EAE models	Induction of SOCS-1?
	Inhibition of EAE by intrathecal IFN-γ administration	Increase in the number of apoptotic CNS-infiltrating cells
IL-6	Stroke animal models	Paracrine neurotrophic effect
	Primary neuronal cultures exposed to glutamate-induced excitotoxicity	Endogenous neuroprotective effect of IL-6
	Animal models of WD after PNS injuries	Promotes axonal growth
M-CSF	Organotypic hippocampal cultures exposed to NMDA	Inhibition of Caspase-3. Over-expression of M-CSFR on microglia cells and neurons
	Ischemic cerebral cortex injury	Upregulation of M-CSFR

Tumor Necrosis Factor-α, one of the central mediators of tissue injury and inflammation, has been implicated in the pathogenesis of many neurological conditions, including MS, GBS, CIDP, AD, AIDS dementia, cerebral ischemia, and brain trauma. Surprisingly, however, systemic blockade of TNF-α in MS patients using a TNF-α receptor fusion protein, or a monoclonal Ab against TNF-α, led to both immune activation and increased disease activity [139–142]. These opposing effects of TNF-α may be partly explained by the existence of two distinct signaling pathways mediated through TNF-α receptor 1 (TNFR1; p55) and TNFR2 (p75), respectively [143]. This further clarifies the unexpected protective role recently proposed for TNF-α in animal models of demyelination, retinal ischemia, and traumatic brain injury [143–145]. Likewise, experiments using cuprizone-treated mice (an animal model for demyelination) and lacking both TNF-α and its associated receptor showed that TNF-α promotes remyelination and OGD regeneration through TNFR2, whereas TNFR1 is involved primarily in initiating demyelination [143,146]. Furthermore, during the course of MOG-induced EAE in the absence of TNF-α, myelin-specific T-cell reactivity fails to regress, and expansion of activated T cells is abnormally prolonged, resulting in the exacerbation of autoimmune demyelination [147]. These data again suggest a dual role for TNF-α: clearly proinflammatory during the acute phase but immunosuppressive during the chronic phase of the disease.

Models of glutamate-induced cell death in primary cortical neurons provide new evidence for the antagonistic function of TNF-α receptors with respect to neuronal survival during exogenous stress signals. Both TNFR1 and TNFR2 induce the NF-κβ pathway, yet with distinguishable kinetics and upstream activating components [148]. Tumor necrosis factor-α receptor 1 induces rapid but transient NF-κβ response, whereas TNFR2 generates a long-term PI3K-dependent NF-κβ activation, which in turn mediates upregulation of the antiapoptotic proteins Bcl-2 and $Bcl\text{-}X_L$ [149]. In an animal model of stroke, TNF-α receptor-deficient mice show enhanced sensitivity to ischemic brain damage compared to wild-type animals [150]. Likewise, after experimental brain injury, TNF-α-deficient mice exhibit significantly reduced deficits in both motor and memory functions compared to brain-injured wild-type mice, though the latter recover faster. Also, histopathology reveals significantly greater cortical tissue loss during the chronic postdamage period in injured TNF-α-deficient mice than in injured wild-type mice [144]. This suggest that in this model, despite a deleterious role during the acute response, TNF-α expression may at a later phase actually enhance repair and recovery in the injured brain. Similar neuroprotective effects have been reported in other *in vitro* models such as amyloid β-peptide toxicity in rat hippocampal cultures [151], and embryonic rat forebrain neurons exposed to glucose deprivation and excitatory amino acid toxicity [152], once again supporting a neuroprotective role for TNF-α in different pathologies.

Like TNF-α, IL-1β is a proinflammatory cytokine associated with a wide spectrum of CNS injuries. This cytokine primarily produced by microglia and macrophages induces IL-6, TNF-α, and nitric oxide production. Contrary to the notion that IL-1β would exacerbate demyelinating diseases, recent reports demonstrate that IL-1β promotes remyelination in the adult CNS. IL-1β-deficient mice, for example, fail to remyelinate properly, apparently due to the correlation with a lack of IGF-1 production by microglia–macrophages and astrocytes, with consequent delay of precursors to differentiate into mature OGDs [153]. Likewise, traumatic CNS injuries in rodents result in rapid IL-1β elevation, followed by upregulation of CNTF [154] and NGF [155]. Both growth factors are of particular interest because they represent survival factors for various neuronal populations. In addition, CNTF has been shown to be an important OGD maturation factor and to protect them from apoptotic death induced by different agents. Increased CNTF and NGF expression in the CNS following injury may therefore be crucial for attenuating neuronal and OGD death. Along the same line, high concentrations of IL-1β induce neuroprotective effects on α-amino-3hydroxy-5methyl-4-isoxazol propionic acid (AMPA)-induced cell death in mouse hippocampal slide cultures when present for a limited period of time before AMPA incubation. This neuroprotection may be attributed to IL-1β-dependent transcription of neurotrophic factors [156]. Similarly, using cultured explants of sciatic nerve, lesion-mediated increase of NGF in the PNS has been shown to be regulated by IL-1β production in activated macrophages [157]. Taken together, these findings provide evidence for the need of IL-1β in the production of different trophic factors following nervous system injury, suggesting that inflammation may be beneficial for recovery.

Extensive evidence incriminates IFN-γ as an autoimmune disease mediator. Surprisingly, however, endogenous and exogenous IFN-γ can also at the same time serve as a protective factor in animal models of autoimmunity. There are experiments that show a significant increase in both EAE morbidity and mortality rates in mice treated with neutralizing mAb against IFN-γ [158,159]. However, other experiments indicate that abrogation of IFN-γ expression by targeted disruption of the IFN-γ gene converts an otherwise EAE-resistant mouse strain into a susceptible phenotype [51]. Furthermore, experiments in mice lacking the gene coding for the ligand-binding chain of the IFN-γ receptor demonstrate that IFN-γ is not essential for the generation or the function of anti-$MOG_{35\text{–}55}$ effector cells. However, this gene does play an important role in downregulating EAE at both the effector and the induction phase of the disease [160]. Finally, intrathecal delivery of IFN-γ inhibits chronic progressive EAE by significantly reducing the neuropathological signs of the disease, namely demyelination and axonal loss [161]. Interestingly, EAE recovery was associated with a significant increase in apoptotic cell number and TNFR1 expression in CNS-infiltrating lymphocytes, suggesting that timing, duration, and location of IFN-γ exposure may cause either tissue damage or preservation, leading to faster clearance of encephalitogenic T cells via this apoptotic pathway [161]. Alternatively, a major immunosuppressive action of IFN-γ is associated with its ability to induce the expression of members of the cytokine-signaling suppressor (SOCS) family, which in turn function as classic feedback inhibitors of cytokine signal transduction pathways. SOCS1 is of particular importance in attenuating IFN-γ effects. SOCS1 gene knockout mice have elevated IFN-γ-inducible gene expression, are hypersensitive to IFN-γ, and have elevated levels of this cytokine [162,163]. Thus, SOCS1 seems to be a key modulator of IFN-γ activity, enabling the protective effect of this cytokine to occur. Although many experiments examining the function of these negative regulatory proteins have been carried out *in vitro*, some of the most informative experiments have been those conducted in animals with mutations engineered in genes encoding for these same molecules. Nevertheless, the clinical relevance of the protective role of IFN-γ observed in murine autoimmune disease models remains to be determined, particularly in EAE, a disease in which the protective effect of IFN-γ is in sharp contrast with that observed in MS, where IFN-γ administration significantly worsens patient condition [164].

Like IL-1 and TNF-α, IL-6 is a key mediator of host–defense responses such as fever, immune activation, endocrine, metabolic, and cardiovascular changes. There are numerous studies indicating that IL-6 may have a deleterious effect on neuronal survival [165]. However, recent *in vitro* studies have shown that IL-6 confers a concentration-dependent neuroprotective effect against excitotoxicity challenge with N-methyl-D aspartic acid (NMDA). IL-6-induced neuroprotection requires activation of the IL-1 receptor and is probably mediated through enhancing NGF activity [166]. Furthermore, in animal models for stroke, both exogenous and endogenous IL-6 expressed in neurons reduce ischemic damage [167,168]. Nevertheless, studies in IL-6 KO mice have provided conflicting data. One study reports a reduced inflammatory response and increased neuronal death after cryoinjury [169], whereas another found no difference between IL-6 KO and wild-type animals in their response to ischemic injury [170]. The levels of IL-6 as well as response duration and the presence of other cytokines in the cerebral environment may represent additional critical factors in the tipping of the balance toward immune system protection or cell death induction.

Macrophage colony-stimulating factor (M-CSF) is a hematopoietic cytokine regulating survival, proliferation, differentiation, and function of mononuclear phagocytes with an important role in innate immunity and in a variety of inflammatory diseases. Macrophage colony-stimulating factor is also expressed in the brain by neurons and glial cells, inducing microglial proliferation and activation, resulting in the expression of a variety of inflammatory molecules [171]. However, it has also been shown that this cytokine can have neuroprotective effects both in animal models and cell cultures. Macrophage colony-stimulating factor applied directly to organotypic hippocampal cultures protects neurons from NMDA-induced apoptosis by inhibiting caspase-3 activation [172]. This protective effect is linked to M-CSF receptor (M-CSFR) overexpression on microglial cells and neurons [173,174]. Interestingly, microglia with increased M-CSFR expression are found surrounding plaques in AD, in mouse models of AD, and after ischemic or traumatic brain injury. Cell response to injury in animals deficient in M-CSF has also been studied. Systemic lack of M-CSF results in a significant increase in neuron vulnerability to ischemic injury, together with marked reduction in microglial activation [175]. Exogenous administration of M-CSF partially restores microglial response to neuron injury and significantly increases neuronal survival in ischemic cortical brain lesions [176], a phenomenon correlating with M-CSFR upregulation in neurons [177].

PNS injury is followed by WD, important for eventual regeneration. Recent observations in animal models indicate that the same cytokines causing detrimental effects during autoimmune demyelination are also expressed in the distal stump of peripheral nerves during WD. Following peripheral nerve axotomy in mice, *in vitro* and *in vivo* experiments show three temporal patterns of cytokine gene induction [31,178]. Injury induces rapid upregulation of IL-1α/β and TNF-α mRNA expression, as well as of macrophage and Schwann cell protein synthesis and secretion, which in turn establishes a cytokine network [179–181]. Both cytokines indirectly regulate PNS growth and survival through NGF production by PNS-resident fibroblasts [157,182], acting in a synergistic manner [183]. Indeed, IL-1α has been seen to act as a comitogen in neonatal rat Schwann cells cultures [184], and IL-1 neutralization using receptor antagonists blocks sciatic nerve regeneration after traumatic injury *in vivo* [185]. This is followed by IL-6 and GM-CSF secretion by fibroblasts and Schwann cells [99,186]. This promotes axonal growth and may be responsible for the reduction of TNF-α production after the first day of WD. Finally, TNF-α and IL-1α/β induce IL-10 production by recruited macrophages, downregulating the production of both inflammatory cytokines and of IL-10 itself, thereby limiting WD [187].

Overall, these data indicate that certain cytokines have the capacity to generate multiple and opposite effects, namely detrimental, protective, and regenerative.

6. SECRETION OF NEUROTROPHIC FACTORS BY INFLAMMATORY CELLS

Three families of proteins supporting proliferation, differentiation, and survival of neurons and glial cells have been described [188]. First, the neurotrophin (NT) family, whose members include NGF, BDNF, NT 3 and NT 4/5, NT 6, and NT 7 [189,190]. The biological activity of these molecules is mediated through interactions with two classes of cell-surface receptors: the low-affinity p75 receptor ($p75^{NTR}$), which is common to all NT, and high-affinity receptors specific for each molecule and which are the members of the tyrosine kinase receptor family (Trk A, Trk B, and Trk C). Second, the family of proteins with neurotrophic activity includes GDNF and three structurally and functionally related proteins: neurturin, artemin, and persephin. All members utilize the GFL/GFR-α receptor complex [191]. Third, the neuropoietic cytokine family includes CNTF and leukemic inhibitory factor (LIF) [192,193]. All members of this group possess two multimeric receptor complexes that share the signal-transducing gp130 receptor as a common subunit. In addition, there are other neurotrophic factors that do not belong to any of these families, for example, IGF-1.

Aside from their effects on the nervous system, neurotrophic factors also play an active role in the immune system. Recently, several laboratories have demonstrated mRNA expression as well as diverse protein secretion of neurotrophic factors by both human and murine immune cells, both *in vitro* and *in vivo* (Table 2) [194–199]. Interestingly, antigen activation significantly increases NT secretion by T and B lymphocytes [197,200]. Furthermore, the presence of Trks and $p75^{NTR}$ on T cells, B cells, and monocytes has also been proven (Table 2).

Increased NT production has been shown in inflammatory infiltrates during EAE: NGF was expressed mainly by macrophages, while high levels of BDNF, NT-3, and GDNF mRNAs were

Table 2. Production of neurotrophic factors and expression of neurotrophic receptors on immune cells

Cells	Neurotrophins	Neurotrophin receptors
Monocytes/macrophages	BDNF, NGF, NT-3	TrkA, TrkB, TrkC, $p75^{NTR}$
Eosinophils	BDNF, NGF	TrkA, TrkB, TrkC, $p75^{NTR}$
Basophils	NGF, NT-3	TrkA
Neutrophils	NGF	
NK cells	BDNF, NGF, NT-4	
B cells	BDNF, NGF, NT-4	TrkA, TrkB, $p75^{NTR}$
T cells	BDNF, NGF, NT-3, NT-4/5, LIF	TrkA, TrkB, TrkC, $p75^{NTR}$

present in T and NK infiltrating cells [201,202]. Brain-derived growth factor immunoreactivity has also been demonstrated in infiltrating immune cells, especially T cells and macrophages, in the brain of patients with acute disseminated encephalomyelitis, and at the actively demyelinating edge of MS plaques at early stages of lesion development [203]. In contrast, in chronic inactive lesions, only very few BDNF+ cells are observed. Also, robust immunoreactivity for Trk B receptor was found in neurons adjacent to BDNF+ inflammatory cells. The finding that certain populations of mature leukocyte populations express Trk A, Trk B, and Trk C receptors, and that $p75^{NTR}$ is expressed in macrophages and microglial cells, particularly in MS lesions, suggests that NT may also exert major autocrine immunomodulatory effects [195]. Essential immunological functions such as B-cell proliferation, immunoglobulin synthesis, antigen presentation by macrophages, and expression of costimulatory molecules are all influenced by NGF [204–211].

Data from the marmoset EAE model provides evidence that after intracranial administration of NGF, CNS-infiltrating mononuclear cells express lower amounts of IFN-γ and higher amounts of IL-10, indicating that CNS-specific immune responses have been switched to the anti-inflammatory Th2 phenotype. Notably, glial cells within the unaffected white matter express higher IL-10 levels even in areas remote from inflammation, suggesting that NGFD induces a CNS-wide immunosuppressive microenvironment [212].

Recent studies have demonstrated that myelin-reactive T cells produce LIF [213]. Interestingly, EAE experiments with double mutant mice with impaired signaling through the LIFR-β/gp130 complex demonstrate enhanced OGD apoptosis, and consequently increased disease severity [214]. Likewise, compared to wild-type animals, CNTF-deficient mice develop more severe clinical deficits, show an earlier onset of disease, and poorer recovery [215]. This pronounced susceptibility to EAE is probably due to increased OGD vulnerability to the inflammatory attack. Thus, LIF and CNTF represent two additional important endogenous protective mediators during the course of experimental neuroinflammation.

Just as in MS, neurotrophic factor secretion by infiltrating immune cells has been described in animal models of ischemic, traumatic, or striatal injury [216–218]. Neurotrophic factors secreted by inflammatory cells may explain the underlying mechanism for CNS neuron protection from the potentially noxious effects of proinflammatory cytokines. Additionally, they may provide trophic support to neurons and OGD in a variety of pathological conditions, through stimulatory mechanisms such as axonal regeneration and remyelination [219–221].

As previously described, studies have shown that passive transfer of autoimmune T cells, specific for myelin proteins, can protect injured neurons in rat CNS from secondary degeneration. Systemic injection of activated T cells causes accumulation of macrophages/microglia as well as B cells in the injured nerves. This is accompanied by the accumulation of large amounts of NGF, BDNF, and NT 3, implying that beneficial injured nerve protection by autoimmune T cells may be mediated at least in part by neurotrophic factors [222]. Supporting this notion, local application of tyrosine kinase-receptor inhibitors partially prevents this neuroprotective effect [223]. Likewise, survival of mechanically injured spinal motoneurons is significantly increased after active immunization with an encephalitogenic MBP peptide. In this animal model, protection was also associated with high levels of NT 3, BDNF, and GDNF, expressed by T and NK cells recruited to the CNS [218].

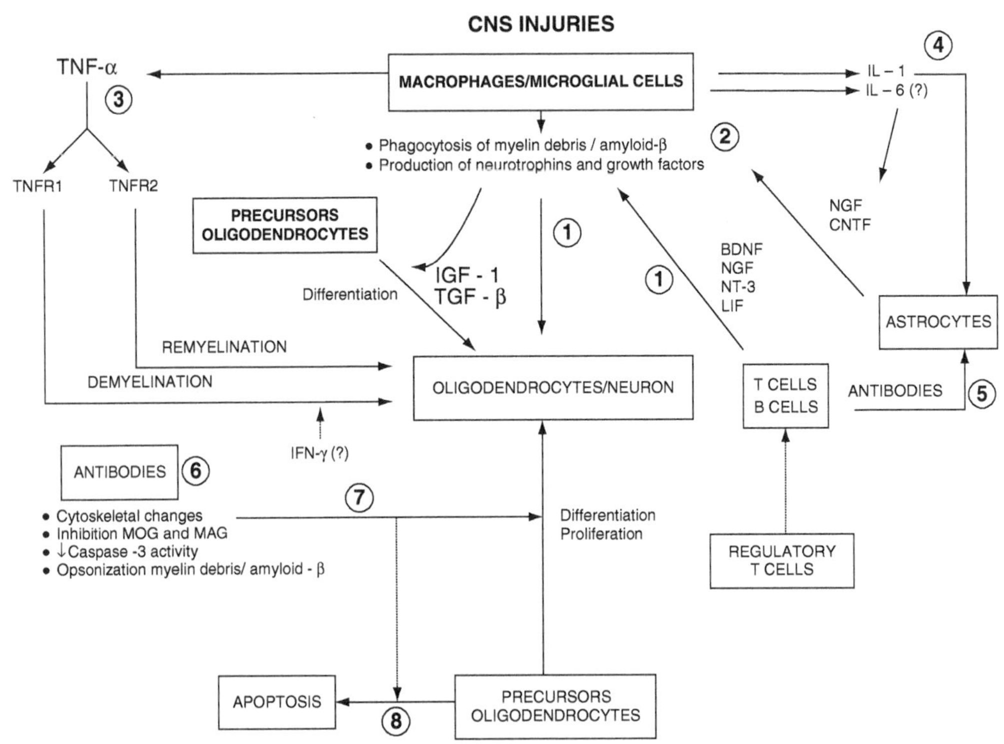

Figure 1. An overview of the immune mechanisms involved in neuroprotection after CNS injuries. Following an injury or a disease, T cells, B cells, and macrophages produce neurotrophins and growth factors that potentially aid injured neurons and OGDs, and thereby contribute to functional repair in the CNS (1). In addition, macrophages are able to remove myelin debris helping remyelinating cells to engage demyelinated axons, as well as fostering maturation of OGD progenitors and axonal regrowth (2). Reactive macrophages and microglial cells also produce cytokines such as TNF-α and IL-1. Once released, these cytokines act on different cell types in the CNS. TNF-α acts through its type 1 receptor (TNFR1) to activate microglial cells further, promoting demyelination. However, TNF-α can also bind to TNFR2 at the surface of OGD progenitors, promoting remyelination and OGD regeneration (3). IL-1β binds to its astrocyte receptor leading to the production of neurotrophins and growth factors (4). Likewise, direct stimulation of astrocytes by remyelination-promoting Abs may induce the release of growth factors (5). It has been proposed that remyelination-promoting Abs bind to receptors on the surface of OGDs and astrocytes, thereby inducing Ca^{2+} signals and subsequent cytoskeletal changes. Moreover, Abs aid in the opsonization of myelin debris, and inhibit caspase activity as well as myelin-associated growth inhibitors such as Nogo and MAG (6). In this fashion, Abs may induce proliferation and/or differentiation of progenitor cells (7), or rescue OGD progenitors from cell death (8). Dashed lines indicate inhibitory pathways.

In conclusion, several different neurotrophic factors produced by activated immune cells may constitute a major mechanism for neuronal protection, either by directly binding to their receptor on neurons or by indirectly modulating the local immune response.

7. CONCLUDING REMARKS

There is substantial evidence that immune-mediated reactions can be harmful to neurons, namely Schwann cells and OGDs, whereas other observations indicate that inflammation exerts beneficial effects by promoting repair and reducing the spread of damage (Figs 1 and 2). However, mechanisms underlying the manifold features of inflammation and their regulation during the course of a wide variety of neurological diseases still remain to be elucidated.

The fact that under certain circumstances inflammation may be beneficial in nervous system maintenance could explain some of the results of failed trials with anti-inflammatory drugs used

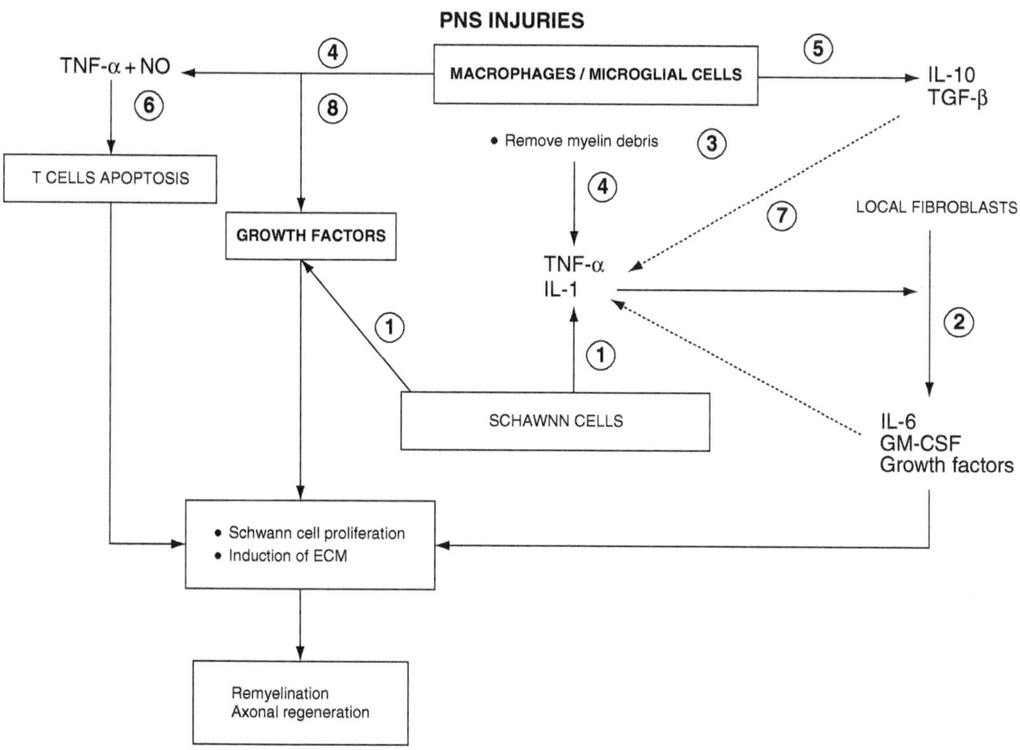

Figure 2. An overview of immune mechanisms involved in neuroprotection after PNS injuries. PNS injuries induce growth factor production, TNF-α, and IL-1 by Schwann cells (1). Thereafter, IL-6, GM-CSF, and growth factors are produced by resident fibroblasts (2). Recruited macrophages are able to remove myelin debris (3) and to produce proinflammatory cytokines such as TNF-α and IL-1 (4), as well as anti-inflammatory cytokines such as IL-10 and TGF-β (5). Macrophages contribute to nervous tissue recovery by promoting T-cell apoptosis (6), inhibiting the inflammatory response through IL-10 and TGF-β (7) and different growth factor secretion (8). ECM, extracellular matrix. Dashed lines indicate inhibitory pathways.

during the course of neurological diseases. It is crucial, therefore, to consider the overall balance of the different mediators of inflammation, because a shift in any given direction in terms of protective versus detrimental effects will have a marked impact on outcome. Sorting out the inflammatory processes that protect versus those that contribute toward neuronal damage is essential in order to design effective new therapies for diseases affecting the nervous system.

ACKNOWLEDGMENTS

The authors are grateful to Adriana Zufriategui for preparation of the figures.

REFERENCES

1. Antel J, Birbbaum G, Hartung HP. Clinical Neuroimmunology. Oxford: Blackwell Science, 1998.
2. Latov N, Wokke JHJ, and Kelly JJ Jr, Eds. Immunology and Infectious Diseases of the Peripheral Nerves. New York: Cambridge University Press, 1998.
3. Antel JP, Becher B. Central nervous system–immune interactions: Contribution to neurologic disease and recovery. In Clinical Neuroimmunology. Antel J, Birbbaum G, and Hartung HP, Eds; Oxford: Blackwell Science, 1998; pp. 26–39.
4. Pachter JS, de Vries HE, Fabry Z. The blood–brain barrier and its role in immune privilege in the central nervous system. J Neuropathol Exp Neurol 2003;62:593–604.
5. Becher B, Antel JP. Comparison of phenotypic and functional properties of immediately *ex vivo* and cultured human adult microglia. Glia 1996;18:1–10.
6. Williams K, Ulvestad E, Antel JP. B7/BB-1 antigen expression on adult microglia studied in vitro and in situ. Eur J Immunol 1994;24:3031–7.
7. Windhangen A, Newcombe J, Dangond F, Strand C, Woodroofe MNB, Cuzner ML, Hafler DA. Expression of costimulatory molecules B7-1 (CD80), B7-2 (CD86), and interleukin 12 cytokine in multiple sclerosis lesions. J Exp Med 1995;182:1985–96.
8. Aloisi F, Ria F, Penna G, Adorini L. Microglia are more efficient than astrocytes in antigen processing and in Th1 but not Th2 cell activation. J Immunol 1998;160:4671–80.
9. Aloisi F, Penna G, Polazzi E, Minghetti L, Adorini L. CD40–CD154 interaction and IFN-gamma are required for IL-12 but not prostaglandin E2 secretion by microglia during antigen presentation to Th1 cells. J Immunol 1999;162:1384–91.
10. Fontana A, Fierz F, Wekerle H. Astrocytes present myelin basic protein to encephalitogenic T-cell lines. Nature 1994;307:–273–6.
11. Massa PT, ter Meulen V, Fontana A. Hyperinducibility of Ia antigen on astrocytes correlates with strain-specific susceptibility to experimental autoimmune encephalomyelitis. Proc Natl Acad Sci USA 1987;84:4219–23.
12. Sedgwick JD, Mossner R, Schwender S, ter Meulen V. Major histocompatibility complex-expressing nonhematopoietic astroglial cells prime only CD8+ T lymphocytes: Astroglial cells as perpetuators but not initiators of CD4+ T cell responses in the central nervous system. J Exp Med 1991;173:1235–46.
13. Matsumoto Y, Ohmori K, Fujiwara M. Immune regulation by brain cells in the central nervous system by microglia but not astrocytes present myelin basic protein to encephalitogenic T cells under in vivo-mimicking conditions. Immunology 1992;76:209–16.

14. Weber F, Meinl E, Aloisi F, Nevinny-Stickel C, Albert E, Wekerle H, Holhfeld R. Human astrocytes are only partially competent antigen presenting cells. Possible implications for lesion development in multiple sclerosis. Brain 2004;117:59–69.
15. Koski CL. Immune interactions in the peripheral nervous system. In Immunology and Infectious Diseases of the Peripheral Nerves. Latov N, Wokke JHJ, and Kelly JJ Jr, Eds; New York: Cambridge University Press, 1998; pp. 1–28.
16. Friedman B, Scherer SS, Rudge JS, Helgren M, Morriscy D, Mclain J, Wang DY, Wiegand SJ, Furth ME, Lindsay RM. Regulation of ciliary neurotrophic factor expression in myelin-related Schwann cells *in vivo*. Neuron 1992;9:295–305.
17. Henderson CE, Phillips HS, Pollock RA, Davies AM, Lemeulle C, Armanini M, Simpson LC, MoVet B, Vandlen RA, Koliatsos VE, Rosenthal A. GDNF: A potent survival factor for motoneurons present in peripheral nerve and muscle. Science 1994;266:1062–4.
18. Yamamoto M, Sobue G, Li M, Arakawa Y, Mitsuma T, Kimata K. Nerve growth factor (NGF), brain derived neurotrophic factor (BDNF) and low-affinity nerve growth factor receptor (LNGFR) mRNA levels in cultures rat Schwann cells; differential time-and-dose–dependent regulation by cAMP. Neurosci Lett 1993;152:37–40.
19. Mancardi GL, Cadoni A, Zicca A, Schenone A, Tabaton A, De Martin A, Zaccheo D. HLA-DR Schwann cell reactivity in peripheral neuropathies of different origins. Neurology 1988;38:848–51.
20. Pollard JD, Baverstocl J, McLeod JG. Class II antigen expression and inflammatory cells in the Guillain-Barre syndrome. Ann Neurol 1987;21:337–41.
21. Armati PJ, Pollard JD, Gatenby P. Rat and human Schwann cells in vitro can synthesize and express MHC molecules. Muscle Nerve 1990;13:106–16.
22. Gold T, Toyka KV, Hartung HP. Synergistic effect of IFN-gamma and TNF-alpha on expression of immune molecules and antigen presentation by Schwann cells. Cell Immunol 1995;165:65–70.
23. Lilje O, Armati PJ. The distribution and abundance of MHC and ICAM-1 on Schwann cells in vitro. J Neuroimmunol 1997;77:75–84.
24. Cowley SA, Butter C, Gschmeissner SE, Curtis J, Turk JL. An immunoelectromicroscopical study of the expression of major histocompatibility complex (MHC) class II antigens in guinea pig sciatic nerves following induction of intraneural mycobacterial granulomas. J Neuroimmunol 1989;23:223–31.
25. Griffin JW, Stoll G, Li CY, Tyor W, Cornblath DR. Macrophage responses in demyelinating neuropathies. Ann Neurol 1990;27:S64–8.
26. Schmidt B, Stoll G, Hartung HP, Heininger K, Schafer B, Toyka KV. Macrophages but not Schwann cells express Ia antigen in experimental allergic neuritis. Ann Neurol 1990;28:70–7.
27. Chaudhry V, Glass JD, Griffin JW. Wallerian degeneration in peripheral nerve disease. Neurol Clin 1992;10:613–27.
28. Brück W, Huitinga I, Dijkstra CD. Liposome-mediated monocyte depletion during Wallerian degeneration defines the role of hematogenous phagocytes in myelin removal. J Neurosci Res 1996;46:477–84.
29. Creange A, Barlovatz-Meimon G, Gherardi RK. Cytokines and peripheral nerve disorders. Eur Cytokine Netw 1997;8:145–51.
30. Fernández-Valle C, Bunge RP, Bunge MB. Schwann cells degrade myelin and proliferate in the absence of macrophages: evidence from in vitro studies of Wallerian degeneration. J Neurocytol 1995;24:667–79.

31. Shamash S, Reichert F, Rotshenker S. The cytokine network of Wallerian degeneration: Tumor necrosis factor-α, and interleukin-1β. J Neurosci 2002;22:3052–60.
32. Allan SM, Rothwell NJ. Cytokines and acute neurodegeneration. Nat Rev Neurosci 2001;2:734–44.
33. Nguyen MD, Rivest S. Innate immunity: The missing link in neuroprotection and neurodegeneration? Nat Rev Neurosci 2002;3:216–27.
34. Profyris C, Cheema SS, Zang DW, Azari MF, Boyle K, Petratos S. Degenerative and regenerative mechanisms governing spinal cord injury. Neurobiol Dis 2004;15:415–36.
35. Iadecola C, Niwa K, Nogawa S, Zhao X, Nagayama M, Araki E, Morham S, Ross ME. Reduced susceptibility to ischemic brain injury and *N*-methyl-D-aspartate-mediated neurotoxicity in cyclooxygenase-2-deficient mice. Proc Natl Acad Sci 2001;98:1294–9.
36. Soriano SG, Lipton SA, Wang YF, Xiao M, Springer TA, Gutierrez-Ramos JC, Hickey PR. Intercellular adhesion molecules-1-deficient mice are less susceptible to cerebral ischemia-reperfusion injury. Ann Neurol 1996;39:618–24.
37. Boutin H. LeFeuvre RA, Horai R, Asano M, Iwakura Y, Rothwell NJ. Role of IL-1 alpha and IL-1 beta in ischemic brain damage. J Neurosci 2001;21:5528–34.
38. Passinetti GM. Cyclooxygenase and inflammation in Alzheimer's disease: Experimental approaches and clinical interventions. J Neurosci Res 1998;54:1–6.
39. Gonzalez-Scarano F, Baltuch G. Microglia as mediators of inflammatory and degenerative diseases. Annu Rev Neurosci 1999;22:219–40.
40. Julien JP. Amyotrophic lateral sclerosis: Unfolding the toxicity of the misfolded. Cell 2001;104:581–91.
41. David S, Aguayo AJ. Axonal elongation into peripheral nervous system "bridges" after central nervous system injury in adult rats. Science 1981;214:931–3.
42. David S, Aguayo AJ. Axonal regeneration after crush injury in rat central nervous system fibers innervating peripheral nerve grafts. J Neurocytol 1985;14:1–12.
43. David S, Bouchard C, Tsatas O, Giftochristos N. Macrophages can modify the nonpermissive nature of the adult mammalian central nervous system. Neuron 1990;5:463–9.
44. Iseda T, Nishio T, Kawaguchi S, Yamanoto M, Kawasaki T, Wakisaka S. Spontaneous regeneration of the corticospinal tract after transection in young rats: A key role of reactive astrocytes in making favourable and unfavourable conditions for regeneration. Neuroscience 2004;126:365–74.
45. Kawaja MD, Gage FH. Reactive astrocytes are substrate for the growth of adult CNS axons in the presence of elevated levels of nerve growth factors. Neuron 1991;7:1019–30.
46. George R, Griffin JW. Delayed macrophage responses and myelin clearance during Wallerian degeneration in the central nervous system: The dorsal radiculotomy model. Exp Neurol 1994;129:225–36.
47. Perry VH, Brown MC, Gordon S The macrophage response to central and peripheral injury: A possible role for macrophages in regeneration. J Exp Med 1987;165:1218–23.
48. Stoll G, Griffin JW, Li CY, Trapp BD. Wallerian degeneration in the peripheral nervous system: Participation of both Schwann cells and macrophages in myelin degradation. J Neurocytol 1989;18:671–83.
49. Chen MS, Huber AB, van der Haar ME, Frank M, Scnell L, Spillman AA, Christ F, Schwab ME. Nogo-A is a myelin-associated neurite outgrowth inhibitor and an antigen for monoclonal antibody IN-1. Nature 2000;403:434–9.
50. Mukhopadhyay G, Doherty P, Walsh FS, Crocker PR, Filbin MT. A novel role for myelin-associated glycoprotein as an inhibitor of axonal regeneration. Neuron 1994;13:757–67.

51. Krakowski M, Owens T. Interferon-gamma confers resistance to experimental allergic encephalomyelitis. Eur J Immunol 1996;26:1641–6.
52. Zeev-Brann AB, Lazarov-Spiegler O, Brenner T, Schwartz M. Differential effects of central and peripheral nerves on macrophages and microglia. Glia 1998;23:181–90.
53. Lazarov-Spiegeler O, Salomon AS, Zeev-Brann AB, Hirschberg DL, Lavie V, Schwartz M. Transplantation of activated macrophages overcomes central nervous system regrowth failure. FASEB J 1996;10:1296–302.
54. Lazarov-Spiegeler O, Salomon AS, Schwartz M. Peripheral nerve-stimulated macrophages simulate a peripheral nerve-like regenerative response in rat transected optic nerve. Glia 1998;24:329–37.
55. Rapalino O, Lazarov-Spiegeler O, Agranov E, Velan GJ, Yoles E, Fraidakis M, Solomon A, Gepstein R, Katz A, Belkin M, Hadani M, Schwartz M. Implantation of stimulated homologous macrophages results in partial recovery of paraplegic rats. Nat Med 1998;4:814–21.
56. Kotter MR, Setzu A, Sim FJ, Van Rooijen N, Franklin RJM. Macrophage depletion impairs oligodendrocyte remyelination following lysolecithin-induced demyelination. Glia 2001;35:204–12.
57. Ousman SS, David S. Lysophosphatidylcholine induces rapid recruitment and activation of macrophages in the adult mouse spinal cord. Glia 2000;30:92–104.
58. Rabchensky AG, Streitt WJ. Grafting of cultured microglia cells into the lesioned spinal cord of adult rats enhances neurite outgrowth. J Neurosci Res 1997;47:34–48.
59. Robinson S, Miller RH. Contact with central nervous system myelin inhibits oligodendrocyte progenitor maturation. Dev Biol 1999;216:359–68.
60. Shields SA, Gilson JM, Blakemore WF, Franklin RJM. Remyelination occurs as extensively but more slowly in old rats compared to young rats following gliotoxin-induced CNS demyelination. Glia 1999;28:77–83.
61. Masson JL, Jones JJ, Taniike M, Morelll P, Suzuki K, Matsushima GK. Mature oligodendrocytes apoptosis precedes IGF-I production and oligodendrocyte progenitor accumulation and differentiation during demyelination/remyelination. J Neurosci Res 2000;61:251–62.
62. Woodruff RH, Franklin RJM. The expression of myelin protein mRNAs during remyelination of lysolecithin-induced demyelination. Neuropathol Appl Neurobiol 1999;25:226–35.
63. Asakura K, Rodriguez M. A unique population of circulating autoantibodies promotes central nervous system remyelination. Mult Scler 1998;4:217–21.
64. Giulian D, Baker TJ, Shih LC, Lachman LB. Interleukin 1 of the central nervous system is produced by amoeboid microglia. J Exp Med 1986;164:594–604.
65. Giulian D, Woodward J, Young DG, Krebs JF, Lachman LB. Interleukin-1 injected into mammalian brain stimulates astrogliosis and neovascularization. J Neurosci 1988;8:2485–90.
66. Aloisi F, Borsellino G, Samoggia P, Testa U, Chelucci C, Russo G, Peschle C, Levi G. Astrocyte cultures from human embryonic brain: Characterization and modulation of surface molecules by inflammatory cytokines. J Neurosci Res 1992;32:494–506.
67. Deller T, Haas CA, Naumann T, Joester A, Faissner A, Frotscher M. Up-regulation of astrocyte-derived tenascin-C correlates with neurite outgrowth in the rat dentate gyrus after unilateral enthorinal cortex lesion. Neuroscience 1997;81:829–46.
68. Poulsen FR, Lagord C, Courty J, Pedersen EB, Barritault D, Finsen B. Increased synthesis of heparin affin regulatory peptide in the perforant path lesioned mouse hippocampal formation. Exp Brain Res 2000;135:319–30.

69. Asada H, Ip NY, Pan L, Razack N, Parfitt MM, Plunkett RJ. Time course of ciliary neurotrophic factor mRNA expression is coincident with the presence of protoplasmic astrocytes in traumatized rat striatum. J Neurosci Res 1995;40:22–30.
70. Hagg T, Varon S. Ciliary neurotrophic factor prevents degeneration of adult rats substantia nigra dopaminergic neurons *in vivo*. Proc Natl Acad Sci 1993;90:6315–9.
71. Ho A, Blum M. Regulation of astroglial-derived dopaminergic neurotrophic factors by interleukin-1beta in striatum of young and middle-aged mice. Exp Neurol 1997;148: 348–59.
72. Haga S, Akai K, Ishii T. Demonstration of microglial cells in and around senile (neuritic) plaques in the Alzheimer brain. An immunohistochemical study using a novel monoclonal antibody. Acta Neuropathol (Berl) 1989;77:569–75.
73. Wegiel J, Imaki H, Wang KC, Rubenstein R. Cells of monocyte/microglial lineage are involved in both microvessel amyloidosis and fibrillar plaque formation in APPsw tg mice. Brain Res 2004;1022:19–29.
74. Bianca VD, Dusi S, Bainchini E, Dal Pra I, Rossi F. β-Amyloid activates the O2 forming NADPH oxidase in microglia, monocytes and neutrophils. A possible mechanism of neuronal damage in Alzheimer's disease. J Biol Chem 1999;274:15493–9.
75. Heneka M, Heneka MT, Loschmann PA, Gleichmann M, Weller M, Schulz JB, Wullner U, Klockgether T. Induction of nitric oxide synthase and nitric-oxide-mediated apoptosis in neuronal PC12 cells after stimulation with tumor necrosis factor-α/lipopolysaccharide. J Neurochem 1998;71:88–94.
76. Siamrd AR, Rivest S. Neuroprotective properties of the innate immune system and bone marrow stem cells in Alzheimer's disease. Mol Psychiatry 2006;11:327–35.
77. Wegiel J, Imaki H, Wang KC, Wronsca A, Osuchowski M, Rubenstein R. Origin and turnover of microglial cells in fibrillar plaques of APPsw transgenic mice. Acta Neuropathol (Berl) 2003;105:393–402.
78. Turrin N, Rivest S. Tumor necrosis factor α but not interleukin 1β mediates neuroprotection in response to acute nitric oxide excitotoxicity. J Neurosci 2006;26:143–51.
79. Weiner HL, Frenkel D. Immunology and immunotherapy of Alzheimer's disease. Nat Rev Immunol 2006;6:404–16.
80. Masliah E, Hansen L, Adame A, Crews L, Bard F, Lee C, Seubert P, Games D, Kirby L, Schenk D. Aβ vaccination effects on plaque pathology in the absence of encephalitis in Alzheimer disease. Neurology 2005;64:129–31.
81. Paresce DM, Ghosh RN, Maxfiled FR. Microglial cells internalize aggregates of the Alzheimer's disease amyloid β-protein via a scavenger receptor. Neuron 1996;17: 553–65.
82. Giulian D, Haverkamp LJ, Yu J, Karshin W, Tom D, Li J, Kazanskaia A, Kirkpatrick J, Roher AE. The HHQK domain of β-amyloid provides a structural basis for the immunopathology of Alzheimer's disease. J Biol Chem 1998;273:29719–26.
83. Xie L, Helmerhorst E, Taddei K, Plewright B, Van Bronswijk W, Martins R. Alzheimer's β-amyloid peptides compete for insulin binding to the insulin receptor. J Neurosci 2002;22:RC221.
84. Qiu WQ, Walsh DM, Ye Z, Vekrellis K, Zhang J, Podlisny MB, Rosner MR, Safavi A, Hersh LB, Selkoe DJ. Insulin-degrading enzyme regulates extracellular levels of amyloid-β protein by degradation. J Biol Chem 1998;273:32730–8.
85. Prineas JW, McLeod JG. Chronic relapsing polyneuritis. J Neurol Sci 1976;27:427–58.
86. Prineas JW. Pathology of Guillain-Barre syndrome. Ann Neurol 1981;9(Suppl):6–19.

87. Griffin JW, Li CY, Ho TW, Tian M, Gao CY, Xue P, Mishu B, Cornblath DR, Macko C, McKhann GM, Asbury AK. Pathology of the motor–sensory axonal Guillain-Barre. Ann Neurol 1996;39:17–28.
88. Jung S, Huitinga I, Schmidt B, Zielasek J, Dijkstra CD, Toyka KV, Hartung HP. Selective elimination of macrophages by dichloromethylene diphosphonate-containing liposomes suppresses experimental autoimmune neuritis. J Neurol Sci 1993;119:195–202.
89. Hartung HP, Schafer B, van der meide PH, Fierz W, Heininger K, Toyka KV. The role of interferon-gamma in the pathogenesis of experimental autoimmune disease of the peripheral nervous system. Ann Neurol 1990;27:247–57.
90. Weishaupt A, Gold R, Hartung H, Gaupp S, Brück W, Toyka KV. Role of TNF-α in high-dose antigen therapy in experimental autoimmune neuritis: Inhibition of TNF-α by neutralizing antibodies reduces T-cell apoptosis and prevents liver necrosis. J Neuropathol Exp Neurol 2000;59:368–76.
91. Zettl UK, Mix E, Zielasek J, Stangel M, Hartung HP, Gold R. Apoptosis of myelin-reactive T cells induced by reactive oxygen and nitrogen intermediates in vitro. Cell Immunol 1997;178:1–8.
92. Bai XF, Zhu J, Zhang GX, Kaponides G, Hojeberg B, van der Meide PH, Link H. IL-10 suppresses experimental autoimmune neuritis and down regulates cytokine mRNA expression of Th1 and macrophage source. Clin Immunol Immunopathol 1997;83:117–26.
93. Kiefer R, Funa K, Schweitzer T, Jung S, Bourde O, Toyka KV, Hartung HP. Transforming growth factor-β1 in experimental autoimmune neuritis: Cellular localization and time course. Am J Pathol 1996;148:211–23.
94. Sindern E, Schweppe K, Ossege LM, Mailin JP. Potential role of transforming growth factor-β1 in terminating the immune response in patients with Guillain-Barré syndrome. J Neurol 1996;243:264–8.
95. Hirata K, Mitoma H, Ueno N, He JW, Kawabuchi M. Differential response of macrophage subpopulations to myelin degradation in the injured rat sciatic nerve. J Neurocytol 1999;28:685–95.
96. Day WA, Kosihi K, McLennan IS. Transforming growth factor beta 1 may regulate the stability of mature myelin sheath. Exp Neurol 2003;184:857–64.
97. Grothe C, Meisinger C, Claus P. In vivo expression and localization of the fibroblast growth factor system in the intact and lesioned rat peripheral nerve and spinal ganglia. J Comp Neurol 2001;434:342–57.
98. Oya T, Zhao YL, Takagawa K, Kawaguchi M, Shirakawa K, Yamauchi T, Sasahara M. Platelet-derived growth factor-b expression induced after rat peripheral nerve injuries. Glia 2002;38:303–12.
99. Reichert F, Levitzky R, Rothshenker S. Interleukin 6 in intact and injured mouse peripheral nerves. Eur J Neurosci 1996;8:530–5.
100. Martin R, McFarland HF, McFarlin DE. Immunological aspects of demyelinating diseases. Annu Rev Immunol 1992;10:153–87.
101. Burns J, Rosenzweig A, Zweiman B, Lisak RP. Isolation of myelin basic protein-reactive T cell lines from normal human blood. Cell Immunol 1983;81:435–40.
102. Correale J, McMillan M, McCarthy K, Le T, Weiner LP. Isolation and characterization of autoreactive proteolipid protein–peptide specific T-cell clones from multiple sclerosis patients. Neurology 1995;45:1370–8.
103. Bieber AJ, Kerr S, Rodriguez M. Efficient central nervous system remyelination requires T cells. Ann Neurol 2003;53:680–4.

104. Yoles E, Hauben E, Palgi O, Agarnov E, Gothilf A, Cohen A, Kuchroo V, Cohen IR, Weiner HL, Schwartz M. Protective autoimmunity is a physiological response to CNS trauma. J Neurosci 2001;21:3740–8.
105. Byram SC, Carson MJ, DeBoy CA, Serpe CJ, Sanders VM, Jones KJ. CD4−positive T cell-mediated neuroprotection requires dual compartment antigen presentation. J Neurosci 2004;24:4333–9.
106. Jones KJ, Serpe CJ, Byram SC, DeBoy CA, Sanders VM. Role of the immune system in the maintenance of mouse facial motoneuron viability after nerve injury. Brain Behav Immun 2005;19:12–9.
107. Kipnis J, Mizrahi T, Hauben E, Shaked I, Shevach E, Schwartz M. Neuroprotective autoimmunity: Naturally occurring CD4+CD25+ regulatory T cells suppress the ability to withstand injury to the central nervous system. Proc Natl Acad Sci 2002;99: 15620–5.
108. Chamorro A, Hallenbeck J. The harms and benefits of inflammatory and immune responses in vascular disease. Stroke 2006;37:291–3.
109. Frenkel D, Huang Z, Maron R, Koldzic DN, Hancock WW, Moskowitz MA, Weiner HL. Nasal vaccination with myelin oligodendrocyte glycoprotein reduces stroke size by inducing IL-10-producing CD4+ T cells. J Immunol 2004;172:6549–55.
110. Becker KJ, McCArron RM, Ruetzler C, Laban O, Sternberg E, Flanders KC, Hallenbeck JM. Immunologic tolerance to myelin basic protein decreases stroke size after transient focal cerebral ischemia. Proc Natl Acad Sci 1997;94:10873–8.
111. Lewitus GM, Kipnis J, Avidan H, Ben-Nun A, Schwartz M. Neuroprotection induced by mucosal tolerance is epitope dependent: Conflicting effects in different strains. J Neuroimmunol 2006;175:31–8.
112. Hauben E, Nevo U, Yoles E, Moaleem G, Agranov E, Mor F, Akselrod S, Neeman M, Cohen IR, Schwartz M. Autoimmune T cells as potential neuroprotective therapy for spinal cord injury. Lancet 2000;355:286–7.
113. Moalem G, Leibowitz-Amit R, Yoles E, Mor F, Cohen IR, Schwartz M. Autoimmune T cells protect neurons from secondary degeneration after central nervous system axotomy. Nature Med 1999;5:49–55.
114. Schwartz M, Hauben E. T-cell based therapeutic vaccination for spinal cord injury. Prog Brain Res 2002;137:401–6.
115. Hauben E, Gothilf A, Cohen A, Butovsky O, Nevo U, Smirnov I, Yoles E, Akselrod S, Schwartz M. Vaccination with dendritic cells pulsed with peptides of myelin basic protein promotes functional recovery from spinal cord injury. J Neurosci 2003;23:8808–19.
116. Correale J, Bassani, Molinas MM. Oligoclonal bands and antibody responses in multiple sclerosis. J Neurol 2002;249:375–89.
117. Rodriguez M, Lennon VA, Benveniste EN, Merril JE. Remyelination by oligodendrocytes stimulated by antiserum to spinal cord. J Neuropathol Exp Neurol 1987;46: 84–95.
118. Asakura K, Miller DJ, Murray P, Bansal R, Pfeiffer SE, Rodriguez M. Monoclonal autoantibody SCH94.03, which promotes central nervous system remyelination, recognizes an antigen on the surface of oligodendrocytes. J Neurosci Res 1996;43: 273– 81.
119. Asakura K, Miller DJ, Pease LR, Rodriguez M. Targeting of IgMκ antibodies to oligodendrocytes promotes CNS remyelination. J Neurosci 1998;18:7700–8.
120. Bieber AJ, Warrington A, Asakura K, Ciric B, Kaveri SV, Pease LR, Rodriguez M. Human antibodies accelerate the rate of remyelination following lysolecithin-induced demyelination in mice. Glia 2002;37:241–9.

121. Warrington AE, Asakura K, Bieber AJ, Ciric B, Van Keulen V, Kaveri SV, Kyle RA, Pease LR, Rodriguez M. Human monoclonal antibodies reactive to oligodendrocytes promote remyelination in a model of multiple sclerosis. Proc Natl Acad Sci 2000;97: 6820–5.
122. Pirko I, Ciric B, Gamez J, Bieber AJ, Warrington AE, Johnson AJ, Hanson DP, Pease LR, Macura SI, Rodriguez M. A human antibody that promotes remyelination enters the CNS and decrease lesion load as detected by T2-weighted spinal cord MRI in a virus-induced murine model of MS. FASEB J 2004;18:1577–9.
123. Warrington AE, Bieber AJ, van Keulen V, Ciric B, Pease LR, Rodriguez M. Neuron-binding human monoclonal antibodies support central nervous system neurite extension. J Neuropathol Exp Neurol 2004;63:461–73.
124. Dyer CA, Benjamins JA. Organization of oligodendrocytes membrane sheets. II. Galactocerebroside: Antibody interactions signal changes in cytoskeleton and myelin basic protein. J Neurosci Res 1998;24:212–21.
125. Dyer CA, Matthieu JM. Antibodies to myelin/oligodendrocyte-specific protein and myelin/oligodendrocyte-specific glycoprotein signal distinct changes in the organization of cultured oligodendroglial membranes. J Neurochem 1994;62: 777–87.
126. Dyer CA, Benjamins JA. Galactocerebroside and sulfatide independently mediate Ca^{2+} responses in oligodendrocytes. J Neurosci Res 1991;30:699–711.
127. Paz Soldan MM, Warrington AE, Bieber AJ, Ciric B, Van Keulen V, Pease LR, Rodríguez M. Remyelination-promoting antibodies activate distinct Ca^{2+} influx pathways in astrocytes and oligodendrocytes: Relationship to the mechanisms of myelin repair. Moll Cell Neurosci 2003;22:14–24.
128. DeJong BA, Smith ME. A role for complement in phagocytosis of myelin. Neurochem Res 1996;22:491–8.
129. Kuhlmann T, Bruck W. Immunoglobulins induced myelin debris clearance by mouse macrophages. Neurosci Lett 1999;19:191–4.
130. Ciric B, Howe CL, Paz Soldan M, Warrington AE, Bieber AJ, van Keulen V, Rodriguez M, Pease LR. Human monoclonal IgM antibody promotes CNS myelin repair independent of Fc function. Brain Pathol 2003;13:608–16.
131. Miller DJ, Bright JJ, Sriram S, Rodriguez M. Successful treatment of established relapsing experimental autoimmune encephalomyelitis in mice with a monoclonal natural antibody. J Neuroimmunol 1997;75:204–9.
132. Ciric B, Van Keulen V, Paz Soldan M, Rodriguez M, Pease LR. Antibody-mediated remyelination operates through mechanism independent of immunomodulation. J Neuroimmunol 2004;146:153–61.
133. Howe CL, Bieber AJ, Warrington AE, Pease LR, Rodriguez M. Antiapoptotic signalling by a remyelination-promoting human antimyelin antibody. Neurobiol Dis 2004;15: 120–131.
134. Irving EA, Vinson M, Rosin C, Roberts JC, Cahpman DM, Facci L, Virley DJ, Skaper SD, Burbidge SA, Walsh FS, Hunter AJ, Parsons AA. Identification of neuroprotective properties of anti-MAG antibody: A novel approach for the treatment of stroke? J Cereb Blood Flow Metab 2005;25:98–107.
135. Karnezis T, Mandemakers W, McQualter JL, Zheng B, Ho PP, Jordan KA, Murray BM, Barres B, Tessier-Levigne M, Bernard CC. The neurite outgrowth inhibitor Nogo A is involved in autoimmune-mediated demyelination. Nat Neurosci 2004;7: 736–44.

136. Mears S, Schachner M, Brushart TM. Antibodies to myelin-associated glycoprotein accelerate preferential motor reinnervation. J Peripher Nerv Syst 2003;8:91–9.
137. Banks WA, Kastin AJ. Blood to brain transport of interleukin links the immune and central nervous systems. Life Sci 1991;48:PL 117–21.
138. Ransohoff RR, Benveniste EN. Cytokines and the CNS. Boca Ratón: CRC Press, 1996.
139. Mohan R, Edwards ET, Cupps TR, Oliverio PJ, Sandberg G, Crayton H, Richert JR, Siegel JN. Demyelination occurring during anti-tumor necrosis factor alpha therapy for inflammatory arthritis. Arthritis Rheum 2001;44:2862–9.
140. Sicotte NL, Voskhul RR. Onset of multiple sclerosis associated with anti-TNF therapy. Neurology 2001;57:1885–8.
141. The Lenercep Group. TNF neutralization in MS: Results of a randomized placebo-controlled multicenter study. Neurology 1999;53:457–65.
142. van Oosten BW, Barkhof F, Truyen L, Boringa JB, Bertelsmann FW, von Blomberg BM, Woody JN, Hartung HP, Polman CH. Increased MRI activity and immune activation in two multiple sclerosis patients treated with the monoclonal anti-tumor necrosis factor antibody cA2. Neurology 1996;47:1531–4.
143. Arnett HA, Masson J, Marino M, Suzuki K, Matsushima GK, Ting JPY. TNFα promotes proliferation of oligodendrocyte progenitors and remyelination. Nat Neurosci 2001; 4:1116–22.
144. Scherbel U, Raghupathi R, Nakamura M, Saatman KE, Trojanowski JQ, Neugebauer E, Marino MW, McIntosh TK. Differential acute and chronic responses of tumor necrosis factor-deficient mice to experimental brain injury. Proc Natl Acad Sci 1999;96:8721–6.
145. Fontaine V, Mohand-Saine S, Hanoteau N, Fuchs C, Pfizenmaier K, Eisel U. Neurodegenerative and neuroprotective effects of tumor necrosis factor (TNF) in retinal ischemia: Opposite roles of TNF receptor 1 and TNF receptor 2. J Neurosci 2002;22:RC216
146. Arnett HA, Wang Y, Matsushima GK, Suzuki K, Ting JPY. Functional genomic analysis of remyelination reveals importance of inflammation in oligodendrocytes regeneration. J Neurosci 2003;23:9824–32.
147. Kassiotis G, Kollias G. Uncoupling the proinflammatory from the immunosuppressive properties of tumor necrosis factor (TNF) at the p55 TNF receptor level: Implications for pathogenesis and therapy of autoimmune demyelination. J Exp Med 2001;193:427–34.
148. Thommesen L, Laegreid A. Distinct differences between TNF receptor-1 and TNF receptor2-mediated activation of NFkappaB. J Biochem Mol Biol 2005;38:281–9.
149. Marchetti L, Klein M, Schletti K, Pfizenmaier K, Eisel ULM. Tumor necrosis factor (TNF)-mediated neuroprotection against glutamate-induced excitotoxicity is enhanced by *N*-methyl-D-aspartate receptor activation. J Biol Chem 2004;279:32869–81.
150. Bruce AJ, Boling W, Kindy MS, Peschon J, Kraemer PJ, Carpenter MK, Holtsberg FW, Mattson MP. Altered neuronal and microglial responses to excitotoxic and ischemic brain injury in mice lacking TNF receptors. Nat Med 1996;2:788–94.
151. Berger SW, Horster D, Furukawa K, Goodman Y, Krieglstein J, Mattson MP. Tumor necrosis alpha and beta protect neurons against amyloid beta-peptide toxicity: Evidence for involvement of a kappa B-binding factor and attenuation of peroxide and Ca^{2+} accumulation. Proc Natl Acad Sci 1995;92:9328–32.
152. Cheng B, Christakos S, Mattson MP. Tumor necrosis factor protects neurons against metabolic–excitotoxic insults and promote maintenance of calcium homeostasis. Neuron 1994;12:139–53.
153. Masson JL, Suzuki K, Chaplin DD, Matsushima GK. Inteleukin-1β promotes repair of the CNS. J Neurosci 2001;21:7046–52.

154. Herx LM, Rivest S, Yong VW. Central nervous system-initiated inflammation and neurotrophism in trauma: IL-1β is required for the production of ciliary neurotrophic factor. J Immunol 2000;165:2232–9.
155. DeKosky ST, Styren SD, O'Malley ME, Gross JR, Kochanek P, Marion D, Evans CH, Robbins PD. Interleukin-1 receptor antagonists suppresses neurotrophin response in injured rat brain. Ann Neurol 1996;39:123–7.
156. Bernardino L, Xapelli L, Silva AP, Jakobsen B, Poulsen FR, Oliveira CR, Vezzani A, Malva JO, Zimmer J. Modulator effects of interleukin-1β and tumor necrosis factor-α on AMPA-induced excitotoxicity in mouse organotypic hippocampal slice cultures. J Neurosci 2005;25:6734–44.
157. Lindholm D, Heumann R, Meyer M, Thoenen H. Interleukin-1 regulates synthesis of nerve growth factor in non-neuronal cells of rat sciatic nerve. Nature 1987;330:658–9.
158. Billiau A, Heremans H, Vandekerckhove F, Dijkmans R, Sobis H, Meulepas E, Carton H. Enhancement of experimental allergic encephalomyelitis in mice by antibodies against IFN-gamma. J Immunol 1988;140:1506–10.
159. Duong TT, St Louis J, Gilbert JJ, Finkelman FD, Strejan GH. Effect of anti-interferon gamma and anti-interleukin-2 monoclonal antibody treatment on the development of actively and passively induced experimental allergic encephalomyelitis in the SJL/J mouse. J Neuroimmunol 1992;36:105–15.
160. Willenborg DO, Fordham S, Bernard CC, Cowden WB, Ramshaw IA. IFN-gamma plays a critical down-regulatory role in the induction and effector phase of myelin oligodendrocyte glycoprotein-induced autoimmune encephalomyelitis. J Immunol 1996;157:3223–7.
161. Furlan R, Brambilla E, Ruffini F, Poliani PL, Bergami A, Marconi PC, Franciotta DM, Penna G, Comi G, Adorini L, Martino G. Intrathecal delivery of IFN-gamma protects C57BL/6 mice from chronic-progressive experimental autoimmune encephalomyelitis by increasing apoptosis of central nervous system-infiltrating lymphocytes. J Immunol 2001;167:1821–9.
162. Alexander WS, Starr R, Fenner JE, Scott CL, Handmann E, Sprigg NS, Corbin JE, Cornish AL, Darwiche R, Owczareck CM, Kay TW, Nicola NA, Hertzog PJ, Metclaf D, Hilton DJ. SOCS1 is a critical inhibitor of interferon gamma signalling and prevents the potentially fatal neonatal actions of this cytokine. Cell 1999;98:597–608.
163. Marine JC, Topham DJ, McKay C, Wang D, Parganas E, Stravopodis E, Yoshimura A, Ihle JN. SOCS1 deficiency causes a lymphocyte-dependent perinatal lethality. Cell 1999;98:609–16.
164. Panitch HS, Hirsch RL, Schindler J, Johnson KP. Treatment of multiple sclerosis with gamma interferon: Exacerbations associated with activation of the immune system. Neurology 1987;37:1097–102.
165. Gruol DL, Nelson TE. Physiological and pathological roles of interleukin-6 in the central nervous system. Mol Neurobiol 1997;15:307–39.
166. Carlson NG, Wieggel WA, Chen A, Bachi A, Rogers SW, Gahring LC. Inflammatory cytokines IL-1α, IL-1β, IL-6, and TNF-α impart neuroprotection to an excitotoxin through distinct pathways. J Immunol 1999;163:3963–8.
167. Yamashita T, Sawamoto K, Suzuki S, Suzuki N, Adachi K, Kawase T, Mihara M, Ohsugi Y, Abe K, Okano H. Blockade of interleukin-6 signaling aggravates ischemic cerebral damage in mice: Possible involvement of Stat3 activation in the protection of neurons. J Neurochem 2005;94:459–68.

168. Herrmann O, Tarabin V, Suzuki S, Attigah N, Coserea I, Schneider A, Vogel J, Prinz S, Schwab S, Monyer H, Brombacher F, Schwaninger M. Regulation of body temperature and neuroprotection by endogenous interleukin-6 in cerebral ischemia. J Cereb Blood Flow Metab 2003;23:406–15.
169. Penkowa M, Penkowa M, Moos T, Carrasco J, Hadberg H, Molinero A, Bluethmann H, Hidalgo J. Strongly compromised inflammatory response to brain injury in interleukin-6 deficient mice. Glia 1999;25:343–57.
170. Clark WM, Rinker LG, Lessov NS, Hazel K, Hill JK, Stenzel-Poore M, Eckenstein F. Lack of interelukin-6 expression is not protective against focal central nervous system ischemia. Stroke 2000;31:1715–20.
171. Hao AJ, Dheen ST, Ling EA. Expression of macrophage colony-stimulating factor and its receptor in microglia activation is linked to teratogen-induced neuronal damage. Neuroscience 2002;112:889–900.
172. Vincent VA, Robinson CC, Simsek D, Murphy GM. Macrophage colony stimulating factor prevents NMDA-induced neuronal death in hippocampal organotypic cultures. J Neurochem 2002;82:1388–97.
173. Wang Y, Berezovskaya O, Fedoroff S. Expression of colony stimulating factor-1 receptor (CSF-1R) by CNS neurons in mice. J Neurosci Res 1999;57:616–32.
174. Mitrasinovic OM, Grattan A, Robinson CC, Lapustea NB, Poon C, Ryan H, Pong C, Murphy GM. Microglia overexpressing the macrophage colony-stimulating factor receptor are neuroprotective in a microglial–hippocampal organotypic coculture system. J Neurosci 2005;25:4442–51.
175. Berezovskaya O, Maysinger D, Fedoroff S. The hematopoietic cytokine colony-stimulating factor1, is also a growth factor in the CNS: Congenital absence of CSF-1 in mice results in abnormal microglial response and increased neuron vulnerability to injury. Int J Dev Neurosci 1995;13:285–99.
176. Fedoroff S, Berezovskaya O, Maysinger D. Role of colony stimulating factor-1 in brain damage caused by ischemia. Neurosci Biobehav Rev 1997;21:187–91.
177. Berezovskaya O, Maysinger D, Fedoroff S. Colony stimulating factor-1 potentiates neuronal survival in cerebral cortex ischemic lesion. Acta Neuropathol (Berl) 1996;92:479–86.
178. Guillen C, Jander S, Stoll G. Sequential expression of mRNA for proinflammatory cytokines and interleukin-10 in the rat peripheral nervous system: Comparison between immune-mediated demyelination and Wallerian degeneration. J Neurosci Res 1998; 51:489–96.
179. Rotshenker S, Aamar S, Barak V. Interleukin-1 activity in lesioned peripheral nerve. J Neuroimmunol 1992;39:75–80.
180. Stoll G, Jung S, Van der Meide P, Hartung HP. Tumor necrosis factor-alpha in immune mediated demyelination and Wallerian degeneration of the peripheral nervous system. J Neuroimmunol 1993;45:175–82.
181. Wagner R, Myers RR. Schwann cells produce tumor necrosis factor alpha: Expression in injured and noninjured nerves. Neuroscience 1996;73:625–9.
182. Heumann R, Korsching S, Bandtlow C, Thoenen H. Changes of nerve growth factor synthesis in non neuronal cells response to sciatic nerve transection. J Cell Biol 1987; 104:1623–31.
183. Hattori A, Iwasaki S, Murase K, Tsujimoto M, Sato M, Hayashi K, Khono M. Tumor necrosis factor is markedly synergistic with interleukin 1 and interferon-γ in stimulating the production of nerve growth factor in fibroblasts. FEBS Lett 1994;340:177–80.

184. Lisak RP, Bealmer B, Ragheb S. Interleukin-1 alpha, but not interleukin 1-beta, is a co-mitogen for neonatal rat Schwann cells in vitro and acts via interleukin-1 receptors. J Neuroimmunol 1994;55:171–7.
185. Guenard V, Dinarello CA, Weston PJ, Aebischer P. Peripheral nerve regeneration is impeded by interleukin-1 receptor antagonist released from a polymeric guidance channel. J Neurosci Res 1991;29:396–400.
186. Saada A, Reichert F, Rotshenker S. Granulocyte macrophage colony stimulating factor produced in lesioned peripheral nerves induces the up-regulation of cell surface expression of MAC-2 by macrophages and Schwann cells. J Cell Biol 1996;133:159–67.
187. Be′eri H, Reichert F, Saada A, Rotshenker S. The cytokine network of Wallerian degeneration: IL-10 and GM-CSF. Eur J Neurosci 1998;10:2707–13.
188. Kerscheinsteiner M, Stadelmann C, Dechant G, Wekerle H, Hohlfeld R. Neurotrophic cross-talk between the nervous and immune systems: Implications for neurological diseases. Ann Neurol 2003;53:292–304.
189. Barde YA, Edgar D, Thoenen H. New neurotrophic factors. Annu Rev Physiol 1983;45:601–12.
190. Lewin GR, Barde YA. Physiology of the neurotrophins. Annu Rev Neurosci 1996; 19:289–317.
191. Baloh RH, Enomoto H, Johnson EM Jr, Milbrandt J. The GDNF family ligands and receptors – Implications for neural development. Curr Opin Neurobiol 2000;10: 103–10.
192. Ip NY. The neurotrophins and neuropoietic cytokines. Two families of growth factors acting on neural and hematopoietic cells. Ann N Y Acad Sci 1998;840:97–106.
193. Ransohoff RM, Howe CL, Rodriguez M. Growth factor treatment of demyelinating disease: At last, a leap into the light. Trends Immunol 2002;23:512–6.
194. Barouch R, Appel E, Kazimirsky G, Brodie C. Macrophages express neurotrophins and neurotrophin receptors. Regulation of nitric oxide production by NT-3. J Neuroimmunol 2001;112:72–7.
195. Besser M, Wang R. Clonally restricted production of the neurotrophin brain-derived neurotrophic factor and neurotrophin-3 mRNA by human immune cells and Th1/Th2-polarized expression of their receptors. J Immunol 1999;162:6303–6.
196. Ehrhard PB, Erb P, Garumann U, Otten U. Expression of nerve growth factor and nerve growth factor receptor tyrosine kinase Trk in activated CD4-positive T cell clones. Proc Natl Acad Sci USA 1993;90:10984–8.
197. Kerscheinsteiner M, Gallmeier E, Behrens L, Vargas Leal V, Misgeld T, Klinkert WEF, Kolbeck R, Hoppe E, Oropeza-Wekerle R, Bartke I, Stadelmann C, Lassmann H, Wekerle H, Hohlfeld R. Activated human T cells, B cells and monocytes produced brain-derived neurotrophic factor in vitro and in inflammatory brain lesions: A neuroprotective role of inflammation?. J Exp Med 1999;189:865–70.
198. Santambrogio L, Benedetti M, Caho MV, Muzaffar R, Kulig K, Gabell N, Hochwald G. Nerve growth factor production by lymphocytes. J Immunol 1994;153:4488–95.
199. Torcia M, Bracci-Laudiero L, Lucibello M, Nencioni L, Labardi D, Rubartelli A, Cozzolino F, Aloe L, Garaci E. Nerve growth factor is an autocrine survival factor for memory B lymphocytes. Cell 1996;85:345–56.
200. Moalen G, Gdalyahu A, Shani Y, Otten U, Lazarovici P, Cohen IR, Schwarz M. Production of neurotrophins by activated T cells: Implications for neuroprotective autoimmunity. J Autoimmun 2000;15:331–45.

201. Micera A, De Simone R, Aloe L. Elevated levels of nerve growth factor in the thalamus and spinal cord of rats affected by experimental allergic encephalomyelitis. Arch Ital Biol 1995;133:131–42.
202. Hammarberg H, Lidman O, Lundberg C, Eltayeb SY, Gielen AW,Muhallab S, Svenningsson A, Linda H, van Der Meide PH, Cullheim S, Olsson T, Piehl F. Neuroprotection by encephalomyelitis: Rescue of mechanically injured neurons and neurotrophin production by CNS-infiltrating T and natural killer cells. J Neurosci 2000;20:5283–91.
203. Stadelmann C, Kerschensteiner M, Misgeld T, Brück W, Hohlfeld R, Lassmann H. BDNF and gp145trkB in multiple sclerosis brain lesions: Neuroprotective interactions between immune and neuronal cells? Brain 2002;125:75–85.
204. Flugel A, Matsumuro K, Neumann H, Klinkert WE, Birnbacher R, Lassmann H, Otten U, Wekerle H. Anti-inflammatory activity of nerve growth factor in experimental autoimmune encephalomyelitis inhibition of monocyte trans-endothelial migration. Eur J Immunol 2001;31:11–22.
205. Lambiase A, Bracci-Laudiero L, Bonini S, Bonini S, Starace G, D'Elios MM, De Carli M, Aloe L. Human CD4+ T cell clones produce and release nerve growth factor and express high-affinity nerve growth factor receptors. J Allergy Clin Immunol 1997;100:408–14.
206. Melamed I, Kelleher CA, Franklin RA, Brodie C, Hempstead B, Kaplan D, Gelfand EW. Nerve growth signal transduction in human B lymphocytes is mediated by gp 140trk. Eur J Immunol 1996;26:1985–92.
207. Neumann H, Misgeld T, Matsumuro K, Wekerle H. Neurotrophins inhibit major histocompatibility class II inducibility of microglia: Involvement of the p75 neurotrophin receptor. Proc Natl Acad Sci 1998;95:5779–84.
208. Otten U, Ehrhard P, Peck E. Nerve growth factor induces growth and differentiation of human B lymphocytes. Proc Natl Acad Sci 1989;86:10059–63.
209. Thorpe LW, Perez-Polo JR. The influence of nerve growth factor on in vitro proliferative response of rat spleen lymphocytes. J Neurosci Res 1987;18:134–9.
210. Villoslada P, Hauser SL, Bartke I, Unger J, Heald N, Rosenberg D, Cheung SW, Mobley WC, Fisher S, Genain CP. Human nerve growth factor protects common marmosets against autoimmune encephalomyelitis by switching the balance of T helper cell type 1 and 2 cytokines within the central nervous system. J Exp Med 2000;191: 1799–806.
211. Wei R, Jonakait GM. Neurotrophins and the anti-inflammatory agents interleukin-4 (IL-4), IL-10, IL-11 and transforming growth factor beta-1 (TGF-beta 1) down-regulate T cell costimulatory molecules B7 and CD40 on cultured rat microglia. J Neuroimmunol 1999;95:8–18.
212. Villoslada P, Genain CP. Role of nerve growth factor and other trophic factors in brain inflammation. Prog Brain Res 2004;146:403–14.
213. Vanderlocht J, Hellings N, Hendriks JJ, Vandenabeele F, Moreels M, Buntinx M, Hoekstra D, Antel JP, Stinissen P. Leukemia inhibitory factor is produced by myelin-reactive T cells from multiple sclerosis patients and protects against tumor-necrosis factor-induced oligodendrocyte apoptosis. J Nuerosci Res 2006;83:763–74.
214. Butzkueven H, Zhang JG, Soilu-Hanninen M, Hochrein H, Chionh F, Shipham KA, Emery B, Turnley AM, Petratos S, Ernst M, Bartlett PF, Kilpatrick TJ. LIF receptor signaling limits immune-mediated demyelination by enhancing oligodendrocyte survival. Nat Med 2002;8:613–9.

215. Linker RA, Maurer M, Gaupp S, Martini R, Holtmann B, Giess R, Rieckmann P, Lassmann H, Toyka KV, Sendtner M, Gold R. CNTF is a major protective factor in demyelinating CNS disease: A neurotrophic cytokine as modulator in neuroinflammation. Nat Med 2002;8:620–4.
216. Albrecht PJ, Dahl JP, Stoltzfus OK, Levenson R, Levinson SW. Ciliary neurotrophic factor activates spinal cord astrocytes, stimulating their production and release of fibroblast growth factor-2, to increase motor neuron survival. Exp Neurol 2002;173:46–62.
217. Batchelor PE, Liberatore GT, Wong JYF, Porrit MJ, Frerichs F, Donnan GA, Howells DW. Activated macrophages and microglia induce dopaminergic sprouting in the injured striatum and express brain-derived neurotrophic factor and glial cell line-derived neurotrophic factor. J Neurosci 1999;19:1708–16.
218. Hammarberg H, Lidman O, Lundberg C, Eltayeb SY, Gielen AW, Muhallab S, Svenningsson A, Lindå H, van der Meide PH, Culheim S, Olsson T, Piehl F. Neuroprotection by encephalomyelitis: Rescue of mechanically injured neurons and neurotrophin production by CNS-infiltrating T and natural killer cells. J Neurosci 2000;20:5283–91.
219. Linker RA, Mäurer M, Gaupp S, Martini R, Holtmann B, Giess R, Rieckmann P, Lassman H, Toyka KV, Sendtner M, Gold R. CNTF is a major protective factor in demyelinating CNS disease: A neurotrophic cytokine as modulator in neuroinflammation. Nature Med 2002;8:620–4.
220. Sobue G, Yamamoto M, Doyu M, Li M, Yasuda T, Mitsuma T. Expression of mRNA for neurotrophins (NGF, BDNF, and NT-3) and their receptors (p75NGFR, trkB, and trkC) in human peripheral neuropathies. Neurochem Res 1998;23:821–9.
221. Yao DL, Liu X, Hudson LD, Webster HD. Insulin-like growth factor I treatment reduces demyelination and up-regulates gene expression of myelin-related proteins in experimental autoimmune encephalomyelitis. Proc Natl Acad Sci USA 1995;92:6190–4.
222. Barouch R, Schwartz M. Autoreactive T cells induce neurotrophin production by immune and neural cells in injured rat optic nerve: Implications for protective autoimmunity. FASEB J 2002;16:1304–6.
223. Nye SH, Squinto SP, Glass DJ, Stitt TN, Hantzopoulos P, Macchi MJ, Lindsay NS, Ip NY, Yancopoulos GD. K-252a and staurosporine selectively block autophosphorylation of neurotrophin receptors and neurotrophin-mediated responses. Mol Biol Cell 1992; 3:677–86.

V. DISEASE

Cytokines and the Brain
Edited by C. Phelps and E. Korneva

Brain Response to Endotoxin

CHRISTOPHER PHELPS AND LI-TSUN CHEN

Department of Pathology and Cell Biology and the Neuroscience Program, College of Medicine, Box 11, University of South Florida, 12901 Bruce B. Downs Blvd., Tampa, FL 33612-4799, USA

ABSTRACT

The cellular and systemic responses of the brain to endotoxin [lipopolysaccharide (LPS)] administration are considered in relation to how the central nervous system receives information about the presence of LPS in blood and extracellular fluids (ECFs). In an attempt to provide an answer to the question of "How does the brain know the body is sick?" we review LPS-induced signaling pathways, reaction of the gastrointestinal (GI) tract epithelium and related vagal sensory nerves, related endothelium of local vascular beds in both the GI tract and the brain, as well as the specialized brain vasculature of circumventricular organs (CVOs) and blood vessels in the brain parenchyma. The featured signal pathways activated by low (subseptic) doses of LPS involve cytokines and prostanoids conveying messages to the brain via perivascular and blood vessel component cells and vagal visceral afferents. Activation of the hypothalamic–pituitary–adrenal (HPA) axis, as it represents a key component of the acute-phase reaction to LPS, is the functional endpoint emphasized in the studies presented. Individual animals were instrumented to permit rapid sampling from ECF in the anterior hypothalamus and concomitant peripheral blood sampling at 15-min intervals before and after intravenous LPS administration. Changes in interleukin (IL)-1β and hormones from pituitary [adrenocorticotropin (ACTH)] and adrenal (corticosterone) components of the HPA axis were studied in individual animals. Intrahypothalamic changes in IL-1β and corticotropin-releasing hormone (CRH) within the first half hour after LPS administration implicate hypothalamic mechanisms contributing to HPA axis activation after endotoxin. The importance of hypothalamus in this acute-phase reaction is examined further in another set of experiments involving excitotoxic chemical lesions of cell bodies in the anterior hypothalamic area (AHA) prior to LPS exposure. When viewed together, the results of these studies emphasize the importance of rapid communication of the periphery with the AHA via multiple routes after peripheral exposure to LPS, which allows the activation of all levels of the HPA axis during the acute-phase reaction.

1. INTRODUCTION

Endotoxin, or lipopolysaccharide (LPS), constitutes a major component of the outer membranes of gram-negative bacteria. Hosts detect the presence of disease-causing organisms by recognizing specific components of the pathogens that are not found in the host. The components of an infectious invader that enable host recognition of this specific pathogen consist of a particular molecular pattern. These pathogen-associated "molecular components" play essential roles in the biology of the disease-causing organism in question, and they are not subject to high mutation rates [1,2]. Lipopolysaccharide is one of these several diverse bacterial cell wall "molecular components" that may also include peptidoglycans, lipopeptides, and teichoic acids [3]. Host exposure to one of these components induces several pathophysiological reactions that are part of the natural or innate immune response, which is the first line of defense against diverse microbial pathogens [1,2].

2. LIPOPOLYSACCHARIDE-INDUCED SIGNALING PATHWAYS

A molecular basis for the innate immune response to pathogens has only recently begun to be understood and mechanisms involved in the first line of response to LPS are understood best. Thus, in the blood stream LPS first binds to a serum protein known as LPS-binding protein (LBP) [4,5]. Studies of LBP-deficient mice revealed that LBP is essential for the rapid induction of an inflammatory response after exposure to small amounts of LPS [6]. Lipopolysaccharide introduction into the mammalian body results in endotoxin binding to LBP, and this complex activates several different populations of cells by binding to its receptor. Over the past decade, several mammalian receptors have been identified (e.g., β_2-integrins, CD11/CD18, the macrophage scavenger receptor for acetylated LDL, L-selectin and CD14). The most important of these is CD14, which exists in two forms: membrane CD14 (mCD14) and soluble CD14 (sCD14) [3]. The mCD14 is found on the surface of bone marrow-derived cells, whereas the sCD14 can bind LPS and then activate cells, which lack mCD14. CD14 functions as a protein ligand for either LPS or LPS–LBP complex, but does not bring about transduction of intracellular signaling [4,7,8]. Finally, in the presence of very high concentrations of LPS, a CD14-independent pathway for signal transduction has also been described, for example, in macrophages of CD14-deficient mice [9, 10].

Myeloid cells have the mCD14 receptor located on their membrane surfaces, which acts as glycophosphatidyl inositol (GPI) – anchored membrane glycoprotein. In contrast, the sCD14 lacks GPI properties, but is able to bind LPS and in doing so, activate cells such as endothelial, epithelial and smooth muscle cells, as well as astrocytes that lack mCD14 [3,7,11]. Activation of the proinflammatory signal transduction pathway occurs after the binding of the LPS–LBP complex to the GPI-anchored mCD14 receptor.

Characterization of human homologues of *Drosophila* Toll proteins has provided significant additional insight toward understanding the mechanisms of LPS signaling [12]. During the past six years, a large group of toll-like receptors (TLRs) in humans have been shown to function as primary innate immune sensors for molecular components of microbes, each responding to specific molecular aspects of microbial origin [13]. Collectively, the TLRs signal the host about the presence of infectious organisms in a matter of minutes. In addition, when the specific microbial molecular component binds to the TLRs, the latter set in motion the systemic inflammatory state that accompanies sepsis and the associated production of tumor necrosis factor-α (TNF-α), interleukin (IL)-1, IL-6, IL-12, type I interferons (INFs), and other related proteins [14]. In order for LPS to exert its pathogenic effects, endotoxin requires one member of the large family of TLRs that is thought to play a critical role in the innate immune response to LPS, TLR4 [14]. TLR4 is present in circulating monocytes/macrophages and other immune cells. Recently, the cellular localization of TLR4 mRNA was studied and localized in the leptomeninges (arachnoid and pia mater) and choroid plexus (CP) of rat brain [15,16]. TLR4 is necessary for signal transduction induced by LPS and the cellular consequences of that exposure, such as expression of reactive oxygen species and various cytokines. Although there is convincing evidence that TLR4 is important for the recognition of LPS and the consequent exertion whereby LBP, CD14, and TLR4 interact to transduce the signal from LPS, the signaling process is not completely known [3].

The transcription factor NFκB mediates LPS-induced release of a number of cytokines and inflammatory signal molecules. The activation of NFκB and translocation to the nucleus results in the transcription of proinflammatory cytokines, including ILs, TNF-α, adhesion molecules and certain enzymes such as cyclooxygenase-2 (COX2), and different isozymes of nitric oxide synthase (NOS). These enzymes have been implicated in the pathogenesis of the systemic inflammatory response syndrome [17]. The activity of NFκB is regulated, in turn, by an inhibitor, IκB. Activation of IκB kinases leads to phosphorylation of IκB, which results in its proteolytic degradation and translocation of NFκB to the nucleus where it is able to regulate the transcription of selected genes by binding to the κB consensus sequence [18]. In the absence of LPS, IκB activity is normally controlled by an NFκB autoregulatory loop [3,19].

To the extent that the molecular and cellular basis for bringing about the innate immune response has only recently begun to become known and is currently understood best for reaction to LPS, this chapter will focus on what is known about not only neuronal and glial (central and peripheral), but also gut mucosal response to LPS exposure. There will also be a consideration of brain endothelial and perivascular cell response to LPS. The area of the brain under consideration includes blood vessels and perivascular cells of the hypothalamus. The hypothalamic region of the brain is emphasized because of the developing knowledge concerning the confluence of messages received at this location: cytokine components of the innate immune response, somatic and visceral (e.g., gut) vagal sensory information, hormones. Furthermore, the grouping together of central neural along with mucosal and vascular cell layers under the rubric of "brain response to endotoxin" is explained as follows: the relationships of traditional brain barriers (meningeal and vascular) and those of model mucosal epithelial barriers (e.g., in this case gastrointestinal (GI)) both represent regions of "initial encounter" for pathogens during disease processes. Moreover, in the case of experimental peripheral LPS administration (or exposure), there can be induction of peripheral and central signals, for example, in the form of proinflammatory cytokines after intravenous injection of LPS. In addition, experimental central injection of specific antagonists for proinflammatory cytokines and other message-bearing molecules are known to attenuate central effects of LPS following peripheral injection such as fever, prolonged slow-wave sleep, hypothalamic–pituitary–adrenal (HPA) axis, and autonomic suppression of peripheral immune responses [20]. Most recently, experimental studies of central-acting antagonists of peripheral LPS effects have focused on locating specific brain region(s) sites of action.

Focus on the hypothalamus in what follows in this chapter will relate to the central regulation of the neuroendocrine system during and after LPS exposure. In this regard, activation of the HPA axis during an immune challenge is considered by most investigators to be a major part of the host–defense response to LPS. Other neuroendocrine mechanisms involving related anterior pituitary hormones will also be briefly discussed [21,22].

3. THE ACUTE INFLAMMATORY RESPONSE

In order for a potentially injurious agent such as LPS to have an effect on a potential host, it must first either cross or destroy the integrity of the primary barriers of the body (epithelial and/or endothelial cells and related specializations). The host responses to this invasion are by definition some aspects of inflammation. The traditional definition of inflammation is the pathophysiological process by which vascularized tissues respond to some form of tissue damage and/or the presence of an invading agent capable of causing tissue and organ damage. Moreover, inflammation has also been further divided into acute and chronic bodily responses to tissue damage. In this context, the designation of "acute" usually refers to host events taking place over minutes to days, whereas "chronic" is described as being of longer duration. Examples of injurious agents that produce acute inflammatory changes include several categories of pathogens (e.g., bacteria, viruses, other pathogens), foreign bodies from exogenous or endogenous sources, as well as physical or chemical agents. The resultant tissue damage and related invasion of pathogens initiates a series of molecular events, which results in the production of soluble proinflammatory mediators that promote the hallmark physical signs of inflammation. These mediators include the plasma protease systems (complement, kinins, clotting, and fibrinolytic proteins), lipid mediators (prostaglandins, leukotrienes, and platelet-activating factor), proinflammatory peptides and amines (neuropeptides such as substance P, vasoactive intestinal peptide and calcitonin gene-related peptide, histamine, serotonin, and nitric oxide), and proinflammatory cytokines (IL-1, IL-4, IL-6, IL-8 and IL-12, TNF-α, and INF-γ) [23].

Two model systems have provided insight into the timing of the appearance and the importance of the mediators in the various processes that characterize inflammation in man and animals: (1) administration of LPS intravenously (iv) and (2) creation of skin blisters by suction. The present review will focus on experimental results obtained in human and laboratory rodents receiving low ($\leq$250 µg/kg body weight) doses of LPS iv.

3.1. Systemic responses to intravenous administration of lipopolysaccharide

There are characteristic changes in both body temperature and peripheral white blood cell count after receiving LPS iv. Body temperature begins to increase after about 1 h and reaches a maximum at about 4 h. During this time, laboratory rodents and humans will exhibit piloerection and experience chills. The peripheral blood leukocyte count shows a characteristic decrease at about 30 min, attributable to neutrophil and monocyte adherence to endothelial cells in the lung and spleen. This is followed by leukocytosis characterized by the presence of immature neutrophils at about 4 h, which can persist through 24 h, gradually returning to baseline by 48 h. The leukocytosis is predominantly attributable to mobilization of immature neutrophils from the bone marrow. Some of the critical components of the acute inflammatory response (fever, neutrophil margination in circulatory vessels, followed by mobilization of neutrophils from the bone marrow) are all associated with detectable changes in circulating mediators of inflammation. For example, TNF-α peaks within 2 h [24] and is likely the predominant pyrogen associated with the febrile response. Early increases in the chemoattractant IL-8, peaking at approximately 4 h after LPS, are major factors in determining the transient decreases in neutrophils at 30 min, based on the results of experimental work demonstrating an increased neutrophil expression of adhesion receptors after iv administration of chemokines like IL-8 [25]. Early increases in plasma IL-1β (over 30–60 min) after iv LPS have been associated with the induction of IL-8 in monocytes, lymphocytes, and neutrophils [21,23]. As mentioned above, acute increases in plasma IL-8 have been correlated with the transient decrease (margination) in circulating neutrophils. Finally, increases in circulating IL-6, with its far-reaching affects on T-and B-lymphocytes, macrophages, and monocyte differentiation, also increase the number of circulating platelets, as well as the synthesis of acute-phase reactant proteins in liver. The latter begins to occur after the onset of blood cytokine level increases, which occur 2–4 h after LPS exposure [26–28].

4. EPITHELIAL AND ENDOTHELIAL BARRIERS

In a general sense, the epithelial barriers to bacteria and LPS are located at the lining surfaces of all body portals to the environment. In this chapter we will focus on (1) the GI tract barriers and their relationship to visceral sensory afferent nerve fibers and (2) the cells lining brain membranes, arteries, veins, and capillaries either surrounding or located within the brain parenchyma. The actual barriers to LPS entry at the epithelial and endothelial locations considered are constituted by cell membranes, in addition to a portion of the intercellular junctional complexes: tight junctions (TJs) (zonula occludens). The latter prevent substances like LPS from moving in between individual barrier cells. Finally, in addition to the brain parenchymal vessels constituting the blood–brain barrier (BBB) [29], there are also vascular locations in the brain and its ventricles where the endothelial capillary barrier is incomplete: the windows on the BBB in circumventricular organs (CVOs) [30] and in the cerebral ventricular CP capillaries [31]. However, it should be pointed out that at the CVOs and the CP, there are additional cell membrane TJs located between the remainder of the brain and related ependymal cells lining the cerebral ventricles at these locations (see below).

4.1. Epithelial barrier function and tight junction structure in gastrointestinal mucosa

4.1.1. Structural considerations

The GI mucosa has three principal functions: protection, absorption, and secretion. The epithelial barrier separates the GI luminal contents from the underlying connective tissue and associated vagal nerves. The TJs between the simple columnar epithelial cells that line most of the GI tract produce a selectively permeable barrier. The TJs represent the most apical component in the epithelial junctional complex. High-resolution transmission electron microscopy (TEM) of an individual TJ (zonula occludens) reveals that this junction is not a continuous seal, but rather a series of focal fusions created by transmembrane proteins of adjoining cells that traverse the cell membrane and join in the intercellular space. The transmembrane protein *occludin* has been identified as the sealing protein. Occludin interacts with the actin cytoskeleton through a TJ protein called ZO-1. A second TJ protein (ZO-2), together with the former, may interact with several pathogenic agents such as cytomegalovirus and cholera toxins causing the TJ to become permeable (see Section 5.1) [32].

4.2. Endothelial blood–brain barrier function and tight junction structure in blood vessels and brain membranes

4.2.1. Blood–brain barrier structural considerations

Endothelial cells are typically flat and elongated, with their long axes oriented to the direction of blood flow within the vessel in question. The endothelia in the brain and the spinal cord are joined by TJs and gap junctions. Endothelial cells also play an important role in blood homeostasis, and their functional properties change in response to various stimuli [29].

Capillaries located in the brain parenchyma (and in muscles and lungs) have endothelia that are described as "continuous" (BBB). Transmission electron microscopy examination of a BBB endothelial cell in cross section reveals two plasma membranes enclosing a very thin cytoplasmic strip. These continuous capillaries have elaborate TJs that are more similar to those found between other types of epithelial cells, as opposed to TJs described between endothelial cells in non-CNS locations. Moreover, TEM studies of BBB capillaries have also revealed a close association of astrocytic end feet processes in contact with the endothelial basal lamina. Studies of BBB capillaries during diseases that involve loss of the BBB effectiveness have revealed that when astrocyte end feet–endothelial relationships are modified, the TJs are also lost [29,33].

The components of the TJs in the BBB have recently been reviewed [29]. Briefly, two transmembrane proteins appear to be integral to TJ function: the *claudins* and the occludin. Other proteins have a peripheral role, as was described earlier for GI TJ structure. The complexity of this TJ, as well as its association with signaling molecules (G proteins and protein C kinase isotypes), suggests that permeability of BBB capillaries may be regulated by a number of factors.

4.2.2. Arachnoid membrane structure

In 1975, Nabeshima et al. [34] described a barrier layer within and between arachnoid membrane cells situated within the membrane proper. This barrier layer consisted of two to three layers of flattened cells, tightly packed and closed by periodic TJs and 20 Å gap junctions. There is an incomplete basement membrane covering the inner aspect of the arachnoid barrier layer. There are no blood vessels in the barrier layer of the arachnoid, whereas the numerous blood vessels located within the subarachnoid space are lined by endothelial cells with TJs. Thus, the arachnoid membrane proper, like the pia mater, is avascular and derives its nutrients by diffusion from the cerebrospinal fluid (CSF). The existence of a barrier between the ECF of the adjacent dura mater and the subarachnoid CSF is assumed based on both (1) the TJs described above and (2) the results of a few existing experimental studies (see below). Moreover, the capillaries of the dura mater are of the highly permeable type that permit rapid penetration of large water-soluble molecules into the dural tissue. If there were no arachnoid barrier, these large molecules (e.g., LPS) would pass rapidly into the subarachnoid CSF and access the brain and the spinal cord [33].

Pape and Katzman [35] studied the functional aspects of the arachnoid barrier in the cat using the penetration of ^{42}K uptake into the brain parenchyma from the subdural perfusion (outside the barrier) was approximately one-third that obtained with a subarachnoid perfusion. More recently, arachnoid barrier TJs have been confirmed in the human arachnoid membrane that was also found to have gap junctions in lower (subarachnoid) cell junctions [36].

4.3. Chroroid plexus epithelium and endothelium

4.3.1. Structural considerations

Each villus of the CP usually has a single capillary located in the center of the villus, near the apical area of the latter. Arterioles and venules with their surrounding connective tissue elements predominate at the base of each villus. In general, capillaries of the villi have extremely large, thin simple endothelium resting on a basement membrane. The endothelia are characterized by many extreme attenuations of their cytoplasm that appear to be pores, 700 Å in diameter. At these sites, the plasma membranes are in close opposition, but have not disappeared. These regions are referred to as fenestrations (windows in the endothelium), and they represent regions of high permeability to fluids and solutes, permitting the rapid formation of a filtrate [33]. The capillary endothelial junctions in the choroid villi are discontinuous and similar to those of venules in the general circulation. These capillary junctions are leaky to molecules up to 20 Å in diameter (e.g., permitting the passage of the tracer microperoxidase). Pre-and postcapillary endothelia are linked together by TJs, providing low permeability. Finally, the junctions between leptomeningeal (pia mater and arachnoid) cell junctions interposed between choroid capillary endothelium at the base of the CP are of the II permeable type [31,33].

The CP epithelium (CPE) is classified as "leaky" in its ability to allow passage to tracer molecules of lanthanum that are ordinarily held-up at TJs of junctional complexes located between endothelia when that substance is administered iv [33]. One implication to be drawn from this observation is that *some* TJs located between the cuboidal to low-columnar CPE intercellular junctions do not have the same degree of barrier restriction as do the TJs of the BBB. The phenomenon of "functional leak" has been described in the CPE. Traces of substances, such as microperoxidase, found their way from blood vessels into the adjacent CSF by crossing the ependymal cell lining of the cerebral ventricle near the base of the CP. In these experimental studies, peroxidase was injected iv and found its way into the stroma of the CP by way of the fenestrated capillaries of the latter. Subsequently, the peroxidase diffused into the CSF by crossing the ependymal cells lining the ventricle and adjacent to the base of the CP [37].

4.4. Endothelial barrier function and related variations in the gastrointestinal tract, liver, and gall bladder

As mentioned earlier in the chapter, although continuous (nonfenestrated) capillaries are typically found in muscles, lungs, and the CNS, fenestrated capillaries are also typically found in endocrine glands, as well as sites of fluid and metabolic absorption such as the intestinal tract and the gall bladder. The fenestrations found in capillaries located within the latter are 80–100 nm in diameter and constitute actual channels across the capillary wall. These fenestrations found in the GI tract and the gall bladder are relatively fewer in number when compared to the other fenestrated capillary locations, and these capillary walls are relatively thicker, as well, when no absorption is taking place in the tract. Discontinuous capillaries (also called sinusoidal capillaries or sinusoids) are found throughout the liver parenchyma, the spleen, and the bone marrow. They are characterized by gaps between neighboring endothelial cells, and the basal lamina underlying the endothelium may be partially or completely absent. Finally, fenestrated capillaries located in the connective tissue of the intestinal tract may also have a thin, nonmembranous diaphragm across its opening. It is thought that the diaphragm may be a remnant of the intestinal epithelial glycocalyx formerly enclosed in a pinocytotic vesicle from which the fenestration channel may have originally formed [38].

All of the locations for fenestrated capillaries described above also have vagal visceral sensory afferent fibers innervating the walls of these organs [38]. Consequently, the walls of the indicated abdominal viscera constitute additional (potential) sites for LPS-induced activation of visceral afferents transmitting impulses to the brainstem.

5. BARRIER RESPONSES TO LIPOPOLYSACCHARIDE

On the basis of the known visceral sensory innervation of abdominal organs, a comprehensive consideration of visceral epithelial, vascular and neural, as well as other related barrier, responses is a necessary contingency of considering the "brain's response" to LPS. Most experimental approaches to this problem administer either septic or subseptic doses of LPS to mammals via either *iv* or *intraperitoneal* routes. Although it is well known that the time course of the body's acute response to low (subseptic) doses of LPS differ according to the route of endotoxin administration, ultimately either route of exposure will impact visceral neural afferents, cranial nerve afferents, and brain barriers (meningeal and vascular) as a result of the systemic presence of LPS. Moreover, the relative influence(s) on brain's response to LPS from these various sources of information continue to be debated and explored [39].

Most studies of immune signaling to the brain from the periphery by multiple routes emphasize cytokines, especially IL-1β and TNF-α, for their roles in the brain's coordination of bidirectional communication between the immune system and the CNS in the maintenance of homeostasis [40]. To begin to develop an accurate view of the acute effects of systemic LPS administration on the brain, there is not only a critical need to inventory all the portals of early interactions of LPS within the nervous system, but more importantly, but also a prerequisite need to understand what is currently known about the different portal responses. The guiding question in this type of inquiry is how does the brain know that the body is sick or that peripheral sepsis has begun? In a "natural" context, the advent of gram-negative sepsis is signaled by the release of endotoxin from the outer membrane of the bacteria. In the featured experimental paradigm, the latter situation is simulated experimentally by the systemic administration of LPS. Finally, evidence is accumulating that the proinflammatory signal transduction pathway sequelae that have been worked out after LPS exposure in peripheral organs and cells also occur within brain barriers and proximal brain cells.

5.1. Intestinal epithelial barrier and related vagal afferent responses to lipopolysaccharide

The normal intestinal epithelium (IE) is not inflamed despite making contact with a high density of commensal bacteria. Ordinarily, intestinal epithelial cells express low levels of TLR4 and MD-2 (a critical TLR4 coreceptor), and the cells are LPS-unresponsive. However, preincubation of either intestinal epithelial cell lines or native colonic epithelial cells with INF-γ and TNF-α sensitizes intestinal epithelial cells to LPS-dependent IL-8 secretion, as well as TLR4 mRNA expression. Introduction of INF-γ into a culture of intestinal cells also upregulates MD-2 promoter activity [41–43]. In healthy humans, a small number of bacteria and related toxins (e.g., LPS) may breach the intestinal epithelial lining. However, further migration is arrested by the gut-associated lymphoid tissue (GALT) [44]. Thus, prior to entry into the GALT, bacteria that have crossed the intestinal epithelial barrier must progress into the loose connective tissue (lamina propria) underlying the gut epithelium where resident tissue macrophages are also well-positioned to interact with pathogens, in conjunction with dendritic cells (DC) residing in epithelia overlying GALT lymph nodules. Moreover, visceral afferent nerve fibers coursing through the vagus nerve have also been traced to innervate the surrounding intestinal submucosa, as well as lamina propria and related epithelium associated with DC-like cells overlying lymphoid nodules [45]. Consequently, vagal sensory fibers are well positioned to be activated by bacteria or LPS invading the body via either the GI portal or related fenestrated blood vessels, and the subsequent cytokine secretion is induced locally as a result of the presence of pathogenic bacteria and related toxins [46].

5.1.1. Effects on proinflammatory cascade

Although there are several known mechanisms by which enteropathogenic *Escherichia coli* or LPS may incite IE inflammation, the end result of the induction is the release of cytokines, chemokines, and recruitment of inflammatory cells. The means by which enteric pathogens initiate inflammatory signals is complex and incompletely understood [32]. NκB clearly plays a pivotal role in the expression of all proinflammatory cytokines, chemokines, and cell surface adhesion molecules, including IL-1β, TNF-α, IL-6, IL-8 and MHC class II molecules (see Section 2). Enteropathogenic *E. coli* and related toxins such as LPS have also been shown to stimulate other transduction pathways (reviewed in [32]).

5.1.2. Effects on immune cascade and vagal afferents

Experimental studies that attempt to understand the role of the vagus nerve in response to endotoxin challenge have revealed that several factors related to LPS administration (dosage, route and time of day for injection) are all important for determining vagal visceral afferent contributions to communication between the body and the brain [39]. Moreover, the experimental approach to understanding vagal response to LPS that is employed by most investigators usually involves subdiaphragmatic vagotomy. However, this experimental manipulation does not permit conclusions to be drawn about potential contributions to experimental outcomes that may involve remaining intact supradiaphragmatic vagal sensory nerves [40].

The subdiaphragmatic branches of the vagus nerve that provide sensory innervation to the intestines extend their afferent neural processes to the submucosal and epithelial (lamina propria) regions that are closely associated with DC-like cells and mast cells [45]. In general, the results of experiments involving subdiaphragmatic vagotomy before the administration of LPS, when compared with sham-operated animals, suggest that the intact vagus nerve may contribute to signaling the brain about the occurrence of a local immune response to LPS (e.g., cytokine release) when low (<1 μg/kg body weight) doses of endotoxin are administered (reviewed in [39,40]). There are also additional potential confounding contributing factors in vagotomy experiments that are involved by necessity in this experimental design. For example, the vagotomy procedure involves a section of both sensory and motor nerve fibers coursing within the vagus nerve below the diaphragm. This nerve surgery may modify the ability of the brain to coordinate "complete" secretomotor and other responses to LPS. Finally, in the experiments where relatively higher (1–50 μg/kg bw) LPS dosages are administered to vagotomized animals, other additional immunosensory neural pathways are known to be activated and have the potential for participating in the brain's response to LPS (see Goehler [46] in this volume for a review).

5.1.3. Summary of intestinal epithelium and vagal responses to lipopolysaccharide

Evidence has been provided in support of the hypothesis that intestinal epithelia and related visceral afferents contribute to the brain's response to low doses of LPS. Examples of graded amounts and routes of LPS administration were described as having the potential for creating a progressive access to the IE either via fenestrated capillaries or through direct contact and disruption of the epithelium (respectively). The corollary hypothesis that GI infection (or another form of injury) alters intestinal permeability to allow further host invasion by enteric pathogens was first proposed over 30 years ago [44]. Since that time, especially during the past 15 years, some of the mechanisms and pathways that contribute to brain activation by, for example, LPS in the gut have begun to be understood [32,40]. Most recently, vagal and nonvagal neural and peripheral neuroimmune pathways for brain activation by LPS have been described [46]. Not only peripheral neural routes contribute important information that helps to enable brain coordination of overall response to LPS (e.g., in low-grade enteropathogenic infection), but cranial and peripheral nerve pathways are also *de facto* portals for information transfer that enable the CNS *coordination* of bodily response to subseptic infections. For example, it has been suggested that in some infections, gut epithelium permeability increases, which may actually allow intestinal bacteria (or endotoxins) to further invade the host. This may provide a greater opportunity for the host's neuroimmune system to amplify its overall response to infection and alter the course of potentially prolonged or life-threatening disease [44]. In order for this type of communication and coordination to occur, it is apparent that the local production of immune messengers in the vicinity of the breached epithelium, for example, lamina propria and submucosa, would include the activation of visceral afferent pathways to the brain.

5.2. Brain vascular and meningeal barrier response to lipopolysaccharide

During the course of infectious disease states associated with inflammatory situations, the production of increasing concentrations of circulating cytokines is known to elicit many of the conditions understood to be adaptive responses: increased sleeping behavior, anorexia, fever, and activation of the hypothalmic–pituitary–adrenal (HPA) axis. Administration of increasing amounts of IL-1 systemically by itself is known to be capable of eliciting all of these responses. However, there is also evidence that indomethacin pretreatment of animals receiving IL-1 lessens the severity of centrally mediated acute-phase response, including activation of the HPA axis [47]. Furthermore, indomethacin administered iv or intracerebrally is also known to block IL-1-induced acute-phase responses such as HPA axis activation. Thus, the drug is apparently able to access the brain via cerebral circulation when administered iv [48]. In the following sections, brain vascular contributions, as well as perivascular brain cells, that participate in brain responses to systemic LPS and induce IL-1 responses will be considered, in addition to a consideration of related prostanoid contributions. Finally, there is also an examination of some of the hypothalamic brain regions activated by systemic LPS and/or IL-1 using the HPA axis and related neuroendocrine acute-phase responses as a functional endpoint.

5.2.1. Vascular and meningeal responses to lipopolysaccharide

In situ expression in brain of mRNA for IL-1β and TNF-α after systemic administration of LPS can be proportionately related to the amount of endotoxin administered to an experimental animal. Thus, subseptic doses of LPS (0.01–10 μg/kg bw) administered iv to rats induce IL-1β and TNF-α expression only in the CP, CVOs, and meninges [49]. In the same type of animal preparation, the expression of the cytokine transcription factor NFκB *inhibitor* (IκBα), the phosphorylation and degradation of which results in the translocation of NFκB to the nucleus (in association with subsequent cytokine expression), is also quite sensitive to subseptic doses of LPS as low as 0.1 μg/kg bw. In some of the CVOs, organum vasculosum of the lamina terminalis (OVLT) and subfornical organ (SFO), and in blood vessels of the brain parenchyma, sensitivity to lower doses of LPS was observed for IκBα mRNA expression, as well as induction of the same mRNA in numerous brain parenchymal cells following higher (1000 μg/kg bw) doses of LPS [49]. When a septic (2.5 mg/kg bw) dose of LPS was injected intraperitoneally in rats, two temporally and spatially distinct waves of IL-1β mRNA induction was observed: (1) cell labeling appeared in the CVOs [OVLT, SFO, median eminence, and area postrema (AP)], CP, leptomeninges and blood vessels: the time course for this first wave was an initial appearance at 0.5 h, peaking at 2.0 h, and declining at 4–8-h postinjection; (2) the second wave of IL-1β mRNA expression occurred 8–12 h postinjection in scattered small cells throughout the brain parenchyma, subsiding by 24 h postinjection. During this second wave, IL-1β mRNA labeling was not observed in neurons [50]. Thus, at earlier time points after LPS injection, IL-1β message production was induced at the BBB and in CVOs where the BBB is "leaky". After a delay of 6–10 h postinjection of a septic dose of LPS, IL-1β mRNA is primarily expressed in non-neural brain cells. These results suggest that the source of initial brain IL-1β activity after peripheral administration of a septic dose of LPS is derived from blood vessels (BBB and in CVOs), whereas more sustained IL-1β activity is derived from glial cells in the brain parenchyma [50].

More recently similar studies undertaken in mice addressed the question of whether or not a septic (1 mg/kg bw) dose of LPS triggers TLR4 receptors in the brain or whether proinflammatory cytokine gene expression is driven by LPS-responsive hematopoietic cells. Chimeric mice were generated that lacked TLR4 receptors in either the nonhematopoietic brain parenchyma or the hematopoietic system [51]. The results of this study revealed that increases in circulating plasma cytokines, IL-1β, TNF-α, IL-12, and INF-γ, required hematopoietic TLR4 signaling at all time points post-LPS injection that were examined. However, prolonged inflammatory cytokine gene induction (at 6 h) occurred within the brain only when nonhematopoietic brain cells carried a functional TLR4 receptor. The results of this group of experiments highlight an unexpected requirement for nonhematopoietic brain cellular TLR4 recognition of systemic LPS in order for persistent innate immune activation of the CNS [51].

The results of the studies of CVOs and adjacent brain structures for post-LPS expression of CD14 mRNA revealed that endotoxin administration caused a rapid expression of the message in both the rich vascular plexus of the OVLT and the brain parenchymal cells surrounding the CVO, followed by the expression of CD14 mRNA in microglia throughout the brain parenchyma [52]. These results are supportive of the sequel in which LPS first reaches structures devoid of the BBB to induce transcription of its own receptor, followed by increased CD14 biosynthesis within the brain parenchyma structures surrounding the OVLT, followed by expression in additional brain microglial cells. However, the interactions between CD14, TLR4, and LBP that may be required in order to bring about activation of NFκB and MAP kinases are not well understood [53]. Structures that are accessible by the bloodstream carrying LPS are known to contain mRNA encoding membrane-bound CD14 and TLR4 (e.g., CVOs, meninges, and CP) [54]. Furthermore, it has been conjectured that TLR4/CD14-bearing cells in the CVOs are activated first, followed by rapid transcription of proinflammatory cytokines in the brain parenchyma, for example, in this case the OVLT, which will lead to subsequent mircoglial activation throughout the brain. This scheme is dependent on the induction of TNF-α, however. Administration of systemic LPS does not always stimulate the gene encoding TLR4 in CVO cells [16], suggesting that TLR4 is located on the surface of the phagocytic cell population of CVOs, CP, and brain meninges [53].

Circulating LPS binds to LBP and to CD14, but in order to exert its pathogenic effects, LPS required TLR4, which is present on circulating monocytes and macrophages [15]. As was mentioned above, in brain, the cellular localization of TLR4 mRNA is reported to be expressed in leptomeninges (pia mater and arachnoid membrane), CP, and in cells resembling microglia [16]. Moreover, TLR4 receptors are necessary for signal transduction induced by LPS and the consequent expression of proinflammatory cytokines and their effects on brain cells. For example, in an *in vitro* brain development model, LPS-stimulated microglial-secreted factors that cause the death of oligodendrocyte precursor cells have no effect on the precursor cells if mixed glial cultures are obtained from mice bearing a loss of function in the TLR4 gene [54].

5.2.2. Perivascular versus vascular cell types in response to lipopolysaccharide

As mentioned above, either systemic administration or intracerebral injection of indomethacin is known to block prostanoid synthesis, as well as a number of other acute-phase reaction responses after IL-1β treatment [47,48,55,56]. Whether or not there is a central or peripheral site of action for IL-1β actions in this context is an unsettled question [57]. For example, IL-1β and LPS are both large enough molecules, so direct access to the brain in functionally significant quantities may not occur. Nevertheless, both substances when administered systemically are capable of inducing acute-phase effects. Given these constraints on both IL-1β and LPS for potential sites of action in brain, there is an increasing interest in the hypothesis that both molecules may bring about their acute-phase effects in brain through second messengers. This hypothesis has, in turn, precipitated a greater focus on cells at known brain barriers, as well as in CVO "windows" on brain barriers. The various vascular sites within the brain that are known to activate a cascade of proinflammatory cytokines upon exposure to LPS were described earlier in this section. Included in the list of cytokines were IL-1β and TNF-α. Both of these cytokines are also known to be independently capable of inducing the synthesis of vascular COX2, the rate-limiting enzyme in prostaglandin synthesis [58–60]. Thus, prostanoids are strong contenders as second messengers to convey cytokine and LPS messages across BBBs, as well as in CVOs. There remain, however, discrepancies in the results of studies designed to identify which of the vascular and/or related cell types demonstrate induced prostanoid synthesis after systemic administration of either IL-1β or LPS. Some of this controversy has begun to be sorted out by the results of a recent report on the effects of different dosage intensities of IL-1β and LPS [57].

Rats were treated with varying iv doses of either recombinant IL-1 (1.87, 10, or 30 μg/kg) or *E. coli* LPS (Sigma serotype 055:B5, 0.1, 2.0, or 100 μg/kg). All injections were per implanted jugular cannulas. Other groups of animals received chronic intracerebroventricle (ivc) cannulas in the lateral ventricle for ivc administration of indomethacin (10 μg in 5 μl of vehicle) 15 min prior to immune challenge. All animals were anesthetized and perfused for histological study at different time points. Immunochemical and hybridization histochemical methods were used to monitor COX2 expression alone or in conjunction with endothelial, perivascular, and glial cell markers in brain [57].

In comparisons of brain sections from IL-1β-challenged animals with those receiving 100 μg/kg of LPS, there was a greater number of COX2-immunoreactive (ir) cells distributed in a greater density over a greater number of blood vessels and capillaries in LPS-treated rats. Moreover, LPS-induced COX2-ir cells exhibited two types of morphologies and cell markers: (1) multipolar/polygonal types as seen after IL-1 treatment with ED2-positive marker for perivascular and meningeal macrophages and (2) distinctly round-shaped nuclear-staining cells costaining with the rat brain endothelial cell marker. The latter were confirmed to be endothelial using immunoelectron microscopic localization [57]. Finally, an LPS dosage study revealed that 0.1 μg/kg of endotoxin resulted in COX2 ir only in the multipolar/polygonal-shaped cells and the 2 μg/kg LPS dose induced COX2 ir in a few weakly stained presumptive endothelial cells, plus multipolar/polygonal cells. In the brains of the largest (100 μg/kg) LPS dose animals, the rounded (endothelial) cells were clearly the predominant cell type stained for COX2, with multipolar/polygonal cells appearing at a density roughly similar to that of the 2 μg/kg LPS dose [57].

The authors concluded that the perivascular multipolar cells exhibited the lower threshold to COX2-ir expression in response to either IL-1β or LPS treatment and that the COX2-ir enzyme expression by endothelial cells required one or more aspects of the more complex immune stimulus of LPS [57]. Although the results of this study support a prominent role for perivascular cells in the transduction of circulating cytokine or LPS "message-bearing" molecules, information about the transduction mechanisms are lacking.

5.3. Hypothalamic neural elements activated by lipopolysaccharide

We have reviewed published reports from several laboratories supporting the view that the non-neuronal cells associated with the cerebral vasculature (parenchymal vessels and within the CVOs), as well as the endothelial cells themselves that are capable of reacting to either cytokines or LPS as a potential critical link in the early steps for signaling the brain of an immune challenge. In addition, we have listed by categories the adaptive responses that are known to occur during infectious and/or inflammatory circumstances, including fever, somnolence, lethargy, anorexia, and activation of the HPA axis. Furthermore, we summarized findings that generally agreed that the cytokines are critically linked with LPS induction of these acute-phase effects via sharp increases in plasma levels of IL-1β and TNF-α after LPS. Finally, there is evidence for elevated intracerebral prostaglandin levels after LPS administration [61]. Moreover, we also noted that there is evidence to show that pharmacological interference with brain prostaglandin synthesis is known to disrupt cytokine-mediated acute-phase effects such as activation of the HPA axis [55,56].

Induced expression of COX2, the rate-limiting enzyme in prostanoid synthesis, is considered by many to be integral to the body's response to immune challenge [58]. Most recently Schiltz and Sawchenko [57] have presented evidence that intracerebral administration of indomethacin 15 min prior to an iv infusion of IL-1β in rats produced a consistent large (83%) reduction in IL-1β-induced Fos expression in the paraventricular nucleus (PVN) of the hypothalamus and the ventrolateral medulla (51%). Similar treatments are also capable of blocking HPA axis activation by IL-1β or LPS. It is generally accepted that stimulation of neurons residing in both the PVN and the ventrolateral medulla is required for complete activation of the HPA axis [62].

5.3.1. Intracerebral and peripheral indicators of hypothalmic–pituitary–adrenal axis stimulus by lipopolysaccharide

Recently, we have described two different experimental approaches for the study of the timing of cytokine contributions to activation of anterior hypothalamic area (AHA) neurons *prior* to HPA axis activation after LPS administration.

The first study was designed to help understand the chronology of intracerebral humoral and hormonal events in relation to peripheral changes associated with HPA axis activation. To this end, release of endogenous IL-1β was studied at two sites in individual adult male rats after a moderate (25 μg/100g bw) iv dose of LPS (*E. coli* 055:B5, Difco) known to activate the HPA axis [21]. To accomplish this goal, individual animals had their peripheral blood sampled via a chronically implanted jugular cannula, coincident with the sampling of their intrahypothalamic ECF through a chronically implanted (7d prior) push–pull cannula (PPC). The PPC guide cannula was stereotaxically positioned so that its inner cannula tip rested lateral to the PVN of the hypothalamus as previously described [63]. Earlier work by us with this type of animal preparation (chronic PPC brain implant) had demonstrated that radiometrically assayable [64] IL-1β concentrations were detectable in hypothalamic ECF withdrawn from freely behaving animals over 10–15-min intervals across several hours, and picogram levels of IL-1β were also detectable in hypothalamic ECF withdrawn from a location lateral to the PVN a few hours after a PPC implant to this region of the AHA [65].

In addition to assaying intracerebral hypothalamic ECF concentrations of IL-1β before and after peripheral iv administration of LPS per the jugular cannula, hypothalamic ECF was also assayed for the HPA axis neuropeptide corticotropin-releasing hormone (CRH). In the same freely behaving animals, peripheral blood samples were also removed, coincident with the intracerebral PPC sampling intervals. Blood plasma was separated for future assays of levels of the HPA axis anterior pituitary hormone adrenocorticotropin (ACTH) and the adrenal cortex hormone corticosterone, in addition to plasma levels of IL-1β. Thus, the data gathered from each individual animal included a profile of (1) hypothalamic ECF concentrations of IL-1β and CRH and (2) peripheral plasma levels of IL-1β, ACTH, and corticosterone at 15-min intervals before and after LPS administration. Baseline concentrations (two 15-min interval samples) were established for all hormones and cytokines when rats freely behaved in their home arenas. The LPS was administered per the cannula (time O) after the withdrawal of blood for future assay of baseline plasma hormone and IL-1β levels. Samples of hypothalamic ECF were also withdrawn via the PPC for baseline determinations of hypothalamic ECF of IL-1β and CRH [21]. Subsequent sampling intervals were at 15, 30, 45, 60, and 180 min post-zero time after LPS injection.

The data obtained from this examination of individual animal intracerebral ECF and peripheral plasma HPA responses to LPS administration demonstrated that an early (15-min after LPS administration) increase in hypothalamic ECF concentration of IL-1β sampled proximal to the PVN *precedes* endotoxin-induced activation of the HPA axis. The measured increase in hypothalamic ECF concentration of IL-1β was statistically significant and progressive over the 3-h post-LPS period of study: beginning at time 0 (77±10 pg/ml) to 393±88 during the first 15 min after LPS, reaching a peak of 867±237 pg/ml) at 45 min, declining to 91± pg/ml at 180-min post-LPS. Corticotropin-releasing hormone concentrations assayed in the same hypothalamic ECF samples (and intervals) as for IL-1β values indicated above were as follows: 41±17 pg/ml at time 0, 97±15 pg/ml at 15 min and increased significantly to 215±56 pg/ml at 30 min post-LPS. These elevated CRH concentrations in hypothalamic ECF remained similarly elevated at 45 min post-endotoxin (226±57 pg/ml), as well as at 60 min, and then declined significantly to basal levels (118±39 pg/ml) at 180 min after LPS administration. The time course for significant post-LPS increases in ACTH and corticosterone in these animals began at 30 min post-LPS. Significant increases in plasma IL-1β in the same samples also began at 30 min post-LPS with a twofold increase from baseline, followed by a more than 30-fold increase at 60 min, increasing again by one-third at the 3-h post-LPS sampling interval [21].

One may conclude that the introduction of LPS into the bloodstream sends a rapid signal to the hypothalamus and during the first 15 min of exposure, an early increase of IL-1β occurs in hypothalamic ECF, which is followed by the rise of CRH in hypothalamic ECF. A significant increase in peripheral blood ACTH and corticosterone follows at 30 min, signifying the full activation of the HPA axis. The increase in plasma Il-1β was delayed approximately by 15 min (reaching its maximum at 30 min) when compared to the rise of intrahypothalamic IL-1β.

5.3.2. *Hypothalamic coordination of neuroendocrine response to lipopolysaccharide*

Our second study was undertaken to further examine the contribution of neurons in the *AHA* to the activation of the HPA axis after LPS administration. Male rats were implanted with chronic jugular cannulas, followed 1 week later by bilateral chemical lesions made in the AHA. The lesions were produced to spare axons of passage, but killing some of the cells that reside in the AHA, including the PVN, which is a site of CRH synthesis, in addition to other neuropeptides [66]. These chemical lesions were accomplished by multiple, bilateral injections of the glutamate analog, *N*-methyl-DL-aspartate (NMA, 0.6 M) in microliter volumes of *artificial CSF (aCSF)*. Sham lesions were produced by bilaterally injecting similar quantities of aCSF alone [66].

The NMA injections destroyed most of the cells residing in the AHA, including the medial and lateral portions of the PVN. Verification of lesion size and location for all animals used in these experiments was accomplished by histological examination of their brains at the end of the experiments. The NMA injections resulted in conical-shaped bilateral lesions, 0.8–1.2 mm from the midline throughout the retrochiasmatic region of the hypothalamus. The anterior–posterior extent of the lesions began with the "base" of the cone at the posterior basal aspect of the preoptic area and extended to the rostral tip of the ventromedial nucleus in the mediobasal hypothalamus. The sham aCSF injections produced only a visible needle track of glial scars, representing the extent of multiple locations of intrahypothalamic needle penetration.

One week after producing the AHA chemical lesions, the animals in this study were evaluated for their HPA axis response to iv LPS (per cannula) over a 3-h period. In addition to plasma levels of ACTH and adrenal corticosterone, plasma prolactin (PRL), considered to be an immunostimulatory hormone, was also measured before and at 30-min intervals after administration of LPS (200 μg/100 g bw).

The results of this experiment provided additional evidence for brain coordination of the HPA axis and immunostimulatory neuroendocrine responses to LPS. Rats that had received either type of AHA injections (NMA or sham aCSF vehicle only) had plasma ACTH secretory responses to LPS that were significantly (twofold) lower than those of control animals (no hypothalamic manipulation prior to LPS) at all intervals of study over 3-h post-LPS. In addition, these differences in plasma ACTH release after LPS were 52–55% less than in controls when total area under the secretory curve (AUC) for ACTH plasma concentrations was calculated. The AUC calculated for hormone secretory release patterns is considered to be a more meaningful functional index when considering the effects of hormones [22]. In contrast, both group mean plasma corticosterone levels and AUC for the latter groups, measured in the same blood samples as for ACTH described above, had equivalent, similar large increases after LPS, in spite of the lower ACTH levels after AHA manipulations.

These findings illustrate several important points regarding the role of the brain in the activation of the HPA axis by LPS. In experimental designs intended to assess the role of hypothalamic contributions to LPS-induced acute-phase neuroendocrine reactions such as the HPA axis, variables selected for the study should be chosen based on their functional *proximity* to the variable under study. For example, in these AHA cell ablation studies designed to destroy neurons making and secreting CRH, the next hormone (most proximal variable) in the HPA axis under study is ACTH (versus corticosterone). If we had measured only corticosterone levels after AHA lesions and LPS administration, we would have concluded that this brain region is not an important site for transducing signals to the adrenal axis received from LPS circulating in systemic blood [22].

Consideration of the PRL data obtained from these animals further emphasizes this point. Unlike ACTH release, PRL synthesis and release by the pituitary is not modulated by a secondary target organ hormone, but instead PRL secretion into systemic blood by the pituitary is regulated by the release of an inhibitory and releasing hypothalamic hormone. Increasing systemic blood levels of PRL are immunostimulatory and are thought to reflect the relative amounts of each hypothalamic regulatory substance reaching the anterior pituitary gland [67]. Given this background, if one considers the amount of PRL release induced by LPS in our control animals (no AHA manipulations) as a reference response, the following "normal" PRL response to LPS (200µg/kg bw) was observed: plasma PRL levels started to rise 30 min after LPS and they peaked at the 1-h post-endotoxin sampling interval. In contrast, rats that had received bilateral AHA cellular damage, caused by NMA injection 1 week prior to LPS, were characterized by having no significant change in plasma PRL at all intervals of study (0, 30, 60, and 180 min after LPS administration). In contrast, rats that received only sham aCSF vehicle injections bilaterally did secrete significant increases in PRL; however, the peak increase in PRL concentration occurred earlier (30 min after LPS administration) when compared with control peaks at 1 hr after endotoxin. However, the total amount of PRL released (AUC) in sham surgery rats was similar to AUC for control PRL release after LPS (42.03 mg/ml per 180 min versus 51.07 µg/ml per 180 min), whereas the total amount of PRL secreted by rats with AHA damage caused by NMA was less than one-third of the above values (27% of control AUC equal to 13.87 µg/ml per 180 min) [22].

6. SUMMARY AND CONCLUSIONS

We have reviewed LPS-induced signaling pathways known to occur in individual cells in an attempt to link the latter to our understanding of brain systemic responses to LPS. Considered in relation to the latter were cell barriers, as well as locations in those body regions where fenestrated capillaries constitute "windows" permitting access to the gut, vagal nerves, and the brain proper. Also considered in this assessment of endotoxin effects on brain were related perivascular and organ parenchymal cells known to be capable of detecting (or transducing related signals of) the presence of LPS in blood and ECF. In addition to the activation of the HPA axis, a systemic acute-phase reaction takes place in response to LPS.

Evidence was presented in support of the hypothesis that intestinal epithelia and related vascular fenestrations and visceral (vagal) afferent nerves all contribute to the overall brain response to LPS in a dose-dependant fashion. Other, nonvagal neural and neuroimmune peripheral pathways for brain activation by LPS were also described. Activation of these peripheral portals for the entry of LPS allows the brain to sense extensively the presence of endotoxin in the body and bodily fluids. It was also suggested that an LPS-induced increase in epithelial permeability in an organ such as the GI tract could also provide for the amplification of the overall host response to LPS as a result of producing an immune cascade of cytokines occurring in the local vicinity of a breached epithelium. This cascade could then activate additional visceral afferent sensory nerves transmitting sensory impulses to the brainstem.

In the brain proper, endothelial cells, perivascular cells, as well as proximal neural and glial cells, are known to react to the presence of endotoxin in a dose-dependent fashion. The reaction included induction of the CD14 receptor, interactions of the latter with LBP and TLR4 by some unknown mechanism that consequently leads to initiation of the proinflammatory signal transduction pathways in a manner similar to activation in peripheral organs. We also reviewed evidence for related LPS induction of brain perivascular cellular cytokine production (IL1-β and TNF-α mRNAs and proteins), as well as COX2, the rate-limiting enzyme for prostanoid synthesis. These substances may complement each other in a time-dependent manner as convergent messengers contributing to the brain's response to LPS. This complementation may take the form of both local (paracrine and autocrine) and more remote messaging (cytokine/prostanoid-neural effects) in brain exposed to LPS via a systemic route. Some of our own evidence for the latter were presented in the form of a rapid (15 min) post-LPS sequel of IL-1β release in the vicinity of the PVN, which lead to activation of the HPA axis. Finally, when the cellular integrity of this brain region was selectively damaged by chemical (NMA) excitotoxic injections, coordination of neuroendocrine immunostimulatory and immunosuppressive responses to LPS were uncoupled in diverse ways. The PVN and AHA regions of the hypothalamus are also neurally linked to brainstem visceral sensory centers by afferent fibers from *the nucleus of the tractus solitarius* [68]. Thus, this region of the hypothalamus is well positioned to receive humoral and visceral afferent messages about the presence of LPS in the periphery, providing for coordination of brain's response to endotoxin.

ACKNOWLEDGMENTS

Research supported by National Institute of Mental Health Grant MH46808.

REFERENCES

1. Medzhitov R, Janeway CA. Innate immunity: The virtues of a nonclonal system of recognition. Cell 1997;91:295–8.
2. Janeway CA. The immune system evolved to discriminate infectious nonself from non-infectious self. Immunol Today 1992;13:11–6.
3. Cadenas S, Cadenas AM. Fighting the stranger-antioxidant protection against endotoxin toxicity. Toxicology 2002;180:45–63.
4. Schumann RR, Leong SR, Flaggs GW, Gray PW, Wright SD, Mathison JC, Tobias PS, Ulevitch RJ. Structure and function of lipopolysaccharide binding protein. Science 1990;249:1429–31.
5. Yu B, Wright SD. Catalytic properties of lipopolysaccharide (LPS) binding protein. Transfer of LPS to soluble CD14. J Biol Chem 1996;271:4100–5.
6. Jack RS, Fan X, Bernheiden M, Rune G, Ehlers M, Weber A, Kirsch G, Mentel R, Furll B, Freudenberg M, Schmitz G, Stelter F, Schutt C. Lipopolysaccharide-binding protein is required to combat a murine gram-negative bacterial infection. Nature 1997;389:742–5.
7. Bazil V, Horejsi V, Bandys M, Kristofova H, Strominger JL, Kostka W, Hilgert I. Biochemical characterization of a soluble form of the 53-kDa monocyte surface antigen. Eur J Immunol 1986;16:1583–9.
8. Ulevitch RJ, Tobias PS. Receptor-dependent mechanisms of cell stimulation by bacterial endotoxin. Annu Rev Immunol 1995;13:437–57.
9. Haziot A, Ferrero E, Kontgen F, Hijiya N, Yamamoto S, Silver J, Stewart CL, Goyert SM. Resistance to endotoxin shock and reduced dissemination of gram-negative bacteria in CD14-deficient mice. Immunity 1996;4:407–14.
10. Perera PY, Vogel SN, Detore GR, Haziot A, Goyert SM. CD14-dependent and CD14-independent signaling pathways in murine macrophages from normal and CD14 knockout mice stimulated with lipopolysaccharide or taxol. J Immunol 1997;158:4422–9.
11. Anderson J, Nagy S, Bjork L, Abrams J, Holin S, Anderson U. Bacterial toxin-induced cytokine production studied at the single-cell level. Immunol Rev 1992;127:69–6.
12. Medzhitov R, Preston-Hurlburt P, Janeway CA. A human homologue of the Drosophila Toll protein signals activation of adaptive immunity. Nature 1997;388:394–7.

13. Hoebe K, Du X, Georgel P, Janssen E, Tabeta K, Kim SO, Goode J, Lin P, Mann N, Mudd S, Crozat K, Sovath S, Han J, Beutler B. Identification of Lps2 as a key transducer of MyD88-independent TIR signaling. Nature 2003;424:743–8.
14. Beutler B. TLR4: Central component of the sole mammalian LPS sensor. Curr Opin Immunol 2000;12:20–6.
15. Zhang G, Ghosh S. Toll-like receptor-mediated NF-kappa β activation: A phylogenetically conserved paradigm in innate immunity. J Clin Invest 2001;107:13–19.
16. Laflamme N, Rivest S. Toll-like receptor 4: The missing link of the cerebral innate immune response triggered by circulating gram-negative bacterial cell wall components. FASEB J 2001;15:155–63.
17. Christman JW, Lancaster LH, Blackwell TS. Nuclear factor-kB: A pivotal role in the systemic inflammatory response syndrome and new target for therapy. Intens Care Med 1998;24:1131–8.
18. Chen Z, Hagler J, Palombella VJ, Melandri F, Scherer D, Ballard D, Maniatis T. Signal-induced site-specific phosphorylation targets Ikβα to the ubiquitin-proteasome pathway. Genes 1995;9:1586–97.
19. Rice NR, Ernst MK. In vivo control of NF-kβ activation by Ikβα. EMBO 1993;12:4685–746.
20. Rothwell NJ, Hopkins SJ. Cytokines and the nervous system: II. Actions and mechanisms of action. Trends Neurosci 1995;18:130–6.
21. Ma XC, Chen LT, Oliver J, Horvath E, Phelps CP. Cytokine and adrenal axis responses to endotoxin. Brain Res 2000;861:135–42.
22. Phelps CP, Dong JM, Chen LT, Menzies RA. Plasma interleukin-1β, prolactin, ACTH and corticosterone responses to endotoxin after damage of the anterior hypothalamic area. Neuroimmunomodulation 2001;9:340–51.
23. Rosenberg HF, Gallin JI. Inflammation. In Fundamental Immunology, 4th Edn. Paul WE, Ed.; Philadelphia, PA: Lippincott–Raven, 1999; pp. 1051–66.
24. Martich GD, Danner RL, Coska M, Suffredini AF. Detection of interleukin-8 and tumor necrosis factor in normal humans after intravenous endotoxin: The effects of anti-inflammatory agents. J Exp Med 1991;173:1021–4.
25. Kishimoto TK, Anderson DC. The role of integrins in inflammation. In Inflammation, 2nd Edn. Gallin JI, Goldstein R, and Snyderman R, Eds.; New York: Raven, 1992; pp. 353–406.
26. Kopf M, LeGros G, Coyle AJ, Kosco-Vilbois M, Brombacher F. Immune responses of IL-4, IL-5, IL-6 deficient mice. Immunol Rev 1995;148:45–69.
27. Shelag PB. Interleukin-6-type cytokines in vivo: Regulated bioavailability. Proc Soc Exp Biol Med 1996;213:238–47.
28. Baatout S. Interleukin-6 and megakaryocytopoiesis and update. Ann Hematol 1996;73:157–62.
29. Keep RF. The blood–brain barrier. In The Neuronal Environment: Brain Homeostasis in Health and Disease. Walz W, Ed.; Totowa, NJ: Humana, 2002; pp. 277–307.
30. Anderson JW, Ferguson AV. Circumventricular organs. In The Neuronal Environment: Brain Homeostasis in Health and Disease. Walz W, Ed.; Totowa, NJ: Humana, 2002; pp. 309–39.
31. Weller RO. Choroid plexus and the cerebrospinal–interstitial fluid systems. In The Neuronal Environment: Brain Homeostasis in Health and Disease. Walz W, Ed.; Totowa, NJ: Humana, 2002; pp. 261–75.
32. Berkes J, Viswanathan VK, Savkovic SD, Hecht G. Intestinal epithelial responses to enteric pathogens: Effects on the tight junction barrier, ion transport, and inflammation. Gut 2003;52:439–51.

33. Davson H, Segal MB. Morphological Aspects of the Barriers. Physiology of the CSF and Blood–Brain Barriers. Boca Raton, FL: CRC, 1996; pp. 93–192.
34. Nabeshima S, Reese TS, Landis DMD, Brightman MW. Junctions in the meninges and marginal glia. J Comp Neurol 1975;164:127–70.
35. Pape LG, Katzman R. ^{42}K distribution in brain during simultaneous ventriculocisternal and subarachnoid perfusion. Brain Res 1972;38:49–69.
36. Alcolado R, Weller RO, Parrish EP, Garrod D. The cranial arachnoid and pia mater in man: Anatomical and ultrastructural observations. Neuropathol Appl Neurobiol 1988;14:1–17.
37. Brightman MW, Klatzo I, Olsson Y, Reese TS. The blood–brain barrier to proteins under normal and pathological conditions. J Neural Sci 1970;10:215–39.
38. Ross MH, Kaye GI, Pawlina W. Digestive System II: Esophagus and Gastrointestinal Tract. Histology. A Text and Atlas. Philadelphia, PA: Lippincott, Williams and Wilkins, 2003; pp. 475–509.
39. Hansen MK, Nguyen KT, Fleshner M, Goehler LE, Gaykema RP. Effects of vagotomy on circulating levels of endotoxin, cytokines, and corticosterone following intraperitoneal lipopolysaccharide. Am J Physiol 2000;278:R331–6.
40. Maier SF, Watkins LR, Nance DM. Multiple routes of action of interleukin-1 on the nervous system. In Psychoneuroimmunology, 3rd Edn. Ader R, et al., Eds.; New York: Academic, 2001; pp. 563–83.
41. Miyake K, Ogata H, Nagai Y, Akashi S, Kimoto M. Innate recognition of lipopolysaccharide by Toll-like receptor 4/MD-2 and RP105/MD-1. J Endotoxin Res 2000;6:389–91.
42. Abreu MT, Arnold ET, Thomas LS, Gonsky R, et al. TLR_4 and MD-2 expression is regulated by immune-mediated signals in human intestinal epithelial cells. J Biol Chem 2002;277:20431–7.
43. Suzuki M, Hisamatsu T, Poldolsky DK. Gamma interferon augments the intracellular pathway for lipopolysaccharide (LPS) recognition in human intestinal epithelial cells through coordinated up-regulation of LPS uptake and expression of the intracellular Toll-like receptor 4-MD-2 complex. Infect Immun 2003;71:3503–11.
44. O'Dwyer ST, Michie HR, Ziegler TR, Revhaug A, et al. A single dose of endotoxin increases intestinal permeability in healthy humans. Arch Surg 1988;123:1459–64.
45. Berthoud H-R, Neuhuber WL. Functional and chemical anatomy of the afferent vagal system. Autonom Neurosci Basic Clin 2000;85:1–17.
46. Goehler LE. Cytokines in neural signaling to the brain. In NeuroImmune Biology (this volume), Cytokines and the Brain. Korneva E and Phelps C, Eds.; Amsterdam: Elsevier, 2007.
47. Lacroix S, Rivest S. Functional circuitry in the brain of immune-challenged rats: Partial involvement of prostaglandins. J Comp Neurol 1997;387:307–24.
48. Ericsson A, Arias C, Sawchenko PE. Evidence for an intramedullary prostaglandin-dependent mechanism in the activation of stress-related neuroendocrine circuitry by intravenous interleukin-1. J Neurosci 1997;17:7166–79.
49. Quan N, Stern EL, Whiteside MB, Herkenham M. Induction of pro-inflammatory cytokine mRNAs in the brain after peripheral injection of subseptic doses of lipopolysaccharide in the rat. J Neuroimmunol 1999;93:72–80.
50. Quan N, Whiteside M, Herkenham M. Time course and localization patterns of interleukin-1β messenger RNA expression in brain and pituitary after peripheral administration of lipopolysaccharide. Neuroscience 1998;83:281–93.

51. Chakravarty S, Herkenham M. How does the brain perceive endotoxin (LPS) challenge? Program No 104.13 2003 Abstract Viewer/Itinerary Planner. Washington, DC: Society for Neuroscience.
52. Lacroix S, Feinstein D, Rivest S. The bacterial endotoxin lipopolysaccharide has the ability to target the brain in upregulating its membrane CD14 receptor within specific cellular populations. Brain Pathol 1998;8:625–40.
53. Rivest S, Nadeau S, Lacroix S, Laflamme N. Proinflammatory signal transduction pathways in the CNS during systemic immune response. In NeuroImmune Biology, Vol. 1, New Foundation of Biology. Berczi I and Gorczynski RM, Eds.; Elsevier, 2001; pp. 163–87.
54. Lehnardt S, Lachance C, Patrizi S, Lefebvre S, et al. The Toll-like receptor TLR4 is necessary for lipopolysaccharide-induced oligodendrocyte injury in the CNS. J Neurosci 2002;22:2478–86.
55. Morimoto A, Murakami N, Nakamori T, Sakata Y, Watanabe Y. Possible involvement of prostaglandin E in development of ACTH response in rats induced by human recombinant interleukin-1. J Physiol (Lond) 1989;411:245–56.
56. Watanabe T, Morimoto A, Sakata Y, Murakami N. ACTH response induced by interleukin-1 is mediated by CRF secretion stimulated by hypothalamic PGE. Experientia 1990;46:481–4.
57. Schiltz JC, Sawchenko PE. Distinct brain vascular cell types manifest inducible cyclooxygenase expression as a function of the strength and nature of immune insults. J Neurosci 2002;22:5606–18.
58. Lacroix S, Rivest S. Effect of acute systemic inflammatory response and cytokines on the transcription of the genes encoding cyclooxygenase enzymes (COX-1 and COX-2) in the rat brain. J Neurochem 1998;70:452–66.
59. Blais V, Rivest S. Inhibitory action of nitric oxide on circulating tumor necrosis factor-induced NF-kappaβ activity and COX-2 transcription in the endothelium of the brain capillaries. J Neuropathol Exp Neurol 2001;60:893–905.
60. Mark KS, Trickler WJ, Miller DW. Tumor necrosis factor-alpha induces cyclooxygenase-2 expression and prostaglandin release in brain microvessel endothelial cells. J Pharmacol Exp Ther 2001;297:1051–8.
61. Sehic E, Szekely M, Ungar AL, Oladehin A, Blatteis CM. Hypothalamic prostaglandin E2 during lipopolysaccharide-induced fever in guinea pigs. Brain Res Bull 1996:39:391–9.
62. Turnbull AV, Rivier CL. Regulation of the hypothalamic–pituitary–adrenal axis by cytokines: Actions and mechanisms of action. Physiol Rev 1999;79:1–71.
63. Kalra PS, Phelps CP, Sahu A, Dube MG. Quantitation of in vivo and in vitro neuropeptide secretion under the influence of steroids. Neuroprotocols 1992;1:87–97.
64. Bristow AF, Mosley S, Poole S. Interleukin-1β production in vivo and in vitro in rats and mice measured using specific immunoradiometric assays. J Mol Endocrinol 1991;7:1–7.
65. Phelps CP, Sahu A, Oliver J, Ma X, Kalra PS, et al. In vivo release of interleukin-1β into hypothalamic extracellular fluid in rats: Effects of repeated sampling. Brain Res 1995;688:193–7.
66. Sisk C, Nunez A, Thebert M. Differential effects of electrolytic and chemical hypothalmic lesions on LH pulses in rats. Am J Physiol 1988;255:E583–90.
67. Dorshkind K, Horseman ND. The role of prolactin, growth hormone, insulin-like growth factor-1 and thyroid hormones in lymphocyte development and function: Insights from genetic models of hormone and hormone deficiency. Endocr Rev 2000;21:292–312.
68. Saper CB. Central Autonomic System. The Rat Nervous System, 2nd Edn. Paxinos G, Ed.; New York: Academic, 1995; pp. 107–35.

Edited by C. Phelps and E. Korneva

Cytokines in Demyelinating Diseases

SERGEY A. KETLINSKIY[1] and NATALIA M. KALININA[2]

[1]*State Research Center of Highly Pure Biopreparations, FMBA, St. Petersburg;*
[2]*All-Russian Center of Emergency and Radiation Medical Research, St. Petersburg, Russia*

ABSTRACT

Among the autoimmune diseases, multiple sclerosis (MS) has been up to now best understood with regard to the pathogenesis of inflammation and demyelination, which affects the white matter of the brain. In MS brain are founded focal infiltrates of T cells and macrophages, axonal injury, and loss of neurological functions [1–3]. The disease typically manifests between the ages of 20 and 40 and affects women twice as often as men. It is the major cause of neurological disability in young people in the Western Hemisphere. MS is generally categorized as being either relapsing–remitting (RR) or primary-progressive (PP) in onset. The RR form of the disease is characterized by a series of attacks that result in varying degrees of disability from which the patients recover partly or completely, usually followed by a remission period of variable duration before another attack. The progressive forms of the disease lack the acute attacks.

Most authors see the cases of MS development as enigmatic, although it seems obvious that immune attack on the white matter of the brain is primarily implicated in MS. The hypothesis that autoreactive type 1 T helpers (Th1) are involved in this attack is based on the identification of these cells in experimental autoimmune encephalitis (EAE). More as well as on the fact that adoptive transfer of autoreactive myelin-specific T lymphocytes to syngeneic mice results in autoimmune damage of their brain. The involvement of pathogenic autoimmune cells in EAE development may be associated with their ability for dichotomy into Th1 cells, which produce Interferon (INF)-γ and induce EAE, and Th2 cells having an antiencephalitis activity.

EAE development is associated with immune inflammation mediated by cytokines produced by Th1 cells [interferon (IFN)-γ, tumor necrosis factor (TNF), granulocyte–macrophage colony-stimulating factor (GM-CSF), interleukin (IL)-17 and antigen-presenting cells (APCs) (IL-1, IL-6, IL-12, and IL-23), which results in myelin and axon degeneration and significant neural cell loss (FasR-FasL, TNF-related apoptosis inducing ligand–TNF-related apoptosis inducing ligand receptor , TRAIL–TRAILR).

However, newer data challenge this paradigm. First of all, these are the data showing that mice with targeted IL12p35, IL-23p19, and tumor necrosis factor-α receptor 1 (TNFR1) gene disruption have an increased susceptibility to EAE. The presence of the above molecules in mice confers resistance to EAE.

Moreover, several murine strains are capable of reparation of EAE lesions, transforming growth factor (TGF)-β being believed to play a major role in murine brain repair. In this review we attempt to analyze the mechanisms of susceptibility and resistance to EAE with emphasis on the cytokine modulation of immune responses and associated inflammation resulting in brain white matter demyelination.

1. CYTOKINE EXPRESSION DURING MULTIPLE SCLEROSIS AND EXPERIMENTAL AUTOIMMUNE ENCEPHALITIS

Data on cytokine level in multiple sclerosis (MS) development are limited by the availability of clinical samples and usually relate to the analysis of cerebrospinal fluid (CSF), brain autopsy, and isolated blood cells from MS patients. Such studies focused on the identification of biological makers for monitoring the activity of the disease, the associated immunological processes, and the efficiency of therapy for MS.

However, it is hard to draw conclusions about the pathophysiological significance of cytokines in autoimmunity basing exclusively on the clinical findings. Therefore, data obtained in both clinical and experimental studies will be discussed.

To study the role of cytokines in the pathogenesis of experimental autoimmune encephalitis (EAE), methods of cytokine blocking by specific antibodies as well as methods of targeting of the genes encoding the studied cytokines are often used (Table 1).

Cytokines play a key role in immune-mediated inflammatory disease of the central nervous system (CNS) such as MS. Furthermore, cytokines play role in brain development and in brain homeostasis. Since it seems that during inflammatory diseases many developmental processes are recapitulated, the role of cytokines may be even more important.

To date, the majority of cytokines and their receptors have been known and found in many CNS cells, including neurons [4] (Table 2). These data were expanded by the discovery of new cytokines, IL-23 and IL-27, which are related to the IL-12 family and the recently found naturally occurring new population of regulatory T cells (Th-1_{17} or Th17). All of the above cytokines and cells are important for the regulation of immunity in MS [5–9].

The transcriptional profiling of 56 relevant genes in brain specimens from eight MS patients and eight normal controls was performed by kinetic reverse transcription polymerase chain reaction (RT-PCR). Significant differences between MS and control samples were observed [10] (Table 3).

The characterization of the transcriptome in MS lesions may provide a better understanding of the mechanisms that generate and sustain the pathogenic immune response in this disease. Key inflammatory type molecules such as interleukin-2 (IL-2), interferon (IFN)-γ, and tumor necrosis factor (TNF)-α did not display consistent and reproducible expression patterns. In contrast, concurrently with elevated expression of IL-5 and IL-6, other cytokines and receptors, such as the IL-6 receptor (IL-6/IL6R), prototypic Th2 molecules like IL-4, IL-10, and IL-13, were undetected [10].

The most important and challenging problem in MS research is to understand the exact role played by the different cytokines in the CNS and the periphery in either the regulatory or the effector mechanisms in the development of inflammatory demyelination.

1.1. Interleukin-12 cytokine family in multiple sclerosis

It is known that peripheral blood monocytes of MS patients produce increased amounts of IL-12, which induces INF-γ in T and natural killer (NK) cells [11] and corresponds to Th1 type

Table 1. Development of EAE in cytokine knockout mice

Knockout mice	EAE inducer	Results	References
IL-6$^{-/-}$	MOG	IL-6$^{-/-}$ mice are protected from EAE	[124]
TNF-α$^{-/-}$, LT$^{-/-}$	EAE	Knockout mice are more susceptible to EAE than wild-type mice are	[125]
TNF-α$^{-/-}$	MOG	Knocking out makes resistant mice susceptible to EAE; TNF-α treatment reduces disease in TNF-α$^{-/-}$ mice	[127]
TNFR1$^{-/-}$	MOG_{35-55} + Ptoxin	Knockout mice do not develop EAE	[58]
IFN-γ$^{-/-}$	MOG	EAE-susceptible B10.PL mice develop a more severe disease; EAE-resistant BALB/c mice become susceptible to EAE	[61,126]
IFN-γ$^{-/-}$	MOG	Anti-IL-12 antibodies prevent mice from EAE	[128]
IFN-γR$^{-/-}$	MOG	Knockout mice are more susceptible to EAE	[59]
IFN-γ R$^{-/-}$	MOG rat, 40–55	Levels of oxidative stress, tissue-protective antioxidant factors, and apoptosis were increased in comparison with immunized wild-type mice	[128]
IL-4$^{-/-}$	MOG	IL-4$^{-/-}$ mice do not develop severe forms of EAE, in comparison with wild-type mice	[88]
IL-12$^{-/-}$	MOG_{35-55}	Susceptibility to EAE is enhanced	[25]
IL-12p35$^{-/-}$	MOG_{35-55}	Susceptibility to EAE is enhanced	[25]
IL-12p40$^{-/-}$	MOG_{35-55}	Deficient mice are resistant to EAE	[25]
IL-18$^{-/-}$	MOG_{35-55}	Mice are resistant to EAE	[122]
IL-17$^{-/-}$	MOG_{35-55}	Mice are resistant to EAE	[131]
IL-23p19$^{-/-}$	MOG_{35-55}	Suppression of EAE development	[6]
IL-27p28$^{-/-}$	MOG_{35-55}	Suppression of ongoing EAE	[40]
Osteopontin$^{-/-}$	MOG_{35-55}	Mice are resistant to EAE	[132]
GM-CSF$^{-/-}$	MOG_{35-55}	Mice are resistant to EAE; antigen-dependent proliferation of splenocytes is decreased	[129]

immune response. Multiple sclerosis patients displayed elevated IL-12 receptor-b1 (IL-12Rβ1) and IL-12Rβ2 expression in phytohemagglutinin (PHA)-activated T cells compared to healthy subjects [12]. The net effect of IL-12 might be augmented in MS by the elevated expression of its receptor. IL-12 mRNA dramatically increases before MS relapses [13]. IL-12 mRNA expressing lymphocyte counts correlates with MS relapses at different stages of the disease [12,13]. The progressive course of MS features stably elevated IL-12 mRNA levels in peripheral blood cells [14,15]. Endogenous IL-10 cannot suppress IL-12 production despite the normal level of IL-10 receptor (IL-10R) mRNA [13,16]. An increase in IL-12 is found in the blood serum of patients with progressive MS [13,17]. Increased level of IL-12p40 in CSF and serum of MS patients is associated with intracephalic inflammation and myelin degradation [18].

IL-12 is expressed in demyelination lesions, the severity of inflammation correlating with the evaluation of local IL-12 production by microglial cells [14,19]. The increasing of IL-12 in the

Table 2. Expression of cytokines and their receptors in the brain [4]

Cytokine	Neurons	Astrocytes	Oligodendrocytes	Microglia
IL-1α, IL-1β	C/R	C/R	C/R	C
IL-1Ra		C		
IL-2	C/R	C	R	R
IL-3	C/R	C/R	R	R
IL-4		R	R	R
IL-5		C		C/R
IL-6	C/R	C/R		C/R
IL-7		R	R	R
IL-8	R	C/R		R
IL-9		R		
IL-10		C/R		C/R
IL-11		C		
IL-12		C	C?	C
IL-13				
IL-14				
IL-15		C		C
IL-16				C
IL-17			C	
IL-18				
TNF	R	C/R	R	C/R
LT				
IFN-γ		C/R		R
IFN-α				
IFN-β				
TGF-β		C/R	C/R	C/R
GM-CSF	R	C/R	R	R
M-CSF		C	R	C/R
IL-23		C		
IL-27				

C, cytokine; R, receptor for cytokine.

CNS results in the penetration of lymphocytes though the blood–brain barrier, Th1-type cells activation, and increased IL-1α and -1β and TNF-α and -β production. Higher percentages of IL-12Rβ1- and IL-12Rβ2-positive T cells in the CSF, compared to those found in blood, were observed in MS [18]. It has been shown that the defective regulation of IFN-γ and IL-12 expression in progressive MS is caused by endogenous IL-10 [16].

IL-23, which is expressed by dendritic cells (DC), may be important for the pathogenesis of autoimmune diseases such as MS. Dendritic cells isolated from MS patient blood have been found to produce significantly more quantity of IL-23 than those from healthy donors. Moreover, T lymphocytes isolated from patient's blood were found to produce increased amounts of IL-17. These data are consistent with the results obtained in the murine MS model (EAE) as far as the former suggest that IL-23, contrary to IL-12, activates Th17 cells to produce IL-17, which has proinflammatory activity and is one of the factors that promote autoimmune diseases. The authors have also shown that the use of antisense oligonucleotides specific for IL-23 and IL-12p40 leads to the suppression of IL-23 and IL-12 gene expression, which correlates with increases of IL-10 production. These data suggest possibilities for immunomodulation in MS [20].

Table 3. Cytokine/chemokine gene expression in brain specimens of MS patients and normal controls [10]

Cytokines and their receptors	Average increase in gene expression for MS samples (relative to normal control)	*P* value
IL-1β	3.07	0.01
IL-1R	1.56	NS
IL2Rγ	4.01	0.01
IL3	2.6	0.05
IL-5	2.22	0.01
IL-6	4.13	0.01
IL-6R	3.51	0.01
IL-8	3.49	NS
IL-9	1.9	0.05
IL-12Rβ1	2.83	0.01
IL-12Rβ2	1.14	NS
IL-15	2.98	0.01
IL-18	2.7	0.01
TGF-β1	2.77	0.05
TGF-β2	1.94	0.05
TGF-β3	2.39	0.05
TNFR1	2.83	0.01
TNFRp55	2.41	0.01
TNFRp75	1.39	NS
IFN-α	1.53	NS

IL-1α, IL-2Rα, IL-4, IL-7, IL-10, IL-11, IL-12p35, IL-12p40, IL-13, IFN-γ, and CCR8 are below detection threshold. NS, no significant difference.

1.1.1. Interleukin-12 in experimental autoimmune encephalitis

IL-12 (IL-12p70) is a heterodimeric protein, which is *N*-glycosylated and consists of two subunits, p40 and p35, linked by disulfide bridges. IL-12p40 cDNA encodes the mature protein consisting of 306 amino acids, 10 cysteine residues, and 4 potential protein glycosylation sites. cDNA of the p35 subunit encodes the mature protein consisting of 197 amino acids, 7 cysteine residues, and 3 glycosylation sites. The subunits contain 20 and 10% of carbohydrates, respectively.

IL-12 receptor consists of two subunits, IL-12Rβ1 and IL-12β2, each showing a high homology to p130, which is the common receptor for IL-6-like cytokine superfamily. Both are type 1 transmembrane glycoproteins. The high-affinity receptor is formed only when both subunits are coexpressed. IL-12R expression is found only in activated T lymphocytes and NK cells, whereas cell types that produce IL-12 are far more numerous and include activated peripheral blood monocytes, macrophages, Langerhans cells, DC isolated from the spleen or peripheral blood and derived from the bone marrow, monocytes, neutrophils, mesoglia, keratinocytes, and endothelial cells.

Signal transduction from IL-12 involves the IL-12Rβ2 subunit of IL-12-receptor, which first activates two phosphokinases, Jak2 and Tyk2, which then phosphorylates and thus activates the transcription factors STAT1, STAT3, STAT4, and STAT-5. The 85-kDa protein associated with IL-12Rβ1 also becomes phosphorylated in response to IL-12 and thus may be another component of IL-12 receptor. Lymphocytes isolated from $STAT4^{-/-}$ mice show impaired responses to IL-12, which results in a shift toward Th2 type immune responses. Specific cellular effects of IL-12 result from STAT4 activation [21].

Early data suggested that IL-12 being the key inducer of Th1 cell development is not involved in MS pathogenesis, because inflammatory demyelination may occur in the absence of either IL-12 or IL-12Rβ2 [22].

Most data suggesting a pathogenetic role for IL-12 in MS and EAE development were published before the discovery of IL-21 and IL-27, which are structurally similar to IL-12 and have many activities in common with IL-12 yet each featuring unique properties. Therefore, data interpretations were not quite adequate. Thus, it was shown that administration of exogenous IL-12 to rats with autoimmune encephalitis increased the number of disease relapses assessed clinically by the number and gravity of paralysis fits. These effects of IL-12 were suggested by the authors to be mediated by reactivation of inflammatory T cells and macrophages in the brain by IL-12 and, also, by initiation of the encephalitogenic responses of Th1 cells derived from the pool of the already sensitized myeline basic protein (MBP)-specific T cells [23].

Following this line of investigation, some authors consider increased rates of EAE complications resulting from several substances is due to the enhancement of IL-12 activity. Vice versa, EAE inhibition may be associated with IL-12 blockage. These conclusions are based on using antibodies that recognize IL-12, but unfortunately no data are provided about the properties of these antibodies. Thus, it is unknown which IL-12 subunits are recognized by the antibodies, whether they can bind the active sites of IL-12, or whether they block signal transduction from IL-12 receptor, among many other issues.

The initially incorrect posing of the problem precludes an understanding of how cyclooxygenase (COX)-2 inhibitors can prevent EAE, while some authors believe that this phenomenon is explained by changes in the signals from IL-12 that block the generation of Th1 cells and, hence, the development of EAE [24]. It remains obscure why inhibition of EAE is dependent on IL-12 expression.

To elucidate the role of IL-12 in the pathogenesis of EAE, the experiments were carried out that involved deleting the genes that code for IL-12p40 and/or IL-12p35 followed by EAE induction using the common technique of immunization with the myelin oligodendrocyte glucoprotein (MOG) peptide (35–55). $IL\text{-}12p40^{-/-}$ mice were resistant to EAE, whereas $IL\text{-}12p35^{-/-}$ mice were susceptible to it. The spinal cord was infiltrated with mononuclear cells in wild-type and $IL\text{-}12p35^{-/-}$ mice and was normal in $IL\text{-}12p40^{-/-}$ mice. $IL\text{-}12p35^{-/-}$ mice developed a weak Th1 response associated with reduced IFN-γ amount. On the contrary, $IL\text{-}12p40^{-/-}$ mice developed the Th2 response to MOG (33–35), which correlated with resistance to EAE. Wild-type and $IL\text{-}12p35^{-/-}$ but not $IL\text{-}12p40^{-/-}$ mice showed TNF production by microglia and CNS infiltration with macrophages and CD4+ lymphocytes [25].

EAE development in $IL\text{-}12p40^{-/-}$ and $IL\text{-}12p35^{-/-}$ mice was similar to the one described above [26]. To make the results more consistent, the authors performed adoptive transfer of encephalitogenic lymphocytes with the objective of identifying the conditions that are needed for EAE development. It was found that cell transfer induced EAE in 40% of $p40^{-/-}$ mice and in 100% of wild-type mice. Experimental autoimmune encephalitis was apparent after 11.5 days in the affected $p40^{-/-}$ mice and after 8.3 days in wild-type mice. It was also found that bone marrow transplantation completely restored Th1 immunity. Using chimeric bone marrow of mice challenged with MOG, it was shown that no Th1-response recovery occurred in $p35^{-/-}$ and $p40^{-/-}$ mice. The final conclusion of the authors was that the heterodimeric IL-12p70 does not contribute to EAE pathogenesis, whereas p40 does play a key role in EAE, probably because p40 is present in other cytokines, for example, IL-23. The absence of p35 confers resistance to EAE by a Th2-dependent mechanism.

It has been shown [27] that astrocytes, which may be APCs, express IL-12p40 and IL23p40 and that their blockage with antibodies prevents EAE development. Signal transduction from IL-12 into cells is known to be mediated by the receptor IL-12β2. To shed light on the role of this receptor in EAE pathogenesis, mice with different degrees of deficiency for IL-12Rβ2 were used. In IL-12Rβ2-deficient mice, susceptibility to EAE was increased. IL-12Rβ2$^{-/-}$ knockouts showed increased inflammation resulting in multiple demyelinization foci. The Th1 response to MOG in deficient mice featured high lymphocyte proliferation and IL-2, IL-17, and IFN-γ production. Interestingly, a high level of IL-23p19 expression was observed in splenocytes of IL-12Rβ2$^{-/-}$ mice suggesting its contribution to increased demyelination foci and mortality rate in deficient mice. It is hypothesized that IL-23p19 compensates for the absence of responsivity in IL-12Rβ2$^{-/-}$ mice affected by EAE and is responsible for the multiplicity of inflammation foci in the CNS [28,29].

To study the other subunit of the receptor, IL-12Rβ1, which binds IL-23 more strongly than IL-12, receptor knockout mice were developed. The most important finding in such mice was that they were completely resistant to MOG-induced EAE, which might be explained by the absence of antibody-specific Th2 response. It should be noted that DC lead to Th2 response in IL12Rβ1$^{-/-}$ mice and to Th1 response in wild-type mice. This may be explained by the fact that, in IL12Rβ1$^{-/-}$ mice, IFN-γ is produced at a low level, which shifts the immune response toward Th2. This explains why wild-type mice develop Th1 response making them more susceptible to EAE [28,29]. Thus, it was suggested [28,29] that EAE and MS may be treated by blocking IL-12Rβ2, IL-12p40, or IL-23p19.

In experiments with the nonhuman primate *Callithrix jacchus*, which is very highly susceptible to EAE, it was shown that human antibodies against human IL-12/23p40 cause a 30- to 65-day delay in EAE development induced with recombinant human MOG. Injecting the antibodies at the advanced stages of EAE was also effective in reducing the accumulation of EAE lesions, which affected 160.4 cm^3 of the brain tissue at the time of antibody injection and 17.0 cm^3 after the injection. Survival time was significantly longer in the monkeys injected with the antibody (12 days) than in the sham-injected monkeys (28 days) [30].

1.1.2. Interleukin-23 in experimental autoimmune encephalitis

Il-23 is a heterodimeric molecule consisting of two subunits, p40 and p19 [31]. The p40 subunit is identical to that included in IL-12. cDNA of the p19 subunit codes for 189 or 196 amino acids in humans or mice, respectively. Both proteins contain five cysteines. N-terminal glycosylation is absent. The homology between murine and human p19 is 70%.

The effects of IL-23 are mediated by the receptor IL-12Rβ1 and of IL-12 by IL-12Rβ2. IL-12 signal transmission involves the activation of the transcription factor STAT-4, that is, it is similar to that of IL-12 [31].

IL-23 is produced in the CNS by resident cells, which regulate the encephalitogenic potential of T cells. It has been detailed [27] that IL-12 is produced by astrocytes constitutively and IL-23, after stimulation with MBP. Astrocytes, which play the role of APCs, produce IL-23 when a peptide from MBP is presented to activated Th1 cells. Antigen-presenting astrocytes activated with INF-γ induce the proliferation of T lymphocytes reactive to MBP. Antibodies to IL-12/-23p40 block this proliferation.

Astrocytes under inflammatory conditions express all of the subunits that comprise IL-12 and IL-23. Their ability to present antigens to encephalitogenic T cells may be blocked. Differences in the expression of IL-12 and IL-23 by microglia of adult mice have been demonstrated. In early EAE, IL-23p19 mRNA expression was found in $CD11b^+$ CNS APCs but not in peripheral macrophages. IL-12p35 and IL-12p40 mRNA were expressed at the peak of EAE [32].

Blockage of IL-23p19 using antibodies or ablation of the gene that codes for this protein results in resistance to EAE because inflammatory cells become unable to enter the CNS [33]. However, this resistance in IL-23p19$^{-/-}$ mice be reversed by the intracerebral administration of IL-23 [34]. The phenotypes of IL-12/IL23p40$^{-/-}$ or p19$^{-/-}$ mice are determined by the absence of IL-23. In IL-12/IL23p40$^{-/-}$ mice, EAE progression is sluggish. Upon encephalitogenic T-cell transfer from wild-type mice with advanced EAE to p40$^{-/-}$ mice, EAE reduction is observed.

Brain infiltration with inflammatory cells is similar in p40$^{-/-}$ and wild-type mice indicating that p40 is not indispensable for infiltration. Most probably, this protein is responsible for the encephalitogenic potential of cells and for the rate of EAE progression [5].

IL-23 is responsible for Th17/Th$_{IL-17}$ cells expansion and differentiation from naïve T cells. Monoclonal antibodies to IL-23p19 were shown to suppress EAE development indicating IL-23 involvement in EAE. Moreover, this is confirmed by the fact that IL-23p19$^{-/-}$ mice are resistant to EAE. Blockage of IL-23 and IL-17 but not of IFN-γ protects from EAE. Anti-p19 antibodies decrease the number of cells that produce IL-17 when they are administered prior to EAE onset (days 7 and 9) and after EAE onset (day 12) [36].

IL-23 is one of the main factors required for the expansion of the pathogenic population of $CD4^+$ cells characterized by IL-17A, IL-17F, IL-6, and TNF, but not IFN-γ and IL-4 production. In addition, the recognized proteolipid EAE inducer, PLP protein, induces IL-23p19 but not IL-12p35 [6].

It has been noted that IL-23 not only induces IL-17 but also increases IL-10 production by CD4 and CD8 lymphocytes. Upon IL-12 addition to cell culture, the cytokines were produced at unaltered rates [36]. Therefore, it may be concluded that the profile of cytokines produced under the effect of IL-23 is significantly different from that of IL-12, irrespective of T-cell activation state.

The effects of IL-23 are mediated by its binding to the receptors, IL-12Rβ1 and IL-23R, which are expressed in Th17 cells. Besides IL-23, other cytokines including IL-2, IL-15, IL-18, and IL-21 can induce IL-17 production. However, IL-12 specifically inhibits IL-17 secretion induced by IL-23 but not by other cytokines. The specificity of inhibition by IL-17 is based on the use of shared receptors and signaling pathways with two other cytokines, IL-12 and IL-23 [33].

The above data raise the hope that antibodies to IL-23 may be used in MS prevention and treatment. Yet, the question remains whether humans and mice are similar in this respect.

In two articles it was shown that IL-23 plays a minimal role (about 7%) in the regulation of IL-17 production by Th17 cells, whereas the main factors of Th17 response are transforming growth factor (TGF)-β and IL-6 [7,37]. The mechanism of this phenomenon is explained by the fact that TGF-b inhibits the effects of IFN-γ and IL-4, which are known to suppress IL-17 production by Th17 cells. In contrast, TGF-β induces Foxp3 expression in $CD4^+CD25^-$ cells thus promoting their differentiation into $CD4^+CD25^+$ cells known as naturally occurring suppressors of the immune response. With regard to these data, it is extremely interesting to know what combinations of the factors that enhance or suppress the immune response would provide for EAE development. Resistance to or development of EAE are suggested to depend not only upon the functional antagonism between the existing Th17 and $CD4^{+}CD25^{+}$ cells, but also upon the dichotomy of their development [7].

1.1.3. Interleukin-27 in experimental autoimmune encephalitis

In 2002, another member of the IL-12 family was discovered, which is a heterodimeric protein differing from the other two members of the family in that while having a high homology to IL-12 it features the unique property to induce the proliferation of naïve T cell and the expression of the IL-12Rb subunit of IL-12 receptors [38].

IL-27 consists of two subunits, EBI3 and IL-27p28, and is a ligand of WSX-1. In humans, the chromosomal localization of p28 gene is 16p11. Human p28 polypeptide does not contain N-glycosylation sites; however, the existence of several O-glycosylation sites is suspected. Murine p28 does contain one N-glycosylation site. The CD loop of the human p28 contains 13 negatively charged glutamate residues, whereas in the murine p28, a series of 14 glutamates interrupted by one lysine is found in this region. Such a highly charged structure was not revealed in any other cytokine comprising a spiral structure. The murine and human IL-27p28 are homologous by 73% [38].

Epstein-Barr virus-inducible protein-3 (EBI3) is a receptor having no known functions. This protein lacks the membrane-anchoring motif and thus has been suggested to be a secretory receptor. EBI3 belongs to the family of receptors that includes IL6Rα, IL-11Rα, CNTFRα (secretory receptor for ciliary neurotrophic factor), receptor for CLC (cardiotrophin-like cytokine), and receptor for CLF-1 (cytokine-like factor-1). EBI3 is homologous to p40 of IL-12, whereas p28 is homologous to p35 of IL-12, which makes the basis to refer IL-27 to the heterodimeric members of the IL-12 family [39].

As early as in the first study of IL-27 significance for EAE, it has been shown that IL-27 acts on the primed T cells including long-living Th1 memory cells implicated in the effector phase of the immune response. IL-27 enhances their antigen-dependent proliferation and IFN-γ production. The use of neutralizing antibodies to IL-27p28 blocks inflammation thus reducing EAE. Blocking of the encephalitogenic T cells does not result in an expressed shift toward Th2 cells that afford protective immunity against EAE. Most likely, a therapy directed against IL-27 will decrease the number of cells that produce INF-γ [40].

IL-27 attracts attention because it is produced by microglia and, taking into account its activity, may be expected to be involved in EAE [41]. IL-27, its subunits, and its receptor WSX-1 are clearly increased in the inflammatory cells in the CNS during EAE. Moreover, the expression of mRNA of EBI3 and p28 of IL-27 was found in APCs from the CNS and lymph nodes at the peak of EAE, whereas WSX-1 mRNA was increased in inflammatory cells at the early stages of inflammation. Thus, IL-27 and its receptor may regulate, in a paracrine fashion, the Th1 response in the course of EAE [42]. However, in spite of such a straightforward explanation, data should be considered indicating that mice deficient in IL-27Ra do not loose the ability to mount Th1 responses. In contrast, upon T-cell activation though T cell receptor (TCR) binding and under highly polarizing conditions, an important role of IL-27/WSX-1 in the suppression of Th1 responses becomes evident. These findings are consistent with the results showing that IL-27/WSX-1 is a negative factor for Th1-type responses [43]. Similar results obtained in mice with EBI3 deficiency show that such mice are able to produce more IFN-γ than wild-type mice are [44].

Mice deficient in ILp27Ra are highly susceptible to EAE. In response to IL-27, their Th17 cells increase the production of IL-17 A and F, IL-6, granulocyte–macrophage colony-stimulating factor (GM-CSF), and TNF more than the cells isolated from wild-type mice. It was found that the suppressive effect of IL-27 is strictly dependent on the presence of IL-7Ra in CD4 cells, whereas APCs are not required. IL-17 suppression by IL-27 requires the activation of the transcription factor STAT1. STAT1-deficient mice have more cells that produce Th17 cytokines and a high susceptibility to EAE. Moreover, INF-γ was shown to reduce the number of Th17 cells. Thus, IL-27 functions are required for Th17 cell differentiation and EAE pathogenesis [8,45].

1.2. Interferon-γ in multiple sclerosis

Expression of IFN-γ and TNF-α mRNA in blood cells correlates with depression scores during an acute attack in patients with MS [47]. A 4.8-fold mean increase in INF-γ mRNA in peripheral blood cells of MS patients was noted [47] when compared to controls, the magnitude of the increase correlating with depression scores of the patients.

In patients with RR MS, during relapses, an increased spontaneous production of IFN-γ and its serum content has been found, whereas upon remission, serum IFN-γ level tends to normalize [48].

In relapse versus remission, the *in vitro*-induced IFN-γ production becomes maximal, even though it is above normal in remission [49,50]. Cytokine production patterns were similar in healthy subjects: T-cell clones produced Th0 and Th1 regulatory cytokines. At the same time, T-cell clones of the MS patients produced IFN-γ and TNF-α at very high levels, and IL-4 at a low level.

Endogenous IL-10 cannot inhibit IFN-γ and IL-12, so high are the levels of their production [51]. In progressive MS, an increased intracellular IFN-γ [52], an increased serum IFN-γ, and spontaneous IFN-γ production as well as decreased induced IFN-γ production are observed [52].

In MS, IFN-γ is found in the cytoplasm of lymphocytes localized in demyelinated lesions perivascularly [52]. IFN-γ increases human leukocyte antigen II (HLAII) expression and induces production of proinflammatory cytokines by microglia and astrocytes [53]. Also, because of Fas-mediated apoptosis, it downregulates the increase in cell number [54]. In patients with RR MS in the acute phase, increased CSF levels of IFN-γ have been found [55]. According to Sindic et al. [56], the expression of IFN-γ mRNA is increased in CSF cells of all MS patients irrespective of the clinical manifestations of the disease. HLA expression in astrocytes induced by IFN-γ makes them APC and activators of T-cells [55].

1.2.1. Interferon-γ in experimental autoimmune encephalitis

IFN-γ is thought to play a pathogenetic role in immunity-mediated demyelination occurring in MS and EAE. Various approaches have been developed to study the mechanisms of IFN-γ involvement in EAE. One of them uses transgenic mice that express INF-γ ectopically and are monitored for the condition of their CNS after induction of EAE or demyelinization with cuprizone. It was shown that in both of the experimental models, INF-γ inhibits remyelinization and hampers remission. These findings correlate with a marked decrease in oligodendroglia cell repopulation in the demyelinized lesions. Moreover, it was found that in cuprizone-treated mice, the adverse effects of INF-γ are associated with stressed endoplasmic reticulum in oligodendrocytes (ODCs) under remyelinization. Thus, these data suggest that IFN-γ can inhibit remyelinization in EAE lesions [57].

Interestingly, IFN-γ expression is increased in TNF-α receptor 1 $(TNFR1)^{-/-}$ mice with EAE, suggesting a correlation between INF-γ level and the multiplicity of brain lesions [58]. Over the last decade, data have been accumulating that suggest that GKO (IFN-γ knockout) mice retain the ability to develop EAE and other autoimmune diseases.

The mechanisms of this effect are probably based on the fact that IFN-γ can induce macrophages to produce nitric oxide (NO), which is known to control EAE [59]. Nitric oxide causes $CD4^+CD25^-$ cell conversion to the $CD4^+CD25^+$ $Foxp3^+$ phenotype, that is, to upregulate naturally occurring suppressors that can inhibit the proliferation of encephalitogenic Th1 lymphocytes. IFN-γ induces the expression of the transcription factor Foxp3 in CD4CD25 Th1 lymphocytes [60].

Mice deficient in IFN-γ acquire susceptibility to EAE, which is similar to that of wild-type mice [61], indicating the absence of the direct effect of IFN-γ on the course of this autoimmune disease. IFN-γ is an important proinflammatory cytokine produced by Th1 cells and is suggested to stimulate encephalitogenic T lymphocytes and to participate in CNS inflammation.

The IFN-γ-inducible chemokine CXCL10 can cause brain tissue infiltration with $CXCR3^+$ T cells and induce EAE irrespective of IFN-γ [62]. T-bet is a new transcription factor for the Th1 cell subset. T-bet expression is restricted to the Th1 subset and is regulated by signals transmitted via the TCR. It is a potent transactivator of the IFN-γ gene, but not IL-2 gene in lymphoid and myeloid cells [63,64]. Retroviral gene transduction of T-bet into primary T cells results in IFN-γ production. T-bet represses IL-4 production in developing Th2 cells and redirects effector Th2 cells into the Th1 pathway. It represses IL-4 and IL-5 production in Th2 cells, in an IFN-γ-independent way [63]. It has been found that IFN-γ induces T-bet expression in monocytes, macrophages, and DC. Moreover, in association with cognate antigen, IFN-γ, in the absence of IL-12, markedly enhances T-bet expression in Th cells. This indicates a role of IFN-γ in evoking the Th1 response through a previously unrecognized, positive feedback loop [64,65].

T-bet, has been shown to promote Th1 development and IFN-γ production. T-bet is expressed preferentially in Th1 cells and NK cells, and its expression correlates with the production of IFN-γ. Overexpression of T-bet in Th2 cells confers a Th1 cytokine production profile on these cells, with the induction of IFN-γ and the repression of IL-4 and IL-5 expression.

1.3. Interleukin-2 in multiple sclerosis

In patients with RR MS during remission, IL-2 synthesis by peripheral blood mononuclear cells (PBMCs) is not different from that of healthy subjects [25]. Before relapses, IL-2 mRNA expression is significantly increased in peripheral blood cells [14]. $CD4^+$ T cells are mainly responsible for this increase. After the acute phase is subsided, IL-2 mRNA expression normalizes. The counts of lymphocytes that express IL-2 mRNA correlate with the severity of MS at its different stages [14,15].

The induced production of IL-2 is above normal in MS patients during remission, which is associated with an increased expression of IL-2 receptor (IL-2R). In relapse versus remission, the *in vitro*-induced IL-2 production is increased [65].

IL-2 is synthesized by astrocytes in the CNS with the intact blood–brain barrier. The expression of IL-2 and its receptor increases in active inflammation foci and adjacent white matter featuring the most pronounced cell infiltration [65]. The increased expression of IL-2 is observed only in developing foci and is transitory.

1.3.1. Interleukin-2 in experimental autoimmune encephalitis

IL-2 has long been known as a growth factor for T lymphocytes, and therefore it was surprising to find out that mice deficient for production of IL-2 or the IL-2Rα or IL-2Rβ subunits of its receptor may develop a lethal lymphoproliferative and autoimmune syndrome referred to as "IL-2 deficiency syndrome." IL-2 administration to the mice or making them transgenic for any of the subunits of IL-2R afforded protection against this syndrome. The most adequate solution of this paradox was provided [66]; it was shown that IL-2 deficiency syndrome is associated with the absence of regulatory $CD4^+CD25^+$ cells, their number being regulated by IL-2. Thus, the main role of IL-2 in EAE is induction of regulatory cell activity. The above data may be useful in developing therapeutical approaches for the treatment of EAE and, possibly, for other autoimmune conditions.

There are but a few authors who do not take into consideration any data about the immunoregulatory functions of neurons. In particular, T lymphocyte interaction with neurons results in that the encephalitogenic T cells become converted into the regulatory $CD4^+CD25^+TGF\text{-}\beta1^+$ $CTLA^+FoxP3^+$ T cells, which inhibit EAE [67].

Chen et al. [35] have suggested an original approach to EAE treatment comprising prophylactic administration of glucocorticoids combined with IL-2, both of which stimulate regulatory cells in the body. The authors have shown the potency of such treatment.

1.4. Interleukin-6 in multiple sclerosis

IL-6 production by peripheral blood monocytes of MS patients is significantly higher than that of healthy subjects [11]. Increased blood counts of monocytes that express IL-6 mRNA are found in all groups of MS patients. Along with an increased expression of IL-6 mRNA by blood mononuclear cells, an increased expression of the mRNA of the receptor for this cytokine is observed [68]. The spontaneous production of cytokines by PBMCs is significantly increased in both remission and relapse in MS patients compared to healthy subjects [2]. According to Chofflon and Juillard [69], the increased spontaneous production of IL-6 in remission is always lower than that observed during relapses in the RR MS.

Patients with RR MS show the maximal spontaneous production of IL-6 during relapses [49]. The induced production of IL-6 in MS patients during remission is above normal [70].

In MS patients with the progressive course of the disease, a significant increase in blood serum IL-6 level, an increase in its spontaneous production, and a decrease in its induced production ($p<0.05$) are observed [49]. It has been suggested [71] that the increase in serum IL-6 levels (112.6 ± 176.2 pg/ml) at the onset of relapse versus remission (7.2 ± 8.6 pg/ml) may be a marker of an approaching relapse.

The expression of IL-6, the factor of the late stages of B-cell differentiation into plasmatic cells, is not related to the penetration of the blood–brain barrier by cells involved in inflammation [72]. Under normal conditions, IL-6 is produced in the CNS by folliculostellate cells and astroglia of the anterior pituitary. During MS development, IL-6 is produced in the CNS by different cells including astrocytes and microglia.

The maximal amounts of IL-6 are produced in inactive demyelinating foci, early remyelination foci, and late active foci [73]. In inflammatory foci, IL-6 is expressed mainly by astrocytes (10–17%) and, to a lesser extent, by macrophages (up to 2%). In brain white matter adjacent to MS lesions, IL-6 is expressed mainly by astrocytes [73]. Several studies suggest a protective role of IL-6 toward ODCs. Thus, a direct correlation between IL-6 expression and the number of surviving ODCs in demyelinating foci has been noted [73].

In *in vitro* cultures of human microglia, a significant production of IL-6 is observed without additional stimulation [74]. *In vitro* studies have shown that the IL-6/IL-6R complex can activate cells including neurons that bear no IL-6R [75]. IL-6/IL-6R addition to cultured glial cells leads to stimulation of the expression of the MBP and its appearance in the cytoplasm of oligodendroglial cells [76].

The levels of IL-6 in the CSF of MS patients are above normal. A positive correlation has been found between IL-6 levels in the CNS and peripheral blood, which is believed by some authors to be the evidence that this cytokine produced in MS lesions can penetrate the blood–brain barrier and enter the peripheral blood [76].

1.4.1. Interleukin-6 in experimental autoimmune encephalitis

IL-6 is a proinflammatory cytokine with two possible effects. On one hand, it inhibits the production of proinflammatory cytokines by macrophages, and on the other hand, it induces acute-phase proteins, activates corticosteroid production, promotes T-cell activation by APCs, increases B-cell proliferation and immunoglobulin induction, and stimulates hematopoiesis and platelet generation.

IL-6 knockout mice are resistant to EAE indicating that IL-6 plays a role in EAE development. Possibly, blocking of proinflammatory cytokines by IL-6 may result in the acute phase of EAE.

Mice lacking the GM-CSF gene are resistant to EAE. GM-CSF is an inflammatory cytokine because it activates macrophages and neutrophils. At present, many pathophysiological conditions are interpreted in terms of Th1 and Th2 responses. The stability of either pattern is based on immunological memory and stability of cytokine gene expression [77].

1.5. Interleukin-1β in multiple sclerosis

A characteristic feature of MS is increased blood counts of lymphocytes that express IL-1β mRNA observed at different stages of the disease and correlated with relapses [14,15]. The spontaneous *in vitro* production of IL-1β during remission remains elevated and increases even more during relapses. The levels of the spontaneous production of IL-1β highly correlate with the clinical activity of the disease: they increase at the onset of a relapse and peak by its 30th day [2]. According to Jong et al. [78], families characterized by a high IL-1β versus IL-1 receptor antagonist (IL-1Ra) ratio (production by PBMCs) compared to those with a low ratio are at a 2.2-fold increased risk to have a relative with RR MS.

IL-1β, a proinflammatory cytokine released by monocytes, microglial cells, astrocytes, and brain endothelial cells, is involved in inflammatory reactions of the CNS. IL-1 species expressed in the CNS under normal conditions contributes to building up interconnection between the nervous and immune systems. In the CNS of healthy subjects, IL-1 mRNA expression is found in microglia and astrocytes in the hippocampus, choroid plexus, cerebellum, and septum [78].

The expression of IL-1 receptors under normal conditions is found in different CNS structures and reaches the greatest density in the hippocampus [79]. Both forms of IL-1α and β are present in the CNS [80]. In MS, IL-1β production is maximal in acute inflammation foci where IL-1β is expressed by cells present in the foci and by cells of the adjacent white matter [80].

1.5.1. Interleukin-1 in experimental autoimmune encephalitis

The role of IL-1 in EAE has been well defined by recent studies. It is believed that IL-1 plays the key role in the induction of IL-17 secretion by Th17 cells. In IL-1R1$^{-/-}$ mice, IL-23 alone does not induce IL-17 production. Moreover, IL-17 production induced by IL-23 may be enhanced by IL-1α and IL-1β. Tumor necrosis factor is synergistic with IL-23 in the induction of IL-17 production in a manner dependent on IL-1. These data suggest that IL-17 acts upstream IL-17 to activate Th7 cells in EAE [9]. However, the latter study does not take into account the fact that IL-1 production is under the control of its natural inhibitor, which is IL-1Ra. The IL-1/IL-1Ra balance is of paramount importance for the development of inflammation including its immune type in demyelination processes [81].

1.6. Tumor necrosis factor-α in multiple sclerosis

In patients with RR MS during remission, the expression of TNF-α mRNA in PBMCs is not different from that of healthy donors [48]. A characteristic feature of MS at different stages of the disease is an increase in blood counts of lymphocytes that express TNF-α mRNA, which correlates with MS relapses [14,15]. Along with the increased expression of TNF-α mRNA in PBMCs, an increased expression of receptors to the cytokine is found in MS [68].

The spontaneous *in vitro* production of TNF-α during remission remains elevated and increases even more during relapses [2,69]. According to Jong et al. [78], families characterized by a high TNF versus IL-10 ratio and a high IL-1β versus IL-1Ra ratio (production by PBMCs) compared to those with a low innate TNF vs. IL-10 and low IL-1β vs. IL-1Ra ratio are at a sixfold increased risk to have a relative with RR MS.

The levels of the induced expression of TNF-α in MS patients during remission are above normal [70]. Tumor necrosis factor-a induction with PHA or myelin basic protein in PBMCs of MS patients 2 weeks before relapse results in an increased expression of the cytokine compared to that observed in remission [69]. Similar results were obtained by other authors [82]. 5.1-fold increase (versus control) in TNF-α mRNA expression in peripheral blood cells of MS patients during relapses and a high correlation between this expression and the clinical scores of depression have been observed [47]. In patients with RR MS, TNF-α production is maximal during relapses [49]. The levels of TNF-α and lymphotoxin-α (LT-a) have been shown to be elevated in plaques and samples of blood and CSFs from MS patients. The expression of MBP is controlled by cytokines, in particular, TNF-α produced by activated T cells. Nuclear factor-κB (NF-κB)-binding cis elements have been found in MBP gene promoter. The data demonstrate that NF-κB is a mediator between TNF-α and MBP expression [83].

TNF-α can damage the myelin sheath and induce ODC apoptosis, suggesting that interactions between TNF-α and TNFR1 have an effector role in the pathogenesis of MS [84]. However, systemic blockade of TNF-α in progressive MS patients led to immune activation and promoted the disease.

The main sources of TNF-α/β are activated macrophages, astrocytes, microglia, and, to a lesser extent, activated T lymphocytes. Tumor necrosis factor-b (LT) is expressed by astrocytes and ODCs. Tumor necrosis factor-a receptor is found in astrocytes [85]. A transitory increase in TNF-α/β expression in reactive inflammation foci has been observed during their development in the CNS [86]. In chronically active and inactive foci and in the adjacent white matter, TNF-α/β expression was similar to control values [85].

TNF-α increases the expression of HLAII and induces the production of other cytokines by microglia and astrocytes [53]. Also, TNF-α induces the expression of Fas and Fas-ligand

(Fas-L) in different cells [55]. The proinflammatory effect of TNF-α on cerebral endothelial cells consists of elevation of the expression of cell adhesion molecules – vascular cell adhesion molecules (VCAM-1) and intercellular adhesion molecule-1 (ICAM-1) – resulting in increased levels of soluble VCAM-1. The soluble VCAM-1 blocks cell adhesion and decreases cellular infiltration [86].

The combined *in vitro* culturing of microglial cells and activated T lymphocytes resulted in a significantly increased TNF-α levels in culture supernatants; the stimulation of TNF-α production required the interaction of costimulatory very late activation protein-4 (VLA-4) molecules on T cells with VCAM-1 molecules on microglial cells [87].

The protective effect of TNF-α against the development of demyelinating lesions observed by some authors may be related to its ability to stimulate the expression of the tissue inhibitor of matrix metalloproteinases (TIMP-1) whose increased levels are found in the CSF of MS patients [88].

A high expression of TNF-α/β mRNA is found in CSF cells of MS patients [87]. Data about the relationships between the clinical manifestation of the diseases and changes in TNF-α/β production by CSF cells of MS patients are conflicting. According to some authors, TNF-α/β concentrations in the CSF correlate with the severity and the type of the disease [89]. However, such correlations were not observed by other authors [90].

1.6.1. Tumor necrosis factor in experimental autoimmune encephalitis

Experiments with $TNF^{-/-}$ mice demonstrated an abnormally prolonged T-cell reactivity to MBP. In $TNF^{-/-}$ mice, at difference from immunized normal mice, MBP-specific T cells accumulated in the spleen. The proliferative response to MBP was less in $TNF^{-/-}$ mice than in control animals. Interestingly, in TNFR $p55^{-/-}$ mice, immune response to MBP is not compromised, in contrast to what occurs in $TNF^{-/-}$ mice. These data show that TNF-neutralizing therapy for autoimmune diseases is not adequate. In $TNFR1^{-/-}$ mice, MOG-induced EAE did not develop. Blocking of TNFR1 in autoimmune demyelination may inhibit the proinflammatory activities of TNF without compromising its immunosuppressive action [61] (Table 1). The resistance to reinduction of EAE depends on TNF but not on TNFR1 p55. It is possible that the immunosuppressive function of TNF that does not involve apoptosis is realized via another cell-signaling pathway of this cytokine (Fig. 1).

Since TNF is produced by both types of T helpers, it can play either a protective or a pathogenic role. The expression of TNF and IFN-γ mRNA in blood cells correlates with clinical scores of depression during acute attacks in patients with MS. Tumor necrosis factor modulates the expression of myelin basic protein gene through NF-κB in human oligodendroglioma [83]. Neutralization of IFN-γ-inducible protein 10/CXC10 exacerbates EAE [91]. However, other authors [92] have obtained conflicting data. They believe that the INF-γ-inducible protein (IP-10) promotes the accumulation of encephalitogenic cells in the CNS during the clinical stage of EAE. Both groups of authors used IP-10-neutralizing antibodies to prove the role of IP-10. The reason for the above discrepancy is unclear.

The majority of studies indicate that TNF-α accelerates EAE in the acute period immediately after the onset of immune response to myelin protein. However, the absence of TNF-α in $TNF^{-/-}$ mice does not lead to the accumulation of myelin-specific T cells in the spleen, which could promote the chronic form of EAE. These results indicate that TNF-α is the main factor in decreasing the number or the activity of myelin-specific T lymphocytes and thus protects animals from the chronic form of demyelinating diseases [93]. Tumor necrosis factor has

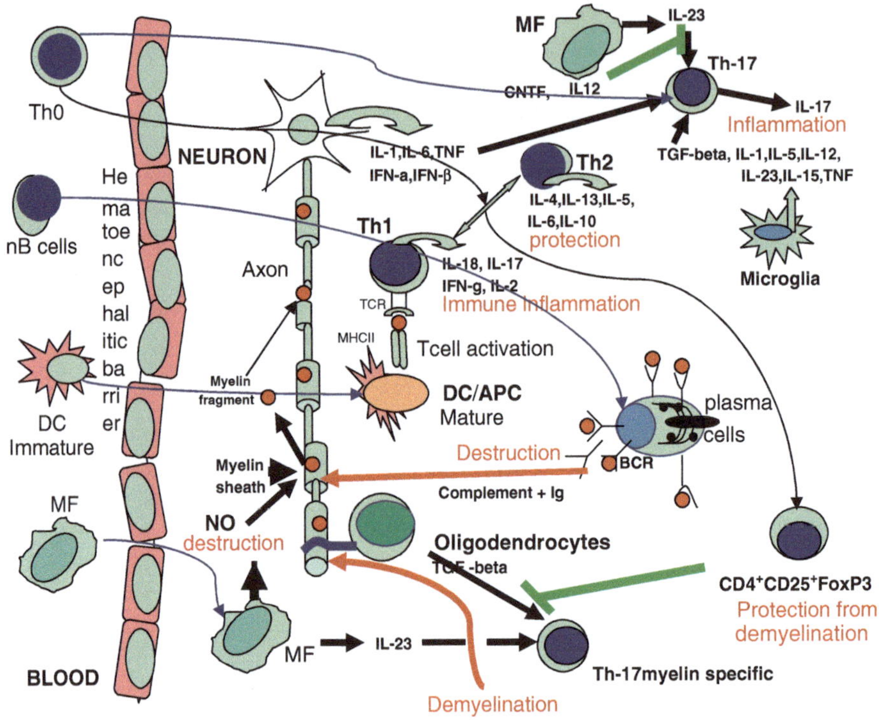

Figure 1. Cytokines in inflammatory and demyelination phases of MS.

been recently shown to be an inhibitor of inflammation, so it may protect against demyelinating diseases [94].

It should be noted that cytokines change their roles in accordance with disease stages. It has been especially emphasized [95] that protection against EAE does not require p55 TNFR, whereas the same receptor is necessary for the detrimental effects of TNF during the acute phase of the disease. According to this logic, one may suggest that the nature of inflammation at the initial and peak stages of a disease is different with regard to the involvement of different cells and the duration of their persistence, different effector cells being able to secrete the same cytokines acting on different target cells. For example, at the initial stage, effector cell involvement in the innate immunity may inhibit the development of encephalitogenic T cells, whereas at the acute phase, their involvement in the autoreactive immunity may result in the killing of encephalitogenic cells by cytotoxic lymphocytes or in chemokine-induced blocking of their migration to the CNS.

1.7. Tumor necrosis factor-α/β receptors in multiple sclerosis

Tumor necrosis factor-a binds to two types of receptors TNFRp55 and TNFRp75, which differ in the molecular weights of their ligand-binding subunits. TNFRp75 is expressed in immune-competent cells. TNFRp55 may be expressed by virtually all human cells including neurons [96,97].

Both TNFRp55 (TNFR1) and TNFRp75 (TNFR2) may be free or membrane-bound. The free form of TNFR at high concentrations inhibits TNF-α effects and at low concentrations, it may increase the duration of the effects.

In comparison with lymphocytes, monocytes express TNFR much more actively and feature the maximal densities of these receptors. ODCs present in MS lesions actively express TNFR, which may result in the apoptosis of these cells [96].

TNFR55 shedding from the surface of cells of MS patients is much less active than that of healthy subjects, whereas the opposite is true for TNFR75 [98]. This may relate to the hyper-production of TNF-α or IL-1, because an increased shedding of TNFR75 (but not of TNFR55) by monocytes is observed *in vitro* in the presence of TNF-α or IL-1. The reduced TNFR55 shedding may explain the significantly increased TNF/sTNFR55 observed in MS patients [98].

Along with an increased expression of TNF-α mRNA by PBMCs, an increased expression of mRNA of receptors to this cytokine [77] as well as a significantly increased serum TNFR1 and TNFR2 levels [99] is observed in MS patients.

The role of sTNFR (soluble TNF receptor) in MS pathogenesis is insufficiently studied. The presence of sTNFR results in the decrease in TNF-α activity because sTNFR competes with the surface receptors for binding to TNF-α. Data about changes in serum sTNFR1 were reported by Rieckmann et al. [99]. The increased serum sTNFR1 in MS patients is more pronounced during remission than during relapse. The maximal level of sTNFR1 was observed by these authors 4 weeks after relapse. The authors concluded that there was a correlation between clinical remission and sTNFR1 level. However, other investigators did not find any relationship between MS course and changes in sTNFR1 [100].

It has been observed [86] that a transitory increase in the expression of TNF receptors (TNFR1 and TNFR2) is present in reactive inflammation foci during their formation in the CNS. The decreased TNFR55 shedding in MS patients may play a significant role in the pathogenesis of the disease because this is the form of TNF receptors that is expressed by neurons [96,97].

1.7.1. Tumor necrosis factor-α receptor 1 in experimental autoimmune encephalitis

Recent studies suggest that apoptosis of ODCs by the FAS-FASL and TNFR1-mediated mechanisms are fundamental in EAE induction, whereas FAS is pivotal in EAE progression. To study this, mice lacking FAS expression in their ODCs were developed and crossed with mice deficient in TNFR1. Inactivation of FAS only or TNF1R only resulted in a partial protection of mice against EAE. The double deficit completely abrogated all manifestations of EAE, brain tissue infiltration with T lymphocytes being reduced but insignificantly. With regard to these data, it may be concluded that such double deficit can afford protection from demyelination and clinical manifestations of autoimmune encephalitis [101].

1.8. Interleukin-15 in multiple sclerosis

It has been suggested [102] that IL-15, a pro-inflammatory cytokine, plays a role in both initiation and progression of MS. It has been shown that IL-15 mRNA expression in stable MS patients is similar to that in healthy subjects. An increased expression of IL-15 mRNA has been found in MS patients during relapses. Moreover, a negative correlation was observed between IL-15 mRNA expression and the duration of the remission period from the last till the next relapse.

1.9. Interleukin-1 receptor antagonist in multiple sclerosis

The ratios of spontaneous production levels of IL-1Ra and IL-1β have been shown to be significantly lower in relapses of RR MS than in remissions (0.14 versus 0.17). The same is true for the induced production [103].

In the CNS, IL-1Ra is expressed even when the blood–brain barrier is intact [104]. The expression is observed in monocytes, macrophages, endothelial cells, glia, fibroblasts, and neurons. Results of experiments about the role of IL-1Ra in EAE is described in Section 1.6.1.

1.10. Interleukin-4 in multiple sclerosis

In MS patients during remission, blood counts of cells that produce IL-4 increase at the same extent as that of cells that produce IFN-γ. However, in no study a relationship between IL-4 production by PBMCs and the clinical manifestation of MS has been revealed [48]. Some investigators note an increase in the soluble form of IL-4 receptor (sIL4-R) in MS patients versus healthy donors [105]. It has shown [49] an increase in peripheral blood B lymphocytes bearing IL-4R (CD20+CD124+) in remitting MS during relapses (18.3%), remission (21.9%), and progression (3.3%) versus control (0.1%). In patients with progressive MS, IL-4 mRNA expression by PBMCs is negligible and does not differ from the norm [103]. *In vitro* cytokine production has been studied [82] after induction by myelin ODC glycoprotein in T-cell clones from MS patients and it was shown that T-cell clones from MS patients produced extremely low levels of IL-4. It was shown [48] that depression scores of MS patients do not correlate with the expression of IL-4 mRNA in PBMCs.

In MS patients' brain, the most active production of IL-4 was found in acute and chronic active lesions where microglia, macrophages, and ODCs located at the edges of the lesions contributed to its expression. *In vitro* experiments showed the ability of IL-4 to inhibit the proliferation of astrocytes that express IL-4R [106].

IL-4 level in the CSF of MS patients is maximal during remission than during relapses and is still somewhat lower than that in the control. During relapses, IL-4 decreases to levels undetectable with enzyme-linked immunosorbent assay (ELISA) [89]. PCR [106] allows detecting of an increased expression of IL-4 mRNA by CSF cells in different groups of MS patients, which does not correlate with the clinical manifestations of the disease.

1.10.1. Interleukin-4 and its transcription factors in experimental autoimmune encephalitis

Differentiation of naive T cells into Th1 and Th2 cells is greatly influenced by the cytokines present during and after antigen presentation. The appearance of IFN-γ-producing cells is promoted by IL-12 signaling, which is mediated by signal transducer and activator of transcription 4 (STAT4). STAT4-deficient mice do not respond to IL-12 normally, having a marked impairment in Th1 differentiation and a propensity for Th2 differentiation. Conversely, IL-4 signals by means of STAT6 to promote Th2 differentiation. STAT6-deficient mice, in turn, have impaired IL-4 signaling and diminished Th2 responses. Moreover, IFN-γ can suppress Th2 differentiation.

To ascertain the roles of STAT4 and STAT6 in EAE development, knockout mice with targeted disruption of these transcription factors were developed. STAT4-deficient mice were shown to be resistant to EAE induction and to develop minimal inflammatory infiltrates in the

CNS. On the contrary, STAT-6 knockouts featured the Th1 phenotype and, as a result, have multiple inflammatory foci. Adaptive transfer of splenocytes isolated from wild-type mice to $STAT4^{-}/^{-}$ recipients results in a less severe disease than that caused by the transfer of $STAT6^{-}/^{-}$ splenocytes [107]. Taken together, the above data unequivocally point at a pathogenic significance of Th1 responses for EAE.

1.11. Interleukin-10 in multiple sclerosis

According to data [108], in patients with RR MS, relapses correlate with a decreased expression of IL-10 mRNA, whereas the transition to remission and remission itself is associated with increased IL-10 mRNA levels in peripheral blood cells. The levels of the induced production of IL-10 by peripheral blood cells of MS patients during remission are above normal [13].

The progressive course of MS is associated with the sustained and significant decrease in IL-10 mRNA expression in peripheral blood cells [10,109]. According to other authors [16], progressive MS is associated with an increased expression of IL-10 mRNA in PBMCs. However, the mechanism of switching to the Th2 response is impaired, and so the endogenous IL-10 cannot inhibit IL-12 and IFN-γ production despite the normal expression of IL-10R mRNA. The expression of IL-10 mRNA is increased in PBMCs [16], mainly in $CD4^{+}$ T cells [51], the proportion of $CD4^{+}$ cells that produce IL-10 being up to 1.2% during relapses in the RR MS, 2.2% during remission, and 1.4% in progressive MS versus 0.69% in healthy subjects [51].

In the CSF of patients with RR MS, high levels of IL-10 are found during remission. High CSF IL-10 associated with the lack of blood–brain barrier defects may be an evidence of its protective role. A sharp decrease in CSF IL-10 and its appearance in the peripheral blood upon activation of the inflammatory process are believed by some authors to be associated with impairments in the penetrability of the barrier [51].

The intracellular production of IL-13 and IL-5 in peripheral blood $CD4^{+}$ and $CD8^{+}$ T cells from MS patients and healthy subjects was analyzed by flow cytometry [106]. IL-13-producing T cells were significantly increased in both T-cell subsets in MS at relapse and returned to normal at remission. IL-5-producing T cells did not vary regardless of the clinical phase or type of MS. A distinct profile of IL-13 production by phase of MS suggests an active involvement of this Th2 cytokine in MS.

It has been suggested [110] that $NO^{\cdot}$ and TNF-α damage myelin sheath, induce macrophages to phagocytize myelin and activate macrophages and T cells to produce osteopontin. Altogether, this results in the induction of Th1 cytokines including INF-γ and IL-12 and in the simultaneous downregulation of the production of Th2 cytokines, for example, IL-10. Th1 cells can aggravate the course of MS, whereas Th2 cells can decrease CNS damage in MS. The concerted attack of the CNS by T and B cells, complement factors, and mediators of inflammation including cytokines, osteopontin, and $NO^{\cdot}$ results in the development of demyelinated foci in brain white matter and impair axonal conductivity thus contributing to MS pathogenesis.

1.11.1. Interleukin-10 in experimental autoimmune encephalitis

In one of the first publications [111] about the role of IL-10 in EAE development, it was shown that this cytokine protects mice against EAE, but its production is controlled by IL-12. The authors suggest that this unique circuit of two cytokines, which are involved, on one hand, in the control of Th1/Th2 balance and, on the other hand, in switching on of the innate immunity, may modulate autoimmune conditions including EAE.

It should be stressed that autoreactive B cells isolated from mice recovered from EAE produce IL-10 in response to autoantigens. IL-10^+ B-cell transplantation to mice with autoimmune encephalitis inhibits its development. IL-10 production by B cells may, at least partly, enhance Th2-type responses. B cells isolated from IL-$10^{-/-}$ mice are unable to inhibit EAE [112]. This is confirmed by *in vivo* experiments showing that IL-10 knockout mice do not recover after autoimmune encephalitis [113]. The use of a congenic murine strain lacking IL-10 made it possible to assess the involvement of this cytokine in EAE development. The lack of protection from EAE in these mice clearly demonstrated the significance of IL-10 in this process [114]. IL-10 is known to be produced by Th2 cells, but it is still unclear whether IL-10 per se or Th2 cells play the key role in the protection. Numerous experiments were carried out to eventually show that the main role is played by IL-10 rather than Th2 cells when such cells are developed by redirecting the immune response to myelin peptide from Th1 to Th2 pathway [115].

1.12. Transforming growth factor-β in multiple sclerosis

TGF-b produced in the CNS participates in ODC differentiation [116]. In MS demyelinating foci, TGF-β is mainly found around blood vessels in the membranes of endothelial cells. TGF-β content is maximal in active foci [117].

In the CSF of MS patients, concentrations of TGF-β1 but not of TGF-β2 are increased compared to those found in healthy donors and subjects suffering from noninflammatory disease of the CNS. The maximal concentrations of TGF-β1 are observed during the relapses of RR MS, and the minimal concentration during progressive MS [105]. Experimental data about TGF-β1 role in EAE is explicated in Section 1.1.2.

1.13. Interleukin-18 in multiple sclerosis

IL-18 contributes to MS pathogenesis as do IL-6, GM-CSF, and osteopontin [17]. Antibodies that neutralize IL-18 prevent EAE [118]. However, this does not mean that IL-18 effects are mediated by INF-γ [130]. IL-18 can induce chronic inflammation not only through INF-γ induction but also through NF-κB activation. Caspase-1 blockage results in the lack of IL-18 production and inflammation [119].

In patients with relapsing MS during relapse versus remission or versus patients with noninflammatory CNS diseases, peripheral blood cells produce significantly more IL-18 (382.0 ± 66.3 pg/ml versus 175.6 ± 58 pg/ml or 112.7 ± 36.6 pg/ml, respectively), which plays an important role in the Th responses because of its ability to induce INF-γ in NK cells [46]. In active MS, macrophages and glial cells in inflamed lesions produce IL-18. In the CSF of patients with relapsing MS during relapse versus remission or versus patients with noninflammatory CNS diseases, significantly more IL-18 is found (653 ± 149.5 pg/ml versus 424.1 ± 136.4 pg/ml or 158 ± 43.5 pg/ml, respectively) [46].

IL-18 production is induced via T-cell receptor activation and IL-18 play a role in INF-γ induction at different MS stages. It was found that in the relapsing–remitting (RR) period, INF-γ production can be neutralized either with anti-IL-12 or anti-IL-18 antibodies, whereas during the progression stage of MS, INF-γ production can be neutralized with both antibodies acting simultaneously. These data show that during MS progression, IL-12 and IL-18 induce INF-γ in the independent manner. Moreover, in the secondary progressive MS, IL-18 levels correlate with disease duration.

1.13.1. Interleukin-18 in experimental autoimmune encephalitis

IL-18 is a potent inducer of proinflammatory cytokines including IFN-γ, GM-CSF, and IL-1. Therefore, the general belief was that IL-18 is implicated in EAE development [120]. Increased IL-18 production and IL-18Ra responsiveness to it may contribute to multiple lesion development. Besides that, increases in IL-18 production and IL-18Ra expression in IL-12Rβ2$^{-/-}$ mice suggest a modulatory effect of IL-12Rβ2 on signal transmission in the IL-18/IL-18R system. Due to IL-18/IL-18R interaction, the production of IL-23 increases [28,29].

IL-18 is secreted by APCs, and IL-18 signals are transmitted via the receptor IL-18R, which is a heterodimer consisting of the ligand-binding subunit IL-18Ra and the signal-transducing subunit IL-18Rb. IL-18R is expressed in cell membranes of lymphocytes and accessory cells [121]. IL-18 binding to the IL-18Ra subunit is characterized by low affinity.

Considering the role of IL-18 in EAE development, it may be concluded that the role is none because IL18$^{-/-}$ mice are susceptible to EAE. At the same time, IL-18Ra$^{-/-}$ mice are completely resistant to EAE, indicating the existence of an alternative ligand having encephalitogenic properties. The lack of IL-18Ra does not disturb priming of T cells nor expansion of antigen-activated T cell. However, increased IL-18Ra expression in APCs is critical for the generation of T helpers that produce IL-17 (Th-17) through a mechanism dependent on IL-23.

Thus, it seems that IL-18 and Th1 cells induced with the participation of IL-18 are irrelevant for EAE, whereas IL-18Ra and Th17 cells, which produce IL-17, are required for inflammation in and autoimmune damage to the CNS [122].

However, there are data that challenge the hypothesis put forward [122]. It was found that the established EAE may be reduced by neutralization of IL-18, which is thought to be associated with the selection of the antigen-specific Th2 cells that cannot transfer the disease. The antigen-specific Th2 cells are selectively accumulated in the CNS in the active phase of EAE, but loose their ability to produce IL-4. This loss may be compensated for by the overexpression of IL-18BP (IL-18-binding protein) at the site of the development of the autoimmune process. In this case, IL-4 production by Th2 cells is sustained after their entry into the CNS. These data may be useful for developing a therapy for MS and, possibly, other autoimmune conditions [123].

It is hard to explain the effect of naturally occurring agent-induced blockage of IL-18, which does not contribute to EAE pathogenesis [122]. Anyway, IL-18 gene disruption or IL-18 suppression is likely to inhibit the development of autoimmune encephalitis or even produce complete resistance to it.

2. CONCLUSIONS

Evidence of the role of cytokines in EAE has generated new models for immune therapy. For example, antibody against IL-23p19, but not against IL-23p35, can be used for protection against MS. They can be used for effective treatment of MS but generalized immunosuppression is not involved in the induction of polarized population of Th1 lymphocytes that mediate MS and EAE. However several of the Th1 cytokines have been shown to exert protective effects in animal models of EAE and participate in remyelinization. Furthermore, antigen-specific Th2 populations can induce EAE (IL-4). It is very important to study the role of new cytokines that play a role in T-cell differentiation and their polarization in Th1 or Th2. Recently, new population of T helpers that increase the development of MS – Th17 cells – have been discovered. It has been shown that both IL-23 and TGF-β1 participate in the differentiation of Th17. Such new data show new roads to the treatment of the CNS autoimmune diseases that tend to spread very quickly worldwide.

REFERENCES

1. Steinman L. Multiple sclerosis: A coordinated immunological attack against myelin in the central nervous system. Cell 1996;85:299–302.
2. Gusev EI. Neurology and neurosurgery. Medicina M. (Russ) 2000, 418 pages.
3. Bisaga GN, Kalinina NM, Akimov SB, Davidova NI. Immunopathogenic differences in remitting and progradient multiple sclerosis. Immunologia (Russ) 2000;3:41–3
4. Sternberg EM. Neural–immune interaction in health and disease. J Clin Invest 1997; 100:2641–7.
5. Becher B, Durell BD, Noele RJ. IL-23 produced by CNS-resident cells controls T-cell encephalitogenicity during the effector phase of experimental autoimmune encephalomyelitis. J Clin Invest 2003;112(8):1186–91.
6. Langrish CL, Chen Y, Bluminschein WM, et al. IL-23 drives a pathogenic T cell population that induces autoimmune inflammation. J Exp Med 2005;201:233–40.
7. Bettelli E, et al. Reciprocal developmental pathways for the generations of pathogenic effector Th17 and regulatory T cells. Nature 2006;441:235–8.
8. Batten M, Li J, Yi S, Kljavin NM, et al. Interleukin 27 limits autoimmune encephalomyelitis by suppressing the development of interleukin 17-producing T cells. Nat Immunol 2006;Advance online publication:1–8.
9. Sutton C, Brereton C, Keoogh B, et al. A crucial role for interleukin (IL)-1 in the induction of IL-17-producing T-cells that mediate autoimmune encephalomyelitis. J Exp Med 2006;203;1685–91.
10. Baranzini SE, Elfstrom C, Chang S-Y, Butunoi C, Murrey R, Higuchi R, Oksenber JR. Transcriptional analysis of multiple sclerosis brain lesions reveals a complex pattern of cytokine expression. J Immunol 2000;165:6576–82.
11. Kouwenhoven M, Teleshova N, Ozenci V, Press R, Link H. Monocytes in multiple sclerosis: Phenotype and cytokine profile. J Neuroimmunol 2001;112:197–205.
12. Őzenci V, Pashenkov M, Kouwenhoven M, Rinaldi L, Link H. IL-12/IL12R system in multiple sclerosis. J Neuroimmunol 2001;114:242–52.
13. van Boxel-Dezaire A, Hoff S, Smits M, et al. Role of IL-10 and IL-12 in progression of multiple sclerosis. Mult Scler 1999;5:17–24.
14. Inoges S, Merino J, Bandres E, et al. Cytokine flow cytometry differentiates the clinical status of multiple sclerosis patients. Clin Exp Immunol 1999;115:521–535.
15. Wang X, Chen M, Wandinger KP, et al. IFN-β-1b inhibits IL-12 production in peripheral blood mononuclear cells in an IL-10-dependent mechanism: Relevance to IFN-beta-1b therapeutic effects in multiple sclerosis. J Immunol 2000;165:548–57.
16. Balashov KE, Comabella M, Ohashi T, et al. Defective regulation of IFNγ and IL-12 by endogenous IL-10 in progressive multiple sclerosis. Neurology 2000;55:192–8.
17. Nicoletti F, Patti F, Cocuzza C, Zaccone P, et al. Elevated serum levels of IL-12 in chronic progressive multiple sclerosis. J Neuroimmunol 1996;70:87–9.

18. Fassbender K, Ragoschke A, Rossol S, Schwartz A, et al. Increased release of IL-12p40 in MS: Association with intracerebral inflammation. Neurology 1998;51:753–6.
19. Ferrante P, Fusi ML, Saresella M, et al. Cytokine production and marker expression in acute and stable multiple sclerosis: Altered IL-12 production and augmented signaling lymphocytic activation molecule (SLAM)-expressing lymphocytes in acute multiple sclerosis. J. Immunol 1998;160:1514–21.
20. Vaknin-Dembinsky A, Balashov K, Weiner HL. IL-23 is increased in dendritic cells in multiple sclerosis and down-regulation of IL-23 by antisense oligos increases dendritic cell IL-10 production. J Immunol 2006;176(12):7768–74.
21. Thierfelder WE, van Deursen JM, Yamamoto K, Tripp RA, Sarawar SR, Carson RT, et al. Requirement for Stat4 in interleukin-12-mediated responses of natural killer and T cells. Nature 1996;382(6587):171–4.
22. Gran B, Zhang GX, Yu S, et al. IL-12p35-deficient mice are susceptible to experimental autoimmune encephalomyelitis: Evidence for redundancy in the IL-12 system in the induction of central nervous system autoimmune demyelination. J Immunol 2002;169(12):7104–10.
23. Ahmed Z, Gveric D, Pryce G, et al. Myelin/axonal pathology in interleukin-12 induced serial relapses of experimental allergic encephalomyelitis in the Lewis rat. Am J Pathol 2001;158:2127–38.
24. Mutchian G, Raikwar HP, Rajasingh J, Bright JJ. 1,25-Dihydrooxyvitamin-D3 modulates JAK-STAT pathway in IL-12/IFNgamma axis leading to Th1 response in experimental allergic encephalomyelitis. J Neurosci Res 2006;83(7):1299–309.
25. Gran B, Zhang GX, Yu S, Chen Xh, Kamoun M, Rostami A. IL-12p35-deficient mice are susceptible to experimental autoimmune encephalomyelitis: Evidence for redundancy in the IL-12 system in the induction of central nervous system autoimmune demyelination. J Immunol 2002;169:7104–10.
26. Becher B, Durell BG, Noelle RJ. Experimental autoimmune encephalitis and inflammation in the absence of interleukin-12. J Clin Invest 2002;110:493–7.
27. Constantinesku CS, Tani M, Ransohoff RM. Astrocytes as antigen-presenting cells: Expression of IL-12/IL-23. J Neurochem 2005;95:331–40.
28. Zhang GX, Gran B, Yu S, Li J, Siglienti I, et al. Induction of experimental autoimmune encephalomyelitis in IL-12 receptor-beta 2-deficient mice: IL-12 responsiveness is not required in the pathogenesis of inflammatory demyelination in the central nervous system. J Immunol 2003;170(4):2153–60.
29. Zhang GX, Yu S, Gran B, et al. Role of IL-12 receptor β1 in regulation of T cell response by APC in experimental autoimmune encephalomyelitis. J Immunol 2003; 171:4485–92.
30. Hart BA, Brok HPM, Remarque E. Suppression of ongoing disease in a nonhuman primate model of multiple sclerosis by a human-anti-human IL-12p40 antibody. J Immunol 2005; 175:4761–8.
31. Oppmann B, Lesley R, Blom B, Timans JC, et al. Novel p19 protein engages IL-12p40 to form a cytokine, IL23, with biological activities similar as well as distinct from IL12. Immunity 2000;13:715–25.
32. Li J, Gran D, Zhang GX, Ventura ES, et al. Differential expression and regulation of IL-23 and IL-12 subunits and receptors in adult mouse microglia. J Neurol Sci 2003; 215:95–103.
33. Hoeve MA, Savage ND, de Boer T, et al. Divergent effects of IL-12 and IL-23 on the production of IL-17 by human T cells. Eur J Immunol 2006;36(3):661–70.

34. Cua DJ, Cherlock J, Chen Y, Murphy CA, et al. Interleukin-23 rather than interleukin 12 is the critical cytokine for autoimmune inflammation of the brain. Nature 2003; 421(6924): 744–8.
35. Chen X, Oppenheim JJ, Wincler-Pickett RT, et al Glucocorticoid amplifies IL-2-depndent expansion of functional FoxP3$^+$ CD4$^+$ CD25$^+$ T regulatory cells in vivo and enhances their capacity to suppress EAE. Eur J Immunol 2006;36:2139–49.
36. Eijnden SV, Goriely S, De Wit D, et al. IL-23 up-regulates IL-10 and induces IL-17 synthesis by polyclonally activated naive T-cells in human. Eur J Immunol 2005;35(2):469–75.
37. Mangan PR, et al. Transforming growth factor-β induces development of the Th17 lineage. Nature 2006;441:231–4.
38. Pflanz S, Timans JC, Cheung J, Rosales R, Kanzler H, Gilbert J, Hibbert L, Churakova T, Travis M, Vaisberg E, et al. IL-27, a heterodimeric cytokine composed of EB13 and p28 protein, induces proliferation of naïve CD4+ T cells. Immunity 2002;16:779–90.
39. Trinchiery G, Pflanz S, Kastelein RA. The IL-12 family of heterodimeric cytokines: New players in the regulation of T cell responses. Immunity 2003;19:641–4.
40. Goldsberg R, Zohar Y, Wildbaum G, et al. Suppression of ongoing experimental autoimmune encephalomyelitis by neutralizing the function of the p28 subunit of IL27. J Immunol 2004;173(10):6465–71.
41. Sonobe Y, Yamata I, Kawanokuchi J, et al. Production of IL27 and other IL-12 family cytokines by microglia and their subpopulations. Brain Res 2005;1040:202–7.
42. Li J, Gran B, Zhang GX, Rostami A, et al. IL-27 subunits and its receptor (WSX-1) mRNAs are markedly upregulated in inflammatory cells in the CNS during experimental autoimmune encephalomyelitis. J Neurol Sci 2005;232(1–2):3–9.
43. Villarino AV, Huang E, Hunter CA. Understanding the pro- and anti-inflammatory properties of IL-27. J Immunol 2004;173:715–20.
44. Nieuwenhuis EE, Neurath MF, Corazza N, et al. Disruption of T helper 2-immune responses in Epstein-Barr virus-induced gene 3-deficient mice. Proc Natl Acad Sci USA 2002;99:16951–6.
45. Veldhoen M, Hocking RJ, Atkins CJ, et al. TGFβ in the context of an inflammatory cytokine milieu supports de novo differentiation of IL-17-producing T cells. Immunity 2006;24:179–89.
46. Losy J, Niezgoda A. Increase of IL-18 in patients with multiple sclerosis and its relation to brain MRI activity. J Neurol Sci 2001;187(Suppl 1):273.
47. Kahl KG, Kruse N, Faller H, Weiss H, Rieckmann P. Expression of TNF-α and IFN-γ mRNA in blood cells correlates with depression scores during an acute attack in patients with multiple sclerosis. Psychoneuroendocrinology 2000;6:671–81.
48. Navicas V, He B, Link J, et al. Augmented expression of TNFa and lymphotoxin in mononuclear cells in MS and optic neuritis. Brain 1996;119:213–20.
49. Kalashnikova AA, Kalinina NM, Bisaga GN, Davidova NM. Variation of immune status in patients with multiple sclerosis depends on activity and form disease. J Allerg Immunol (Russ) 2001;2:274–5.
50. Poehlau D, Rieks M, Alberti T, et al. Th-1 and Th-2 type cytokine producing peripheral blood mononuclear cells of patients with relapsing–remitting multiple sclerosis. Mult Scler 1997;2:313–20.
51. Balashov KE, Rottman JB, Weiner HL, Hancock WW. CCR5+ and CXCR3+ T cells are increased in multiple sclerosis and their ligands MIP-1a and IP-10 are expressed in demyelinating brain lesions. Proc Natl Acad Sci 1999;96:6873–8.

52. Shrikant P, Benveniste EN. The central nervous system as an immune competent organ: Role of glial cells in antigen presentation. J Immunol 1996;156:1819–22.
53. Nakamura Y. Regulating factors for microiglial activation. Biol Pharm Bull 2002;25:945–53.
54. Sellebjerg F, Christiansen M, Nielsen PM. Cytokines and apoptosis in patients with multiple sclerosis. Mult Scler 1998;4:475–9.
55. Lindberg R. Matrix metalloproteinases in the brain tissue and CSF of MS. J Neurol Sci 2001;187(Suppl 1):262–9.
56. Sindic CJ, Van Laere V, Monteyne P. Cytokine mRNA expression in CSF and peripheral blood mononuclear cells in multiple sclerosis. Mult Scler 1997;3:32–9.
57. Lin W, Kemper A, Dupree JL, et al. Interferon-gamma inhibits CNS remyelination through a process modulated by endoplasmic reticulum stress. Brain 2006;129:1306–18.
58. Wheeler RD, Zehntner SP, Kelly LM. Elevated interferon-gamma expression in the CNS of tumor necrosis factor receptor 1-deficient mice with experimental autoimmune encephalomyelitis. Immunology 2006;118:527–38.
59. Willenborg DO, Fordham S, Bernard CCA, Cowden WB, Remshow IA. IFN-γ plays a critical down-regulatory role in the induction and effector phase of myelin oligodendrocyte glycoprotein-induced autoimmune encephalitis. J Immunol 1996;157:3223–30.
60. Wang Z, Hong J, Sun W, Yu G, et al. Role of IFN-γamma in induction of Foxp3 and conversion of CD4+CD25− T cells to CD4+ Tregs. J Clin Invest 2006;116(9):2434–41.
61. Krakowski M, Owens T. Interferon-γ confers resistance to experimental allergic encephalomyelitis. Eur J Immunol 1996;26:1641–6.
62. Klein RS, Izikson L, Means T, et al. IFN-inducible protein 10/CXC chemokine ligand 10-independent induction of experimental autoimmune encephalomyelitis. J Immunol 2004;172(1):550–9.
63. Lighvani AA, Frucht DM, Jankovic D, Yamane H, Aliberti J, Hissong BD, Nguyen BV, Gadina M, Sher A, Paul WE, O'Shea JJ.T-bet is rapidly induced by interferon-γ in lymphoid and myeloid cells. Proc Natl Acad Sci USA 2001;98:15137–42.
64. Mullen AC, High FA, Hutchins AS, Lee HV, Villarino AV, Livingstone DM, Kung AL, Cereb N, Yao T-P, Yang S-Y, Reiner SL. Role of T-bet in commitment of Th1 cells before IL-12-dependent selection. Science 2001;292:1907–10.
65. Malek TR. T helper cells, IL-2 and the generation of cytotoxic T cell responses. Trends Immunol 2002;23:465–7.
66. Furtado G, Currotto de Lafaille M, Kutchukhidze N, Lafaille J. Interleukin-2 signalling is required for CD-4 regulatory T cell function. J Exp Med 2002;196:851–7.
67. Liu L, Huang D, Matsui M. Severe disease, unaltered leukocyte migration and reduced IFN-γ production in CXCR3-/- mice with experimental autoimmune encephalomyelitis. J Immunol 2006;176:4399–409.
68. Owens T, Wekerie H, Antel J. Genetic models for CNS inflammation. Nat Med 2001;7:161–6.
69. Chofflon RS, Juillard C. Tumor necrosis factor production capacity in multiple sclerosis. Eur Cyt Netw 1997;8:253–7.
70. Achiron A, Miron S, Zeitun I, et al. Cytokine production is increased in multiple sclerosis patients. Eur Cyt Netw 1998;9:501.
71. Miron S, Zeitun I, Barak Y, et al. TNF-α and IL-6 serum levels as markers of relapse detection in multiple sclerosis patients. Eur Cytokine Netw 1998;9:337–40.
72. Campbell LL, Stalder AK, Pagenstecher YAA, Asensio VC. Transgenic models to study the actions of cytokines in the CNS. Neuroimmunomodulation 1998;5:126–35.

73. Schonrock LM, Galowski G, Rruck W. IL-6 expression in early multiple sclerosis lesions. Mult Scler 1999;5:4–8.
74. Chabot S, Yong VW. Interferon beta-1b increases interleukin-10 in a model of T cell – Microglia interaction. Neurology 2000;55:1497–505.
75. Rose-John S, Marz P, Ficher M, et al. IL-6 and soluble IL-6 receptor are potent stimulators of hematopoietic and neuronal cells in vitro and in vivo. Eur Cyt Netw 1998;9:328–36.
76. Haggiag S, Chebat J, Revel M. Induction of myeline gene expression in cell culture by IL-6Receptor-IL-6 chimera. Mult Scler 1999;5:53–9.
77. Perez-Vicente JA, Minguela A, Alvares-Lopez MR, et al. Th2 cytokine patterns in multiple sclerosis. Eur J Immunol 1996;3:40–7.
78. Jong B, Huizinga T, Bollen E, Uitdehaag B, Bosma G, et al. Production of IL-1β and IL-1Ra as risk factors for susceptibility and progression of relapse-onset multiple sclerosis. J. Neuroimmunol 2002;126:172–9.
79. Lindberg R, De Groot CJA, Montagne L, et al. Matrix Metalloproteinases, but not their endogenous inhibitors TIMPs and CSF cells of MS patients. J Neurol Sci 2001;187(Suppl 1):263.
80. Schneider H, Prossi F, Balschun D, et al. A neuromodulatory role of interleukin-1β in hippocampus. Proc Natl Acad Sci USA 1998;95:7778–83.
81. Matsuki T, Nakae S, Sudo K, et al. Abnormal T cell activation caused by imbalance of the IL1/IL1R antagonist system is responsible for the development of experimental autoimmune encephalomyelitis. Int Immunol 2006;18(2):399–407.
82. Koehler NK, Genain CP, Giesser B, Hauser SL. The human T cell response to myelin oligodendrocyte glycoprotein: A multiple sclerosis family-based study. J Immunol 2002;168:5920–7.
83. Huang C, Nazarian R, Lee J, Espinosa-Jeffery A, Vellis J. Tumor necrosis factor modulates transcription of myelin basic protein gene through nuclear factor kappa B in a human oligodendroglioma cell line. Int J Neurosci 2002;20:289.
84. Eugster HP, Frei K, Bachmann R, Lassman H, Fontana A. Severity of symptoms and demyelination in MOG-induced EAE depends on TNFR1. Eur J Immunol 1999;29:626–32.
85. Raine CS, Bonetti B, Cannella B. Multiple sclerosis: Expression of molecules of the tumor necrosis factor ligand and receptor families in relationship to demyelinated plaque. Rev Neurol (Paris) 1998;154:577–85.
86. Lindberg R, De Groot CJA, Montagne L, et al. Matrix metalloproteinases, but not their endogenous inhibitors TIMPs and CSF cells of MS patients. J Neurol Sci 2001;187(Suppl 1):263.
87. Steinman L. Some misconceptions about understanding autoimmunity through experiments with knockouts. J Exp Med 1997;185:2039–46.
88. Jovicic A, Djordjevic D, Kataranovski M. Possible role of tumor necrosis factor-α (TNF-α), interleukin-4 (IL-4), and transforming growth factor-β (TGF-β) in multiple sclerosis patients. Mult Scler 1997;3:313–20.
89. Haviv R, Stein R. Neuronal expression of TNF-R. J Neurosci Res 1998;52:380–9.
90. Bongioanni P, Lombardo F, Moscato G, et al. T-cell interleukin-6 receptor binding in interferon-beta-1b-treated multiple sclerosis patients. Eur J Neurol 2000;7:647–53.
91. Kassiotis G, Pasparakis M, Kollias G, Probert L. TNF accelerates the onset but does not alter the incidence and severity of myelin basic protein-induced experimental autoimmune encephalomyelitis. Eur J Immunol 1999;29:774–80.

92. Lui J, Marino MW, Wong G, Grail D, Dunn A, Bettadapura J, Slavin AJ, Old L, Bernard CC. TNF is a potent antiinflammatory cytokine in autoimmune-mediated demyelination. Nat Med 1998;4:78–83.
93. Miura M, Zhu H, Rotello R, Hartweig EA,Yuan J. Induction of apoptosis in fibroblasts by IL-1β converting enzyme, a mammalian homolog of the *C. elegans* cell death gene ced-3. Cell 1993;75:653–8.
94. Furlan R, Martino G, Galbiati F, Poliani PL, Smiroldo S, Bergami A, Desina G, Comi G, Flavell R, Su MS, Adorini L. Caspase-1 regulates the inflammatory process leading to autoimmune demyelination. J Immunol 1999;163:2403–9.
95. Kassiotis G, Kollias G. Uncoupling the proinflammatory from the immunosuppressive properties of tumor necrosis factor (TNF) at the p55 TNF receptor level: Implications for pathogenesis and therapy of autoimmune demyelination. J Exp Med 2001;193: 19–26.
96. Sipe KJ, Srisawasdi D, Dantzer R, Kelley KW, Weyhenmeyer JA. An endogenous 55 kDa TNF receptor mediates cell death in neural cell line. Brain Res Mol Brain Res 1998;38:222–32.
97. Jurewicz A, Matysiak M, Walczak A, et al. Signal transduction pathways in glial cells after TNF stimulation. Mult Scler 1999;5:24–31.
98. Kraus J, Fischer O, Kuehne BS, et al. Diurnal changes of common immunological markers in the serum of MS patients and healthy donors – Preliminary results. Abstract of the XVII world congress of neurology. London J Neurol Sci 2001;187(Suppl 1): 273.
99. Rieckmann P, Albrecht M, Kitze B. Cytokine mRNA levels in mononuclear blood cells from patients with multiple sclerosis. Neurology 1994;44:1523–8.
100. Spuler S, Yousry T, Scheller A, et al. Multiple sclerosis: Prospective analysis of TNF-α and 55kDa TNF receptor in CSF and serum in correlation with clinical and MRI activity. Neuroimmunology 1996;66:57–64.
101. Hovelmeyer N, Hao Z, Kranidioti K, Kassiotis G, et al. Apoptosis of oligodendrocytes via Fas and TNF-R1 is a key event in the induction of experimental autoimmune encephalomyelitis. J Immunol 2005;175(9):5875–84.
102. Blanco-Jeres C, Plaza JF, Masjuan J, Orensanz LM, Alvares-Cermeno JC. Increased levels of IL-15 mRNA in relapsing-remitting multiple sclerosis attacks. J Neuroimmunol 2002;128:90–4.
103. Kalinina NM, Solnceva O, Bisaga G, Golovkin V. TNF-α, IL-1β, IL-2 & apoptosis in patients with multiple sclerosis. Eur Cytokine Netw 1998;9:501–2.
104. Femstein GZ, Wang X, Barone FC. IL-1ra production in central nervous system in multiple sclerosis. Mult Scler 1998;4:313.a.
105. Sellebjerg F, Bendtzen K, Christiansen M, Frederiksen J. Cytokines and soluble IL-4 receptor in patients with acute optic neuritis and multiple sclerosis. Eur J Neurol 1996;3(Suppl):31.
106. Ochi H, Osoegawa M, Wu XM, Minocjhara M, Horiuchi I, Murai H, Furuya H, Kira J. Increased IL-13 but not IL-5 production by CD4-positive T cells and CD-8-positive T-cells in multiple sclerosis during relapse phase. J Neurol Sci 2002;201:45–51.
107. Chitnis T, Najafian N, Benou C, Salama AD, Grusby MJ, Sayegh MH, Khoury SJ. Effect of targeted disruption of STAT4 and STAT6 on the induction of experimental autoimmune encephalomyelitis. J Clin Invest 2001;108:739–47.
108. van Boxel-Dezaire AHH. Differences in cytokine expression between relapsing-remitting and SP multiple sclerosis patients. Mult Scler 1997;2:34–40.

109. Djordjevic D, Jovicic A, Kataranovski M, et al. Interleukin-10 possible immunomodulatory role in multiple sclerosis patients. Mult Scler 1997;3:314–20.
110. Steinman L. Multiple sclerosis: Two-stage disease. Nat Immunol 2001;2:762–4.
111. Segal BM, Dwyer BK, Shevach EM. An interleukin(IL)-10/IL12 immunoregulatory circuit controls susceptibility to autoimmune disease. J Exp Med 1998;187(4):537–46.
112. Fillatreau S, Sweenie CH, McGeachy MJ, Gray D, Anderton SM. B cells regulate autoimmunity by provision of IL-10. Nat Immunol 2002;3:944–50.
113. Yang J-S, Xu L-Y, Huang Y-M, van der Meide PH, Link H, Xiao BG. Adherent dendritic cells expressing high levels of interleukin-10 and low levels of interleukin-12 induce antigen-specific tolerance to experimental autoimmune encephalomyelitis. Immunology 2000;101:397–403.
114. O'neill EJ, Day MJ, Wraith DC. IL-10 is essential for disease protection following intranasal peptide administration in the C57BL/6 model of EAE/. J Neuroimmunol 2006;Jul 25 [Epub ahead of print].
115. Mekala DJ, Alli RS, Geiger TL. IL-10-Dependent suppression of experimental allergic encephalomyelitis by Th2-differentiated, anti-TCR redirected T lymphocytes. J Immunol 2005;174:3789–97.
116. Eugster HP. IL-6 deficient mice resist myelin oligodendrocyte glycoprotein-induced autoimmune encephalomyelitis. Eur J Immunol 1998;28:2178–87.
117. Cannella B, Raine CS. The adhesion molecule and cytokine profile of multiple sclerosis lesions. Ann Neurol 1995;37:424–35.
118. Yu S, Chen Z, Mix E, Zhu SW, Winbland B, Ljunggren HG, Zhu J. Neutralizing antibodies to IL-18 ameliorate experimental immune neuritis by counter-regulation of autoreactive Th1 responses to peripheral myelin antigen. J Neuropathol Exp Neurol 2002;61:614–22.
119. Dinarello CA. IL-18: A Th1-inducing, proinflammatory cytokine and new member of the IL-1 family. J Allergy Clin Immunol 1999;103:11–5.
120. Shi FD, Takeda K, Akira S, et al. IL-18 directs autoreactive T cells and promotes autodestruction in the central nervous system via induction of IFN-γ by NK cells. 2000;165:3099–104.
121. Kaser A, Kaser S, Kaneider NC. Interleukin-18 attacts plasmocytoid dendritic cells (DC2s) and promotes Th1 induction by DC2s through IL-18 receptor expression. Blood 2004;103:648–55.
122. Gutcher I, Urich E, Wolter K, Prinz M, Becher B. Interleukin 18-independent engagement of interleukin 18 receptor-α is required for autoimmune inflammation. J Immunol 2006;7(9):946–53.
123. Schif-Zuck S, Westermann J, Netzer N, et al. Targeted overexpression of IL-18 binding protein at the central nervous system overrides flexibility in functional polarization of antigen-specific Th2 cells. J Immunol 2005;174:4307–15.
124. Okuda Y. IL-6-deficient mice are resistant to the induction of autoimmune encephalomyelitis provoked by myelin oligodendrocyte glycoprotein. Int Immunol 1998;10:703–8.
125. Frei K, Eugaster H-P, Bopst M, Constantinesku CS, Lavi E, Fontana A. Tumor necrosis factor α and lymphotoxin α are not required for induction of acute experimental autoimmune encephalitis. J Exp Med 1997;185:2177–82.
126. Ferber IA, Brocke S, Taylor-Edwards C, et al. Mice with disrupted IFN-γ gene are susceptible to the induction of experimental autoimmune encephalomyelitis. J Immunol 1996;156:5–7.

127. Segal BM, Cross AH. FasL track to apoptosis in multiple sclerosis: TNF receptor may suppress or potentiate CNS demyelination. Neurology 2000;55:906–7.
128. Espejo C, Penkowa M, Saez-Torres I, Hidalgo J, Garcia A, Montalban X, Martinez-Caceres EM. Interferon-gamma regulates oxidative stress during experimental autoimmune encephalomyelitis. Exp Neurol 2002;177:21–31.
129. McQuatler JL, Darwich R, Ewing C, Onuki M, Kay T, Hamilton J, Reid H, Bernard C. Granulocyte macrophage colony-stimulating factor: A new putative therapeutic target in multiple sclerosis. J Exp Med 2001;194:873–81.
130. Shi F-D, Takeda K, Akira S, Sarvetnick N, Ljunggren H-G. IL-18 directs autoreactive T cells and promotes autodestruction in the central nervous system via induction of IFN-γ by NK cells. J Immunol 2000;165:3099–104.
131. Aggarwal S, Ghilardi N, Xie M-H, de Sauvage FJ, Gurney AL. Interleukin 23 promotes a distinct CD4 T cell activation state characterized by the production of interleukin-17. J Biol Chem 2003;278:1910–14.
132. Chabas D, Baranzini SE, Mitchell D, Bernard CC, Rittling C, Denhardt DT, Sobel R, Lock C, Karpuj M, Pedott R, Heller R, Oksenberg JR, Steinman L. The influence of the proinflammatory cytokine, osteopontin, on autoimmune demyelinating disease. Science 2001;294:1731–5.

Cytokines and the Brain
Edited by C. Phelps and E. Korneva

The Cytokines and Depression Hypothesis: An Evaluation

ADRIAN J. DUNN

Department of Pharmacology, Toxicology and Neuroscience, Louisiana State University Health Sciences Center, Shreveport, LA, USA

ABSTRACT

Recently, there has been great interest in the possibility that cytokines are involved in the genesis of depressive illness. The primary stimulus for this concept derived from similarities between the symptoms of depression and the behavioral responses known as sickness behavior observed in animals following the administration of interleukin-1 (IL-1) or endotoxin [lipopolysaccharide (LPS)]. This concept was bolstered by the observation that certain cytokine therapies can induce depression, and that abnormalities in immune function have long been believed to be more common in humans suffering from psychiatric illness than in the general population. This chapter will review the evidence for the cytokine hypothesis of depression and evaluate it critically. It is concluded that although cytokines can indeed induce some symptoms of depression and may be responsible for some depressive symptoms in some individuals, it is unlikely that cytokines are the essential or the normal mediators of depression.

1. INTRODUCTION

The initial concept for a cytokine model of depression derived from the observation that the cytokine, interleukin (IL)-1, induced a behavioral pattern known as sickness behavior. In a classic review, Benjamin Hart [1] argued that "the behavior of a sick individual is not a maladaptive and undesirable effect of illness but rather a highly organized strategy that is at times critical to the survival of the individual if it were living in the wild state." Smith [2] was the first to suggest that depression might be associated with increased secretion of cytokines. He proposed that IL-1, secreted by macrophages, could cause depression. He noted that administration of IL-1 not only mimicked some of the behavioral characteristics of depression, but also activated the hypothalamic–pituitary–adrenocortical axis (HPAA). This "macrophage hypothesis of depression" attracted significant attention, because it was consistent with a number of facets of depressed patients. Subsequently, Stephen Kent coined the term "sickness behavior" in an article in which the similarities between the behavioral responses elicited by the administration of IL-1 to rodents were compared to those characteristic of sickness [3]. Kent et al. [3] went so far as to propose that antagonists of IL-1 might be used to treat depression. These articles thus conceived the cytokine hypothesis of depression.

2. ORIGINS OF THE CYTOKINE HYPOTHESIS OF DEPRESSION

Almost all of the evidence for the cytokine hypothesis of depression depend on experiments performed with IL-1, although there is some limited evidence of such actions for IL-2, IL-6, and interferon (IFN)-α in humans (see below).

An obvious potential complicating factor is that an infection or other disease may induce cytokine production as a side effect, whereas the depression may result from manifestations of the disease state itself, the recognition by the individual that they have a disease, or the treatments implemented. Thus depression may coexist with the disease, but it is not necessarily the case that the body's immune response to the disease causes the depression.

The evidence for the IL-1 hypothesis of depression is based on the following assertions (Table 1): treatment of patients with certain cytokines induces symptoms of depression; activation of the immune system is observed in many depressed patients; depression occurs more frequently in those with medical disorders associated with immune dysfunction; activation of the immune system and administration of lipopolysaccharide (LPS) or IL-1 to animals induce sickness behavior, which resembles depression; chronic treatment with antidepressants inhibits sickness behavior induced by LPS; IL-1, IL-6, tumor necrosis factor (TNF)-α and IFN-α stimulate the HPAA, which is commonly activated in depressed patients; IL-1 activates cerebral noradrenergic systems, also commonly observed in depressed patients; and IL-1 and IL-6 activate brain serotonergic systems, which have been implicated in major depressive illness and its treatment. A discussion of the evidence for each of these tenets is as follows (see also Dunn [4]).

2.1. Immune abnormalities in depressed patients

Relationships between immune abnormalities and psychiatric disease have been discussed for more than half a century. Historically, the psychiatric disease most commonly associated with immune abnormalities was schizophrenia. However, although it appeared that immune abnormalities were more common in psychotic patients than in the general population, not all psychotic patients exhibited immune abnormalities and no specific immune parameter or parameters have been

Table 1. Experimental evidence cited to support the cytokine hypothesis of depression

Evidence	Cytokines implicated
1. Treatment of patients with cytokines induces symptoms of major depressive disorder	IL-2, IFN-α
2. Activation of the immune system is observed in many depressed patients. This may be reflected in elevated circulating concentrations of cytokines	IL-6
3. Depressive disorders occur more frequently in medical disorders involving immune dysfunction	Various
4. Infections and the administration of LPS or cytokines to animals induce sickness behavior, which bears some resemblance to major depressive disorder. Chronic treatment with antidepressants inhibits LPS-induced sickness behavior	IL-1
5. Cytokines activate the HPA axis. Elevated plasma cortisol is commonly observed in depressed patients	IL-1, IL-6, TNF-α, IFN-α
6. Cytokines activate brain noradrenergic systems. Elevated NE and its catabolites are observed in the CSF of depressed patients	IL-1, TNF-α
7. Cytokines activate brain serotonergic systems. Abnormalities of brain serotonin have been implicated in major depressive disorder and its treatment	IL-1, IL-6, TNF-α

implicated. It is relevant that psychosis may result from brain damage, which in turn may result in immune activation. Immune abnormalities in depressed patients have been widely reported for more than 30 years. The earliest studies of immune function in depressed patients suggested a deficiency in immune function, but the results were very inconsistent. Subsequent studies have indicated that depressed patients more often exhibit immune activation. However, the results have been highly variable; immune function is normal in many depressed patients, but when it is abnormal, it is decreased in some aspects, but activated in others (see reviews by Weisse [5], Kronfol [6], and Irwin and Miller [7]). It was also noted that depression was more frequent in patients with diseases having an immunological component. However, once again the results were highly variable and it was clear that not all patients with immune abnormalities were depressed [6,8].

The cytokine hypothesis of depression also derives support from clinical observations that patients treated with IFNs (IFN-α and IFN-β) or IL-2 often display influenza-like symptoms and nonspecific neuropsychiatric symptoms, some of which are characteristics of depression.

3. EXPERIMENTAL EVIDENCE FOR THE CYTOKINE HYPOTHESIS OF DEPRESSION

3.1. Responses to cytokine treatments in humans

Cytokine administration has been found to be effective in the treatment of many medical conditions, such as hepatitis C, multiple sclerosis, certain infections, leukemia, Kaposi's sarcoma, melanoma, myeloma, renal carcinoma, and other cancers (see Table 1). The cytokines most commonly used are IFN-α, IFN-β, and IL-2. These treatments produce side effects such as asthenia, myalgia, confusion, and influenza-like symptoms. Depression is most commonly associated with treatment with IFN-α and IL-2, and occasionally with IFN-β, but not with IFN-γ [8]. Many other neurobehavioral symptoms are observed, the most common of which is fatigue [3,9,10]. The incidence of depression associated with cytokine therapy is highly variable, ranging from 0 to 45% in different studies. The variability is in part related to the specific disease under study, as well as the cytokines used and the doses. However, the assessment measures, the patient's psychiatric history, and the choice of comparison (control) subjects are also important factors [11]. Frequently, the depressive symptoms can be treated effectively with standard antidepressant drugs. Interestingly, antidepressant treatments are most effective on the mood symptoms, and do not typically affect the neurovegetative symptoms, such as anorexia, psychomotor retardation, and sleep disorders [12–14]. By contrast, most of the animal studies have focused on the neurovegetative symptoms.

The administration of IFNs to rodents does not induce classical sickness behavior. There may be some behavioral symptoms, but they appear to be less pronounced and the literature is inconsistent (see Dunn [4]). It is notable that IFN-α does not typically induce fever or HPAA activation in rats and mice as it does in humans (see Dunn [4]).

3.2. Immune activation in depressed patients

Major depression has been associated with increased plasma concentrations of positive acute-phase proteins (haptoglobin, ceruloplasmin, C-reactive protein, hemopexine, β_1-antitrypsin, and β_1-acid glycoprotein) and lower plasma concentrations of negative acute-phase proteins (transferrin, albumin, and retinol-binding protein) [6,15]. Increases in acute-phase proteins in combination with decreases in negative acute-phase proteins are considered to indicate an inflammatory state, leading Maes [15] to suggest that depression is associated with chronic inflammation. He reported that depressed patients display elevated concentrations of inflammatory mediators, such as the prostaglandins [16] and complement [17]. He specifically proposed that the chronic inflammatory state he believes underlies depression can be explained by increased production of cytokines by circulating monocytes and macrophages [15]. Some support for this concept has been provided by reports of increased circulating leukocytes and activated T cells in depressed patients [6,18].

More recent reports have assessed the ability of immune cells obtained from patients to produce cytokines *in vitro* [7,17,19]. The results have been quite variable, but there is some evidence for increased production of IL-1 and IL-6 from lymphocytes of depressed patients [20]. Such assays have been argued to provide functional measures of immune activity, but the biological significance of such *ex vivo* measures and their relevance to *in vivo* immune function are far from clear. Neopterin, considered to be a marker of activation of cell-mediated immunity, was reported to be increased in the plasma and urine of depressed patients [19,21]. However, other studies have failed to find evidence for immune activation in depressed patients [22,23].

Most recent studies have focused on assessing the concentrations of circulating cytokines in depressed patients. IL-1β, IL-6, and the IFNs have all been reported to be increased in the plasma of depressed patients [24–26]. Increases in plasma IL-1β were reported in a small number of studies [27,28], but such findings were not replicated in others (e.g., [29]). Musselman et al. [30] reported elevated IL-6 concentrations in cancer patients with depression, but others observed normal concentrations of circulating cytokines in patients diagnosed with depression [31]. In a relatively small group of depressed patients, Levine et al. [32] reported increases in the concentrations of IL-1β in the cerebrospinal fluid (CSF), which were accompanied by decreases in IL-6. In this study, the increases in IL-1β correlated with the severity of the depression. However, to date, no replication (or refutation) of this study has been reported.

Zorrilla et al. [33] performed a meta-analysis of the existing clinical data on immune abnormalities in depressed patients. This analysis showed that patients with major depression exhibited an overall leukocytosis, manifested as a relative neutrophilia and lymphopenia; as well as increased CD4/CD8 ratios; increased circulating haptoglobin, prostaglandin E2 (PGE_2), and IL-6 concentrations; reduced natural killer (NK)-cell cytotoxicity; and reduced lymphocyte proliferative responses to mitogens. Plasma concentrations of IL-1, the only cytokine that clearly induces depression-like symptoms, were not consistently altered. Zorrilla et al. [33] commented that "the degree of heterogeneity of the studies' results raises questions about their robustness." Pollmächer et al. [34] pointed out that certain cytokines (particularly IL-1β) are undetectable in human plasma under normal physiological conditions as well as during experimental endotoxemia [34]. They also noted that the physiological fluctuations of the detectable cytokines (IL-6 and TNF-α) are very poorly characterized in normal and pathological states, and that the alterations of circulating cytokines observed in the depressed and immune-related medical conditions are extremely modest compared with the concentrations of cytokines that occur in the circulation during cytokine treatments. Another important caveat is that depression may occur in response to a medical condition (perhaps not detected) that is responsible for the immune activation.

To summarize, the evidence for an activation of immune responses in patients with major depressive disorder is not particularly strong or consistent [11,33], although there are trends to leukocytosis, increased CD4/CD8 ratios, elevated plasma concentrations of haptoglobulin, IL-6 and PGE_2, and decreased NK cell activity and lymphocyte proliferative responses. However, immune activation is not observed in all depressed patients.

3.3. Depression in medical conditions that display immune activation

If immune activation can cause depression, then depression should be more common in patients that suffer from medical conditions that exhibit immune activation, and there should be a correlation between the symptoms of depression and the immune activation (e.g., increased circulating concentrations of cytokines) in such conditions.

Depression has been reported to be more common in noninfectious diseases associated with a chronic activation of the immune system, such as multiple sclerosis [35], allergy [36], rheumatoid arthritis [37], and stroke [38]. In these diseases, the immune abnormalities precede the development of depression [39]. Nevertheless, data on the correlations between depression and immune activation or plasma concentrations of cytokines in patients with autoimmune diseases or chronic infections or inflammation are few and conflicting (see [34]).

3.4. Does sickness behavior resemble clinical depression?

The Diagnostic and Statistical Manual (DSM-IVR) criterion for a major depressive episode requires either depressed mood or anhedonia (the inability to experience pleasure), plus any four of the following symptoms: significant loss of weight or appetite, insomnia, psychomotor agitation or retardation, fatigue or loss of energy, feelings of worthlessness, diminished ability to think or concentrate, and indecisiveness, recurrent thoughts of death, or suicidal ideation for a 2-week period. Weight gain and hypersomnia are regarded as symptoms of atypical depression. In Table 2, these symptoms are compared with those observed in patients treated with IFN-α or IL-2, and with sickness behavior in rodents. Table 3 compares the various physiological responses in the same way.

Sickness behaviors induced by IL-1 and LPS include hyperthermia, hypomotility, hypophagia, hyperalgesia, decreased interest in exploring the environment, decreased libido in males, and increased time spent sleeping (see [3,40]. Although there are some similarities between sickness behavior and depression, the analogies are far from perfect. Biologically, sickness behavior is considered to be a coping strategy that bestows benefits on the individual and has evolutionary

Table 2. Comparison of the behavioral symptoms or concomitants of major depressive disorder and animal models of depression

Humans		Animals
DSM-IV symptoms of depression	Cytokine therapy	LPS or IL-1 administration
1. Depressed mood	+	Indeterminate
2. Anhedonia	+	+
3. Significant loss of weight or appetite (weight gain in atypical)	Anorexia	Hypophagia
4. Insomnia (hypersomnia in atypical)	+	Slow-Wave Sleep ↑
5. Psychomotor agitation or retardation	+	Locomotor activity ↓
6. Fatigue or loss of energy	+	Locomotor activity ↓ Motivation ↓
7. Feelings of worthlessness	0	Indeterminate
8. Diminished ability to think or concentrate, or indecisiveness	+	Indeterminate
9. Recurrent thoughts of death or suicidal ideation	+/−	Indeterminate

A diagnosis of a depressive disorder requires symptoms 1 or 2, plus four of 3–9 for a period of 2 weeks or more.

Table 3. Comparison of physiological changes in human depression, sickness behavior, and animal models

	Humans	Animals
Physiological parameter	Depression	LPS or IL-1 injection
HPAA (plasma corticosteroids)	+	+
Plasma catecholamines	+	+
Plasma tryptophan	(−)	
Brain NE activity	+	+
Brain 5-HT activity		+
Brain tryptophan		+
Brain CRF	+/−	+/−
Plasma IL-1β	0	+
Plasma IL-6	+	+
DST	+	
CRF test	+	

advantages [1]. Whether or not psychiatric states have adaptive value is controversial, and the benefit of depression for the suffering patient is highly questionable [41,42].

3.4.1. *Hyperthermia (fever)*

Fever is almost universally associated with infection with pathogens and many other pathologies, but is not a characteristic of depression, although there is a slight trend to nocturnal hyperthermia in depressed patients [43].

3.4.2. *Hypomotility*

Hypomotility is a marked characteristic of sickness behavior in animals, and this is consistent with the psychomotor retardation observed in depressed patients, although depressed patients are often agitated.

3.4.3. *Feeding: hypophagia*

IL-1, LPS, and infections induce marked reductions in feeding, and consequent loss of body weight [44,45]. However, depressed patients may exhibit either hypo- or hyperphagia. Hypophagia is characteristic of typical depression, whereas hyperphagia is a defining characteristic of atypical depression.

3.4.4. *Pain: hyperalgesia*

IL-1 and LPS have been reported to induce hypersensitivity to pain [46]. However, depressed patients typically do not exhibit hypersensitivity; more often they exhibit hypoalgesia.

3.4.5. *Sleep: hypersomnia*

IL-1 and LPS administration to rodents increases sleep time, specifically an increase in the duration of slow-wave sleep [47]. But, hypersomnia is not typical of depression; depressed patients are often insomniac, although increased sleep may occur in atypical depression.

4. ARE SYMPTOMS OF DEPRESSION OBSERVED IN SICK ANIMALS?

It is not currently possible to assess mood in animals, so we cannot evaluate this symptom in studies of sickness behavior. However, the second key symptom, anhedonia, can be studied in animals, albeit not directly. It has been argued that the consumption of palatable foods, such as solutions of sucrose or saccharine, or sweetened milk may reflect hedonia, so that reduced consumption reflects anhedonia, although it is difficult experimentally to distinguish the hedonic from the appetitive effects (see [48]). This may be confounded further if studies are performed in food-deprived animals. The most popular models use saccharine, which it is argued appeals to the taste buds, but because it lacks caloric value, it is not considered to reflect feeding [49,50]. Nevertheless, IL-1 and LPS decrease ingestion of sucrose or saccharine consistent with the anhedonia in humans, although hyperphagia is more common in depressed patients.

Hedonia can also be assessed using intracranial self-stimulation (ICSS). Rats implanted with electrodes in certain regions of the brain, most notably in the medial forebrain bundle (MFB), will press a bar if doing so results in electrical stimulation of the electrodes. In a series of studies, Anisman's group [51,52] has shown that peripheral (ip) injection of IL-2 altered self-stimulation via electrodes in the MFB. There was relatively little effect immediately after the IL-2 injections; only a modest increase in the threshold was observed at intermediate currents. But over the next several days, more substantial increases in the ICSS threshold were observed. These effects were interpreted to indicate that IL-2 induced a long-lasting anhedonia. No such effects were observed following administration of IL-1β or IL-6 [52]. Because of the well-accepted role of dopamine (DA) as a mediator of reward, the effect of IL-2 would be consistent with a reduction in DA release, although the latter was observed only after chronic injection of IL-2 [51]. Similar effects of IL-2 were observed in mice stimulated in the ventral tegmental area. In this case, intracerebroventricular (icv) application of IL-2 (5 ng) elevated the frequency required to elicit self-stimulation from the dorsal, but not the ventral, A10 region [53]. These results suggest that IL-2, but not IL-1 and IL-6, can have long-lasting anhedonic effects in rats and mice.

There has been relatively little research on cognitive behavior. Sick animals may show cognitive deficits, for example, in learning and memory tasks (see reviews [40,54]). Needless to say, some of the above manifestations of sickness behavior could explain the cognitive deficits. Major cognitive deficits are not often observed in depressed patients, and when they are, they are generally mild, normally restricted to some cognitive bias [55]. However, clinical research has shown that cytokine administration (IFN-α and IL-2) in cancer patients can affect cognitive functions, with effects reminiscent of the dysfunctions observed in neurodegenerative diseases [9].

5. THE POTENTIAL ROLE IN SICKNESS BEHAVIOR OF CYTOKINES IN THE BRAIN

The most recent versions of the cytokine hypothesis of depression propose a role for cytokines within the brain. In this model, immune activation in the periphery can induce the synthesis or appearance of cytokines in the brain parenchyma, and it is these cytokines that are directly responsible for inducing sickness behavior and the symptoms in depressed patients [56–58]. The literature to date has focused more or less exclusively on IL-1 [58]. The initial impetus for the proposal was the observation that intracerebral injection of IL-1 induced symptoms of sickness behavior like those of ip injection and at lower doses.

The experimental data cited to support this hypothesis are the induction of the appearance of IL-1 within the brain and the ability of intracerebrally administered IL-1 antagonists to prevent the effect of the peripheral treatments. This is an interesting hypothesis, but it is not at all clear why an involvement of brain cytokines is necessary to initiate sickness behavior. It is also unclear what functions would be associated with IL-1 if it appears in the brain. The major route by which LPS and IL-1 in the periphery are thought to signal the brain is by IL-1 stimulating afferents of the vagus nerve, because lesioning the vagus can substantially diminish sickness behaviors and other responses induced by intraperitoneal injection of IL-1 and LPS [56,57]. Lipopolysaccharide appearing in the peritoneum (which may occur following the ingestion of pathogenic bacteria) can induce the production of IL-1 by Kupfer cells (considered to be a specialized form of macrophage) in the liver. The IL-1 can then bind to paraganglion cells associated with the vagus nerve, and subsequently activate vagal afferents that signal the brain [56,59]. The vagal afferents within the brain can activate various brain stem cell groups, including noradrenergic neurons in the brain stem, which in turn appear to be involved in the activation of the HPAA [60,61]. However, norepinephrine (NE) does not appear to be involved in the hypophagia and hypomotility associated with ip IL-1 and LPS administration [62]. An alternative route is direct action of IL-1 on the receptors of endothelial cells, especially in the anterior hypothalamus [63]. These receptors appear to be coupled to cyclooxygenase (COX) enzymes, the activation of which results in the production of prostaglandins, including PGE_2. There is strong evidence that this mechanism is important for the early stages of IL-1-induced fever [64], and reason to believe that this may be a mechanism for IL-1-induced hypophagia [65].

The concentrations of IL-1 present in the normal brain are very low indeed and whether or not active IL-1 occurs in the normal brain is controversial. An early immunohistochemical study indicated the presence of IL-1β in postmortem human brain, apparently associated with neurons [66]. Subsequent immunohistochemical studies have reported IL-1 in neurons in the hypothalamus of the rat [67] and the pig [68]. In Breder's study, the anatomical distributions of the IL-1β immunoreactivity differed somewhat among the individual human brains, and there was little concordance among the anatomical distributions observed in the brains of the human, the rat, and the pig.

A more recent study performed in the brains of multiple sclerosis patients and controls observed IL-1β-like immunoreactivity primarily in microglia, especially in the sclerotic plaques [69]. However, IL-1β-like immunoreactivity was found not only in neuronal cell bodies in the hypothalamus particularly in the paraventricular nucleus (PVN), but also in several other hypothalamic nuclei and in certain neuronal fibers [69]. The neuronal IL-1β immunoreactivity was dramatically reduced in the brains of multiple sclerosis patients. Curiously, all of the IL-1β-positive neurons also contained oxytocin, although only about one-third of the oxytocin-containing neurons contained IL-1β. None of the corticotropin-releasing factor (CRF)- or vasopressin-positive neurons contained IL-1β immunoreactivity. Huitinga et al. [69] were aware that the association of IL-1β with oxytocin neurons could be artifactual and reflect another protein with similar antigenic properties, although it was observed using three different antibodies to IL-1β. If the immunoreactivity is truly for IL-1β, then the association between oxytocin-containing neurons and sickness behavior should be explored.

Most of the studies indicating IL-1 induction in the brain have employed ip administration of high or very high doses of LPS. Many studies reported only the induction of mRNA for IL-1 and/or the IL-1 Type I receptor using nonquantitative techniques [70–73]. Quan et al. [73] administered 2.5 mg/kg LPS (ip) to rats, which induced IL-1 mRNA in the pituitary and the circumventricular organs (CVOs), that is, outside the blood–brain barrier. Small amounts of mRNA for

IL-1β were detected in the brain, mostly in microglia thought to be derived from macrophages that had penetrated the endothelia and/or accessed the brain via the CVOs. A demonstration of the presence of mRNA indicates only the propensity for the synthesis of the proteins, and in the case of IL-1β, only the synthesis of its precursor, pro-IL-1β, which must be cleaved by IL-1-converting enzyme (caspase 1) to produce active IL-1β. However, Quan et al. [74] detected IL-1β using an immunoassay after injection of 2.5 mg/kg LPS ip. Subsequently, Van Dam et al. [75] observed immunoreactivity for IL-1β in the rat brain hypothalamus in microglia and sparsely in astroglia after ip administration of the same dose of LPS (ip or intravenously).

Several laboratories have used an IL-1β-specific ELISA to demonstrate the presence of IL-1 in the brain, especially in the hippocampus, following various treatments including LPS and various stressors (see [56]). To obtain measurable quantities of IL-1β protein, it is necessary to homogenize the whole hypothalamus of a rat in a small volume of fluid, which is then centrifuged to produce a supernatant that is used undiluted for the ELISA. It is not clear whether the assays used are specific for IL-1 in the presence of such high concentrations of other proteins. However, in none of these studies was the purported IL-1 purified, even partially, prior to the assay, so we cannot be certain that the apparent immunoreactivity truly reflected IL-1, especially because the amounts of IL-1 reported were around the detection limit of the assay.

Classical criteria for the identification of hormones and neurotransmitters require more rigorous techniques, involving extensive purification of the candidate molecule and demonstration of its identity using both immunological and bioassays. However, to date, only one report included a Western blot to support the existence of IL-1 in the rat hypothalamus [67]. And, Quan et al. [76] employed a bioassay to demonstrate the activity of partially purified IL-1 in the brains of untreated rats.

Nevertheless, IL-1 is likely to be induced in the brain under pathological conditions. When a cell dies, IL-1 will almost certainly be involved in the immunological reaction that clears the dead tissue, and this may account for much of the IL-1 measured in samples of brain tissue. This phenomenon could be regarded as a local inflammation within the brain, and will undoubtedly occur associated with any form of cannulation or intracerebral injections. Thus it cannot be excluded that the presence of the limited amounts of IL-1β in some of the reports reflects local pathologies. The IL-1 may well have been present in microglia involved in the phagocytosis of dead cells or cellular components.

If IL-1β functions in the brain, it needs receptors on which to act. An early study suggested widespread binding of ^{125}I-IL-1 (mouse) to slices of rat brain [77]. Subsequent careful studies by Haour et al. [78] and Takao et al. [79] found that the only binding of radiolabeled IL-1 detected in the brains of rats and mice was associated with endothelial cells, which may explain the earlier results of Farrar et al. [77]. Neither group found any evidence for binding of IL-1 in the brain parenchyma of the rat, although both observed limited binding in the hippocampus of the mouse. It is possible that there are so few IL-1 receptors in the rat and mouse brain that they could not be detected by the binding techniques used. The presence of a very small number of IL-1 receptors on a cell may be sufficient to induce biological responses [80], so the failure to demonstrate IL-1 binding in the brain does not necessarily preclude the presence of functional receptors. Nevertheless, it is clear that the existence of IL-1 receptors in the brain parenchyma is limited at best. Other researchers have identified mRNA for the IL-1 Type 1 receptor in the brain [71,81,82]. However, the presence of mRNA is insufficient to prove the existence of functional receptors, for which other proteins, such as the IL-1 receptor accessory protein, are required.

The major evidence cited to indicate a role for cytokines in the brain is based on the use of intracerebral injections of IL-1ra [83–85]. Such data are useful and suggestive of a role for IL-1, but much more evidence will be needed to define and establish a clear physiological role for IL-1 in the brain, not the least in depression.

6. EFFECTS OF ANTIDEPRESSANT TREATMENTS ON SICKNESS BEHAVIOR

Treatment of rats with LPS decreased the frequency with which rats pressed a bar to obtain a drop of saccharine solution [86]. This response was considered to reflect anhedonia, a cardinal symptom of depression. When the rats were treated chronically (for 3–5 weeks) with the tricyclic antidepressant, imipramine, which inhibits the reuptake of both NE and 5-HT, the treatment prevented the induction of "anhedonia" by LPS in the rats [49,86]. Similar results were obtained by Shen et al. [87] using the NE-selective reuptake inhibitor, desmethylimipramine, and by Yirmiya et al. [88] using the serotonin-selective reuptake inhibitor (SSRI), fluoxetine, although in the latter case, the prevention was less complete. However, Shen et al. [87] observed no such effect of the SSRI, paroxetine, or of the 5-HT and NE reuptake inhibitor, venlafaxine. Thus the effect of antidepressants appears to be less evident with the SSRIs than with the tricyclic antidepressants [49,87]. The atypical antidepressant, tianeptine, was reported to be effective chronically but not acutely [89]. We failed to observe any effect of chronic imipramine or venlafaxine on the inhibitory effect of LPS on the drinking of sweetened milk in mice [90], and Yirmiya et al. [49] have also indicated a failure to observe an effect of antidepressants on LPS-induced saccharin drinking in mice. Based on these studies, it appears that the effect of chronic antidepressant treatment is observed more clearly with LPS than with IL-1, and the effects of the antidepressants are apparent in rats but not mice. Other evidence suggests that antidepressants may affect the induction of IL-1 production by LPS [49,89,91] and reflect a peripheral rather than a central mechanism. Thus these observations have failed to provide strong support for the IL-1 hypothesis of depression.

Thus there are some similarities between sickness behavior and the behavior of patients suffering from major depressive disorder, but there are many aspects in which sickness behavior does not resemble depressive illness in humans. It has also been shown that in experimental endotoxemia, fever and corticosterone responses occur at a time when there are no detectable increases in circulating endotoxin or cytokines [92]. These discrepancies suggest that cytokines may not be necessary for the activation of the endocrine and sickness responses, nor for major depressive disorders.

7. CYTOKINE ACTIVATION OF THE HYPOTHALAMIC–PITUITARY–ADRENOCORTICAL AXIS: RELATIONSHIP TO DEPRESSION

Smith [2] cited the ability of IL-1 to activate the HPAA [93] as support for the cytokine hypothesis of depression. Activation of the HPAA is perhaps the most consistent biological marker of depression, yet it occurs in only 50–70% of patients. Similar proportions of depressed patients are abnormal in the dexamethasone suppression (DST) and CRF challenge tests [94]. The HPAA-activating effect of IL-1 is shared by certain other cytokines, such as IL-6 and TNF-β, but both cytokines are far less potent than IL-1 and may have limited physiological significance, except perhaps in the absence of IL-1 [95,96]. Pretreatment with an antibody to IL-6 and studies in IL-6-knockout mice indicate that IL-6 contributes only modestly to the elevation of plasma adrenocorticotropic hormone (ACTH) and corticosterone by LPS in mice [97,98]. But, because plasma concentrations of IL-6 can be elevated dramatically by infections and other pathologies in which the HPAA is activated [99], IL-6 may contribute to this activation. Depending on the physiological conditions, IL-6 has the ability to activate the HPAA via the hypothalamus, the pituitary and the adrenal cortex [96,99].

Administration of IFN-α and IFN-γ, but not IFN-β, markedly activates the HPAA in humans [100,101]. However, curiously, IFN-γ has been reported to elevate cortisol, but has little effect on ACTH [102]. There may be species differences because in rats, both ip and icv administration of hIFN-α decreased plasma ACTH and corticosterone [103,104]. However, another study showed modest increases in ACTH and corticosterone following iv injection of rat IFN-α [105]. We have not observed increases in plasma corticosterone following administration of a range of doses of IFN-α (human or mouse) in mice and rats [96,106].

The HPAA activation by IL-1 can be dissociated from its behavioral effects. Inhibitors of COX more or less prevent several behavioral responses to IL-1 [45,107], but the HPAA activation is affected only after iv injection of IL-1 and only in the early phase after ip injections [108]. The role of CRF also differs in the HPAA and the behavioral responses. It appears to be involved in the HPAA responses to IL-1, because antibodies to CRF prevent IL-1-induced HPAA activation in rats and mice [109,110]. Also, HPAA activation was minuscule in CRF knockout mice [111,112], although the hypophagic responses to IL-1 and LPS were normal [112].

Smith [2] and Maes et al. [25] argued that IL-1 may be responsible for the increased HPAA activation in depressed patients. However, support for IL-1 or the IFNs as mediators of the HPAA activation in depression is diminished because neither IL-1 nor the IFNs have been shown to be elevated consistently in depressed patients (see above).

8. CYTOKINES AND BRAIN CATECHOLAMINES

In mice and rats, central noradrenergic systems are markedly activated by IL-1 as indicated by classical neurochemical studies showing increases in NE catabolites [113,114]. The activation of brain noradrenergic systems has been confirmed by many microdialysis studies (see [114]). This noradrenergic activation is not uniform; the ventral system is activated substantially more than the dorsal system. IL-6 administration does not induce noradrenergic activation [115], but TNF-β has modest effects at high doses in mice [116].

IL-1, IL-6, and TNF-β do not induce consistent effects on DA metabolism, although small increases are occasionally observed with IL-1 [117]. However, IL-2 (1 μg ip) induced a marked reduction in microdialysate concentrations of DA from the nucleus accumbens of the rat [51]. The reported effects of IFNs on brain catecholamines have been quite varied, but we have not observed consistent effects of IFN-α or IFN-β on brain DA or NE in rats or mice (see review by Dunn [114]).

Many clinical studies have indicated apparent hypersecretion of brain NE in depression, assessed by increased CSF or urinary concentrations of the NE catabolite, 3-methoxy, 4-hydroxyphenylethyleneglycol (MHPG). More recently, such findings have been extended by direct measurement of CSF NE (e.g., [118]). Therefore, IL-1-mediated elevation of cerebral NE activity could be taken as evidence for a role of IL-1 in depression, although again the failure to observe consistent elevations of plasma IL-1 in depressed patients tempers this evidence.

8.1. Cytokines and brain serotonin

Current hypotheses of depression are focused on brain serotonin, based largely on the ability of drugs that inhibit the reuptake of serotonin to act as antidepressants. Thus it is posited that depression may be caused by some abnormality in brain serotonergic neurotransmission. Consistent with this, serotonin receptor abnormalities have been reported in the brains of deceased depressed patients (for reviews, see [119–122]). Antidepressants are presumed to work by normalizing a defect in brain serotonergic transmission through plastic changes or neurotropic effects [122].

IL-1, IL-6, and TNF-α have all been shown to activate brain serotonergic systems, increasing brain tryptophan concentrations and the metabolism of 5-HT as indicated by increases in the brain concentrations of its catabolite, 5-hydroxyindoleacetic acid (5-HIAA) [117,123,124]. Increased 5-HT release by peripherally administered IL-6 is supported by studies employing microdialysis and *in vivo* chronoamperometry in rats [125]. Acutely, IL-1 and IL-6 increase free tryptophan in the brain, which has the potential to enhance serotonergic activity, because tryptophan can be rate-limiting for serotonin synthesis (see below). The problem is that this action of the cytokines should cure depression, not cause it. This could perhaps be explained if the chronic effects of the cytokines differed from the acute ones. For example, the chronic elevation of cytokines might downregulate 5-HT secretion. The effects of chronic administration of the cytokines on serotonin metabolism have not been studied extensively, but one report indicated that repeated TNF-α administration increased 5-HT metabolism [126]. IL-1β has also been reported to activate the serotonin transporter, which could decrease extracellular 5-HT [127].

An interesting possibility for the mechanism by which infections might induce depression relates to the metabolism of tryptophan and the consequences for 5-HT. Infections, especially viral infections, induce IFN-γ, which is a potent inducer of indoleamine 2,3-dioxygenase (IDO) in macrophages and certain other cells. IL-2 and IFN-α have similar effects on IDO, but to a lesser extent. IDO degrades tryptophan, and this probably accounts for the frequent association of infections with decreases in plasma tryptophan. This may be a defense mechanism designed to limit the production of new proteins by pathogens (tryptophan is the rate-liming amino acid for protein synthesis). IDO enables the conversion of tryptophan to kynurenine and the production of neopterin, both of which are elevated in the plasma of infected subjects. The resulting catabolism of tryptophan decreases its circulating concentrations.

Decreases in the plasma concentrations of tryptophan have been reported in some depressed patients [120], but low plasma tryptophan is not a reliable biological marker for depressive illness [119]. Low concentrations of plasma tryptophan limit the availability of tryptophan to the brain, thus limiting serotonin synthesis, which might induce depression [121]. Such a mechanism for the induction of depression has not been adequately demonstrated, but lowering plasma tryptophan can precipitate a relapse in depressed patients treated with SSRIs [128,129], although not in those treated with desmethylimipramine (whose primary mechanism of action is thought to be inhibition of NE uptake) [129] or cognitive therapy [130]. Tryptophan depletion may also induce depression in susceptible individuals [119]. There is also evidence that individuals that exhibit mood changes in response to rapid tryptophan depletion may be at risk for depression [131].

Animal studies have provided some limited support for this mechanism. Repeated tryptophan depletion in rats increased anxiety-like behavior in the open field and immobility in the forced swim test (thought to reflect depression-like behavior), but did not alter behavior in the Morris water maze [132]. However, acute tryptophan depletion did not alter the behavior of rats in the open-field test, the home cage emergence test, and the forced swim test, but decrements were observed in an object recognition test [133].

Therefore, there are several hypothetical mechanisms by which cytokines could act on brain serotonergic systems and thus contribute to the pathophysiology of depression. However, any relationship between these potential mechanisms and depression remains to be demonstrated in depressed patients.

9. THE VALUE OF CYTOKINE-INDUCED SICKNESS BEHAVIOR AS A MODEL FOR THE STUDY OF DEPRESSION

The value of animal models for human diseases is to provide insight into the mechanisms involved in the diseases, which may suggest potential therapies that can then be tested in the models. Because such models will normally involve artificial interventions, it is important to create them intelligently. It is unlikely that treatments effective in the model will work if the method used to induce the model is unrelated to the basic mechanism(s) of the disease. Thus if cytokine administration is used as a model for depression, then the introduction of appropriate cytokine antagonists should prevent the symptoms in the model. However, unless cytokines are the cause (or a cause) of a major depressive disorder, cytokine antagonists are unlikely to be effective clinically. Moreover, provided they are safe and nontoxic, the cytokine antagonists could be tested directly in patients, negating the need for the model. The model could be used to assess the toxicity of the treatments or the efficacy of chronic treatment, but such tests would normally employ standard tests for toxicity and not use the sickness behavior model. If cytokine antagonists are not effective in the clinical situation, there would be no need to test them in the animal model, and the model would have little value. Models are most useful if they are based on the etiology of the human disease, or to test a proposed underlying mechanism.

We do not know the value of the cytokine-induced sickness behavior model of depression yet, but we do have some insight into what will probably not work. For example, there is a general agreement that many behavioral responses to IL-1 are largely prevented by inhibition of COX, a key enzyme for the production of prostaglandins and other eicosanoids [45,134,135]. This observation led Charlton [136] (and subsequently others) to propose that COX inhibitors [the nonsteroidal anti-inflammatory drugs (NSAIDs)], such as aspirin, indomethacin and ibuprofen, should have antidepressant activity. However, depressed patients have frequently taken NSAIDs, but there is no good evidence for antidepressant activity. Reference to the animal data indicates something of the problem. Although COX inhibitors more or less prevent several sickness behaviors induced by IL-1, they are much less effective against LPS and have virtually no effect against more complex immune challenges, such as influenza virus infection or tumors [45,137, 138].

The results from the animal tests used to assess antidepressant activity have not been particularly useful. We have studied the behavior of mice in the Porsolt forced swim test [139,140], and the tail suspension test [141]. The results showed that ip administration of IL-1β (30–300 ng) or LPS (1–5 μg/mouse) induced depression-like activity. However, the same doses of IL-1β and LPS induced substantial decreases in locomotor activity [142]. This latter effect generally appeared at doses lower than those necessary to induce statistically significant results in both tests. Thus it is difficult to dissociate these behavioral effects of IL-1 in the forced swim and tail suspension tests from the general reduction in activity. These results are generally consistent with the literature [138,143].

10. CONCLUSIONS

The foregoing review indicates some association between the appearance of cytokines and major depressive disorder, but no clear causal relationship has been established. Administration of certain cytokines (e.g., IFN-α and IL-2) to humans induces symptoms of depression in some patients, but these responses occur in only a proportion of patients, and many other neuropsychiatric symptoms may also be induced. Immune activation appears more frequently in depressed patients than in the general population, but is not observed in all depressed patients. The immune activation may reflect other medical conditions that independently induce depression. Alternatively, depression may ensue when a patient learns of a diagnosis with a poor prognosis. Nevertheless, elevated concentrations of some cytokines may induce symptoms of depression in some patients.

There is some evidence that depressed patients exhibit elevated plasma concentrations of cytokines. However, the only cytokine consistently elevated in the plasma of depressed patients is IL-6. Plasma concentrations of IL-6 are frequently elevated during infections and other pathological conditions (and associated with stress), but not all such occurrences are associated with depression. Thus, there is unlikely to be a specific direct rclationship between plasma IL-6 and depression. Also, in studies in which IL-6 was administered to rats and mice, it failed to elicit sickness behavior [98]. IL-1 is the only cytokine that has been shown to induce depression-like symptoms, but there is very little evidence for elevations of this cytokine in depressed patients (except for one unconfirmed study of CSF).

Sickness behavior induced by LPS or IL-1 has some similarities with the behavior of depressed patients, but there are important differences, for example, in the effects on body temperature, in the sensitivity to pain, and in the specific changes in sleep patterns. Therefore it may be premature to consider experimentally induced sickness behavior as a useful model for depressive illness. Experiments in which animals were treated chronically with antidepressant drugs have failed to provide strong support for the hypothesis that such treatments work by antagonizing the actions of cytokines.

The cytokine hypothesis is not in conflict with earlier hypotheses of depression, such as those involving hyperactivity of the HPAA, of CRF, or of noradrenergic systems, because IL-1 activates the HPAA and brain noradrenergic systems. Also, CRF appears to be involved in the HPAA responses to IL-1, because antibodies to CRF prevent IL-1-induced HPAA activation in rats and mice, and HPA responses to IL-1 are very small in CRF knockout mice. The cytokine hypothesis could also be consistent with the serotonin hypothesis of depression, because IL-1, IL-6 and TNF-α each have been shown to affect brain serotonergic transmission. However, IL-1, IL-6, and TNF-α administered acutely increase brain 5-HT release, whereas the most commonly prescribed treatments for depression have similar effects. It is possible that the chronic elevation of circulating cytokine concentrations would downregulate 5-HT release, which might contribute to depressive illness.

Drugs that inhibit serotonin reuptake are currently the most useful for the treatment of depression. Research on the effects of antidepressants on the immune system would be useful. In particular, the associations of abnormalities in tryptophan and serotonin with pathogen infection and with immune activation are potentially very significant and need further investigation.

These observations do not exclude a role for cytokines in inducing depression. It is certainly possible that increased cytokine production may induce depression in some patients, and some cytokines may contribute to various neuropsychiatric symptoms. Nevertheless, we can conclude that the actions of cytokines are unlikely to account for all depressive illness.

It is also possible that by activating brain CRF, the HPAA, and noradrenergic and serotonergic mechanisms, cytokines may complement (or even synergize with) other factors that induce depression.

ACKNOWLEDGMENTS

The author's research included or referred to in this article was supported by the US National Institutes of Health (MH46261 and NS35370).

REFERENCES

1. Hart BL. Biological basis of the behavior of sick animals. Neurosci Biobehav Rev 1988;12:123–37.
2. Smith RS. The macrophage theory of depression. Med Hypotheses 1991;35:298–306.
3. Kent S, Bluthé R-M, Kelley KW, Dantzer R. Sickness behavior as a new target for drug development. Trends Pharmacol Sci 1992;13:24–8.
4. Dunn AJ, Swiergiel AH, de Beaurepaire R. Cytokines as mediators of depression: What we can learn from animal studies? Neurosci Biobehav Rev 2005;29:891–909.
5. Weisse CS. Depression and immunocompetence: A review of the literature. Psychol Bull 1992;111:475–89.
6. Kronfol Z. Immune dysregulation in major depression: A critical review of existing evidence. Int J Neuropsychopharmacol 2002;5:333–43.
7. Irwin M, Miller AH. Depressive disorders and immunity: 20 years of progress and discovery. Brain Behav Immun 2007;21:374–83.
8. Valentine AD, Meyers CA, Kling MA, Richelson E, Hauser P. Mood and cognitive side-effects of interferon-α therapy. Semin Oncol 1998;25:S39–47.
9. Meyers CA. Mood and cognitive disorders in cancer patients receiving cytokine therapy. Adv Exp Med Biol 1999;461:75–81.
10. Cleeland CS, Bennett GJ, Dantzer R, Dougherty PM, Dunn AJ, Meyers CA, Miller AH, Payne R, Reuben JM, Wang X, Lee B-N. Are the symptoms of cancer and cancer treatment due to a shared biologic mechanism? A cytokine-immunologic model of cancer symptoms. Cancer 2003;97:2919–25.
11. de Beaurepaire R, Swiergiel AH, Dunn AJ. Neuroimmune mediators: Are cytokines mediators of depression. In Biology of Depression. Licinio J and Wong M-L, Eds; Weinheim:Wiley, 2005; pp. 557–81.
12. Miyaoka H, Otsubo T, Kamijima K, Ishii M, Onuki M, Mitamura K. Depression from interferon therapy in patients with hepatitis C. Am J Psychiatry 1999;156:1120.
13. Musselman DL, Lawson DH, Gumnick JF, Manatunga AK, Penna S, Goodkin RS, Greiner K, Nemeroff CB, Miller AH. Paroxetine for the prevention of depression induced by high-dose interferon alfa. N Engl J Med 2001;344:961–6.
14. Capuron L, Ravaud A, Neveu P, Miller AH, Maes M, Dantzer R. Association between decreased serum tryptophan concentrations and depressive symptoms in cancer patients undergoing cytokine therapy. Mol Psychiatry 2002;7:468–73.
15. Maes M. Evidence for an immune response in major depression: A review and hypothesis. Prog Neuropsychopharmacol Biol Psychiatry 1995;19:11–38.

16. Maes M, Lambrechts J, Bosmans E, Jacobs J, Suy E, Vandervorst C, de Jonckheere C, Minner B, Raus J. Evidence for a systemic immune activation during depression, results of leukocyte enumeration by flow cytometry in conjunction with monoclonal antibody staining. Psychol Med 1992;22:45–53.
17. Maes M. Major depression and activation of the inflammatory response system. In Cytokines, Stress, and Depression. Dantzer R, Wollman EE, and Yirmiya R, Eds; New York: Kluwer Academic/Plenum Publishers, 1999; pp. 25–46.
18. Kronfol Z, House JD. Lymphocyte mitogenesis, immunoglobulin and complement levels in depressed patients and normal controls. Acta Psychiatr Scand 1989;80:142–7.
19. Anisman H, Ravindran AV, Griffiths J, Merali Z. Endocrine and cytokine correlates of major depression and dysthymia with typical or atypical features. Mol Psychiatry 1999;4:182–8.
20. Seidel A, Arolt V, Hunstiger M, Rink L, Behnisch A, Kirchner H. Cytokine production and serum proteins in depression Scand J Immunol 1995;41(6):534–8.
21. Dunbar PR, Hill J, Neale TJ, Mellsop GW. Neopterin measurement provides evidence of altered cell-mediated immunity in patients with depression, but not with schizophrenia. Psychol Med 1992;22:1051–7.
22. Landmann R, Schaub B, Link S, Wacker HR. Unaltered monocyte function in patients with major depression before and after three months of antidepressive therapy. Biol Psychiatry 1997;15:675–81.
23. Natelson BH, Denny T, Zhou XD, LaManca JJ, Ottenweller JE, Tiersky L, et al. Is depression associated with immune activation? J Affect Disord 1999;53:179–84.
24. Maes M, Scharpé S, Meltzer H, Bosmans E, Suy E, Calabrese J, Cosyns P. Relationships between interleukin-6 activity, acute phase proteins, and function of the hypothalamic–pituitary–adrenal axis in severe depression. Psychiatry Res 1993;49:11–27.
25. Maes M, Bosmans E, Meltzer HY, Scharpé S, Suy E. Interleukin-1β: A putative mediator of HPA axis hyperactivity in major depression? Am J Psychiatry 1993;150:1189–93.
26. Maes M, Meltzer HY, Bosman E. Psychoimmune investigation in obsessive–compulsive disorder: Assays of plasma transferrin, IL-2 and IL-6 receptor, and IL-1 beta and IL-6 concentrations. Neuropsychobiology 1994;30:57–60.
27. Griffiths J, Ravindran AV, Merali Z, Anisman H. Neuroendocrine measures and lymphocyte subsets in depressive illness: Influence of a clinical interview concerning life experiences. Psychoneuroendocrinology 1997;22:225–36.
28. Owens MJ, Nemeroff CB. Physiology and pharmacology of corticotropin-releasing factor. Pharmacol Rev 1991;43:425–73.
29. Brambilla F, Maggioni M. Blood levels of cytokines in elderly patients with major depressive disorder. Acta Psychiatr Scand 1998;97:309–13.
30. Musselman DL, Miller AH, Porter MR, Manatunga A, Gao F, Penna S, Pearce BD, Landry J, Glover S, McDaniel JS, Nemeroff CB. Higher than normal plasma interleukin-6 concentrations in cancer patients with depression: Preliminary findings. Am J Psychiatry 2001;158:1252–7.
31. Haack M, Hinze-Selch D, Fenzel T, Kraus T, Kuhn M, Schuld A, Pollmacher T. Plasma levels of cytokines and soluble cytokine receptors in psychiatric patients upon hospital admission: Effects of confounding factors and diagnosis. J Psychiatr Res 1999;33:407–18.
32. Levine J, Barak Y, Chengappa KN, Rapoport A, Rebey M, Barak V. Cerebrospinal cytokine levels in patients with acute depression. Neuropsychobiology 1999;40:171–6.
33. Zorrilla EP, Luborsky L, McKay JR, Rosenthal R, Houldin A, Tax A, McCorkle R, Seligman DA, Schmidt K. The relationship of depression and stressors to immunological assays: A meta-analytic review. Brain Behav Immun 2001;15:199–226.

34. Pollmächer T, Haack M, Schuld A, Reichenberg A, Yirmiya R. Low levels of circulating inflammatory cytokines – Do they affect human brain functions? Brain Behav Immun 2002;16:525–32.
35. Minden SL, Schiffer RB. Affective disorders in multiple sclerosis. Review and recommendations for clinical research. Arch Neurol 1990;47:98–104.
36. Marshall PS. Allergy and depression, a neurochemical threshold model of the relation between the illnesses. Psychol Bull 1993;113:23–43.
37. Dickens C, McGowan L, Clark-Carter D, Creed F. Depression in rheumatoid arthritis, a systematic review of the literature with meta-analysis. Psychosom Med 2002;64:52–60.
38. Schwartz JA, Speed NM, Brunberg JA, Brewer TL, Brown M, Greden JF. Depression in stroke rehabilitation. Biol Psychiatry 1993;33:694–9.
39. Foley FW, Traugott U, LaRocca NG, Smith CR, Perlman KR, Caruso LS, Scheinberg LC. A prospective study of depression and immune dysregulation in multiple sclerosis. Arch Neurol 1992;49:238–44.
40. Dantzer R, Bluthé R-M, Castanon N, Chauvet N, Capuron L, Goodall G, Kelley KW, Konsman J-P, Layé S, Parnet P, Pousset F. Cytokine effects on behavior. In Psychoneuroimmunology, 3rd Edn. Ader R, Felten D, and Cohen N, Eds; San Diego, CA: Academic Press, 2001; pp. 703–27.
41. Nesse RM. Is depression an adaptation? Arch Gen Psychiatry 2000;57:14–20.
42. Dubrovsky B. Evolutionary psychiatry. Adaptationist and nonadaptationist conceptualizations. Prog Neuropsychopharmacol Biol Psychiatry 2002;26:1–19.
43. Dietzel M, Saletu B, Lesch OM, Sieghart W, Schjerve M. Light treatment in depressive illness: Polysomnographic, psychometric, and polyendocrinal findings. Eur Neurol 1986; 25(Suppl 2):93– 103.
44. McCarthy DO, Kluger MJ, Vander AJ. Effect of centrally administered interleukin-1 and endotoxin on food intake of fasted rats. Physiol Behav 1986;36:745–9.
45. Swiergiel AH, Smagin GN, Dunn AJ. Influenza virus infection of mice induces anorexia: Comparison with endotoxin and interleukin-1 and the effects of indomethacin. Pharmacol Biochem Behav 1997;57:389–96.
46. Watkins LR, Wiertelak EP, Goehler LE, Smith KP, Martin D, Maier SF. Characterization of cytokine-induced hyperalgesia. Brain Res 1994;654:15–26.
47. Krueger JM, Majde JA. Microbial products and cytokines in sleep and fever regulation. Crit Rev Immmunol 1995;14:355–79.
48. Merali Z, Brennan K, Brau P, Anisman H. Dissociating anorexia and anhedonia elicited by interleukin-1b: Antidepressant and gender effects on responding for "free chow" and "earned" sucrose intake. Psychopharmacology 2003;165:413–8.
49. Yirmiya R, Weidenfeld J, Pollak Y, Morag M, Morag A, Avitsur R, Barak O, Reichenberg A, Cohen E, Shavit Y, Ovadia H. Cytokines, "Depression due to a general medical condition," and antidepressant drugs. Adv Exp Med Biol 1999;461:283–316.
50. De La Garza R. Endotoxin- or pro-inflammatory cytokine-induced sickness behavior as an animal model of depression: Focus on anhedonia. Neurosci Behav Rev 2005;29:761–70.
51. Anisman H, Kokkinidis L, Merali Z. Interleukin-2 decreases accumbal dopamine efflux and responding for rewarding lateral hypothalamic stimulation. Brain Res 1996; 731:1–11.
52. Anisman H, Kokkinidis L, Borowski T, Merali Z. Differential effects of interleukin (IL)-1β, IL-2 and IL-6 on responding for rewarding lateral hypothalamic stimulation. Brain Res 1998;779:177–87.

53. Hebb ALO, Zacharko RM, Anisman H. Self-stimulation from the mesencephalon following intraventricular interleukin-2 administration. Brain Res Bull 1998;45:549–56.
54. Larson SJ, Dunn AJ. Behavioral effects of cytokines. Brain Behav Immun 2001;15:371–87.
55. McKenna PJ, McKay AP, Laws K. Memory in functional psychosis. In Memory Disorders in Psychiatric Practice. Berrios GE and Hodges JR, Eds; Cambridge: Cambridge University Press, 2000.
56. Watkins LR, Maier SF, Goehler LE. Cytokine-to-brain communication: A review and analysis of alternative mechanisms. Life Sci 1995;57:1011–26.
57. Bluthé R-M, Michaud B, Kelley KW, Dantzer R. Vagotomy attenuates behavioural effects of interleukin-1 injected peripherally but not centrally. NeuroReport 1996;7:1485–8.
58. Licinio J, Wong M-L. The role of inflammatory mediators in the biology of major depression: Central nervous system cytokines modulate the biological substrate of depressive symptoms, regulate stress-responsive systems, and contribute to neurotoxicity and neuroprotection. Mol Psychiatry 1999;4:317–27.
59. Goehler LE, Relton JK, Dripps D, Kiechle R, Tartaglia N, Maier SF, Watkins LR. Vagal paraganglia bind biotinylated interleukin-1 receptor antagonist (IL-1ra) in the rat: A possible mechanism for immune-to-brain communication. Brain Res Bull 1997;43:357–64.
60. Wieczorek M, Pournajafi-Nazarloo H, Swiergiel AH, Dunn AJ. Physiological and behavioral responses to interleukin-1b and LPS in vagotomized mice. Physiol Behav 2005;84:500–11.
61. Wieczorek M, Dunn AJ. Effect of subdiaphragmatic vagotomy on the noradrenergic and HPA axis activation induced by intraperitoneal interleukin-1 administration in rats. Brain Res 2006;1101:73–84.
62. Swiergiel AH, Burunda T, Patterson B, Dunn AJ. Endotoxin- and interleukin-1-induced hypophagia are not affected by noradrenergic, dopaminergic, histaminergic and muscarinic antagonists. Pharmacol Biochem Behav 1999;63:629–37.
63. Quan N, Herkenham M. Connecting cytokines and brain: A review of current issues. Histol Histopathol 2002;17:273–88.
64. Blatteis CM, Li S, Li Z, Feleder C, Perlik V. Cytokines, PGE2 and endotoxic fever: A re-assessment. Prostaglandins Other Lipid Mediat 2005;76:1–18.
65. Dunn AJ, Swiergiel AH, Zhang H, Quan N. The hypophagia induced by interleukin-1 and lipopolysaccharide is associated with induction of cyclooxygenase-2 in brain endothelia. Neuroimmunomodulation 2006;13:96–104.
66. Breder CD, Dinarello CA, Saper CB. Interleukin-1 immunoreactive innervation of the human hypothalamus. Science 1988;240:321–4.
67. Lechan RM, Toni R, Clark BD, Cannon JG, Shaw AR, Dinarello CA, et al. Immunoreactive interleukin-1β localization in the rat forebrain. Brain Res 1990;514:135–40.
68. Molenaar GJ, Berkenbosch F, van Dam A-M, Lugard CMJE. Distribution of interleukin 1β immunoreactivity within the porcine hypothalamus. Brain Res 1993;608:169–74.
69. Huitinga I, van der Cammen M, Salm L, Erkut Z, van Dam A-M, Tilders F, Swaab D. IL-1β immunoreactive neurons in the human hypothalamus: Reduced numbers in multiple sclerosis. J Neuroimmunol 2000;107:8–20.
70. Layé S, Parnet P, Goujon E, Dantzer R. Peripheral administration of lipopolysaccharide induces the expression of cytokine transcripts in the brain and pituitary of mice. Mol Brain Res 1994;27:157–62.
71. Wong M-L, Licinio J. Localization of interleukin-1 type I receptor mRNA in rat brain. Neuroimmunomodulation 1994;1:110–5.

72. Wong ML, Bongiorno PB, Rettori V, McCann SM, Licinio J. Interleukin (IL) 1β, IL-1 receptor antagonist, IL-10, and IL-13 gene expression in the central nervous system and anterior pituitary during systemic inflammation: Pathophysiological implications. Proc Natl Acad Sci 1997;94:227–32.
73. Quan N, Whiteside M, Herkenham M. Time course and localization patterns of interleukin-1β messenger RNA expression in brain and pituitary after peripheral administration of lipopolysaccharide. Neuroscience 1998;83:281–93.
74. Quan N, Sundar QN, Weiss JM. Induction of interleukin-1 in various brain regions after peripheral and central injections of lipopolysaccharide. J Neuroimmunol 1994; 49:125–34.
75. Van Dam AM, Bauer J, Tilders FJH, Berkenbosch F. Endotoxin-induced appearance of immunoreactive interleukin-1b in ramified microglia in rat brain: A light and electron microscopic study. Neuroscience 1995;65:815–26.
76. Quan N, Zhang Z, Emery M, Bonsall R, Weiss JM. Detection of interleukin-1 bioactivity in various brain regions of normal healthy rats. Neuroimmunomodulation 1996;3:47–55.
77. Farrar WL, Kilian PL, Ruff MR, Hill JM, Pert CB. Visualization and characterization of interleukin 1 receptor in brain. J Immunol 1987;139:459–63.
78. Haour F, Ban E, Marquette C, Milon G, Fillion G. Brain interleukin-1 receptors: Mapping, characterization and modulation. In Interleukin-1 in the Brain. Rothwell NJ and Dantzer RD, Eds; Oxford: Pergamon Press, 1992; pp. 13–25.
79. Takao T, Culp SG, Newton RC, De Souza EB. Type I interleukin-1 receptors in the mouse brain–endocrine–immune axis labelled with [^{125}I] recombinant human interleukin-1 receptor antagonist. J Neuroimmunol 1992;41:51–60.
80. Krakauer JV, Oppenheimer JJ. Pro-inflammatory cytokines. TNF and IL-1 families, chemokines, TGF-beta and others. In Fundamental Immunology. Paul WE, Ed.; Philadelphia, PA: Lippincott–Raven, 1999; pp. 775–811.
81. Yabuuchi K, Minami M, Katsumata S, Satoh M. Localization of type 1 interleukin-1 receptor mRNA in the rat brain. Mol Brain Res 1994;27:27–36.
82. Ericsson A, Liu C, Hart RP, Sawchenko PE. Type 1 interleukin-1 receptor in the rat brain: Distribution, regulation, and relationship to sites of IL-1-induced cellular activation. J Comp Neurol 1995;361(4):681–98.
83. Dantzer R, Aubert A, Bluthé R-M, Gheusi G, Cremona S, Layé S, Konsman J, Parnet P, Kelley KW. Mechanisms of the behavioural effects of cytokines. Adv Exp Med Biol 1999;461:83–105.
84. Pugh CR, Nguyen KT, Gonyea JL, Fleshner M, Watkins LR, Maier SF, Rudy JW. Role of interleukin-1 beta in impairment of contextual fear conditioning caused by social isolation. Behav Brain Res 1999;106:109–18.
85. Borsody MK, Weiss JM. The effects of endogenous interleukin-1 bioactivity on locus coeruleus neurons in response to bacterial and viral substances. Brain Res 2004;1007: 39–56.
86. Yirmiya R. Endotoxin produces a depressive-like episode in rats. Brain Res 1996; 711:163–74.
87. Shen Y, Connor TJ, Nolan Y, Kelly JP, Leonard BE. Differential effect of chronic antidepressant treatments on lipopolysaccharide-induced depressive-like behavioural symptoms in the rat. Life Sci 1999;65:1773–86.
88. Yirmiya R, Pollak Y, Barak O, Avitsur R, Ovadia H, Bette M, Weihe E, Weidenfeld J. Effects of antidepressant drugs on the behavioral and physiological responses to lipopolysaccharide (LPS) in rodents. Neuropsychopharmacology 2001;24:531–44.

89. Castanon N, Konsman J-P, Medina C, Chauvet N, Dantzer R. Chronic treatment with the antidepressant tianeptine attenuates lipopolysaccharide-induced Fos expression in the rat paraventricular nucleus and HPA axis activation. Psychoneuroendocrinology 2003; 28:19–34.
90. Dunn AJ, Swiergiel AH. The reductions in sweetened milk intake induced by interleukin-1 and endotoxin are not prevented by chronic antidepressant treatment. Neuroimmunomodulation 2001;9:163–9.
91. Connor TJ, Harkin A, Kelly JP, Leonard BE. Olfactory bulbectomy provokes a suppression of interleukin-1b and tumour necrosis factor-a production in response to an in vivo challenge with lipopolysaccharide: Effect of chronic desipramine treatment. Neuroimmunomodulation 2000;7:27–35.
92. Campisi J, Hansen MK, O'Conner KA, Biedenkapp JC, Watkins LR, Maier SF, Fleshner M. Circulating cytokines and endotoxin are not necessary for the activation of the sickness or corticosterone response produced by peripheral *E. coli* challenge. J Appl Physiol 2003;95:1873–82.
93. Besedovsky HO, del Rey A, Sorkin E, Dinarello CA. Immunoregulatory feedback between interleukin-1 and glucocorticoid hormones. Science 1986;233:652–4.
94. Strohle A, Holsboer F. Stress responsive neurohormones in depression and anxiety. Pharmacopsychiatry 2003;36(Suppl 3):S207–14.
95. Silverman MN, Pearce BD, Miller AH. Cytokines and HPA axis regulation. In Cytokines and Mental Health. Kronfol Z, Ed.; Norwell, MA: Kluwer Academic Publishers, 2003; pp. 85–122.
96. Dunn AJ. Cytokine activation of the hypothalamo–pituitary–adrenal axis. In Handbook of Stress and the Brain Part 2: Stress: Integrative and Clinical Aspects. Steckler T, Kalin N, and Reul JMHM, Eds; Amsterdam: Elsevier, 2005; pp. 157–74.
97. Wang JP, Dunn AJ. The role of interleukin-6 in the activation of the hypothalamo–pituitary–adrenocortical axis induced by endotoxin and interleukin-1b. Brain Res 1999; 815:337–48.
98. Swiergiel AH, Dunn AJ. Feeding, exploratory, anxiety- and depression-related behaviors are not altered in interleukin-6-deficient male mice. Behav Brain Res 2006;171:94–108.
99. Silverman MN, Miller AH, Biron CA, Pearce BD. Characterization of an interleukin-6 and adrenocorticotropin-dependent, immune-to-adrenal pathway during viral infection. Endocrinology 2004;145:3580–9.
100. Shimizu H, Ohtani K-I, Sato N, Nagamine T, Mori M. Increase in serum interleukin-6, plasma ACTH and serum cortisol levels after systemic interferon-α administration. Endocr J 1995;42:551–6.
101. Capuron L, Raison CL, Musselman DL, Lawson DH, Nemeroff CB, Miller AH. Association of exaggerated HPA axis response to the initial injection of interferon-alpha with development of depression during interferon-alpha therapy. Am J Psychiatry 2003;160:1342–5.
102. Holsboer F, Stalla GK, von Bardeleben U, Hammann K, Müller H, Müller OA. Acute adrenocortical stimulation by recombinant gamma interferon in human controls. Life Sci 1988;42:1–5.
103. Saphier D. Neurophysiological and endocrine consequences of immune activity. Psychoneuroendocrinology 1989;14:63–87.
104. Saphier D, Welch JE, Chuluyan HE. α-Interferon inhibits adrenocortical secretion via μ_1-opioid receptors in the rat. Eur J Pharmacol 1993;236:183–91.
105. Menzies RA, Phelps CP, Wiranowska M, Oliver J, Chen LT, Horvath E, Hall NRS. The effect of interferon-alpha on the pituitary–adrenal axis. J Interferon Cytokine Res 1996;16:619–29.

106. Dunn AJ. The role of interleukin-1 and tumor necrosis factor a in the neurochemical and neuroendocrine responses to endotoxin. Brain Res Bull 1992;29:807–12.
107. Dunn AJ, Swiergiel AH. The role of cyclooxygenases in endotoxin- and interleukin-1-induced hypophagia. Brain Behav Immun 2000;14:141–52.
108. Dunn AJ, Chuluyan H. The role of cyclo-oxygenase and lipoxygenase in the interleukin-1-induced activation of the HPA axis: Dependence on the route of injection. Life Sci 1992;51:219–25.
109. Dunn AJ. Role of cytokines in infection-induced stress. Ann N Y Acad Sci 1993;697:189–202.
110. Turnbull AV, Rivier C. Regulation of the hypothalamic–pituitary–adrenal axis by cytokines: Actions and mechanisms of action. Physiol Rev 1999;79:1–71.
111. Muglia LJ, Bethin KE, Jacobson L, Vogt SK, Majzoub JA. Pituitary–adrenal axis regulation in CRH-deficient mice. Endocr Res 2000;26:1057–66.
112. Dunn AJ, Swiergiel AH. Behavioral responses to stress are intact in CRF-deficient mice. Brain Res 1999;845:14–20.
113. Dunn AJ. Systemic interleukin-1 administration stimulates hypothalamic norepinephrine metabolism parallelling the increased plasma corticosterone. Life Sci 1988; 43:429–35.
114. Dunn AJ. Effects of cytokines and infections on brain neurochemistry. Clin Neurosci Res 2006;6:52–68.
115. Wang JP, Dunn AJ. Mouse interleukin-6 stimulates the HPA axis and increases brain tryptophan and serotonin metabolism. Neurochem Int 1998;33:143–54.
116. Ando T, Dunn AJ. Mouse tumor necrosis factor-α increases brain tryptophan concentrations and norepinephrine metabolism while activating the HPA axis in mice. Neuroimmunomodulation 1999;6:319–29.
117. Dunn AJ. Effects of cytokines and infections on brain neurochemistry. In Psychoneuroimmunology, 3rd Edn. Ader R, Felten DL, and Cohen N, Eds; New York: Academic Press, 2001; pp. 649–66.
118. Wong ML, Kling MA, Munson PJ, Listwak S, Licinio J, Prolo P, Karp B, McCutcheon IE, Geracioti TD, DeBellis MD, Rice KC, Goldstein DS, Veldhuis JD, Chrousos GP, Oldfield EH, McCann SM, Gold PW. Pronounced and sustained central hypernoradrenergic function in major depression with melancholic features: Relation to hypercortisolism and corticotropin-releasing hormone. Proc Natl Acad Sci 2000;97:325–30.
119. Bell C, Abrams J, Nutt D. Tryptophan depletion and its implications for psychiatry. Br J Psychiatry 2001;178:399–405.
120. Coppen A, Wood K. Tryptophan and depressive illness. Psychol Med 1978;8:49–57.
121. Wichers MC, Maes M. The role of indoleamine 2,3-dioxygenase (IDO) in the pathophysiology of interferon-alpha-induced depression. J Psychiatry Neurosci 2004;29:11–7.
122. Manji HK, Drevets WC, Charney DS. The cellular neurobiology of depression. Nat Med 2001;7:541–7.
123. Zalcman S, Green-Johnson JM, Murray L, Nance DM, Dyck D, Anisman H, Greenberg AH. Cytokine-specific central monoamine alterations induced by interleukin-1, -2 and -6. Brain Res 1994;643:40–9.
124. Clement HW, Buschmann J, Rex S, Grote C, Opper C, Gemsa D, Wesemann W. Effects of interferon-γ, interleukin-1β, and tumor necrosis factor-α on the serotonin metabolism in the nucleus raphe dorsalis of the rat. J Neural Transm. 1997;104:981–91.
125. Zhang J-J, Terreni L, De Simoni M-G, Dunn AJ. Peripheral interleukin-6 administration increases extracellular concentrations of serotonin and the evoked release of serotonin in the rat striatum. Neurochem Int 2001;38:303–8.

126. Hayley S, Wall P, Anisman H. Sensitization to the neuroendocrine, central monoamine and behavioural effects of murine tumor necrosis factor-α: Peripheral and central mechanisms. Eur J Neurosci 2002;15:1061–76.
127. Morikawa O, Sakai N, Obara H, Saito N. Effects of interferon-α, interferon-γ and cAMP on the transcriptional regulation of the serotonin transporter. Eur J Pharmacol 1998;349:317–24.
128. Delgado PL, Price LH, Miller HL, Salomon RM, Aghajanian GK, Heninger GR, Charney DS. Serotonin and the neurobiology of depression – Effects of tryptophan depletion in drug-free depressed patients. Arch Gen Psychiatry 1994;51:865–74.
129. Delgado PL, Miller HL, Salomon RM, Licinio J, Krystal JH, Moreno FA, Charney DS. Tryptophan-depletion challenge in depressed patients treated with desipramine or fluoxetine: Implications for the role of serotonin in the mechanism of antidepressant action. Biol Psychiatry 1999;46:212–20.
130. O'Reardon JP, Chopra MP, Bergan A, Gallop R, DeRubeis RJ, Crits-Christoph P. Response to tryptophan depletion in major depression treated with either cognitive therapy or selective serotonin reuptake inhibitor antidepressants. Biol Psychiatry 2004;55:957–9.
131. Moreno FA, Heninger GR, McGahuey CA, Delgado PL. Tryptophan depletion and risk of depression relapse: A prospective study of tryptophan depletion as a potential predictor of depressive episodes. Biol Psychiatry 2000;48:327–9.
132. Blokland A, Lieben CKJ, Deutz NEP. Anxiogenic and depressive-like effects, but no cognitive deficits, after repeated moderate tryptophan depletion in the rat. J Psychopharmacol 2002;16:39–49.
133. Lieben CKJ, van Oorsouw K, Deutz NEP, Blokland A. Acute tryptophan depletion induced by a gelatin-based mixture impairs object memory but not affective behavior and spatial learning in the rat. Behav Brain Res 2004;151:53–64.
134. Uehara A, Ishikawa Y, Okumura T, Okamura K, Sekiya C, Takasugi Y, Namiki M. Indomethacin blocks the anorexic action of interleukin-1. Eur J Pharmacol 1989;170:257–60.
135. Bluthé RM, Crestani F, Kelley KW, Dantzer R. Mechanisms of the behavioral effects of interleukin 1. Role of prostaglandins and CRF. Ann N Y Acad Sci 1992;650:268–75.
136. Charlton BG. The malaise theory of depression: Major depressive disorder is sickness behavior and antidepressants are analgesic. Med Hypotheses 2000;54:126–30.
137. McCarthy DO, Daun JM. The effects of cyclooxygenase inhibitors on tumor-induced anorexia in rats. Cancer 1993;71:486–92.
138. Jain NK, Kulkarni SK, Singh A. Lipopolysaccharide-mediated immobility in mice: Reversal by cyclooxygenase enzyme inhibitors. Meth Find Exp Clin Pharmacol 2001; 23:441–4.
139. Porsolt RD, Bertin A, Jalfre M. Behavioural despair in mice: A primary screening test for antidepressants. Arch Int Pharmacodyn 1977;229:327–36.
140. Porsolt RD, Le Pichon M, Jalfre M. Depression: A new animal model sensitive to antidepressant treatments. Nature 1977;266:730–2.
141. Steru L, Chermat R, Thierry B, Simon P. The tail suspension test: A new method for screening antidepressants in mice. Psychopharmacology (Berl) 1985;85:367–70.
142. Dunn AJ, Swiergiel AH. Effects of interleukin-1 and endotoxin in the forced swim and tail suspension tests in mice. Pharmacol Biochem Behav 2005;81:688–93.
143. Deak T, Bellamy C, D'Agostino LG, Rosanoff M, McElderry NK, Bordner KA. Behavioral responses during the forced swim test are not affected by anti-inflammatory agents or acute illness induced by lipopolysaccharide. Behav Brain Res 2005;160:125–34.

Cytokines and the Brain
Edited by C. Phelps and E. Korneva

Clinical Relevance: Cytokines in Alzheimer's Disease

WILLIAM K. SUMMERS

President, Solo Research Non-Profit Ltd, 3118 Doolittle Avenue, Arcadia, CA 91006
President, Alzheimer's Corporation, 6000 Uptown Blvd, Suite 308 Albuquerque, NM 87110, Telé 505.878.0192

ABSTRACT

Alzheimer's disease (AD) is a complex interplay between various possible brain insults and the cytokine response resulting in a microlocalized inflammation in predictable areas of the central nervous system (CNS). The process of Alzheimer's is like 'rheumatoid arthritis of the brain'. Progression of AD leads to a total body phenomenon, which leads to death in 8 years on average. Because the illness is multifaceted, treatment and prevention should be as well. Prevention may be possible through the use of daily oral antioxidants to reduce vulnerability to AD. Chronic use of cyclooxygenase (COX)-2 inhibitors could be supported with the onset of pre-Alzheimer's forgetfulness. The use of sex steroids could be considered at about this time. With the onset of illness, anticholinesterase inhibitor is the principal therapy. Selective use of antidepressants should be considered as depression is so frequent in AD. Curiously, there appears to be a role for psychostimulants to treat the atypical depressions seen in AD patients while slowing the progress of AD.

1. INTRODUCTION

Alzheimer's disease (AD) is the most common cause of dementia in those over the age of 60. The frequency of common dementias are Lewy body dementia (7%), mixed Lewy body with Alzheimer's (5%) and mixed vascular with AD (10%), and AD (65%). This means 87% of dementias in the elderly have a form of treatment with currently available medications [1].

The major neuropathological features include astrogliosis with neuronal and synaptic loss, accompanied by the presence of abnormal intra- and extraneuronal proteinaceous deposits. The intracellular protein deposits, called neurofibrillary tangles, are tau protein. Tau is a microtubule-associated protein consisting of paired helical filaments (PHF). The extracellular deposits are the lipoprotein, amyloid. When accumulated around cerebral blood vessels, the deposit is called amyloid angiopathy. In the neuropil, the extracellular deposits are called amyloid plaques. Astrocytes, microglia, and neuritic processes surround the amyloid deposits and leave the clue that the process is inflammatory and involves cytokines.

For over two decades, research in AD focused on β-amyloid (Aβ) in various forms as causal to AD. This not only has resulted in much interesting material, but also has misdirected much of the research efforts in the discovery of treatments for AD.

Figure 1 presents the pathophysiology of AD in year 2003:

1) Insult (infection, anoxia, toxin exposure, aging, trauma, metabolic crisis, etc.)
2) Microlocalized inflammation with upregulation of cytokines, complement, and other inflammatory mediators
3) Microlocalized free radical production and oxidative injury
4) Microlocalized pathological protein accumulation
5) Microlocalized apoptosis with further accumulation of compact β-pleated amyloid material into plaques

Although the process is microlocalized, AD is a total body disease [2,3]. Recently autopsy-derived mRNAs for complement markers, C-reactive protein, and amyloid-P were found to be significantly elevated in AD heart, spleen, and kidney compared to age-matched controls [4]. The question is whether an insult to the brain results in the total body disease of AD or the converse.

Various central nervous system (CNS) insults are associated with AD. Head trauma, myocardial infarcts, hypertension, metabolic X syndrome, other cardiovascular risk factors, and diabetes have been suspected as risk factors [5–7]. Infectious agents such as *Chlamydia pneumoniae* and various viruses such as herpes simplex are also suspected as potential precipitants of AD [8,9]. Even aging itself is now considered an insult associated with AD [10].

Genetic vulnerabilities affect the CNS cell response to physiological insults. The current state of knowledge dictates that AD and other neurodegenerative disease have multifactorial involvement with an unknown number of genes and an unknown number of environmental factors. The most

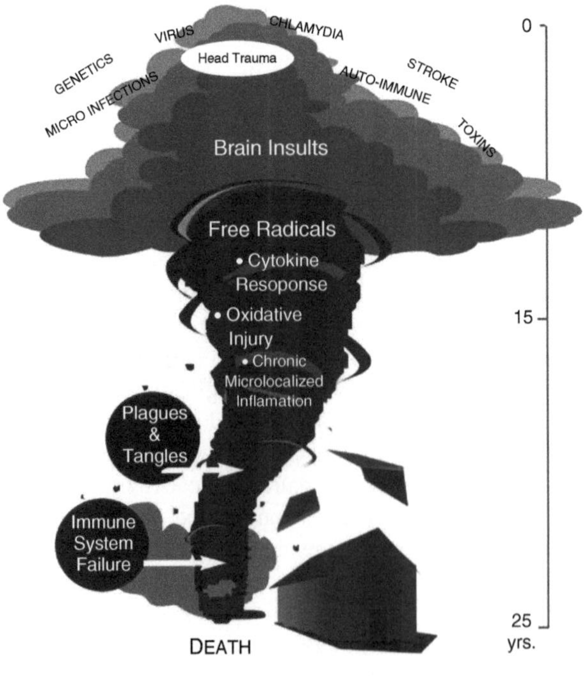

Figure 1. Pathophysiology of Alzheimer's disease.

common of the genetic risks in late-onset AD is the fourth allelic form of apolipoprotein E (apo ε4) [11]. Mutations in genes that produce amyloid protein precursor (APP), a 110–130-kDa glycoprotein, are associated with early-onset familial AD [12]; however, these are rare.

The genetic theory of Aβ causing AD does not seem to be substantiated. Identical twins have only a 30% degree of concordance for AD. Of the concordant monozygotic twins, only 50% of the none-index cases develop dementia within 5 years of their twin [13]. Certainly there are other genetic effects, say on the immune system, that influence the presence or the progression of AD.

2. CYTOKINES

In 1985, Korneva et al. established that immune homeostasis depends largely on the influence of the nervous system [14]. A brain insult, such as a pathogen, results in the following:

1) A local response at the blood–brain (epithelial) barrier (BBB) with local macrophages and complement systems
2) An early systemic response by recruiting inflammatory cells
3) Late response with antigen-specific inflammatory cells and antibodies

At each level of response, the threat may be overcome, the brain returns to immune homeostasis, and cytokines are involved in each of these steps.

Cytokines are regulatory proteins secreted by cells to modulate the organism's immune response. They are low molecular weight (generally <200 amino acid) nonantibody proteins that are made of a wide range of cells. In non-CNS tissue, cytokines are produced by macrophages, B lymphocytes, T lymphocytes, granulocytes, and endothelial cells. In the CSN, cytokines are produced by astrocytes, microglia, and neurons [15–17]. About 80 cytokines are currently recognized. They are synthesized as required or stored in intracellular granules and released after reception of a specific signal. Cytokines have very short half-lives and they are the ideal candidates for causing localized microinflammatory phenomena, such as seen in AD.

In vitro cytokines work in groups or in networks that have synergistic and/or antagonistic interactions [18]. One cytokine may increase (cascade) or decrease (truncate) the production of other cytokines The result(s) of cytokine signaling depends on a complex network of feedback loops. There is a functional *pleiotrophy* of cytokines, in that each cytokine has multiple target cells and multiple actions. There is also *redundancy* as different cytokines have similar actions.

There are six structural superfamilies of cytokines [19]. The first four superfamilies are principally produced by lymphoid, endothelial, and microglial cells. *Cell adhesion molecules* are a special class of regulators. Adhesion molecules allow circulating macrophages and leukocytes to bind to the blood vessel endothelium. Adhesion molecules may also allow lymphocytes to enter the brain parenchyma [20]. Cell adhesion molecules may also allow embryos to adhere to the endometrium [21].

The *chemokine superfamily* has almost 60 members, which are classified by the number of N-terminal cysteine pairs ("– C–") [22]. Chemokines, which serve to attract migration of specific inflammatory cells, may be produced by any cell in the body [23]. Chemokines are low molecular weight proteins with 40–80 amino acids that have an affinity for heparin. They induce migration by specific inflammatory cells to the site of signaling.

Chemokines in the CNS are produced by astrocytes and microglia. These include Macrophage Inflammatory Protein (MIP)1α, MIP-1β, RANTES, Monocyte Chemotactic Protein (MCP)-1, and Interferon-gamma Inducible Protein (IP)-10 [24]. Mononuclear cells exist in the CNS and they also migrate into the CNS after chemokine stimulus. Mononuclear cells signal with interleukin (IL)-1β or tumor necrosis factor (TNF)-α to activate astrocytes. These same cytokines stimulate microglia chemotaxis. Experimental studies of anoxic and inflammatory stimuli have demonstrated that key cytokines, TNFα, IL-1β, and IL-1ra, are produced by glial cells, CNS endothelial cells, and neurons [25]. It appears that the brain is immunocompetent and is capable of summoning assistance from beyond the BBB.

Cytokines can be loosely thought of as proinflammatory or anti-inflammatory. IL-1β, for example, is proinflammatory and is produced by microglia, astrocytes, neurons and endothelium after CNS insults [15,25]. IL-1 is neurotoxic and is associated with apoptosis [15]. IL-6 is produced by astroglia and microglia, which induce acute-phase inflammatory proteins, increase vascular permeability, and activate lymphocytes by chemokine and adhesion molecule upregulation. Chronic IL-6 exposure promotes N-*methyl-D-aspartate (NMDA)* excitotoxin-mediated injury. IL-6 also causes increased release of norepinephrine and serotonin and may have an inhibitory effect on acetylcholinesterase (*AChE*).

Tumor necrosis factor-α is another pro-inflammatory cytokine produced by astroglia and microglia. Tumor necrosis factor-α induces ICAM-1 and other adhesion molecules. Tumor necrosis factor-α administration causes increased catecholamine neuronal transmission, fever, hypersomnia, and demyelination. The levels of this cytokine are known to rise with the progression of AD [25].

Tumor necrosis factor-β is known to be neuroprotective, especially in the case of dopaminergic neurons. Tumor necrosis factor-β is produced by astroglia and microglia and is also known to induce IL-1ra (interleukin-1 receptor antagonist). IL-1ra is produced in cerebral endothelial cells, neurons, and glial cells and has protective effects in anoxic or traumatic brain injury. IL-6 is also reported to be anti-inflammatory. IL-6 inhibits TNF-α and induces IL-1ra [26].

3. CYTOKINE RECEPTORS

There are eight superfamilies of cytokine receptors [18]. Cytokines can upregulate or downregulate receptors on or within target cells. Details of these receptors are covered in other chapters of this volume. There are two special forms of cytokine receptors: soluble cytokine receptors (sCRs) and receptors bound to cell membranes (MRs). The sCRs can be generated from any of the seven superfamilies by proteolytic cleavage of the transmembrane receptors (receptor shedding) or by *de novo* synthesis of sCRs. The sCRs can prevent or augment cytokine binding to typical membrane receptors. MRs bind cytokines on the cell surface. Adhesion molecules mediate cell-to-cell interaction and play a major role in the function of leukocytes and mast cells. The presence of recruited leukocytes at the site of inflammation in the CNS is critically dependent upon the coordinated expression of adhesion molecules (ligands and receptors) on the inflammatory cells with subsequently activated cerebral capillary endothelium. Timed release of adhesion molecules facilitates "rolling and docking" of activated immune cells on the endothelium, allowing them to pass through the BBB to the site of CNS inflammation that is signaling them. Among these adhesion molecules, P-selectin, L-selectin, integrin, trophinin, ICAM-1, ELAM-1, and their respective adhesion ligand receptors play predominant roles.

4. INTRACELLULAR SIGNALING AND CYTOKINES

The flow of intracellular signaling mechanisms has recently been described [27]. Briefly, a cytokine (stimulus) combines with a specific cell surface receptor (activator) to initiate a complex and predetermined series of protein kinase enzymatic reactions (a cascade). The cascade sends specific signals from the cell surface to the nucleus. Cascade sequences follow a format seen in mitogen-activated protein kinase (MAPK) pathways [28]. This cascade is as follows:

1) MKKK (MAPK kinase kinase), which phosphorylates and activates
2) MKK (MAPK kinase or MEKK), which phosphorylates and activates
3) MAPKs, which act on
4) Nuclear substrates to effect gene expression

The MAPKs stimulate DNA production of proteins, affect cell mitosis, cell movement, metabolism, and apoptosis. The intracellular messengers are protein kinases. Genomic sequencing has determined that there are 518 such putative protein kinase genes, many of which are still undiscovered in the cell [29]. It is significant that the cascade can be 'played' by a diverse set of stimuli. These stimuli include growth factors, cytokines, oxidative stressors (anoxia, aging, etc.), viral infections, sCRs, and carcinogens.

5. NEUROTRANSMITTERS AND CYTOKINES

Most neurons have a cell body, multiple dendrites and a single axon. Ninety-nine percent of the total cell volume may be associated with the dendritic arbor. Synapses involving one neuron communicating with other nerve cells can occur at any point on the cell membrane. Neurotransmitters are packets of small molecular weight signaling chemicals that create stimulatory or inhibitory effect on adjacent neurons by way of synapses. Traditional neurotransmitters are norepinephrine, histamine, serotonin, dopamine, and acetylcholine. More recently accepted neurotransmitters are amino acids (glutamate, γ-aminobutyric acid, and glycine), peptides (enkephalins, endorphins, cholecystokinin, and substance P), and purines (adenosine triphosphate and adenosine) [29].

Additionally, there are substances with important neurotransmitter-like effects. These putative neurotransmitters include steroids (pregnenalone, dehydroepiandrosterone), nitric oxide, and eicosanoids (prostaglandins). It is beyond the scope of this chapter to provide detail about neurotransmitters, but rather to present the linkage between neurotransmitters and cytokines. For additional information the reader is referred to neurotransmitter reviews elsewhere [29].

There is substantial interplay between the neurotransmitter systems and cytokines. The release of norepinephrine reduces glial inflammatory responses by downregulating IL-1β, IL-6, and inducible nitric oxide synthase (iNOS) [30]. In *in vitro* studies, IL-1β was reported to increase the release of the neuropeptide–neurotransmitter, substance P, whereas *in vivo* IL-1β inhibited somatostatin release [31].

Traditional neurotransmitters have been associated with the enzymes that break down and allow reuptake of the neurotransmitters by the neurons. For example, *AChE* breaks down the neurotransmitter acetylcholine. Acetylcholinesterase is produced by neurons and is "tethered" to the synaptic membrane. *Butyrylcholinesterase (BuChE),* a similar enzyme, is produced by activated astrocytes. Acetylcholinesterase and BuChE appear to have activity beyond the neurotransmitter function. For example, the cytokine Nerve Growth Factor (NGF), which is produced by neurons, is upregulated by AChE [32]. NGF is a cystine-knot superfamily cytokine. NGF is principally active in cholinergic subcortical neurons. NGF is produced in cortical cholinergic neurons of the hippocampus and other locations [33]. Transport to subcortical receptors is regulated by the acetylcholine neurotransmitter system through the actions of AChE and BuChE. Furthermore, intraventricular infusion of NGF will prevent the loss of cholinergic hippocampal or nucleus basalis neurons induced by trans-section or by ibotenic acid injection.

Acetylcholinesterase also downregulates glutamatergic synaptetic development and transmission in hippocampal neurons by reducing the surface insertion of NMDA receptors [34]. As AD progresses, AChE produced by neurons decreases while BuChE produced by astrocytes increases. This increase may allow for a change in acetylcholine levels in the brain. However, it seems likely that upregulated BuChE could also prevent NMDA-related apoptosis of neurons. There is additional theory that BuChE is involved in the Aβ cross-linking between tyrosines at position 10. This results in proteolytic-resistant precipitation of Aβ. Butyrylcholinesterase colocalizes with final stage of Aβ-pleated deposits in plaques [35].

Brain-derived neurotropic factor (BDNF) offers another example of the interplay between cytokines and neurotransmitter systems. Brain-derived growth factor is reported to increase the activity of the NMDA class of glutamate receptors [36]. Brain-derived growth factor actually enhances the neurotransmitter release in hippocampal nerve terminals and upregulates the release of dopamine and glutamate [37]. Brain-derived growth factor is upregulated in inflammation and nerve injuries and acts through the receptor tyrosine kinases (RTKs). Brain-derived growth factor supports the function of hippocampal, cortical, and basal forebrain cholinergic neurons, which loose function in AD, and BDNF is downregulated in AD [38].

6. OXIDATIVE FREE RADICALS, AGING, AND INFLAMMATION IN THE CENTRAL NERVOUS SYSTEM

Evidence is accumulating that oxidative injury and inflammation, which originate from increases in brain free radicals, set the stage for the development of AD. This free radical accumulation appears to precede pathological accumulations of the protein called Aβ. Elevated tau and Aβ42 concentrations in CSF are reported to be present in pre-AD [39].

Aging alone is associated with the appearance of increased markers for oxidative injury to DNA. DNA is particularly vulnerable to free radical damaged proteins, lipids, and nucleic acids. A reliable marker of this oxidative DNA stress is *8-hydroxy-2′deoxyguanosine*. The levels of this compound in the CNS and the peripheral blood rise with aging, but even more with AD [40].

The human brain is uniquely vulnerable to oxidative injury for numerous reasons [41]. First, high oxygen consumption is one contributing factor. The brain receives 20% of cardiac output, a high percent given the relatively small mass of the brain. Indeed, in stressful circumstances, the body will divert blood flow to the brain from the periphery or abdominal organs. Second, neurons have to maintain a relatively large surface area due to their long axons. ATP production is required in order to maintain ion gradients (high intracellular K^+, low intracellular Na-, very low 'free' Ca^{++}) across these surface areas. This requires high oxygen consumption in the mitochondria, which are stationed throughout the neuron. Moreover, mitochondria are particularly vulnerable to production of free radicals. Third, neurons depend upon glucose (glycolysis and Krebs cycle) for production of energy. The normal brain (10^{11}–10^{12} neurons) requires 4×10^{21} ATP molecules constantly. This is a massive energy requirement. Glycogen is not stored in the brain, thus interruptions of glucose *cause reactive oxygen species* (ROS) and *reactive nitrogen species* (RNS), which can create oxidative injury, and (through actions of a protein kinase cascade ROS and RNS accumulation) can result in apoptosis.

Fourth, high Ca^{++} traffic across neuronal membranes causes rapid increases in intracellular free Ca^{++}, which can lead to the accumulation of ROS/ RNS. In addition, common neurotransmitters such as dopamine, serotonin, and noradrenaline are auto-oxidizable. They react with O_2 to generate $O_2^{\bullet -}$ (superoxide radical). Thus, neuronal function can lead to the accumulation of free radicals. Metabolism of *monoamine neurotransmitters* (R-CH_2NH_2) by mitochondrial *monoamine oxidases* (MAO) can also result in the formation of hydrogen peroxide (H_2O_2) and ammonia (NH_3) in the following reaction:

$$R - CH_2NH_2 + O_2 + H_2O \Rightarrow RCHO\,(\text{aldehyde}) + H_2O_2 + NH_3$$

It is well known that amino acids can act as neurotransmitters. *Glutamate* and *aspartate* are examples of amino acids that, when they are released in excess amounts, can act as excitotoxins and result in intracellular signaling leading to apoptosis. Iron ions are found throughout the brain in necessary proteins such as cytochromes, ferritin, tyrosine hydroxylase, and tryptophan hydroxylase. Copper ions are also important constituents of *metalloproteins*. Traumatic, anoxic, or other injury result in the release of iron and copper, which form free radical reactants such as the hydroxyl radical ($OH^{\bullet}$), H_2O_2, and lipid peroxidation. Generally iron salts are excluded from the brain by the BBB. However, injections of iron salts directly into brain can result in convulsions, dopamine depletion, lipid peroxidation, neuronal apoptosis, and $OH^{\bullet}$ production. Finally, neuronal cell membrane lipids have a high percentage of *polyunsaturated side chain fatty acids*. Fifty to eighty percent of neuron membranes consist of lipids. Thus, neuron membranes are subject to oxygen-dependent deterioration, known as rancidity or *lipid peroxidation*. The irony of brain structure is that oxygen can result in the destruction of cell membranes, yet high oxygen consumption is required. Substances such as iron accelerate lipid peroxidation. Once initiated, lipid peroxidation tends to propagate in a domino-like effect. $H^{\bullet}$ (free hydrogen) combines with peroxyl radicals to produce lipid hydroperoxide, which precipitates a chain reaction of membrane lipid disruption.

The ravages of aging alone may contribute to the etiology of AD. Aging is associated with *oxidative injury* to the CNS. Oxidative injury then triggers a cytokine cascade response. Oxidative insults also result from CNS viral infection, bacterial infection, or head trauma. Anoxia, which creates free radicals, can occur from microembolic cerebral infarct, cerebral hemorrhage, or even cardiac arrhythmia. It is not surprising that all of these CNS insults are associated with AD.

7. NEUROPATHOPHYSIOLOGY OF ALZHEIMER'S DISEASE

The neuropathophysiology of AD, that has not been worked out in full, is presented in Fig. 2. The initial thought that genetics alone could account for the etiology of AD is now discredited. However, there are enough known pieces of the puzzle to enable educated speculation about completing the puzzle.

7.1. Insults to the central nervous system

In an aging brain, oxidative injury may trigger AD. Infections, such as herpes virus and chlamydia, may be a trigger [9,42]. Other CNS insults are seizures, ischemia, hypoglycemia,

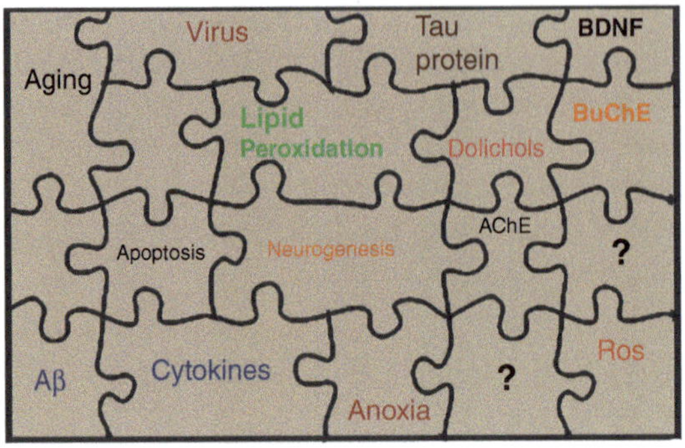

Figure 2. Puzzle Pieces: Aging, CNS Insults, Cytokines, Amyloid β, Neurofibrillary tangles, Neurotransmitters, ROS, Lipid peroxidation, Apoptosis, Neurogenesis, Dolichols.

and traumatic injury. Ischemia can come from cardiac, embolic, hemorrhagic, or thrombotic sources. All such stressors may result in local CNS inflammatory response [43].

7.2. Reactive oxygen species (ROS)

Reactive oxygen species are largely generated by mitochondria. Mitochondria create intracellular energy in the form of ATP. Hydrogen ion gradients are created across the inner oxidative mitochondrial membranes by five membrane-associated protein complexes. The H^+ generated drives the phosphorylation of ADP to ATP. However, a superoxide anion radical is formed during the oxidative phosphorylation process. *Superoxide dismutases* within the mitochondria (SOD and manganese-SOD) and cytoplasm (Cu/Zn-SOD) convert superoxide to hydrogen peroxide. Location of the ROS is mobile because mitochondria are mobile. Mitochondria are not fixed within the cell. They move along microtubules and actin filaments to where energy production is needed [44]. Catalase and glutathione peroxidase convert hydrogen peroxide to water. Presumably, these means of ROS removal travel with the mitochondria. Failing this, vitamins E, C, and A are capable of removing hydroxyl radicals.

7.3. Cyclooxygenase-2 inhibitors

Cyclooxygenase-2 (COX-2) inhibitors have been suggested to slow or prevent AD [45]. Figure 3 displays the metabolism from cell membrane lipids to active prostaglandins [46]. Cyclooxygenase-2 is also called *prostaglandin G/H synthase II*. Here, arachidonic acid is metabolized into prostaglandin PGG_2, then into other prostaglandins. Arachidonic acid is provided by membrane lipids. The products of COX-1 or COX-2 peroxidation cascade depend upon the tissue, but are generally highly reactive and unstable. For example, platelets form TXA_2, which is a vasoconstrictor, and promotes platelet aggregation. Small doses of aspirin prevent TXA_2 formation. Vascular endothelium favors production of prostacyclin (PGI_2), which is a potent vasodilator and platelet disperser. Thus PGI_2 is an antagonist of TXA_2. Here aspirin would have an opposite effect. In the brain, prostaglandin metabolism is unique [47]. Cyclooxygenase-1 functions in both glia and neurons in housekeeping functions in the brain. Cyclooxygenase-2 is primarily expressed in neurons and functions in inflammation and cellular differentiation. The

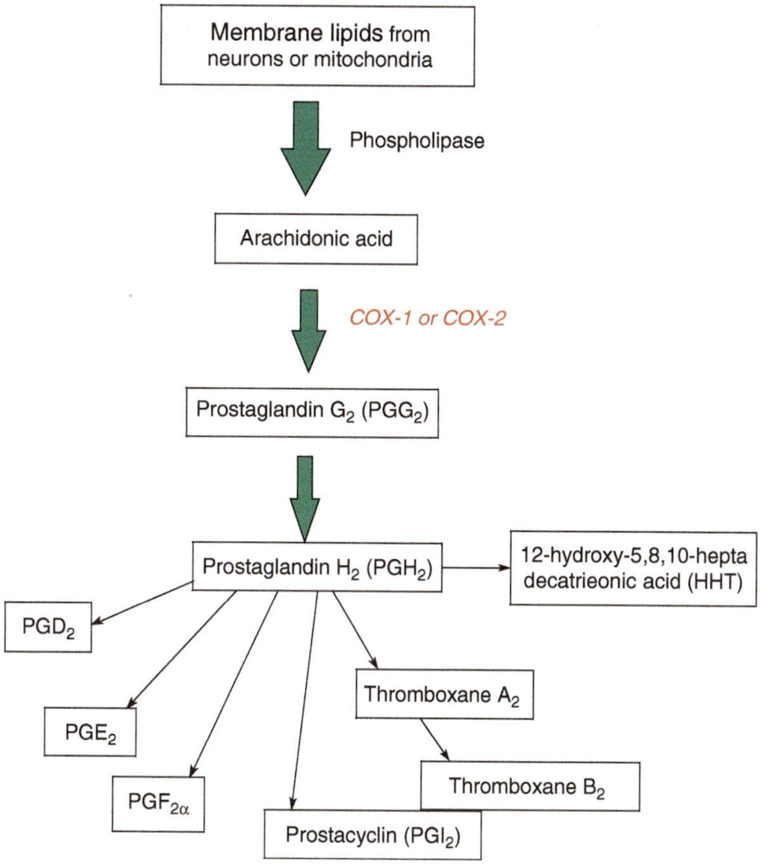

Figure 3. Prostaglandin metabolism from arachidonic acid.

principal product in neurons is prostacyclin, which accentuates local inflammation. Thus, the use of COX-2 inhibitors would reduce CNS inflammation and hence slow the progression of AD.

As cognitive function falls, neuronal COX-2 concentrations rise [47]. Speculation has been that the rise in COX-2 is related to its function in moving the neurogenesis cell cycle from G_0 phase (resting) to G_1 phase (growth), which would lead to S phase (DNA synthesis), then G_2 phase (growth for cell division), and M phase (mitosis). It is true that neurogenesis is functional and occurs in humans [48]. It is also probable that a homeostatic response to AD would be neurogenesis. However, it is more probable that the elevation of COX-2 is related to the increased availability of membrane lipids from lipid peroxidation, which would elevate arachidonic acid.

7.4. Lipid peroxidation

Lipid peroxidation is an important phenomenon that has received little attention in AD research to date. This is a complex field of study as it is hard to visualize the structures involved. Figure 4 [adapted from Halliwell and Gutteridge, 49] provides some insights into the problem. Membrane lipids are amphipathic molecules because they have polar hydrophobic regions and

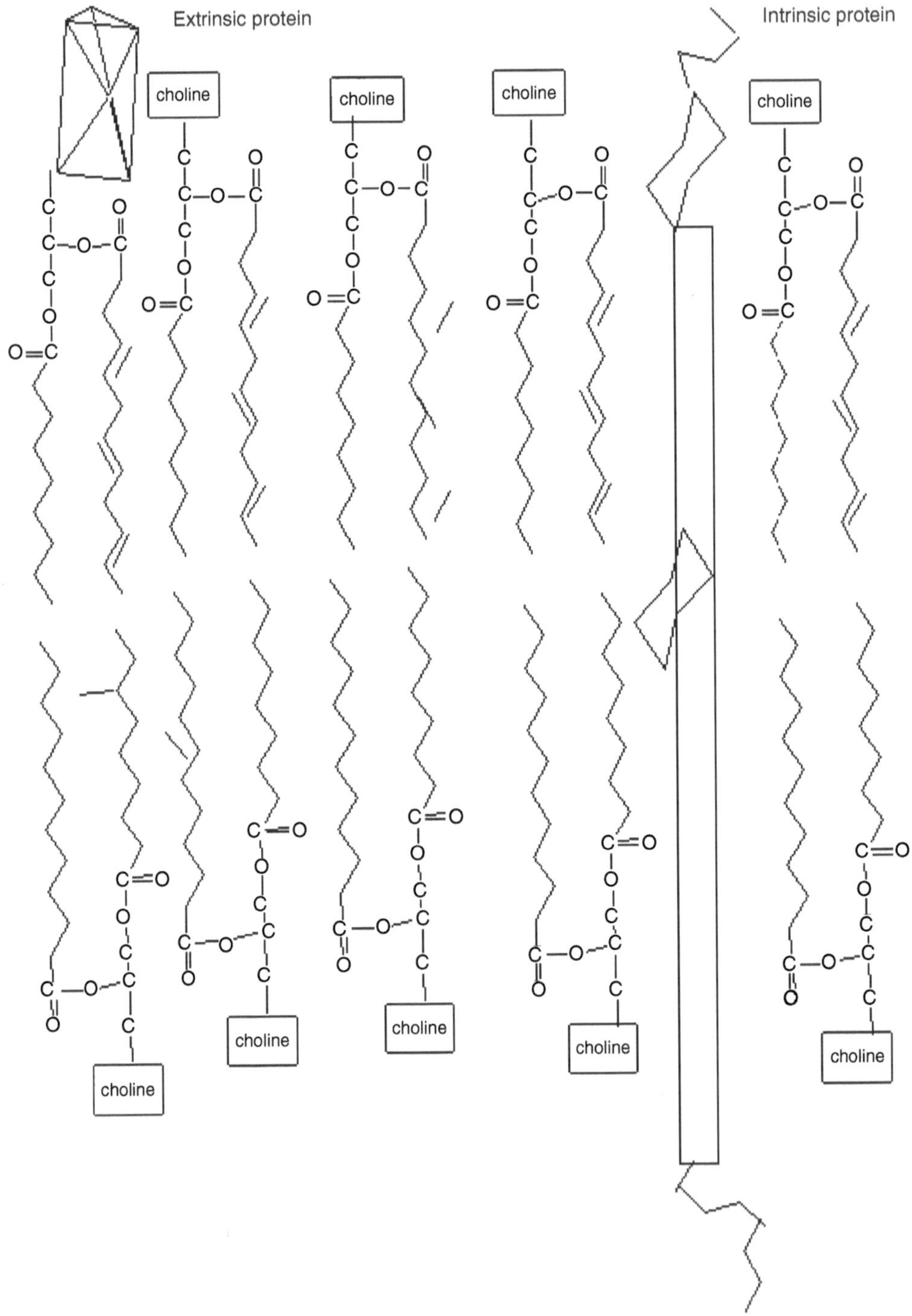

Figure 4. Bilayer lipid membrane with proteins.

nonpolar lipophilic regions. In neuronal plasma membranes and mitochondrial membranes, the dominant lipid is *phosphatidylcholine*. Mitochondria additionally use *cardiolipin* (phosphatidylethanolamine) in the inner membrane. The hydrophilic heads (choline) of these floppy molecules line up in a bilayer. One set of heads toward the intracellular space and one toward the cytosol (see Fig. 4). The lipid bilayer also hosts proteins. These are divided into extrinsic (attached to the external surface) and intrinsic (embedded in the membrane or through the membrane) proteins.

Cell membranes are very sensitive to peroxidation. This is an oxygen-dependent deterioration, also known as rancidity. The fatty acid side chains of phosphatidylcholine and phosphatidylethanolamine influence the susceptibility of peroxidation. The greater the number of carbon-to-carbon double bonds (C=C) in the fatty acids, the greater the instability of the molecule. When there are more than two double bonds, the lipids are called polyunsaturated fatty acids (PUFAs). The membranes are somewhat protected from ROS by free radical scavengers such as mannitol, vitamin E, and glutathione peroxidase. However, hydroxyl radicals ($-OH^{\bullet}$) and protonated oxygen ($O_2^{\bullet}$) tend to break through this protection. Both will strip hydrogen off lipids to generate H_2O and H_2O_2, respectively. The effect on the membrane lipid is similar to cracking an egg, the propagation of lipid peroxidation begins. The $H^{\bullet}$ stripped from the carbon on the lipid has an unpaired electron ($-C^{\bullet}H-$). In the presence of oxygen, this can become a *peroxyl radical* ($ROO^{\bullet}$). Oxygen in its molecular form tends to be hydrophobic and hence often lurks within the cell membranes. Now the new peroxyl radicals are highly reactive and extract hydrogen from the neighboring lipid molecule (especially from double bonds of PUFAs). A chain reaction follows. By metaphor, it is as if the phospholipids were two packed layers of eggs lined up like dominos. Cracking one egg leads to instability of the one next to it and so on.

Lipid peroxidation is, of course, more complex than dominos. Lipid peroxidation often involves iron chelation, copper interactions, Ca^{2+} shifts, membrane protein damage, and other nuances. Without repair, apoptosis is the outcome.

Repair of lipid peroxidation is the rule. Phospholipid hydroperoxide glutathione peroxidases convert lipid peroxides to alcohols, which are cleaved by phospholipase A2 in the presence of Ca^{2+}. Repair is completed by reacylation with *fatty acyl-coenzyme A*. The rising Ca^{2+} levels caused by rents in the membrane actually upregulate and activate phospholipase A2. The fatty acids are recruited from available extracellular and intracellular phospholipids, cholesterol, and serum lipids. The *acyl-coenzyme A* is *vitamin* B_5 (pantothenate) and provided by diet.

7.5. Amyloid-β

Aβ has several forms such as Aβ protein precursor (AβPP), subfragment $A\beta_{1-40}$, and subfragment $A\beta_{1-42}$. The AβPP fragments are accumulated in the microscopic amyloid plaques of AD. Curiously, AβPP is upregulated after CNS stress [50,51]. By definition, AβPP is one of the acute-phase proteins that arise early in the microlocalized inflammation. AβPP is a cytokine.

In small amounts, AβPP maintains homeostasis and is beneficial.

AβPP is physiologically produced and protects from transient ischemic episodes and excitotoxic injury [52,53]. Soluble AβPP stimulates NGF increasing neuronal survival, synaptic density, and neuron growth [54,55]. Nanomolar amounts AβPP have significant antioxidant activity and prevent apoptosis of neurons [56,57]. Soluble Aβ inhibits lipoprotein auto-oxidation and lipid peroxidation [58,59].

In the extracellular space, small amounts of AβPP exert beneficial effects by complexing copper (Cu) and zinc (Zn) released from metalloproteins in the face of oxidative insult. Both

Cu and Zn exist in the brain in high concentration [60]. Cu is required for essential intracellular enzymes such as SOD, cytochrome oxidase, and dopamine-β-hydroxylase. Zinc is also involved in essential intracellular enzymes such as Cu–Zn-SOD, zinc-finger proteins, and matrix metalloproteinases (MMPs). Insults to CNS result in the release of Cu and Zn into the extracellular matrix [9,61]. In inflammatory microenvironments, the pH becomes acidic. Aβ is the only known protein that binds toxic Cu and Zn in an acidic extracellular matrix. In head trauma and stroke, where Cu and Zn are released locally, there is rapid deposition of Aβ [62].

However, higher concentrations of AβPP are associated with stimulating inflammation. AβPP can upregulate cell adhesion molecules, cytokine production, complement activation, and microglia activation [63]. Activated microglia are a major source of reactive oxidative species [64]. What happens next is that AβPP becomes overwhelmed in attempting to control ROS and becomes insoluble. This would explain why activated microglia cluster at sites of precipitated Aβ in the AD brain [65]. The excess Aβ directly activates NADPH-oxidase complexes of activated microglia resulting in O_{2-} and H_2O_2. These ROS are further associated with a respiratory burst of proteins, such as myeloid-specific enzyme myeloperoxidase [66]. MyeloPerOxidase (MPO)–H_2O_2 is designed to kill pathogens. MPO–H_2O_2 tends to accelerate the formation of RNS. However, MPO–H_2O_2 also tends to cross-link proteins such as Aβ. In physiological circumstances, Aβ is a soluble and constitutive protein. In AD, Aβ is insoluble and aggregates with other proteins such as apoE, MPO, and amyloid P components [67,68].

The pathological effect of Aβ is associated with cross-linking of tyrosine at position 10 resulting in dimeric and trimeric Aβ [69,70]. Tyrosine cross-linkage of proteins is a mechanism that increases the structural strength of cell membranes against proteolysis and physical trauma. Thus, cross-linked Aβ in amyloid plaques is very resistant to resolubilization.

Curiously, Aβ also directly effects neurotransmitter systems such as the cholinergic system. Acetylcholinesterase and BuChE are both related to amyloid plaques. In transgenic mice and *in vitro*, high concentration Aβ appears related to absorption of AChE activity, downregulating membrane-bound AChE [71,72]. Acetylcholinesterase is rather diverted to intracellular lysosomal degradation. Thus, Aβ excess, by loss of AChE expression, causes upregulation of NMDA receptors related to apoptosis.

7.6. Apoptosis

Apoptosis stands for *programmed cell death* precipitated by cysteine proteases of the *caspase* family with increased permeability of mitochondrial membranes and release of cytochrome *c* from mitochondria [73]. In physiological circumstances, apoptosis is useful in eliminating neurons that do not successfully integrate into functional neuronal circuits. There is significant interaction between apoptotic mechanism in the mitochondria and synaptic function. Activation of glutamate receptors of the types α-amino-3-hydroxy-5-methylisoxazole-4-propionate (AMPA) and NMDA is associated with calcium influx and precipitation of intracellular signaling leading to apoptotic pathways. In AD, Aβ-induced reduction of AChE results in an increase in excitotoxic glutamate receptors, which may be instrumental in localized apoptosis. There are compensating homeostatic mechanisms at play that limit the extent of the localized apoptosis. For example, microglia surrounding precipitated Aβ in transgenic mouse models upregulate macrophage colony-stimulating factor receptor (M-CSFR) to a level of over expression [74]. Macrophage colony-stimulating factor receptor is a microglial cytokine that apparently protects against excitotoxicity. Further, the microglial production of BuChE would protect against excitotoxic apoptotic triggering.

7.7. Cytokine cascades and truncations

Cytokine action and reaction dominate the progression from an acute CNS microinflammation to a perpetuating chronic inflammation. In AD, there is elevation of cytokines and acute-phase proteins in peripheral blood. These include IL-1β, IL-6, and TNF-α [25,75,76]. Other acute-phase molecules elevated in AD blood are cell adhesion molecules (ICAM-1), M-CSF, C-reactive protein, serum amyloid A, transthyretin, and 8-oxy-dG [40,76]. This occurs early in the process of AD and appears to be in response to CNS cytokine production.

Excessive inflammatory response results in a chronic activation of microglia with oxidative injury and precipitation of Aβ. Proinflammatory cytokines (IL-1α, ILl-1β, IL-2, IL-6, etc.) produced by microglia, astrocytes, and neurons attract other local and systemic inflammatory cells. Activated astrocytes line up at the precipitated Aβ, and homeostatic mechanisms come into play. Anti-inflammatory cytokines are produced, such as IL-1ra, IL-6, IFN-α, IFN-γ, and TNFα. There is interplay between cytokines and neurotransmitters. For example, systemic IL-6 increases noradrenergic and serotonergic neurotransmission [14]. Noradrenaline in turn reduces proinflammatory cytokines produced by microglia responding to aggregated A$\beta_{1\text{-}42}$[77]. IL-6 also upregulates cholinergic transmission.

7.8. Neurofibrillary tangles

These tangles are seen by ordinary histological examination if silver staining is used and was part of the original observation of AD described by Alois Alzheimer himself [78]. As neurons are lost to apoptosis, cellular contents are spilled into the intracellular space. *Tau proteins*, which are members of microtubule-associated proteins (MAP) are among the cellular contents released. The normal role of tau is to stabilize microtubules, which are the highways of the intraneuronal transport systems. Physiological conditions do not allow tau to precipitate into the PHF-tau that form the neurofibrillary tangles characteristic of AD. PHF-tau seen in AD is pathologically hyperphosphorylated. The role of PHF-tau in AD remains speculative, but it *destabilizes intracellular microtubules*. Tau pathology is seen in more than 20 different diseases from normal aging to myotonic dystrophy [79]. If nothing else, tau is a tombstone of neuron death.

8. SPECULATIONS ON THE CAUSE, TREATMENT, AND PREVENTION OF ALZHEIMER'S DISEASE

8.1. The cause

As the brain ages, it becomes more vulnerable. Homeostatic mechanisms initially manage the free radicals created by aging. Intracellular antioxidants assisting in this process include vitamin E (mitochondrial), glutathione (cytosol), and CoQ10 (cytosol). Extracellular antioxidants are present include vitamin C.

One or more localized brain insults may trigger localized inflammation. These insults include infections, toxins, anoxia, and trauma. Cytokines are summoned to contain the injury. Soluble Aβ is an early cytokine that inhibits lipid peroxidation and excitotoxin injury. As neurons die and leak their contents into the extracellular space, Aβ absorbs the toxic Cu and Zn. As the inflammation progresses, Aβ upregulates other cytokine production, microglia activation, complement activation, and the expression of cell adhesion molecules. The latter allows circulating leukocytes to participate in the local defense.

As the microlocalized inflammation continues, it becomes embedded or chronic. The Aβ precipitates due to cross-linkage between tyrosines at position 10. This sulfur dimeric and trimeric linkage of Aβ oligomers makes the precipitant insoluble and resistant to proteolysis. Intercellular BuChE and released MPO–H_2O_2 free radicals create these modifications of Aβ. In doing so, Aβ becomes neurotoxic. This makes the local inflammation self-generating. As the contents of neurons spill into the intercellular space, tau is converted to neurofibrillary tangles. With the CNS sustaining progressive injury, the AD patient becomes *immunocompromised* and dies of infections such as pneumonia or urosepsis. On the basis of this neuropathophysiology, it would appear that combination treatment is necessary for best results.

8.2. Treatment and prevention of Alzheimer's disease

Mainstream treatment of AD is AChE inhibitors. Tacrine and rivastigmine both have shown long-term benefit with economic savings. Both increase cerebral blood flow. Both slow Aβ deposition into amyloid plaques. Both are potent inhibitors of both AChE and BuChE, and thus may impair the progression of illness.

There are many speculations on the future of therapy for AD. Variations on anticholinesterase therapies, low molecular weight peptide preparations from purified brain proteins, androgen treatments, hypoxanthine derivatives, vaccination of N-terminal regions of Aβ, statin therapy, and common antibiotics such as tetracycline are under consideration [80,81].

The use of nonsteroidal anti-inflammatory drugs (NSAIDs), especially the newer COX-2 inhibitors, has been proposed as treatment for AD. The logic is that inhibition of the COX-2 enzyme will slow the progression of AD. So far the results have been mixed [47]. Furthermore, the potential for toxicity in the lifetime use of NSAIDs in a population, which is known for inability to report changes such as gastric pain, makes this approach questionable. Perhaps a better COX-2 solution is the preventative use of herbal preparations such as green tea, basil, turmeric, rosemary, ginger, and oregano. These compounds have COX-2 inhibition, are available, and have few known adverse effects from long-term use [82].

Vitamin use, especially vitamin E, has been proposed. Although mixed favorable results in AD treatment are reported, the effect is not very robust. This may be due to the use of only one of the eight forms of vitamin E. Exploration of other potentially useful vitamins has not been done. These would include vitamin C (ascorbic acid), vitamin B_2 (riboflavin), vitamin B_5 (panthothenate), vitamin B_6 (pyridoxine), vitamin B_{12} (methylcobalamin), CoQ_{10} (ubiquitin), folic acid (pteroylglutamic acid), α-lipoic acid, and phosphatidylcholine (lecithin). All are effective antioxidants that work synergistically. For example, vitamin C is the principal extracellular antioxidant. Vitamin C in combination with vitamin B_2 upregulates glutathione reductase. Glutathione is the principal antioxidant defense in the cytosol of the neuron. Glutathione is the most important means of correcting lipid peroxidation. Vitamin B2 can be toxic at relatively low doses, yet when coadministered with vitamin E, the toxicity is markedly reduced. And vitamin E is better absorbed and more active when combined with phosphatidylcholine. Furthermore, phosphatidylcholine, a lipid, is more stable and less likely to become rancid if combined with vitamin E. To fully appreciate the potential for simple vitamin effect on the prevention of AD, a well thought-out combination must be employed. Recently, the benefit of vitamin combinations has received some attention [83].

"Poor circulation" was the common reason given for dementia in the elderly 30 years ago. Today most physicians scoff at this explanation. Yet, the importance of cerebrovascular pathology in AD is now receiving renewed attention. de la Torre and Hachinski's symposium on the topic stimulates the imagination on possible therapeutic alternatives [84]. The role of circulation and possibly circulating cytokines is best dramatized by a neurosurgical paper where the highly vascular omentum was freed from the intestine and passed subcutaneously up to the scalp where a burr hole allowed the omentum to be gently laid on the surface of brain [85]. This is a radical procedure and not a practical therapy. However, the results in small number of cases were startling. There was a dramatic reduction in amyloid plaques without a change in neurofibrillary tangles. The insoluble Aβ became surprisingly soluble. This indicates that improved vascular flow and addressing brain microvascular changes may be preventative and therapeutic.

The alkaloids seem to offer more than standard pharmaceuticals in improving cerebral circulation. *Ginkgo biloba*, *gotu kola*, *ginger* (zingiber officinale), and *bioflavonoids* (e.g., citrus lemon) have circulatory benefit and can be beneficial to the endothelial cells, which constitute the BBB [86,87]. Among current treatments of AD, both tacrine and rivastigmine have been shown to significantly increase cerebral blood flow. *Ergotamine alkaloids* and *vinpocetine* (extract of periwinkle) may yet become part of the treatment for AD.

Vaccines and infectious diseases are now becoming a focus of treatment for AD. Studies of amyloid precursor protein in transgenic mice suggested that immune responses to Aβ may remove plaques from the brain. But clinical human trial of the Aβ vaccine resulted in serious neurological complications. Indeed, the immune response to Aβ indicated increased neurotoxicity of Aβ [88]. The concept of such a vaccine is flawed because it assumes that Aβ has no useful physiological function. In contrast, development of pre-exposure vaccination against possible viral vectors and *Chlamydia pneumoniae* might prove useful. Justification would be difficult, as such vaccinations would occur in early childhood to prevent an illness six or more decades later. The use of antibiotics in those who have evidence of *Chlamydia infection* or *herpes virus* involvement in their pathology could be justified at this time.

The benefit of *sex steroids* is suggestive in delaying AD for estrogens and theoretically beneficial for androgens. A specific effect of CNS androgen receptors is postulated as a mechanism for male androgens. What might unify both approaches is the steroid modulation of the cytokine inflammatory response. In this case, lack of consistent dose and timing of sex steroid intervention would explain the lack of consistent findings in this field of research. Discussion of the inhibition of secretases and the use of cholesterol-reducing statins is premature at this time.

REFERENCES

1. Small GW, Rabins PV, Barry PP, Buckholtz NS, DeKosky ST, Ferris SH, Finkel ST, Gwyther LP, Khachaturian ZS, Lebowitz BD, McRae TD, Morris JC, Oakley F, Schneider LS, Streim JE, Sunderland T, Teri LA, Tune LE. Diagnosis and treatment of Alzheimer disease and related disorders: Consensus statement of the American Association for Geriatric Psychiatry, the Alzheimer's Association, and the American Geriatrics Society. JAMA 1997;278:1363–71.
2. Terry RD. Morphological changes in Alzheimer's disease-senile dementia: Ultrastructural changes and quantitative studies. Res Publ Assoc Res Nerv Ment Disord 1979;57:99–105.
3. Jarvik LF, Perl M. Overview of psychological dysfunctions related to psychiatric problems in the elderly. In Neuropsychiatric Manifestations of Physical Disease in the Elderly. Levenson AJ and Hall RCW, Eds; New York: Raven Press, 1981; pp. 1–15.

4. McGeer PL, Yasojima K, McGeer EG. Innate immunity, autotoxicity and degenerative neurologies. In Alzheimer's Disease: Advances in Etiology, Pathogenesis and Therapeutics. Iqbal K, Sisodia, SS, and Winblad, B, Eds; New York: John Wiley & Sons Ltd, 2001; pp. 331–9.
5. Guo Z, Cupples LA, Kurz A, Auerbach SH, Volicer L, Chui H, Green RC, Sadovnick AD, Duara A, DeCarli C, Johnson K, Go RC, Growdon JH, Haines JL, Kukull WA, Farrer LA. Head injury and the risk of AD in the MIRAGE study. Neurology 2000;54:1316–23.
6. Aronson MK, Ooi WL, Morgenstern H, Hafner A, Masur D, Crystal H, Frishman WH, Fisher D, Katzman R. Women, myocardial infarction and dementia in the very old. Neurology 1990;40:1102–6.
7. Ott A, Solk RP, van Harskamp F, Pols HA, Hofman A, Breteler MM. Diabetes mellitus and the risk of dementia: The Rotterdam study. Neurology 1999;53:1937–42.
8. Appelt DM, Hammond CJ, MacIntyre A, Balin BJ. A model for sporadic Alzheimer's disease: Detection of AD-like pathology in the brains of BALB/c mice following infection with *Chlamydia pneumoniae*. 32nd Neuroscience Meetings (Orlando, FL) 2002; Program No 18.10: November 3.
9. Martinez-Torres FJ, Wagner S, Kluge B, Hass J, Kehm R, Sellner J, Hacke W, Meyding-Lamadé U. Expression and activity of matrix metalloproteinases (MMPS) 2 and 9 in a murine model of herpes simplex virus encephalitis. 32nd Neuroscience Meetings (Orlando, FL) 2002; Program No 395.12: November 4.
10. Perry G, Nunomura A, Avila J, Smith MA. Oxidative damage and antioxidant responses in Alzheimer's disease. In Alzheimer's Disease: Advances in Etiology, Pathogenesis and Therapeutics. Iqbal K, Sisodia SS, and Winblad B, Eds; New York: John Wiley & Sons Ltd., 2001; pp. 371–8.
11. Roses AD. Apolipoprotein E affects the rate of Alzheimer disease expression: β-amyloid burden is a secondary consequence dependent on APOE genotype and duration of disease. J Neuropathol Exp Neurol 1994;53:429–37.
12. Hardy J. Amyloid, the presenilins and Alzheimer's disease. Trends Neurosci 1997;20:154–9.
13. Finch CE, Tanzi RE. Genetics of aging. Science 1997;278:407–11.
14. Korneva EA, Klimenko VM, Shkhinek EK. Neurohumoral Maintenance of Immune Homeostasis. Chicago: University of Chicago Press, 1985; pp. 1–14.
15. Wilson CJ, Finch CE, Cohen HJ. Cytokines and cognition – The case for a head-to-toe inflammatory paradigm. J Am Geriatr Soc 2002;50:2041–56.
16. Dantzer R. Sickness behavior: A neuroimmune-based response to infectious disease. In Psychoneuroimmunology: An Interdisciplinary Introduction. Schedlowski M and Tewes U, Eds; Dordrecht: Kluwer Academic/Plenum Publishers, 1999; pp. 235–40.
17. Campbell IL, Stalder AK, Akwa Y, Pagenstecher A, Asensio VC. Transgenic models to study in the actions of cytokines in the central nervous system. Neuroimmunomodulation 1998;5:126–35.
18. Vilcek J. The cytokines. In The Cytokine Handbook, 3rd Edn. Thomson AW, Ed.; New York: Academic Press, 1998; pp. 1–20.
19. Geoffrey R, Pugh-Humphreys P, Thomson AW. Cytokines and their receptors as potential therapeutic targets. In The Cytokine Handbook, 3rd Edn. Thomson AW, Ed.; London: Academic Press, 1998; pp. 885–951.
20. Alon R, Feigelson S. From rolling to arrest on blood vessels: Leukocyte tap dancing on endothelial integrin ligands and chemokines at sub-second contacts. Semin Immunol 2002;14:93–104.

21. Genbacev OD, Prakobphol A, Foulk RA, Krtolica AR, Ilic D, Singer MS, Yang ZQ, Kiessling LL, Rosen SD, Fisher SJ. Trophoblast L-selectin mediated adhesion at the maternal–fetal interface. Science 2003;299:405–8.
22. Bacon KB, Greaves DR, Dairaghi DJ, Schall TJ. The expanding universe of C, CX3 and CC chemokines. In The Cytokine Handbook, 3rd Ed. Thomson AW, Ed.; London: Academic Press, 1998; pp. 752–75.
23. Van Damme J. Interleukin-8 and related chemotatic cytokines. In Cytokine Handbook, 2nd Edn. Thomson AW, Ed.; London: Academic press, 1994; pp. 185–208.
24. Sun D, Hu X, Liu X, Whitaker JN, Walker WS. Expression of chemokine genes in rat glial cells: The effect of myelin basic protein-reactive encephalitogenic T-cells. J Neurosci Res 1997;48:192–200.
25. Bruunsgaard H, Andersen-Ranberg K, Jeune B, Pedersen AN, Skinhøj P, Pedersen BK. A high plasma concentration of TNF-α is associated with dementia in centenarians. J Gerontol A Biol Sci Med Sci 1999;54A:M357–64.
26. Feuerstein GZ, Wang X, Barone FC. The role of cytokines in the neuropathology of stroke and neurotrauma. Neuroimmunomodulation 1998;5:143–59.
27. Johnson GL, Lapadat R. Mitogen-activated protein kinase pathways mediated by ERK, JNK and p38 protein kinases. Science 2002;298:1911–2.
28. Manning G, Whyte DB, Martinez R, Hunter T, Sudarsanam S. The protein kinase complement of the human genome. Science 2002;298:1912–34.
29. Webster RA, Ed. Neurotransmitters, Drugs, and Brain Function. New York: John Wiley & Sons Ltd, 2000.
30. Feinstein DL, Galea E, Gavrilyuk V, Landreth GE, O'Banion MK, Heneka MT. Noradrenergic depletion potentiates β-amyloid induced cortical inflammation. 32nd Neuroscience Meetings (Orlando, FL) 2002; Program No 123.2: November 3.
31. Li X, Duflo F, Eisenach JC. Clonidine and interleukin-1-Beta modulate somatostatin release in cultured rat DRG cells. 32nd Neuroscience Meetings (Orlando, FL) 2002; Program No 46.23: November 3.
32. Deschenes-Furry JL, Belanger G, Perrone-Bizzozero NI, Jasmin BJ. Nerve growth factor regulates acetylcholinesterase expression in PC12 cells via posttranscriptional mechanisms. 32nd Neuroscience Meetings (Orlando, FL) 2002; Program No 30.17: November 3.
33. Webster RA. Alzheimer's disease. In Neurotransmitters, Drugs, and Brain Function. Webster RA, Ed. New York: John Wiley & Sons Ltd, 2000; pp. 391–2.
34. Dong H, Wu Y, Yang BB, Sheng W, Lee V, Lu WY. Non-cholinergic trophic action of acetylcholinesterase on glutamatergic synaptic transmission. 32nd Neuroscience Meetings (Orlando, FL) 2002; Program No 747.3: November 6.
35. Guillozer AL, Smiley JF, Mash DC, Mesulam MM. Butyrylcholinesterase in the life cycle of amyloid plaques. Am Neurol Assoc 1997;42:909–18.
36. Mizuno M, Yamada K, He J, Nakajima A, Nawa H, Nabeshima T. Interaction of BDNF and NMDA receptor in spatial memory formation. 32nd Neuroscience Meetings (Orlando, FL) 2002; Program No 81.04: November 3.
37. Granholm AE, Paredes DA, Bickford PC. *In vivo* effects of BDNF and NGF on dopamine and glutamate release in the hippocampus. 32nd Neuroscience Meetings (Orlando, FL) 2002; Program No 336.12: November 3.
38. Garzon DJ, Yu G, Fahnestock M. A new human BDNF transcript and decrease in BDNF transcripts 1, 2, and 3 in Alzheimer's disease in parietal cortex. 32nd Neuroscience Meetings (Orlando, FL) 2002; Program No 30.11: November 3.

39. Riemenschneider M, Lautenschlager N, Wagenpfeil S, Diehl J, Drzezga A, Kurz A. Cerebrospinal fluid tau and β-amyloid 42 proteins identify Alzheimer disease with mild cognitive impairment. Arch Neurol 2002;59:1729–34.
40. Mecocci P, Polidori MC, Ingegni T, Cherubini A, Chionne F, Cecchetti R, Senin U. Oxidative damage to DNA in lymphocytes from AD patients. Neurology 1998;51:1014–7.
41. Halliwell B, Gutteridge JMC. Free Radicals in Biology and Medicine, 3rd Edn. Oxford: Oxford University Press, 2001; pp. 721–51.
42. Hudson AP, Gйrard HC, Whittum-Hudson JA, Clayburne G, Schummacher HR. *Chlamydia pneumoniae*, APOE genotype, and Alzheimer's disease. In *Chlamydia pneumoniae* and Chronic Diseases. L'age-Stehr J, Ed. Berlin: Springer-Verlag, 2000; pp. 121–37.
43. Xiong Y, Gu Q, Peterson PL, Muizelaar JP, Lee CP. Mitochondrial dysfunction and calcium perturbation induced by traumatic brain injury. J Neurotrauma 1997;14:23–34.
44. Brodin L, Bakeeval L, Shupliakov O. Presynaptic mitochondria and the temporal pattern of neurotransmitter release. Philos Trans R Soc Lond B Biol Sci 1999;354:365–72.
45. McGeer PL, Schulzer M, McGeer EG. Arthritis and anti-inflammatory agents as possible protective factors for Alzheimer's disease: A review of 17 epidemiologic studies. Neurology 1996;47:425–32.
46. Halliwell B Gutteridge JMC. Free Radicals in Biology and Medicine. Oxford: Oxford University Press, 2001; pp. 471–81.
47. Pasinetti GM. Cyclooxygenase (COX)-2 and clinical progression of Alzheimer's disease dementia: Implications in the role of neuronal COX-2 in cell cycle. In Alzheimer's Disease: Advances in Etiology, Pathogenesis and Therapeutics. Iqbal K, Sisodia, SS, and Winblad, B, Eds; New York: John Wiley & Sons Ltd, 2001; pp. 379–92.
48. Barinaga M. Newborn neurons search for meaning. Science 2003;299:32–4.
49. Halliwell B, Gutteridge JMC. Free Radicals in Biology and Medicine, 3rd Edn. Oxford: Oxford University Press, 2001; pp. 284–313.
50. Shi J, Yang SH, Stubley L. Hypoperfusion induces over expression of beta-amyloid precursor protein mRNA in a focal ischemia rodent model. Brain Res 2000;853:1–4.
51. Atwood CS, Huang X, Moir RD, Smith MA, Tanzi RE, Roher AE, Bush AI, Perry G. Neuroinflammatory responses in Alzheimer's disease brain promote the oxidative post-translational modification of amyloid deposits. In Alzheimer's Disease: Advances in Etiology, Pathogenesis and Therapeutics. Iqbal K, Sisodia, SS, and Winblad, B, Eds; New York: John Wiley & Sons Ltd, 2001;331–9.
52. Smith-Swintosky Vl, Pettigrew LC, Craddock SD, Culwell AR, Rydell RE, Mattsch MP. Secreted forms of beta-amyloid precursor protein protect against ischemic brain injury. J Neurochem 1994;63:781–4.
53. Masliah E, Westland CE, Rockenstein EM, Abraham CR, Mallory M, Veinberg I, Sheldon E, Mucke L. Amyloid precursor proteins protect neurons of transgenic mice against acute and chronic excitotoxic injuries *in vivo*. Neuroscience 1997;78: 135–46.
54. Akar CA, Wallace WC. Amyloid precursor protein modulates the interaction of nerve growth factor with p75 receptor and potentiates its activation of trkA phosphorylation. Brain Res Mol Brain Res 1998;56:125–32.
55. Roch JM, Mashlia E, Roch-Levecq AC, Sundsmo MP, Otero DA, Veinberg I, Saitoh T. Increase of synaptic density and memory retention by a peptide representing the trophic domain of amyloid beta/A4 protein precursor. Proc Natl Acad Sci 1994;91:7450–4.

56. Bush AI, Lynch T, Cherny RA, Smith MJ, Masters CL. Alzheimer Aβ functions as a superoxide antioxidant *in vitro* and *in vivo*. Soc Neurosci Abstr 1999;25:14.
57. Chan CW, Dharmarajan A, Atwood CS. Anti-apoptotic action of Alzheimer Aβ. Alzheimer's Rep 1999;2:1–6.
58. Kontush A, Berndt C, Weber W, Akopyan V, Arlt S, Schippling S, Beisiegel U. Amyloid-β is an antioxidant for lipoproteins in cerebrospinal fluid and plasma. Free Rad Biol Med 2001;30:119–28.
59. Andorn AC, Kalaria RN. Factors affecting pro- and anti-oxidant properties of fragments of the β-protein precursor (βPP): Implication for Alzheimer's disease. J Alzheimer's Dis 2000:2;69–78.
60. Lovell MA, Robertson JD, Teesdale WJ, Campbell JL, Markesbery WR. Copper, iron and zinc in Alzheimer's disease senile plaques. J Neurol Sci 1998;158:47–52.
61. Atwood CS, Huang X, Khatri A, Scarpa RC, Kim YS, Moir RD, Tanzi RE, Roher AE, Bush AI. Copper catalyzed oxidation of Alzheimer Aβ. Cell Mol Biol 2000;46:777–83.
62. Roberts GW, Gentleman SM, Lynch A, Murray L, Landon M, Graham DI. Beta amyloid protein deposition in the brain after severe head injury implications for the pathogenesis of Alzheimer's disease. J Neurol Neurosurg 1994;57:419–25.
63. Eikelenboom P, Veerhuis R. The role of complement and activated microglia in the pathogenesis of Alzheimer's disease. Neurobiol Aging 1996;17:673–80.
64. Colton CA, Chernyshev ON, Gilbert DL, Vitek MP. Microglial contribution to oxidative stress in Alzheimer's disease. Ann NY Acad Sci 2000;899:292–307.
65. McGeer PL, Akiyama H, Itagaki S, McGeer EG. Activation of the classical complement pathway in brain tissue of Alzheimer patients. Neurosci Lett 1989;107:341–6.
66. Jacob JS, Cistola DP, Hsu FF, Muzaffar S, Mueller DM, Hazen SL, Heinecke JW. Human phagocytes employ the myeloperoxidase–hydrogen peroxide system to synthesize dityrosine, trityrosine pulcherosine, and isodityrosine by a tyrosyl radical-dependent pathway. J Biol Chem 1996;271:19950–6.
67. Glenner GG, Wong CW. Alzheimer's disease: Initial report of the purification and characterization of a novel cerebrovascular amyloid protein. Biochem Biophy Res Commun 1984;120:885–90.
68. Reynolds WF, Rhees J, Maciejewski D, Paladio T, Seiburg H, Maki RA, Masliah E. Myeloperoxidase polymorphism is associated with gender specific risk for Alzheimer's disease. Exp Neurol 1999;155:31–41.
69. Atwood CS, Scarpa RC, Huang X, Moir RD, Jones WD, Fairlie DP, Tanzi RE, Bush AI. Characterization of copper interactions with Alzheimer amyloid beta peptides: Identification of an attomolar-affinity copper binding site on amyloid beta 1–42. J Neurochem 2000;75:1219–33.
70. Hensley K, Maidt ML, Yu Z, Sang H, Markesbery WR, Floyd RA. Electrochemical analysis of protein nitrotyrosine and dityrosine in the Alzheimer brain indicates region-specific accumulation. J Neurosci 1998;18:8126–32.
71. Dumont M, Lalonde R, Fukuchi K, Strazielle C. Acetylcholinesterase activity and regional mitochondrial function in the brains of transgenic mice expressing human C99-Beta-APP. 32nd Neuroscience Meetings (Orlando, FL) 2002; Program No 191.8: November 2.
72. Hu TW, Gray N, Brimijoin S. Amyloid-beta induces alteration in acetylcholinesterase expression due to delayed degradation. 32nd Neuroscience Meetings (Orlando, FL) 2002; Program No 688.18: November 6.
73. Gilman CP, Mattson MP. Do apoptotic mechanisms regulate synaptic plasticity and growth-cone motility? Neuromol Med 2002;2:197–214.

74. Mitrasinovic OM, Rhee J Grattan A, Robinson CC, Phong C, Murphy GM. Amyloid beta phagocytosis, neuroprotection and signal transduction by microglia over expressing the macrophage colony stimulating factor receptor. 32nd Neuroscience Meetings (Orlando, FL) 2002; Program No 123.10: November 3.
75. Singh VK, Guthikonda P. Circulating cytokines in Alzheimer's disease. J Psychiatr Res 1997;31:657–60.
76. Neuroinflammation Working Group. Inflammation and Alzheimer's disease. Neurobiology 2000;21:383–421.
77. Feinstein DL, Galea E, Gavrilyuk V, Landreth GE, O'Banion MK, Heneka MT. Noradrenergic depletion potentiates β-amyloid induced cortical inflammation. 32nd Neuroscience Meetings (Orlando, FL) 2002; Program No 123.2: November 3.
78. Alzheimer A. Uber eine eigenartige erkrankung der hirnrinde. Allg Z Psychiatr 1907;64:146–8.
79. Delacourte A. Neurofibrillary degeneration: patterns of tau isoform expression. In Alzheimer's Disease: Advances in Etiology, Pathogenesis and Therapeutics. Iqbal K, Sisodia, SS, and Winblad, B, Eds; New York: John Wiley & Sons Ltd, 2001;pp. 331–9.
80. Vellas B and Fitten LJ, Eds. Research and Practice in Alzheimer's Disease. New York: Springer Publishing Co, 2000; pp. 197–305.
81. Iqbal K, Sisodia, SS, and Winblad, B, Eds. Alzheimer's Disease: Advances in Etiology, Pathogenesis and Therapeutics. New York: John Wiley & Sons, Ltd, 2001;pp. 705–821.
82. Newmark TM, Schulick P. Beyond Aspirin. Prescott, AZ: Hohm Press, 2000.
83. Fairfield KM, Fletcher RH. Vitamins for chronic disease prevention in adults. JAMA 2002;287:3116–26.
84. de la Torre JC, Hachinski V, Eds. Cerebrovascular Pathology in Alzheimer's Disease. Ann N Y Acad Sci 1997;826:1–521.
85. Goldsmith HS. Omental transposition to the brain for Alzheimer's disease. In Cerebrovascular Pathology in Alzheimer's Disease. de la Torre JC, Hachinski V, Eds; Ann N Y Acad Sci 1997;826:323–36.
86. Chevallier A. Encyclopedia of Herbal Medicine. London: Dorling Kindersley, 2000.
87. Gruenwald J, Brendler T, Jaenicke C (Eds). PDR for Herbal Medicines. Montvale, NJ: Medical Economics Company Inc., 2000.
88. Nath A, Hall, Toxova M, Dobbs M, Jones, Anderson C, Woodward J, Kryscio R, Guo Z, Wekstein D, Smith C, Markesbery W. Amyloid Beta-peptide autoantibodies are increased in Alzheimer's disease and enhance peptide neurotoxicity: Implications for disease pathogenesis and vaccine development. 32nd Neuroscience Meetings (Orlando, FL) 2002; Program No 686.17: November 6.

VI. CYTOKINES AND BEHAVIOR

Cytokines and the Brain
Edited by C. Phelps and E. Korneva

Cytokines and Immune-Related Behaviors

ARNAUD AUBERT and JULIEN RENAULT

Faculté des Sciences Parc de Grandmont, 37200 Tours, France

ABSTRACT

Theodosius Dobzhansky noted that "nothing in biology makes sense except in the light of evolution". The success of species evolution owes much to the complex set of adaptive responses, which they are able to mount when confronted with environmental changes. These changes are perceived through various sensory receptors and trigger a complicated array of physiological responses. As animals have always lived surrounded by pathogenic microorganisms and will continue to do so, invasion of tissues by either bacteria or viruses similarly initiates a set of adaptive responses involving not only the immune system but also the functionally linked nervous system. These responses include behavioral changes including sleepiness, hypophagia, and general withdrawal, and have been found to potentiate the efficiency of the immune system to clear off pathogens. These behavioral changes have been for long commonly considered as the inability to achieve normal activities resulting from a debilitation of the general physical state of the sick individual. On the contrary, sick animals retain the ability to express complex behaviors under specific circumstances. The expression of the behavioral repertoire remains flexible and therefore depends on changes in the situation the sick individual is facing. This flexibility can be interpreted in motivational terms and therefore open the gate to the further study of behavioral dynamics induced by inflammatory processes.

1. INTRODUCTION

Adaptation has several meanings in biology. It can mean the adjustment of living matter to environmental conditions and to other living things either in an organism's lifetime (physiological adaptation) or in a population over many generations (evolutionary adaptation). The ability to adapt is a fundamental property of life and constitutes a basic difference between living and nonliving matter. Thus, sickness can be easily understood as a form of adaptation and has long been considered as such. Indeed, sickness has been formerly defined by Emile Littré, a famous French lexicographer and philosopher of the nineteenth century, as "*a reaction of life, either local or general, either immediate or mediate, against an obstacle, a trouble, a lesion.*"

Animals acutely sick from systemic protozoan, bacterial or viral infections, commonly display a whole set of nonspecific symptoms. Among the most reliable symptoms are fever, sleepiness, uneasiness, reduced sexual activity, reduced exploration, and reduction in food and water intake [1–4]. Animals have always lived surrounded by pathogenic microorganisms and will continue to do so. However, for most physicians and veterinarians, these nonspecific symptoms, and in particular the behavioral changes, are traditionally interpreted as the inability to achieve normal activities resulting from a debilitation of the general physical state of the sick individual.

This conception of infectious diseases has led to the neglect of the fact that evolution has selected animals for increased fitness in their natural environments. Over thousands of generations and throughout their evolutionary history, all species have been exposed to and have successfully resisted the noxious effects of various invading microorganisms and are the products of an "arms race" [5,6]. Indeed, for the most part, scientific literature dealing with behavioral adaptation, including ultimate and proximate causation, mainly focuses attention on healthy individuals. In their natural environment, animals are challenged by numerous pathogens that can originate from infectious or inflammatory illnesses it seems. Therefore, it is reasonable to hypothesize that such animals would have developed various physiological and behavioral mechanisms to fight illness and promote recovery.

In the context of the host–pathogen relationship, the results of various studies have pointed out the emergence of immune strategies in mammals in order to fight infection [7,8]. Moreover, the large amount of data indicating a dense cross-talk between the neuroendocrine and immune systems also suggest the evolution of nonimmune (e.g., behavioral) strategies to serve the same purpose. This supposition is reinforced for at least one author by the fact that sick animals are submitted to a situation in which their integrity is heavily compromised, either through the bypass of their immune resources or by delaying the expression of basic life-supporting behaviors [1]. However, despite the prominence of published data supporting bidirectional immune–neuroendocrine communications, the study of behavioral patterns in sick animals has only received attention in the literature for a few years. Almost all behavioral studies of the neurobiology of cytokines use behavior as a dependent variable in order to determine which cytokines are neurologically active and what underlying receptor mechanisms may be involved. The approach that is proposed here is different. It considers sickness behavior as an independent variable that deserves to be studied as such, in order to determine the way it is organized and regulated at the behavioral level. Furthermore, a better understanding of sickness behavior *per se* should also provide additional basic knowledge on a little-studied natural animal behavior. It is our starting assumption that sickness behavior pattern is an orchestrated attempt to produce energy for fighting infection and to preserve energy through behavioral changes [1]. Information will be assembled to support the proposition that the behavioral modifications accompanying a course of infection or inflammation (e.g., lethargy, adipsia, hypophagia) reflect the onset of motivational reorganization, specifically devoted to counter the pathogenic microorganisms, while expressly preserving behavioral flexibility.

2. HOST RESPONSE TO INFECTION AND INFLAMMATION

The acute-phase response of the organism is a reaction to disturbances of its homeostasis during viral, bacterial, or parasitic infection and tissue injury [9]. It consists of both a local and a systemic reaction initiated within an hour or two after infection. The local reaction is dominated by inflammation. It involves many different cell types, including phagocytic cells, lymphocytes, and endothelial cells. Interactions between these different cell types are mediated by cytokines.

The systemic reaction is mediated by the action of these cytokines on distant cellular targets and includes neurological, endocrine, and metabolic changes.

Proinflammatory cytokines, such as interleukin-1 (IL-1), IL-6, and tumor necrosis factor (TNF), are proteins that are synthesized and liberated by immunocompetent cells (activated monocytes and macrophages) in response to the presence of a pathogen. As information-bearing molecules, cytokines play a major role in the coordination of the immune response. Besides this key role, these soluble proteins trigger numerous peripheral effects and have potent neurotropic effects as well.

In the periphery, they initiate an immune reaction by activating leukocytes and acting on numerous organs and tissues. They induce profound proprioceptive changes, for example, in producing hyperalgesic effects and joint pain, as well as a decrease in intestinal motility. In addition, they also act on the liver, which then liberates other immune factors known as the acute-phase proteins [10].

Central effects of cytokines are numerous. One of the most striking effect is fever, which does not result from the direct action of pathogens, but from a rise in the hypothalamic thermal set point by a change in the neuronal activity of the preoptic area [11]. In addition, cytokines have been shown to alter monoaminergic transmission mainly in the hypothalamus, the hippocampus, and the nucleus of the tractus solitarius [12–14]. Finally, cytokines are also well known for their ability to activate the hypothalamic–pituitary–adrenal (HPA) axis, the increase in the release of corticotropin-releasing hormone (CRH), adrenocorticotropin hormone (ACTH), α-melanocyte-stimulating hormone (α-MSH), β-endorphin, and glucocorticoids exerting, in turn, a negative effect upon further cytokine production. Activation of the HPA axis constitutes a regulatory loop for reducing fever and inflammatory processes [15].

In essence, the immune system uses cytokines to convey information to the brain about the level of peripheral immunological activity. By the use of recombinant cytokines and their antagonists, it has been demonstrated that behavioral effects of infection and inflammation (e.g., increased slow sleep, hypophagia, reduced social investigation) involve the central effects of proinflammatory cytokines that are released by activated accessory immune cells during infection and inflammation [2,16]. Indeed, the brain contains microglial cells that play a role analogous to the peripheral macrophages. Moreover, numerous studies have demonstrated the central synthesis of cytokines as well as the presence of receptors for these molecules in brain [17]. It is now well established that the transmission of peripheral immune messages and the production and action of cytokines in the brain are mediated by the convergence of humoral and neural pathways [18]. However, if it is correct to assume that the behavior of sick individuals depends on the sum total of peripheral and central effects of proinflammatory cytokines, then the precise mechanisms by which these molecules influence behavior are far from being fully understood.

A valuable tool for investigating the behavioral changes in response to inflammatory processes is bacterial endotoxin. Indeed, lipopolysaccharide (LPS) is a potent cytokine inducer [19], and when injected intraperitoneally, LPS reproduces nonspecific symptoms of infection and inflammation [2,20]. Lipopolysaccharide is the active fragment of the cell wall of gram-negative bacteria. Its use provides a valid simulation of acute sickness since it does not alter the sequence of events taking place during the acute phase. Moreover, the use of LPS is often selected as less stressful for the general welfare of experimental subjects. Other models, such as infection with viral or complete bacterial pathogens, while more natural, will lead to significantly more short- and long-term suffering and lethality, as well as specific pathogenic effects that could hide host nonspecific responses.

3. NONSPECIFIC SYMPTOMS AS ADAPTIVE RESPONSES TO INFECTION AND INFLAMMATION

3.1. Reduced food intake and diet selection

Hart [1] convincingly argued that the behavioral features of sick animals are not maladaptive responses or side effects of pathogen-induced debilitation. Rather, he proposed that these changes form a coordinated ensemble termed "sickness behavior." This concept refers to a behavioral strategy intended to support the metabolic and physiological changes that occur in the infected organism and that help to fight the pathogen. For example, cytokine-induced hypophagia can be interpreted as a way to reduce ingestion of some micronutrients (e.g., iron or zinc) necessary for pathogen multiplication. Support for this interpretation can be found in the positive relationship that has been observed in mice between the extent of hypophagia in response to *Listeria monocytogenes* infection and the survival rate of animals [21]. In essence, Murray and Murray [21] experimentally infected mice with *L. monocytogenes* (LD50) and let consume some food *ad libitum*, while others were intubated and force fed to the level of free-feeding, noninfected controls. Mice allowed to consume food *ad libitum* ate 58% of the controls and were much more likely to survive than those force-fed. Furthermore, there was a positive relationship between weight loss and survival for the infected mice with *ad libitum* access to food. Therefore, survival appears to be positively related to anorexia and weight loss, at least in the short term. Of course, if anorexia and weight loss caused by the degradation of body protein and fat persists, a condition known as cachexia or wasting develops. A positive relationship between loss of lean body mass and mortality in a number of diseases has been reported.

In addition to the well-established decrease in food intake in sick animals, qualitative changes in food consumption have been found in LPS- and IL-1β-treated rats when given a free choice of fat-, protein- or carbohydrate-rich food [22]. Indeed, when both LPS and IL-1β decreased total caloric intake, rats that were made sick with such treatments also increased their relative intake of carbohydrates. Moreover, since carbohydrates are calorie-rich macronutrients offering easy energy compared to fats, this result is in accordance with the pyrogenic and metabolic effects of cytokines and, thus, with the key role of fever in the recuperative process [22]. In addition to stimulation of lipolysis, infection is accompanied by an increased uptake of glucose by various tissues [23]. Administration of LPS and IL-1β induces the same effects in rats [24,25]. As a result, plasma glucose concentrations are increased or decreased depending on the dose and time since treatment, but in all cases these changes are independent of circulating concentrations of insulin. The enhanced intake of carbohydrate by LPS- and IL-1β-treated rats in the self-selection paradigm is likely to reflect the increased demand of the host for immediately assimilated energy.

These results reinforce the idea that sickness-related behaviors correspond to the expression of a behavioral homeostasis contributing to the holistic maintenance of the body. Indeed, it has been amply demonstrated that when rats are given the opportunity to select components of their diets, their selection patterns reflect the organism's nutritional and energetic requirements [26]. But perhaps the best known of all the studies revealing the expression of adaptive appetite was the one conducted by Curt Richter [27]. The rats developed "specific hungers" for substances missing in their diets, such as salt, protein, or fat. Animals deprived of vitamins in their diets would specifically seek them out when given the choice. Similarly, operations that modified the rat's ability to taste or synthesize various minerals prompted corrective ingestive behaviors.

Removing the rat's adrenal glands not only promoted sodium excretion, but also increased the salt appetite, helping to preserve the total amount available for use by the body. Finally, it has been recently shown that malaria-infected mice displayed less parasitemia and mortality when given access to a chloroquine solution [28]. This sampling has been argued to represent a possible mean to therapeutically absorb beneficial compounds (which is referred to as chemoprophylaxis) that are also associated with a bitter taste [29,30]. Nevertheless, these authors did not find an increased consumption of chloroquine in infected mice relative to controls.

3.2. Fever

Overwhelming evidence now indicates that pathogens induce fever and sickness-related behavior by stimulating leukocytes to produce cytokines [31]. Moreover, it has been shown that fever plays a critical role in recovery from pathogenic infection. A temperature ranging from 38 to 40°C potentiates the functional activity of immune cells while it consistently diminishes the growth rate of numerous viral or bacterial pathogens [32,33]. Fever also increases bacterial killing by neutrophils [34], enhances lymphocyte proliferative response to antigens and mitogens [35], enhances antibody synthesis [36,37], and potentiates IL-1 induction of T-cell proliferation [38]. The inhibition of pathogens' growth by the febrile response has been reported in pneumococci [33], in herpes virus [39], and in pathogens responsible for syphilis and gonorrhea [40]. It should be noted that this suppression in bacterial growth is attributed not only to the febrile response but also to the withholding of plasma iron [41].

Circumstantial evidence on elderly human patients reveals that their commonly blunted fever response is associated with more severe illness and higher mortality from infectious diseases when compared with younger patients [42]. Vaughn et al. [43] described an experiment in which desert iguanas *Dipsausorus dorsalis* injected with *Aeromonas hydrophyllia*, a bacterial agent, displayed maximum survival rates when exposed to ambient temperatures ranging from 38 to 42°C. Like all ectotherms, lizards regulate body temperature by behavioral thermoregulation. Thus, when placed in an experimental chamber where one end was kept at 50°C (a lethal temperature) and the other end at 30°C, lizards regulated body temperature by shuttling from one end to the other. The body temperature at which lizards moved from the hot end to the cooler end represented the high set point, and the body temperature at which lizards moved from the cooler end to the hot end represented the low set point. Interestingly, lizards challenged with killed bacteria had a higher high set point temperature and a higher low set point temperature than control had [11]. Thus, immune-challenged lizards "chose" to develop a fever, which in this case was behaviorally mediated. The importance of the behavioral response was later revealed when lizards were inoculated with live bacteria, but kept at a constant 34, 36, 38, 40, or 42°C so as to prevent behavioral thermoregulation. The results showed a high positive correlation between higher body temperatures (i.e., environmental temperature) and survival [44]. A few years later, similar results were reported in mammals [45]: rabbits infected with *Pasteurella multocida* have an optimal survival rate when their fever is about 2°C. Moreover, rabbits in which fever was suppressed by an antipyretic were much more likely to die than control subjects [46].

Generally, reduced activity and increased sleepiness are considered as a strategy of energy conservation in order to allow the full development of a fever. Indeed, the obligatory acceleration in metabolic rate during a fever ranges from around 30 to 50%, amounting to an estimated metabolic cost of about 13% for each 1°C body temperature rise [47]. From an evolutionary standpoint, it has been argued that, since fever is very costly metabolically, there must be potential benefits in the survival of animals for the response to have persisted all along the evolution of vertebrates. Thus, the energetic cost of fever is very high and could have contributed to the inhibition of energetically expensive behaviors, thereby minimizing heat loss. In a wild animal, a high febrile temperature that produces some tissue damage in the central nervous system, liver, and heart (i.e., above 41°C) still may prove adaptive if the fever is effective in combating infectious disease in an animal that would otherwise succumb to the disease [7].

4. SICKNESS AS A MOTIVATIONAL STATE: THE FLEXIBILITY OF IMMUNE-RELATED BEHAVIORS

If behavioral changes that accompany infectious illness are the expression of an organized strategy critical to the survival of the organism, then it follows that sick animals should be able to reorganize their behavioral goals depending on both the internal and external constraints they are exposed to. This flexibility is characteristic of what psychologists now call goal-directed behavior.

Although drive states or motivation takes a central position in psychology, its definition (or the range of phenomenon it accounts for) never ceases to undergo constant change according to behavioral theories. Under this single term, a set of notions have been assembled, as disparate as instincts, drives, needs, incentives, arousal, or impulses. If the emerging vagueness of the whole concept appears to render it quite useless, there is a more restricted view of motivation that is useful when engaging in characterization of a behavioral phenomenon. This view defines motivation as a "central state that organizes both perception and action" [48]. Defined as such, motivation is a modulating and coordinating influence on the direction, strength, and composition of behavioral goals. Interestingly, this view also reduces the traditional dichotomy between emotional and cognitive processes, and implies that the specific state of central drive (i.e., motivation) enables the uncoupling of perception from action and therefore the selection of the most appropriate strategy in relation to the eliciting situation [48].

The first evidence that sickness behavior may be the expression of a motivational state rather than the consequence of weakness was provided by Neal Miller [49]. Although Miller was engaged in the search for the motivational determinants of thirst, he observed that the bacterial endotoxin suppressed bar pressing in rats trained to obtain water in an operant task. However, motivation to drink was not suppressed since when endotoxin-treated rats were given free access to water, they drank although to a lesser extent than normal. This effect has been replicated in studies where specific cytokines modulated responses to rewarding intracranial self-stimulation, thus suggesting a specific action of cytokines upon motivational arousal [50]. Although providing important clues, it is difficult to draw specific conclusions on the basis of these studies, since the reduced activity observed does not provide enough specificity and could be explained by a large range of different factors.

An important feature of a particular motivational state is to compete with other motivational states for behavioral output. As a typical example, it is difficult, if not impossible, to search for food and at the same time court a sexual partner, since the behavioral patterns of foraging and courtship are incompatible. The normal expression of behavior therefore requires a hierarchical structure of motivational state, which is continuously updated according to stimulus input. When an infection occurs, the sick individual may be at a life or death juncture and its physiology and behavior must be altered to overcome the disease. However, this is a relatively long-term process that, if necessary, needs to make room to more urgent needs. It is easy then to imagine that if a sick person lying in bed hears a fire alarm ringing in his house and sees flames and smoke coming out of the basement, he will be able to momentarily overcome his sickness behavior in order to escape danger. Translated into motivational terms, this means that fear competes with sickness, and fear-motivated behaviors take precedence over sickness-motivated behaviors.

A demonstration of the motivational processes induced by sickness shows that this motivational processing is rather specific and affects behavior not necessarily as a whole but differentially relative to the considered component. This conclusion has emerged from a study on the effects of LPS on pup retrieving and nest building in lactating dams. In rodents, neonates are poïkilotherms and need the regulatory action of the mother. Maternal behavior is therefore critical for the survival of the progeny. In accordance with the motivational priority of maternal behavior, lactating mice that were made sick by the administration of LPS (400 μg/kg ip) were still able to retrieve pups that had been removed from the nest, although they did not engage in nest building. However, when the ambient temperature (22°C) was lowered to 6°C prior to testing, LPS-treated dams not only retrieved their whole litter, but also engaged in nest building as efficiently as did controls [51]. Thus, LPS (which is likely to be experienced as pain because of its effects on joints and muscles) triggers a competitive motivational system that makes the animal tend to lay down and be lethargic. In standard thermal conditions, this motivational system is stronger than the maternal motivation elicited by nest materials, so that no nest is built up. When the environmental temperature drops, the nest-building motivation increases in the survival hierarchy and bypasses the sickness motivation, so that sick dams engage in nest building.

To show that sickness does not interfere with the subject's ability to adjust his/her survival behavior strategies with regard to his/her needs and capacities, the effects of LPS (250 μg/kg ip) on food hoarding and food consumption in rats were compared [52]. Briefly, rats were trained to get food for 30 min in an apparatus consisting of a cage connected to an alley in which food was freely available at the distal end. Tested rats were separated into two groups. The first group received a food supplement comprising the difference between the amount of food hoarded and consumed during the daily 30-min test period and a fixed amount of food necessary to achieve body weight stability. Rats in the second group did not receive this supplement and had to rely on their own behavior in order to maintain their body weight. In response to LPS, food intake was decreased to the same extent irrespective of whether rats received the food supplement or not. However, food hoarding was much less affected in rats that were left to maintain body weight through their own behavior. These results are important since they indicate that the sickness-induced internal state is more effective in suppressing immediate responses to food than the anticipatory response to future needs. The differential effect of LPS on hoarding and eating can be interpreted within the context of Whishaw and Kornelson's [53] concept of dissociated motivations for different components of feeding behavior. Hoarding behavior is an anticipatory response to future needs and is mainly under the influence of secondary incentive motivation, whereas food consumption is a response to primary food cues. LPS, in the serotype and the dose used in the mentioned experiment, would impair primary motivation to a larger extent than secondary motivation.

Another contribution to the understanding of motivational processes in sick animals has assessed the influence of ongoing behavior on cytokine-induced behavioral changes [54]. It was found that IL-1β (100 ng/mouse ip) decreased home cage consumption of sweetened milk to a greater extent in *ad libitum*-fed mice than in mice that were food-restricted to maintain 85–90%

of their free-feeding body weight. Moreover, when operant responding for milk was maintained under a fixed-ratio 10 response schedule (FR10) of milk delivery, IL-1 significantly decreased milk-maintained responding in *ad libitum*-fed mice, but not in food-restricted mice. Finally, when food-restricted mice were trained under either a low- or a high-requirement (i.e., FR4 or FR32, respectively) response schedule of milk delivery, IL-1 produced significant decreases in the higher but not in the lower responding schedule. These findings provide further support for the motivational conceptualization of sickness, because the behavior of cytokine-treated mice is shaped by the dynamic trade-off between the food restriction and the ratio requirement.

Several years ago, Michael Fanselow and Robert Bolles [55] proposed a motivational interpretation of pain. They invested in the fine analysis of the interactions between different motivational states, using laboratory procedures drawn from behavioral ecology principles. According to their view, pain is part of a sensation but the consequence of a motivational system. They proposed the "perceptual–defensive–recuperative" model in which pain and fear are two antagonist motivational systems. This model promoted the idea that endogenous analgesia plays a role in supporting an individual's antipredatory defensive behavior, while nociceptive stimuli promotes a recuperative model. The antagonism of pain (analgesia) is proposed to prevent disruptive influences of nociception on coordinated defensive activities. In the formulation of their theoretical model, Bolles and Fanselow [55] noted that sickness should constitute a potentially important setting for the emergence of recuperative behaviors [55].

Indeed, the results presented above clearly show that sick animals are not incapacitated or debilitated and are still able to express complex behaviors, but they can also accurately evaluate the situation they are exposed to. The results confirm that the establishment of new priorities by the induction of sickness is expressed by a general decrease in behavioral activities. The animals are tested under the "optimal" conditions of laboratory housing (e.g., freely available food and water, warm ambient temperature, absence of stressful events), and the sick individual remains an open system still able to respond to changes in the immediate environment. These environmental cues may be evaluated as relevant to new priorities. For instance, to examine the increase in the efficiency of recuperative processes versus cues from scattered pups or cues from nest materials, when the ambient temperature is low. In this case, the sick individual interrupts its "sickness behavior" in order to respond specifically to the particular cues (e.g., retrieving of the pups or nest building) [51]. Once this response has occurred and external threats have reduced, the animal reengages in previously interrupted recuperative behavior.

5. SICKNESS AND INFORMATION PROCESSING: THE SICK ANIMAL IN ITS ENVIRONMENT

5.1. Learning and cognitive processes

As pointed out by Jensen and Toates [56], interesting mechanistic aspects of motivation take place in the brain. In motivational research, behavioral scientists seek to understand the processes by which an animal integrates information from different stimuli, both internal and external to the body, and subsequently translates into quantifiable behavior. Therefore, if we admit that sickness corresponds to a motivational state, this then poses the critical problem of how sick animals perceive their environment.

A first approach consist of exploring the effects of either cytokines or other immunogens on learning and cognitive functions. Three lines of evidence tend to suggest that learning is disrupted by injections of cytokines. Clinical experience suggests that some forms of cognitive reasoning are less efficient during an illness [57–59]. More precisely, it appears that influenza virus induces a disturbance in the execution of psychomotor tasks requiring visual and motor coordination [60–62]. The same type of infection does not seem to bring about deterioration in the performance of simple learning tasks such as free recall. However, it should be noted that when presented with more complex information, the sick patients display a greater total capacity (i.e., able to memorize more details of a given story) but have more difficulty when asked to restore the narrative logic [60]. Moreover, different pathogens exert variable influence in cognitive processes. Contrary to illnesses caused by rhinovirus, infection by the influenza virus increased the general vigilance and the distractibility of patients [63,64]. Finally, neither the influenza nor the rhinoviruses affected the performances of the patients in complex cognitive tasks such as tests of logical reasoning [63].

A second result was reported by Katsuki et al. [65] who demonstrated that IL-1β inhibited long-term potentiation (thought to be a correlate of memory) in hippocampal slice preparations.

Finally, a direct test for cognitive effects of IL administration was reported by Oitzl et al. [66]. They reported that centrally administrated IL-1β, but not IL-6, induced a transient performance deficit in rats tested in a spatial memory task (i.e., Morris water maze). Although the results of this study are highly suggestive of a specific spatial memory effect, another aspect of their data that these authors report, but do not comment upon, is the extremely labile nature of their effect. Their graphs suggest that performance was disrupted 24 h after the IL-1β injection, but that acquisition was not, because the performance of the treated group was equal to that of the controls after the first trial of the test session. However, a similar deficit in spatial learning was also observed following *Legionella pneumophila* treatment in mice [67].

Cognitive effects of cytokines have also been investigated in an autoshaping task in which rats were presented with a stimulus (introduction of a retractable lever in an operant chamber) that predicted food delivery [68]. Control rats quickly learned to press the lever, although this response did not influence the probability of food delivery. When pyrogens (250 μg/kg ip LPS; 4 μg/rat IL-1β or 300 mg/rat yeast) were injected into rats during acquisition of this task, the pyrogens severely disrupted task acquisition, while the pyrogen was active. The same treatments were however without effect on performance when injected later (i.e., when performance had stabilized). Many aspects of this autoshaping procedure have been discussed and analyzed [69,70]. It has been argued, for example, that the contact of the response lever during the intertrial interval (ITI) revealed that there were attentional components in the task. Therefore, the fact that in this study ITI lever presses were reduced following pyrogen injection is consistent with an attention deficit.

5.2. Reactivity and emotional processes

Several recent studies point to the possible existence of a relationship between stress, depression, and cytokines. In humans, a positive correlation has been observed between the development of an infectious episode and a transient depressive state [71–74]. A chronic activation of the cytokine network (e.g., sclerosis, rheumatoid arthritis) has also been correlated with a propensity to develop a depressive mood [75]. Finally, a correlation has also been shown in women between the increased liberation of cytokines occurring at childbirth and a postpartum depressive mood [76]. These results form the basis for the hypothesis of a dysthymic symptomatology resulting from an overexpression of cytokines [77–80].

Further studies have demonstrated that the administration of IL-2 or interferon (IFN) induces depressive symptoms in cancer patients [17,81]. The hyperactivity of the HPA axis that is also observed in response to a systemic increase in cytokines is reminiscent of the activation of this axis during some depressive states [17,82]. Experimentally, depressive-like effects in animals are mainly assessed through anhedonic responses (i.e., the incapacity to experience pleasure), and the administration of cytokines has been found to induce such effects in rats [79,83]; these effects are blocked by chronic treatment with imipramine, a tricyclic antidepressant [79]. However, recent findings have revealed that chronic antidepressant treatment was unable to prevent reduced intake of a highly palatable solution (e.g., sweetened milk) induced by IL-1 or LPS [84].

The major difficulty arising in the human studies was that the most significant correlation between immune processes and mood disorders has been obtained in patients suffering from major illnesses (e.g., AIDS, cancer) or in particularly emotionally laden situations (e.g., giving birth), which have associated psychological burdens. It is, therefore, unclear whether the related immune activation is purely coincidental. Although results from animal studies are still too few in number, they nevertheless raise important questions concerning the relationship between emotional and adaptive processes.

A series of experiments were undertaken to further test the relationships between sickness and hedonic processes [4,85–87]. Hedonism in animals is commonly assessed through preference and/or consumption of sweet solutions. To overcome possible biases due to the effect of LPS on fluid intake, a taste reactivity paradigm was used in freely drinking rats [88]. In this test, the influence of LPS (250 μg/kg ip) was assessed on orofacial reactivity patterns (e.g., specific tongue and mouth movements) to threshold and standard concentrations of saccharin, quinine, and sucrose. Reactivity patterns to quinine and sucrose were unaltered by LPS treatment, for both threshold and standard concentrations. In particular, a standard concentration of quinine (0.1 mM) elicited the same intensity of aversive reaction, and sucrose (90 mM), the same intensity of hedonic reaction, regardless of treatment. In contrast, however, LPS treatment altered the reactivity pattern to saccharin: LPS-treated rats displayed less ingestive and more aversive responses to a standard concentration of saccharin (5 mM) when compared to controls [85–87]. At a threshold concentration of saccharin, LPS- and saline-injected rats did not differ from each other [85–87]. Analogous results were reported when LPS (200 μg/kg ip) treatment by itself did not alter the palatability of a sucrose solution (0.3 M) infused intraorally in rats, as evidenced by continuous high levels of ingestion responses [89].

Although these results support the idea that there are some consistent relationships between systemic cytokine levels and sensory pleasure, they argue more in favor of producing an increased finickiness than anhedonia. This interpretation is based on the fact that a given taste stimulus does not always elicit the same behavioral response. In this regard, such stimuli may elicit ingestion or rejection, or be rated as pleasant or unpleasant, depending on the physiological state of the organism. In this context, changes in taste reactivity have been noted in various models of depression in animals. In a different experimental situation, rats submitted to unsignaled, inescapable shocks were found to drink less of a weak quinine solution than control rats that consumed the solution as though it was plain water. This result was not observed in rats treated with LPS. Furthermore, LPS-treated rats displayed the same hedonic patterns in response to sucrose as controls, whereas animal models of depression were characterized by a decrease in sucrose preference and consumption. In this experimental model,

LPS-induced taste reactivity changes can be interpreted as reflecting alliesthesia, wherein a given stimulus, for example, saccharin, is perceived as pleasant or unpleasant depending on the internal state of the subject [90,91]. In this case, an appetitive concentration of saccharin would be perceived in LPS-treated rats through a negative alliesthesia process, resulting in an increase in its aversive properties and a decrease in its hedonic aspects. This interpretation was confirmed by a series of taste-reactivity tests in which an increasing concentration of quinine (0, 0.07, and 0.1 mM) was added to an appetitive concentration of sucrose (90 mM). As it was predicted, the greater the quinine concentration, the greater the shift from appetitive to aversive reactions, but this shift appeared earlier for LPS-treated rats [87].

To account for such an alliesthesic change and its occurrence for saccharin (but not for quinine or sucrose), it is important to keep in mind that sensory processes derive from the evolutionary history of the species under consideration. Feeding is regulated on the basis of the taste and composition of nutrients [92]. The taste of a nutrient is a consistent indicator of the composition of this nutrient, and the taste system of a given species has been shaped by its dietary history. For example, a significant correlation has been found between the sweet taste of a "natural" nutrient and its caloric density, and between the bitter taste of a food and its toxicity. Thus, sugars prevalent in fruits are potent oral ingestion stimuli for omnivores such as rodents. In contrast, alkaloids found in plants are potent aversive stimuli for rodents, thus fostering the notion that "bitter tastes" signal the presence of toxins. The spontaneous attraction for sweet nutrients and aversion for bitter food in rats therefore has adaptive value. The fact that LPS-treated rats respond in the same way as controls to quinine could be interpreted to suggest that sick rats retain all their capacities for bitter rejection, thus all their oral capacities to reject potentially toxic food. In a similar way, the fact that LPS-injected rats still respond appetitively to sucrose suggests that sickness does not interfere with the hedonic (ingestion) value of this nutritive compound, such that sick animals can continue to ingest high-benefit food. Saccharin has mixed gustatory properties (i.e., a mixture of sweet- and quinine-like properties) [93]. Therefore, the negative alliesthesia (i.e. increased finickiness) occurring for a normally appetitive concentration of saccharin in LPS-treated rats, corresponds to the increased sensitivity of these animals to the aversive component of saccharin and a decreased responsiveness to its hedonic component. This could be considered as a decreased rejection threshold, preventing sick animals from absorbing any toxic compound that (although normally tolerated in healthy animals) could compete with the health-restoring processes in sick animals. This adaptive interpretation fits with Cabanac's hedonic theory of motivation [90,91], hedonism supporting behavioral responses to useful stimuli, and displeasure facilitating avoidance of potential dangerous stimuli.

The question that arises from previous data would be whether this negative alliesthesia can be generalized in sick subjects or not. Indeed these changes in finickiness are related to the modification of emotional processes (pleasure versus displeasure) occurring during an immune challenge, and it seems interesting to ask how an immune challenge can affect emotional reactivity to external stimuli. A recent study by Renault and Aubert [94] tried to investigate the modification of LPS-treated subjects' emotional reactivity using the "forced-swimming test" procedure (FST). In this test, rodents, forced to swim in a narrow inescapable tank of water, will develop a characteristic immobility that has been argued to represent "behavioral despair" [95], which is usually regarded as a depressive-like symptom. Authors used a two-session procedure using a modified version of the test adapted for mice [96], and they focused not only on the immobility time induced by the exposition to the water tank but also on the defensive behaviors developed by the subject. Although the first exposition to the water focuses on immediate effects of the treatment on defensive behaviors (climbing and swimming), the second exposition (24 h later) mainly assesses the delayed effects of the first one without the direct effects of endotoxin (no more inflammation process 24 h after an LPS injection). The results of this study showed that LPS-treated mice increase their defensive behaviors during the first exposition while decreasing them in the second one (higher immobility time). In contrast, a subchronic treatment with a tricyclic antidepressant (imipramine) blocked the enhancement of active defensive behaviors during the first exposition in LPS-treated group and decreased the "despair" in the second exposition. The conclusion of this study is that immune activation, by enhancing reactivity to the negative features of a given situation (i.e., the exposition to water), increases defensive motivation of subjects, but makes them more vulnerable to the deleterious emotional consequences of failure in defensive strategies (i.e., incapacity to escape). This comforts the likeness between immune activation and depression but not a direct relationship. Indeed, the activation of the immune system produces depressive-like symptoms in this study but the explanation comes from a modification of emotional processes rather than a direct causality. Authors hypothesize the existence of an "excitation transfer" occurring during an immune challenge that will induce these emotional disturbances. This theory comes from previous findings by Zillman [97], who described how a situation can modify the immediate responses in a subsequent one because they share the same physiological arousal. In other words, when an animal is confronted with an external threat, it will first evaluate the threat (cognitive process), then it will develop an emotional state that will be sustained by physiological arousal (activation of the HPA axis, for example) and provoke the emergence of defensive behaviors. In the case of this study, the immune activation follows the same physiological event and will induce the activation of the HPA axis that will enhance the emotional arousal provoked by the cognitive evaluation of the threat, thus increasing the defensive behaviors of the subjects. Because of the persistence of the threat, the failure of the defense against the threat will potentate the deleterious effects on the subjects and provoke increased depressive-like symptoms.

6. EVOLUTIONARY ROOTS OF IMMUNE-RELATED BEHAVIORS

In recent years, studies of innate immunity have led to the discovery of common molecular mechanisms used for host defense in plants, insects, and mammals [98]. The recent progress in the field of innate immunity shows that nature, despite its enormous diversity of species, very often uses the same basic principles for solving related problems in defense against microorganisms [99,100]. Certainly, plants, insects, and mammals are organized very differentially in order to carry out their essential functions, but many similarities exist in the way they resist attacks to microorganisms. This includes the various receptors (i.e., Toll-like receptors) that recognize classes of microbial cell-surface molecules, the similar signal transduction pathways (i.e., NF-κB pathway) that activate the transcription of genes related to host defense, and the ubiquitous cationic peptides (i.e., cathelicidins and defensins) that act as antimicrobial effectors [101]. For example, the study of *Drosophila melanogaster* has established this insect as an interesting model for the study of innate immunity. In particular, the molecular characterization of the regulatory pathway controlling the antifungal peptide drosomycin has revealed the importance of Toll receptors in innate immunity. Furthermore, it has been reported that injection of LPS into flies induces an immune response, suggesting that LPS receptors are used in Drosophila to detect gram-negative bacterial infection [102].

Despite the recent interest in vertebrate immunobehavioral cross-links and the relations that have been pointed out between the innate immune system of invertebrates and mammals [103], little effort has been invested to explore such phenomena in insects. Until now, attention has been essentially focused on behavioral defensive strategies against parasites and parasitoids as revealed by the extensive literature addressing this topic [104]. However, recent studies have disclosed the existence of neuroimmune bidirectional communications in invertebrates also as existing in mammals. Indeed, the invertebrate immune system shares signaling features with the nervous system (e.g., the immunosuppressive action of the opioid system), parasites taking advantage of these processes in their cross-talk with their host [105].

Recent studies suggest that invertebrates share the same kind of behavioral modifications during an immune challenge as those traditionally observed in vertebrates. Here are two examples of the recent advancement using insect as a model for studying the communication between immune and nervous systems. The first example concerns the presence of reduced associative learning abilities that has been demonstrated in insects and more precisely in honeybees, after the activation of the immune system by LPS [106]. In another study by the same authors [107], bumblebee workers, previously starved with no protein in their food, showed reduced score in the proboscis extension reaction test. This test is based on a simple classical conditioning using a sugar solution as the unconditioned stimulus and an odor as the conditioned one, the response is the extension of the worker's proboscis. The reason for using starved workers instead of *ad libitum*-fed ones is to avoid the synthesis of new proteins and especially octopamine (OA). These results call for more investigation about the mechanisms involved in such learning disruption, and it has been proposed that the octopaminergic system is implicated in the link between immune and nervous systems [107]. Indeed, OA in invertebrates is analogous to norepinephrine in vertebrates and is known to be involved in many physiological processes including the immune response and is a modulator of many different kinds of behaviors [see review 108]. OA may play a key role in the interaction between the immune system and the nervous system in insects. The activation of immune processes by LPS will deplete the OA available for learning processes in the nervous system, which cannot be synthesized again because of the protein deprivation induced before the experiment. In the same way, a study by Adamo [109] discussed about the possible relationship between an "immune-activated anorexia" and the octopaminergic system. In that study, Adamo shows that the host of a parasitic wasp displays reduced feeding behavior when the wasp's larvae are about to emerge from the host's cuticle. The experiment shows a positive relationship between immune activation, OA concentration in hemolymph, and reduction of feeding in the host.

These two studies show that invertebrates, and more precisely insects, display – at least two – the same type of behavioral changes when facing an immune challenge. Furthermore, as invertebrates' innate immune response shares the same basic features with vertebrates; the examination of immune-related behaviors in invertebrates should provide important new information on their evolutionary history.

7. CONCLUSIONS

The commonly observed nonspecific symptoms accompanying inflammation and infection (e.g., fever, hypophagia, adipsia, sleepiness) have been identified by multiple investigators and called sickness behavior. It is an evolutionarily ancient physiological response to infection, in which cognitive structures are invoked. Several investigators have provided convincing experimental evidence to support the theoretical position of sickness behavior as sustaining a meaningful pattern that can be described as a central motivational state that reorganizes the perceptions and actions of the sick individual. If sickness as a motivational state organizes new behavioral priorities geared to increase the efficiency of the individual's recuperative processes, this behavioral expression is flexible and sensitive to changes in the situation context. These changes in motivational priorities correspond to a decrease in responsiveness to various external or internal cues normally triggering activities that would compete with the efficacy of the organism to clear the invading pathogen. If the setup of these new priorities is expressed by a general decrease in behavioral activities (e.g., immobility, sleepiness), the sick individual nevertheless retains the capacity to respond to environmental stimuli. In some cases, motivational conflicts may arise, the resulting behavior expressing a trade-off between divergent activities. This theoretical framework offers new insights for further exploration of the cognitive processes that assist the subjects in adapting to their actual environment. Sickness behavior, when viewed as triggering a specific central state, should provide a potent tool to further approach the adaptive organization of cognition in animals and also in humans.

REFERENCES

1. Hart BL. Biological basis of the behaviour of sick animals. Neurosci Biobehav Rev 1988;12:123–37.
2. Dantzer R, Bluthé RM, Aubert A, Goodall G, Bret-Dibat JL, Kent S, Goujon E, Layé S, Parnet P, Kelley KW. Cytokines actions on behaviour. In Cytokines in the Nervous System. Rothwell NJ, Ed.; Austin: Landes, 1996.
3. Maier SF, Watkins LR. Bidirectional communication between the brain and the immune system: Implications for behaviour. Anim Behav 1999;57:741–51.
4. Aubert A. Sickness and behaviour in animals: A motivational perspective. Neurosci Biobehav Rev 1999;23:1029–36.
5. Anderson RM, May RM. Regulation and stability of host parasite population interactions. J Anim Ecol 1979;47:219–47.
6. Dawkins R, Krebs JR. Arms race between and within species. Proc R Soc B 1979;205:489–511.
7. Ewald EW. Evolutionary biology and the treatment of signs and symptoms of infectious disease. J Theor Biol 1980;86:169–76.
8. Templeton JW, Smith R, Adams LG. Natural disease resistance in domestic animals. J Am Vet Med Assoc 1988;1992:1306–15.
9. Kusher I. The acute phase response: From hippocrates to cytokine biology. Eur Cyt Netw 1991;2:75–80.
10. Hendler N. Infectious mononucleosis and psychiatrics disorders. In Viruses, Immunity and Mental Disorders. Kurstak E, Lipowski ZJ, and Morozov PV, Eds; New York: Plenum Press, 1987; pp. 81–94.
11. Kluger MJ. Fever: Role of pyrogens and cryogens. Physiol Rev 1991;71:93–127.
12. Dunn AJ. Systemic interleukin-1 administration stimulates hypothalamic norepinephrine metabolism paralleling the increased plasma corticosterone. Life Sci 1988;43:429–35.
13. Dunn AJ. Endotoxin-induced activation of cerebral catecholamine, serotonin metabolism: Comparison with interleukin-1. J Pharmacol Exp Ther 1992;261:964–9.

14. Dunn AJ. Effects of cytokines and infections on brain neurochemistry. In Psychoneuroimmunology, 3rd Edn. Ader R, Felten DL, and Cohen N, Eds; New York: Academic Press, 2001; pp. 649–66.
15. Berczi I. Neuroendocrine response to endotoxin. Ann N Y Acad Sci 1998;851:411–5.
16. Meijer A, Zakay-Rones Z, Morag A. Post-influenzal psychiatric disorders in adolescents. Acta Psychiatr Scand 1988;78:176–81.
17. Maes M, Sharpé S, Bosmans E, Vanderwoude M, Suy E, Calabrese J, Cosnys P. Relationships between inteleukin-6 activity, acute phase proteins, and function of the hypothalamic–pituitary–adrenal axis in severe depression. Psychiatr Res 1993;49:11–27.
18. Maier SF, Watkins LR. The role of the vagus nerve in cytokine-to-brain communication. Ann N Y Acad Sci 1998;840:1004–17.
19. Zuchermann SH, Shellhass J, Butler LD. Differential regulation of lipopolysaccharide-induced interleukin-1 and tumor necrosis factor synthesis: Effects of endogenous and exogenous glucocorticoids and the role of the pituitary–adrenal axis. Eur J Immunol 1989;19:301–5.
20. Kent S, Bluthé RM, Kelley KW, Dantzer R. Sickness behavior as a new target for drug development. Trends Pharmacol Sci 1992;3:24–8.
21. Murray MJ, Murray AB. Anorexia infection as a mechanism of host defense. Am J Clin Nutr 1979;32:593–6.
22. Aubert A, Goodall G, Dantzer R. Compared effects of cold ambient temperature and cytokines on macronutrient intake in rats. Physiol Behav 1995;57:869–73.
23. Beisel WR. Metabolic responses to infection. Ann Rev Med 1975;26:9–20.
24. Lang CH, Bagby GJ, Spitzer JJ. Glucose kinetics following lethal and sublethal doses of endotoxin. Am J Physiol 1985;248:R471–8.
25. Del Rey A, Besedovsky H. Interleukin-1 affects glucose homeostasis. Am J Physiol 1987;253:R794–8.
26. Collier GH, Leshner AI, Squibb RL. Dietary self-selection in active and non active rats. Physiol Behav 1969;4:79–82.
27. Richter CP. Total self regulatory functions in animals and human beings. Harvey Lect Ser 1943;38:63–103.
28. Vitazkova SK, Long E, Paul A, Glendinning JI. Mice suppress malaria infection by sampling a 'bitter' chemotherapy agent. Anim Behav 2001;61:887–94.
29. Johns T. With Bitter Herbs They Shall Eat It: Chemical Ecology and the Origins of Human Diet and Medicine. Tucson: University of Arizona Press, 1990.
30. Budavari S, Smith A, Heckelman P, Kinneary J, O'Neill MJ. The Merck Index: An Encyclopedia of Drugs, Chemicals, and Biologicals, 12th Edn. Rathway: Merck Inc, 1996.
31. Larson SJ, Dunn AJ. Behavioral effects of cytokines. Brain Behav Immun 2001;15:371–87.
32. McKowiack PA. Direct effects of hyperthermia on pathogenic microorganisms: Teleologic implications with regard to fever. Rev Infect Dis 1981;3:508–20.
33. Small PM, Tauber MG, Hackbarth CJ, Sandie MA. Influence of body temperature on bacterial growth rates in experimental pneumococcal meningitis in rabbits. Infect Immun 1986;52:484–7.
34. Sebag J, Reed WP, Williams RC. Effect of temperature on bacterial killing by serum and by polymorphonuclear leukocytes. Infect Immunol 1977;10:947–54.
35. Manzella JP, Roberts NJ Jr. Human macrophage and lymphocyte responses to mitogen stimulation after exposure to influenza virus, ascorbic acid, and hyperthermia. J Immunol 1979;123:1940–44.

36. Banet M, Fisher D, Hartmann KU, Hilling U. The effect of whole body heat exposure and of cooling the hypothalamus on antibody titre in the rat. Pflugers Arch 1981;391:25–7.
37. Janpel HD, Duff GW, Gershon RK, Atkins E, Durum SK. Fever and immunoregulation: 3. Hyperthermia augments the primary in vitro humoral immune response. J Exp Med 1983;157:1229–38.
38. Duff GW, Durum SK. Fever and immunoregulation: Hypothermia, interleukin 1 and 2 and T-cell proliferation. Yale J Biol Med 1982;55:437–42.
39. Carmichael J, Rizzoli R, Dayer JM, et al. Interleukin-1 decreases renal sodium reabsorption: Possible mechanism of indotoxin-induced natriuresis. J Infect Dis 1969;120:664.
40. McKowiack PA. Direct effects of hyperthermia on pathogenic microorganisms: Teleologic implications with regard to fever. Rev Infect Dis 1981;3:508–20.
41. Weinberg ED. Iron withholding: A defense against infection and neoplasia. Physiol Rev 1984;64:65.
42. Norman DC, Grahn D, Yoshikawa TT. Fever and aging. J Am Geriatr Soc 1985;33:859–63.
43. Vaughn LK, Bernheim HA, Kluger MJ. Fever in the lizard *Dipsosaurus dorsalis*. Nature 1974;252:473–4.
44. Kluger MJ. Fever: Its Biology, Evolution and Function. Princeton, NJ: Princeton University Press, 1979.
45. Kluger MJ, Vaughn LK. Fever and survival in rabbits infected with *Pasteurella multicida*. J Physiol 1978;282:243–51.
46. Kluger MJ. Fever: Role of pyrogens and cryogens. Physiol Rev 1991;71:93–127.
47. Hensel H, Bruck K, Raths P. Homeothermic organisms. In Temperature and Life. Precht H, Christopherson J, Hensel H, Larcher W., Eds; Berlin: Springer-Verlag, 1973; pp. 503–761.
48. Bolles RC. Theory of Motivation. New York: Harper & Row, 1967.
49. Miller NE. Some psychophysiological studies of motivation and of the behavioural effects of illness. Bull Psychol Soc 1964;17:1–20.
50. Anisman H, Kokkinidis L, Borowski T, Merali Z. Differential effects of interleukin (IL)-1β, IL-2 and IL-6 on responding for rewarding lateral hypothalamic stimulation. Brain Res 1998;779:177–87.
51. Aubert A, Goodall G, Dantzer R, Gheusi G. Differential effects of lipopolysaccharide on pup retrieving and nest building in lactating mice. Brain Behav Immun 1997;11:107–18.
52. Aubert A, Kelley KW, Dantzer R. Differential effect of lipopolysaccharide on food hoarding behaviour and food consumption in rats. Brain Behav Immun 1997;11:229–38.
53. Whishaw IQ, Kornelson RA. Two types of motivation revealed by ibotenic acid nucleus accumbens lesions: Dissociation of food carrying and hoarding and the role of primary and incentive motivation. Behav Brain Res 1993;55:283–95.
54. Larson SJ, Romanoff RL, Dunn AJ, Glowa JR. Effects of interleukin-1beta on food-maintained behavior in the mouse. Brain Behav Immun 2002;16:398–410.
55. Bolles RC, Fanselow MS. A perceptual-defensive recuperative model of fear and pain. Behav Brain Sci 1980;3:291–323.
56. Jensen P, Toates FM. Stress as a state of motivational systems. Appl Anim Behav Sci 1997;53:145–56.
57. Tye J. The Invisible Factor – An Inquiry into the Relationship between Influenza and Accidents. London: British Safety Council, 1960.
58. Grant J. Postinfluenzal judgement deflection among scientific personnel. Asian J Med 1972;8:535–9.

59. Heazlett M, Whaley RF. The common cold: Its effect on perceptual ability and reading comprehension among pupils of a seventh grade class. J School Health 1976;46:145–7.
60. Smith AP, Tyrell DAJ, Coyle K, Willman JS. Selective effects of minor illnesses on human performance. Br J Psychol 1987;78:183–8.
61. Smith AP, Tyrell DAJ, Al-Nakib W. The effects of experimentally induced respiratory virus infection on performance. Psychol Med 1988;18:65–71.
62. Smith AP. Respiratory virus infections and performance. Philos Trans R Soc Lond 1989;327:519–28.
63. Smith AP. Effects of influenza and the common cold on the Stroop color-word test. Percept Mot Skills 1990;74:668–70.
64. Smith AP, Tyrell DAJ, Barrow GI. Mood and experimentally induced respiratory virus infections and illnesses. Psychol Health 1992;6:205–12.
65. Katsuki H, Nakai S, Hirai Y, Akaji K, Kiso Y, Satoh M. Interleukin-1α inhibits long-term potentiation in the CA3 region of mouse hippocampal slices. Eur J Pharmacol 1990;181:323–6.
66. Oitzl MS, Vanoers H, Schobitz B, Dekloet ER. Interleukin-1-beta, but not interleukin-6, impairs spatial navigation learning. Brain Res 1993;613:160–3.
67. Gibertini M, Newton C, Friedman H, Klein TW. Spatial learning impairment in mice infected with *Legionella pneumophilia* or administered exogenous interleukin-1β. Brain Behav Immun 1995;9:113–28.
68. Aubert A, Vega C, Dantzer R, Goodall G. Pyrogens specifically disrupt the acquisition of a task involving cognitive processing in the rat. Brain Behav Immun 1995;9:129–48.
69. Sparber, SB, Bollweg GL, Messing RB. Food deprivation enhances both autoshaping and autoshaping impairment by a latent inhibition procedure. Behav Proc 1991;23:59–74.
70. Messing RB, Kleven MS, Sparber SB. Delaying reinforcement in an autoshaping task generates adjunctive and superstitious behaviors. Behav Proc 1986;13:327–38.
71. Brown GR, Rundell JR, McManis SE, Kendall SN, Zachary R, Temoshok L. Prevalence of psychiatric disorders in early stages of HIV. Psychosom Med 1992;54:588–601.
72. Greenwood R. Residual mental disorders after herpes virus infections. In Viruses, Immunity and Mental Disorders. Kurstak E, Lipowski ZJ, and Morozov PV, Eds; New York: Plenum Press, 1987; pp. 65–80.
73. Hendler N. Infectious mononucleosis and psychiatric disorders. In Viruses, Immunity and Mental Disorders. Kurstak E, Lipowski ZJ, and Morozov PV, Eds. New York: Plenum Press, 1987; pp. 81–94.
74. Meijer A, Zakay-Rones Z, Morag A. Post-influenza psychiatric disorders in adolescents. Acta Psychiatr Scand 1988;78:176–81.
75. Marshal PS. Allergy and depression: A neurochemical threshold model of the relation between the illnesses. Psychol Bull 1993;113:23–43.
76. Parry BL. Mood disorders linked to the reproductive cycle in woman. In Psychopharmacology: The Fourth Generation of Progress. Bloom FE and Kupfer DJ, Eds; New York: Raven Press, 1995; pp. 1029–42.
77. Hall S, Smith A. Investigation of the effects and aftereffects of naturally occurring upper respiratory tract illnesses on mood and performance. Physiol Behav 1996;59:56977.
78. Malek-Ahmadi P. Neuropsychiatric aspects of cytokines research: An overview. Neurosci Biobehav Rev 1996;20:359–65.
79. Yirmiya R. Endotoxin produces a depressive-like episode in rats. Brain Res 1996;711:163–74.

80. Maes M, Sharpé S, Bosmans E, Vanderwoude M, Suy E, Calabrese J, Cosnys P. Relationships between inteleukin-6 activity, acute phase proteins, and function of the hypothalamic–pituitary–adrenal axis in severe depression. Psychiatr Res 1993;49:11–27.
81. Sluzewska A, Rybakowski J, Bosmans E, Sobieska M, Berghmans R, Maes M, Wiktorrowicz K. Indicators of immune activation in major depression. Psychiatr Res 1996;64:161–7.
82. Linkowski P, Mendelwicz J, LeClerq R, Brasseur M, Hubain P, Golstein J, Copinschi G, VanCauter E. The 24-hour profile of adrenocorticotropin and cortisol in major depressive illness. J Clin Endocrinol Metab 1985;61:429–38.
83. Anisman H, Kokkinidis L, Borowski T, Merali Z. Differential effects of interleukin (IL)-1β, IL-2 and IL-6 on responding for rewarding lateral hypothalamic stimulation. Brain Res 1998;779:177–87.
84. Dunn AJ, Swiergiel AH. The reductions in sweetened milk intake induced by interleukin-1 and endotoxin are not prevented by chronic antidepressant treatment. Neuroimmunomodulation 2001;9:163–9.
85. Aubert A. Cytokines and anhedonia: Evidence from animal studies. Eur Neuropsychopharmacol 1998;8:S103.
86. Aubert A, Dantzer R. Effects of lipopolysaccharide on taste reactivity in rats. Neuroimmunomodulation 1998;5:63.
87. Aubert A, Dantzer R. The taste of sickness: Lipopolysaccharide-induced finickiness in rats. Physiol Behav 2005;84:437–44.
88. Grill HJ, Berridge KC. Taste reactivity as a measure of the neural control of palatability. Prog Psychobiol Physiol Psychol 1985;11:1–61.
89. Cross-Mellor SK, Kent WD, Ossenkopp KP, Kavaliers M. Differential effects of lipopolysaccharide and cholecystokinin on sucrose intake and palatability. Am J Physiol 1999;277:R705–15.
90. Glendinning JI. Is the bitter rejection response always adaptive? Physiol Behav 1994;56:1217–27.
91. Dess NK. Saccharin's aversive taste in rats: Evidence and implications. Neurosci Biobehav Rev 1993;17:359–72.
92. Cabanac M. Physiological role of pleasure. Science 1971;173:1103–7.
93. Cabanac M. Sensory pleasure. Q Rev Biol 1979;54:1–29.
94. Renault J, Aubert A. Immunity and emotions: Lipopolysaccharide increases defensive behaviours and potentiates despair in mice. Brain Behav Immun 2006;20:517–26.
95. Porsolt RD, Le Pichon M, Jalfre M. Depression: A new animal model sensitive to antidepressant treatments. Nature 1977;266:730–2.
96. Lucki I. The forced swimming test as a model for core and component behavioral effects of antidepressant drugs. Behav Pharmacol 1997;8:523–32.
97. Zillman D. Transfer of excitation in emotional behaviour. In Social Psychophysiology: A Sourcebook. Cacciopo TJ and Petty RE, Eds; New York: Guilford Press, 1983; pp. 215–40.
98. Magor BG, Magor KE. Evolution of effectors and receptors of innate immunity. Dev Comp Immunol 2001;25:651–82.
99. Ottaviani E, Franceschi C. The neuroimmunology of stress from invertebrates to man. Prog Neurobiol 1996;48:421–40.
100. Ottaviani E, Franceschi C. A new theory on the common evolutionary origin of natural immunity, inflammation and stress response: The invertebrate phagocytic immunocyte as an eye-witness. Domest Anim Endocrinol 1998;15:291–6.

101. Borregaard N, Elsbach P, Ganz T, Garred P, Svejgaard A. Innate immunity: From plants to humans. Immunol Today 2000;21:68–70.
102. Imler JL, Tauszig S, Jouanguy E, Forestier C, Hoffmann JA. LPS-induced immune response in Drosophila. J Endotoxin Res 2000;6:459–62.
103. Vilmos P, Kurucz E. Insect immunity: Evolutionary roots of the mammalian innate immune system. Immunol Lett 1998;62:59–66.
104. Schmid-Hempel P. Parasites in Social Insects. Princeton, NJ: Princeton University Press, 1998.
105. Salzet M. Invertebrate molecular neuroimmune processes. Brain Res Rev 2000;34:69–79.
106. Mallon EB, Brockmann A, Schmid-Hempel P. Immune response inhibits associative learning in insects. Proc Biol Sci 2003;270:2471–3.
107. Riddell CE, Mallon EB. Insect psychoneuroimmunology: Immune response reduces learning in protein starved bumblebees (*Bombus terrestris*). Brain Behav Immun 2006;20:135–8.
108. Roeder T. Tyramine and octopamine: Ruling behavior and metabolism. Annu Rev Entomol 2005;50:447–77.
109. Adamo SA. Parasitic suppression of feeding in the tobacco hornworm, *Manduca sexta*: Parallels with feeding depression after an immune challenge. Arch Insect Biochem Physiol 2005;60:185–97.

Cytokines and the Brain
Edited by C. Phelps and E. Korneva

The Production and Effects of Cytokines Depend on Brain Lateralization

PIERRE J. NEVEU

Neurobiologie Integrative, INSERM U 394, Institut François Magendie, Rue Camille Saint-Saëns, 33077 Bordeaux, France

ABSTRACT

Brain asymmetry, ascertained by anatomical, neurochemical, and functional data, is involved in the regulation of physiological systems including the immune system. Two different approaches have been used to demonstrate the influence of brain asymmetry on immune functional activity and cytokine production. The first one, formerly studied in laboratory animals, consists of measuring the immune effects of left or right ablations of the brain cortex. T-cell functions are depressed after left-side lesions but enhanced after symmetrical right-side lesions. These results have been recently confirmed in patients after stroke or surgical ablations for epilepsy. These data lead to the conclusion that in normal conditions the right hemisphere has a depressive effect on T-cell functions whereas the left hemisphere has a stimulating effect.

The second approach to the study of brain asymmetry results from clinical observations where left-handedness has been associated with a high incidence of immune disorders. An association between functional lateralization and immune reactivity has been clearly demonstrated in animals, mainly using paw preference as an index of lateralization. This experimental approach, which avoids the phenomenon of neural plasticity observed after brain lesions, enables the study of mechanisms whereby the brain can asymmetrically modulate the activity of the immune system. Preliminary experiments suggest that the roles of different brain hemispheres on immune responses result from an asymmetrical brain control of the sympathetic nervous system.

Cytokines produced in the periphery by the immune system function as mediators of an afferent pathway to the brain and are considered part of a neuroendocrine–immune signaling system. It is well known that cytokines can modulate the neuroendocrine systems and brain metabolism and can induce behavioral alterations. Interestingly, all these effects of cytokines depend on lateralization.

Key words: brain asymmetry, functional lateralization, immune reactivity, cytokine production, brain metabolism, HPA axis, sympathetic nervous system, stress, sickness behavior.

1. INTRODUCTION

Modulation of the immune system by the brain is a well-known phenomenon that has been demonstrated in various clinical and experimental situations. Stressful events or lesions of brain structures induce various alterations of immune functional activity. Unfortunately these alterations are not readily predictable, possibly because, in a particular situation, the respective implications of specific central nervous structures, neuroendocrine pathways, and the immune targets are usually not known. This implies that there are many potential contributing causes of variability, and *hemispheric asymmetry* is one of them.

Anatomical, neurochemical, and functional studies clearly demonstrate that the human brain is lateralized and that each hemisphere can be functionally linked to the other (for review see [1]). Morphological brain asymmetries mainly involve the cortical areas located around the Sylvian fissure, and in most instances, the number of cortical asymmetries favor the left hemisphere. Asymmetries in the distribution of various neurotransmitters have also been described. There is also a widespread agreement that the two hemispheres contribute differently to the regulation of behavior. In right-handers, the left hemisphere is specialized for speech and handedness and the right hemisphere for spatial ability and the expression of affect. Brain asymmetry may result from the differential effects of sex hormones on the right and left hemispheres during intrauterine development. Testosterone has been proposed to slow down the development of the left hemisphere, and this has been related to an increased frequency of left-handedness, learning disorders, and immune diseases in males [2]. Functional brain asymmetries are not restricted to human beings and have also been extensively studied in animals [3]. Since the brain is involved in the regulation of the neuroendocrine and immune systems in rats and mice, brain asymmetry has certainly important functional consequences [4,5].

The immune alterations associated with either stress or brain lesions are multiple. In this regard, recent advances in immunology offer the possibility to have a more rational approach to neural modulation of the immune system. The Th1/Th2 model provides a valuable framework for investigating and explaining many immune reactions. T lymphocytes can be classified into two groups, Th1 and Th2, based on the cytokine production and their functional activities. Th1 cells are defined by their production of interleukin (IL)-2, interferon-γ (IFN-γ), and tumor necrosis factor-β (TNF-β), while Th2 cells are characterized by their production of IL-4, IL-5, IL-6, IL-10, and IL-13; both cell types produce IL-3, TNF-α, and granulocyte–macrophage colony-stimulating factor (GM-CSF) [6]. Functionally, Th1 lymphocytes mediate cellular immunity and cytotoxicity, whereas Th2 cells are involved in humoral responses and antibody production. The Th1/Th2 balance is known to be affected by stress. Usually, stress and glucocorticoids induce a shift toward Th2-like response [7]. Therefore, brain–immune interactions may be productively studied by analyzing the effects of brain lateralization on the production of cytokines by the immune system.

Cytokines produced at the periphery by the immune system function as mediators of an afferent pathway to the brain and can therefore be considered as part of a neuroendocrine–immune signaling system. A large body of literature clearly demonstrates that cytokines modulate the neuroendocrine systems and brain metabolism and induce behavioral alterations (for review see [8]). Cytokine activities that may depend on brain lateralization and behavioral lateralization have been shown to be related to interindividual differences in stress responses [5]. Therefore, hemispheric asymmetry is likely to be part of the variability among members of a population in their responses to various insults including psychological stressors and infections, which involve cytokines.

The influence of *brain lateralization* on cytokine production can be demonstrated using two different approaches, a neuroanatomical one and a functional one. Using the neuroanatomical approach, we will assess whether asymmetrical brain lesions have different effects on cytokine production in experimental animals or in patients. Using the functional approach, we will search for a possible association between functional lateralization and the profile of cytokine production in animals and humans. As the brain–immune pathways mainly involve the neuroendocrine and sympathetic nervous systems, differences in the functioning of these systems will be discussed in relation to lateralization. Finally, we will examine the possibility that brain lateralization modulates the production and effects of cytokines in diseases like depression.

2. EFFECTS OF UNILATERAL BRAIN LESIONS ON IMMUNITY AND CYTOKINE PRODUCTION

The effects of unilateral brain lesions on various immune functions where cytokines would be involved have been extensively studied, especially in laboratory animals. However, very few data specifically dealing with the effects of lateralized brain lesions on cytokine production are available.

The effects of brain lesions on immunity depend on the size and location of the lesions, reflecting that each hemicortex contains both activating and suppressing areas that can interact within a hemisphere and between hemispheres. Furthermore, subcortical structures are asymmetrically implicated in cortical immunomodulation, and immune consequences of brain lesions depend on the neuronal reorganization that occurs mainly during the early postoperative period [9]. In laboratory animals, the contrasting effects of right and left large cortical lesions on immune reactivity are obvious when measured 6–10 weeks after surgery. In mice, mitogen-induced T-cell proliferation is decreased relative to controls after ablation of the left frontoparietooccipital cortex. After a symmetrical ablation of the right cortex, mitogenesis is relatively enhanced [10,11]. Similar results were observed in female Sprague–Dawley and in male Wistar rats [12,13]. Additionally, natural killer cell activity is impaired after left cortical ablation but is unaffected by lesions of the right cortex [14,15]. The production of antibodies of the IgG isotype is depressed after left lesions, whereas IgM antibody production remains unaffected [11]. Brain modulation of macrophage activation also demonstrates an asymmetry. The intraperitoneal injection of Camette-Guerin bacillus is known to induce an accumulation of activated macrophages in the peritoneum. Such an accumulation is not observed after a left cortical ablation. Moreover, the oxidative metabolism of macrophages is decreased after left, but not right, cortical lesions [16]. Damaging the cortex does not alter the function of nonactivated resident macrophages. Although it has been shown that the neocortex modulates the activity of both T and B lymphocytes, as well as macrophages, the mechanisms of immunomodulation are not yet known.

Cortical lesions do not induce a lymphocyte redistribution similar to that observed during stress [17]. Furthermore, the effects of cortical lesions on mitogen-induced lymphoproliferation were similar in the spleen and lymph nodes [18]. Therefore, the cortex modulation of the immune system does not appear to occur directly at the level of secondary lymphoid organs, but possibly at the level of the bone marrow or the thymus. If this were the case, the neocortical damage could affect immune functions in two nonexclusive manners. As both lymphocytes and macrophages are affected after cortical lesions, it is possible that the primary target is a hematopoietic stem cell at the bone marrow level as previously postulated [16]. Alternatively, the neocortex may first alter the functions of T lymphocytes in the thymus and only secondarily modulate B-cell and macrophage functions through the production of cytokines produced by T lymphocytes.

Studying the effects of brain lesions on cytokine production and especially on the Th1/Th2 balance would help to answer the question of mechanisms of cortical lesion effects.

Unfortunately until now, very few data are available. Renoux and Bizière [19] showed that left cortical lesions decrease the production and/or the release of serum factor(s) involved in T-cell maturation, but these factor were not identified. IL-1 production by macrophages stimulated by lipopolysaccharide (LPS) was depressed after left lesions, but enhanced after right lesions [20]. Likewise, *in vitro* production of IL-2 by lymphocytes stimulated by concanavalin A was depressed after left lesions [21]. In a related experimental design for rats, the production of IL-10 and IFN-γ by ConA-stimulated splenocytes was increased after left hemisphere seizures, but not after right seizures [22]. Further experiments, measuring the *in vivo* production of both Th1 and Th2 cytokines following unilateral brain manipulations is needed to address this question.

Findings in humans are in accordance with the data obtained in laboratory animals. Indeed, immune defects observed after brain lesions in humans depend on the side of the lesion. However, literature on clinical data is still sparse and, as in the case of animal studies, various immune functions involving cytokines have been studied. However, very few data specifically dealing with the effects of brain lesions on cytokine production are available.

The perturbations of blood lymphocyte subsets observed 3 weeks after a stroke in humans [23] are in agreement with the modifications of mitogenesis observed in mice 2 weeks after brain cortex ablation. One month poststroke patients displayed enhanced antigen-specific T-cell reactivity on the paretic side of the body when the lesions were on the right but not the left side [24]. These authors have further suggested that the frontal cortex–putamen are key brain structures in regulating the magnitude of immune responses. Their results indicate a similar specialization in humans for the regulation of immune responses between the right and left brain hemispheres, as was described in mice. In patients with stroke or cerebral palsy, immune alterations are more important after left than after right hemispheric lesions [25]. When studying the effects of surgical resections for epilepsy on T-cell indices, Meador et al. [26] show differential immunological responses to focal cerebral lesions as a function of cerebral lateralization. Absolute lymphocyte count, total T cells, helper T cells, and cytotoxic/suppressor cells were reduced following language-dominant resections, but were increased following nondominant resections. Finally, preliminary data on elderly patients with cerebrovascular diseases reveal a greater incidence of severe infection, mainly pneumonia, after left lesions [27].

As expected from known brain–immune pathways, stroke and related lateralized damage has been shown to affect cytokine production. For example, the *in vitro* production of IL-6 and TNF-α by peripheral blood cells increased after acute stroke [28]. Plasma levels of anti-inflammatory cytokines increased after intracerebral hemorrhage [29]. However, these effects of stroke and hemorrhage on cytokine production were not studied in relation to lateralization. Only very recently, Dzledzic et al (data to be published) reported that plasma levels of IL-10, but not that of IL-6, were related to lateralization; patients with left hematoma had higher levels of IL-10 than patients with right hematoma had.

3. EFFECTS OF FUNCTIONAL LATERALIZATION ON IMMUNITY AND CYTOKINE PRODUCTION

The effects of *functional lateralization* on immune reactivity were first observed in humans and thereafter in laboratory animals. The observations in humans were carried out in the context of immune disorders like allergy and autoimmune diseases in which the Th1/Th2 balance is known to play a crucial role. Unfortunately, few data dealing with the measures of immune variables are available. Geschwind and Behan [2] were the first to describe a higher incidence of immune diseases in *left-handers* as compared to *right-handers*. Following this first observation, an association between left-handedness and autoimmune diseases has been found by other authors for certain disorders such as Crohn's disease, ulcerative colitis, and diabetes type I [30], but not for others like lupus erythematosus [31]. Similarly, for allergic diseases whose ethiopathogenesis is better known, the association remains controversial. Strong associations are found in some studies. Geschwind and Behan [2] have reported a significant increase in allergies in strongly left-handed individuals. Smith [32] has found that among patients attending an allergy clinic, there are significantly more left-handers when compared to control subjects; the proportion of left-handedness was largest in patients suffering from rhinitis, asthma, eczema, and urticaria. Lelong et al. [33] have reported that the frequency of nonright-handedness is significantly greater in children allergic to mites than in a control population of school children of the same age. In a study of mathematically precocious youths at the Johns Hopkins University, Benbow and Benbow [34] found a marked predominance of males in this group and 20% of these students were left-handed or ambidextrous and 56% had symptomatic atopic disease. However, other authors have been unable to demonstrate an association between left-handedness and allergies. Bishop [35], in a study of a population of British children, did not find allergies (usually hay fever), such as eczema, asthma, or psoriasis, to be associated with left-handedness. Although Pennington et al. [36] found a significant elevation of both autoimmune and allergic disorders in *dyslexic patients*, there was no association between left-handedness and immune disorders. In our study, we have compared the distribution of right- and left-handers in a population of patients consulting an allergy clinic and a control group with a similar sex and age distribution [37]. There was no overall association between left-handedness and allergy. However, in accordance with Geschwind's theory, we found a tendency toward left-handedness in patients whose allergic symptoms began before puberty, indicating that left-handers may have an increased predisposition to allergic disease that manifests itself during early life. The possible association between handedness and immune functions is poorly documented. Burke et al. [38] found that handedness was not correlated with immune measures including delayed hypersensitivity, mitogen-induced lymphoproliferation, plasma levels of autoantibodies, and lymphocyte blood counts. The percentage of various T-cell subpopulations was similar in left- and right-handed individuals with the exception of the percentage of suppressor inducer cells that were lower in left-handers [39]. Chengappa et al. [40] reported a lower *in vitro* production of IL-2 by blood cells from non-right-sided as compared to right-sided individuals. The influence of lateralization on immune reactivity, and especially on cytokine production, should now be studied in humans since it is clearly established in laboratory animals.

In laboratory animals, the influence of *lateralization* on immunity was first demonstrated by studying immune functions in which cytokines are known to play a role. The influence of lateralization on the production of cytokines *per se* was demonstrated only recently. In rodents, functional lateralization was assessed by measuring postural/motor asymmetries in various tests (see [41]). Animals usually display a consistent lateralized response when tested with the same protocol, but they do not show a consistent bias across the different tests. Indeed, studies on the relationship between different measures of behavioral asymmetries have shown that the direction of one postural/motor asymmetry does not predict the direction of another one [42–44]. This shows that in rodents as in human beings, different measures of behavioral lateralization may also reflect different brain asymmetries. In our experiments, lateralization was tested by paw preference in a food-reaching task, as previously described by Collins [45]. Paw preference is stable over time and results obtained in test sessions separated by intervals of several weeks are highly correlated [45,46]. According to paw preference distributions, a mouse population may be divided into subpopulations of right-pawed, left-pawed and ambidextrous animals.

An association between paw preference and immune reactivity was first described in C3H mice. In female, left-pawed animals exhibited higher mitogen-induced T-lymphocyte proliferation than did right-pawed mice. However, B-lymphocyte mitogenesis did not vary according to behavioral lateralization [46]. Such an association between paw preference and T-cell mitogenesis was not found in males of the same strain. A depression of T-cell induced mitogenesis in response to LPS was observed in right-pawed mice, but not in left-pawed [47]. In males, but not in females, natural killer cell activity was associated with paw preference. Left-pawed males exhibited lower natural killer cell activity than did right-pawed animals [15]. Recently, using another criterion of lateralization, Kim et al. [48] showed that Balb/c mice with right-turning preference had a higher host resistance to an intracellular bacteria, *Listeria monocytogenes*. Mice with right-turning preference also exhibited a higher primary antibody response and a higher delayed-type hypersensitivity to a protein antigen. However, the secondary humoral response was similar in right- and left-turners. An association between paw preference and immune reactivity was also demonstrated for the production of autoantibodies [49]. New Zealand Black mice spontaneously develop autoimmune disorders such as lupus-like glomerulonephritis and hemolytic anemia related to anti-DNA and antierythrocyte antibodies, respectively. No correlation between paw preference and autoantibody production was observed in males. However, in females, both antierythrocyte and anti-DNA antibodies of the IgG isotype appeared earlier in left-pawed animals. The production of autoantibody production of the IgM isotype was not influenced by lateralization. Assuming that IgG, in contrast to IgM, are strongly T-cell-dependent, it can be proposed that paw preference is mainly associated with T-cell functions. The available data concerning the association between behavioral lateralization and immune reactivity clearly show that this association depends on the *sex* of animals, suggesting an involvement of sex. This is not surprising because sex hormones are involved in both brain development [50] and immune reactivity [51]. Finally, the fact that the importance of the link between behavioral lateralization and immunity depends on the *strain* under test [51,52,53] indicates that this association could be influenced by *genetic factors*, which are still unknown. This point has been specifically questioned using a genetic approach. The association between lateralization and immune reactivity also depends on the *immune parameters* tested. This suggests that not all the different components of the immune system are under the influence of lateralization. The T-cell lineage appears to be more susceptible than the B-cell lineage. No experiments have yet tested the possible influence of behavioral lateralization on macrophage functions.

Very recently, the influence of lateralization on cytokine production was studied under either *psychological* or *immune stress* conditions. Various stressors can increase cytokine production. In rats, stress has been shown to increase central IL-1β, IL-1 mRNA, and plasma IL-6 [54–58]. Usually, stress able to induce cytokines is severe. Mild stress, like 1-h restraint, did not alter plasma levels of IL-1 and IL-6 in mice [59]. The lack of effect could be due to a negative action of glucocorticoids. In support of this interpretation, brain IL-1β increased in *adrenalectomized* mice, but not in intact rats, in response to inescapable shocks [60]. Restraint induced an increase in plasma concentrations of IL-1, which depended on lateralization, since it occurred in right-but not in left-pawed mice [61]. In a similar manner, plasma levels of IL-1 were higher in right-than in left-pawed mice 2 h after peripheral administration of LPS [62]. By contrast, there was no effect of lateralization on restraint-induced increase in plasma IL-6 or in the LPS-induced increase in plasma IL-10. Further experiments, including time response studies, are needed to demonstrate a possible influence of lateralization on the production of cytokines other than IL-1. In Balb/c mice, the production of cytokines in response to LPS also depended on lateralization [63]. Likewise, Kim et al. [48] showed that the production of cytokines (IL-6 and IFN-γ) in response to *L. monocytogenes* was lower in right-turners. These preliminary results clearly show that lateralization influences cytokine production. However, its role on the Th1/Th2 balance remains to be studied.

4. INFLUENCE OF LATERALIZATION ON THE EFFECTS OF CYTOKINES

Very few experiments have been performed to answer the question of whether or not lateralization influences responses to cytokines. In our experiments, we first studied the role of lateralization in the effects of cytokines using mycobacteria that are known to activate cellular immune functions and especially the production of IL-1 [64]. The injection of mycobacteria is associated with long-lasting changes in the distribution of brain monoamines [65,66]. Interestingly, modifications of norepinephrine (NE) and serotonin (5-HT) concentrations observed 8 weeks after mycobacteria administration were found mainly in right-pawed mice [67]. In further experiments, we used LPS as a cytokine inducer. Lipopolysaccharide increased central NE and serotonin (5-HT) turnover and stimulated adrenocorticotropin (ACTH) and corticosterone production [68]. These effects of LPS could be possibly mediated by the cytokines produced in response to LPS, especially IL-1. Indeed, IL-1 mediates the enhanced ACTH secretion that follows administration of LPS [8,69]. Furthermore, direct administration of IL-1 induces changes in central monoamine concentrations that are quite similar to those observed after LPS [69,70]. Among the neurochemical alterations induced by LPS, the increase in serotonin turnover in the medial hypothalamus was observed only in right-pawed animals. Likewise, the elevation of plasma ACTH and the decrease of T-lymphocyte proliferation induced by LPS were observed in right-pawed and ambidextrous but not in left-pawed animals [47]. These results clearly show that neurochemical, neuroendocrine, and immune responses to LPS depend on lateralization. Although cytokine levels were not measured in this study, these effects of LPS are likely mediated by cytokines.

The effect of lateralization on the effects of cytokines was directly assessed by studying IL-1-induced sickness behavior in animals selected for paw preference. In rats and mice, sickness behavior is characterized by a decrease in food intake, body weight, social interaction, and motility [71]. We showed that IL-1-induced sickness was more pronounced in right-pawed than in left-pawed mice [72]. The influence of lateralization in other effects of IL-1 remains to be determined. Furthermore, the influence of lateralization in the effects of other cytokines, especially TH1 versus Th2 cytokines, has not yet been studied.

5. BRAIN–IMMUNE PATHWAYS INVOLVED IN LATERALIZED CONTROL OF CYTOKINE PRODUCTION

Because the communication pathways between the brain and the immune system involve the neuroendocrine system and, especially, the hypothalamic–pituitary–adrenal (HPA) axis [73], as well as the autonomic nervous system [74], it may be hypothesized that the asymmetrical brain organization influences the reactivity of these systems.

Only a few differences have been observed in the activity of the HPA axis in basal conditions. The *in vitro* production of ACTH by the hypothalamus and corticosterone by adrenals is similar irrespective of whether tissues are taken off from right- or left-pawed mice [75]. As the activity of the *HPA axis* is regulated by the *hippocampus*, we searched for a possible difference in the distribution of corticoid receptors in relation to behavioral lateralization. In the hippocampus, the binding capacity of glucocorticoid receptors was distributed symmetrically and did not depend on the behavioral lateralization of animals. By contrast, the percentage of right over total *mineralocorticoid receptors* was inversely correlated with individual paw preference scores of animals [76]. Since the inhibitory function of the hippocampus on the HPA axis is mainly mediated by mineralocorticoid receptors, differences in hippocampal mineralocorticoid receptor-binding activity may explain interindividual differences in the corticoid stress response [77,78], and by extension, the above-described findings.

The influence of lateralization in the activation of the HPA axis may be easily studied during the stress response. As already mentioned, a stress-like response can be induced by an injection of LPS. This response includes an increase in ACTH and corticosterone plasma levels. Interestingly, plasma corticosterone levels increased from the baseline to the same extent in both left- and right-pawed animals, but *plasma ACTH* levels increased only in right-pawed and ambidextrous animals [45]. After a 1-h restraint stress, the increase in plasma corticosterone varied according to lateralization [79]. Besides the HPA axis, several neuroendocrine systems, including the hypothalamic–thyroid axis and the hypothalamic–gonadal axis, can modulate immune responses [51]. Functioning of these axes has been shown to be lateralized [4]. Furthermore, during stress, the production of prolactin, which is known to modulate immune reactivity [80], depends on behavioral lateralization [81]. Although it is clear that lateralization modulates both cytokine production and neuroendocrine activity, it is not yet known whether lateralization necessarily modulates cytokine production through its effects on the neuroendocrine system.

Some experiments, performed in stress conditions, favor the hypothesis that the lateralization effect on cytokine production is at least partly mediated by the sympathetic nervous system. Stimulation of the sympathetic nervous system during stress can be responsible for immune alterations associated with stress and especially for the production of cytokines. Catecholamines are potent modulators of cytokine production through β2- and α-adrenergic receptors. Most of the data on the immunomodulating effects of catecholamines concern the activation of β2 receptors in *in vitro* conditions. Catecholamines upregulate the production of IL-10 and IL-8 and downregulate TNF-α [82–84]. Cytokine production by murine Th1 lymphocytes was inhibited by β2 adrenergic stimulation, whereas production of cytokines by Th2 clones remained unchanged [85]. The immunomodulating role of α-adrenoreceptors is less well known. However, catecholamines can augment the production of TNF-α via α-adrenoreceptors that are expressed on macrophages [86]. *In vivo*, α-adrenergic drugs can modulate immune reactivity. For example, prazosin, an α1/α2-adrenoreceptor antagonist, suppressed experimental autoimmune encephalomyelitis in rats [87] and modulated cytokine production in response to LPS in mice [88]. However, these *in vivo* effects could be indirect. Liberation of catecholamines in the vicinity of immune cells depends on the presence of presynaptic adrenoreceptors, which regulate the release of noradrenaline [89].

The role of the sympathetic nervous system in lateralization effects on cytokine production was previously suggested. LPS-induced immune depression and activation of the HPA axis, which have been previously demonstrated to be lateralized, were partly or totally suppressed by pharmacological depletion of peripheral catecholamines using 6-hydroxydopamine [90]. Administration of an α1/α2 adrenergic receptor antagonist, prazosin, reduced plasma levels of IL-1 and abolished the effect of lateralization observed after LPS alone [62]. This indicates that modulation of IL-1 production via α-adrenoreceptors could depend on lateralization, perhaps because of differences in the distribution and/or the affinity of α1/α2 adrenoreceptors on macrophages that produced cytokines or on presynaptic endings. Further experiments are needed to elucidate these points. The sympathetic nervous system could be indirectly involved in lateralized control of cytokine production via the production of heat-shock proteins (HSP). The HSP response to restraint appeared to be under α1-adrenergic control [91], and HSP, when released by cells, can stimulate cytokine production. In fact, the induction of HSP70 mRNA was increased in the cerebral cortex and stomach after restraint water immersion in rats [92]. Likewise, restraint has been shown to induce a high expression of HSP70 in the adrenal gland and the aorta, but not in any of a dozen other tissues tested [91]. Therefore, stress could induce necrosis resulting in the release of HSP that in turn stimulates the production of inflammatory cytokines. Indeed, exogenous HSP70 activates NF-κB and upregulates the expression of proinflammatory cytokines in human monocytes [92,93]. In accordance with this hypothesis, Campisi and Fleshner [94] recently reported that an acute stress – inescapable tail shocks – induced the release of extracellular HSP72 and inflammatory cytokines. According to this hypothesis, the lateralization effect in cytokine production should be related, at least partly, to the activity of the sympathetic nervous system. Whether the stress-induced inflammatory cytokine production results from the direct effects of catecholamines on immune cells or the production of extracellular HSP, the lateralization effect should result from asymmetrical brain control of the sympathetic nervous system.

6. CONCLUSION AND PERSPECTIVES

The consistent observation of an association between lateralization and immune functional activity has led to important hypotheses. Lateralization could represent a neurobehavioral trait linked to the activity of physiological systems involved in the response to external aggressions (either intra- or interspecific). Therefore, the description of these associations should be a first step for delineating the personality of individual (taken as the whole of its characteristics) including various aspects of behavior, brain metabolism, neuroendocrine, and immune responsiveness in order to predict the responsiveness to stressful conditions. Only few of the data already available on the role of lateralization on immune functional activity are related to cytokine production. Furthermore, the Th1/Th2 balance, which is relevant in physiopathology, has not yet been studied in relation with lateralization. Several experimental studies indicate that the sympathetic nervous system is involved in the effects of lateralization on cytokine production, but this needs further investigation.

In clinical conditions, several studies might follow from these experimental observations. For studying the association between handedness and immunity, groups of subjects must be better defined and the various components of the immune responses, especially cytokine production, must be studied instead of the clinical course of disease. In fact, experimental models have shown that this association can be observed for some immune parameters but not for others. As the immune processes involved in the pathogenesis of various diseases may be different, the association may or may not be relevant for predictability related to the disease process. The association between lateralization and cytokine production can also exist in other unexpected clinical situations. For example, depressive symptomatology has been shown to depend on lateralization [95]. Recent work shows that depression is accompanied by an increased production of cytokines [96], which may have a pathogenic role in the activation of the HPA axis and in the depression of brain serotonin [97]. As animal studies have shown that the production of cytokines in response to an inflammatory stimulus depends on lateralization, the latter could be used as a behavioral marker to predict the severity and evolution of depression. This example may be generalized to inflammatory diseases like multiple sclerosis or arthritis in which stressful situations are associated with relapses, since lateralization is involved in both stress response and cytokine production. These considerations may open a large field of new investigations.

REFERENCES

1. Geschwind N, Galaburda AM. Cerebral Lateralization: Biological Mechanisms, Associations, and Pathology. London: The MIT Press, 1987.
2. Geschwind N, Behan P. Left-handedness: Association with immune disease, migraine and developmental learning disorders. Proc Natl Acad Sci 1982;79:5097–100.
3. Glick SA. Cerebral Lateralization in Nonhuman Species. New York: Academic Press, 1985.
4. Gerendai T, Halasz B. Neuroendocrine asymmetry front. Neuroendocrinology 1997;18:354–81.
5. Neveu PJ. Lateralization and stress responses in mice: Interindividual differences in the association of brain, neuroendocrine, and immune responses. Behav Genet 1996;26:373–7.
6. Kelso A. Th1 and Th2 subsets: Paradigms lost? Immunol Today 1995;16:374–9.
7. Elenkov IJ, Chrousos GP. Stress hormones, Th1/Th2 patterns, pro/anti-inflammatory cytokines and susceptibility to disease. Trends Endocrinol Metab 1999;10:359–68.
8. Besedovsky HO, Del Rey A. Immune–neuro–endocrine interactions: Facts and hypotheses. Endocrine Rev 1996;17:64–102.
9. Neveu PJ. Cerebral lateralization and the immune system. Int Rev Neurobiol 2002; 52:303–23.
10. Neveu PJ, Taghzouti K, Dantzer R, Simon H, Le Moal M. Modulation of mitogen-induced lymphoproliferation by cerebral neocortex. Life Sci 1986;38:1907–13.
11. Renoux G, Bizière K, Renoux M, Guillaumin JM, Degenne D. A balanced brain asymmetry modulates T cell-mediated events. J Neuroimmunol 1983;5:227–38.
12. Barnéoud P, Neveu PJ, Vitiello S, Le Moal M. Brain neocortex immunomodulation in rats. Brain Res 1988;474:394–8.
13. Lahoste GJ, Neveu PJ, Mormède P, Le Moal M. Hemispheric asymmetry in the effects of cerebral cortical ablations on mitogen-induced lymphoproliferation and plasma prolactin levels in female rats. Brain Res 1989;483:123–9.
14. Bardos P, Degenne D, Lebranchu Y, Bizière K, Renoux G. Neocortical lateralization of NK activity in mice. Scand J Immunol 1981;13:609–11.
15. Betancur C, Neveu PJ, Vitiello S, Le Moal M. Natural killer cell activity is associated with brain asymmetry in male mice. Brain Behav Immunol 1991;5:162–9.
16. Neveu PJ, Barnéoud P, Georgiades O, Vitiello S, Vincendeau P, Le Moal M. Brain neocortex influence of the mononuclear phagocytic system. J Neurosci Res 1989; 22:188–93.
17. Dhabhar FS, Miller AH, McEwen BS, Spencer RL. Effects of stress on immune cell distribution. Dynamics and hormonal mechanisms. J Immunol 1995;154:5511–27.

18. Barnéoud P, Neveu PJ, Vitiello S, Le Moal M. Lymphocyte homing after left or right brain neocortex ablation. Immunol Lett 1990;24:31–6.
19. Renoux G, Bizière K. Asymmetrical involvement of the cerebral neocortex on the response to an immunopotentiator, sodium diethyldithiocarbamate. J Neurosci Res 1987;18:230–8.
20. Li QS, Yang GS. Immunoregulatory effect of neocortex in mice. Immunol Invest 1987;16:87–96.
21. Neveu PJ, Barnéoud P, Vitiello S, Kelley KW, Le Moal M. Brain neocortex modulation of mitogen-induced interleukin-2 but not interleukin-1 production. Immunol Lett 1989;21:307–10.
22. Goldstein KR, Bhatt R, Barton BE, Zalcman SS, Rameshwar P, Siegel A. Effects of hemispheric lateralization and site specificity on immune alterations induced by kindled temporal lobe seizures. Brain Behav Immun 2002;16:706–19.
23. Czolonkowska A, Korlak J, Kuczynska-Zarzewialy A. Lymphocyte subsets after stroke (abstract). J Neuroimmunol 1987;16:40.
24. Tarkowski E, Jensen C, Ekholm S, Ekelund P, Blomstrand C. Localization of the brain lesion affects the lateralization of T-lymphocyte dependent cutaneous inflammation. Evidence for an immunoregulatory role of the right frontal cortex-putamen region. Scand J Immunol 1998;47:30–6.
25. Rogers SL, Coe CL, Karaszewski JW. Immune consequences of stroke and cerebral palsy in adults. J Neuroimmunol 1998;91:113–20.
26. Meador KJ, DeLecuona JM, Helman SW, Loring DW. Differential immunologic effects of language-dominant and nondominant cerebral resections. Neurology 1999;52:1183–7.
27. Kawaharda M, Urasawa K. Immunological functions and clinical course of elderly patients with cerebrovascular diseases. Jpn J Geriatr 1992;29:652–60.
28. Ferrase C, Mascarucci P, Zola C, Cavaretta R, Frigo M, Begni B, Sarinella F, Frattola L, De Simoni MG. Increased cytokine release from peripheral blood cells after acute stroke. J Cereb Blood Flow Metab 1999;19:1004–9.
29. Dziedzic T, Bartus S, Klimkowicz A, Motyl M, Slowik A, Szczudlik A. Intracerebral hemorrhage triggers interleukin-6 and interleukin-10 release in blood. Stroke 2002;33:2334–5.
30. Searleman A, Fugali AK. Suspected autoimmune disorders and left-handedness. Evidence for individuals with diabetes, Crohn's disease and ulcerative colitis? Neuropsychologia 1987;25:367–74.
31. Salcedo JR, Spiegler BJ, Gibson E, Magilavy DB. The autoimmune disease systemic lupus erythematosus is not associated with left-handedness. Cortex 1985;21:645–7.
32. Smith J. Left-handedness: Its association with allergic disease. Psychopharmacologica 1987;25:665–74.
33. Lelong M, Thelliez F, Thelliez P. Les gauchers sont-ils plus souvent des allergiques? Allergie et Immunologie 1986;18:10–3.
34. Benbow CP, Benbow RM. Biological correlates of high mathematical reasoning ability. Prog Brain Res 1984;61:469–90.
35. Bishop DVM. Is there a link between handedness and hypersensitivity? Cortex 1986;22:289–96.
36. Pennington BF, Smith SD, Kimberling WJ, Green PA, Haith MM. Handedness and immune disorders in familial dyslexics. Arch Neurol 1987;44:634–9.
37. Betancur C, Velez A, Cabanieu G, Le Moal M, Neveu PJ. Association between left-handedness and allergy: A reappraisal. Neuropsychologia 1990;28:223–7.
38. Burke HL, Yeo RA, Vranes L, Garry PJ, Goodwin JS. Handedness, developmental disorders, and in vivo and in vitro measurements of immune responses. Dev Neuropsychol 1988;4:103–15.

39. Yokoyama MM, Hara A, Shiotsuki K. Lymphocyte subsets of left-handers. Brain Behav Immun 1987;1:36–9.
40. Chengappa KNR, Ganguli R, Yang ZW, Schurin G, Cochran J, Brar JS, Rabin B. Non-right sideness; an association with lower IL-2 production. Life Sci 1992;51:1843–9.
41. Betancur C, Neveu PJ, Le Moal M. Strain and sex differences in the degree of paw preference in mice. Behav Brain Res 1991;45:97–101.
42. Camp DM, Robinson TE, Becker JB. Sex differences in the effects of early experience on the development of behavioral asymmetries in rats. Physiol Behav 1984;33:433–9.
43. Castellano MA, Diaz-Palarea MD, Rodriguez M, Barroso J. Lateralization in male rats and dopaminergic system: Evidence of right-side population bias. Physiol Behav 1987;40: 607–12.
44. Myslobodsky MS, Braun H. Postural asymmetry and directionality of rotation in rats. Pharmacol Biochem Behav 1980;13:743–5.
45. Collins RL. On the inheritance of direction and degree of asymmetry. In Cerebral Lateralization in Non-Human Species. Glick SD, Ed.; New York: Academic Press, 1985; pp. 41–71.
46. Neveu PJ, Barnéoud P, Vitiello S, Betancur C, Le Moal M. Brain modulation of the immune system: Association between lymphocyte responsiveness and paw preference in mice. Brain Res 1988;457:392–4.
47. Delrue C, Deleplanque B, Rouge-Pont F, Vitiello S, Neveu PJ. Brain monoaminergic, neuroendocrine and immune responses to an immune challenge in relation to brain and behavioral lateralization. Brain Behav Immun 1994;8:137–52.
48. Kim D, Carlson JN, Seegal RF, Lawrence DA. Differential immune responses in mice with left- and right-turning preference. J Neuroimmunol 1999;93:164–71.
49. Neveu PJ, Betancur C, Barnéoud P, Preud'homme JL, Aucouturier P, Le Moal M, Vitiello S. Functional brain asymmetry and murine systemic lupus erythematosus. Brain Res 1989;498:159–62.
50. McLusky NJ, Naftolm F. Sexual differentiation of the central nervous system. Science 1981;211:1294–303.
51. Comsa J, Leonhardt H, Wekerle H. Hormonal coordination of the immune response. Rev Physiol Biochem Pharmacol 1982;92:115–91.
52. Fride E, Collins RL, Skolnick P, Arora PK. Strain-dependent association between immune function and paw preference in mice. Brain Res 1990;522:246–50.
53. Neveu PJ, Betancur C, Vitiello S, Le Moal M. Sex-dependent association between immune function and paw preference in two substrains of C3H mice. Brain Res 1991;559:347–51.
54. Minami M, Kuraishi Y, Nakai S., Hirai Y, Satoh M. Immobilization stress induces interleukin-1β mRNA in the rat hypothalamus. Neurosci Lett 1991;123:245–56.
55. Zhou D, Kusnekov AW, Shurin MR, De Paoli M, Rabin BS. Exposure to physical and psychological stressors elevates plasma interleukin-6: Relationship to the activation of the hypothalamic–pituitary–adrenal axis. Endocrinology 1993;133:2523–30.
56. Takaki A, Huang QH, Somogyvari-Vigh A, Arimura A. Immobilization stress may increase plasma interleukin-6 via central and peripheral catecholamines. Neuroimmunomodulation 1994;1:335–42.
57. Shintini F, Nakaki T, Kanba S, Sato K, Yagi G, Shiozawa M, Aiso S, Kato R, Asai M. Involvement of interleukin-1 in immobilization stress-induced increase in plasma adrenocorticotropic hormone and in the release of hypothalamic monoamines in the rat. J Neurosci 1995;15;1961–70.

58. Shizuya K, Komori T, Fujiwara R, Miyahara S, Ohmori M, Nomura J. The influence of restraint stress on the expression of mRNA for IL-6 and the IL-6 receptor in the hypothalamus and midbrain of the rat. Life Sci 1997;61:PL135–40.
59. Li KS, Liège S, Moze E, Neveu PJ. Plasma corticosterone and immune reactivity in restrained C3H mice. Stress 2000;3:285–98.
60. Nguyen KT, Deak T, Owens SM, Kohno T, Fleshner M, Watkins LR, Maier SF. Exposure to acute stress induces brain interleukin-1β protein in the rat. J Neurosci 1998;18:2239–46.
61. Merlot E, Moze E, Dantzer R, Neveu PJ. Suppression of restraint-induced plasma cytokines in mice pretreated with LPS. Stress 2002;5:131–5.
62. Dong J, Mrabet O, Moze E, Li KS, Neveu PJ. Lateralization and catecholaminergic neuroimmunomodulation: Prazosin, an α1:α2 adrenergic receptor antagonist, suppresses interleukin-1 and increases interlukin-10 production induced by lipopolysaccharides. Neuroimmunomodulation 2002;10:163–8.
63. Gao MX, Li KS, Dong J, Liège S, Jiang B, Neveu PJ. Strain-dependent association between lateralization and LPS-induced IL-1β and IL-6 production in mice. Neuroimmunomodulation 2000;8:78–82.
64. Oppenheim JJ, Mizel SB, Melzer MS. In Biology of the Lymphokines. Cohen S, Pick E, and Oppenheim JJ, Eds.; New York: Academic Press, 1979; pp. 291–323.
65. Barnéoud P, Rivet JM, Vitiello S, Le Moal M, Neveu PJ. Brain norepinephrine levels after BCG-stimulation of the immune system. Immunol Lett 1988;18:201–4.
66. Deleplanque B, Neveu PJ. Brain regions involved in modulation of immune responses. In Brain Control of Responses to Trauma. Rothwell N and Berkenbosch F, Eds; Cambridge: Cambridge University Press, 1994; pp. 108–122.
67. Deleplanque B, Delrue C, Vitiello S, Neveu PJ. Distribution of brain monoamines in left- and right-handed mice injected with bacillus Calmette-Guérin. Int J Neurosci 1993;72:287–98.
68. Dunn AJ, Welch J. Stress and endotoxin-induced increases in brain tryptophan and serotonin depend on sympathetic nervous system activity. J Neurochem 1991;57:1615–22.
69. Rivier C, Chizzonite R, Vale W. In the mouse, the activation of the hypothalamic–pituitary–adrenal axis by a lipopolysaccharide (endotoxin) is mediated through interleukin-1. Endocrinology 1989;125:2800–5.
70. Berkenbosch F, de Goeij DEC, Del Rey A, Besedovsky HO. Neuroendocrine, sympathetic and metabolic responses induced by interleukin-1. Neuroendocrinology 1989;50:570–6.
71. Kent S, Bluthé RM, Kelley KW, Dantzer R. Sickness behavior as a new target for drug development. Trends Pharmacol Sci 1992;13:24–8.
72. Neveu PJ, Bluthé RM, Liège S, Michaud B, Moya S, Dantzer R. Interleukin-1 induced sickness behavior depends on behavioral lateralization in mice. Physiol Behav 1998;63: 587–90.
73. Bateman A, Singh A, Kral T, Solomon S. The immune–hypothalamic–pituitary–adrenal axis. Endocr Rev 1989;10:92–112.
74. Ader R, Felten D, Cohen N. Interactions between the brain and the immune system. Annu Rev Pharmacol Toxicol 1990;30:561–602.
75. Betancur C, Sandi C, Vitiello S, Borrell J, Guaza C, Neveu PJ. Activity of the hypothalamo–pituitary–adrenal axis in mice selected for left-or right-handedness. Brain Res 1992; 589:302–6.
76. Neveu PJ, Liège S, Sarrieau A. The asymmetrical distribution of hippocampal mineralocorticoid receptors depends on lateralization in mice. Neuroimmunomodulation 1998; 5:16–21.

77. Schöbitz B, Sutanto W, Carey MP, Holsboer F, DeKloet ER. Endotoxin and interleukin-1 decrease the affinity of hippocampal mineralocorticoid (type I) receptor in parallel to activation of the hypothalamic–pituitary–adrenal axis. Neuroendorinology 1994;60:124–33.
78. Oitzl MS, VanHaarst AD, Sutanto W, DeKloet ER. Corticosterone, brain mineralocorticoid receptors (MRs) and the activity of the hypothalamic–pituitary–adrenal (HPA) axis: The Lewis rat is an example of increased central MR capacity and a hyporesponsive HPA axis. Psychoneuroendocrinology 1995;20:655–75.
79. Neveu PJ, Moya S. In the mouse, the corticoid stress response depends on lateralization. Brain Res 1997;749:344–6.
80. Matera L. Endocrine, paracrine and autocrine actions of prolactin on immune cells. Life Sci 1996;59:599–614.
81. Waters NS, Badura LL, Denenberg VH. Differential prolactin responsiveness to stress in left- and right-pawed mice. Brain Res 1996;724:112–6.
82. Kavelaars A, van de Pol M, Zijlstra J, Heijnen, CJ. Beta-2-adrenergic activation enhances interleukin-8 production by human monocytes. J Neuroimmunol 1997;77:211–6.
83. Van der Poll T, Jansen J, Endert E, Sauerwein HP, Deventer, SJH. Noradrenaline inhibits lipopolysaccharide-induced tumor necrosis factor and interleukin 6 production in human whole blood. Infect Immun 1994;62:2046–50.
84. Lombardi MS, Kavelaars A, Schedlowski M, Bijlsma JW, Okihara, KL, van de Pol M, Ochsmann S, Pawlak C, Schmidt RE, Heijnen CJ. Decreased expression and activity of G-protein-coupled receptor kinases in peripheral blood mononuclear cells of patients with rheumatoid arthritis. FASEB J 1999;13:715–25.
85. Ramer-Quinn DS, Baker RA, Sanders VM. Activated T helper 1 and 2 cells differentially express the β2-adrenergic receptor. J Immunol 1997;159:4857–67.
86. Spengler RN, Allen RM, Remick DG, Strieter RM, Kunkel SL. Stimulation of α-adrenergic receptor augments the production of macrophage-derived tumor necrosis factor. J Immunol 1990;145:1430–4.
87. Brosnan CF, Goldmuntz EA, Cammer W, Factor SM, Bloom BR, Norton WT. Prazosin, an α1-adrenergic receptor antagonist, suppresses experimental autoimmune encephalomyelitis in the Lewis rat. Proc Natl Acad Sci USA 1985;82:5915–9.
88. Szelenyi J, Kiss JP, Puskas E, Selmeczy Z, Szelenyi M, Vizi, ES. Opposite role of α2- and β-adrenoreceptors in the modulation of interleukin-10 production in endotoxaemic mice. Neuroreport 2000;11:3565–8.
89. Elenkov IJ, Wilder RL, Chrousos GP, Vizi ES. The sympathetic nerve – an integrative interface between two super systems: The brain and the immune system. Pharmacol Rev 2000;52:595–638.
90. Delrue-Perollet C, Li KS, Neveu PJ. Peripheral catecholamines are involved in the neuroendocrine and immune effects of LPS. Brain Behav Immun 1995;9:149–62.
91. Udelsman R, Blake MJ, Stagg CA, Holbrook NJ. Endocrine control of stress-induced heat shock protein 70 expression in vivo. Surgery 1994;115:611–6.
92. Fukudo S, Abe K, Hongo M, Utsumi A, Itoyama Y. Brain–gut induction of heat shock protein (HSP) 70 mRNA by psychophysiological stress in rats. Brain Res 1997;757:146–8.
93. Asea A, Rehli M, Kabingu E, Boch JA, Baré O, Auron PE, Stevenson AM, Calderwood SK. Novel signal transduction pathway utilized by extracellular HSP 70. J Biol Chem 2002;277:15028–34.
94. Campisi J, Fleshner M. Extracellular HSP72 contributes to stress-induced facilitation of innate immunity. Brain Behav Immun 2002;16:174, abstract.

95. Flor-Henry P. Lateralized temporal-limbic dysfunction and psychopathology. Ann N Y Acad Sci 1976;280:777–95.
96. Capuron L, Ravaud A, Gualde N, Bosmans E, Dantzer R, Maes M, Neveu PJ. Association between immune activation and early depressive symptoms in cancer patients treated with interleukin-2-based therapy. Psychoneuroendocrinology 2001;26:797–808.
97. Capuron L, Ravaud A, Neveu PJ, Miller AH, Maes M, Dantzer R. Association between decreased serum tryptophan concentrations and depressive symptoms in cancer patients undergoing cytokine therapy. Mol Psychiatry 2002;7:468–73.

VII. CONCLUSIONS

Cytokines and the Brain
Edited by C. Phelps and E. Korneva

Concluding Remarks

The first phase of Neuroimmune Biology (immunophysiology, neuroimmunomodulation, phychoneuroimmunology, etc.) involved investigations into the role of the nervous and endocrine systems in the regulations of the immune system. These seminal studies discovered many mechanisms by which the neuroendocrine system regulates the immune system (I. Berczi, A. Szentivanyi). The involvement of the immune system and of cytokines in brain physiology and pathophysiology has been considered only recently.

Cytokines and the Brain is one of the first books dedicated to addressing this comprehensive research area. This volume contains contributions from well-known scientists dealing with the basic and applied aspects of these multifaceted problems. Reviews supplemented with original findings provide an integrated presentation of advances and current objectives of biology and medicine related to this field. The titles of the chapters in this book testify to the breadth of current approaches to the problem and to its maturity.

An important part of neuroimmunophysiology – the blood–brain barrier (BBB) and cytokines – is also illuminated (W. Banks). Recent results indicate that most BBB–cytokine interactions are important for physiological neuroimmune interactions. Many cytokines are transported across the BBB by saturable or nonsaturable transport pathways; others are presented on the membrane of endothelial and other cells, which form the BBB. Cytokine–BBB interactions are not static; they respond to physiological and pathological events. The cytokines that are transported across the BBB or secreted by brain endothelial cells are interleukin (IL)-1, IL-3, IL-6, IL-8, tumor necrosis factor (TNF), etc. Transport/secretion may be constitutive or stimulated. As it becomes clear, cytokine transport/release is selective. For example, normally IL-2, IL-10, TNF-α and some others are normally not transported from the blood to the brain. These mechanisms might change under pathological conditions, which is of special importance.

A significant part of the book discusses the question of cytokine involvement in brain functions, that is, the influence of cytokines on the activities of the neurons and glial cells. Real progress has been made in the understanding of the role of cytokines in neuronal interaction, including the regulation of synaptic functions (Spengler). Many cytokines are shown as signaling molecules for neurons of the central nervous system (CNS). Each cytokine can influence many cell types and exert many different effects, while acting by autocrine, paracrine, or endocrine mechanisms.

For example, TNF plays an important role in the nervous system by regulating the synaptic release of neurotransmitters: it inhibits nerve growth factor (NGF) release physiologically, which is altered under pathological conditions. Collectively, the findings indicate that TNF regulates synaptic plasticity in regions of the brain associated with learning and memory. IL-1, depending on its concentration, can either promote or inhibit synaptic plasticity. IL-6 predominantly plays a protective role. It improves the survival of neurons and attenuates pain signals.

The involvement of interferons (IFNs) and chemokines in the modulation of neuronal functions is also demonstrated. These data are especially important, because they show principally new mediators and mechanisms of neuroimmune interactions, disorders of which play a role in the development of many neurological diseases (Schettini et al.).

The chapters discussing cytokine receptors in this volume provide the most extensive information. A number of important aspects of the overall problem are discussed, including cytokine receptors of brain cells, the role of the receptors, the significance of their density and affinity for neuronal functions. The authors (Conti et al., Takao et al., Rybakina and Korneva, Konsman) provide evidence for significant progress in our understanding of the localizations and the role of the receptor systems in the brain.

It has become increasingly clear that cytokine receptors in the brain underlie previously unrecognized mechanisms of neural and immune dysfunctions, including those developing during stress. In particular, studies of the receptor apparatus of neurons, lymphoid cells, and mechanisms for signal transduction from the cell membrane to the nucleus have been clarified. This information allowed the elucidation of the mechanisms of immunostimulation and inhibition during various forms of stress, which results in increased blood glucocorticoids and IL-1 (Rybakina and Korneva).

The localization of cytokine receptors in different brain structures formed the basis for studies of receptor roles in the functions of brain neurons and brain nuclei (Goehler, Prinz, Konsman, Hanisch).

Recent observations show that cytokines, particularly IL-1α and IL-1 receptors, play an important role in coordinating hypothalamic–pituitary–adrenal (HPA) reactions during stress. The fact that CRF is participating in modulating IL-1 receptor affinity under stress also must be taken into consideration (Takao et al.). These pioneering studies provide evidence for cytokine involvement in the regulation of brain functions. Now it is also apparent that stromal elements contribute to the regulation of brain cells in a major way (Wiranowska).

This volume of NIB enables the reader to obtain insight not only into the changes in the functional activity of neurons influenced by cytokines, but also into their effects on some CNS functions including sleep (Krueger)-feeding behavior, memory, and involvement in biological defense system (Oomura), cytokine effects on animal behavior (Aubert), and the role of lateralization of cytokine production in these effects (Neveu).

Of special interest is the contribution of K. Bulloch's laboratory dealing with the role of sex hormones, in particular estrogen, in brain tissue regeneration and brain functional recovery, as well as with the involvement of cytokines in these processes (Gottfied-Blackmore et al.), which is important in view of the ever increasing interest toward attempts to use stem cells in the therapy for brain injury. Data available in the literature and the author's own results and position make it possible to suggest that estrogen and the cytokines stimulated by this hormone may be useful for the optimization of treatment of patients provided with stem cell therapy.

Thus, the *Cytokines and Brain in Stress and Pathology* part of the book represents a broad spectrum of lines, along which developments currently occur in the field of studies of the role of cytokines in brain functions. On the whole, it provides evidence that investigations into the roles and mechanisms of cytokine involvement in brain functions are very promising and that the work carried out so far is only the beginning, but a very significant one.

It is now clear that chemoattractant cytokines participate in the CNS not only in pathological situations but also in physiological functions. Fascinating new data concern the influence of chemokines on neural and glial cell proliferation and migration during the process of CNS development (Shettini et al.).

Although glial cell functions in brain as related to immunity have been studied for quite a long time, recent developments in methodology and understanding make it reasonable to revisit the field in this book. Importantly, the brain has been found to be relatively privileged only with regard to immunity, since molecules of the major and minor histocompatibility complexes are present even in neurons. Not only glial cells but also neurons produce cytokines, their repertoire being quite extended. Moreover, immunocompetent cells, including T lymphocytes, can enter the brain under pathological conditions. In contrast, antigen presentation to brain tissue does not elicit an adequate adaptive immune response, but rather evokes reactions, including cytokine production, as a part of innate immunity. IL-1β is present in the normal brain, and its expression increases upon parenteral administration of lipopolysaccharide (LPS). IL-1β regulates neuronal survival and contributes to the realization of the neuroendocrine response to stress, the latter being brought about by IL-1β expression in the neuroendocrine nuclei of the hypothalamus (Phelps and Chen). *De novo* IL-1β synthesis in the BBB upon infections contributes to behavioral responses to infections. Especially interesting are data about IL-1β involvement in normal brain physiology (Konsman).

Cytokines produced by astrocytes have been found to activate information processing in neurons and their synaptic integration (Penkova et al.). Moreover, a role for astroglia and its cytokine products in the development of neural diseases and neuroimmune interactions has also been demonstrated (Suzumura).

Cytokines participate in neuroimmune interactions as signal molecules, for example, IL-1β and IFN-α stimulate the electrical activity of the efferent nerves of the spleen and reduce natural killer (NK)-cell activity. Data on the production of these cytokines by hypothalamic cells during stress and in other situations suggest that these cytokines produced in the brain may be key factors of stress-induced immunosuppression (Katafuchi). Several mechanisms have been described for the participation of the sympathoadrenal system and norepinephrine in the interaction of the nervous and immune systems during inflammation. For example, a role of norepinephrine released by lymphocytes in the modulation of immunological processes has been demonstrated (Kannan et al.).

Different stress and pathological conditions are associated with cytokine production in the brain, which play significant roles in disease development and outcome. A new discovery in this area is the involvement of IL-2 in the mediation of brain responses to stressors and to antigenic stimuli. The IL-2 gene, similar to immediate response genes, is expressed at a low level in brain neurons and glial cells under normal conditions. Increases in IL-2 mRNA occur under stress or after antigenic stimulation, which may be modulated by therapeutic drugs. These findings warrant further studies of IL-2 functions in the brain (Korneva and Kazakova).

The valuable data, reported by J. Correale and colleagues, support the position that some of the inflammatory cells in the CNS and the peripheral nervous system (PNS) have beneficial effects, promote repair and reduce the extent of damage of nerve cells after insult and injury. This information may serve for the development of novel therapeutic approaches for neurological diseases.

The development of autoimmune pathological processes in the brain, particularly in the case of multiple sclerosis, is associated with increased blood and brain levels of proinflammatory cytokines (IL-1, IL-2, TNF-α, and IFNs). Further studies of the positive role of IL-6, IL-18, and osteopontin are warranted. The recently discovered interleukins IL-17, IL-23, and IL-27 are already under investigation with regard to their role in the pathogenesis of multiple sclerosis, since there are reasons to believe that these cytokines are also involved in the development of this disease. Imbalances in the production and levels of the above cytokines may also result in shifts in the Th1/Th2 balance and CD122/CD28B7 ratio, which, under certain conditions, may lead to the development of autoimmune diseases, including those associated with demyelinization (Ketlinskiy and Kalinina).

Much effort has been devoted to the studies of changes in cytokine imbalances in Alzheimer's disease. It should be stressed that modern theories of this disease take into account the importance for correction of disturbances in cytokine expression in the pathological process (Summers).

The role of cytokines in the development of depression is complex and controversial. Some experimental and clinical observations do not assign a significant role to cytokines in the pathogenesis of depression. Moreover, it is not possible to exclude the participation of cytokines in depression (Dunn). This is an important conclusion, which calls for future investigations.

Brain cytokines represent a novel area of Neuroimmune Biology, which deals with the role of these important biological mediators in the physiology and pathophysiology of the central nervous system. This volume of Neuroimmune Biology provides a testimony for the significance and potential of this new area of Biology.

Despite the intensification of studies carried out in this field, it is quite evident that there is a lot to be discovered yet in this area. Nevertheless, the direction and the framework of these studies have already been determined. In our opinion, one of the objectives and merits of this book is to aid the development of this field further. Hopefully, further advances in studies within this area will lead to the discovery of still unknown mechanisms of disturbances in neuroimmune interactions and to a better insights into triggers for the development of diseases and thus expanding the possibilities of their treatment. That is why this book is expected to attract the attention of a wide audience of biologists and health care specialists.

Elena A. Korneva

Subject Index

Kolibri 360 GmbH
Ferdinand-Jühlke-Straße 7
99095 Erfurt, Deutschland
produktsicherheit@kolibri360.de